Paul McKenzie
2011

Urban Remote Sensing

Urban Remote Sensing

Monitoring, Synthesis and Modeling in the Urban Environment

Xiaojun Yang

A John Wiley & Sons, Ltd., Publication

This edition first published 2011 © 2011 by John Wiley & Sons, Ltd.

Wiley-Blackwell is an imprint of John Wiley & Sons, formed by the merger of Wiley's global Scientific, Technical and Medical business with Blackwell Publishing.

Registered office: John Wiley & Sons, Ltd, The Atrium, Southern Gate, Chichester, West Sussex, PO19 8SQ, UK

Editorial offices: 9600 Garsington Road, Oxford, OX4 2DQ, UK
The Atrium, Southern Gate, Chichester, West Sussex, PO19 8SQ, UK

111 River Street, Hoboken, NJ 07030-5774, USA

For details of our global editorial offices, for customer services and for information about how to apply for permission to reuse the copyright material in this book please see our website at www.wiley.com/wiley-blackwell.

The right of the author to be identified as the author of this work has been asserted in accordance with the UK Copyright, Designs and Patents Act 1988.

All rights reserved. No part of this publication may be reproduced, stored in a retrieval system, or transmitted, in any form or by any means, electronic, mechanical, photocopying, recording or otherwise, except as permitted by the UK Copyright, Designs and Patents Act 1988, without the prior permission of the publisher.

Designations used by companies to distinguish their products are often claimed as trademarks. All brand names and product names used in this book are trade names, service marks, trademarks or registered trademarks of their respective owners. The publisher is not associated with any product or vendor mentioned in this book. This publication is designed to provide accurate and authoritative information in regard to the subject matter covered. It is sold on the understanding that the publisher is not engaged in rendering professional services. If professional advice or other expert assistance is required, the services of a competent professional should be sought.

Library of Congress Cataloging-in-Publication Data applied for.

A catalogue record for this book is available from the British Library

This book is published in the following electronic format: ePDF:9780470979570, Wiley Online: 9780470979563, ePub: 9780470979891

Typeset in 9/11pt Minion by Laserwords Private Limited, Chennai, India

Printed and bound in Malaysia by Vivar Printing Sdn Bhd

First Impression 2011

This book is dedicated to the memory of Dr. Chor-Pang (C.P.) Lo (1939–2007), whose creative and groundbreaking work in urban remote sensing inspires a new generation of scholars working for a better understanding of the complex, dynamic urban environment.

CONTENTS

List of Contributors xiii
Author's Biography xvi
Preface xix

PART 1 INTRODUCTION 1

1 What is urban remote sensing? 3
Xiaojun Yang

1.1 Introduction 4
1.2 Remote sensing and urban studies 5
1.3 Remote sensing systems for urban areas 6
1.4 Algorithms and techniques for urban attribute extraction 7
1.5 Urban socioeconomic analyses 7
1.6 Urban environmental analyses 8
1.7 Urban growth and landscape change modeling 8
Summary and concluding remarks 9
References 10

PART 2 REMOTE SENSING SYSTEMS FOR URBAN AREAS 13

2 Use of archival Landsat imagery to monitor urban spatial growth 15
Xiaojun Yang

2.1 Introduction 16
2.2 Landsat program and imaging sensors 16
2.3 Mapping urban spatial growth in an American metropolis 18
　　2.3.1 Research design 18
　　2.3.2 Data acquisition and land classification scheme 19
　　2.3.3 Image processing of remotely sensed data 20
　　2.3.4 Change detection 21
　　2.3.5 Interpretation and analysis 25
　　2.3.6 Summary 27
2.4 Discussion 27
　　2.4.1 A generic urban growth monitoring workflow 27
　　2.4.2 Image resolution and land use/cover classification 27
　　2.4.3 Image preprocessing 28
　　2.4.4 Change detection methods 29
Summary and concluding remarks 30
References 30

3 Limits and challenges of optical very-high-spatial-resolution satellite remote sensing for urban applications 35
Paolo Gamba, Fabio Dell'Acqua, Mattia Stasolla, Giovanna Trianni and Gianni Lisini

3.1 Introduction 36
3.2 Geometrical problems 36
3.3 Spectral problems 38
3.4 Mapping limits and challenges 38
3.5 Adding the time factor: VHR and change detection 39
3.6 A possible way forward 39
3.7 Building damage assessment 43
Conclusions 46
References 47

4 Potential of hyperspectral remote sensing for analyzing the urban environment 49
Sigrid Roessner, Karl Segl, Mathias Bochow, Uta Heiden, Wieke Heldens and Hermann Kaufmann

4.1 Introduction 50
4.2 Spectral characteristics of urban surface materials 50
　　4.2.1 Categories of interest for material mapping 50
　　4.2.2 Establishment of urban spectral libraries 52
　　4.2.3 Determination of robust spectral features 52
4.3 Automated identification of urban surface materials 54
　　4.3.1 State of the art of automated hyperspectral image analysis 54
　　4.3.2 Processing system for automated material mapping 57
4.4 Results and discussion of their potential for urban analysis 58
References 60

5 Very-high-resolution spaceborne synthetic aperture radar and urban areas: looking into details of a complex environment 63
Fabio Dell'Acqua, Paolo Gamba and Diego Polli

- 5.1 Introduction 64
- 5.2 Before spaceborne high-resolution SAR 64
- 5.3 High-resolution SAR 66
  - 5.3.1 Extraction of single buildings 66
  - 5.3.2 Damage assessment with VHR SAR data 67
  - 5.3.3 Vulnerability mapping with VHR SAR data 69
- Conclusions 70
- Acknowledgments 70
- References 70

6 3D building reconstruction from airborne lidar point clouds fused with aerial imagery 75
Jonathan Li and Haiyan Guan

- 6.1 Lidar-drived building models: related work 76
  - 6.1.1 Building detection 76
  - 6.1.2 Building reconstruction 76
- 6.2 Our building reconstruction method 77
  - 6.2.1 Our strategy using fused data 77
  - 6.2.2 Building detection 78
  - 6.2.3 Building reconstruction 81
- 6.3 Results and discussion 85
  - 6.3.1 Datasets 85
  - 6.3.2 Results 85
- Concluding remarks 89
- Acknowledgments 90
- References 90

PART 3 ALGORITHMS AND TECHNIQUES FOR URBAN ATTRIBUTE EXTRACTION 93

7 Parameterizing neural network models to improve land classification performance 95
Xiaojun Yang and Libin Zhou

- 7.1 Introduction 96
- 7.2 Fundamentals of neural networks 96
  - 7.2.1 Neural network types 96
  - 7.2.2 Network topology 98
  - 7.2.3 Neural training 98
- 7.3 Internal parameters and classification accuracy 100
  - 7.3.1 Experimental design 100
  - 7.3.2 Remotely sensed data and land classification scheme 101
  - 7.3.3 Network configuration and training 101
  - 7.3.4 Image classification and accuracy assessment 103
  - 7.3.5 Interpretation and analysis 103
  - 7.3.6 Summary 105
- 7.4 Training algorithm performance 105
  - 7.4.1 Experimental design 105
  - 7.4.2 Network training and image classification 105
  - 7.4.3 Performance evaluation 106
- 7.5 Toward a systematic approach to image classification by neural networks 107
- Future research directions 108
- References 108

8 Characterizing urban subpixel composition using spectral mixture analysis 111
Rebecca Powell

- 8.1 Introduction 112
- 8.2 Overview of SMA implementation 112
  - 8.2.1 SMA background 112
  - 8.2.2 Endmember selection 114
  - 8.2.3 SMA models 116
  - 8.2.4 Mapping fraction images 117
  - 8.2.5 Model complexity 118
  - 8.2.6 Accuracy assessment 118
- 8.3 Two case studies 118
  - 8.3.1 Evolution of urban morphology on a tropical forest frontier 119
  - 8.3.2 Discriminating urban tree and lawn cover in a western US city 122
- Conclusions 124
- Acknowledgments 126
- References 126

9 An object-oriented pattern recognition approach for urban classification 129
Soe W. Myint and Douglas Stow

- 9.1 Introduction 130
- 9.2 Object-oriented classification 130
  - 9.2.1 Image segmentation 130
  - 9.2.2 Features 131
  - 9.2.3 Classifiers 132
- 9.3 Data and study area 133
- 9.4 Methodology 134

9.4.1 Rule-based detection of swimming pools 134
9.4.2 Nearest neighbor classifier to extract urban land covers 136

9.5 Results and discussion 137
9.5.1 Decision rule set to extract pool 137
9.5.2 Nearest neighbor classifier to extract urban land covers 138

Conclusion 139
References 140

10 Spatial enhancement of multispectral images on urban areas 141

Bruno Aiazzi, Stefano Baronti, Luca Capobianco, Andrea Garzelli and Massimo Selva

10.1 Introduction 142
10.1.1 Component substitution fusion methods 142
10.1.2 Multiresolution analysis fusion methods 142
10.1.3 Injection model of details 143
10.1.4 Quality assessment 143
10.2 Multiresolution fusion scheme 144
10.3 Component substitution fusion scheme 144
10.4 Hybrid MRA – component substitution method 146
10.5 Results 147

Conclusions 152
References 152

11 Exploring the temporal lag between the structure and function of urban areas 155

Victor Mesev

11.1 Introduction 156
11.2 Micro and macro urban remote sensing 156
11.3 The temporal lag challenge 157
11.4 Structural–functional links 157
11.5 Temporal–structural–functional links 159
11.6 Empirical measurement of temporal lags 159

Conclusions 161
References 161

PART 4 URBAN SOCIOECONOMIC ANALYSES 163

12 A pluralistic approach to defining and measuring urban sprawl 165

Amnon Frenkel and Daniel Orenstein

12.1 Introduction 166

12.2 The diversity of definitions of sprawl 166
12.2.1 Definitions describing an urban spatial development phenomenon 167
12.2.2 Definitions based on consequences of sprawl; sprawl is as sprawl does 167
12.2.3 Definitions according to the social and/or economic processes that give rise to particular urban spatial development patterns 168
12.2.4 Sprawl redux: focusing on the concerns of remote sensing experts 168
12.3 Historic forms of "urban sprawl" 168
12.4 Qualitative dimensions of sprawl and quantitative variables for measuring them 169
12.4.1 Criteria for a good sprawl measurement variable 170
12.4.2 What shall we measure? 170
12.4.3 Choosing among the sprawl measures 173

Conclusion 178
References 178

13 Small area population estimation with high-resolution remote sensing and lidar 183

Le Wang and Jose-Silvan Cardenas

13.1 Introduction 184
13.2 Study sites and data 185
13.3 Methodology 186
13.3.1 Data preprocessing 186
13.3.2 Building extraction 186
13.3.3 Land use classification 186
13.3.4 Population estimation models 187
13.3.5 Accuracy assessment 187
13.4 Results 187
13.4.1 Building detection results 187
13.4.2 Land use classification results 189
13.4.3 Population estimation results 189

Discussion and conclusions 192
Acknowledgments 192
References 192

14 Dasymetric mapping for population and sociodemographic data redistribution 195

James B. Holt and Hua Lu

14.1 Introduction 196
14.2 Dasymetric maps, dasymetric mapping, and areal interpolation 196
14.2.1 Ancillary data 197
14.2.2 Dasymetric mapping 197

- 14.2.3 Origins 197
- 14.2.4 Dasymetric mapping variations 198

14.3 Application example: metropolitan Atlanta, Georgia 200
- 14.3.1 Data 200
- 14.3.2 Dasymetric maps 201
- 14.3.3 Areal interpolation 203

Conclusions 205

Acknowledgments 208

References 208

15 Who's in the dark—satellite based estimates of electrification rates 211
Christopher D. Elvidge, Kimberly E. Baugh, Paul C. Sutton, Budhendra Bhaduri, Benjamin T. Tuttle, Tilotamma Ghosh, Daniel Ziskin and Edward H. Erwin

15.1 Introduction 212

15.2 Methods 212
- 15.2.1 Data sources 212
- 15.2.2 Data processing 213

15.3 Results 213

15.4 Discussion 214

Conclusion 223

Acknowledgments 223

References 223

16 Integrating remote sensing and GIS for environmental justice research 225
Jeremy Mennis

16.1 Introduction 226

16.2 Environmental justice research 226

16.3 Remote sensing for environmental equity analysis 227

16.4 Integrating remotely sensed and other spatial data using GIS 229

16.5 Case study: vegetation and socioeconomic character in Philadelphia, Pennsylvania 230

Conclusion 234

References 235

PART 5 URBAN ENVIRONMENTAL ANALYSES 239

17 Remote sensing of high resolution urban impervious surfaces 241
Changshan Wu and Fei Yuan

17.1 Introduction 242

17.2 Impervious surface estimation 242
- 17.2.1 Pixel-based models 242
- 17.2.2 Object-based models 243

17.3 Pixel-based models for estimating high-resolution impervious surface 243
- 17.3.1 Introduction 243
- 17.3.2 Study area and data 243
- 17.3.3 Methodology 244
- 17.3.4 Results 248

17.4 Object-based models for estimating high-resolution impervious surface 249
- 17.4.1 Study area and data preparation 249
- 17.4.2 Object-oriented classification 249
- 17.4.3 Results 251

Conclusions 252

References 252

18 Use of impervious surface data obtained from remote sensing in distributed hydrological modeling of urban areas 255
Frank Canters, Okke Batelaan, Tim Van de Voorde, Jarosław Chormański and Boud Verbeiren

18.1 Introduction 256

18.2 Spatially distributed hydrological modeling 256

18.3 Impervious surface mapping 257

18.4 The WetSpa model 258
- 18.4.1 Surface runoff 258
- 18.4.2 Flow routing 260
- 18.4.3 Water balance 261

18.5 Impact of different approaches for estimating impervious surface cover on runoff calculation and prediction of peak discharges 261
- 18.5.1 Study area and data 261
- 18.5.2 Impervious surface mapping 262
- 18.5.3 Impact of land-cover distribution on estimation of peak discharges 264

Conclusions 270

Acknowledgments 270

References 270

19 Impacts of urban growth on vegetation carbon sequestration 275
Tingting Zhao

19.1 Introduction 276

19.2 Vegetation productivities and estimation 276
- 19.2.1 Vegetation productivities 276
- 19.2.2 Estimation of vegetation productivities 276

19.3 Data and analysis 277

- 19.3.1 Identifying urban growth 277
- 19.3.2 Preparing vegetation maps and light-use efficiency parameters 279
- 19.3.3 Estimating APAR, GPP and changes in GPP 279

19.4 Results 280

19.5 Discussion 283
- 19.5.1 Urban growth in the South Atlantic division 283
- 19.5.2 Impacts of urban growth on vegetation productivities 283

Conclusions 284

Acknowledgments 284

References 285

20 Characterizing biodiversity in urban areas using remote sensing 287
Marcus Hedblom and Ulla Mörtberg

20.1 Introduction 288

20.2 Remote sensing methods in urban biodiversity studies 288
- 20.2.1 Direct approaches 289
- 20.2.2 Indirect approaches 289

20.3 Hierarchical levels and definitions of urban ecosystems 292
- 20.3.1 Flora and fauna along urban gradients 292
- 20.3.2 Using remote sensing to quantify urbanization patterns 293

20.4 Using remote sensing to interpret effects of urbanization on species distribution 294

20.5 Long-term monitoring of biodiversity in urban green areas – methodology development 295

20.6 Applications in urban planning and management 296

Conclusions 297

Acknowledgments 300

References 300

21 Urban weather, climate and air quality modeling: increasing resolution and accuracy using improved urban morphology 305
Susanne Grossman-Clarke, William L. Stefanov and Joseph A. Zehnder

21.1 Introduction 306

21.2 Physical approaches for the representation of urban areas in regional atmospheric models 306
- 21.2.1 Roughness approach 307
- 21.2.2 Single-layer urban canopy approaches 307
- 21.2.3 Multilayer urban canopy approaches 307

21.3 Remotely sensed data as input for regional atmospheric models 307
- 21.3.1 Urban land use and land cover data 308
- 21.3.2 Building data 310

21.4 Case studies investigating the effects of urbanization on weather, climate and air quality 311
- 21.4.1 Studies investigating effects of urban land use and land cover on meteorology and air quality 311
- 21.4.2 Case study for Phoenix 312

Conclusions 316

Acknowledgments 316

References 316

PART 6 URBAN GROWTH AND LANDSCAPE CHANGE MODELING 321

22 Cellular automata and agent base models for urban studies: from pixels to cells to hexa-dpi's 323
Elisabete A. Silva

22.1 Introduction 324

22.2 Computation: the raster–pixel aproach 324

22.3 Cells: migrating from basic pixels 324

22.4 Agents: joining with cells 327

22.5 Cells and agents in a computer's "artificial life" 328

22.6 The hexa-dpi: closing the cycle in the digital age 330

Conclusions 332

References 332

23 Calibrating and validating cellular automata models of urbanization 335
Paul M. Torrens

23.1 Introduction 336

23.2 Calibration 336
- 23.2.1 Conditional transition rules 336
- 23.2.2 Weighted transition rules 337
- 23.2.3 Seeding and initial conditions 337
- 23.2.4 Specifying the value of calibration parameters 338
- 23.2.5 Coupling automata and exogenous models 338
- 23.2.6 Automatic calibration 339

23.3 Validating automata models 339
- 23.3.1 Pixel matching 339
- 23.3.2 Feature and pattern recognition 340
- 23.3.3 Running models exhaustively 341

Conclusions 341
Acknowledgments 342
References 342

24 Agent-based urban modeling: simulating urban growth and subsequent landscape change in suzhou, china 347
Yichun Xie and Xining Yang

24.1 Introduction 348
24.2 Design, construction, calibration, and validation of ABM 348
24.3 Case study – desakota development in Suzhou, China 350
24.4 The Suzhou Urban Growth Agent Model 351
- 24.4.1 The model design 351
- 24.4.2 The model construction 352
- 24.4.3 The model calibration 352
- 24.4.4 The model validation 353

Summary and conclusion 354
References 355

25 Ecological modeling in urban environments: predicting changes in biodiversity in response to future urban development 359
Jeffrey Hepinstall-Cymerman

25.1 Introduction 360
- 25.1.1 Using urban remote sensing to develop land cover maps for ecological modeling 360
- 25.1.2 One example of ecological modeling: modeling species habitat 360
- 25.1.3 Predicting future land use and land cover 361
- 25.1.4 Integrating models to predict future biodiversity 362

25.2 Predicting changes in land cover and avian biodiversity for an area north of Seattle, Washington 362
- 25.2.1 Land cover maps 362
- 25.2.2 Land use change model 362
- 25.2.3 Land cover change model 364
- 25.2.4 Avian biodiversity model 365
- 25.2.5 Predicted land cover change for study area 365
- 25.2.6 Predicted changes in avian biodiversity for study area 365

Conclusions 365
Acknowledgments 367
References 368

26 Rethinking progress in urban analysis and modeling: models, metaphors, and meaning 371
Daniel Z. Sui

26.1 Introduction 372
26.2 Pepper's world hypotheses: the role of root metaphors in understanding reality 373
26.3 Progress in urban analysis and modeling: metaphors urban modelers live by 373
- 26.3.1 Cities as forms – the spatial morphology tradition 374
- 26.3.2 Cities as machines – the social physics tradition 375
- 26.3.3 Cities as organisms – the social biology tradition 375
- 26.3.4 Cities as arenas – the spatial event tradition 376

26.4 Models, metaphors, and the meaning of progress: further discussions 377

Summary and concluding remarks 377
Acknowledgments 378
Notes 378
References 378

Index 383

LIST OF CONTRIBUTORS

Bruno Aiazzi
Institute of Applied Physics "Nello Carrara"
CNR Research Area of Florence
10 Via Madonna del Piano, 50019 Sesto F.no (FI), Italy
E-mail: b.aiazzi@ifac.cnr.it

Stefano Baronti
Institute of Applied Physics "Nello Carrara"
CNR Research Area of Florence
10 Via Madonna del Piano, 50019 Sesto F.no (FI), Italy
E-mail: s.baronti@ifac.cnr.it

Okke Batelaan
Department of Hydrology and Hydraulic Engineering
Vrije Universiteit Brussel
Pleinlaan 2, 1050 Brussels, Belgium
E-mail: okke.batelaan@vub.ac.be

Kimberly E. Baugh
Cooperative Institute for Research in Environmental Sciences
University of Colorado
Boulder, Colorado 80309, USA
E-mail: kim.baugh@noaa.gov

Budhendra Bhaduri
Oak Ridge National Laboratory
Oak Ridge, Tennessee 37831, USA
E-mail: bhaduribl@ornl.gov

Mathias Bochow
Helmholtz Centre Potsdam – GFZ
German Research Centre for Geosciences
Section 1.4 Remote Sensing
Telegrafenberg, 14473 Potsdam, Germany
E-mail: mathias.bochow@gfz-potsdam.de

Frank Canters
Cartography and GIS Research Unit
Department of Geography
Vrije Universiteit Brussel
Pleinlaan 2, 1050 Brussels, Belgium
E-mail: fcanters@vub.ac.be

Luca Capobianco
Department of Information Engineering
University of Siena
Via Roma 56, 53100 Siena, Italy
E-mail: capobianco@dii.unisi.it

Jose-Silvan Cardenas
Department of Geography
University at Buffalo, State University of New York
Buffalo, NY 14261, USA
E-mail: jlsilvan@buffalo.edu

Jarosław Chormański
Department of Hydraulic Structures and Environmental Restoration
Warsaw University of Life Sciences
ul. Nowoursynowska 159, 02–776 Warsaw, Poland
E-mail: j.chormanski@levis.sggw.pl

Fabio Dell'Acqua
Department of Electronics
University of Pavia, via Ferrata 1
27100 Pavia, Italy
E-mail: fabio.dellacqua@unipv.it

Christopher D. Elvidge
Earth Observation Group
NOAA National Geophysical Data Center
Boulder, Colorado 80303, USA
E-mail: chris.elvidge@noaa.gov

Edward H. Erwin
NOAA National Geophysical Data Center
Boulder, Colorado 80303, USA
E-mail: Edward.H.Erwin@noaa.gov

Amnon Frenkel
Center for Urban and Regional Studies
Faculty of Architecture and Town Planning
Technion – Israel Institute of Technology
Haifa 32000, Israel
E-mail: amnonf@tx.technion.ac.il

Paolo Gamba
Department of Electronics
University of Pavia, via Ferrata 1
27100 Pavia, Italy
E-mail: paolo.gamba@unipv.it

Andrea Garzelli
Department of Information Engineering
University of Siena
Via Roma, 56, 53100 Siena, Italy
E-mail: garzelli@dii.unisi.it

Tilotamma Ghosh
Cooperative Institute for Research in Environmental Sciences
University of Colorado
Boulder, Colorado 80309, USA
E-mail: Tilottama.Ghosh@noaa.gov

Susanne Grossman-Clarke
Global Institute of Sustainability
Arizona State University
Tempe, AZ, 85287-5302, USA
E-mail: sg.clarke@asu.edu

Haiyan Guan
Department of Geography and Environmental Management
University of Waterloo
200 University Avenue West
Waterloo, Ontario, Canada N2L 3G1
E-mail: h6guan@uwaterloo.ca

Marcus Hedblom
Department of Ecology
Swedish University of Agricultural Sciences
PO Box 7044
SE-750 07, Uppsala, Sweden
E-mail: marcus.hedblom@ekol.slu.se

Uta Heiden
German Aerospace Center (DLR)
German Remote Sensing Data Center (DFD)
Münchener Strasse 20, 82234 Wessling, Germany
E-mail: Uta.Heiden@dlr.de

Wieke Heldens
German Aerospace Center (DLR)
German Remote Sensing Data Center (DFD)
Münchener Strasse 20, 82234 Wessling, Germany
E-mail: wieke.heldens@dlr.de

Jeffrey Hepinstall-Cymerman
Warnell School of Forestry and Natural Resources
University of Georgia
Athens, Georgia 30602, USA
E-mail: jhepinstall@warnell.uga.edu

James B. Holt
Division of Adult and Community Health
Centers for Disease Control and Prevention
Mailstop K-67, 4770 Buford Highway, NE
Atlanta, Georgia 30341, USA
E-mail: jgh4@cdc.gov

Hermann Kaufmann
Helmholtz Centre Potsdam – GFZ
German Research Centre for Geosciences
Section 1.4 Remote Sensing
Telegrafenberg, 14473 Potsdam, Germany
E-mail: hermann.kaufmann@gfz-potsdam.de

Jonathan Li
Department of Geography and Environmental Management
University of Waterloo
200 University Avenue West
Waterloo, Ontario, Canada N2L 3G1
E-mail: junli@uwaterloo.ca

Gianni Lisini
IUSS, Centre for Risk and Security
Via le Lungo Ticino Sforza 56
27100 Pavia, Italy
E-mail: gianni.lisini@unipv.it

Hua Lu
Division of Adult and Community Health
Centers for Disease Control and Prevention
Mailstop K-67, 4770 Buford Highway, NE
Atlanta, Georgia 30341, USA
E-mail: hlu1@cdc.gov

Jeremy Mennis
Department of Geography and Urban Studies
Temple University
Philadelphia, PA 19122, USA
E-mail: jmennis@temple.edu

Victor Mesev
Department of Geography
Florida State University
Tallahassee, FL 32306, USA
E-mail: vmesev@fsu.edu

Ulla Mörtberg
Department of Land and Water Resources
  Engineering
Royal Institute of Technology
SE-100 44 Stockholm, Sweden
E-mail: mortberg@kth.se

Soe W. Myint
School of Geographical Sciences and Urban Planning
Arizona State University
Tempe, AZ 85287, USA
E-mail: soe.myint@asu.edu

Daniel Orenstein
Center for Urban and Regional Studies
Faculty of Architecture and Town Planning
Technion – Israel Institute of Technology
Haifa 32000, Israel
Email: daniel.orenstein@gmail.com

Diego Polli
Department of Electronics
University of Pavia, via Ferrata 1
27100 Pavia, Italy
E-mail: diegoaldo.polli@unipv.it

Rebecca Powell
Department of Geography
University of Denver
Denver, Colorado 80208, USA
E-mail: rebecca.l.powell@du.edu

Sigrid Roessner
Helmholtz Centre Potsdam – GFZ
German Research Centre for Geosciences
Section 1.4 Remote Sensing
Telegrafenberg, 14473 Potsdam, Germany
E-mail: sigrid.roessner@gfz-potsdam.de

Karl Segl
Helmholtz Centre Potsdam – GFZ
German Research Centre for Geosciences
Section 1.4 Remote Sensing
Telegrafenberg, 14473 Potsdam, Germany
E-mail: karl.segl@gfz-potsdam.de

Massimo Selva
Institute of Applied Physics "Nello Carrara"
CNR Research Area of Florence
10 Via Madonna del Piano, 50019 Sesto F.no (FI), Italy
E-mail: m.selva@ifac.cnr.it

Elisabete A. Silva
Department of Land Economy and Fellow of Robinson College
University of Cambridge
19 Silver Street, Cambridge CB3 9EP, UK
E-mail: es424@cam.ac.uk

Mattia Stasolla
Department of Electronics
University of Pavia, via Ferrata 1
27100 Pavia, Italy
E-mail: mattia.stasolla@unipv.it

William L. Stefanov
Image Science & Analysis Laboratory/ESCG
NASA Johnson Space Center
Houston, TX 77058, USA
E-mail: william.l.stefanov@nasa.gov

Douglas Stow
Department of Geography
San Diego State University
San Diego, California 92182-4493, USA
E-mail: stow@mail.sdsu.edu

Daniel Z. Sui
Center for Urban & Regional Analysis
Department of Geography
The Ohio State University
Columbus, OH 43210, USA
E-mail: sui.10@osu.edu

Paul C. Sutton
Department of Geography
University of Denver
Denver, Colorado 80208, USA
E-mail: psutton@du.edu

Paul M. Torrens
Geosimulation Research Laboratory
School of Geographical Sciences & Urban Planning
Arizona State University
Tempe, AZ 85287-5302, USA
E-mail: torrens@geosimulation.com

Giovanna Trianni
Joint Research Centre, via Enrico Fermi 2749
21027 Ispra, Italy
E-mail: giovanna.trianni@jrc.ec.europa.eu

Benjamin T. Tuttle
Cooperative Institute for Research in Environmental Sciences
University of Colorado
Boulder, Colorado 80309, USA
E-mail: Ben.Tuttle@noaa.gov

Boud Verbeiren
Department of Hydrology and Hydraulic Engineering
Vrije Universiteit Brussel
Pleinlaan 2, 1050 Brussels, Belgium
E-mail: bverbeir@vub.ac.be

Tim Van de Voorde
Cartography and GIS Research Unit
Department of Geography
Vrije Universiteit Brussel
Pleinlaan 2, 1050 Brussels, Belgium
E-mail: tvdvoord@vub.ac.be

Le Wang
Department of Geography
University at Buffalo, State University of New York
Buffalo, NY 14261, USA
E-mail: lewang@buffalo.edu

Changshan Wu
Department of Geography
University of Wisconsin-Milwaukee
Milwaukee, WI 53201, USA
E-mail: cswu@uwm.edu

Yichun Xie
Department of Geography and Geology
Eastern Michigan University
Ypsilanti, Michigan 48197, USA
E-mail: yxie@emich.edu

Xiaojun Yang
Department of Geography
Florida State University
Tallahassee, FL 32306, USA
E-mail: xyang@fsu.edu

Xining Yang
Department of Geography
The Ohio State University
Columbus, OH 43210, USA
E-mail: xyang5@emich.edu

Fei Yuan
Department of Geography
Minnesota State University, Mankato
Mankato, Minnesota 56001, USA
E-mail: fei.yuan@mnsu.edu

Tingting Zhao
Department of Geography
Florida State University
Tallahassee, FL 32306, USA
E-mail: tzhao@fsu.edu

Joseph A. Zehnder
Department of Atmospheric Sciences
Creighton University
Omaha, NE 68178, USA
E-mail: zehnder@creighton.edu

Libin Zhou
Department of Geography
Florida State University
Tallahassee, FL 32306, USA
E-mail: lz06c@fsu.edu

Daniel Ziskin
Cooperative Institute for Research in Environmental Sciences
University of Colorado
Boulder, Colorado 80309, USA
E-mail: Daniel.Ziskin@noaa.gov

AUTHORS' BIOGRAPHY

Bruno Aiazzi is a Researcher of "Nello Carrara" Institute of Applied Physics, IFAC-CNR, Italy. His research interests include image processing of remote sensor data and environmental applications. He has published over 30 journal articles.

Stefano Baronti is a Researcher of "Nello Carrara" Institute of Applied Physics, IFAC-CNR, Italy. His research interests include image compression, processing of optical and SAR images, and image fusion. He has published nearly 50 journal articles.

Okke Batelaan is an Associate Professor of Eco-hydrology/GIS at the Vrije Universiteit Brussel, Belgium and part-time Associate Professor of Hydrogeology at the K.U.Leuven. His research centers on distributed hydrological modeling of shallow subsurface and surface hydrological processes using GIS and remote sensing. He has published more than 100 papers.

Kimberly E. Baugh is an Associate Scientist with the Cooperative Institute of Research in the Environmental Sciences at the University of Colorado, Boulder. Her research focuses on processing and calibration of the night-time data from the DMSP OLS sensor.

Budhendra Bhaduri leads the Geographic Information Science & Technology group at Oak Ridge National Laboratory. His research centers on novel implementation of geospatial science and technology for sustainable development. He has published over 50 papers. He currently serves on the Mapping Sciences Committee of the National Academy of Sciences/National Research Council.

Mathias Bochow is a Research Scientist at the Remote Sensing Section, the GFZ German Research Centre for Geosciences. His research interests include imaging spectroscopy, image classification, and applied remote sensing.

Frank Canters is an Associate Professor and Head of the Department of Geography at the Vrije Universiteit Brussel, Belgium. Since 2001 he has been a Visiting Professor in Geomatics at Ghent University. His main research interests are urban remote sensing, multisensor/multiresolution image analysis, modeling of spatial data uncertainty, and map projection design.

Luca Capobianco is with the Department of Earth Science at the University of Florence, Italy. His research focuses on several areas in remote sensing, especially kernel-based machine learning methods, information mining, multispectral and hyperspectral data analysis, SAR data processing, and data fusion.

Jose Luis Silvan-Cardenas was a Postdoctoral Scholar with the State University of New York at Buffalo. He is currently with the Geography and Geomatic Research Center in Mexico. His research centers on subpixel remote sensing and lidar data analysis.

Jarosław Chormański is an Assistant Professor in Hydrology and Water Resources, Warsaw University of Life Science, Poland. His research emphasizes the applications of geographic information systems and remote sensing in hydrology and hydrological modeling. He has published over 40 papers.

Fabio Dell'Acqua is an Assistant Professor of Remote Sensing at the University of Pavia, Italy. His research interests include synthetic aperture radar (SAR) data processing, earthquake damage assessment, and seismic vulnerability evaluation. He has published over 30 journal articles. He is an Associate Editor of the *Journal of Information Fusion*.

Christopher D. Elvidge leads the Earth Observation Group at NOAA's National Geophysical Data Center. He and his team have been developing the algorithms for constructing global maps of satellite observed nighttime lights since 1994. His current projects include satellite estimation of gas flaring volumes at oil production facilities in 60 countries and global mapping of the density of constructed surfaces.

Edward H. Erwin is a Physical Scientist with NOAA's National Geophysical Data Center. He processes and archives DMSP data and is responsible for the collection and historical preservation of various types of space weather data.

Amnon Frenkel is chair of the Graduate Program for Urban and Regional Planning at the Faculty of Architecture and Town Planning at the Technion – Israel Institute of Technology. He served as the secretary of the Israel Association of Planners and he is the Chair of the Israeli Section of the European Regional Science Association. His research interests include issues of urban and regional planning and technology policy with an emphasis on land use, urban sprawl, and diffusion of innovation in space.

Paolo Gamba is an Associate Professor of Telecommunications at the University of Pavia, Italy. His research centers on urban remote sensing. He has played a key role in organizing the Joint Urban Remote Sensing Event (JURSE). He has been invited to give keynote lectures in many international conferences. He has published more than 70 journal articles. He has served as Editor-in-Chief of the *IEEE Geoscience and Remote Sensing Letters* since 2009.

Andrea Garzelli is an Associate Professor of Telecommunications in the Department of Information Engineering at the University of Siena, Italy. His research interests are signal and image analysis, processing, and classification, including filtering, SAR image analysis, and image fusion for optical and radar remote-sensing applications. He has published over 150 papers.

Tilottama Ghosh is with NOAA National Geophysical Data Center (NGDC). Her research interests include human geography, remote sensing, and GIS. At NGDC, she is responsible for generating global mosaics of nighttime lights and performing socioeconomic analyses. She documents the DMSP algorithms and accomplishments through conference proceedings and journal submissions.

Susanne Grossman-Clarke is an Assistant Research Professor at Arizona State University. Her research emphasizes the improvement of the representation of urban areas in atmospheric models as well as the application of the latter to study the influence of urbanization on weather, climate, air quality and human comfort and health.

Haiyan Guan is a Postdoctoral Fellow at the University of Waterloo, Waterloo, Ontario, Canada. Her research interests include lidar remote sensing, mobile mapping, and spatial modeling. She has published over 10 articles.

Marcus Hedblom leads the Swedish national monitoring program of urban landscapes, Urban NILS, in the Department of Ecology at the Swedish University of Agricultural Sciences. His research concerns the development of methods to monitor biodiversity in urban areas, human perception of biodiversity, bird abundance in urban woodlands, and green corridors as movement conduits for butterflies.

Uta Heiden is a Research Scientist with the Department of Land Applications of the German Aerospace Center (DLR). Her research centers on the application of imaging spectroscopy for urban areas and brown fields. Currently, she is involved in the development of the ground segment for the forthcoming hyperspectral EnMAP satellite mission.

Wieke Heldens is a Research Scientist with the Department of Land Applications of the German Aerospace Center (DLR). Her research focuses on the application of hyperspectral remote sensing data to support urban planning and urban microclimate analysis.

Jeffrey Hepinstall-Cymerman is an Assistant Professor in the Warnell School of Forestry and Natural Resources at the University of Georgia. His research centers on using geospatial data to model urban development and ecological phenomena including wildlife habitat and the effects of future urban development and climate change upon ecological systems.

James B. Holt is a Geographer at the Centers for Disease Control and Prevention in Atlanta, Georgia. His research focuses on spatial analysis of public health data for epidemiology, public health policy, and program planning. He was instrumental in establishing the CDC Geography and Geospatial Science Working Group.

Hermann Kaufmann is Head of Department 1 – Geodesy and Remote Sensing of the GFZ German Research Centre for Geosciences and holds a chair at the University of Potsdam. His major scientific experiences are in the fields of data processing, sensor definition and applications dedicated to various disciplines. He is the scientific leader of the forthcoming EnMap hyperspectral satellite program.

Jonathan Li is Professor of Geomatics at the University of Waterloo, Canada. His research interests include remote sensing, mobile mapping, and geographic information systems. He has published five books and over 150 papers. He is Vice Chair of ICA Commission on Mapping from Satellite Imagery and Chair of ISPRS Intercommission Working Group V/I on Land-based Mobile Mapping Systems.

Gianni Lisini is with IUSS, Centre for Risk and Security, Pavia, Italy. His research centers on high-resolution SAR remote sensing of urban areas and the development of methods to extract different kinds of objects. He has published more than 50 articles.

Hua Lu is a Geographer at the Division of Adult and Community, Centers for Disease Control and Prevention (CDC). Her research interest centers on spatial analysis of public health data.

Jeremy Mennis is an Associate Professor of Geography and Urban Studies at Temple University. His research interests are in geographic information science and its application in the social and health sciences. He has served as Chair of AAG Geographic Information Systems and Science Specialty Group and on the UCGIS Board of the Directors.

Victor Mesev is a Professor of Geography at Florida State University. His research focuses on the analytical interface between geographic information systems and remote sensing, particularly for measuring and modeling urban growth and density patterns. He is author of over 50 publications.

Ulla Mörtberg is an Assistant Professor in Land and Water Resources Engineering, Department of Land and Water Resources Engineering, Royal Institute of Technology, Sweden. Her research concerns urban landscape ecology, environmental systems analysis, and GIS-based spatial modeling.

Soe W. Myint is an Associate Professor in the School of Geographical Sciences and Urban Planning at Arizona State University. His research interests include remote sensor data analysis, geostatistical modeling, data mining, and pattern recognition. He currently serves as Chair of the Remote Sensing Specialty Group (RSSG) at the Association of American Geographers (AAG).

Daniel Orenstein is a Researcher with the Center for Urban and Regional Planning at the Technion – Israel Institute of Technology. His research interests include population and environment interactions, environmental implications of urban spatial growth, and interdisciplinary approaches to long-term socioecological research.

Diego Aldo Polli is a doctoral student at the University of Pavia, Italy. His research interests include SAR data processing, earthquake damage assessment, and seismic vulnerability evaluation.

Rebecca L. Powell is an Assistant Professor of Geography at the University of Denver. Her research interests include applications of remote sensing to quantitatively assess ecological properties of land cover and to characterize the physical transformation of landscapes through time. In particular, her work has focused on characterizing urban ecosystems and vegetation structure in tropical savannas.

Sigrid Roessner is a Senior Research Scientist in the Remote Sensing Section at the GFZ German Research Centre for Geosciences. Her research interests include hyperspectral remote sensing of the urban environment and satellite remote sensing for natural hazard assessment.

Karl Segl is a Senior Research Scientist in the Remote Sensing Section at the GFZ German Research Centre for Geosciences. His research centers on methodological developments for hyperspectral data analysis, sensor design and validation.

Massimo Selva is with the Institute of Applied Physics "Nello Carrara" (IFAC-CNR) in Florence, Italy. His main scientific interests include multi-resolution image analysis, data fusion and image quality assessment.

Elisabete A. Silva is a University Lecturer in the Department of Land Economy and Fellow of Robinson College at the University of Cambridge. Her research centers on the application of new technologies to spatial urban planning. She is a Fellow of the Royal Institution of Chartered Surveyors (FRICS).

Mattia Stasolla is a Research Engineer at the Microwave Laboratory, University of Pavia, Italy. Her research interests include radar and optical data processing, mathematical morphology, fuzzy rule-based classifiers, neural networks, and applied remote sensing for risk and crisis management.

William L. Stefanov is a Senior Geoscientist with the Image Science & Analysis Laboratory at NASA Johnson Space Center. His research interest is with the application of remote sensing in geological and ecological studies, with a particular focus on urban areas. He has published over 40 articles.

Douglas Stow is a Professor of Geography at San Diego State University (SDSU). His research centers land cover change analyses, particularly for Mediterranean-type and Arctic tundra ecosystems, and major cities of developing countries. He is the Co-Director of the Center for Earth Systems Analysis Research. He has published over 110 refereed articles.

Daniel Sui is a Professor of Geography, Distinguished Professor of Social & Behavioral Sciences and Director of the Center for Urban & Regional Analysis (CURA) at the Ohio State University. His research interests include GIS-based spatial analysis and modeling, volunteered geographic information, legal and ethical issues of using geospatial technologies in society. He has published 4 books and over 100 articles. He was a 2009 Guggenheim Fellow. He is also a current member of the US National Mapping Science Committee and serves as Editor-in-Chief for *GeoJournal*.

Paul C. Sutton is an Associate Professor in the Department of Geography at the University of Denver. His research centers on the human–environment-sustainability problem. He has worked to demonstrate the potential of nighttime satellite imagery as a spatially explicit proxy measure of various human impacts on the environment.

Paul M. Torrens is an Associate Professor in the School of Geographical Sciences & Urban Planning at the Arizona State University. His research centers on GISci, development of geosimulation and geocomputation tools, applied modeling of complex urban systems, and new emerging cyberspaces. He earned a Presidential Early Career Award for Scientists and Engineers in 2008.

Giovanna Trianni is with the Institute for the Protection and Security of the Citizen, Joint Research Centre of the European Commission, Ispra, Italy. Her research centers on the use of optical and SAR satellite data to study the links between natural resources and armed conflicts and to analyze the damage caused by natural disasters.

Benjamin T. Tuttle is a doctoral student in the Department of Geography at the University of Denver. His research interests include human–environment interactions, the Geoweb, Cyberinfrastructure, and nighttime lights. His research has been published in various journals.

Boud Verbeiren is a doctoral researcher in the Department of Hydrology and Hydraulic Engineering at the Vrije Universiteit Brussel, Belgium. His research centers on the use of GIS and remote sensing in hydrological modeling.

Tim Van de Voorde is a Researcher in the Department of Geography at the Vrije Universiteit Brussel, Belgium. His research emphasizes the development of remote sensing for the study of urban land-use dynamics and environmental impacts of urbanization.

Le Wang is an Associate Professor of Geography at the State University of New York, Buffalo. His research centers on the use of remote sensing for population estimation, coastal mangrove mapping, and the study of the spread of invasive species. He has published more than 30 referred articles. He is the recipient of the 2008 Early Career Awards from the AAG Remote Sensing Specialty.

Changshan Wu is an Associate Professor of Geography at the University of Wisconsin-Milwaukee. His research interests include geographic information science and remote sensing with applications in urban development, population estimation, housing studies, and transportation analysis. He is the author of more than 20 papers.

Yichun Xie is a Professor of Geography and Founding Director of Institute for Geospatial Research and Education at Eastern Michigan University. His research centers on GISci, dynamic urban modeling, spatial decision support system, and China. He is the author of 1 book and over 80 papers. He is recipient of One Hundred Distinguished Overseas Scholars from the Chinese Academy of Sciences.

Xiaojun Yang, Editor of this volume, is with the Department of Geography at Florida State University. His research focuses on the development of geospatial science and technologies to support geographic inquiries in urban and environmental domains. He has published over 80 articles. He currently serves as Chair of the Commission on Mapping from Satellite Imagery, International Cartographic Association.

Xining Yang is a PhD student in the Department of Geography, Ohio State University. His research interests include geography, computer science and statistics.

Fei Yuan is an Associate Professor at the Minnesota State University, Mankato. Her research interests include land use/cover change, urban growth monitoring, urban impervious mapping, and urban environmental analysis. Her research has been published in various journals.

Joseph A. Zehnder is a Professor of Atmospheric Science at Creighton University. His research centers on dynamic meteorology. He has published widely on the formation and motion of tropical cyclones, energetics of the air–sea interface and the urban boundary layer, and the transition from shallow to deep convection in continental tropical cumulus.

Tingting Zhao is an Assistant Professor in the Department of Geography at Florida State University. Her research centers on spatial inventory of carbon emissions and vegetation carbon sinks; and assessment of human carbon impacts, especially from settlement development and energy consumption. Her research has been published in various journals.

Libin Zhou is a doctoral student in the department of Geography at Florida State University. Her research interests include GIS and remote sensing with applications in the urban environment.

Daniel C. Ziskin is a Research Associate with the Cooperative Institute for Research in Environmental Sciences at the University of Colorado. He is in the Earth Observing Group of the NOAA National Geophysical Data Center, enhancing the scientific value of the DMSP Nighttime Lights data set.

PREFACE

Remote sensing has traditionally been the colony of earth scientists and national security communities, and urban questions have been largely marginalized. With the recent innovations in data, technologies, and theories in the broad arena of Earth Observation, urban remote sensing, or urban applications of remote sensing, has rapidly gained the popularity across a wide variety of communities, such as urban planners, geographers, environmental scientists, and global change researchers. This surge of interest in urban remote sensing has been predominately driven by the need to derive critical urban information from remote sensing in support of various scientific inquiries and urban management activities.

The development of urban remote sensing has prompted much interest from the academics, and dedicated scholarly forums on urban remote sensing began to appear in 1995 when the European Science Foundation sponsored a specialist meeting on remote sensing and urban analysis. This meeting featured the research conducted by 16 invited scholars mostly from Europe, with a clear focus on interpreting urban physical structure and land use. This European-style urban remote sensing research framework has dominated the two subsequent major urban remote sensing forums: International Symposia on Remote Sensing of Urban Areas sponsored by the International Society for Photogrammetry and Remote Sensing (ISPRS) and Workshops on Remote Sensing and Data Fusion over Urban Areas jointly sponsored by Geoscience and Remote Sensing Society and ISPRS. Since 2005, the two forums have colocated to form a joint event that was officially named "Joint Urban Remote Sensing Event" in 2007.

In the United States, I began to organize special paper sessions on remote sensing and geographic information systems (GIS) for urban analysis at the annual meetings of the Association of American Geographers (AAG) since 2000. In addressing the multidisciplinary needs, several major areas have been identified as the session themes, including remote sensor data requirements for urban areas, development of digital image processing techniques for urban feature extraction, deriving urban socioeconomic indicators by remote sensing and spatial analysis, assessment of environmental consequences of urbanization by remote sensing, urban and landscape modeling using remote sensor data, urban change case studies, interface between remote sensing and urban geography, and urban remote sensing education. Sponsored by AAG Remote Sensing, GIS and Urban Geography Specialty Groups, these urban remote sensing conference sessions have been well received. More than 100 papers have been presented during the past 10 years, which featured the research conducted by some well-established urban remote sensing scholars, quite a few rising stars in urban remote sensing and GIS, as well as a large number of doctoral students predominately from U.S. universities. The Remote Sensing and GIS for Urban Analysis Special Paper Session has therefore become a major urban remote sensing forum in the United States.

The above forums have led to the publication of at least eight theme issues on urban remote sensing by virtually all major remote sensing journals during the last decade, along with at least ten books with urban remote sensing as the subject. While urban remote sensing is rapidly emerging as a major field of study receiving more attention than ever, there was no book with a broad vision on urban remote sensing research that resembles the themes formulated by myself for the urban analysis special paper sessions. Most of the published books were restricted on extracting urban features and interpreting land use using various remote sensing systems and digital image processing techniques. They offer little insights on the synergistic use of remote sensing and relevant geospatial techniques for deriving socioeconomic and environmental indicators in the urban environment and for modeling the spatial consequences of past, current and future urban development.

Within the above context, a broad-vision book on urban remote sensing is timely. This book examines how the modern concepts, technologies, and methods in remote sensing can be effectively used to solve problems relevant to a wide range of topics extending beyond urban feature extraction into urban socioeconomic and environmental analyses and predictive modeling of urbanization. The book is divided into six major parts. The first part introduces a broad vision of urban remote sensing research that draws upon a number of disciplines to support monitoring, synthesis and modeling in the urban environment. The second and third parts review the advances in remote sensors and image processing techniques for urban attribute information extraction. The fourth and fifth parts showcase some latest developments in the synergistic use of remote sensing and relevant geospatial techniques for developing urban socioeconomic and environmental indicators. The last part examines the developments of remote sensing and dynamic modeling techniques for simulating and predicting urban growth and landscape changes.

This book is the result of extensive research by interdisciplinary experts, and will appeal to students, researchers and professionals dealing with not only remote sensing, geographic information systems and geocomputation but also urban planning, geography, environmental science and global change science. The Editor is grateful to all of those who contributed papers and revised their papers one or more times and those who reviewed papers according to my requests and timelines. The group of reviewers who contributed their time, talents, and energies is listed here: John Agnew, Li An, Gilad Bino, Alexander Buyantuyev, Jin Chen, Mang Lung Cheuk, Galina Churkina, Joshua Comenetz, Helen Couclelis, Mike de Smith, Manfred Ehlers, Michael Einede, Thomas Gillespie, Jack Harvey, John E. Hasse, Gary Higgs, Zhirong Hu, Minhe Ji, Xiaoyan Jiang, Byong-Woon Jun, Niina Käyhkö, Verda Kocabas, Mike Lackner, Chun-Lin Lee, Alexandre Leroux, Noam Levin, Peijun Li, Arika Ligmann-Zielinska, Yangrong Ling, Xiaohang Liu, Dengsheng Lu, Yasunari Matsuno, Xueliang Meng, David O'Sullivan, Fabio Pacifici, Amy Pocewicz, Ruiliang Pu, Dale Quattrochi, Tarek Rashed, Andrea Sarzynski, Conghe Song, Haider Taha, Junmei Tang, Céline Tison, Tim van de Voorde, Uwe Weidner, Cédric

Wemmert, Alan Wilson, Bev Wilson, Changshan Wu, Zhixiao Xie, and Weiqi Zhou. This book project would not have been completed without the help and assistance from several staff members at John Wiley & Sons Ltd, especially Liz Renwick, Fiona Woods, Izzy Canning, and Sarah Karim. Acknowledgements are due to Ting Liu and Daniel Sui for their help, to my wife Xiaode Deng and my son Le Yang for their patience and love, to Dr. James O. Wheeler (1938–2010) who inspired me to pursue my passion on the urban environment, and to Dr. Chor-Pang (C.P.) Lo (1939–2007) who offered brilliant guidance and boundless encouragement over many years of my professional career.

Xiaojun Yang
Tallahassee, Florida, USA
2011

i

INTRODUCTION

This introductory part discusses the rationale and motivation leading to the development of remote sensing for urban studies, emphasizing the need to adopt a broad vision on urban remote sensing research. It discusses some major benefits and possible challenges of using remote sensing for urban studies, and provides an overview on the book structure and a topic-by-topic preview. It also identifies several conceptual or technical areas that need further attentions.

1
What is urban remote sensing?

Xiaojun Yang

This introductory chapter defines the scope of urban remote sensing research. It begins with a discussion on the rationale leading to the development of remote sensing for urban studies and the motivation behind this book project emphasizing the need to adopt a broad vision on urban remote sensing research. It then discusses the benefits and possible challenges of using remote sensing for urban studies, followed by an overview of the major topics discussed in the book. Finally, the chapter highlights several areas that need further attention.

Urban Remote Sensing: Monitoring, Synthesis and Modeling in the Urban Environment, First Edition. Edited by Xiaojun Yang.
© 2011 John Wiley & Sons, Ltd. Published 2011 by John Wiley & Sons, Ltd.

1.1 Introduction

Remote sensing is the art, science and technology of acquiring information about physical objects and the environment through recording, measuring and interpreting imagery and digital representations of energy patterns derived from noncontact sensors (Colwell, 1997). Remote sensing has traditionally been the colony of earth scientists and national security communities and urban questions have been largely marginalized (Sherbinin et al., 2002).

With recent innovations in data, technologies, and theories in the wider arena of Earth Observation, urban remote sensing, or urban applications of remote sensing, has rapidly gained the popularity among a wide variety of communities. First, urban and regional planners are increasingly using remote sensing to derive information on the urban environment in a timely, detailed and cost-effective way to accommodate various planning and management activities (e.g., Sugumaran, Zerr and Prato, 2002; Alberti, Weeks and Coe, 2004; Mittelbach and Schneider, 2005; Santana, 2007; Bhatta, 2010). Second, more urban researchers are using remote sensing to extract urban structure information for studying urban geometry, which can help develop theories and models of urban morphology (e.g., Batty and Longley, 1994; Longley, 2002; Herold, Scepan and Clarke, 2002; Yang, 2002; Lo, 2004, 2007; Rashed et al., 2005; Batty, 2008; Schneider and Woodcock, 2008). Third, environmental scientists are increasingly relying upon remote sensing to derive urban land cover information as a primary boundary condition used in many spatially distributed models (e.g., Lo, Quattrochi and Luvall, 1997; Lo and Quattrochi, 2003; Arthur-Hartranft, Carlson and Clarke, 2003; Carlson, 2004; Stefanov and Netzband, 2005; Hepinstall, Alberti and Marzluff, 2008). Lastly, the global change community has recognized remote sensing as an enabling and acceptable technology to study the spatiotemporal dynamics and consequences of urbanization as a major form of global changes (e.g., Bartlett, Mageean and O'Connor, 2000; Small and Nicholls, 2003; Auch, Taylor and Acededo, 2004; Small, 2005; Turner, Lambin and Reenberg, 2007; Grimm et al., 2008), given the facts that more than half of the global population are now residing in cities (UN-HABITAT, 2010) and urban areas are the home of major global production and manufacture centers (Kaplan, Wheeler and Holloway, 2009).

The development in urban remote sensing has prompted much interest from academics, and dedicated scholarly forums on urban remote sensing began to appear in 1995 when the European Science Foundation (ESF) sponsored a specialist meeting on remote sensing and urban analysis as part of its GISDATA Programme. This meeting featured the research conducted by 16 invited scholars mostly from Europe, with the exception of Michael Batty and C.P. Lo. Batty, a British scholar and an urban modeling pioneer, was with the State University of New York at Buffalo (USA) during 1990–1995; Lo, a British-trained scholar and a pioneer in urban remote sensing, was with the University of Georgia (USA) from 1984 to 2007. The papers presented at the ESF-sponsored event largely centered on interpreting urban physical structure and land use (see Donnay, Barnsley and Longley, 2001). This European-style urban remote sensing research framework has dominated the two subsequent major urban remote sensing forums: International Symposia on Remote Sensing of Urban Areas (since 1979) sponsored by the International Society for Photogrammetry and Remote Sensing (ISPRS) and Workshops on Remote Sensing and Data Fusion over Urban Areas (since 2001) jointly sponsored by Geoscience and Remote Sensing Society (GRSS) and ISPRS. Since 2005 the two forums have colocated to form a joint event that was officially named the "Joint Urban Remote Sensing Event (JURSE)" in 2007.

In the United States, the author began to organize special paper sessions on remote sensing and geographic information systems (GIS) for urban analysis at the annual meetings of the Association of American Geographers (AAG) since 2000. In addressing the multidisciplinary needs, several major areas have been identified as the session themes, which include remote sensor data requirements for urban areas, development of digital image processing techniques for urban feature extraction, deriving urban socioeconomic indicators by remote sensing and spatial analysis, assessment of environmental consequences of urbanization by remote sensing, urban and landscape modeling using remote sensor data, urban change case studies, interface between remote sensing and urban geography, and urban remote sensing education.

Sponsored by AAG's Remote Sensing, GIS and Urban Geography Specialty Groups, these urban remote sensing conference sessions have been well received. More than 100 papers have been presented during the past 10 years, which featured the research conducted by some well-established urban remote sensing scholars, quite a few rising stars in urban remote sensing and GIS, as well as a large number of doctoral students predominately from American universities. The Remote Sensing and GIS for Urban Analysis Special Paper Session has therefore become a major urban remote sensing forum in the United States.

The above forums have led to the publication of at least eight theme issues on urban remote sensing by virtually all major remote sensing journals during the last decade, along with at least ten books with urban remote sensing as the subject (Yang, 2009). While urban remote sensing is rapidly emerging as a major field of study receiving more attention than ever, there was no any book with a broad vision on urban remote sensing research that resembles the themes formulated by the author for the urban analysis special paper sessions. Most of the published books were restricted to extracting urban feature and interpreting land use using various remote sensing systems and digital image processing techniques. They offer little insights on the synergistic use of remote sensing and spatial data analysis techniques for deriving socioeconomic and environmental indicators in the urban environment and for modeling the spatial consequences of past, current and future urban development.

Within the above context, a broad vision book on urban remote sensing research is timely. Designed for both the academic and business sectors, this book examines how the modern concepts, technologies and methods in remote sensing can be creatively used to solve problems relevant to a wide range of topics extending beyond urban feature extraction into two core inquiring areas in urban studies, i.e., urban socioeconomic and environmental analyses and predictive modeling of urbanization. Specifically, the book covers the following major aspects (Fig. 1.1):

- Introduces a broad vision of urban remote sensing research that draws upon a number of disciplines to support monitoring, synthesis and modeling in the urban environment;

- Reviews the advances in remote sensors and image processing techniques for urban attribute information extraction;

- Examines some latest developments in the synergistic use of remote sensing and other types of geospatial information for

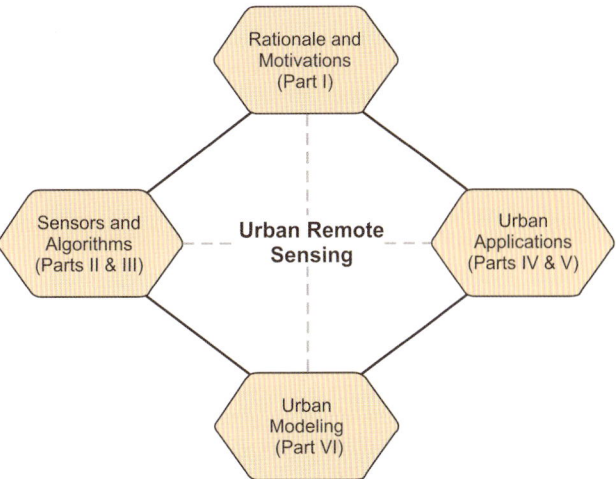

FIGURE 1.1 A graphic overview of the book structure.

developing urban socioeconomic and environmental indicators; and

- Examines the developments of remote sensing and dynamic modeling techniques for simulating and predicting urban growth and landscape changes.

In addition to scientific research, the book has incorporated a management component that can be particularly found in the chapters discussing urban socioeconomic and environmental analyses and predictive modeling or urbanization. Cutting-edge remote sensing research helps improve our understanding of the status, trends and threats in the urban environment; such knowledge is critical for formulating effective strategies towards sustainable urban planning and management.

Unlike most edited books with a contributing author pool from a single event, this book is written by a carefully selected group of interdisciplinary scholars:

- Researchers who presented a scholarly paper in an urban remote sensing session the author has organized at the annual meetings of the Association of American Geographers (AAG) since 2000;
- Researchers who recently presented a scholarly paper at a Joint Urban Remote Sensing Event;
- Some active researchers largely identified from their recent presentations at several other remote sensing conferences (e.g., annual meetings of American Society for Photogrammetry and Remote Sensing or International Geoscience and Remote Sensing Symposium); and
- A small number of other world-class scholars in remote sensing, geocomputation, urban studies, geography, and environmental science.

A total of 59 authors from Belgium, Canada, Germany, Israel, Italy, Poland, Sweden, the United Kingdom, and the United States contribute to this book. Although this book is authored by US and European scholars with case studies predominately drawn from North America and Europe, the knowledge gained from these two regions can be applied to other urban areas globally.

The sections to be followed will discuss the benefits and possible challenges of using remote sensing for urban studies, provide an overview of the major topics discussed in the book, and highlight several areas that need further research.

1.2 Remote sensing and urban studies

The technology of modern remote sensing began with the invention of the camera more than 150 years ago, and by now a wide variety of remote sensing systems has been developed to detect and measure energy patterns from different portions of the electromagnetic spectrum. Remote sensing can help improve our understanding of urban areas in several ways, although the realistic potential for making these improvements is often challenged by the complexity in the urban environment.

Remote sensing provides several major benefits for urban studies. First, perhaps the largest benefit of remote sensing is its capability of acquiring photos or images that cover a large area, providing a synoptic view that allows identifying objects, patterns, and human-land interactions. This unique perspective is highly relevant to the interdisciplinary approach we advocate to study the urban environment in this volume since many urban processes are operating over a rather large area; failure in observing the entire mosaic of an urban phenomenon may hinder our ability to understand the potential processes behind the observed patterns.

Second, remote sensing provides additional measures for urban studies. Urban researchers frequently use data collected from field surveys and measurements. This way of data collection is considered to be accurate but can introduce potential errors due to the bias in sampling design (Jensen, 2007). Field measurements can become prohibitively expensive over a large area. Remote sensing can collect data in an unbiased and cost-effectiveness fashion. Moreover, remote sensors can measure energy at wavelengths which are beyond the range of human vision; remote sensor data collected from the ultraviolet, infrared, microwave portions of the electromagnetic spectrum can help obtain knowledge beyond our human visual perception. For example, thermal remote sensing can measure spatially continuous surface temperature that is useful to examine the urban heat island effect (e.g., Lo, Quattrochi and Luvall, 1997). Data fusion from different sensors can improve urban mapping and analysis (see Ch. 10).

Third, remote sensing allows retrospective viewing of Earth's surface, and time-series of remote sensor data can be used to develop a historical perspective of an urban attribute or process, which can help examine significant human or natural processes that act over a long time period. Examples in this volume include time-series land use/cover data that have been used to examine the suburbanizing process in the Atlanta metropolitan area over nearly the past four decades (Ch. 2); increasing gross primary production (GPP) that may be linked with vegetation carbon sequestration due to urban growth in the eastern United States (Ch. 19); historical land use changes affecting upon near-surface air temperature during recent extreme heat events in the Phoenix metropolitan area (Ch. 21); and urban growth and landscape changes affecting biodiversity in northern Washington (Ch. 25).

Fourth, remote sensing can help make connections across levels of analysis for urban studies. Urban science disciplines and subdisciplines have their own preferred levels of analysis and normally do not communicate across these levels. For example, urban planners tend to work at street and neighborhood levels; regional planners deal with a larger environment such as several counties, one or more metropolitan areas, or even a whole state; urban meteorologists and ecologists tend to work at levels defined by physiographical features or ecological units; and urban geographers tend to work at various levels depending upon specific topics under investigation. On the other hand, the temporal scales used by these different urban researchers vary greatly, from hourly, daily, weekly, monthly, seasonally, to annual or decadal basis. Remote sensing provides essentially global coverage of data with individual pixels ranging from submeters to a few kilometers and with varying temporal resolutions; such data can be combined to allow work at any scales or levels of analysis, appropriate to the urban phenomenon or process being examined. Therefore, remote sensing offers the potential for promoting urban researchers to think across levels of analysis and to develop theories and models to link these levels.

Last, remote sensing integrated with relevant geospatial technologies, such as geographic information systems, spatial analysis and dynamic modeling, offers an indispensible framework of monitoring, synthesis and modeling in the urban environment. Such frameworks support the development of a spatio-temporal perspective of urban processes or phenomena across different scales and the extension of historical and current observations into the future. They can also be used to relate different human and natural variables for developing an understanding of the indirect and direct drivers of urban changes and the potential feedbacks of such changes on the drivers in the urban environment.

Nevertheless, urban environments are complex by nature, challenging the applicability and robustness of remote sensing. The presence of complex urban impervious materials, along with a variety of croplands, grasslands and vegetation cover, causes substantial interpixel and intrapixel scenic changes, thus complicating the classification and characterization of urban landscape types. Moreover, it is always difficult to integrate remote sensor data with other types of geospatial data in urban social or environmental analyses because of the fundamental differences in data sampling and measurement. Some additional challenges will be addressed in the sections to be followed.

1.3 Remote sensing systems for urban areas

Remote sensor data used for urban studies should meet certain conditions in terms of spatial, spectral, radiometric, and temporal characteristics (Jensen and Cowen, 1999). There is a wide variety of passive and active remote sensing systems acquiring data with various resolutions that can be useful for urban studies. Medium-resolution remote sensor data have been used to examine large-dimensional urban phenomena or processes since early 1970s when Landsat-1 was successfully launched. With the launch of IKONOS, the world's first commercial, high-resolution imaging satellite, on September 24 1999, very-high-spatial-resolution satellite imagery became available, which allow detailed work concerning the urban environment. Independent of weather conditions, active remote sensing systems, such as airborne or space-borne radar, can be particularly useful for such applications as housing damage assessment or ground deformation estimation in connection to some disastrous events in urban areas. Another active sensor system, similar in some respects to radar, is lidar (light detection and ranging), which can be used to derive height information useful for reconstructing three-dimensional city models.

With five major chapters, Part II of this volume reviews some major advances in remote sensors that are particularly relevant for urban studies. It begins with a chapter (Ch. 2) discussing the utilities of medium-resolution satellite remote sensing for the observation and measurement of urban growth and landscape changes, emphasizing the use of the data from the Landsat imaging sensors. Over a period of nearly four decades, the Landsat program has acquired a scientifically valuable image archive unmatched in quality, details, coverage, and length, which has been the primary source of data for urbanization studies at the regional, national and global scales. The chapter comprises a moderate review on the past, present and future of the Landsat program and its imaging sensors, a case study focusing on a rapidly suburbanizing metropolis, and an extended discussion on some conceptual and technical issues emerging when using archival satellite images acquired by different sensors and perhaps during different seasons.

The other four chapters within Part II review the utilities of high-resolution optical and radar remote sensing, hyperspectral remote sensing, and lidar remote sensing for urban feature extraction. Chapter 3 discusses some major challenges and limitations when using very-high-resolution optical satellite imagery for monitoring human settlements, including geometric, spectral, classification, and change detection problems. Then, the authors propose an integrated spatial approach to deal with some of these problems, which is followed by a discussion of some interesting results using very-high-resolution satellite imagery for building damage assessment in connection to major earthquake events. Chapter 4 reviews the methodological development of urban hyperspectral remote sensing emphasizing the progress in developing an automated system for mapping urban surface materials. This system comprises an iterative procedure that involves field- and image-based spectral investigations to automatically derive quantitative spectral features that serve as the input information for a multi-step processing system. It allows detailed mapping of urban surface materials at a sub-pixel level and provides area-wide information about the fractional coverage of surface materials for each pixel. Chapter 5 discusses some new possibilities and challenges when using very-high-resolution spaceborne radar data for urban feature extraction. The authors compare airborne versus spaceborne radar data in terms of image geometry and other aspects that have been elaborated in connection to single building extraction, building damage assessment, and vulnerability mapping. They also discuss the suitability of adopting the algorithms and methods originally developed for processing high-resolution airborne radar data to spaceborne radar data. The last chapter (Ch. 6) included in Part II discusses the use of lidar remote sensing for three-dimensional building reconstruction. The chapter comprises a moderate review on lidar-based building extraction techniques and a detailed discussion on a comprehensive approach for automated creation of three-dimensional building models from airborne lidar point cloud data fused with aerial imagery.

1.4 Algorithms and techniques for urban attribute extraction

The urban environment is characterized by the presence of heterogeneous surface covers with large interpixel and intrapixel spectral variations, thus challenging the applicability and robustness of conventional image processing algorithms and techniques. Largely built upon parametric statistics, conventional pattern classifiers generally work well for medium-resolution scenes covering spectrally homogeneous areas, but not in heterogeneous regions such as urban areas or when scenes contain severe noises due to the increase of image spatial resolution. Developing improved image processing algorithms and techniques for working with different types of remote sensor data has therefore become a very active research area in urban remote sensing. For years, various strategies have been developed to improve urban mapping, and some of the most exciting developments are discussed in Part III.

The first three chapters in Part III are dedicated to a set of image processing techniques that can be used to improve urban mapping performance at the per-pixel, sub-pixel, or object levels. The first chapter (Ch. 7) discusses some algorithmic parameters affecting the performance of artificial neural networks in image classification at the per-pixel level. The chapter comprises a moderate review on the basic structure of neural networks, two focused studies with a satellite image covering an urban area to assess the sensitivity of image classification by neural networks in relation to various internal parameter settings and the performance of several training algorithms in image classification, and a discussion on a generic framework that can guide the use of neural networks in remote sensing. Chapter 8 reviews the spectral mixture analysis (SMA) technique that allows the decomposition of each pixel into independent endmembers or pure materials to map urban subpixel composition. It then discusses two case studies highlighting the flexibility of multiple endmember spectral mixture analysis (MESMA) to map vegetation, impervious and bare soil components. Chapter 9 provides an overview on the principles of object-based image analysis (OBIA) and demonstrates how the OBIA can be applied to achieve satisfactory urban mapping accuracy. Two case studies are conducted with Quickbird data to demonstrate two object-based analysis procedures, namely, decision rule and nearest neighbor classifiers.

The last two chapters included in Part III deal with two important aspects for urban mapping: image fusion technique (Ch. 10) and temporal lag between urban structure and function (Ch. 11). Chapter 10 reviews some advanced pan-sharpening algorithms and discusses their performance in terms of objective and visual quality. Chapter 11 examines the issue of temporal lag between when decisions are made to change a city to when these changes actually physically materialize. This seems to be an important issue for urban mapping. Yet it has been largely neglected in urban remote sensing literatures. The author explores the temporal lag largely from a conceptual perspective.

It should be noted that there are some other urban mapping techniques or methods that have been discussed in other chapters of this volume. For example, Chapter 2 (Part II) discusses a hybrid approach combining unsupervised classification and spatial reclassification that has been successfully used to produce accuracy-compatible land use/cover maps from a decades-long time series of satellite imagery acquired by the three Landsat imaging sensors. Chapter 3 discusses a filtering step built upon the use of some operators of mathematical morphology as part of an integrated adaptive spatial approach that can be used to improve urban mapping from very-high-resolution remote sensor data. Finally, due to the space limit, we are not able to cover some other pattern classification techniques that can be used to improve urban mapping, such as expert systems (*e.g.*, Stefanov, Ramsey and Christensen, 2001), support vector machines (e.g., Yang, 2011), or a fuzzy classifier (e.g., Shalan, Arora and Ghosh, 2003). Readers who are interested in learning more about these methods should refer to the references provided.

1.5 Urban socioeconomic analyses

Applying remote sensing to urban socioeconomic analyses has been an expanding research area in urban remote sensing. There are two major types of such analyses. The first type centers on linking socioeconomic data to land use/cover change data derived from remote sensing in order to identify the drivers of landscape changes (e.g., Lo and Yang, 2002; Seto and Kaufmann, 2003). The other type of analyses focuses on developing indicators of urban socioeconomic status by combined use of remote sensing and census or field-survey data (e.g., Lo and Faber, 1997; Yu and Wu, 2006). While some aspects relating to the first type of analyses will be addressed in Part VI, here we focus on some latest developments in the second type of urban socioeconomic analyses.

Part IV examines some latest developments in the synergistic use of remote sensing and other types of geospatial information for developing urban socioeconomic indicators. It begins with a chapter (Ch. 12) discussing a pluralistic approach to defining and measuring urban sprawl. This topic is included as part of urban socioeconomic analyses because defining urban sprawl involves not only urban spatial characteristics but also socioeconomic conditions such as population density and transportation. The chapter reviews the literature and debates on the definition of urban sprawl, emphasizing common themes in definitions and those measurable spatial characteristics that would be of specific interest to the remote sensing community. It shows that sprawl can be described by multiple quantitative measures but different sprawl measures can yield contradictory outcomes. The chapter suggests that sprawl should be best defined for a given case study and quantified using different indicators that can accommodate the researcher's definition of sprawl, spatial scale of analysis, and specific characteristics of the study site.

The remaining four chapters in Part IV focus on population estimation (Ch. 13), dasymetric mapping (Ch. 14), electrification rate estimation (Ch. 15), and environmental justice research (Ch. 16). Chapter 13 discusses a method for small area population estimation by combined use of high-resolution imagery with lidar data. This type of information is critical for decision-making by both public and private sectors but is only available for one date per decade. The work has provided an alternative that can be used to derive reliable population estimation in a timely and cost-effective fashion. Chapter 14 reviews various areal interpolation techniques emphasizing dasymetric mapping, followed by an example in which population estimates and sociodemographic data are derived for different spatial units by using dasymetric mapping methods that rely upon ancillary data from a variety of sources including satellite imagery. Chapter 15 discusses a method that has been developed to estimate the global percent population having electric power access based on the presence of satellite detected night-time lighting. The satellite-derived results are pretty close to the reported electrification rates by the International Energy Agency. The last chapter (Ch. 16) included in Part IV discusses the role of remote sensing for urban environmental justice research. The chapter comprises a review on the principles and issues in environmental justice research, a case study investigating the relationship between a satellite-derived vegetation index and indicators of race and socioeconomic status in Philadelphia, and a discussion on some issues that need further research.

1.6 Urban environmental analyses

Although urban areas are quite small relative to the global land cover, they significantly alter hydrology, biodiversity, biogeochemistry, and climate at local, regional, and global scales (Grimm *et al.*, 2008). Understanding environmental consequences of urbanization is a critical concern to both the planning (Alberti, Weeks and Coe, 2004) and global change science communities (Turner, Lambin and Reenberg, 2007). Urban environmental analyses can help understand the status, trends, and threats in urban areas so that appropriate management actions can be planed and implemented. This is a research area in which remote sensing can play a critical role.

Part V (Chs 17–21) reviews the latest developments in the synergistic use of remote sensing and relevant geospatial techniques for urban environmental analyses. Chapter 17 discusses a remote sensing approach to high-resolution urban impervious surface mapping. This topic is included as part of urban environmental analyses because landscape imperviousness has recently emerged as a key indicator being used to address a variety of urban environmental issues such as water quality, biodiversity of aquatic systems, habitat structure, and watershed health (Yang and Liu, 2005). The chapter reviews some major pixel-based and object-based techniques for impervious surface estimation, and compares the performance of the two groups of methods with case studies.

The other chapters included in Part V deal with urban hydrological processes (Ch. 18), vegetation carbon sequestration (Ch. 19), biodiversity (Ch. 20), and air quality and climate (Ch. 21). Chapter 18 discusses the impact of different remote sensing methods for characterizing the distribution of impervious surfaces on runoff estimation, and how this can affect the assessment of peak discharges in an urbanized watershed. The study shows that detailed information on the spatial distribution of impervious surfaces strongly affects local runoff estimation and has a clear impact on the modeling of peak discharges. Chapter 19 reviews the light-use efficiency (LUE) models and applied them to estimate gross primary production (GPP) in the eastern United States in two different years. The estimated GPP was associated with various settlement densities. The LUE-based vegetation productivity estimates may be integrated with carbon emissions data, thus providing a comprehensive view of net carbon exchange between land and the atmosphere due to urban development. Chapter 20 discusses the utilities of remote sensing for characterizing biodiversity in urban areas, how urbanization affects biodiversity, and how remote sensing-based biodiversity research can be integrated with urban planning and management for biodiversity conservation. The last chapter (Ch. 21) in Part V reviews the existing literature concerning the influence of urban land use/cover changes on urban meteorology, climate and air quality. This is followed by a case study focusing on the Phoenix metropolitan area to demonstrate how remote sensing can be used to study the effect of historic land use changes on near-surface air temperature during recent extreme heat events.

1.7 Urban growth and landscape change modeling

A group of important activities in urban studies is to understand urban dynamics and to assess future urban growth impacts on the environment. There are two major types of models that can be used to support such activities: analytical models that are useful to explain urban expansion and evolving patterns as well as dynamic models that can be used to predict future urban growth and landscape changes. Here we direct our attention to the second type of models because of their predictive power that can be used to imagine, test and assess the spatial consequences of urban growth under specific socioeconomic and environmental conditions. The role of remote sensing is indispensable in the entire model development process from model conceptualization to implementation that includes input data preparation, model calibration, and model validation (Lo and Yang, 2002; Yang and Lo, 2003).

Part VI (Chs 22–26) examines the developments of remote sensing and dynamic modeling techniques to simulate and predict urban growth and landscape changes. The first four chapters deal with the three major types of dynamic modeling techniques, i.e., cellular automata modeling, agent-based modeling, and ecological modeling, while the last chapter shifts the discussion from technical aspects to the underlying root metaphors embedded in various modeling efforts. Chapter 22 explores how the developments in remote sensing, together with advances in physics, mathematics, chemistry and computer science, contributed to the exploration of urban complexity. It discusses the evolution from pixel-matrix structures towards cell and agent-based models and the challenge of integrating spatial and a-spatial data structures and models. The chapter then proposes a new data structure and modeling approach in which remote sensing can play an important role. Chapter 23 reviews the developments in calibration and validation of urban cellular automata models emphasizing on models tasked with simulating human–environment interactions. It discusses calibration mechanisms and the derivation of calibration parameter values. It then moves to the consideration of model validation routines and various procedures for sweeping the parameter space of models. Chapter 24 introduces the agent-based urban modeling technique, followed by a case study to demonstrate the utilities of this type of modeling technique for urban growth and landscape change simulation. Chapter 25 discusses the utilities of ecological modeling for predicting changes in biodiversity in response to future urban development, emphasizing an integrated modeling environment that can predict future land cover, estimate biodiversity, and link the output from land cover change models into models estimating biodiversity.

The last chapter in Part VI (Ch. 26) provides a comprehensive review on the progress in urban modeling during the past 50 years, emphasizing the underlying root metaphors embedded in various modeling efforts. It considers the four major urban modeling traditions, i.e., spatial morphology, social physics, social biology, and spatial events. The chapter argues that the root metaphors embedded in these traditions correspond to those in Pepper's world hypotheses – the world as forms, machines, organism, and arenas. The author believes that urban modeling progress is actually a shift of metaphors used for conceptualizing cities, and we need to pay attention on the process whereby meaning is produced from metaphor to metaphor, rather than between model and the world. In this regard, we should not only check the validity of our models from the technical perspective but also examine the driving conceptual metaphors deeply embedded in the models. Only then can we weave the insights gained from the urban modeling efforts with other urban narratives to have a more sensible urban future.

Summary and concluding remarks

This chapter discusses the rationale and motivation leading to the development of remote sensing for urban studies emphasizing the need to adopt a broad vision on urban remote sensing research in order to address the multidisciplinary needs. It discusses some major benefits and possible challenges in urban remote sensing research. Moreover, the chapter provides an overview on the book structure and a topic-by-topic preview.

While many exciting progresses have been made in urban remote sensing, as discussed in this volume, there are several major conceptual or technical issues that deserve further attentions. First, while the development of urban remote sensing has been largely technology driven, urban remote sensing professionals should be equipped with not only solid technical skills but also essentials of intellectual knowledge on the urban environment, including relevant core concepts, theoretical debates, and emerging methods; such knowledge can help better plan and implement an urban remote sensing project, as indicated by some recent literatures (e.g., Yang and Lo, 2003; Lo, 2004; Dietzel *et al.*, 2005) and several chapters included in this volume (Ch. 2, Ch. 11, Ch. 12, and Ch. 26).

Second, although the issue of remote sensor data resolutions has been extensively discussed in various remote sensing literatures, there is no consistent guidance on the choice of image resolutions. While the current literature overwhelmingly focusses on the issue of image spatial resolution, recent studies suggest the importance of image spectral characteristics in urban feature mapping (e.g., Herold *et al.*, 2004; Ch. 2, Ch. 4). Continuing research is needed to help acquire good and sufficient *in situ* data for building comprehensive spectral libraries of different urban features that can help improve urban feature mapping accuracy and to develop practical guidance on the choice of image resolutions that should not only consider the spatial component but also the spectral, radiometric, and temporal characteristics.

Third, an emerging research effort is needed to balance the different needs by remote sensing and urban planning communities. Within the remote sensing community, there is an increasing research demand to develop improved methods and techniques for working with medium-resolution images covering spectrally heterogeneous areas (such as urban areas) and with high-resolution images; some exciting developments in these aspects have been reported in this volume. Within the urban and regional planning community, on the other hand, there is an urgent need to operationalize the advanced information extraction techniques or procedures that have been recently developed by the remote sensing community so that they can be widely used to support various urban applications.

Fourth, data and technological integration play a key role in urban remote sensing research, particularly for urban socioeconomic and environmental analyses and predictive modeling of urban growth and landscape changes. More efforts are needed to develop innovative data models used for representing dynamic processes, to identify improved methods and techniques that can be used to deal with data incompatibility in terms of parameter measuring and sampling schemes, and to develop more realistic predictive models that can be used to support various urban and regional planning activities.

Finally, with a broad vision on urban remote sensing research, this book advocates an interdisciplinary approach to the study of urban environments. We need to understand not only urban structure and patterns but also the underlying processes, consequences, and possible feedbacks. To this end, conceptualizing cities as a complex ecosystem can be very helpful. The success of implementing this approach depends upon not only technological soundness in remote sensing but also intensive research collaboration from interdisciplinary experts and broad partnerships including virtual communities as well.

References

Alberti, M., Weeks, R., and Coe, S. (2004) Urban land-cover change analysis in Central Puget Sound. *Photogrammetric Engineering and Remote Sensing*, **70**, 1043–1052.

Arthur-Hartranft, S.T., Carlson, T.N., and Clarke, K.C. (2003) Satellite and ground-based microclimate and hydrologic analyses coupled with a regional urban growth model. *Remote Sensing of Environment*, **86**, 385–400.

Auch, R., Taylor, J. and Acededo, W. (2004) *Urban Growth in American Cities: Glimpses of US Urbanization*. USGS Circular 1252. 52p.

Batty, M. (2008) The size, scale, and shape of cities. *Science*, **319**, 769–771.

Batty, M. and Longley, P. (1994) *Fractal Cities: A Geometry of Form and Function*. Academic Press, London.

Bartlett, J.G., Mageean, D.M., and O'Connor, R.J. (2000) Residential expansion as a continental threat to US coastal ecosystems. *Population and Environment*, **21**, 429–468.

Bhatta, B. (2010). *Analysis of Urban Growth and Sprawl from Remote Sensing Data*, Springer-Verlag, Berlin, Heidelberg.

Carlson, T.N. (2004) Analysis and prediction of surface runoff in an urbanizing watershed using satellite imagery. *Journal of the American Water Resources Association*, **40**, 1087–1098.

Colwell, R.N. (1997) History and place of photographic interpretation, in *Manual of Photographic Interpretation* (2nd) (ed W.R. Phillipson), ASPRS, Bethesda, pp. 33–48.

Dietzel, C., Herold, M., Hemphill, J.J., and Clarke, K.C. (2005) Spatio-temporal dynamics in California's central valley: Empirical links to urban theory. *International Journal of Geographical Information Science*, **19**, 175–195.

Donnay, J.P., Barnsley, M.J. and Longley, P.A. (eds) (2001) *Remote Sensing and Urban Analysis*. CRC Press, Boca Raton.

Grimm, N.B., Faeth, S.H., Golubiewski, *et al.* (2008) Global change and ecology of cities. *Science*, **319**, 756–760.

Herold, M., Roberts, D.A., Gardner, M.E., and Dennison, P.E. (2004) Spectrometry for urban area remote sensing – Development and analysis of a spectral library from 350 to 2400 nm. *Remote Sensing of Environment*, **91**, 304–319.

Herold, M., Scepan, J., and Clarke, K.C. (2002) The use of remote sensing and landscape metrics to describe structures and changes in urban land uses. *Environment and Planning A*, **34**, 1443–1458.

Hepinstall, J.A., Alberti, M., and Marzluff, J.M. (2008) Predicting land cover change and avian community responses in rapidly urbanizing environments. *Landscape Ecology*, **23**, 1257–1276.

Jensen, J.R. (2007) *Remote Sensing of the Environment: an Earth Resource Perspective*, Prentice Hall, New Jersey.

Jensen, J.R. and Cowen, D.C. (1999) Remote sensing of urban suburban infrastructure and socio-economic attributes. *Photogrammetric Engineering and Remote Sensing*, **65**, 611–622.

Kaplan, D.H., Wheeler, J.O., and Holloway, S.R. (2009) *Urban Geography*, 2nd edition, John Wiley & Sons, Inc., New York.

Lo, C.P. (2004) Testing urban theories using remote sensing. *GIScience & Remote Sensing*, **41**, 95–115.

Lo, C.P. (2007) The application of geospatial technology to urban morphological research. *Urban Morphology*, **11**, 81–90.

Lo, C.P. and Faber, B.J. (1997) Integration of Landsat Thematic Mapper and census data for quality of life assessment. *Remote Sensing of Environment*, **62**, 143–157.

Lo, C.P., Quattrochi, D.A., and Luvall, J.C. (1997) Application of high-resolution thermal infrared remote sensing and GIS to assess the urban heat island effect. *International Journal of Remote Sensing*, **18**, 287–304.

Lo, C.P. and Quattrochi, D.A. (2003) Land-use and land-cover change, urban heat island phenomenon, and health implications: A remote sensing approach. *Photogrammetric Engineering and Remote Sensing*, **69**, 1053–1063.

Lo, C.P. and Yang, X. (2002) Drivers of land-use/land-cover changes and dynamic modeling for the Atlanta, Georgia Metropolitan Area. *Photogrammetric Engineering and Remote Sensing*, **68**, 1073–1082.

Longley, P.A. (2002) Geographical information systems: will developments in urban remote sensing and GIS lead to 'better' urban geography? *Progress in Human Geography*, **26**, 231–239.

Mittelbach, F.G. and Schneider, M.I. (2005) Remote sensing: With special reference to urban and regional transportation. *Annals of Regional Science*, **5**, 61–72.

Rashed, T., Weeks, J.R., Stow, D. and Fugate, D. (2005) Measuring temporal compositions of urban morphology through spectral mixture analysis: toward a soft approach to change analysis in crowded cities. *International Journal of Remote Sensing*, **26**, 699–718.

Santana, L.M. (2007) Landsat ETM+ image applications to extract information for environmental planning in a Colombian city. *International Journal of Remote Sensing*, **28**, 4225–4242.

Schneider, A. and Woodcock, C.E. (2008) Compact, dispersed, fragmented, extensive? A comparison of urban growth in 25 global cities using remotely sensed data, pattern metrics and census information. *Urban Studies*, **45**, 659–692.

Seto, K.C. and Kaufmann, R.K. (2003) Modeling the drivers of urban land use change in the Pearl River Delta, China: Integrating remote sensing with socioeconomic data. *Land Economics*, **79**, 106–121.

Shalan, M.A., Arora, M.K. and Ghosh, S.K. (2003) An evaluation of fuzzy classifications from IRS 1C LISS III imagery: a case study. *International Journal of Remote Sensing*, **24**, 3179–3186.

Sherbinin, A.D., Balk, D., Jaiteh, M. *et al.* (2002) A CIESIN Thematic Guide to Social Science Applications of Remote Sensing. Available http://sedac.ciesin.columbia.edu/tg/guide_frame.jsp?rd=RS&ds=1 (accessed 10 July 2010).

Small, C. (2005) Urban remote sensing: Global comparisons. *Architectural Design*, **178**, 18–23.

Small, C. and Nicholls, R.J. (2003) A global analysis of human settlement in coastal zones. *Journal of Coastal Research*, **18**, 584–599.

Stefanov, W.L. and Netzband, M. (2005) Assessment of ASTER land cover and MODIS NDVI data at multiple scales for ecological characterization of an urban center. *Remote Sensing of Environment*, **99**, 31–43.

Stefanov, W.L., Ramsey, M.S. and Christensen, P.R. (2001) Monitoring urban land cover change: An expert system approach to land cover classification of semiarid to arid urban centers. *Remote Sensing of Environment*, **77**, 173–185.

Sugumaran, R., Zerr, D. and Prato, T. (2002) Improved urban land cover mapping using multi-temporal IKONOS images for local government planning. *Canadian Journal of Remote Sensing*, **28**, 90–95.

Turner, B.L., Lambin, E.F., Reenberg, A. (2007) The emergence of land change science for global environmental change and sustainability. *Proceedings of the National Academy of Sciences, USA*, **104**, 20666–20671.

UN-HABITAT (2010) *State of the World's Cities 2010/2011 – Cities for All: Bridging the Urban Divide*. Available http://www.unhabitat.org/content.asp?cid=8051&catid=7&typeid=46&subMenuId=0 (accessed 10 July 2010).

Yang, X. (2002) Satellite monitoring of urban spatial growth in the Atlanta metropolitan region. *Photogrammetrical Engineering and Remote Sensing*, **68**, 725–734.

Yang, X. (2009) *Characterizing Urban Ecosystems: A Research Agenda*. Presented at the Annual Meeting of the Chinese Academy of Sciences International Partnership Project 'Ecosystem Process and Services', 29 December 2009, Beijing, China.

Yang, X. (2011) Parameterizing support vector machines for land cover classification. *Photogrammetric Engineering and Remote Sensing*, **77**, 27–37. (in press).

Yang, X. and Liu, Z. (2005) Use of satellite-derived landscape imperviousness index to characterize urban spatial growth. *Computers, Environment and Urban Systems*, **29**, 524–540.

Yang, X. and Lo, C.P. (2003) Modelling urban growth and landscape changes in the Atlanta metropolitan area. *International Journal of Geographical Information Science*, **17**, 463–488.

Yu, D.L. and Wu, C.S. (2006) Incorporating remote sensing information in modeling house values: A regression tree approach. *Photogrammetric Engineering and Remote Sensing*, **72**, 129–138.

REMOTE SENSING SYSTEMS FOR URBAN AREAS

Part II (Ch. 2–Ch. 6) reviews the latest developments in remote sensing systems that are particularly relevant to urban studies. It covers a group of imaging systems that has been in use for quite a long period and another group of relatively new systems with advanced capabilities. The first group focuses on the Landsat imaging systems that have acquired a scientifically valuable image archive unmatched in quality, details, coverage, and length, which has been the primary source of data for urban studies at the regional, national and global scales. Chapter 2 discusses the utilities of archival Landsat imagery for the observations and measurement of urban spatial growth and landscape changes. The other group includes several advanced systems acquiring data with very-high resolutions over the optical or microwave portion of the electromagnetic spectrum. Chapter 3 examines some major challenges, limitations, and possible solutions when using the data from very-high-resolution optical satellite remote sensing systems for monitoring human settlements. Chapter 4 discusses the development of an automated system for mapping urban surface materials from hyperspectral remote sensor data. Chapter 5 discusses some new possibilities and challenges when using very-high-resolution space-borne radar data for urban feature extraction. Finally, Chapter 6 details a comprehensive approach for automated creation of three-dimensional building models from airborne lidar point cloud data fused with aerial imagery.

2

Use of archival Landsat imagery to monitor urban spatial growth

Xiaojun Yang

The Landsat program has provided the longest continuous observations of Earth's surface, and the freely available Landsat image archive has been an invaluable resource for examining natural and anthropogenic changes in the environment. This chapter discusses the utilities of satellite remote sensing for the observation and measurement of urban spatial growth emphasizing upon the use of archival Landsat data. We begin our discussion with an overview of the past, present and future of the Landsat program and its imaging sensors, which is tied with various inventorying and mapping activities in the urban environment. Then, we present a case study focusing on a rapidly suburbanizing American metropolis to demonstrate the usefulness of time-sequential Landsat imagery for monitoring urban growth and landscape changes over nearly the past four decades. Lastly, based on this case study and other literatures, we identify a generic workflow for urban growth monitoring and discuss several conceptual and technical issues emerging when using archival satellite images acquired by different sensors and perhaps during different seasons, which include image resolution and information contents, image pre-processing, and change detection methods. We believe such discussions can help identify the outstanding issues that must be addressed in order to implement an urban growth monitoring protocol effectively.

Urban Remote Sensing: Monitoring, Synthesis and Modeling in the Urban Environment, First Edition. Edited by Xiaojun Yang.
© 2011 John Wiley & Sons, Ltd. Published 2011 by John Wiley & Sons, Ltd.

2.1 Introduction

Over the past several decades, humans have substantially altered Earth's surface, predominately through agriculture, deforestation, and urbanization (Kondratyev, Krapivin and Phillips, 2002; Foley *et al.*, 2005; Turner, Lambin and Reenberg, 2007). Rates of deforestation and the dedication of marginal lands to high-impact agriculture have varied widely across the world, but there has been a consistent world-wide increase in the number of people residing in cities (Kaplan, Wheeler and Holloway, 2009). In 1950, only one-third of the world's 2.5 billion were urban dwellers. In 2010, more than half of the 6.9 billion people of our planet live in cities. At the global scale, the growth of urban areas shows no signs of slowing down and likely continues unabated into the next several decades (UN-HABITAT, 2010). Urban growth has frequently been viewed as a sign of the vitality of a regional economy, but it has rarely been well planned, thus provoking concerns over the degradation of our environment and ecological health (Lo and Quattrochi, 2003; Carlson, 2004; Grimm *et al.*, 2008). Monitoring urban growth and landscape change is critical both to those who study urban dynamics and those who must manage resources and provide services in the urban environment (Yang, 2002, 2007, 2010; Alberti, Weeks and Coe, 2004; Auch, Taylor and Acededo, 2004).

Assessment of urban growth and landscape change involves the procedures of inventorying and mapping that require a reliable information base and robust techniques (Yang, 2003). Urban and landscape patterns are observable and therefore can be mapped through ground surveys or remote sensing. While ground surveys are often limited by logistical constraints and largely localized by nature, remote sensing makes direct observations across large areas of the land surface, thus allowing urban and landscape patterns to be mapped in a timely and cost-effective mode (Lindgren, 1985; Lo, 1986; Jensen and Cowen, 1999; Mittelbach and Schneider, 2005). Using archival remote sensor data, a spatio-temporal assessment of urban growth and landscape changes can be obtained (e.g., Yang, 2002; Herold, Goldstein and Clarke, 2003; Seto and Fragkias, 2005; Bhatta, 2010). Evaluation of both static and dynamic attributes of land surface extracted from remote sensor data may allow the types of changes to be characterized and the proximate sources of change to be identified or inferred (Lo and Yang, 2002). Such information is useful for the evaluation of interactions among the various driving forces that can further help develop computer-based models to predict future urban growth and landscape changes (Kline, Moses and Alig, 2001; Yang and Lo, 2003).

Over the past several decades, data from various remote sensors have been used to map urban growth and landscape changes. Before the advent of satellite remote sensing, aerial photography played a key role in producing urban land use/cover maps, and remained critical for such a purpose (e.g., Lindgren, 1985; LaGro and DeGloria, 1992) until the late 1990s when high-resolution satellite imagery became available (e.g., Herold, Goldstein and Clarke, 2003; Ellis *et al.*, 2006). The high-resolution remote sensor data allow a substantial proportion of the basic land use/cover units to be distinguished, yet they are constrained by the high cost of data acquisition and the technical difficulties in data processing when the study area under investigation is quite large. The acquisition of information on regional, national and global urban land use/cover has been the subject of numerous studies and evaluations since the early 1970s (e.g., Gaydos and Newland, 1978; Jensen, 1981; Haack, Bryant and Adams, 1987; Kwarteng and Chavez, 1998; Yang and Lo, 2002; Auch, Taylor and Acededo, 2004; Seto and Fragkias, 2005; Small, 2005; Schneider and Woodcock, 2008), which was largely stimulated by the launch of ERTS-1 (Earth Resources Technology Satellite-1; later renamed as Landsat) in 1972. Images acquired by the US Landsat program and French SPOT satellites are the principal sources of data. Additionally, large volumes of valuable data have been acquired by the Indian remote sensing satellites (IRS), the NASA Terra satellite, the China–Brazil Earth resources satellites (CBERS), and several European, Canadian, and Japanese satellites carrying active imaging devices.

This chapter will focus on the use of the data acquired by the Landsat program that provides the longest continuous observations of Earth's surface from space. The Landsat system is the only satellite system designed and operated to repetitively observe Earth's landmass at moderate resolution. It offers a rich archive of highly calibrated, multispectral data of global coverage that recently becomes available at no charge from the USGS EROS Data Center. The Landsat data set has been an invaluable resource for examining natural and anthropogenic changes on Earth's surface. This chapter will specifically discuss the utilities of archival Landsat data for the observation and measurement of urban spatial growth and landscape changes. It comprises three major components. First, we provide an overview of the past, present and future of the Landsat program and its imaging sensors, which will be tied with various inventorying and mapping activities in the urban environment. Second, we present a case study focusing on a rapidly suburbanizing American metropolis to demonstrate the usefulness of time-sequential Landsat imagery for monitoring urban growth and landscape changes over nearly the past four decades. Last, based on this case study and other literature, we further identify a common workflow for urban growth monitoring, and discuss some conceptual and technical issues emerging when using archival satellite images acquired by different sensors and perhaps during different seasons.

2.2 Landsat program and imaging sensors

The Landsat (originally Earth Resources Technology Satellite-ERTS) program was initiated in 1966, which was largely inspired by the success of the early meteorological satellites and the orbital photography of the Earth's surface taken during the manned spacecraft missions in early 1960s. The program has resulted in the successful launch of six land satellites, with the first one on 23 July 1972, signaling a new age of terrestrial remote sensing from space. The first three satellites were launched with the Return Bean Vidicon (RBV) and the Multispectral Scanner (MSS) onboard, the fourth and fifth include the MSS and the Thematic Mapper (TM), the sixth with the Enhanced Thematic Mapper (ETM) onboard failed upon launch, and the seventh carries the Enhanced Thematic Mapper Plus (ETM+). These Landsat sensors were designed with moderate resolution that is strategically important because it is coarse enough for a global coverage, yet fine enough to capture human-dimension processes such as deforestation and urban growth (Paulsson, 1992). Over a period of nearly four decades, the Landsat program has acquired a scientifically valuable image archive unmatched in quality, details, coverage, and length, which has become freely available since 9 January 2009. This visually stunning data set has supported a wide variety of applications such as natural resource assessment and management, urban and regional planning, surveying and mapping, global change studies, among others.

As the first generation space-borne multispectral imaging system, the Landsat MSS sensor began to acquire data in 1972, with a spatial resolution approaching that of medium-scale aerial photographs. The MSS system acquires images over 4 spectral bands with the first two (green and red bands) suitable for detecting cultural features such as urban areas, roads, new subdivisions, gravel pits, and quarries. The image size is 185 × 185 km that allows some large natural or human-dimensional patterns or processes to be examined within a single scene. In addition, the MSS data set has been well archived and maintained at the USGS EROS Data Center and more than one dozen of international Landsat ground stations (Draeger et al., 1997). It is the only digital data set from operational satellites available for the period of 1972 to early 1982. The Landsat data were made available at an affordable price before the entire data set is open at no charge since early 2009. Given the above considerations, the MSS data have the unique values and thus have been widely used in connection with urban and landscape change analysis. Some examples include: mapping urban change in Atlanta (Todd, 1977), Denver and Richmond (Toll et al., 1980); monitoring anthropogenic land use as well as albedo changes in the Montréal, Ottawa, and Québec regions (Royer, Charbonneau and Bonn, 1988); and mapping urban spatial extent in the Bombay metropolitan region (Pathan et al., 1993).

The TM data have become available since 1982 when Landsat-4 was successfully launched, prompting a surge of interests to evaluate the utilities of this type of data for urban and landscape change analysis (Jensen et al., 1989). Comparing to the MSS, the TM sensor acquires data over seven carefully designed bands, including new bands in the visible (blue), mid-infrared, and thermal portions of the spectrum, which have helped improve the spectral differentiability of major land surface features, particularly vegetation whose proportion is a criterion for the discrimination of different types of urban land use such as commercial, industrial, or residential use. With the improved spatial, spectral, and radiometric resolutions, the Landsat-5 TM sensor has become the Landsat workhorse imager, providing invaluable data for urban land use/cover mapping with Anderson Level I&II classification (Anderson et al., 1976) and for detecting finer urban changes with an improved accuracy with some advanced algorithms or techniques (see the extended discussions in Part III of this book). For example, Gomarasca et al. (1993) used two TM images in conjunction with other digitized data to assess one century of land use modifications driven by rapid agricultural transformation and urban development in the Milan metropolitan area, Italy. Green, Kempka and Lackey (1994) used two TM images to detect land cover/use change in the Portland metropolitan area through the image differencing technique. Hill and Hostert (1996) employed a multitemporal TM data set to monitor the growth of Greater Athens, Greece by using the spectral mixing technique. Masek, Lindsay and Goward (2000) used three TM images, together with one MSS scene, to detect urban growth in the Washington DC area and found that the urban physical growth can be reasonably correlated with regional and national economic patterns. Seto and Fragkias (2005) used two TM images to quantify spatio-temporal patterns of urban land-use changes in four Chinese cities through the landscape metrics technique.

Landsat-7 was successfully launched on 15 April 1999, with the ETM+ sensor onboard. This latest Landsat multispectral sensor includes two important updates when comparing to the TM sensor: a new panchromatic band with 15 m spatial resolution and an improved thermal band with 60 m resolution. These improvements allow more details in urban land use/cover changes to be detected. On the other hand, there was a significant change in Landsat-7 ETM+ data delivery and distribution policies according to the 1992 Land Remote Sensing Policy Act. With a free license, the initial US price for a single ETM+ scene was at $600 from the USGS EROS Data Center, compared to $4400 for one TM scene ordered from the Earth Observation Satellite Company (EOSAT), a commercial company that was contracted to run the Landsat program during 1985–2001 according to the 1984 Land Remote Sensing Commercialization Act. This substantially low price and free licensing allowed the ETM+ data to be widely used for urban and landscape change analysis (e.g., Yang, 2002; Lo and Choi, 2004; Doygun and Alphan, 2006; Millward, Piwowar and Howarth, 2006). Unfortunately, the Scan Line Corrector (SLC) in the ETM+ instrument, which compensates for the along-track forward motion of the spacecraft, failed permanently on 31 May 2003. This failure forces the ETM+ instrument to image the Earth in a zigzag fashion, resulting in some areas that are imaged twice and approximately one-fourth of an ETM+ scene that is not imaged at all. Although the Landsat-7 data acquired after this failure are available with the data gaps optionally filled in by using other data, recent studies indicate that the SLC failure has undermined the role of ETM+ data as a complete land use/cover inventory dataset (Trigg, Curran and McDonald, 2006; Wulder et al., 2008).

The technical problems with the Landsat-7 instrument, the operation of Landsat-5 substantially beyond its designed life, and delays in the development and launch of a successor have increased the likelihood that a gap in the Landsat data archive may occur. And a Landsat Data Gap Study team formed by USGS and NASA in 2005 found that there are no other systems in orbit or planned for launch in the short term that can supply data to fill the Landsat gap should system failures occur to Landsat-5 and -7 (Wulder et al., 2008). Considering the scientific, environmental, economic, and social benefits offered by the Landsat program, the Bush Administration decided in December 2005 to continue Landsat instruments in the form of a free-flyer spacecraft, which has resulted in a NASA and USGS joint initiative called the Landsat Data Continuity Mission (LDCM). Under this initiative, a new satellite named LDCM-1 with the Operational Land Imager (OLI) and the Thermal Infrared Sensor (TIRS) onboard is scheduled to launch in December 2012. This plan, once implemented, would ensure the continuity of the Landsat data acquisition that has become unique and indispensable for monitoring, management, and scientific activities.

2.3 Mapping urban spatial growth in an American metropolis

2.3.1 Research design

In this study, we focus on the utilities of archival Landsat imagery for urban growth and landscape change analysis with the Atlanta metropolitan area as a case. The study site includes the 10 counties under the Atlanta Regional Commission (ARC) as well as three additional counties, i.e., Coweta, Forsyth, and Paulding, which have shown a similar growth pattern to the ARC counties (Fig. 2.1). For the past four decades, Atlanta has been one of the fastest growing metropolises in the United States as it emerged to become the premier commercial, industrial, and transportation urban center of the southeast. Population increased 27% between 1970–1980, 33% between 1980–1990, 40% between 1990–2000, and 36% between 2000–2010. The city has expanded greatly as suburbanization consumes large areas of agricultural and forest land adjacent to the city, pushing the periurban fringe farther and farther away from the original urban boundary. This rampant suburban sprawl has provoked concerns over the losses of large areas of primary forests, inadvertent climate repercussions, and the degradation of the quality of life in this region (Bullard, Johnson and Torres, 2000; Lo and Quattrochi, 2003). By using population trends, land use, traffic congestion, and open space loss, Sierra Club's 1998 Annual Report ranked Atlanta as America's most sprawl-threatened large city (Sierra Club, 1998).

On the other hand, because of the significant physical growth, Atlanta's urban spatial structure and pattern have changed dramatically, making the city a 'hot spot' in urban studies. Urban geographers have recognized Atlanta as one of the few typical postmodern metropolises in North America (Hall, 2001). Soja (1989) first used the term "postmodern" to describe cities that have undergone restructuring in the United States after the rise of post-Fordist industrial organization, which is characterized by a flexible subcontracted production system based on small-size and small-batch units. The postmodern city exhibits unique urban forms, architectural styles, and socioeconomic characteristics. Architecturally, it has multicentered business districts with office glass towers and stylish buildings. Socially, it is increasingly minoritized and polarized along class, income, racial and ethnic lines (Dear and Flusty, 1998). Los Angeles is in fact a typical postmodern city, and many cities in the United States Sunbelt are

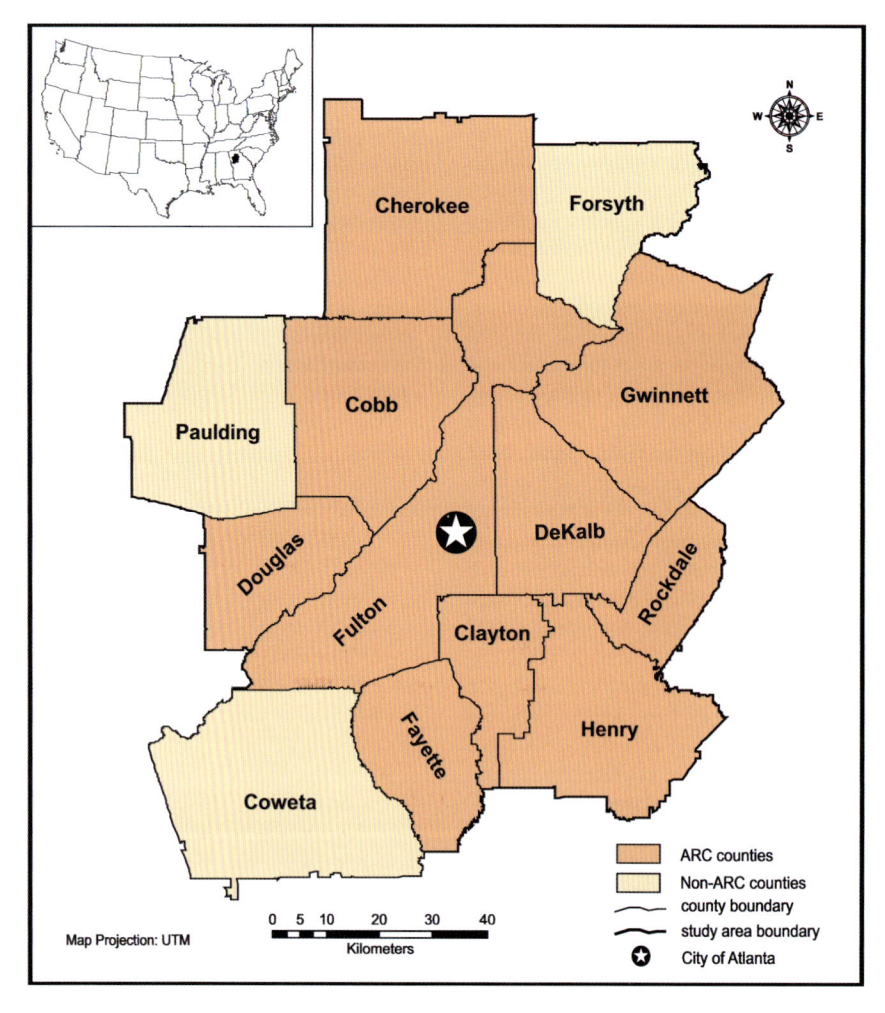

FIGURE 2.1 Location of the study site. It comprises the 10 counties under the Atlanta Regional Commission (ARC) plus three additional counties (Forsyth, Paulding, and Coweta) that show a similar growth pattern. The city of Atlanta is shown.

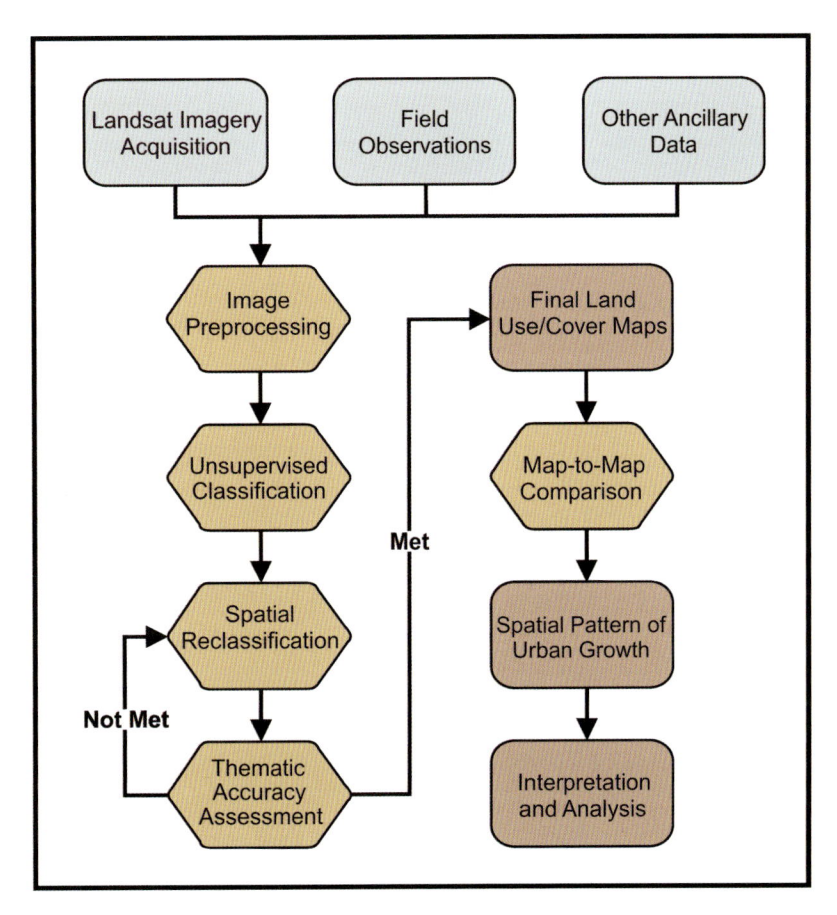

FIGURE 2.2 Working procedural route adopted, with several major components: primary and secondary data acquisition, image preprocessing, image classification and thematic accuracy assessment, and change detection and analysis.

growing in the fashion of Los Angeles. Atlanta is a good example of such a city because it exhibits all the characteristics of the postmodern city described above.

Research on Atlanta's internal structure has led to the formulation of the urban realms model to depict the multi-nuclei nature of the city in contrast to the conventional single-core urban form (Hartshorn and Muller, 1989). Thus, Atlanta is an ideal city to study postmodern urban dynamics and environmental consequences of accelerating urban growth. During the past 15 years, the author has been involved various research projects aiming to understand the dynamics of change in Atlanta by using remote sensing. The case study examines urban spatial growth and landscape change along the outskirts of the Atlanta metropolitan area by using a time series of Landsat data covering a period of nearly four decades. The research methodology identified here includes several major components: primary and secondary data acquisition, image processing of remote sensor data, change detection, and interpretation and analysis (Fig. 2.2).

2.3.2 Data acquisition and land classification scheme

A time series of Landsat images was used as the primary data to detect and measure the spatio-temporal urban growth pattern at 6–8-year intervals, beginning in 1973 when Landsat MSS data became available. This series comprises 13 predominantly cloud-free images acquired by the three Landsat sensors for Atlanta during 1973–2007 (Table 2.1). Note that because of the permanent SLC failure in the ETM+ instrument on 31 May 2003, two cloud-free TM scenes were acquired for 2007. Most of the scenes were acquired during the spring or earlier summer seasons, when vegetation is in the stage of vigorous growth. The 1998 and 1999 scenes are the two exceptions. The 1998 TM scene, acquired in the winter season, was used to improve vegetation mapping. The ETM+ scenes were acquired in late summer because they are the only scenes free from clouds available between April and September 1999. Because of good image quality, the 1988 MSS scene was mainly used as the reference image for relative radiometric normalization of other MSS images.

To facilitate satellite image-based change mapping, a variety of ancillary data have been collected, including digital images of Advanced Thermal and Land Applications Sensor (ATLAS) acquired on 11 May 1997, contact prints of aerial photographs for 1986–1988, 1993 USGS digital orthophotos, the 1988–1990 land-cover classification from Georgia Department of Natural Resources, and high-resolution satellite imagery from Google Earth. In addition, a GPS-guided field survey was conducted to help establish the relationship between image signals and ground conditions. Fieldwork also helped obtain first hand information about suburban sprawl throughout the study area, which can be useful for understanding the dynamics of change.

TABLE 2.1 List of the Landsat images used.

Date	Landsat No.	Sensor Type	Scene Location	Nominal IFOV (m)	Sun Elevation (degree)	Sun Azimuth (degree)	RMSE (Control Point No.)	RRN[c]	Purpose[d]
4-13-73	1	MSS	north Atlanta	57 × 79	54.16	129.37	0.58 (13)	Yes	1
4-13-73	1	MSS	south Atlanta	57 × 79	54.83	127.39			
6-11-79	3	MSS	north Atlanta	57 × 79	61.65	105.65	0.46 (13)	Yes	1
6-11-79	3	MSS	south Atlanta	57 × 79	61.74	102.77			
6-29-87	5	TM	center-shifted[a]	28.5 × 28.5	61.84	103.71	0.22 (13)	No	1
5-14-88	5	MSS	center-shifted	57 × 79	61.61	115.51	0.51 (12)	ref.	4
7-31-93	5	TM	center-shifted	28.5 × 28.5	57.00	110.00	0.40 (15)	No	1
7-10-97	5	TM	center-shifted	28.5 × 28.5	61.00	106.00	ref.	No	2 and 4
1-02-98	5	TM	center-shifted	28.5 × 28.5	27.00	150.00	0.27 (14)	No	3
9-09-99	7	ETM+	north Atlanta	28.5 × 28.5[b]	53.8	140.10	0.52 (15)	No	1 and 4
9-09-99	7	ETM+	south Atlanta	28.5 × 28.5[b]	54.8	138.40	0.44 (13)		
5-19-07	5	TM	north Atlanta	30 × 30	65.65	121.91	0.42 (14)	No	1
5-19-07	5	TM	south Atlanta	30 × 30	66.17	118.68			

[a]The panchromatic band has a nominal IFOV (instantaneous field of view) of 15 m.
[b]The center of the north scene has been shifted by 50%. The scene size is approximately 185 × 185 km^2.
[c]RRN = Relative radiometric normalization.
[d]Code: 1 = land use/cover mapping; 2 = reference data for geometric correction; 3 = reference data for ground truth in the Winter season; and 4 = land use/cover mapping accuracy assessment.

Based on the Landsat images, the ancillary data, and the knowledge gained during the field survey, we carefully designed a mixed Anderson Level I/II land use/cover classification scheme (Anderson *et al.*, 1976), with the following six major classes:

- high-density urban use: mostly large commercial and industrial buildings, large transportation facilities, and high-density residential areas in the city cores;
- low-density urban use: mostly single/multiple family houses, apartment complexes, yards, local roads, and small open spaces;
- cultivated/exposed land: mainly non-impervious areas with sparse vegetation, such as clear-cuts, quarries, barren rock or sand along river/stream beaches;
- cropland/grassland: crop fields and pasture as well as cultured grasses (such as golf courses, lawns, city parks);
- forest land: deciduous, coniferous, and mixed forest land; and
- water: streams, rivers, lakes, and reservoirs.

2.3.3 Image processing of remotely sensed data

The design of the image processing procedures to be used in this study was based on a thorough understanding of the research objective, image characteristics, landscape complexity, and the status of technological development. The image processing procedures identified here involved a variety of remote sensing and relevant geospatial techniques, including data preprocessing, image classification, spatial reclassification, and thematic accuracy assessment.

2.3.3.1 Image preprocessing

Both geometric rectification and radiometric normalization were conducted. The geometric correction strategy used is actually an image-to-image rectification. The reference image was the 1997 Landsat TM image (see Table 2.1) that has been orthorectified by EOSAT to UTM projection (Zone 16), NAD83 horizontal datum, and GRS80 ellipsoid. All other scenes were rectified to the TM image by using a first-degree polynomial transformation, given the relatively even terrain relief in the study site. The number of control points and RMSEs (root mean squared errors) are summarized in Table 2.1.

The two MSS images used for image classification were acquired by Landsat-1 and Landsat-3, and are very different from each other in contrast although they were processed identically. To help restore a common radiometric response among them, the relative radiometric normalization method developed by Hall *et al.* (1991) was applied to the 1973 and 1979 MSS images by using the 1988 MSS image as the reference (Yang and Lo, 2000). Note that the RRN method developed by Hall *et al.* (1991) is based on the use of radiometric control sets that should have little or no variation through time. These control sets were extracted by using the two non-vegetated extremes of the Kauth–Thomas (KT) greenness-brightness scattergram which was constructed using the first two bands of a Tasseled Cap transformation of the raw image. Radiometric normalization was not attempted to the TM and ETM+ images because they have been processed to good radiometric quality.

2.3.3.2 Unsupervised classification

The 1973, 1979, 1987, 1993, 1999, and 2007 images were classified by using a two-step unsupervised method. Firstly, the ISODATA (Iterative Self Organizing Data Analysis) algorithm was used to identify spectral clusters from image data, excluding the thermal bands for the TM and ETM+ images because of their coarse spatial resolutions. It was implemented without assigning predefined signature sets as starting clusters to avoid the impacts of sampling characteristics (Vanderee and Ehrlich, 1995). Some important clustering parameters specified include number of classes, convergence value, and maximum number of iterations. To determine the optimum class number, we tested 20, 40, 60, and 80, and found that 60 for the TM images or 80 for the ETM+ image allows the resultant clusters to be better interpreted in relation to the classification scheme. The convergence value was specified as 0.990 for all types of data. Finally, for the TM or MSS data, 60 was used for the maximum number of iterations while 80 for the ETM+ scene.

The resultant clusters were assigned into one of the six land use/cover classes through visual inspection of the original images in relation to appropriate ancillary data. To label the clusters, the original image and the clustered map were displayed side by side and then spatially linked. The class assignment for specific clusters was based on an examination of the cluster at the two different detail levels. At the large-scale level, image color was mainly used in decision making; at the small-scale level, however, additional image elements such as association and site were combined to improve classification quality. Also a note was taken for any spectral cluster containing more than one land use/cover type. This happens when spectral contents of more than two different land use/cover classes are similar. Whenever this occurred, the cluster was initially labeled as one of the most likely land use/cover classes. And at a later stage, these clusters were split into smaller clusters for the correct land use/cover labels using the spatial reclassification procedures described below.

2.3.3.3 Spatial reclassification

Spatial reclassification was used to reduce the two types of misclassification errors on the initial maps produced through the unsupervised classification. The first type was the boundary error due to the occurrence of spectral mixing within a pixel, which is small relative to the areas of correct classification. Within a class there are some anomalous pixels representing the noise in the data. A modified 3×3 modal filter with the four corner neighbors disabled was used to suppress the boundary errors.

The second type of misclassification errors was the spectral confusion error due to the spectral similarity between different land use/cover classes, which is inevitable for an image acquired with a broad-band sensor and tends to be more perceptible in an urban scene than in a rural one. Defining the spectral confusion involves the use of spatial and contextual properties through an image interpretation method that can be incorporated effectively into a digital classification procedure with on-screen digitizing, multiple zooming, Area Of Interest (AOI) functionality, and other relevant spatial analysis tools. In addition, several image processing programs permit advanced tools for geoprocessing through which some "manual" operations can be implemented automatically. With the use of this method, four major types of spectral confusion were identified: low-density urban (mostly residential) versus forest (clearcuts, sparse forest, and wetlands); low-density urban (sparse residential) versus cropland or grassland; forest (sparse forest and shrubs) versus cropland/grassland; and high-density urban (large open roof buildings, air fields, and multilane highways) versus cultivated or exposed land (large barren landmass, river beach, fallowed land). Spectral confusions were found to be generally more serious for the MSS data than for the TM and ETM+ data. Spectrally confused clusters were first identified, and AOI layers were created by on-screen digitizing. The AOI layers served as masks for splitting confused clusters. Finally, GIS-based reclassification functionality was employed to recode the split clusters into correct land classes. This was an interactive process until an acceptable accuracy was obtained.

2.3.3.4 Thematic accuracy assessment

Due to the limited availability of ground truth data, it is impossible to perform accuracy assessment for each map exhaustively. The strategy adopted here was to assess the maps produced from each type of imagery covering the study area. Three maps were selected for the thematic accuracy assessment: the 1988 map from MSS data, the 1997/1998 map from a summer 1997 and a winter 1998 TM images, and the 1999 map from ETM+ data. The first two were produced by the author for other projects using the same method described in this chapter. The accuracy assessment was carried out by using a standard method recommended by Congalton (1991). Results revealed that the land use/cover maps based on TM or ETM+ images yielded a slightly better accuracy than that from MSS data (Table 2.2). The maps derived from TM or ETM+ data show a slightly higher kappa index for low-density urban, cropland/grassland, and forest than that from MSS data. This could be the result of higher spatial, spectral, and radiometric resolution of the TM or ETM+ data. However, the map derived from MSS data is compatibly accurate in every respect to the ones from TM or ETM+ data. Overall, all these maps meet the minimum 85 percent accuracy stipulated by the Anderson classification scheme (Anderson et al., 1976), indicating that the image processing procedures adopted here have been effective in producing compatible land use/cover data over time, despite the differences in spatial, spectral, and radiometric resolution of the three generations of Landsat data used in this project. Other maps were produced using the same procedures. It is anticipated that the same level of accuracy should be maintained. Land use/cover classification maps for six dates are shown in Fig. 2.3. The statistics of each classification map are summarized in Table 2.3 and the land use/cover changes during different periods are shown in Table 2.4.

2.3.4 Change detection

Change detection was used to examine urban growth and the nature of change. To analyze urban growth, the spatial distribution of each urban land class was extracted from each map in the time series. The change in urban land was summarized by using the GIS minimum dominate overlay method, which allows the smallest amount of high-density or low-density urban use in the earliest year to show up fully, while only the net addition

TABLE 2.2 Results of thematic accuracy assessment for the land use/cover maps produced with three different types of Landsat data.

Land use/cover[a]	Map from MSS Data (1988)			Map from TM Data (1997/1998)			Map from ETM + Data (1999)		
	Producer's accuracy (%)	User's accuracy (%)	Conditional K index	Producer's accuracy (%)	User's accuracy (%)	Conditional K index	Producer's accuracy (%)	User's accuracy (%)	Conditional K index
High-density urban	93.24	87.34	0.86	96.43	83.31	0.81	88.00	84.62	0.82
Low-density urban	81.63	85.71	0.81	93.44	78.08	0.75	81.33	87.14	0.84
Cultivated/exposed land	95.38	83.78	0.82	85.42	82.00	0.80	86.84	82.50	0.80
Cropland/grassland	85.92	77.22	0.74	81.36	96.00	0.95	86.79	82.14	0.79
Forest land	85.88	91.25	0.85	94.23	98.00	0.98	90.29	93.00	0.90
Water	94.74	94.74	0.95	100	98.00	0.98	95.12	92.85	0.92
Overall	87.00		0.83	90.10		0.88	87.78		0.85

[a] At least 50 sample points for each class were used in the thematic accuracy assessment.

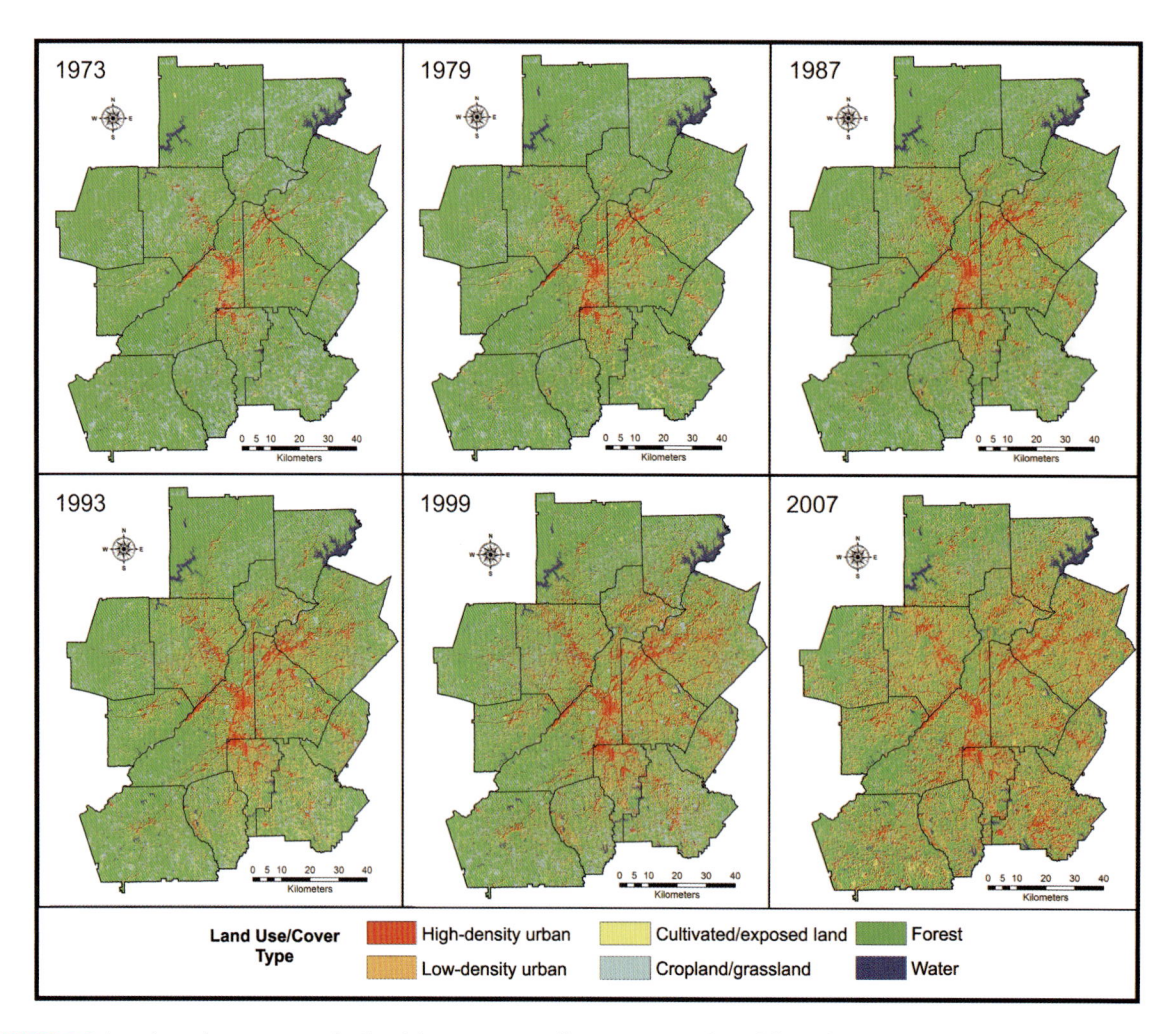

FIGURE 2.3 Land use/cover maps in the Atlanta metropolitan area, produced from Landsat images. The boundary of 13 counties in Atlanta is also shown.

TABLE 2.3 Land use/cover statistics for the 13 counties in Atlanta: 1973–2007.

Years	High-density urban Area (ha)	%	Low-density urban Area (ha)	%	Cultivated/exposed land Area (ha)	%	Cropland/grassland Area (ha)	%	Forest land Area (ha)	%	Water Area (ha)	%	Total Area (ha)	%
1973	30 112	2.88	76 861	7.35	14 507	1.39	159 366	15.25	750 880	71.84	13 491	1.29	1 045 217	100
1979	38 388	3.67	129 092	12.35	20 570	1.97	117 419	11.23	725 511	69.41	14 236	1.36	1 045 216	100
1987	54 629	5.23	177 757	17.01	15 506	1.48	117 737	11.26	664 204	63.55	15 383	1.47	1 045 216	100
1993	67 631	6.47	214 629	20.53	21 143	2.02	96 721	9.25	626 652	59.95	18 439	1.76	1 045 216	100
1999	87 446	8.37	283 169	27.09	5034	0.48	101 470	9.71	545 773	52.22	22 325	2.14	1 045 216	100
2007	106 860	10.22	373 685	35.75	17 489	1.67	69 608	6.66	454 962	43.53	22 613	2.16	1 045 217	100

TABLE 2.4 Land use/cover changes for 13 counties in Atlanta during different periods.

Period	High-density urban Area (ha)	%	Low-density urban Area (ha)	%	Cultivated/exposed land Area (ha)	%	Cropland grassland Area (ha)	%	Forest land Area (ha)	%	Water Area (ha)	%
1979–1973	8276	27.48	52 231	67.96	6063	41.79	−41 947	−26.32	−25 369	−3.38	745	5.52
1987–1979	16 242	42.31	48 665	37.70	−5064	−24.62	318	0.27	−61 307	−8.45	1147	8.06
1993–1987	13 002	23.80	36 872	20.74	5637	36.35	−21 016	−17.85	−37 552	−5.65	3057	19.87
1999–1993	19 815	29.30	68 540	31.93	−16 110	−76.19	4749	4.91	−80 879	−12.91	3886	21.07
2007–1999	19 414	22.20	90 516	31.97	12 456	247.45	−31 862	−31.40	−90 811	−16.64	287	1.29
2007–1973	76 748	254.88	296 824	386.19	2982	20.56	−89 758	−56.32	−295 918	−39.41	9122	67.61

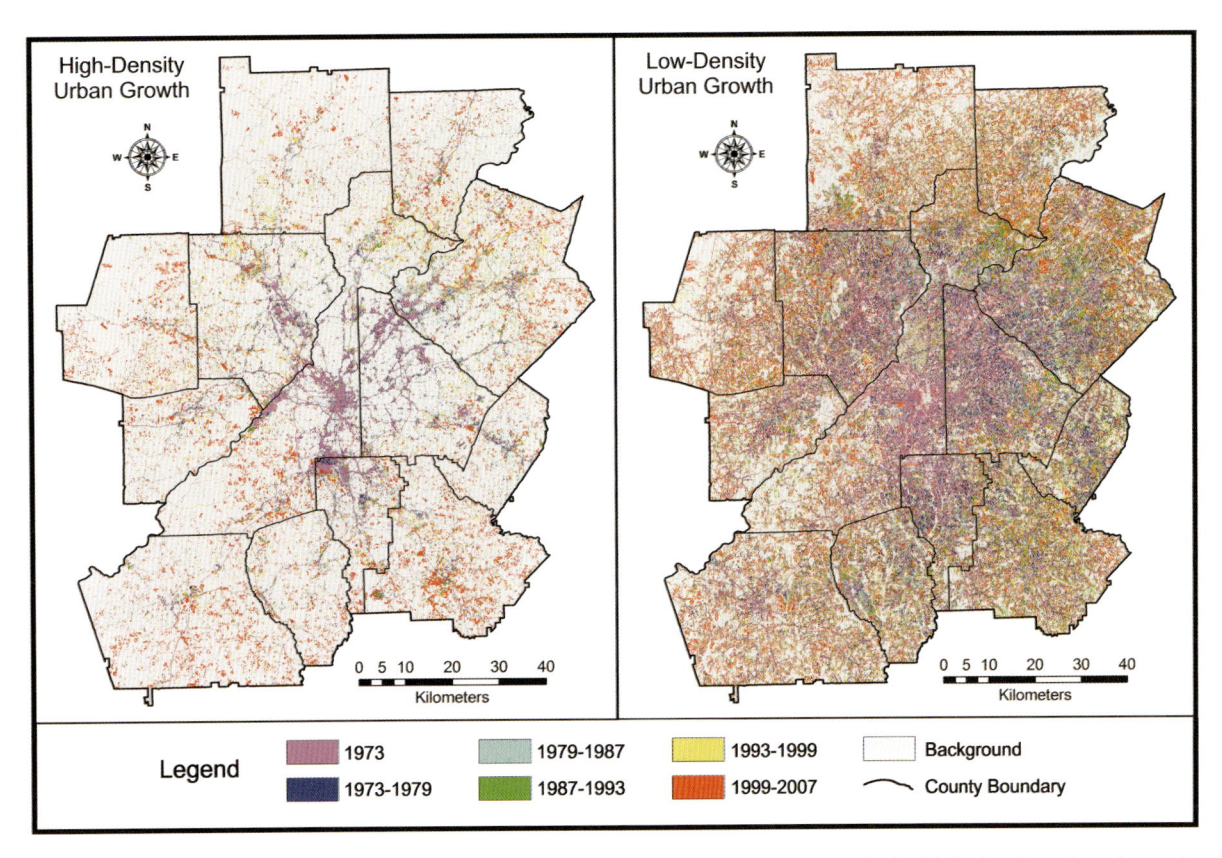

FIGURE 2.4 Urban spatial growth in the Atlanta metropolitan area during different periods: high-density urban (upper) and low-density urban (lower). The boundary for 13 counties in Atlanta is shown.

in the following years in the time series is shown. In this way, the urban extent of six dates was summarized on one map. By assigning a unique color to the net addition of each year on the combined map, the progressive growth of high-density or low-density urban land can be perceived (Fig. 2.4).

The nature of change is to analyze land conversions using cross-tabulation or matrix analysis that assigns a unique class for each coincidence in two input layers, thus capturing different combinations of change. This method was used to characterize the land conversions for different time periods. Given the total number of land classes, there are 36 possible combinations for each period. Because this study focused on the conversion of forest, cropland/grassland, or cultivated/exposed land into urban uses, only six combinations were selected for further analysis, while the other were merged into a single unit. The conversion coding system is shown in Table 2.5, along with the conversion statistics for different periods.

2.3.5 Interpretation and analysis

Based on Fig. 2.4 (left), the spatial expansion of high-density urban use is clearly visible. In 1973, the high-density urban use was small, occupying only 2.88% of the total land area for Atlanta (Table 2.3). The outward spread is quite clear in the 1979, 1987 and 1993 patterns, following the major transportation routes. The net addition was 8276 hectares (or 27.48% increment) between 1973–1979, 16 242 hectares (or 42.31% increment) between 1979–1987, and 13 002 hectares (or 23.80% increment) between 1987–1993. These additions were primarily concentrated in several inner counties, such as Fulton, Cobb, DeKalb, and Clayton. Significant growth took place by 1999 and 2007, with more new development areas concentrating in northern and northeastern areas. The linearly concentrated pattern became more multinucleated. The 1999 and 2007 distributions show further enhancement of this transition as the spread took place largely along transportation routes and around urban centers, particularly in some peripheral counties, such as Coweta, Cherokee, Forsyth, Paulding, and Fayette. In 2007, the high-density urban use occupied 106 860 hectares, or 10.22% of the total land area, which was about 254.88% increase in land when comparing to 1973. The daily increment was approximately 6 hectares or 15 acres between 1973–2007.

The evolution of spatial pattern of low-density urban use, mainly residential, is clearly perceived in Fig. 2.4 (right). In 1973, the low-density urban use occupied 76 861 hectares, or 7.35% of the total area (Table 2.3). Although the low-density urban land shows signs of spreading outward, its large share was clearly concentrated in the inner city core and several inner counties. A somewhat linear pattern can also be seen along several major transportation routes. Thus, the spatial pattern of low-density urban use in 1973 can be described as a form of concentration mixed with some degree of dispersal. Significant growth occurred in 1979 and 1987, with net addition of 52 231 and 48 665 hectares, respectively (Table 2.4). Most of the new additions occurred outside the central city core, concentrating in four inner counties of DeKalb, Clayton, Fulton, and Cobb, and in three exterior counties of Gwinnett, Rockdale, and Fayette. Growth in the northern, northwestern, and northeastern directions was quite

TABLE 2.5 Land use/cover conversion statistics for 13 counties in Atlanta during different periods.

Maps		Nature of change code[a]	1973–1979 Area (ha)	%	1979–1997 Area (ha)	%	1987–1993 Area (ha)	%	1993–1999 Area (ha)	%	1999–2007 Area (ha)	%	1973–2007 Area (ha)	%
Year A	Year B													
3	1	C1	344	0.03	1064	0.10	894	0.09	623	0.06	1455	0.14	3874	0.37
3	2	C2	1464	0.14	1544	0.15	2199	0.21	1073	0.10	1490	0.14	6936	0.66
4	1	C3	2017	0.19	2281	0.22	2146	0.21	3161	0.30	19072	1.82	27423	2.62
4	2	C4	21550	2.06	11484	1.10	13852	1.33	10260	0.98	49567	4.74	84417	8.08
5	1	C5	2657	0.25	5661	0.54	4025	0.39	16522	1.58	25407	2.43	47988	4.59
5	2	C6	33014	3.16	42683	4.08	28976	2.77	56154	5.37	114576	10.96	223800	21.41
All other combinations		C7	984170	94.16	980499	93.81	993123	95.02	957422	91.60	833647	79.76	650776	62.26

[a]Note: C1, converted from cultivated/exposed land into high-density urban; C2, converted from cultivated/exposed land into low-density urban; C3, converted from cropland/grassland into high-density urban; C4, converted from cropland/grassland into low-density urban; C5, converted from forest into high-density urban; C6, converted from forest into low-density urban; and C7, all other combinations which are not considered here.

clear. The spatial distribution pattern became more dispersed. Low-density urban use continued to grow after 1987. Most of the new development, however, took place in the exterior metropolis. This widely spread-out pattern is a major indicator of suburbanization. Quantitatively, the low-density urban use occupied 373 685 hectares or 35.75% of the total area in 2007, indicating a 386.19% increase between 1973 and 2007. The daily increment was about 24 hectares or 59 acres for the same period.

Tables 2.3 and 2.4 also indicate that the continuing decline in cropland/grassland and forest land. The shrinking pattern of these two classes was proportional to the growth of the two urban classes. In general, the decline of cropland/grassland and forest land predominately took place in the interior metropolis before 1987 but in the exterior after 1987. Quantitatively, cropland/grassland occupied 159 366 hectares or 15.25% of the total study area in 1973. It declined to 69 608 hectares (or 6.66%) by 2007. This represents a decrease of 56.32%, or a daily rate of approximately 7 hectares (17 acres). Similarly, forest declined from 750 880 hectares (or 71.84%) in 1973 to 454 962 hectares (or 43.53%) in 2007, thus representing a decrease of 39.41%, or a daily rate of about 24 hectares (59 acres) in land area.

The nature of change is quite clear from Table 2.5. From 1973 to 2007, the loss of forest land contributed to 68.90% of the urban growth while the loss of cropland/grassland accounted for 28.35%. The high-density urban use had a net addition of 79 285 hectares, among which 60.53% came from the loss in forest land (C5) and 34.59% resulted from the loss in cropland/grassland (C3). The loss in cultivated/exposed land only contributed to 4.89% of the increase in high-density urban use (C1). For the low-density urban use, 70.01% (C6) and 26.79% (C4) of the increase came from the loss in forest land and cropland/grassland, respectively. The loss of cultivated/exposed land only accounted for 2.20% (C2) of the net addition in low-density urban use.

2.3.6 Summary

By using Atlanta as a case, this study has demonstrated the usefulness of satellite remote sensing for urban change mapping and measurement. Central to this study was a time series of satellite images acquired by three Landsat sensors, namely, MSS, TM and ETM+, which has been used to produce land use/cover maps through image classification. The combined use of unsupervised classification and GIS-based spatial reclassification procedures has been quite effective to resolve the spectral confusion among different land classes, a typical problem with broad-band sensors and within an urban image scene. The time series of land classification maps was further used to analyze urban spatial growth and the nature of change. This was built upon the combined use of post-classification comparison and GIS-based overlay techniques that made possible the production of single-theme change maps, which emphasize spatial dynamics. On the other hand, this study has established a well-documented regional case study focusing on Atlanta, a typical postmodern American metropolis having undergone rapid demographic and economic growth during the past several decades. This study reveals a rampant growth in urban land during 1973–2007, which substantially outpaced the rate of population growth in Atlanta. Urban spatial form experienced a transition from linear concentration to multinucleated pattern for high-density urban use (mainly commercial, industrial and transportation) and from concentration mixed with some degree of dispersal to more dispersed pattern for low-density urban use (mainly residential) in Atlanta.

2.4 Discussion

2.4.1 A generic urban growth monitoring workflow

Urban spatial growth can be mapped and measured by using time-sequential satellite imagery such as archival Landsat data, as demonstrated in our case study and some other studies (e.g., Pathan *et al.*, 1993; Gomarasca *et al.*, 1993; Green, Kempka and Lackey, 1994; Hill and Hostert, 1996; Masek, Lindsay and Goward, 2000; Seto and Fragkias, 2005; Liu *et al.*, 2010). Based on these studies, a generic workflow for urban growth monitoring by remote sensing can be summarized in Fig. 2.5, which includes several major components: defining research questions, data acquisition and collection, image preprocessing, change detection, and interpretation and analysis. Appropriate handling each of these components is not a trivial task by any means. Defining research questions involve some non-technical issues such as the purpose and scope of research as well as the physical and cultural characteristics of a study site, which will ultimately determine specific technical procedures to be adopted. While we will not delve into these non-technical issues, a moderate understanding of the physical and cultural characteristics of the area under investigation can improve the design and implementation of subsequent technical procedures (Yang and Lo, 2002). Moreover, some essential knowledge on urban geography and landscape ecology can help identify appropriate remote sensor data or information extraction techniques (Herold, Scepan and Clarke, 2002; Longley, 2002; Yang and Lo, 2003; Lo, 2004). For example, urban physical composition varies greatly across the world, which is linked with underlying social, economic, and cultural circumstances (Kaplan, Wheeler and Holloway, 2009). Compared to American cities, most non-American counterparts are compact and clustered by nature, which may need to use data with much higher spatial resolution (Welch, 1982; Jensen and Cowen, 1999). Our further discussion will focus on some issues relating to the three technical components in the generic workflow (Fig. 2.5).

2.4.2 Image resolution and land use/cover classification

Among the technical components (see Fig. 2.5), acquiring an appropriate remote sensor data set can be quite challenging. Theoretically, data used for urban studies must meet certain conditions in terms of spatial, spectral, radiometric, and temporal characteristics (Lo, 1986; Jensen and Cowen, 1999; Jensen, 2007). However, the choice of available data sets can be quite limited for a study covering a decades-long period. For many such studies, Landsat data become the only choice, particularly during the period of early 1970s to early 1980s when there was no any alternative available because other Earth resource satellite programs had not been initiated. Although within the range of

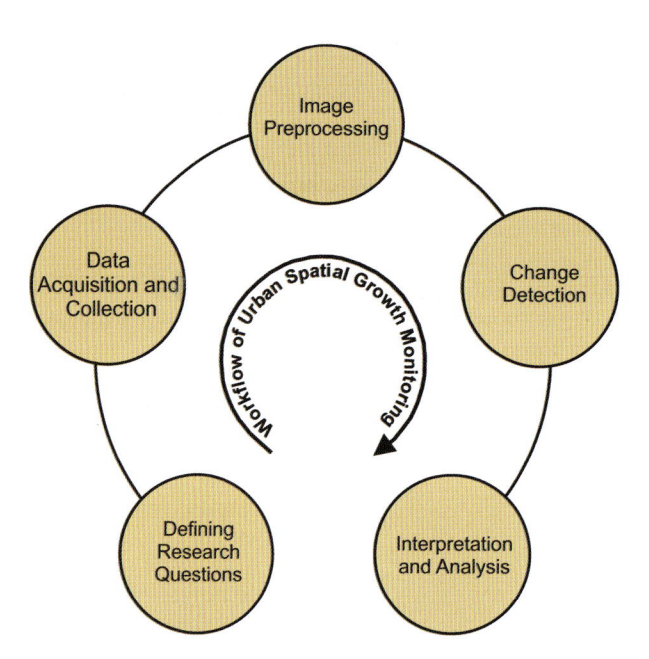

FIGURE 2.5 A generic workflow for urban growth monitoring by remote sensing. Note that defining research questions is critical for designing specific technical procedures.

medium resolution, the three generations of Landsat sensors provide multispectral images with spatial resolution varying from 30–80 m, excluding the thermal bands. When a study must combine the use of data acquired by these sensors, caution should be taken as the information contents revealed by the images with changing resolutions may vary, as indicated by several empirical studies (e.g., Markham and Townshend, 1981; Townshend and Justice, 1981; Mumby and Edwards, 2002).

In a related project, the author evaluated the impacts of image resolutions upon land use/cover classification by using three images of the same day in 1987 for the Atlanta metropolitan area: a MSS image, a TM image (excluding the thermal band), and a degraded TM image (Table 2.6). The degraded TM image comprises bands 2, 3, and 4 of the original TM image but the radiometric quantification has been downgraded to 7 bits. Thus, the degraded TM image has similar spectral and radiometric resolutions to the MSS image but the spatial resolution is identical to that of the original TM image. All three images were classified using unsupervised method described in Section 2.3, excluding the spatial reclassification procedures. The thematic accuracy for each classified map is evaluated and the area for each land use/cover category is compared. Results indicate that the improved spectral and radiometric resolutions are quite helpful in achieving better classification accuracies. But the improved spatial resolution does not help much in this regard. This is not a surprise because improved spatial resolution can lead to an increase not only in the interclass variability but also in the intraclass variability, which can undermine classification accuracies if a classic pixel-by-pixel classification method is used without refinements (Markam and Townshend, 1981; Williams *et al.*, 1984; Haack, Bryant and Adams, 1987).

When looking at the specific statistics in Table 2.6, the original TM image consistently gives a better classification accuracy for each category than that of the MSS, especially for the low-density urban use. If the land use/cover category area statistics extracted from the original TM image are used as the reference, the MSS data tend to greatly overestimate the area of low-density urban use, a suburban housing category associated with woodland, lawns, and open space. Therefore, "low-density urban use" clusters of pixels produced by unsupervised classification tend to exhibit mixed reflectance from residential use, forest land, and exposed/cultivated land. Because of the coarser spatial resolution, the MSS image mixes the reflectance of these component land cover types per pixel more than that of the TM image. This explains why the areas for "exposed/cultivated land" and "forest" were underestimated from the MSS image relative to the original TM image.

The observations from our image resolution impact assessment experiment can be used to guide the design and implementation of digital image processing strategies for ensuring reliable results being derived from multidate, multiresolution Landsat data. First, because the spectral and radiometric characteristics of an image can be significant for digital land use/cover mapping, any enhancements to spectral and radiometric resolutions should be pursued. Second, appropriate remedial procedures, such as image preprocessing, should be used to improve the quality of Landsat MSS data. Finally, proper image processing procedures should be identified to ensure improved classification accuracies being achieved from the Landsat TM data with higher resolutions.

2.4.3 Image preprocessing

As a prerequisite for change detection, geometric rectification makes possible the production of spatially corrected maps of land use/cover through time. Ideally, sufficient ground control data should be collected to rectify at least one good-quality image within a satellite time series, which can be further used as the 'master' image to rectify other "slave" images. While this preprocessing procedure has been conducted in virtually every change detection project, the need for radiometric correction among a satellite time series has recently been stressed (e.g., Cihlar, 2000; Yang and Lo, 2000, 2002; Du, Teillet and Cihlar,

TABLE 2.6 Image resolution impacts upon land use/cover classification.

Land use/cover class	Area statistics[a] (ha)			Kappa index		
	Original TM image	Degraded TM image	MSS image	Original TM image	Degraded TM image	MSS image
High-density urban use	50 112	49 917	52 086	0.70	0.66	0.60
Low-density urban use	180 153	214 312	256 103	0.72	0.63	0.47
Exposed/cultivated land	34 083	35 164	5767	0.47	0.31	0.36
Cropland or grassland	140 499	136 692	130 494	0.68	0.49	0.61
Forest	619 024	587 587	581 032	0.88	0.79	0.77
Water	20 212	20 351	18 605	0.93	0.82	0.72
Overall Kappa index	NA			0.71	0.65	0.63
Overall accuracy				75.19%	69.63	68.15%

[a] The total area is approximately 1 044 030 ha.

2002; Schroeder et al., 2006; Vicente-Serrano, Perez-Cabello and Lasanta, 2008; Yang and Chen, 2010). The latter becomes particularly relevant when using satellite data spanning several decades, such as archival Landsat imagery acquired by different sensors.

The differences in radiometry among a satellite time series, caused by such factors as sensor variations, atmospheric properties, and sensor-target-illumination geometry, must be eliminated or minimized so that a common radiometric response among the data set can be restored. To this end, the relative radiometric normalization (RRN) method is preferred over the absolute radiometric correction method as no *in situ* atmospheric data at the time satellite overpasses are necessary (Hall et al., 1991). There are some RRN methods that have been proposed, such as pseudoinvariant features, radiometric control set, image regression, no change set determined from scattergrams, histogram matching, among others. A comprehensive review on these methods was given by Yang and Lo (2000). These methods are based on different assumptions and hence can perform differently with respect to the variations in land use/cover distribution, water-land proportion, topographic relief, similarity between the reference and the subject scenes, and sample size. Caution must be taken when using a RRN method being visually and statistically robust since it can substantially reduce the magnitude of meaningful spectral differences. Because of the space limit, we are not able to delve into specific RRN procedures, and readers can always refer to the work published by Yang and Lo (2000) for some further discussions.

2.4.4 Change detection methods

After data acquisition and preprocessing, the next step in the generic urban spatial growth monitoring workflow is to identify a digital change detection method. Either image-to-image comparison or map-to-map comparison can be used for this purpose. General reviews of different algorithms under these two approaches are given elsewhere (e.g., Singh, 1989; Lu et al., 2004; Radke et al., 2005). The effectiveness of these techniques for urban land use/cover change detection was assessed by Kam (1995) and Ridd and Liu (1998). The image-to-image comparison can be conducted by using raw images (e.g., Green, Kempka and Lackey, 1994), a composite image (e.g., Liu and Lathrop, 2002), or derived indexes (e.g., Yang and Liu, 2005a). Being generally accurate, the image-to-image comparison approach provides detailed information on intra-class land use/cover modification or intensification but suffers from the inability to provide detailed information of how various urban land use/cover categories change (i.e., inter-class change) (Singh, 1989; Kam, 1995; Ridd and Liu, 1998; Yang and Liu, 2005a). On the other hand, the map-to-map comparison can be completed by using two classified maps, providing a full matrix of change, as demonstrated in the case study reported in this chapter. However, the effectiveness of the map-to-map comparison approach is highly dependent upon the assumptions and techniques used to produce maps of the same area at different times. Conventional pattern classifiers are largely built upon parametric statistics. They generally work well for medium-resolution scenes covering spectrally homogeneous areas, but not in heterogeneous regions or when scenes contain severe noises due to the increase of image spatial resolution.

For years substantial research efforts have been made to improve the performance of image pattern classification for working with different types of remote sensor data, and some strategies have been developed as a result of such efforts. Examples include: (i) the identification of various hybrid approaches that combine two or more classifiers, or incorporate pre- and postclassification image transformation and feature extraction techniques (e.g., Yang and Liu, 2005b); (ii) the development of 'soft' classifiers by introducing partial memberships for each pixel to accommodate the heterogeneous and imprecise nature of the real world (e.g., Shalan, Arora and Ghosh, 2003); (iii) the decomposition of each pixel into independent endmembers or pure materials to conduct image classification at sub-pixel level (e.g., Verhoeye and Wulf, 2002; Chapter 9); (iv) the incorporation of the spatial characteristics of the neighboring (contextual) pixels to develop object-oriented classification (e.g., Walker and Briggs, 2007; Chapter 7); (v) the fusion of multisensor, multitemporal, or multisource data for combining multiple spectral, spatial, and temporal features and ancillary information in the image classification (e.g., Tottrup, 2004; Chapter 10); and (vi) the use of artificial intelligence technology, such as rule-based classifiers (e.g., Schmidt *et al.*, 2004), artificial neural networks (see Chapter 8) and support vector machines (Yang, 2011), for pattern classification. These developments are quite promising. Nevertheless, further efforts will certainly be maintained and will probably intensify in order to adopt these techniques to solve practical problems in a productive fashion.

Summary and concluding remarks

In this chapter, we discussed the utilities of satellite remote sensing for the observation and measurement of urban spatial growth and landscape changes emphasizing the use of archival Landsat data. We targeted the data acquired by the three generations of Landsat imaging sensors because they provide the principal source of data for urbanization studies at the regional, national and global scales. After a review on the past, present and future of the Landsat program and its imaging sensors, we present a case study focusing on a rapidly suburbanizing metropolis to demonstrate the usefulness of time-sequential Landsat imagery for monitoring decades-long urban spatial growth. The Atlanta metropolitan area as the case study site is an ideal city to study the postmodern urban dynamics and environmental consequences of accelerating urban growth. The technical procedures used were designed and implemented after a careful examination of the image characteristics, landscape complexity, and the status of technological development. Central to this study was a time series of satellite imagery acquired by the three Landsat imaging sensors, which has been used to produce accuracy-compatible land use/cover maps through unsupervised classification and spatial reclassification procedures. The time series of classified maps was further used to analyze urban spatial growth and the nature of change. This study reveals a rampant urban land growth during a period of nearly four decades. Urban spatial form experienced a transition from linear concentration to multinucleated pattern for high-density urban use and from concentration mixed with some degree of dispersal to more dispersed pattern for low-density urban use in Atlanta.

Lastly, based on the case study and other literature, we further formulated a generic workflow for urban spatial growth monitoring, and discussed some conceptual and technical issues emerging when using archival satellite images acquired by different sensors and perhaps during different seasons. Conceptually, we strongly suggested remote sensing professionals should be equipped with some essential knowledge on urban geography and landscape ecology that can help identify appropriate remote sensor data or information extraction techniques in an urban study project. Technically, we emphasized a thorough understanding of the spatial, spectral, radiometric and temporal characteristics of the satellite time series being used, a mandatory radiometric normalization procedure to help restore a common radiometric response among the multi-date, multi-sensor data set, and a carefully-tailored change detection procedure with the research objectives and scope in mind. We believe such discussions can help identify the outstanding issues that must be addressed in order to implement an urban growth monitoring protocol effectively.

References

Alberti, M., Weeks, R. and Coe, S. (2004) Urban land-cover change analysis in Central Puget Sound. *Photogrammetric Engineering and Remote Sensing*, **70**, 1043–1052.

Anderson, J.R., Hardy, E.E., Roach, J.T. and Witmer, R.E. (1976) *A Land Use and Land Cover Classification System for Use with Remote Sensor Data*. USGS Professional Paper 964. Available http://landcover.usgs.gov/pdf/anderson.pdf (accessed 10 July 2010).

Auch, R., Taylor, J. and Acededo, W. (2004) *Urban Growth in American Cities: Glimpses of U.S. Urbanization*. USGS Circular 1252. Available http://pubs.usgs.gov/circ/2004/circ1252/| (accessed 10 July 2010).

Bhatta, B. (2010). *Analysis of Urban Growth and Sprawl from Remote Sensing Data*, Springer-Verlag, Berlin, Heidelberg.

Bullard, R.D., Johnson, G.S. and Torres, A.O. (eds.) (2000). *Sprawl City: Race, Politics, and Planning in Atlanta*, Island Press, Washington, D.C.

Carlson, T.N. (2004) Analysis and prediction of surface runoff in an urbanizing watershed using satellite imagery. *Journal of the American Water Resources Association*, **40**, 1087–1098.

Cihlar, J. (2000). Land cover mapping of large areas from satellites: status and research priorities. *International Journal of Remote Sensing*, **21**, 1093–1114.

Congalton, R.G. (1991) A review of assessing the accuracy of classifications of remotely sensed data. *Remote Sensing of Environment*, **1**, 35–46.

Dear, M. and Flusty, S. (1998) Post modern urbanism. *Annals of the Association of American Geographers*, **88**, 50–72.

Doygun, H. and Alphan, H. (2006) Monitoring urbanization of Iskenderun, Turkey, and its negative implications. *Environmental Monitoring and Assessment*, **114**, 145–155.

Draeger, W.C., Holm, T.M., Lauer, D.T. and Thompson, R.J. (1997) The availability of Landsat data: past, present, and future. *Photogrammetric Engineering and Remote Sensing*, **63**, 869–875.

Du, Y., Teillet, P.M., and Cihlar, J. (2002) Radiometric normalization of multitemporal high-resolution satellite images with quality control for land cover change detection. *Remote Sensing of Environment*, **82**, 123–134.

Ellis, E.C., Wang, H.Q., Xiao, H.S. *et al.* (2006). Measuring long-term ecological changes in densely populated landscapes using current and historical high resolution imagery. *Remote Sensing of Environment*, **100**, 457–473.

Foley, J.A., DeFries, R., Asner, G.P. *et al.* (2005) Global consequences of land use. *Science*, **309**, 570–574.

Gaydos, L. and Newland, W.L. (1978) Inventory of land-use and land cover of Puget Sound region using Landsat digital data. *Journal of Research of the US Geological Survey*, **6**, 807–814.

Gomarasca, M.A., Brivio, P.A., Pagnoni, F., and Galli, A. (1993) One century of land use changes in the metroplitan area of Milan (Italy). *International Journal of Remote Sensing*, **14**, 211–223.

Green, K., Kempka, D. and Lackey, L. (1994). Using remote sensing to detect and monitor land cover and land use change. *Photogrammetric Engineering and Remote Sensing*, **60**, 331–337.

Grimm, N.B., Faeth, S.H., Golubiewski, N.E. et al. (2008) Global change and ecology of cities. *Science*, **319**, 756–760.

Haack, B., Bryant, N. and Adams, S. (1987) An assessment of Landsat MSS and TM data for urban and near-urban land-cover digital classification. *Remote Sensing of Environment*, **21**, 201–213.

Hall, F.G., Strebel, D.E., Nickeson, J.E. and Goetz, S.J. (1991) Radiometric rectification: toward a common radiometric response among multidate, multisensor images. *Remote Sensing of Environment*, **35**, 11–27.

Hall, T. (2001). *Urban Geography*, 2nd edition, Routledge, London.

Hartshorn, T. and Muller, P. (1989) Suburban downtown and the transformation of metropolitan Atlanta's business landscape. *Urban Geography*, **10**, 375–395.

Herold, M., Goldstein, N.C. and Clarke, K.C. (2003) The spatiotemporal form of urban growth: measurement, analysis and modeling. *Remote Sensing of Environment*, **86**, 286–302.

Herold, M., Scepan, J. and Clarke, K.C. (2002) The use of remote sensing and landscape metrics to describe structures and changes in urban land uses. *Environment and Planning A*, **34**, 1443–1458.

Hill, J. and Hostert, P. (1996) Monitoring the growth of a Mediterranean metropolis based on the analysis of spectral mixtures a case study on Athens (Greece), in *Progress in Environmental Remote Sensing Research and Applications* (ed. E. Parlow), Balkema, Rotterdam, pp. 21–30.

Jensen, J.R. (1981) Urban change detection mapping using Landsat digital data. *American Cartographer*, **8**, 127–147.

Jensen, J.R. (2007) *Remote Sensing of the Environment: An Earth Resource Perspective*, 2nd edition, Prentice Hall, New Jersey.

Jensen, J.R. and Cowen, D.C. (1999) Remote sensing of urban suburban infrastructure and socio-economic attributes. *Photogrammetric Engineering and Remote Sensing*, **65**, 611–622.

Jensen, J.R., Campbell, J., Dozier, J. et al. (1989) Remote sensing, in *Geography in America* (eds G.L. Gaile and C.J. Willmott), Merrill, Columbus, pp. 746–775.

Kaplan, D.H., Wheeler, J.O., and Holloway, S.R. (2009) *Urban Geography*, 2nd edition, John Wiley & Sons, Inc., New York.

Kam, T.S. (1995) Integrating GIS and remote sensing techniques for urban land-cover and land-use analysis. *Geocarto International*, **10**, 39–49.

Kline, J.D., Moses, A. and Alig, R.J. (2001) Integrating urbanization into landscape-level ecological assessments. *Ecosystems*, **4**, 3–18.

Kondratyev, K.Y., Krapivin, V.F. and Phillips, G.W. (2002) *Global Environmental Change: Modelling and Monitoring*, Springer-Verlag, Berlin, Heidelberg.

Kwarteng, A.Y. and Chavez, P.S. (1998) Change detection study of Kuwait City and environs using multi-temporal Landsat Thematic Mapper data. *International Journal of Remote Sensing*, **19**, 1651–1662.

LaGro, J. and Degloria, S. (1992) Land use dynamics within an urbanizing non metropolitan county in New York State (USA). *Landscape Ecology*, **7**, 275–289.

Lindgren, D.T. (1985) *Land Use Planning and Remote Sensing*, Martinus Nijhoff, Boston.

Liu, X. and Lathrop, R.G. (2002) Urban change detection based on an artificial neural network. *International Journal of Remote Sensing*, **23**, 2513–2518.

Liu, Y.X., Zhang, X.L., Lei, J. and Zhu, L. (2010) Urban expansion of oasis cities between 1990 and 2007 in Xinjiang, China. *International Journal of Sustainable Development and World Ecology*, **17**, 253–262.

Lo, C.P. (1986) *Applied Remote Sensing*, Burnt Mill, England.

Lo, C.P. (2004) Testing urban theories using remote sensing. *GIScience & Remote Sensing*, **41**, 95–115.

Lo, C.P. and Choi, J. (2004) A hybrid approach to urban land use/cover mapping using Landsat 7 Enhanced Thematic Mapper Plus (ETM+) images. *International Journal of Remote Sensing*, **25**, 2687–2700.

Lo, C.P. and Quattrochi, D.A. (2003) Land-use and land-cover change, urban heat island phenomenon, and health implications: A remote sensing approach. *Photogrammetric Engineering and Remote Sensing*, **69**, 1053–1063.

Lo, C.P. and Yang, X. (2002). Drivers of land-use/land-cover changes and dynamic modeling for the Atlanta, Georgia Metropolitan Area. *Photogrammetric Engineering and Remote Sensing*, **68**, 1073–1082.

Longley, P.A. (2002). Geographical Information Systems: will developments in urban remote sensing and GIS lead to 'better' urban geography? *Progress in Human Geography*, **26**, 231–239.

Lu, D., Mausel, P., Brondizio, E. and Moran, E. (2004) Change detection techniques. *International Journal of Remote Sensing*, **25**, 2365–2407.

Markham, B.L. and Towshend, R.G. (1981) Land cover classification accuracy as a function of sensor spatial resolution, in *Proceedings of the Fifteenth International Symposium on Remote Sensing of Environment*, Ann Arbor, Michigan, pp. 1075–1090.

Masek, J.G., Lindsay, F.E. and Goward, S.N. (2000). Dynamics of urban growth in the Washington DC metropolitan area, 1973–1996, from Landsat observations. *International Journal of Remote Sensing*, **21**, 3473–3486.

Millward, A.A., Piwowar, J.M. and Howarth, P.J. (2006) Time-series analysis of medium-resolution, multisensor satellite data for identifying landscape change. *Photogrammetric Engineering and Remote Sensing*, **72**, 653–663.

Mittelbach, F.G. and Schneider, M.I. (2005) Remote sensing: With special reference to urban and regional transportation. *Annuals of Regional Science*, **5**, 61–72.

Mumby, P.J. and Edwards, A.J. (2002) Mapping marine environments with IKONOS imagery: enhanced spatial resolution can deliver greater thematic accuracy. *Remote Sensing of Environment*, **82**, 248–257.

Pathan, S.K., Sastry, S.V.C., Dhinwa et al. (1993) Urban-growth trend analysis using GIS techniques: a case-study of the Bombay metropolitan region. *International Journal of Remote Sensing*, **14**, 3169–3179.

Paulsson, B. (1992) *Urban Applications of Satellite Remote Sensing and GIS Analysis*, The World Bank, Washington, DC. Available http://ww2.unhabitat.org/programmes/ump/documents/UMP9.pdf (accessed 10 July 2010).

Radke, R.J., Andra, S., Al-Kofahi, O., and Roysam, B. (2005) Image change detection algorithms: A systematic survey. *IEEE Transactions on Image Processing*, **14**, 294–307.

Ridd, M.K. and Liu, J.J. (1998) A comparison of four algorithms for change detection in an urban environment. *Remote Sensing of Environment*, **63**, 95–100.

Royer, A., Charbonneau, L. and Bonn, F. (1988) Urbanization and Landsat MSS albedo change in the Windsor Québec corridor since 1972. *International Journal of Remote Sensing*, **9**, 555–566.

Schneider, A. and Woodcock, C.E. (2008) Compact, dispersed, fragmented, extensive? A comparison of urban growth in 25 global cities using remotely sensed data, pattern metrics and census information. *Urban Studies*, **45**, 659–692.

Schroeder, T.A., Cohen, W.B., Song, C.H., Canty, M.J. and Yang, Z.Q. (2006) Radiometric correction of multi-temporal Landsat data for characterization of early successional forest patterns in western Oregon. *Remote Sensing of Environment*, **103**, 16–26.

Schmidt, K.S., Skidmore, A.K., Kloosterman, E.H. *et al.* (2004) Mapping coastal vegetation using an expert system and hyperspectral imagery. *Photogrammetric Engineering and Remote Sensing*, **70**, 703–715.

Seto, K.C. and Fragkias, M. (2005) Quantifying spatiotemporal patterns of urban land-use change in four cities of China with time series landscape metrics. *Landscape Ecology*, **20**, 871–888.

Shalan, M.A., Arora, M.K. and Ghosh, S.K. (2003) An evaluation of fuzzy classifications from IRS 1C LISS III imagery: a case study. *International Journal of Remote Sensing*, **24**, 3179–3186.

SIERRA CLUB (1998) Sprawl: The Dark Side of the American Dream. Available http://www.sierraclub.org/sprawl/report98/report.asp (accessed 10 July 2010).

Singh, A. (1989) Digital change detection techniques using remotely-sensed data. *International Journal of Remote Sensing*, **10**, 989–1003.

Small, C. (2005) Urban remote sensing: Global comparisons. *Architectural Design*, **178**, 18–23.

Soja, E. (1989) *Postmodern Geographies: The Reassertion of Space in Critical Social Theory*, Verso, London.

Todd, W.J. (1977) Urban and regional land-use change detected by using Landsat data. *Journal of Research of the US Geological Survey*, **5**, 529–534.

Toll, D.L., Royal, J.A. and Davis, J.B. (1980) Urban area up date procedures using Landsat data, in *Proceedings of the Fall 1980 Technical Meeting of the American Society of Photogrammetry*, ASP, Falls Church, Virginia, pp. RS-E1–17.

Tottrup, C. (2004) Improving tropical forest mapping using multi-date Landsat TM data and pre-classification image smoothing. *International Journal of Remote Sensing*, **25**, 717–730.

Townshend, J. and Justice, C. (1981) Information extraction from remotely sensed data, a user view. *International Journal of Remote Sensing*, **2**, 313–329.

Trigg, S.N., Curran, L.M. and McDonald, A.K. (2006) Utility of Landsat 7 satellite data for continued monitoring of forest cover change in protected areas in Southeast Asia. *Singapore Journal of Tropical Geography*, **27**, 49–66.

Turner, B.L., Lambin, E.F. and Reenberg, A. (2007) The emergence of land change science for global environmental change and sustainability. *Proceedings of the National Academy of Sciences, USA*, **104**, 20666–20671.

UN-HABITAT (2010) State of the World's Cities 2010/2011 – Cities for All: Bridging the Urban Divide. Available http://www.unhabitat.org/content.asp?cid=8051&catid=7&typeid=46&subMenuId=0 (accessed 10 July 2010).

Vanderee, D. and Ehrlich, D. (1995) Sensitivity of ISODATA to changes in sampling procedures and processing parameters when applied to AVHRR time series NDVI data. *International Journal of Remote Sensing*, **16**, 673–686.

Verhoeye, J. and Wulf, R.D. (2002) Land cover mapping at sub-pixel scales using linear optimization techniques. *Remote Sensing of Environment*, **79**, 96–104.

Vicente-Serrano, S.M., Perez-Cabello, F. and Lasanta, T. (2008) Assessment of radiometric correction techniques in analyzing vegetation variability and change using time series of Landsat images. *Remote Sensing of Environment*, **112**, 3916–3934.

Walker, J.S. and Briggs, J.M. (2007) An object-oriented approach to urban forest mapping in Phoenix. *Photogrammetric Engineering and Remote Sensing*, **73**, 577–583.

Welch, R. (1982) Spatial resolution requirements for urban studies. *International Journal of Remote Sensing*, **3**, 139–146.

Williams, D.L., Irons, J.R., Markham, B.L. *et al.* (1984) A statistical evaluation of the advantages of Landsat Thematic Mapper data in comparison to multispectral scanner data. *IEEE Transactions on Geoscience and Remote Sensing*, **GE-22**, 294–301.

Wulder, M.A., White, J.C., Goward, S.N. *et al.* (2008) Landsat continuity: Issues and opportunities for land cover monitoring. *Remote Sensing of Environment*, **112**, 955–969.

Yang, X. (2002) Satellite monitoring of urban spatial growth in the Atlanta metropolitan region. *Photogrammetrical Engineering and Remote Sensing*, **68**, 725–734.

Yang, X. (2003) Remote sensing and GIS for urban analysis: An introduction. *Photogrammetrical Engineering and Remote Sensing*, **69**, 937, 939.

Yang, X. (2007) Integrating remote sensing, GIS and dynamic modeling for sustainable urban growth management, in *Integration of GIS and Remote Sensing* (ed V. Mesev), John Wiley & Sons, Inc., New York.

Yang, X. (2010) Integration of remote sensing with GIS for urban growth characterization, in *Geospatial Analysis and Modeling of Urban Structure and Dynamics* (eds B. Jiang and X. Yao), Springer, Dordrecht.

Yang, X. (2011) Parameterizing support vector machines for land cover classification. *Photogrammetric Engineering and Remote Sensing*, **77**, 27–37. (in press).

Yang, X. and Chen, L. (2010) Using multi-temporal remote sensor imagery to detect earthquake-triggered landslides. *International Journal of Applied Earth Observation and Geoinformation*, **12**, 487–495.

Yang, X. and Liu, Z. (2005a) Use of satellite-derived landscape imperviousness index to characterize urban spatial growth. *Computers, Environment and Urban Systems*, **29**, 524–540.

Yang, X. and Liu, Z. (2005b). Using satellite imagery and GIS to characterize land use and land cover changes in an estuarine watershed. *International Journal of Remote Sensing*, **26**, 5275–5296.

Yang, X. and Lo, C.P. (2000) Relative radiometric normalization performance for change detection from multi-date satellite images. *Photogrammetric Engineering and Remote Sensing*, **66**, 967–980.

Yang, X. and Lo, C.P. (2002) Using a time series of normalized satellite imagery to detect land use/cover change in the Atlanta, Georgia metropolitan area. *International Journal of Remote Sensing*, **23**, 1775–1798.

Yang, X. and Lo, C.P. (2003) Modelling urban growth and landscape changes in the Atlanta metropolitan area. *International Journal of Geographical Information Science*, 17, 463–488.

3

Limits and challenges of optical very-high-spatial-resolution satellite remote sensing for urban applications

Paolo Gamba, Fabio Dell'Acqua, Mattia Stasolla, Giovanna Trianni and Gianni Lisini

This chapter discussed the major challenges and limits when using very-high-spatial-resolution (VHR) optical remote sensor data for monitoring human settlements. Specifically, the discussion centers on the spectral aspects, mapping limits and challenges in connection to urban change analysis. Then, an integrated adaptive spatial approach is proposed to help deal with the challenges, followed by an example of the major research lines still to be pursued in this active field. Finally, the chapter discusses some interesting results using VHR data for damage mapping in connection to major events.

3.1 Introduction

The use of very-high-spatial-resolution (VHR) satellite imagery is steadily increasing in many fields from precision mapping to location-based businesses. The trend toward finer spatial resolution is pushed by peculiar applications, whose spatial requirements cannot be met even by high resolution sensors. Urban-related applications are among the front-runners. However, it is true that VHR data come with equally numerous limits and challenges than advantages and improvements. Of course, this is true for both radar and optical images, but in this chapter we will focus on the optical data, while in a companion chapter radar images will be considered.

Indeed, VHR optical imagery is nowadays offered by many sensors, from Quickbird to Worldview-1 and -2, from Ikonos II to Geoeye-1, from Cartosat-1 to the EROS (Earth Remote Observation Satellite) constellation. Moreover, future systems such as the French Pleiades constellation will provide faster revisit times, thus enhancing the timeliness of the data for urban disaster management and similar applications. A very detailed analysis of the current situation of VHR sensors for urban applications may be found in Ehlers (2009) and won't be repeated here. However, following and updating that source, Table 3.1 summarizes the main information. It is worth noting that less than 1 m spatial resolution and less than 1 day revisit images are to be considered a very close goal.

The question we would like to address here is whether the users (all, and not only "end-users") are really able to exploit the wealth of information coming from these sensors. As researchers in remote sensing, and especially in urban remote sensing, we are aware that the answer is generally negative. The reasons are not only, however, the lack of knowledge by the users of the potentials of these data, but also the problems and various issues related to interpretation and (semi)automatic analysis of VHR imagery. According to what is available in technical literature, we will attempt in the following sections to address to what extent these issues are critical, and which is their impact on urban remote sensing and urban area applications.

To this aim, we will follow this list:

- issues related to geometrical problems of optical VHR data: all the challenges coming from the geometric accuracy of the data and its positioning in a common reference system;
- issues related to spectral problems, especially the lack of discrimination capability of current VHR sensors and the need of a compromise between VHR in the spectral and the spatial sense;
- issues related to mapping problems, i.e. the need to analyze the scene no more using a per-pixel approach, but gradually shifting to a per-object approach;
- issues related, finally, to multitemporal analysis, e.g. for change detection, whose validity is strongly connected to the ability to correlate features and objects more than isolated pixels.

The final part of the chapter will provide an example coming from our experience. The approach proposed in those paragraphs is meant to provide a possible way to overcome some of the problems discussed in the first sections. Although it is not the "best" available methodology, it might be useful to highlight one or more interesting and possible research paths and thus invite the interested readers to find their own way to solve any specific urban remote sensing problem of their interest.

3.2 Geometrical problems

A first group of issues peculiar to VHR images are due to geometric accuracy. In images whose pixels correspond to a ground resolution of less than 1 m, a correct georeferencing may be a real problem. There are a few examples aimed at discussing how precise the standard products provided by the data providers are. For instance, in Davis and Wang (2001a) many Ikonos scenes have revealed a consistent RMS (root mean square) error of nearly 2 m between the referenced and the actual position of the images. The reasons for these currently challenging precision values are the reduced availability of same resolution digital elevation models (DEM) and the approximations in the satellite orbital parameters. Although the effects of the second problem may be reduced by accurate post analysis of the satellite position at the data acquisition time, the lack of accurate DEM is still a problem. Lidar and aerial systems have been extensively used to obtain DEMS in urban areas (e.g., Davis and Wang, 2001b; Liu et al., 2007), but their data is often limited to the city centre and do not cover the extensive areas depicted by VHR optical sensors. The availability of InSAR (Interferometric SAR) DEMs (Gamba and Houshmand, 2000) is increasingly being considered as one way to reduce the problem. Similar results are also obtained using stereo pair of VHR optical satellite images (Baltsavias, 2005).

A second problem, more peculiar to urban areas, is the need for orthorectification of VHR images. The fine spatial resolution in fact causes geometric distortion of the three-dimensional urban landscape, which must be compensated when using the images for mapping purposes. There are many examples of approaches developed for orthorectification of VHR optical images. One example is Volpe and Rossi (2003), where the problem is put in the framework of the overall exploitation of Quickbird data.

Finally, the big improvement in new-generation VHR sensors, which is the capability to steer the sensor, brings further limits to the usability of these data. In fact, the possibility of different viewing angles for the same scene is an advantage to reduce revisit rime (the current 3-day lower limit of many sensors is obtained due to this capability). It is, however, a further challenge in order to design analysis approaches that can exploit scenes taken in different dates with different acquisition geometries and obtain consistent results. For instance, different viewing angles do not guarantee the availability of comparable orthorectified products for the same area. Indeed, occlusions, especially in dense urban areas, prevent achieving mapping products with similar accuracy. In turn, this results in erroneous or incomplete change maps, when these acquisitions are used for change detection after classification.

Just to provide an example of problems due to slanted viewing angles, in Fig. 3.1 a Quickbird image covering the downtown of San Francisco is shown, made available from Digitalglobe after the 2007 Joint Urban Remote Sensing Event conference (the complete image set is available at http://tlc.unipv.it/urban-remote-sensing-2007. In the same figure a very simple classification map is also provided, and even by a visual inspection it is clear how the geometric distortion due to the lateral viewing angle with respect to some of the high-rise building in the area affects the mapping results.

TABLE 3.1 Current very-high-spatial-resolution satellite systems (modified from Ehlers 2009).

Company/institution	GeoEye				Digital Globe				KARI		ImageSat
System	Ikonos II		GeoEye-1		QuickBird 2		WorldView-1	WorldView-2		KOMPSAT 2		EROS B
Sensors	Pan	Multisp.	Pan	Multisp.l	Pan	Multisp.	Pan	Pan	Multisp.	Pan	Multisp.	Pan
Geometric resolution	1 m	4 m	0.41 m	1.65 m	0.61 m	2.44 m	0.5 m	0.46 m	1.84 m	1 m	4 m	0.7 m
Spectral resolution (nm)	525–929	445–516 B 506–595 G 632–698 R 767–853 N-d	450–900	450–520 B 520–600 G 625–695 R 760–900 N-d	450–900	450–520 B 520–600 G 630–690 R 760–900 N-d	450–900		400–450 630–690 450–510 705–745 510–580 770–895 585–625 860–1040	500–900	450–520 520–600 630–690 760–900	500–900
Image size	11 × 11 km²		8 × 8 km²		16.5 × 16.5 km²		17.6 × 17.6 km²			15 × 15 km²		7 × 7 km²

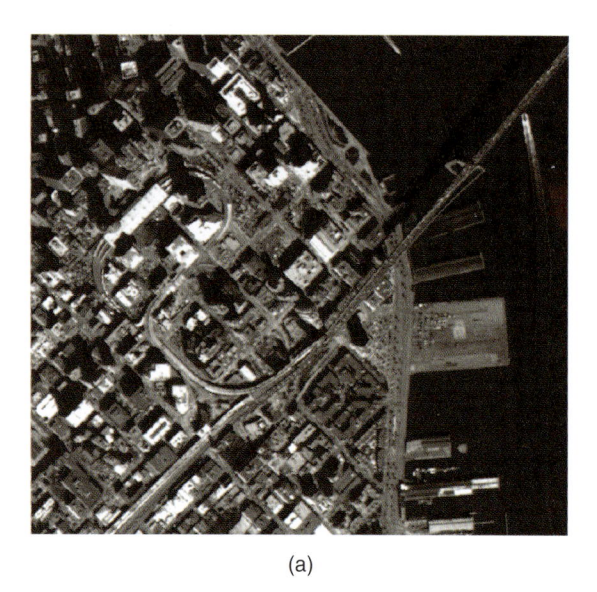

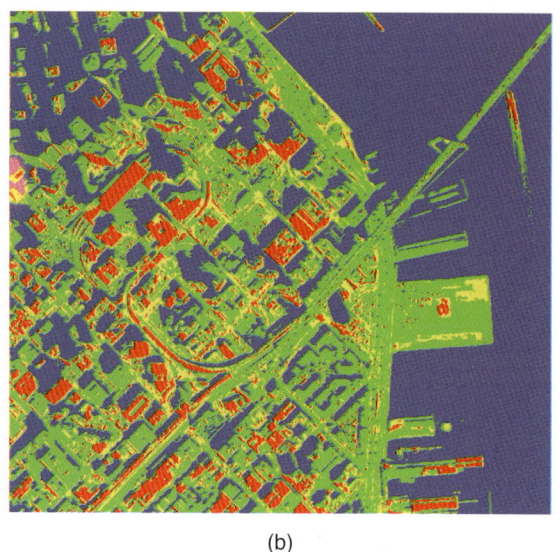

FIGURE 3.1 A sample of multispectral Quickbird data covering the downtown of San Francisco, USA: (a) the original image; (b) the output after a supervised classification.

3.3 Spectral problems

VHR images in urban areas may be used for many applications, some of them connected to environmental monitoring. Mapping of dangerous materials (Marino *et al.*, 2000), "urban forest" monitoring (Xiao *et al*, 1999) and sealed part detection (Segl, Roessner and Heiden, 2000) are only examples. In these applications, however, the stress on VHR resolution is due to the need to detect the exact location of the above mentioned features, which are of course at the level of single objects in urban areas, and thus require a sub-meter spatial resolution. It is however extremely important for these applications the availability of VHR imagery also with very high spectral resolution. This is immediately obtained by observing that most of the above mentioned examples refer to airborne hyperspectral imagery, which pair spectral and spatial fine resolution. The availability of specific spectral bands for urban roof material mapping have been discussed for instance in Herold, Gardner and Roberts (2003) showing that there are specific frequencies in the 2 m wavelength range that are extremely useful to map urban materials.

Generally, the current operation VHR optical sensors from satellite platform do not match the requirements in term of spectral bands for many urban applications. Their use beyond basic mapping is hampered by the lack of important bands in the medium infrared, which can be used to characterize building materials and surface properties (e.g., for road asphalt status monitoring as in Herold and Roberts, 2005). WorldView-2 is somehow the next move in this direction, as it couples VHR spatial resolution in the panchromatic sensor with eight bands in the multispectral one.

A different problem, which has not been addressed so far in urban research, is the need of a complete characterization of the bidirectional reflectance function for urban materials. In fact, the dependence of the material spectral signature from the orientation between the sensor and the observed specimen is an issue when spatial and spectral fine resolution are paired. For this kind of sensors, per-pixel analysis must take into account the effect or the classification will results in errors because of differences in the apparent spectra. The effect is well-known in hyperspectral data analysis (see for instance the work by Schiefer, Hostert and Damm, 2005) but still to be considered for sensors with a smaller set of wavelengths.

3.4 Mapping limits and challenges

As a consequence of the challenges related to the fine spatial resolution and the relatively coarse spectral resolution of current optical VHR data from spaceborne platforms, urban mapping is not an easy task. Although there are studies focused on the trade-off between these two characteristics (comparing airborne and satellite sensors, see Gamba and Dell'Acqua, 2006), they are limited to the specific considered legend and still need to be validated on large and various data sets.

As a result of the geometrical and spectral inconsistencies of single pixels, the stress in urban mapping using optical VHR sensor is on using information from more than one pixel at a time. Spatial aggregations of pixels are considered by means of textural (Yashon and Tateishi, 2008) or geometrical features (Lisini *et al.*, 2005), or by tightening the class assignment for a pixel to the class membership of its neighbors (e.g. by means of a Markov Random Field approach, as in Gamba and Trianni, 2007). The most exploited technique, however, is the so called object-based image analysis (OBIA, see Blaschke, Lang, and Hay, 2008). The idea is that objects (i.e. portions of the scene that are statistically consistent with respect to a set of spectral and spatial features) are to be searched for. They define a higher level of interpretation of the scene beyond the land cover obtainable from the single pixel. By aggregating the features at the pixel level it is possible to assign each of these objects to a class, with either a soft or a hard decision. Eventually, both the pixel-based and the segment-based decisions can be profitably used together to characterize the scene.

OBIA or spatiospectral analysis of the data at a single spatial scale must however be complements with the ability to adapt to data with different spatial resolutions and different viewing angles (here called *spatial adaptiveness*). To this aim, most of the proposed and more advanced methodologies are based on adaptive combined spectral and spatial processing of the whole scene. For instance, in Bruzzone and Carlin (2006) an initial segmentation of the scene at multiple resolutions is performed, to adapt the data interpretation procedure to the different scales of the scene objects.

Moreover, interpretation algorithms may fully exploit the available information in the scene by exploiting *class adaptiveness*. The concept behind this term is that different classes require different processing steps, or different parameters for the same sequence of processing steps, and tailoring general algorithms to these requirements (in addition to different scales) is a way to further improve their effectiveness for each class.

By combining properties at different scales, spatial characteristics and class-tailored processing, mapping algorithm for VHR data are able to go beyond basic land cover mapping. However, to this aim, and especially in urban areas, they need to accommodate the *a-priori* information by the human interpreter or the ancillary data that often are available for these areas. So far, one of the still open and more interesting challenges of VHR optical urban remote sensing is to define a strategy useful to mapping land use with a consistent legend in many parts of the world. The challenge is the development of these methodologies, robust to different geographical locations, and able to capture the variable spatial scales of land use classes in this environment, is one of the most important to move from "academic exercises" to the operational use of VHR optical data.

3.5 Adding the time factor: VHR and change detection

One additional set of challenges and problems of VHR optical images is related to the exploitation of the time factor, i.e. the use of multitemporal data. In order to find changes or to highlight persistent features in coarse and high resolution optical images it is often enough to require the same viewing angle (which for most sensors is fixed) and ascending or descending orbit. This is no more the case for VHR images, which might be affected by (co-)registration errors due to the uncertainties in DEMS or orbital measures, as discussed in the previous sections. There is therefore the need for methodologies meant to cope with the so called "registration noise" and provide robust change detection at the pixel level. As an example, we refer here to Bruzzone and Cossu (2003) and Bovolo, Bruzzone and Marchesi (2008), both very effective approaches. Additionally, as an example of the problems researchers are called to face, Fig. 3.2 shows pre- and post-earthquake images taken over the town of L'Aquila, central Italy. The earthquake struck the town in April 2009, causing death and damages. A pixel-by-pixel post-classification comparison of the center of the town in both images is shown in the figure. It is clearly visible that most of the changes highlighted between the two classifications are simply due to coregistration errors, while the real changes are "lost" among the false positives.

The trend in multitemporal urban remote sensing using VHR optical data is, as already mentioned for single date classification, to exploit OBIA or feature based comparison to avoid these problems. Object-based comparison was proposed by means of the exploitation of a scene model by Hazel (2001) and by Carlotto (1997). In the latter paper non-linear prediction technique for measuring changes between images and temporal segmentation and filtering techniques for analyzing patterns of change over time are used. Other examples of object-based change detection can be found in Dekker (2004), Walter (2004), Niemeyer and Nussbaum (2006), Blaschke (2003), and Benz et al. (2004). As a different approach, geometrical features may be exploited, often in addition to pixel-based change maps, to detect important elements of the scene (Lisini *et al.* 2005) and changes (Dell'Acqua, Gamba and Lisini, 2006). No ultimate approach has been designed so far, but many *ad hoc* solutions for specific problems have been devised.

3.6 A possible way forward

Summarizing the analysis of the previous sections, it is expected that a possible way forward, able to design efficient and effective ways to exploit VHR optical data in urban areas, is to match spatially-aware and adaptive classifiers to object-based analysis and change detection. In the following we will show one possible way to implement these ideas together with some results to validate it.

We will start first by the above mentioned notion of *class adaptiveness* to design an example of such a scene classifiers. As discussed in the introduction, spatial feature are often peculiar to specific classes, especially in anthropogenic environments. Furthermore, this idea could be widened to spatial relationships among elements of the same class or of different classes in the same environment. Therefore, the idea is to exploit these characteristics by means of a class-specific spatial post-processing of a basic pixel-based classification map. The key is to define and apply a set of rules in a different and adaptive way to each of classes of the original classification map, as depicted in Fig. 3.3. This set needs to be designed including flexible and efficient spatial operators, and must able to reflect the known information about the spatial features of the objects belonging to a class. The operators need to be designed to reduce misclassifications and discard wrong associations between scene objects and mapping classes due to errors in the first classification step. According to this description, the best set that we could think of is composed by some of the operators of the so called "mathematical morphology".

Mathematical morphology is a theory for the analysis of spatial structures, since it aims at analyzing the shape and form of objects. The most important operations can be grouped into morphological and geodesic transformations, and are defined on single-band grey level image $X = X(i,j)$. For multi-spectral images, operators are computed separately for each component and then combined to obtain a scalar value. Usually the maximum of the magnitude among the bands is considered. In the following, the main morphological transformations will be quickly introduced. A longer and more precise description can be found in Soille (1999).

The main morphological transformations are:

- Erosion. The erosion of a set X by a structuring element B is defined as the locus of points x such that B is included in X

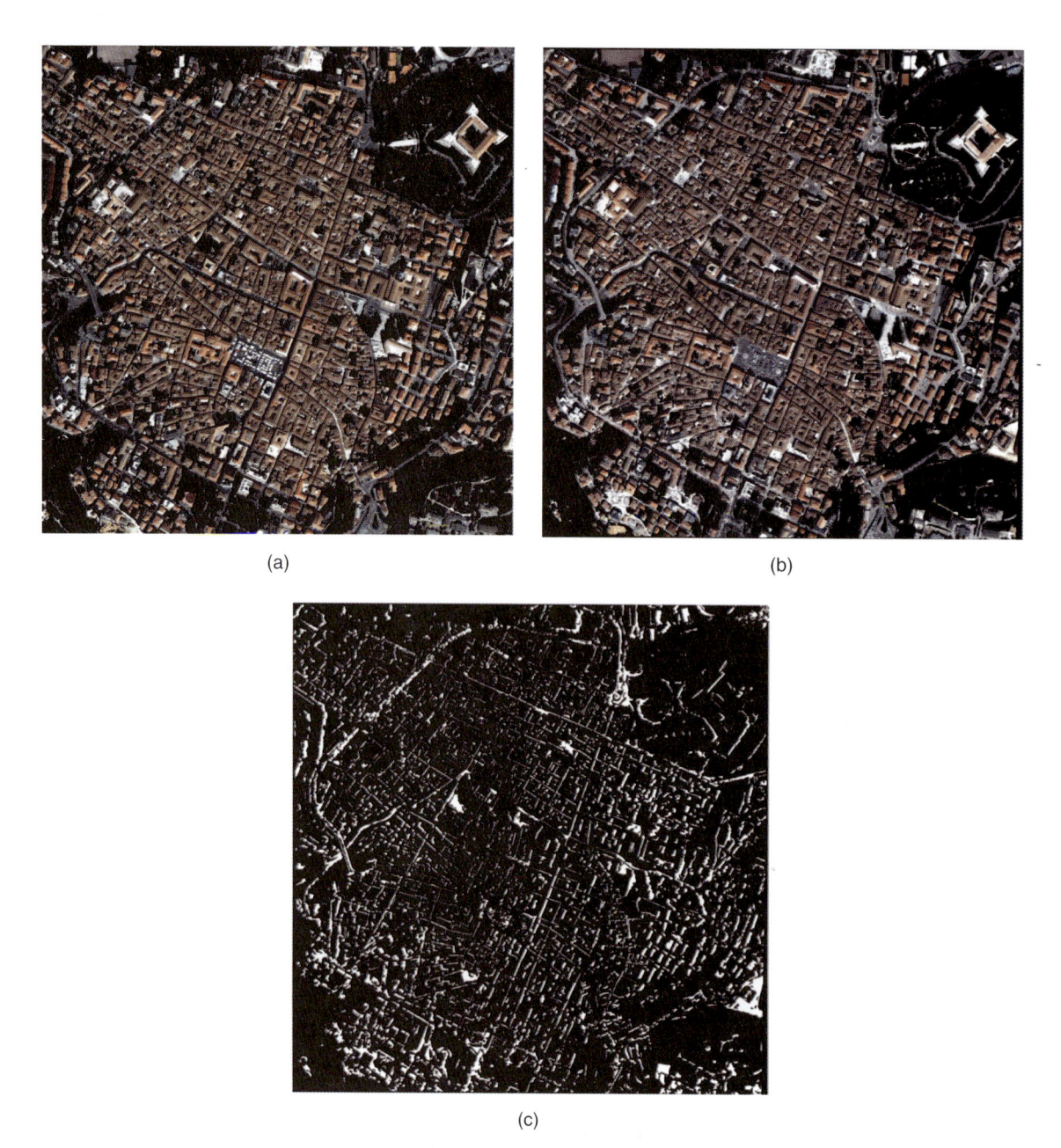

FIGURE 3.2 A small sample of two Quickbird images over L'Aquila in central Italy: (a) and (b) multitemporal VHR optical dataset over the city centre; (c) change detection through post-classification comparison of the pixels belonging to the "building" class.

when its origin is placed at x:

$$\varepsilon_B(X) = \{x \mid B_x \subseteq X\} \quad (3.1)$$

- Dilation. The dilation of a set X by a structuring element B is defined as the locus of points x such that B hits X when its origin coincides with x:

$$\delta_B(X) = \{x \mid B_x \cap X \neq \emptyset\} \quad (3.2)$$

Morphological transformations involve combinations of one input image with specific structuring elements (or SEs). The approach for geodesic transformations is to consider two input images, so that a morphological transformation is applied to the first image, which must remain greater or lower than the second one.

- Geodesic dilation. The geodesic dilation of size 1 of the marker image X with respect to a mask image Y with the same domain, and such as $X \leq Y$, is defined as the point-wise minimum between the mask and the elementary dilation of the marker:

$$\delta_Y^{(1)}(X) = \delta^{(1)}(X) \wedge Y \quad (3.3)$$

- Geodesic erosion. The geodesic erosion is the dual transformation of geodesic dilation with respect to set complementation:

$$\varepsilon_Y^1(X) = \varepsilon^{(1)}(X) \vee Y \quad (3.4)$$

- Reconstruction by dilation. The reconstruction by dilation of a mask image Y from a marker image X is defined as the

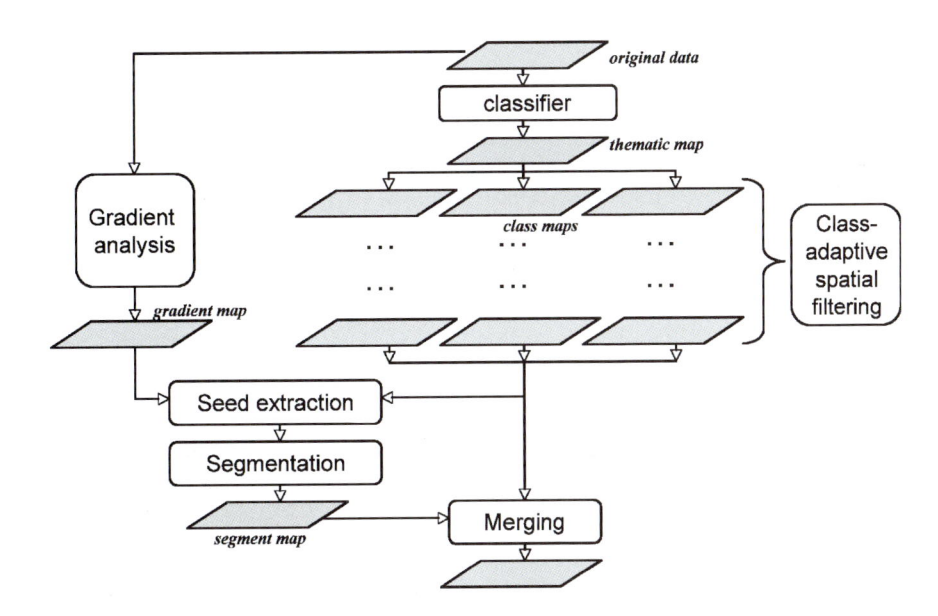

FIGURE 3.3 The processing chain proposed.

geodesic dilation of X with respect to Y until stability:

$$R_Y(X) = \delta_Y^{(i)}(X), \quad (3.5)$$

where i is such that $\delta_Y^{(i)}(X) = \delta_Y^{(i+1)}(X)$.

- Reconstruction by erosion. The reconstruction by erosion is similarly defined as:

$$R_Y^*(X) = \varepsilon_Y^{(i)}(X), \quad (3.6)$$

where i is such that $\varepsilon_Y^{(i)}(X) = \varepsilon_Y^{(i+1)}(X)$.

- Fillhole. The holes of a binary image correspond to the set of its regional minima that are not connected to the image border. Thus, filling the holes of a grayscale image means to remove all minima not connected to the border:

$$FILL(X) = R_X^*(X_m), \quad (3.7)$$

where $X_m(x)$ is equal to $X(x)$ if x belongs to the border, and to $t_{\max}$ otherwise.

It is clear that a way to implement a class-dependent procedure based on morphological operators and with large robustness and suitability is guaranteed by the possibility to choose the size and shape of the structuring elements in each of the previously mentioned steps in accordance with the mean size and shape of the objects in the class. All these possibilities ensure to match the scale of the spatial analysis performed by the filters with the scales of the objects in each mapped class. Indeed, this has been proposed by Soille (1996) and also exploited in a number of stimulating papers, where the size and shape of different elements of the scenes under test have been exploited to improve substantially the mapping results. This is the case for instance for Soille and Pesaresi (2002), where different examples are proposed. A multi-scale approach using these operators, able to adapt to different resolution and capture the characteristics of multiple objects, is the so called Differential Morphological Profile (DMP), introduced in Pesaresi and Benediktsson (2001). Since this publication, DMPs have been considered as one of the most promising approaches in spatial processing of urban scenes.

In fact, no spatial post-processing based on spectral classification is able to capture all the details of a complex scene. This is due to the very different scales of the objects. It is instead mandatory to include spatial features within the actual classification process. Thus, one proposed way forward is to insert the results of the class-adaptive filtering step into a spatial analysis performed using a segmentation algorithm, which could be scaled according to the requirements of the scene and the sensor resolution. To this aim, the seeds of the segmentation are extracted from the refined class maps. Following this idea, in Fig. 3.3 it is shown that the class-adaptive filtering step is followed by a seed extraction and then fed into a segmentation algorithm. Seed extraction can be performed using morphological operators as well, using a minima imposition procedure. Similarly, the gradient information required by the segmentation algorithm is provided by a gradient computation based again on morphological operators.

In particular, minima imposition is performed in two steps. First the pointwise minimum between the input image and the marker image is computed. Then a reconstruction by erosion is applied:

$$MI(X) = R_{(X+1)}^* \wedge X_m(X_m), \quad (3.8)$$

where $X_m(x)$ is equal to 0 if x belongs to a marker, and to $t_{\max}$ otherwise. Gradient values are computed as the arithmetic difference between the dilation and the erosion by the elementary SE:

$$\rho_B(X) = \delta_B(X) - \varepsilon_B(X) \quad (3.9)$$

The segmentation algorithm exploited in this work is the watershed segmentation algorithm. As is well known, the basic idea of watershed segmentation is to consider the regions to be extracted as catchment basins in topography (Lin et al., 2006). The watershed lines are then the boundaries of these basins. If one applies the watershed segmentation method directly to the intensity image, we may not obtain meaningful segmentation results for most images. Except for a few simple cases where the target object is brighter than the background or vice versa, watershed segmentation cannot be applied directly to raw images, because of a severe over-segmentation.

On the one hand this is due to the presence of irrelevant local minima and local maxima in the image. On the other hand, the reason is the presence of texture effects derived by the spatial interaction between the size of the object in the scene and the spatial resolution of the sensor (Pesaresi and Benediktsson, 2001). This is the reason why the input image is chosen to be the adaptive filtered one.

The final step of the procedure is then a recombination of the classification and the segmentation maps. This is accomplished by a assigning to each segment a unique label according to the label of the majority of the pixels in it ("majority voting algorithm"). However, to preserve situations where the majority is extremely challenging, a *stability threshold* is introduced. If the votes of the most abundant class in a segment do not overcome the ones of the second most voted by more than this threshold, the segment is considered a mixed one, and no assignment to a unique class is performed. In order to show very quickly the different phases of the procedure, Figure 3.4 provides the outputs

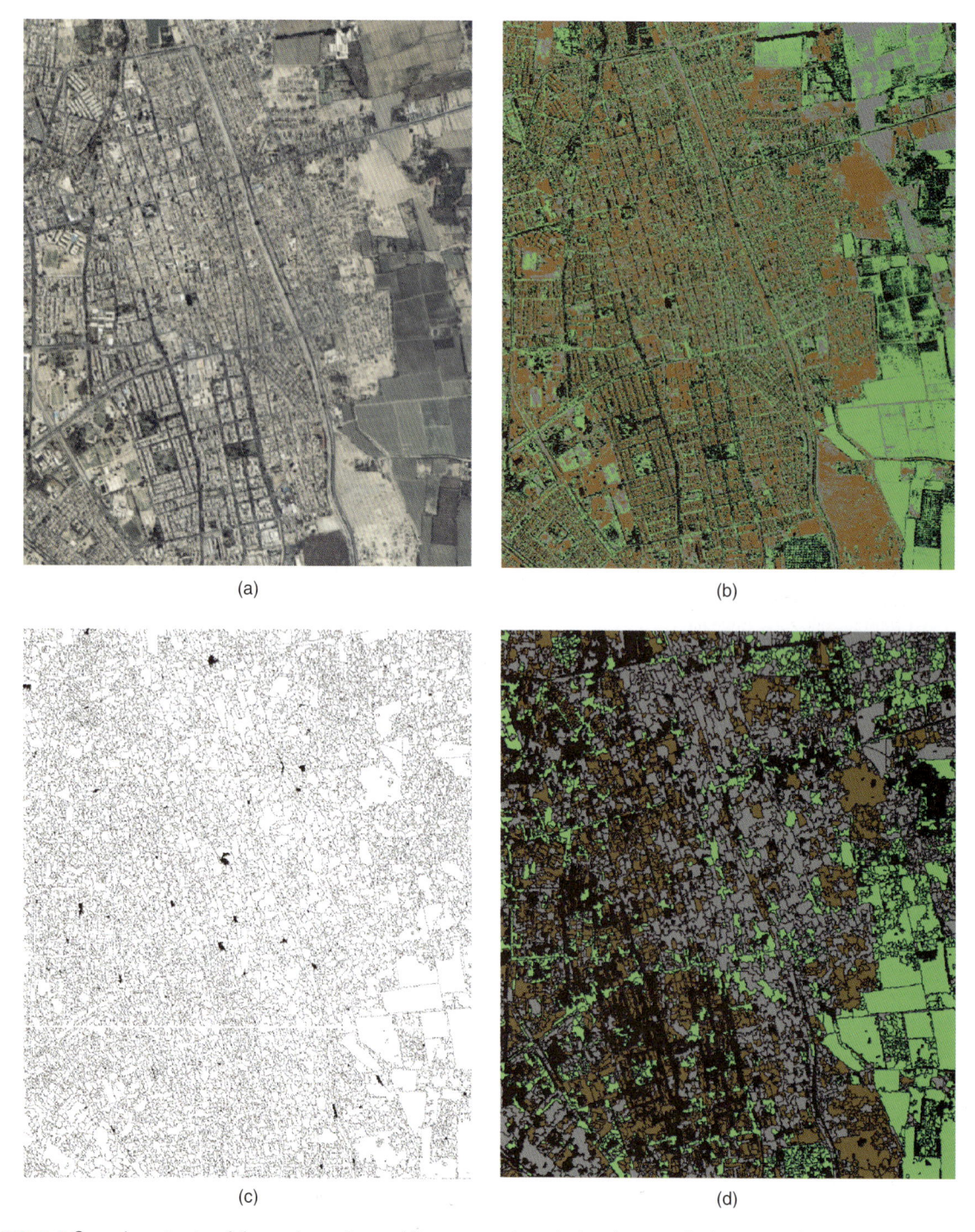

FIGURE 3.4 Sample outputs of the various steps of the processing chain when applied to part of an image over Ica, Peru: (a) original image (© Geoeye, 2007); (b) the initial classification; (c) the output after the watershed segmentation algorithm; (d) the final classification map.

of the various steps starting from a sample of an Ikonos II image of the town of Ica (Peru). The image was made available by GeoEye to the press after the August 2007 earthquake event in the area. The classification results in Fig. 3.4 are meaningless from a quantitative point of view, because the image details were degraded by GeoEye before making it public and are useful only to understand the procedure and see how robust is. Indeed, one might easily note that the recognition of the various elements of the scene is improved by comparing Fig. 3.4(b) with Fig. 3.4(d).

3.7 Building damage assessment

A very interesting application of the procedure proposed in the previous section is related to built-up object detection for building inventories, cadastral map updates and, for civil protection purposes, damage assessment after a major disaster. The examples in the following paragraphs refer to earthquakes because of the specific research interests of the authors, but the same approach may be used for any other situation involving rapid change detection in urban areas. Moreover, two examples are shown because the areas under test are different, both in the sense that the two considered urban areas are in different part of the globe, but also have very different structures and building typologies.

The first site is Bam (Iran) and the dataset is formed of a couple of Quickbird images acquired on 30 September 2003 and 3 January 2006, respectively before and after a major earthquake occurred in the area (26 December 2003). To exploit the high spatial resolution of the panchromatic data and high spectral resolution of the multi-resolution data, two pan-sharpened images have been generated and coregistered. For testing our algorithm, only a limited 2000 × 2000 subsample has been used.

As stated in previous sections, the first step of the algorithm is a classification of the raw image in order to extract relevant spectral classes. In this example, the quick and simple ISODATA algorithm was considered, looking for four different land cover categories, labeled as "houses," "soil," "trees," and "shadow/roads". The choice of this classifier has all those drawbacks derived from using unsupervised algorithms, but basically matches the requirement of the application. In particular, it provides rapid output maps, even in case of unknown areas. A sample of the resulting classification map is shown in Fig. 3.5(a), with an intuitive color legend, with buildings highlighted in white.

Following the analysis procedure, the four classes must be separately processed by the morphological "regularization" step. Strictly speaking, the only meaningful class for the application is the one referring to built-up structures ("houses"). It is however easy to understand that the improvement of the whole map yields a segmentation function that enhances the watershed algorithm performance.

Therefore, first of all *salt and pepper* classification noise is reduced by filling holes in "shadows/roads," "trees," and "houses" classes. Then, specific steps for each class are applied:

- Trees: noise blobs are removed after considering some geometric information about trees. Basically, they have a circular form and at least a 2 m diameter. Thus, erosion by a 1 × 1 disc and then reconstruction by dilation are performed.

- Houses: houses are eroded and reconstructed as well. The SE element is a 5 × 5 square, since we may assume that blobs smaller then 3 × 3 m are not buildings. In fact, besides noisy pixels, there are also cars on the streets. At the same time, blobs bigger than 6000 pixels (2000 m^2) are discarded, assuming they are very large bright soil areas.

- Shadows/roads: blob areas smaller than 15 pixels are considered noise. No assumptions about geometric features are made.

All classes are finally merged into a unique image and eroded by a 2 × 2 square element. This marker image is then used to impose minima on the multispectral gradient image (using a 1 × 1 square SE) of the original scene to obtain the final segmentation function. The watershed algorithm can now be applied avoiding over-segmentation. Actually, noise could be still present, thus is preferable to remove isolated watershed lines shorter than 40 pixels at the end of the segmentation procedure.

Spectral and spatial features are merged via the majority voting step described in the previous section, and the final regularized map is proposed in Fig. 3.5(b). A comparison of the original and final maps shows that the ISODATA classifier has quite good performances, even though no parameter tuning has been done. The main drawbacks are, on the one hand, the very similar spectral responses of buildings and bright soil, on the other hand, the presence of a few roofs made of different materials, which are not correctly classified as "houses". A very accurate spectral classification was not the aim of this work, thus there are margins for improvement. The final map is definitely more accurate and precise than the ISODATA output. Besides detecting more homogeneous and better defined areas, this process reduces some misclassification errors, like the presence of huge bright soil areas, and even removes ambiguities, for example when the same roof has different cover materials.

Once the segmentation scheme has been applied to both pre- and post-images, it was also tried to evaluate any difference in building features. The basic assumption is that all undamaged buildings must be found in both the images. In case of damage, instead, post-event buildings must have a different shape and presumably a smaller footprint than pre-event ones. These considerations do not take into account all possible changes (e.g., the presence of new constructed buildings or bright debris areas larger than original buildings), but they cover the most part of cases. Damage mapping is thus shifted into the evaluation of changes in the buildings' shape.

To this aim, common areas of the buildings are detected by doing the difference between pre- and post-classified maps. Then, specifically for building damage evaluation, and following the above mentioned assumptions, a *building damage index* (BDI) is introduced, defined as the ratio between the building area in the difference image, *common building area*, and its area in the pre-event image, *pre-building area*. By choosing an appropriate threshold, it is possible to get an output grayscale map in which buildings can be separated into two classes: slightly damaged (including undamaged ones), and heavily damaged. According to the intuitive idea that a building which is heavily damaged is more than half destroyed, the above mentioned threshold can be set to 0.6.

The validation of the results is made by a direct comparison with the visual inspection in Yamazaki, Yano and Matsuoka (2005) and shown in Fig. 3.5(d). These validation

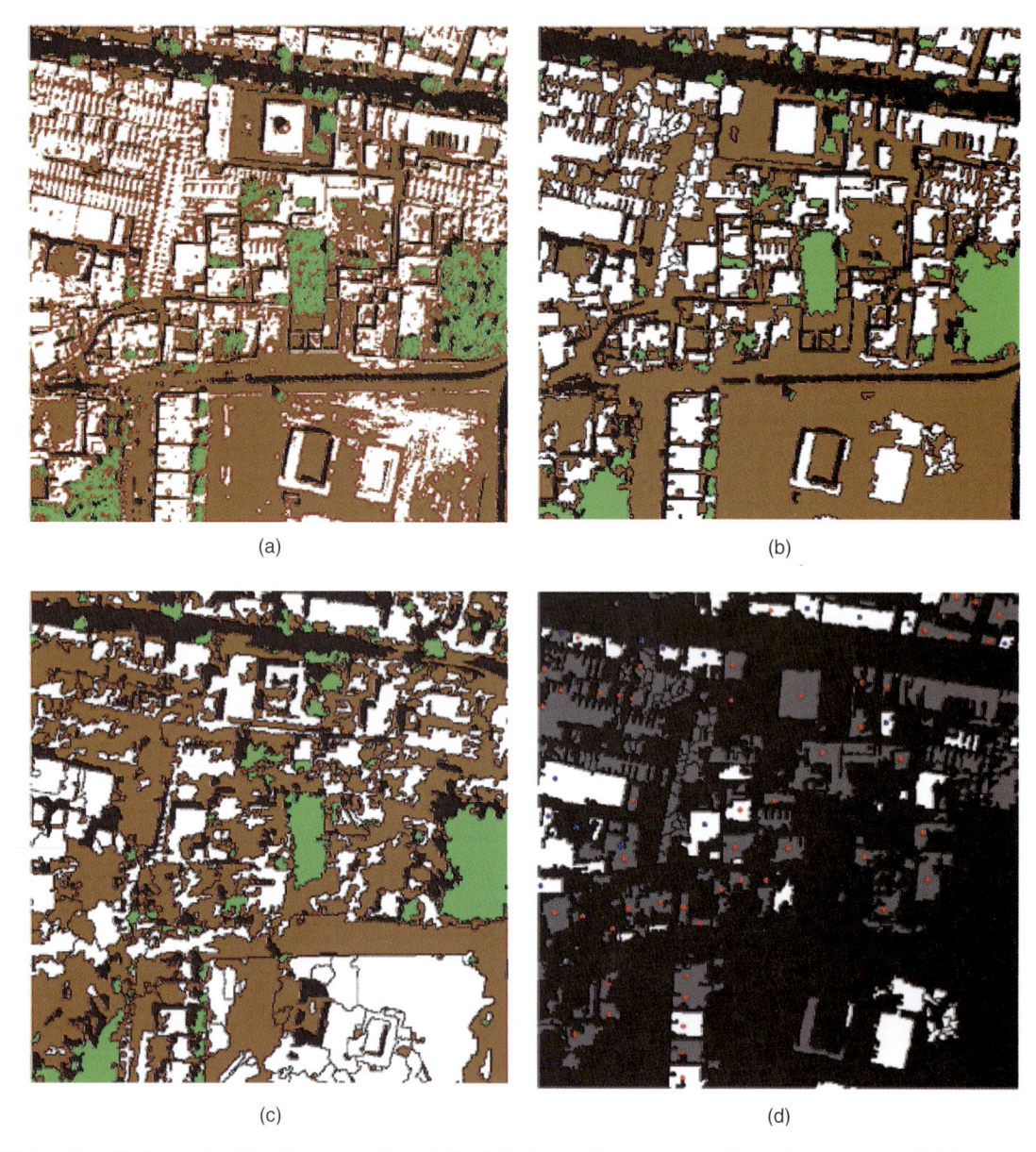

FIGURE 3.5 Results for part of the Bam test site: (a) the initial classification maps from the pre-event Quickbird image; (b) the final classification map from the pre-event scene; (c) the final classification map from the post-event scene; (d) visual representation of the thresholded BDI values, compared with the validation data that are in blue and red dots (see the relevant text for the explanation).

data has been made available as a vector file containing the center location of each building, with an attribute defining its damage level according to the five grades of the European Macroseismic Scale (EMS), from "undamaged" to "heavily damaged". For this work, building damage classes were reduced to "strongly damaged" (red dots, EMS classes 4 and 5) and "lightly damaged" (blue dots, EMS class 1 to 3). The confusion matrix between the reference and the BDI-derived damage assessment in Table 3.2 shows a really good accuracy (nearly 80%), with greater precision, as expected, for strongly damaged buildings. The BDI seems then to be a good measure of damage extent, even in presence of four main limiting factors that do not directly depend on the index itself:

- some buildings could not be selected because of a lack of precision in classification (there are also multiple objects classified as a unique building);

- performances are limited by differences in atmospheric and acquisition conditions, since they increase the number of false alarms;

- the confusion matrix takes into account just true negatives and not false positives;

- some objects are considered errors because of little co-registration misplacements between the output map and the ground truth.

TABLE 3.2 Confusion matrix for damage evaluation of the detected buildings in the Bam pre- and post-event Quickbird image pair.

	78.71%	Strongly damaged	Lightly damaged
BDI $\leq$ 0.6	90.48%	608	140
BDI $>$ 0.6	51.05%	64	146

The second experimental results refer to the magnitude 6.8 earthquake occurred in northern Algeria on 21 May 2003. Centered on the Boumerdes province (Fig. 3.6) some 50 km east of Algiers, the worst-affected urban areas included the cities of Boumerdes, Zemmouri, Thenia, Belouizdad, Rouiba, and Reghaia. For this event also two Quickbird images (one pre- and one post-event) were selected.

The images were acquired on 22 April 2002 and 23 May 2003, respectively (see Fig. 3.7). In this case, with respect to the first test site, there is a high variability in building spectral features. Roofs have different responses and it is not easy to isolate a "building" class, especially by means of unsupervised classifiers. On the other hand is very time-consuming to specify manually all the possible classes. As it is usual in many other cities, there is also an intrinsic confusion between roads and buildings with bitumen roofs.

The processing scheme in this case exploits the obvious difference between these two categories: the height of the objects. Therefore, after ISODATA classification, "shadow" and "tree" classes are isolated. No assumptions are made on the others. After some morphological operations, such as noise removal and hole filling, from these classes all the objects that cast shadows are selected. All the residuals are left to the bare soil class.

Alternatively, we have also considered a simplified version of the classification approach (Fig. 3.8) presented in Gamba and Dell'Acqua (2006). Again, the first step is a spectral classification of the pre-event image, focusing on four different roof types, roads, vegetation, bare soil and shadows for discrimination purposes and performed through the neuro-fuzzy network. From that classification map the building footprints were extracted and used to focus the post event damage analysis only on buildings. In order to correct some errors due to misclassified pixels in the urban area, a dilate filter has been applied to the area of interest extracted from the classification. Then, the urban area of the post event image has been classified in damaged and non-damaged buildings; then each building of the available GIS has been assigned following a majority rule. In Fig. 3.9 the result

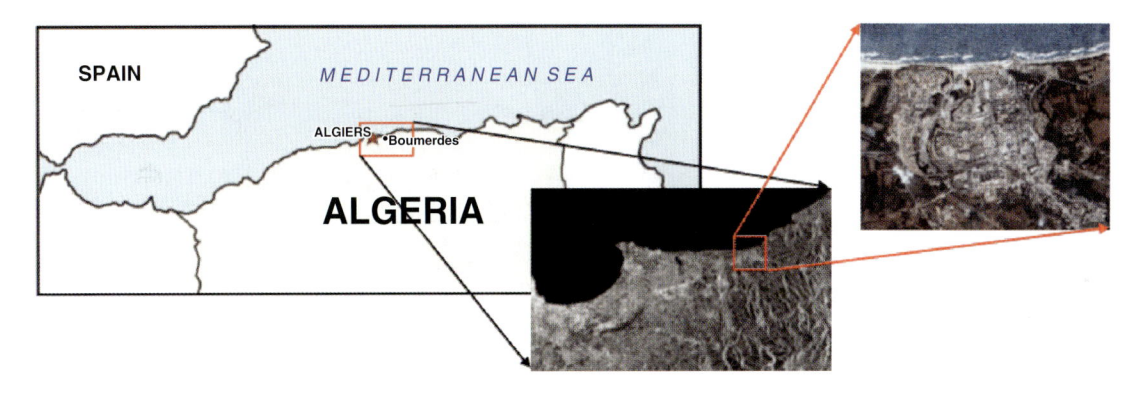

FIGURE 3.6 Location of the urban area of Boumerdes in Algeria.

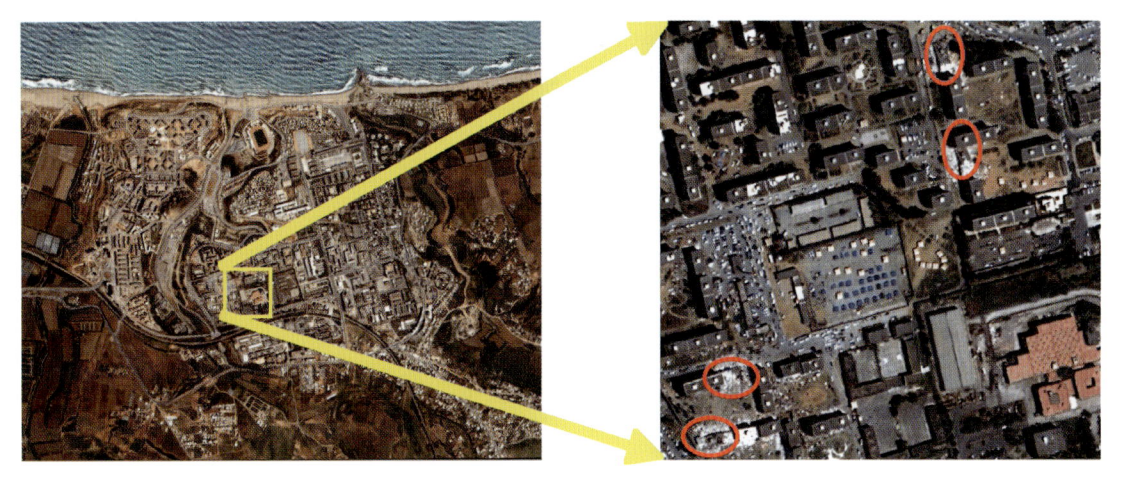

FIGURE 3.7 Post event pan-sharpened image of Boumerdes (left) and a small area of interest (right), with the collapsed buildings highlighted in red.

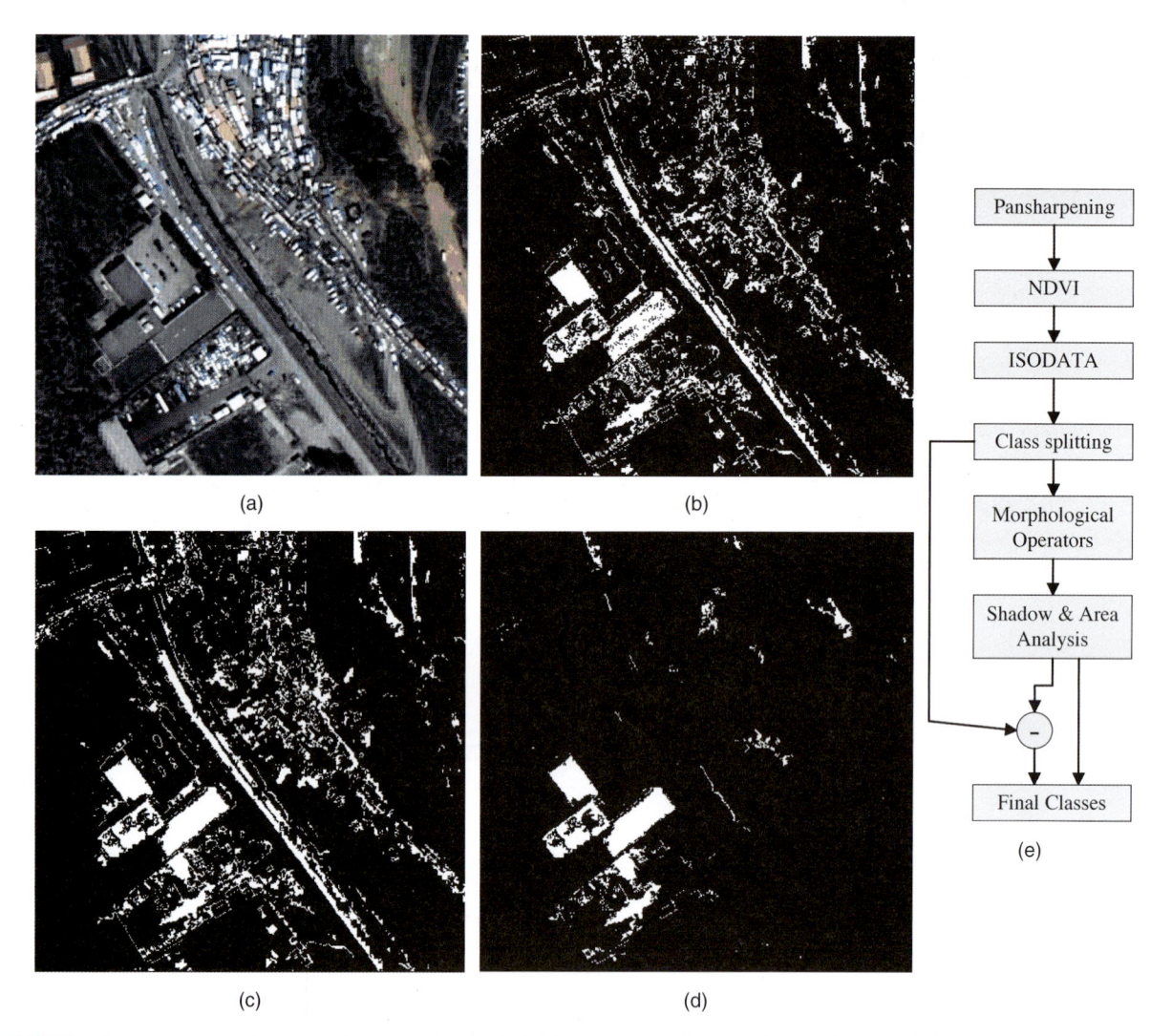

FIGURE 3.8 The proposed approach applied to the VHR images over Boumerdes to detect building and damages. (a) A sample of the original image through which the buildings are classified (b) and refined (c) by using morphological operators. Adding shadow information helps reduce the buildings to the actual situation (d).

of this last method is presented, and a comparison with the GIS manually extracted from the pre-event image is shown.

The results obtained show that all the buildings present in the image have been identified, and the same can be said of the damaged buildings, while only 3 out of 73 buildings have been misclassified as damaged, giving rise to an overall accuracy of 96%.

Conclusions

In this chapter we have discussed multiple limits and challenges of VHR optical imagery for urban-related applications. We also reviewed some algorithms suitable to reduce some of the problems and cope with some of the challenges related to urban mapping and change detection at the building level.

Using several examples, it has been shown that per-pixel analysis is no more the right way to exploit the information in the data when the spatial resolution is finer than the relevant elements in a scene. Per-object or, more generally, spatially-aware and context-sensitive approaches are required and should rely not only on remotely sensed data, but on the *a priori* knowledge about the test site and its environment.

As an example of this way forward, change detection in case of severe earthquakes using mathematical morphology operators to refine pre- and post-event maps and combine their results in an effective way has been proposed. Results are encouraging from a quantitative point of view. However, the main aim of this chapter is to provide some ideas and a general framework, leaving to the many researchers working in the active field of urban remote sensing the freedom to develop their own specialized procedures.

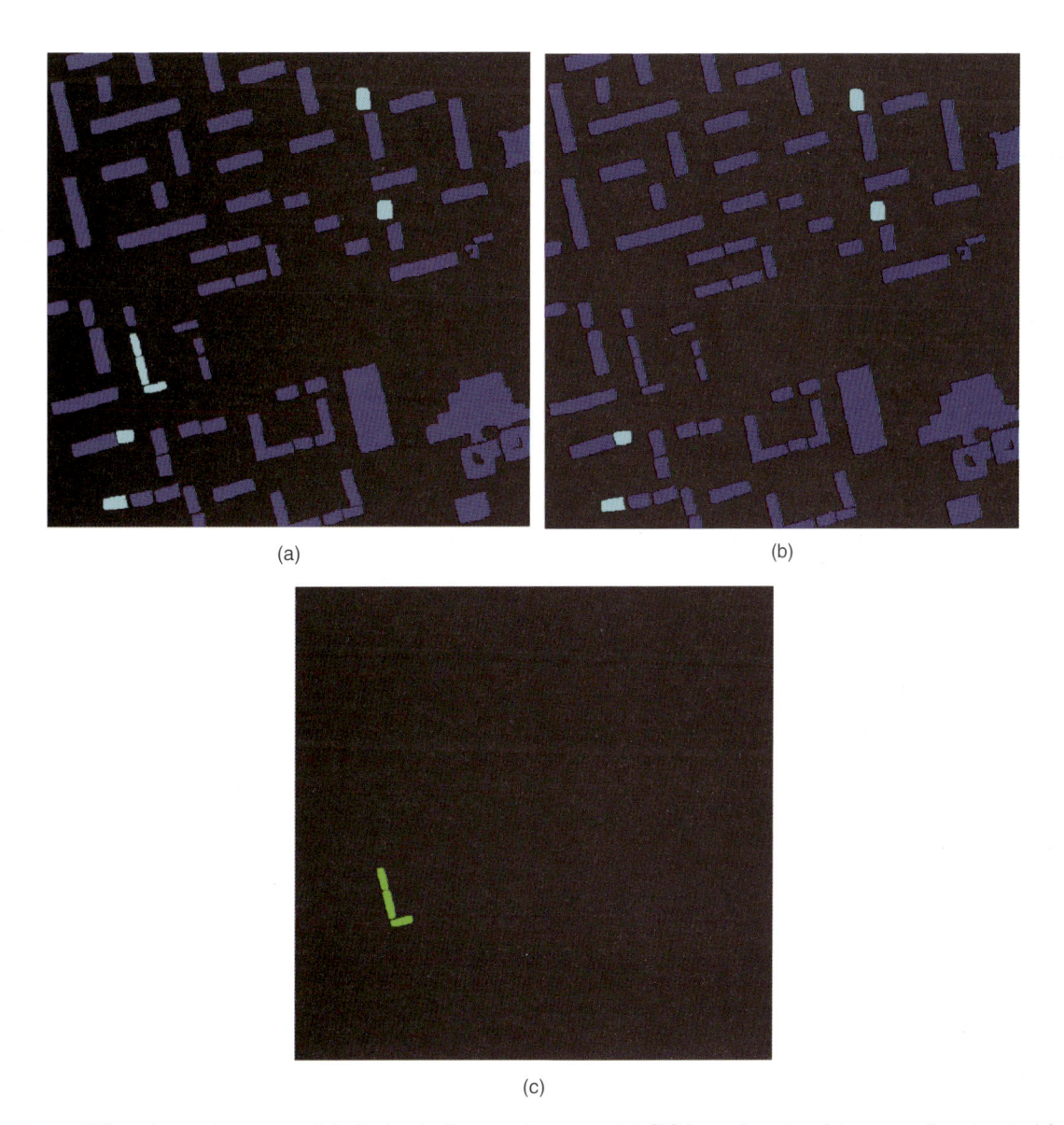

FIGURE 3.9 GIS and remote sensor data fusion in Boumerdes area: (a) GIS-based vector data manually extracted from a pre-event image where undamaged buildings are in blue while collapsed buildings are in light blue (b); the difference between (a) and (b).

References

Baltsavias, E. (2005) Potential of optical satellite sensors for urban mapping. *SIFET and AIT workshop on Urban Remote Sensing*, Mantova, Italy, unformatted CD-ROM.

Blaschke, T., Lang, S. and Hay, G.J. (eds.) (2008) *Object-Based Image Analysis*, Springer Verlag, Berlin.

Blaschke, T. (2003) Object-based contextual image classification built on image segmentation, Proc. of *IEEE Workshop on Advances in Technology and Analysis of Remote Sensed Data*, Washington DC, pp. 113–119.

Benz, U.C., Hofmann, P., Willhauck, G., Lingenfelder, I., and Heynen, M. (2004) Multi-resolution, object-oriented fuzzy analysis of remote sensing data for GIS-ready information, *ISPRS Journal of Photogrammetry and Remote Sensing*, **58**(3/4), 239–258.

Bovolo, F., Bruzzone, L. and Marchesi, S. (2008) A context-sensitive technique robust to registration noise for change detection in very high resolution multispectral images. *Proceedings of IEEE International Geoscience and Remote Sensing Symposium, IGARSS'08*, Boston, Massachusetts, III, pp. 150–153.

Bruzzone, L. and Cossu, R. (2003) An adaptive approach for reducing registration noise effects in unsupervised change detection, *IEEE Transactions on Geoscience and Remote Sensing*, **41** (11), 2455–2465.

Bruzzone, L. and Carlin, L. (2006) A Multilevel context-based system for classification of Very High Spatial resolution images. *IEEE Transactions on Geoscience and Remote Sensing*, **44** (9), pp. 2587–2600.

Carlotto, M.J. (1997) Detection and analysis of change in remotely-sensed imagery with application to wide area surveillance. *IEEE Transactions on Image Processing*, **6** (1), 189–202.

Davis, C.H. and Wang, X. (2001a) Planimetric Accuracy of IKONOS 1-m Panchromatic Image Products, *Proc. of the 2001 ASPRS Annual Conference*, St. Louis, MI, USA, unformatted CD-ROM.

Davis, C.H. and Wang, X. (2001b) High-resolution DEMs for urban applications from NAPP photography, *Photogrammetric Engineering and Remote Sensing*, **67**, 585–592.

Dekker, R.J. (2004) Object-based updating of land-use maps of urban areas using satellite remote sensing, *Proc. of the 12th International Conference on Geoinformatics*, University of Gävle, Sweden, 7–9 June 2004, unformatted CD-ROM.

Dell'Acqua, F., Gamba, P., and Lisini, G. (2006) Change detection of multi-temporal SAR data in urban areas combining feature-based and pixel-based techniques, *IEEE Transactions on Geoscience and Remote Sensing*, **44** (10), 2820–2827.

Ehlers, M. (2009) Future EO sensors of relevance – integrated perspective for global urban monitoring, in *Global Mapping of Human Settlement: Experiences, Datasets, and Prospects* (eds. P. Gamba and M. Herold), Taylor & Francis, New York.

Gamba, P, & Houshmand, B. (2000) Digital surface models and building extraction: a comparison of IFSAR and LIDAR data, *IEEE Transactions on Geoscience and Remote Sensing*, **38** (4), 1959–1968.

Gamba, P. and Dell'Acqua, F. (2006) Spectral resolution in the context of Very High Resolution urban remote sensing, in *Urban Remote Sensing* (eds. Q. Weng and D. Quattrochi), CRC Press, New York.

Gamba, P. and Trianni, G. (2007) Boundary-adaptive MRF classification of VHR images. *Proceedings of IGARSS'07*, Barcelona, Spain, pp. 1493–1496.

Gamba, P., Dell'Acqua, F., Lisini, G. and Trianni, G. (2007) Exploiting spectral and spatial analysis for urban area data interpretation. *Proceedings of the 27th EARSEL Symposium*, Bolzano, Italy, pp. 117–124.

Hazel, G.G. (2001) Object-level change detection in spectral imagery. *IEEE Transactions on Geoscience and Remote Sensing*, **39**(3), 553–561.

Herold, M. Gardner, M.E. and Roberts, D.A. (2003) Spectral resolution requirements for mapping urban areas. *IEEE Transactions on Geoscience and Remote Sensing*, **41**(9), 1907–1919.

Herold, M. and Roberts, D.A. (2005) Mapping asphalt road conditions with hyperspectral remote sensing. *Proceedings of the 5th International Symposium on Remote Sensing of Urban Areas*, unformatted CD-ROM.

Lin, Y.C., Tsai, Y.P., Hung, Y.P. and Shih, Z.C. (2006) Comparison between immersion-based and toboggan-based watershed image segmentation, *IEEE Transactions on Image Processing*, **15**(3), 632–640.

Lisini, G., Gamba, P., Dell'Acqua, F. and Trianni, G. (2005) Image interpretation through problem segmentation for very high resolution data. *Proceedings of IGARSS'05*, Seoul (Korea), pp. 5634–5637.

Liu, X., Zhang, Z., Peterson, J. and Chandra, S. (2007) LiDAR-Derived high quality ground control information and DEM for image orthorectification, *Geoinformatica*, **11**(1), 37–53.

Marino, C.M., Panigada, C., Busetto, L., Galli A. and Boschetti, M. (2000) Environmental applications of airborne hyperspectral remote sensing: asbestos concrete sheeting identification and mapping. *Proceedings of the 14th International Conference and Workshops on Applied Geologic Ramote Sensing*, August 2000.

Niemeyer, I. and Nussbaum, S. (2006) Change detection – the potential for nuclear safeguards, in *Verifying Treaty Compliance. Limiting Weapons of Mass Destruction and Monitoring Kyoto Protocol Provisions* (eds. R. Avenhaus, N. Kyriakopoulos, M. Richard and G. Stein), Springer, Berlin.

Pesaresi, M. and Benediktsson, J.A. (2001) A new approach for the morphological segmentation of high-resolution satellite imagery. *IEEE Transactions on Geoscience and. Remote Sensing*, **39**(2), 309–320.

Schiefer, S., Hostert, P. and Damm, A. (2005) Correcting brightness gradients in hyperspectral data from urban areas. *Remote Sensing of the Environment*, **101**(1), 25–37.

Segl, K., Roessner, S. and Heiden, U. (2000) Differentiation of urban surfaces based on hyperspectral image data and a multitechnique approach. *Proceedings of the IEEE Geoscience and Remote Sensing Symposium*, Honolulu, 4, pp. 1600–1602.

Soille, P. (1996) Morphological partitioning of multispectral images. *Journal of Electronic Imaging*, **5**(3), 252–265.

Soille, P. (1999) *Morphological Image Analysis*, Springer-Verlag, Berlin.

Soille, P. and Pesaresi, M. (2002) Advances in mathematical morphology applied to geoscience and remote sensing, *IEEE Transactions on Geoscience and Remote Sensing*, **40**(9), 2042–2055.

Volpe, F. and Rossi, L. (2003) Quickbird high resolution satellite data for urban applications. *Proceedings of the 2nd GRSS/ISPRS Joint Workshop on Remote Sensing and Data Fusion over Urban Areas*, Berlin, Germany, unformatted CD-ROM.

Walter, V. (2004) Object-based classification of remote sensing data for change detection, *ISPRS Journal of Photogrammetry & Remote Sensing*, **58**, 225–238.

Xiao, Q., Austin, S.L., McPherson, E.G. and Peper, P.J. (1999) Characterization of the structure and species composition of urban trees using high resolution AVIRIS data. *Proceedings of the AVIRIS Workshop*, Pasadena, CA.

Yamazaki, F., Yano, Y. and Matsuoka, M. (2005) Visual damage interpretation of buildings in Bam City using Quickbird images following the 2003 Bam, Iran, earthquake. *Earthquake Spectra*, **21**(S1), S329–S336.

Yashon, O. and Tateishi, R. (2008) Urban-trees extraction from Quickbird imagery using multiscale *spectex*-filtering and nonparametric classification. *ISPRS Journal of Photogrammetry and Remote Sensing*, **63**(3), 333–351.

4

Potential of hyperspectral remote sensing for analyzing the urban environment

Sigrid Roessner, Karl Segl, Mathias Bochow, Uta Heiden, Wieke Heldens and Hermann Kaufmann

In contrast to widely used multispectral data, hyperspectral imagery resolves material-specific spectral reflection and absorption features making them especially suitable for detailed and comprehensive mapping of urban surface materials. However, this requires development of automated methods for efficient information extraction that take into account the special conditions in urban areas characterized by a big variety of materials and a large heterogeneity of small-sized urban structures. This chapter deals with the current state of methodological developments in urban hyperspectral remote sensing emphasizing the research of the authors towards the development of an automated system for comprehensive mapping of urban surface materials. This includes field- and image-based spectral investigations aiming at the automated derivation of robust quantitative spectral features. They serve as input information for the developed multi-step processing system allowing detailed mapping of urban surface materials. In this context, the iterative procedure is capable of analyzing urban structures at a subpixel level and provides area-wide information about the fractional coverage of surface materials for each pixel. Thus, the obtained results are characterized by a new level of thematic and spatial detail which significantly increases their suitability for subsequent modeling, evaluation and monitoring of the urban environment.

Urban Remote Sensing: Monitoring, Synthesis and Modeling in the Urban Environment, First Edition. Edited by Xiaojun Yang.
© 2011 John Wiley & Sons, Ltd. Published 2011 by John Wiley & Sons, Ltd.

4.1 Introduction

Remote sensing technologies have been used for analyzing the urban environment covering a wide range of applications in the fields of urban land cover/land use mapping, urban planning, urban growth monitoring, urban hazard and risk analysis as well as urban ecology studies including investigations of urban climate and hydrology. They are mainly driven by the need for up-to-date information about surface conditions in the rapidly changing urban environment whereas special emphasis is put on the analysis of the spatially dominating manmade materials/structures since they significantly determine the ecological, climatologic and human living conditions and are most vulnerable to natural and man-made hazards. These studies have been carried out at different spatial and temporal scales based on a variety of sensor systems whereas medium and high spatial resolution multispectral data, such as Landsat-(E)TM, SPOT and IKONOS have played the biggest role. For this purpose a wide range of automated methods for land cover/land use mapping and change detection has been developed. Despite the demonstrated high potential of these approaches, there have been limitations in automated multispectral analysis of material properties mainly for man-made urban surfaces restricting consequential material mapping which is required for physically-based modeling of environmental conditions and for the derivation of quantitative indicators allowing objective evaluation of urban areas.

In this context, advances in imaging spectroscopy (Schaepman et al., 2009) have opened up new opportunities for spatially and thematically detailed analysis of urban structures. Hyperspectral image data are especially suitable for material mapping due to their high spectral resolution allowing detailed per pixel assessment of spectral reflectance characteristics which are determined by the chemical composition of the materials. The resulting material-specific spectral absorption and reflection features form the basis for identification and characterization of urban surface materials. The potential of these data has already been investigated in a number of studies (see Section 4.2), which is still small in comparison to studies based on multispectral data. In the near future new opportunities will open up with the operational availability of satellite-based hyperspectral remote sensing systems, such as EnMAP (Environmental Mapping and Analysis Program) (Kaufmann et al., 2006), PRISMA (PRecursore Iper-Spettrale della Missione Operativa) (Sacchetti et al., 2010) and HyspIRI (Hyperspectral Infrared Imager) (Green et al., 2008). However, the existing studies have already shown the high spectral information content of hyperspectral data. At the same time they have revealed the challenges in automated information extraction mainly consisting of the big variety of urban surface materials and the small-sized urban structures leading to a large spectral variability and a high number of mixed pixels. This requires the development of specific methods for automated information extraction which are capable of meeting the thematic needs for detailed urban material mapping.

This chapter discusses the current state of the art of urban hyperspectral remote sensing with emphasis on comprehensive and quantitative assessment of spectral characteristics of urban surface materials (Section 4.2) as the main prerequisite for their automated identification (Section 4.3). In this connection the methodological experiences of the authors in urban hyperspectral remote sensing are presented. They have led to the development of an automated system for urban material mapping which is described in Section 4.3.2. Exemplary results are discussed and evaluated in their potential for thematic applications in Section 4.4. This way, the complete chain of urban hyperspectral data analysis ranging from primary spectral investigations to derivation of higher-level information products is covered demonstrating the high potential of hyperspectral remote sensing for analyzing the urban environment.

4.2 Spectral characteristics of urban surface materials

This section deals with comprehensive assessment of spectral characteristics of urban surface materials using hyperspectral imagery. In contrast to natural materials, such as soil, rock and vegetation, the number of investigations for man-made materials dominating the urban environment has been fairly limited (e.g., Ben-Dor, Levin and Saaroni, 2001; Herold et al., 2004; Heiden et al., 2007). Motivated by the big variety of urban surface materials discussed in more detail in Section 4.2.1, the authors have carried out systematic research on spectral characteristics of these surfaces in major German cities (Berlin, Dresden, Munich, Potsdam) using field and airborne hyperspectral imaging spectroscopy. The investigations have resulted in field and image spectral libraries of urban surface materials widespread in German cities (Section 4.2.2). They form the basis for the derivation of robust spectral features as an important prerequisite for automated mapping of urban surface materials (Section 4.2.3).

4.2.1 Categories of interest for material mapping

Spatial inventories of surface materials are required for a wide range of thematic applications ranging from mapping of imperviousness to characterizing thermal conditions (Section 4.4). Therefore, several classification schemes have been developed whereas the majority is structured in a hierarchical way (e.g., Roessner et al., 2001; Herold et al., 2004; Heiden et al., 2007). Most of them can be traced back to the land cover/land use classification scheme of Anderson et al. (1976) allowing a flexible and application-oriented categorization of the earth surface at different hierarchical levels based on the main land cover classes, vegetation, water, built up/man-made areas, water and bare ground. This scheme has been adapted by the authors to the needs of comprehensive mapping of urban surface materials with special emphasis on urban ecological aspects leading to primary differentiation between manmade/artificial and natural surfaces at level 1 of Table 4.1.

At level 2 main land cover types are distinguished whereas man-made surfaces are further subdivided into buildings and open spaces. Natural surfaces are categorized into vegetation, bare ground and water bodies. In level 3 these categories are further specified according to material types determining the appearance of surfaces in hyperspectral imagery. Man-made roofing materials are grouped into mineral, metallic, hydrocarbon, and biomass ones. Artificial open spaces are distinguished according to imperviousness. Vegetation is further distinguished into trees, shrubs/bushes and meadow/lawn. Bare ground is categorized into soil, sand and rock and water bodies in ocean/sea, lakes, ponds and rivers. These material types represent the main categories and thus the framework for comprehensive spectral analysis of urban surface materials described in Section 4.2.2.

Level 4 contains a compilation of urban surface materials occurring in Europe and North America whereas emphasis has been put on man-made materials. Despite their dominance in the urban environment, man-made surface materials have been neglected to a large extent in spectral investigations. Although the resulting list of Table 4.1 is not meant to be exhaustive,

TABLE 4.1 Categories for comprehensive urban material mapping based on Anderson scheme (Anderson et al., 1976).

Level 1	Level 2: Land cover types	Level 3: Material types	Level 4: Surface materials
Man-made/artificial surfaces	Buildings/roofs	Mineral materials	Asbestos
			Bitumen roof sheeting
			Clay tiles
			Concrete slabs
			Concrete tiles
			Fiber cement
			Glass
			Gravel
			Slate
		Metallic materials	Aluminum
			Copper
			Zinc
			Steel with protective coating
			Corrugated metal sheet
			Lead
			Gold leaf
			Tin
		Hydrocarbon materials	Coated corrugated metal sheet (PVC, Polyethylene, coating color)
			Polyvinylchloride (PVC)
			Polyethylene (PE)
			Polyisobutylene (PIB)
			Plexiglas
			Tar Paper
		Biomass materials	Green roof
			Thatched roof
			Wood shingles
	Artificial open spaces	Partially impervious surfaces	Cinder
			Clay-baked paving stones
			Cobblestone pavement
			Concrete paving stones
			Gravel
			Grass pavers
			Loose chippings
			Railway tracks
		Fully impervious surfaces	Asphalt
			Concrete
			Flagstone (Granite)
			Synthetic turf
			Tartan
		Water bodies with artificial bottom	Pool
			Garden pond

TABLE 4.1 (continued).

Level 1	Level 2: Land cover types	Level 3: Material types	Level 4: Surface materials
Natural surfaces	Vegetation	Trees	*Coniferous*
			Deciduous
		Shrubs/bushes	
		Meadow/lawn	*Meadow - dry*
			Meadow - fresh
			Ornamental lawn
			Sports turf
	Bare ground	Soil	*Soil - dark*
			Soil - bright
		Sand	*Sand - coarse*
			Sand - fine
		Rock	
	Water bodies	Ocean/sea	
		Inland waters	*Lakes*
			Ponds
			Rivers

Surface materials (level 4) marked in italics have been spectrally investigated by the authors.

the hierarchical structure of the developed scheme allows for successive extensions at levels 3 and 4. The surface materials listed at level 4 are characterized by a large number of spectral variations due to varying material properties and bidirectional reflectance distribution function (BRDF) effects resulting from the imaging process. In case of man-made surfaces material-inherent differences are caused by color, coating, weathering and usage. For natural surfaces possible variations are the distinction between green and non-photosynthetic vegetation, the determination of main plant communities, rock and soil types as well as the determination of the eutrophic state of water bodies. Since the resulting number of categories is very large, only selected ones are included in Table 4.1. However, comprehensive material mapping requires systematic investigation of such variations for the materials of interest in order to determine their effects on spectral reflectance (Section 4.2.2).

4.2.2 Establishment of urban spectral libraries

In spectral libraries detailed knowledge about spectral material characteristics is stored in form of spectra. They are recorded by laboratory, field and imaging spectroscopy allowing the analysis of materials at different spatial scales, spectral resolutions and environmental conditions. In hyperspectral remote sensing they serve as important input information for pre-processing and classification procedures and contribute to the general spectral understanding of the phenomena of interest. Spectral libraries presented in this paper have been established during the last 10 years based on field and imaging spectroscopy for various test sites in the German cities of Berlin, Dresden, Munich and Potsdam.

Assessment of field spectra was performed using a field spectrometer (Analytical Spectral Device (ASD) Field SpecPro FR) recording data with 1 nm spectral sampling distance (SSD) in the wavelength range between 350 nm and 2500 nm. In the result, field spectra for about 80 spectral targets were stored in the *field spectral library* comprising about 50 targets for roofing and 30 targets for open surface materials. All of these targets were measured under open sky and direct sun illumination in nadir position. A detailed methodological description of spectral data acquisition assessed material variations and description of the spectral behavior of selected man-made materials in relation to their spectrally relevant chemical compounds can be found in Heiden *et al.* (2007).

Since the establishment of a comprehensive field spectral library is very time consuming and does not cover the large spectral variety occurring in hyperspectral imagery, an *image spectral library* has been built using the field spectral library to assign surface materials to the image spectra. They were identified interactively in the hyperspectral image database comprising 11 flight lines acquired for test sites in four cities between 1999 and 2007 (Table 4.2). The data were recorded by the hyperspectral HyMap sensor operated by the German Aerospace Center (DLR) and the obtained radiances have been transformed into reflectances for normalization purposes (Heiden *et al.*, 2007). For each material the number of spectra varies between a few hundreds and several thousands depending on its frequency of occurrence and spectral variability. This way a total of about 60 000 image spectra has been stored in the image spectral library representing those surface materials of Table 4.1 marked in italic print. Subsequently, this database has been used for determination of robust spectral features allowing discrimination between these materials (Section 4.2.3).

4.2.3 Determination of robust spectral features

A spectral feature represents a specific spectral behavior within two wavelength positions in the reflectance spectrum. It can be

TABLE 4.2 Hyperspectral flight lines used for establishment of the image spectral library.

Date	Time (UTC)	Test site	Flight heading (°)	Flight altitude (m)	Resampled pixel size	No. of bands
19.05.1999	10:46	Dresden	179	2540	6.0	128
19.05.1999	12:50	Potsdam	180	2560	6.0	128
01.08.2000	11:50	Dresden	−6	1660	3.0	126
20.07.2003	10:20	Dresden	175	1680	3.0	126
07.07.2004	9:39	Dresden	−4	2050	4.0	126
07.07.2004	10:29	Potsdam	0	1950	3.5	126
20.06.2005	9:35	Berlin	95	1960	3.5	126
17.06.2007	9:30	Munich	0	2520	4.0	126
17.06.2007	9:45	Munich	180	2520	4.0	126
17.06.2007	10:15	Munich	180	2520	4.0	126
25.06.2007	11:05	Munich	0	2520	4.0	126

transformed into a single numerical value for further quantitative analysis using feature functions, such as absorption position and depth. Incorporation of such feature values in automated classification procedures considerably reduces the dimension of the feature space (Hughes phenomenon) and the variability of input parameters in comparison to using original reflectance values as input information for hyperspectral classification methods. The occurrence of such features is caused by differences in the intensity of absorption and reflection throughout the reflective part of the electromagnetic spectrum. The type and wavelength position of spectral features provide valuable information about the chemical composition of a material (Clark, 1999). One of the most important feature types is the absorption band represented by a local reflectance minimum. The spectrum shown in Fig. 4.1(d) depicts the absorption band at 2200 nm which is characteristic for clay minerals, such as kaolinite. Another commonly used spectral feature type is the increase in reflectance which can be observed amongst others between the red and near-infrared wavelength ranges. This so-called "red edge" is caused by the chlorophyll content of green vegetation and described by

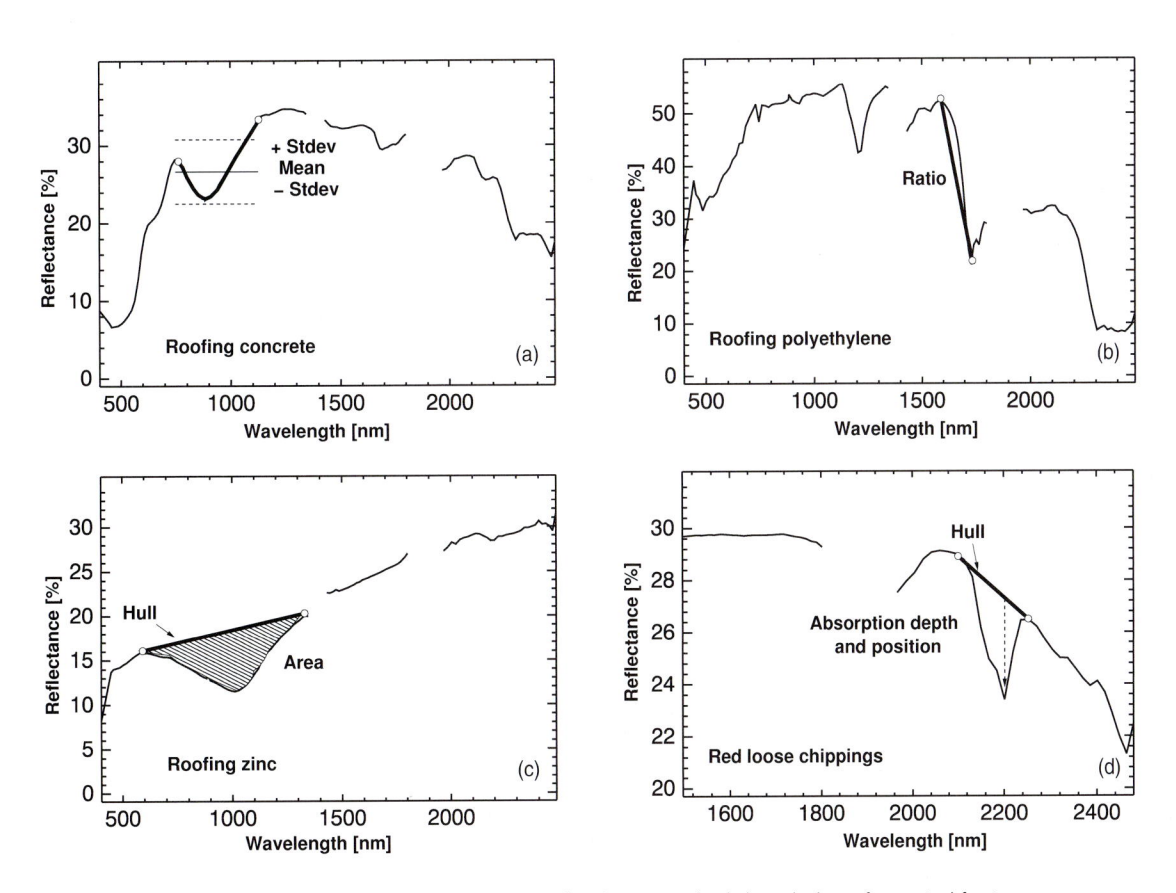

FIGURE 4.1 Exemplary feature functions for the numerical description of spectral features.

several functions, such as ratios and vegetation indices (Bannari et al., 1995).

Figure 4.1 depicts exemplary feature functions for numerical description of spectral features based on spectra of different man-made materials. The functions mean and standard deviation (Fig. 4.1a) characterize the albedo and are mostly used for identification of materials with typical brightness and flat curve progression and weak or no spectral absorption features, such as concrete and a number of dark materials. Another feature function is the ratio describing increase and decrease of reflectance occurring in a specific wavelength range. The example shown in Fig. 4.1b depicts the use of this function for describing the absorption feature of polyethylene near 1700 nm. Broad absorption bands which are typical for aluminum and zinc are well expressed by the feature function of the area that is enclosed by the reflectance curve and the hull function between two wavelength positions (Fig. 4.1c). The spectrum of red loose chippings contains the above mentioned kaolinite absorption band between 2134 and 2272 nm with the maximum absorption depth at 2200 nm (Fig. 4.1d). The applied feature function calculates the maximum of the difference between the reflectance curve and the hull function within a defined wavelength range.

So far, most of the approaches for hyperspectral image analysis only consider one spectral feature type for a certain analytical task. Due to the wide range of surface materials and their spectral variations occurring in the urban environment there is the need for techniques which are capable of simultaneously analyzing different spectral feature types in order to derive the most characteristic combination for a certain material. For this purpose the authors have developed an interactive approach for identification of robust spectral features based on image spectra since they contain the most comprehensive variability of spectral characteristics for the analyzed surface materials. The approach is described in detail in Heiden et al. (2007). It has led to the identification of distinct spectral features for 21 manmade surface materials. Figure 4.2 shows the results for roofing polyethylene (a), and the open surface materials red loose chippings (b) and asphalt (c). For these examples representative image spectra obtained from different flight lines are plotted against a single field spectrum. Additionally, for each material the number of analyzed image spectra, the specific feature types, the used feature functions and the wavelength ranges where the features occur are specified, whereas the wavelength values correspond to the center wavelength of the HyMap bands. In case of these three materials the number of identified spectral features varies between nine for roofing polyethylene and three for asphalt reflecting the overall finding that the biggest number of spectral features can be identified for bright materials, such as roofing polyethylene and roofing tiles. In contrast, some of the dark materials, such as asphalt and roofing tar paper do not show distinct absorption features. Their separation from other materials had only been possible based on brightness represented by the feature functions mean and standard deviation of all reflectance values between two wavelength positions.

All of the interactively identified spectral features have been validated with regard to their robustness determining the separability. The developed approach and the obtained results are discussed in detail in Heiden et al. (2007). They show that the identified spectral features have resulted in overall good separation between materials. Best separability has been achieved for bright materials with distinct spectral features (e.g., polyethylene). Problems have occurred for dark materials with low reflectance and weak spectral characteristics. These results confirm the capability of hyperspectral imagery to resolve robust spectral features for most of the investigated urban surface materials. However, the investigations have also revealed that the interactive approach is very time consuming and thus can only be applied in test areas with a limited number of surface materials. Hence, there is the goal to develop methods which are capable of automatically determining such robust spectral features as part of an automated material mapping system.

4.3 Automated identification of urban surface materials

Automated mapping approaches for the urban environment have to be capable of fully exploiting the spectral information content of the hyperspectral imagery and related expert knowledge in form of spectral libraries. Furthermore, such approaches need to be able to handle a large number of possible surface materials including their spectral variations and a high percentage of mixed pixels in the image data due to the dominance of heterogeneous and small-sized structures. In the following, standard approaches of automated hyperspectral image analysis are briefly discussed and evaluated in regard to their suitability for automated mapping of urban surface materials (Section 4.3.1). The obtained findings have formed the basis for the development of a multi-step hyperspectral processing system for urban material mapping which is presented in Section 4.3.2.

4.3.1 State of the art of automated hyperspectral image analysis

Most of the classification and spectral unmixing approaches follow a three-step processing scheme. In the first step, the spectral dimensionality of the reflectance imagery is reduced (Section 4.3.1.1). In the second step, key endmembers are identified that are represented by spectrally pure image pixels (Section 4.3.1.2). In the third step, these endmembers are used to perform automated classification and spectral unmixing of surface materials for the whole image data set. This inversion process results in classification maps or in case of spectral unmixing in surface abundance maps for each material. In the following, the steps of this overall approach are discussed in more detail focusing on the suitability of existing methods for analyzing the urban environment.

4.3.1.1 Dimensionality reduction

Dimensionality reduction aims at eliminating redundant information from a dataset because many algorithms perform better in a low-dimensional and uncorrelated subspace. In case of hyperspectral data, dimensionality reduction is a requirement for many endmember detection methods that rely on convex geometry concepts (Section 4.3.1.2). In this context, methods, such as principal components analysis (PCA), the minimum

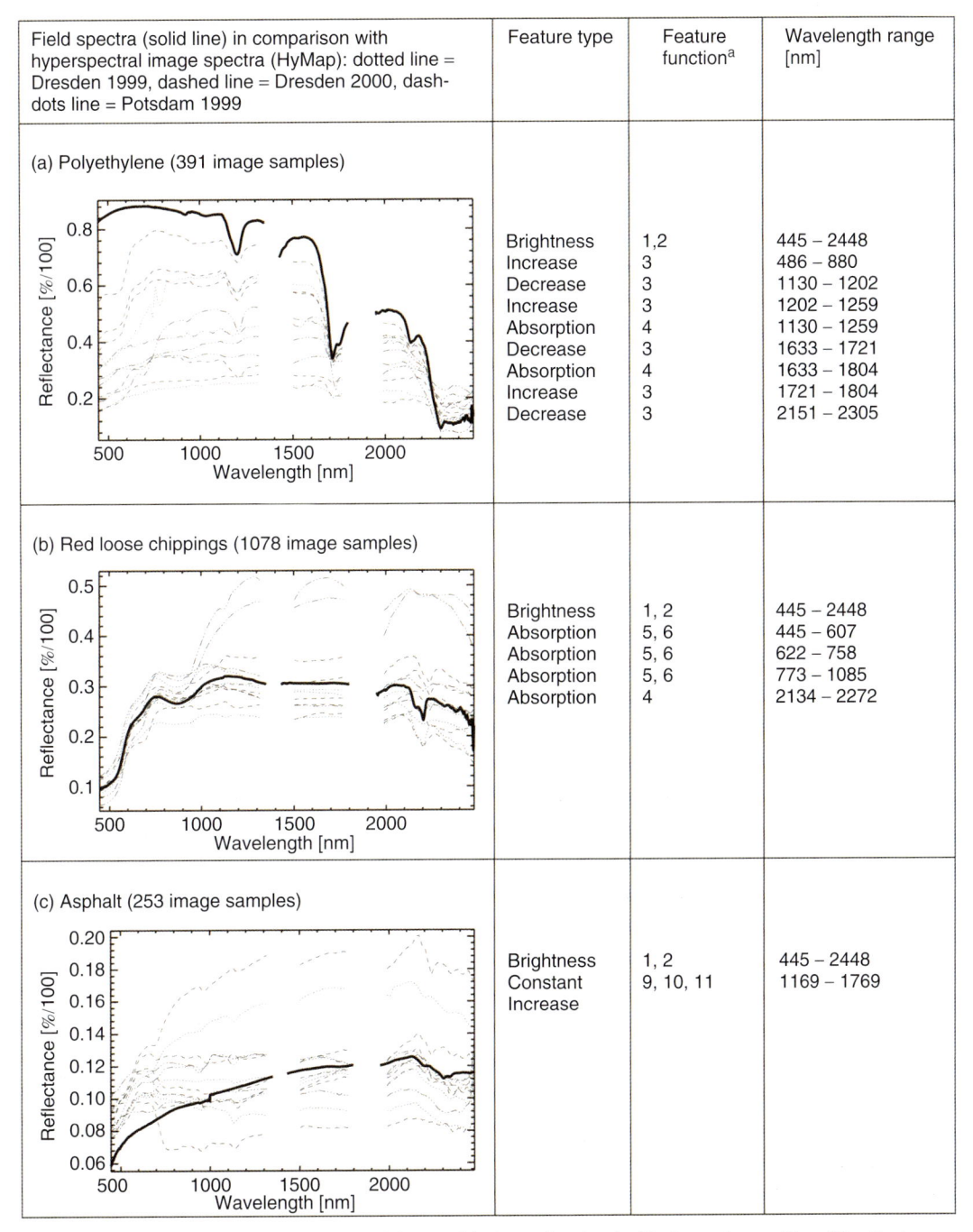

Field spectra (solid line) in comparison with hyperspectral image spectra (HyMap): dotted line = Dresden 1999, dashed line = Dresden 2000, dash-dots line = Potsdam 1999	Feature type	Feature function[a]	Wavelength range [nm]
(a) Polyethylene (391 image samples)	Brightness Increase Decrease Increase Absorption Decrease Absorption Increase Decrease	1, 2 3 3 3 4 3 4 3 3	445 – 2448 486 – 880 1130 – 1202 1202 – 1259 1130 – 1259 1633 – 1721 1633 – 1804 1721 – 1804 2151 – 2305
(b) Red loose chippings (1078 image samples)	Brightness Absorption Absorption Absorption Absorption	1, 2 5, 6 5, 6 5, 6 4	445 – 2448 445 – 607 622 – 758 773 – 1085 2134 – 2272
(c) Asphalt (253 image samples)	Brightness Constant Increase	1, 2 9, 10, 11	445 – 2448 1169 – 1769

[a](1) mean, (2) standard deviation, (3) ratio, (4) area, (5) absorption depth, (6) absorption position, (7) reflectance height, (8) reflectance position and (9) offset, (10) gain and (11) RMS of a regression line

FIGURE 4.2 Description of interactively identified spectral features for selected urban surface materials. The small step at about 1000 nm in the reflectance signature of asphalt results from a spectral jump between the VNIR and SWIR I detectors which is typical for the ASD field spectrometers.

noise fraction (MNF) transformation (Green et al., 1988), and the singular value decomposition (SVD) (Golub and van Loan, 1996: chapter 12.4) are widely used. The PCA seeks for principal components that explain most of the variance of the dataset. The MNF transformation optimizes the SNR whereas the SVD results in the projection that leads to the best representation of the data in the maximum power sense. All of them reduce the feature space to a significantly smaller number of bands allowing the determination of the inherent dimensionality of the data as a prerequisite for optimal endmember detection.

4.3.1.2 Endmember detection

Endmembers are represented by distinct spectral signatures of surface materials. The correct determination of such endmembers is very important in classification and unmixing of hyperspectral image data since they have a direct influence on the quality of the result (Theseira et al., 2003). The uniqueness of such endmembers is assured by two requirements. First, the data recording instrument has to be able to resolve spectrally distinct features of the respective surface materials allowing their identification and discrimination between them. Second, these endmembers have to be spectrally pure in the sense of not being influenced by spectral signals of other surface materials caused by neighborhood effects or mixing of materials. There are several approaches for the identification of such endmembers. They can be measured directly using field or laboratory spectrometers. Another opportunity is interactive selection of endmembers based on image data. In general, image based endmember detection leads to more accurate results in subsequent image classification and unmixing procedures (Song, 2005). This is due to the fact that they originate from the same data source and thus, reflect image-data-specific aberrations caused by sun illumination, weathering and preprocessing. However, both methods are very time consuming. Hence, most approaches automatically extract endmembers from the image data.

For this purpose, several techniques have been developed based on convex geometry concepts applied to the feature space. They assume that the endmembers are part of the convex hull of the point cloud. In this context, the well-known Pixel-Purity-Index method represents a semiautomatic empirical approach. The first generation of fully automatic methods has been detecting endmembers by fitting a simplex to the point cloud of the dataset in the feature space. These model-based approaches include minimum volume transforms (Craig, 1994) and the N-FINDR. Additional optimization techniques have been used in a number of other methods, such as iterative error analysis (IEA), Optical Real-time Adaptive Spectral Identification System (ORASIS) and Automated Morphological Endmember Extraction (AMEE). These methods are discussed and compared in detail by Plaza et al. (2004). However, all of these approaches are unsupervised and thus, they require subsequent assignment of the identified endmembers to thematic categories of interest. Additionally, all of them are based on convex geometry concepts causing the problem that only endmembers characterized by an extreme spectral shape are detected. Thus, they do not consider the full spectral variability of the materials of interest in the process of endmember selection (Bateson, Asner and Wessman, 2000). In the result, the missing spectral endmember variations lead to imprecise surface abundance estimations (Theseira et al., 2003).

4.3.1.3 Classification and spectral unmixing

One of the main goals of hyperspectral image analysis is the development of automated methods for area-wide material mapping. The high spectral resolution and – in case of airborne hyperspectral sensors – high spatial detail of hyperspectral imagery required new mapping approaches since traditional methods, such as maximum likelihood classification cannot deal with the high dimensionality of the data and are limited to analyzing whole pixels. However, in urban areas there has been a big need for spectral analysis at the subpixel level due to the small-sized and heterogeneous structures causing a high percentage of mixed pixels in the image data. The big variety of urban surface materials also requires the development of approaches allowing the incorporation of expert knowledge into the identification process. Therefore, supervised methods are preferably used relying on material-specific assessment of spectral training and reference information. Commonly used mapping approaches can be grouped into classification and unmixing techniques which are discussed in more detail in the following.

Supervised classification approaches

Image classification is the process of assigning a specific category of interest to each image pixel. Parametric methods, such as the maximum-likelihood classification rely on second order statistics (variance of classes) requiring a much larger number of training pixels than spectral bands for each class. Therefore, these methods can only be applied to high-dimensional spectral data if a dimensionality reduction is performed prior to classification. In order to use the full information content of the hyperspectral data, special classification techniques have been developed based on first order statistics or spectral features, e.g. spectral angle mapper (SAM) (Kruse et al., 1993) and spectral feature fitting (SFF) (Clark, Gallagher and Swayze, 1990). These methods only require a single reference spectrum per material class. For this purpose, automatically detected endmembers (Section 4.3.1.2) can be used. Besides these parametric approaches, non-parametric methods, such as artificial neural networks (e.g., Multilayer Perzeptron – MLP) and support-vector-machines (SVM) can also be used for classification. These techniques allow the most detailed endmember-based modeling of classes in the feature space. However, they require representative and comprehensive training information. All of these methods only assign one material class to each pixel and thus, do not consider the possible occurrence of more than one material per pixel. In order to deal with such mixed-pixel situations the spectral mixture analysis approach can be used.

Spectral mixture analysis

The high spectral dimensionality of hyperspectral image data allows the application of linear spectral mixture models resulting in estimation of abundance for each endmember class within one pixel. They are stored in abundance layers that contain the proportional coverage of the respective endmember for each pixel. A wide range of unmixing methods has been developed which can be grouped into full and partial unmixing techniques. The application of *full spectral unmixing* to an image dataset requires the knowledge of all endmembers contained in the dataset. Examples of full spectral unmixing methods are linear spectral unmixing (LSU) – also referred to as spectral mixture analysis (SMA) (Adams, Smith and Gillespie, 1993), non-negative least-squares estimation (NNLS) (Lawson and Hanson, 1974), and multiple endmember spectral mixture analysis (MESMA) (Roberts et al., 1998). Most of these methods only yield meaningful results if the endmembers are spectrally distinct and their number is fairly limited. Therefore, they are not applicable for urban material mapping comprising a large number of spectrally similar endmembers. In order to avoid this problem, advanced methods for endmember selection are required. One example is the MESMA approach (Roberts et al., 1998) which is capable of minimizing the number of possible endmember combinations within each pixel. *Partial spectral unmixing* aims at selected spectral targets of interest and is used to minimize the output of undesired signatures in order to further improve target discrimination. Representatives are matched filtering (MF) (Chen and Reed, 1987), or constraint energy minimization (CEM) (Resmini et al., 1997). However, partial unmixing can only be applied for a limited number of materials of interest. Thus, these techniques are not suitable for comprehensive urban material mapping.

4.3.2 Processing system for automated material mapping

A complex processing system optimized for automated identification of urban surface materials has been developed at the Remote Sensing Section of the GFZ Potsdam, Germany. Standard hyperspectral image analysis techniques discussed in Section 4.3.1 have been combined, modified and extended in order to meet the special needs of hyperspectral material mapping in the urban environment described in Section 4.3. The resulting approach (Fig. 4.3) has reached a high degree of automation and consists of three major steps. First, endmember identification is carried out in the hyperspectral image data using an automated spectral-feature based approach. Second, spectrally pure pixels are identified by a maximum likelihood classification whereas the previously identified endmembers serve as training information. Third, the remaining unclassified pixels are analyzed by neighborhood-oriented iterative linear spectral unmixing (Roessner et al., 2001; Segl et al., 2003). The approach also includes the possibility of integrating additional object height information at all processing steps reducing confusion between materials (Sections 4.3.2.1 and 4.3.2.2). A more detailed overview of the processing steps and their relation to each other is given in the flowchart depicted in Fig. 4.3.

4.3.2.1 Feature-based identification of endmembers

The developed method automatically extracts representative endmembers and their spectral variations from the image data. For this purpose, the existing comprehensive urban image spectral library (Section 4.2.2) has been used to build a classifier that – in contrast to the methods discussed in Section 4.3.1.2 – enables supervised endmember detection.

The developed classifier is based on spectral features (Section 4.2.3) and consists of $\binom{n}{2}$ pairwise maximum likelihood classifiers (n = total number of surface materials). The feature space of each pairwise classifier is optimized for each two-class problem using an automated feature selection algorithm (sequential forward selection). This way, a class-specific dimensionality reduction is achieved that is

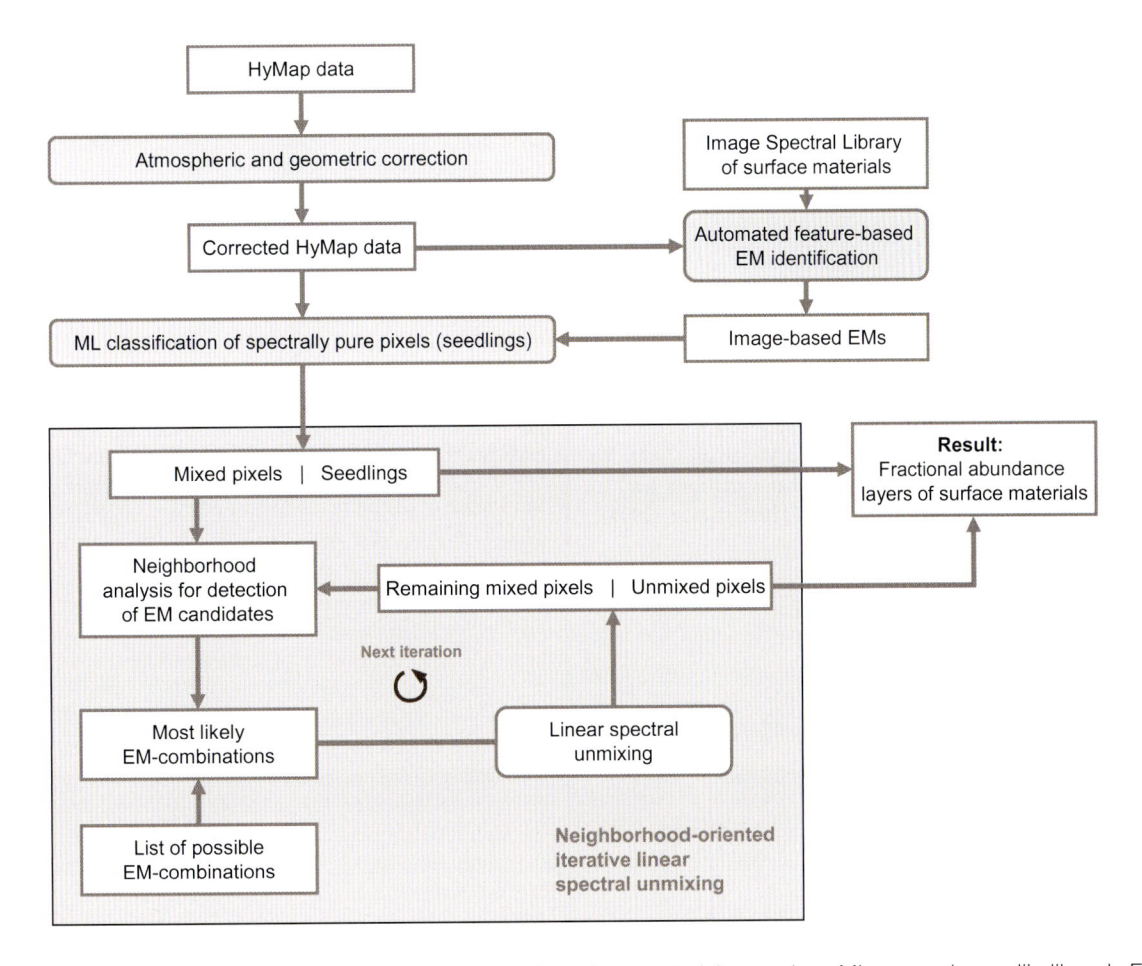

FIGURE 4.3 Automated multi-step processing system for urban material mapping. ML = maximum likelihood, EM = endmember.

superior to unsupervised projections based on global statistics (Section 4.3.1.1). For feature selection thousands of spectral features are calculated based on all spectra contained in the image spectral library. Since the pairwise maximum likelihood classification uses covariance statistics, it is capable of integrating features of different scales. These features include band ratios, absorption depths as well as means, standard deviations and coefficients of polynomial fits.

The feature selection process results in a set of spectral features for each class pair allowing its optimal separation. After parameterizing the pairwise classifiers, the final material decision is made by applying a MIN-MAX operation. First, the minimum probability is selected for each class and second, the pixel is assigned to the material characterized by the highest minimum probability. Once the classifier has been established, it can be applied to any HyMap-like hyperspectral image dataset in order to identify the specific endmembers. For this purpose a user-defined threshold is applied to the obtained class probabilities. This way, the developed endmember detection method reduces the effort for collecting ground truth information.

4.3.2.2 Classification of spectrally pure pixels

The endmember pixels identified in the first processing step only represent a small percentage of all image pixels. The remaining majority can be subdivided into spectrally pure and mixed pixels whereas it is assumed that the spectral signal of the pure ones is caused by a single material covering the whole pixel. In case of mixed pixels the spectral signal results from a combination of several materials occurring within one pixel. The second processing step aims at optimal separation between these two groups including the identification of the respective surface material for the spectrally pure ones. For this purpose an algorithm has been developed which is based on the maximum likelihood classification. The already identified endmembers serve as training information for the classifier after they have been further subdivided into representative subclasses by a cluster algorithm based on reflectance data in order to accommodate the high spectral variability of urban surface materials in the image data. In the result, the classifier is parameterized with the mean and covariance statistics of Gaussian distributed subclasses. The separation between pure and mixed pixels builds on the fact that in feature space mixed pixels are located further away from the class centers compared with spectrally pure pixels. Hence, the Mahalanobis distance threshold is used to exclude mixed pixels from the classification process. This threshold is calculated for each subclass based on the training pixels and one global user-defined parameter for the rejection of a certain percentage of distant pixels. The combined identification results of the first and second processing step represent the total number of spectrally pure pixels. They serve as seedling pixels for the third processing step analyzing the remaining mixed pixels.

4.3.2.3 Neighborhood-oriented iterative linear spectral unmixing

A special spectral mixture analysis similar to the MESMA approach has been developed that includes additional optimization strategies in order to reduce the number of possible endmembers per pixel. For this purpose spectrally pure seedlings and a list of possible endmember combinations are introduced into neighborhood-oriented iterative linear spectral unmixing (Roessner et al., 2001). The iterative unmixing procedure starts at pixels that are adjacent to the previously identified spectrally pure pixels (seedlings). These seedlings represent potential endmembers forming the mixed pixel spectrum. Depending on their number and spatial configuration in relation to unknown mixed pixels (neighborhood analysis), different combinations are tested successively according to their likelihood based on a list of possible endmember combinations. During further unmixing iterations the endmember information of previously unmixed pixels is also considered in the analysis of neighboring mixed pixels. This way, the iterative unmixing procedure leads to spatial growing of the unmixing results around seedling pixels. The process is repeated until no more pixels are left which can be unmixed in a meaningful way. The result of the spectral mixture analysis consists of a multi-band image whereas each band represents one surface material containing the fractional abundance of this endmember for each pixel.

4.4 Results and discussion of their potential for urban analysis

Figure 4.4(b–d) depicts exemplary results of the developed automated urban material mapping system in comparison to a HyMap true color composite (Fig. 4.4a) for a 2 × 2.5 km subset containing a part of the inner city of Berlin, Germany. The area is characterized by different urban structure types dominated by industrial areas, perimeter block development, single family homes and mid-rise dwellings development. For a classification-like representation of the results the dominating endmember (Fig. 4.4b) has been derived for each pixel based on the 45 abundance layers representing the surface materials automatically detected in the study area. This map is complemented by Fig. 4.4c showing the endmember with the second highest abundance for each mixed pixel. For visualization purposes these surface materials have been further aggregated into the mapping categories contained in the legend for the maps in Fig. 4.4(b and c). The numbers in brackets indicate the number of surface materials (level 4 of Table 4.1) aggregated into the respective categories. These mapping results are supplemented by a grayscale image depicting the fraction of the dominating endmember ranging between 50 and 100% (Fig. 4.4d).

Despite thematic aggregation of the mapping categories, the large spatial heterogeneity of surface materials is well visible in the mapping result. Although the developed approach operates on a per-pixel basis, the shapes of the real-world objects are well preserved. This is due to the large number of material classes and the low number of unclassified pixels (coded black) amounting to less than three percent. The grayscale image of Fig. 4.4d indicating the fractional abundance of the dominating endmember shows that about 60% of the 3–4 m resolution image data of the test area are covered by mixed pixels. Areas dominated by small-sized objects are depicted in various shades of gray indicating a large number of mixed pixels. They can clearly be distinguished from areas characterized by large homogeneous objects consisting of several pure pixels framed by mixed pixels (e.g., industrial

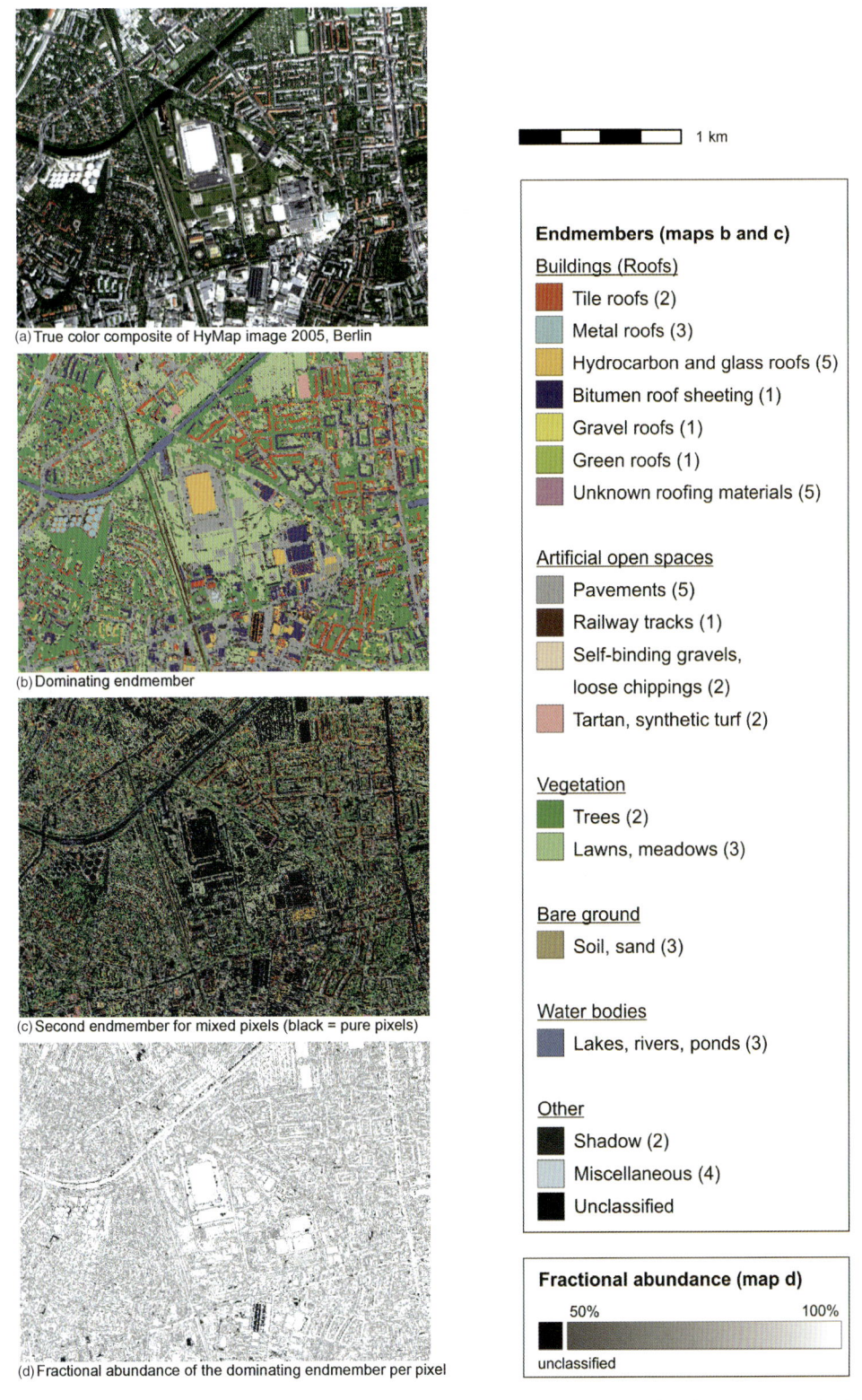

FIGURE 4.4 Exemplary results of neighborhood-oriented iterative linear spectral unmixing for a test area in the city of Berlin, Germany. The numbers in brackets indicates the number of surface materials (level 4 of Table 4.1) aggregated into the respective mapping categories.

sites in the center of the test area). In case of low abundances for the dominating endmember, the material coverage of the second endmember provides valuable information analyzing the occurrence of surface materials at a sub-pixel level. Additionally, a few smaller black areas can be identified representing unclassified pixels. They show the ability of the developed processing system to reject pixels if they are covered by materials which have not been introduced yet to the supervised classification and unmixing procedure. Such areas require subsequent spectral analysis leading to an extension of the existing image spectral library. This way, successive completion of the spectral knowledge base can be achieved resulting in a comprehensive automated mapping system covering the complete variety of urban surface materials occurring in the region of interest.

The obtained results represent a new level of physically based material mapping derived from automated hyperspectral image analysis. The high spatial and thematic degree of detail of the mapped surface materials opens up new opportunities for a wide range of applications. One example is the assessment of the imperviousness of surfaces. So far, simplified approaches, such as the Vegetation-Impervious-Soil (V-I-S) model have been used to derive the degree of imperviousness from multispectral data (Ridd, 1995). Originally, these models were developed for medium-resolution satellite imagery, such as Landsat-(E)TM and later on they also were applied to high-resolution IKONOS data (Small, 2003). However, they are based on the assumption that all impervious materials are fully impervious and that they only comprise one or two spectral classes. This represents an oversimplification preventing application of these models to spatially detailed analysis of urban structures required in ecologically oriented urban planning. Heiden *et al.* (2003) have shown that the presented approach allows material-based quantitative assessment of imperviousness by linking the spatial coverage of each material within a mapping unit to an empirically measured infiltration rate. In this connection the use of hyperspectral image data allows overcoming problems of insufficient differentiation between materials, such as soil and paved surfaces which have been occurring in studies based on multispectral data (Lu and Weng, 2006). Additionally, semi-permeable surfaces, such as gravel or cobblestones can be distinguished leading to a more realistic quantitative assessment of imperviousness. These results can serve as input information for spatially distributed modeling of hydrological processes characterizing the water cycle in urban areas (Gill *et al.*, 2007). Another potential application of the presented material mapping results is the analysis of urban microclimate using energy flux models in order to identify the drivers of local heating and cooling. Hoyano *et al.* (1999) have modeled the sensible heat flux of different urban structure types showing that the size and arrangement of buildings as well as the material coverage of the surfaces have a great influence on the resulting air temperatures.

In summary, this chapter has demonstrated that high spectral resolution hyperspectral imagery yields the potential for detailed and comprehensive mapping of urban surface materials which goes far beyond the widely known landcover classification schemes derived from multispectral data. In order to assess the full information content of these data for automated mapping of a big variety of surface materials, specialized approaches are needed which are capable of simultaneously analyzing different spectral feature types. Within the developed multi-step processing system supervised feature-based endmember identification forms the basis for automated classification and iterative neighborhood-oriented linear spectral unmixing. The resulting surface materials can serve as input information for detailed large-scale modeling. They also allow quantitative derivation of a multitude of ecological indicators (e.g. Krause, 1989; Pauleit and Duhme, 2000; Whitford, Ennos and Handley, 2001) and automated identification of of areas of similar spatial structure, such as urban biotopes (Bochow *et al.*, 2010). This way, hyperspectral remote sensing can significantly contribute to a more objective characterization of urban areas and the development of automated monitoring systems based on efficient and detailed change detection in a rapidly changing urban environment.

References

Adams, J. B., Smith, M. O. and Gillespie, A. R. (1993) Imaging spectroscopy: Interpretation based on spectral mixture analysis, in *Remote Geochemical Analysis: Elemental and Mineralogical Composition* (eds C. M. Pieters and P. Englert), Cambridge University Press, New York, 145–166.

Anderson, J. R., Hardy, E. E., Roach, J. T. and Witmer, R. E. (1976) A land use and land cover classification scheme for use with remote sensor data, US Geological Survey, Reston VA.

Bannari, A., Morin, D., Bonn, F. and Huete, A. R. (1995) A review of vegetation indices. *Remote Sensing Reviews*, **13**, 95–120.

Bateson, A., Asner, G. P. and Wessman, C. A. (2000) Endmember bundles: A new approach to incorporating endmember variability into spectral mixture analysis. *IEEE Transactions on Geoscience and Remote Sensing*, **38** (2), 1083–1094.

Ben-Dor, E., Levin, N. and Saaroni, H. (2001) A spectral based recognition of the urban environment using the visible and near-infrared spectral region (0.4–1.1 µm). A case study over Tel-Aviv, Israel. *International Journal of Remote Sensing*, **22** (11), 2193–2218.

Bochow, M., Peisker, T., Roessner, S., Segl, K. and Kaufmann, H. (2010) Towards an automated update of urban biotope maps using remote sensing data: What is possible? *Urban Biodiversity and Design*. N. Müller, P. Werner and J. G. Kelcey Conservation Science and Practice Series, no. 7 Wiley-Blackwell, Oxford 255–272.

Chen, J. Y. and Reed, I. S. (1987) A detection algorithm for optical targets in clutter. *IEEE Transactions on Aerospace and Electronic Systems*, **AES-23** (1), 46–59.

Clark, R. N. (1999) Spectroscopy of rocks and minerals, and principles of spectroscopy, in *Manual of Remote Sensing* (ed. A. N. Rencz), John Wiley & Sons, Inc., New York, pp. 3–58.

Clark, R. N., Gallagher, A. J. and Swayze, G. A. (1990) Material absorption band depth mapping of imaging spectrometer data using a complete band shape least-squares fit with library reference spectra, in *2nd AVIRIS Earth Science Workshop*, Pasadena, JPL.

Craig, M. D. (1994) Minimum Volume Transforms for Remotely Sensed Data. *IEEE Transactions on Geoscience and Remote Sensing*, **32**, 542–552.

Gill, S. E., Handley, J. F., Ennos, A. R. and Pauleit, S. (2007) Adapting cities for climate change: the role of the green infrastructure. *Built Environment*, **33** (1), 115–133.

Golub, G. H. and van Loan, C. F. (1996) *Matrix Computations*, Johns Hopkins University Press, Baltimore, MD, USA.

Green, A. A., Berman, M., Switzer, P. and Craig, M. D. (1988) A transformation for ordering multispectral data in terms of image quality with implications for noise removal. *IEEE Transactions on Geoscience and Remote Sensing*, **26** (1), 65–74.

Green, R. O., Asner, G., Ungar, S. and Knox, R. (2008) NASA mission to measure global plant physiology and functional types, in *Proceedings of the 2008 IEEE Aerospace Conference*. Big Sky, Montana.

Heiden, U., Segl, K., Roessner, S. and Kaufmann, H. (2003) Ecological evaluation of urban biotope types using airborne hyperspectral HyMap data, in *Proceedings of the 2nd GRSS/ISPRS Joint Workshop on Remote Sensing and Data Fusion over Urban Areas*, Berlin.

Heiden, U., Segl, K., Roessner, S. and Kaufmann, H. (2007) Determination of robust spectral features for identification of urban surface materials in hyperspectral remote sensing data. *Remote Sensing of Environment*, **111** (4), 537–552.

Herold, M., Roberts, D. A., Gardner, M. E. and Dennison, P. E. (2004) Spectrometry for urban area remote sensing – development and analysis of a spectral library from 350 to 2400 nm. *Remote Sensing of Environment*, **91** (3–4), 304–319.

Hoyano, A., Iino, A., Ono, M. and Tanighchi, S. (1999) Analysis of the influence of urban form and materials on sensible heat flux – a case study of Japan's largest housing development "Tama New Town". *Atmospheric Environment*, **33** (24–25), 3931–3939.

Kaufmann, H., Segl, K., Chabrillat, S. *et al.* (2006) A Hyperspectral Sensor for Environmental Mapping and Analysis, in *IEEE International Geoscience and Remote Sensing Symposium (IGARSS 2006) & 27th Canadian Symposium on Remote Sensing*, Denver, Colorado, USA.

Krause, K.-H. (1989) Zur Erfassung der Oberflächenarten für eine stadtökologische Zustandsbeschreibung. *Hallesches Jahrbuch für Geowissenschaften*, **14**, 124–130.

Kruse, F. A., Lefkoff, A. B., Boardman, J. W. *et al.* (1993) The spectral image processing system (SIPS) – interactive visualization and analysis of imaging spectrometer data. *Remote Sensing of Environment*, **44** (2–3), 145–163.

Lawson, C. L. and Hanson, R. J. (1974) *Solving Least Squares Problems*, Prentice Hall, Englewood Cliffs, NJ.

Lu, D. S. and Weng, Q. H. (2006) Use of impervious surface in urban land-use classification. *Remote Sensing Of Environment*, **102** (1–2), 146–160.

Pauleit, S. and Duhme, F. (2000) Assessing the environmental performance of land cover types for urban planning. *Landscape and Urban Planning*, **52** (1), 1–20.

Plaza, A., Martinez, P., Perez, R. and Plaza, J. (2004) A quantitative and comparative analysis of endmember extraction algorithms from hyperspectral data. *IEEE Transactions on Geoscience and Remote Sensing*, **42** (3), 650–663.

Resmini, R. G., Kappus, M. E., Aldrich, W. S. *et al.* (1997) Mineral mapping with Hyperspectral Digital Imagery Collection Experiment (HYDICE) sensor data at Cuprite, Nevada, USA. *International Journal of Remote Sensing*, **18**, 1553–1570.

Ridd, M. K. (1995) Exploring a V-I-S (vegetation-impervious surface-soil) model for urban ecosystem analysis through remote sensing: comparative anatomy for cities. *International Journal of Remote Sensing*, **16** (12), 2165–2185.

Roberts, D. A., Gardner, M. E., Church, R. *et al.* (1998) Mapping chaparral in the Santa Monica Mountains using multiple endmember spectral mixture analysis. *Remote Sensing of Environment*, **65**, 267–279.

Roessner, S., Segl, K., Heiden, U. and Kaufmann, H. (2001) Automated differentiation of urban surface based on airborne hyperspectral imagery. *IEEE Transactions on Geoscience and Remote Sensing*, **39** (7), 1523–1532.

Sacchetti, A., Cisbani, A., Babini, G. and Galeazzi, C. (2010) The Italian precursor of an operational hyperspectral imaging mission, in *Small Satellite Missions for Earth Observation* (eds R. Sandau, H.-P. Roeser and A. Valenzuela), Springer, Berlin, 73–81.

Schaepman, M. E., Ustin, S. L., Plaza, A. J. *et al.* (2009) Earth system science related imaging spectroscopy – an assessment. Session on the State of Science of Environmental Applications of Imaging Spectroscopy held in honor of Alexander FH Goetz, *Remote Sensing of Environment*, **113**, S123–S137.

Segl, K., Roessner, S., Heiden, U. and Kaufmann, H. (2003) Fusion of spectral and shape features for identification of urban surface cover types using reflective and thermal hyperspectral data. *ISPRS Journal of Photogrammetry and Remote Sensing*, **58**(1–2), 99–112.

Small, C. (2003) High spatial resolution spectral mixture analysis of urban reflectance. *Remote Sensing of Environment*, **88**(1–2), 170–186.

Song, C. H. (2005) Spectral mixture analysis for subpixel vegetation fractions in the urban environment: How to incorporate endmember variability? *Remote Sensing of Environment*, **95**(2), 248–263.

Theseira, M. A., Thomas, G., Taylor, J. C. *et al.* (2003) Sensitivity of mixture modelling to end-member selection. *International Journal of Remote Sensing*, **24**(7), 1559–1575.

Whitford, V., Ennos, A. R. and Handley, J. F. (2001) "City form and natural process" – indicators for the ecological performance of urban areas and their application to Merseyside, UK. *Landscape and Urban Planning*, **57**(2), 91–103.

5

Very-high-resolution spaceborne synthetic aperture radar and urban areas: looking into details of a complex environment

Fabio Dell'Acqua, Paolo Gamba and Diego Polli

The late years of the last decade marked a turnover from typical ground resolutions around ten meters to one meter for images acquired from space-borne radar sensors. When complex environments such as urban areas are concerned, this big step forward means more than just more visibility of details: it implies a change of approach to the problem of information extraction, both in terms of goals and of techniques. With this level of very-high-resolution it makes sense to turn one's attention from general, extensive inspection of urban areas (e.g., land cover & urban sprawl, density of buildings) to more intensive extraction of specific features such as individual building characteristics. Meanwhile, new problems are encountered. For example, geometric distortions inherent to SAR images become evident and need to be specifically addressed if building feature extraction is to be carried out; for some land cover classes, some previously developed speckle models may not be valid any more. This chapter deals with the new possibilities opened by the increased resolution and the challenges when using such new satellite data in some urban applications, including building extraction and disaster management.

5.1 Introduction

The peculiar characteristics of the urban environment were generally reflected in somehow complex characteristics of remotely sensed data acquired over it. In optical images, for example, the many different materials found within a small parcel (roof, vegetation, building materials, and so on) turn into a wide variability of spectral responses over a small spatial stretch.

In radar images, the influence of material is limited, and geometry of the target becomes more critical. Just to give an example, Fig. 5.1 shows an example of backscattering range profiles of a simple flat-roof building model. The building is represented by a box with width w and height h, uniform surfaces and flat surroundings, viewed by a synthetic aperture radar (SAR) sensor with incidence angle θ.

In Fig. 5.1 the return from the ground is represented by letter a, while b highlights the double bounce caused by the corner reflection created by the intersection of the building vertical wall and the surrounding ground, c indicates single backscattering from the front wall, d is the return caused by the building roof and e represents the shadow area with absence of any return from the building and from the ground. Symbols l [$l = h^*\cot(\theta)$] and s [$s = h^*\tan(\theta)$] represent the areas affected by layover and shadow in the ground-projected image space, respectively.

The scattering effects of a gable-roof building and a flat-roof building are different. The major difference is that a gable-roof building has a second bright scattering feature, appearing closer to the sensor than the double bounce, resulting from direct backscattering caused by the part of the roof, which is oriented toward the sensor.

Clearly, the complexity of the urban environment as seen in the above examples, places several requirements on remote sensors to be used. Spatial resolution, for example, is crucial, because of the considerable variability and non-homogeneity of urban areas even at small scales.

While optical satellite sensors have acquired data with as high as 1-meter resolution in 1999 with the launch of IKONOS, only several years later, satellite radar sensors were able to obtain data with the comparable spatial resolution since the launches of the German TerraSAR-X (Roth, 2003), the Italian COSMO/SkyMed (Impagnatiello et al., 1998), the Canadian Radarsat-2 (Morena, James and Beck, 2004) in 2007. Such data are becoming more widely available while projects on TerraSAR-X announcement of opportunity (AO) are running, projects on COSMO/SkyMed AOs are about to start, and the Science and Operational Applications Research for RADARSAT-2 Program (SOAR) is also on its way. The timescale is summarized in Table 5.1, where the main features of some among the most important spaceborne sensors are reported, along with their launch year. It can be noted that a switch to the X band marked the achievement of 1-m resolution.

Scientific research over data provided by the above new generation sensors is also fully active, as shown by the increasing number of publications. In this chapter, we will first see what sort of information can be extracted on urban areas at the resolution typical for older sensors, and then will move on to the new generation, to highlight what new types of information can be obtained. The focus will naturally be on the benefits of such information extraction for urban area analysis. Some examples of applications will be given and the benefits of SAR remote sensing will be highlighted.

5.2 Before spaceborne high-resolution SAR

A SAR sensor generates the electromagnetic waves that illuminate the target in a coherent manner, and collects the backscattered field thus estimating the radar backscattering features of the observed scene through complex processing of the collected signal. This adds up to the inherent insensitivity to the presence of water drops in the atmosphere, due to the long wavelength, to form two clear advantages for radar data with respect to optical data.

Notwithstanding these nice features, SAR sensors have not been considered suitable for precise characterization of the urban environment over a quite long period; until very recently, very-high-resolution (VHR) optical data have been considered somehow mandatory due to their efficient and reliable mapping capabilities.

This does not mean that SAR sensors were not considered in literatures. In fact, remote sensing papers explicitly considering urban areas appeared as early as in 1990 (Bender, Gilg and Ottl,

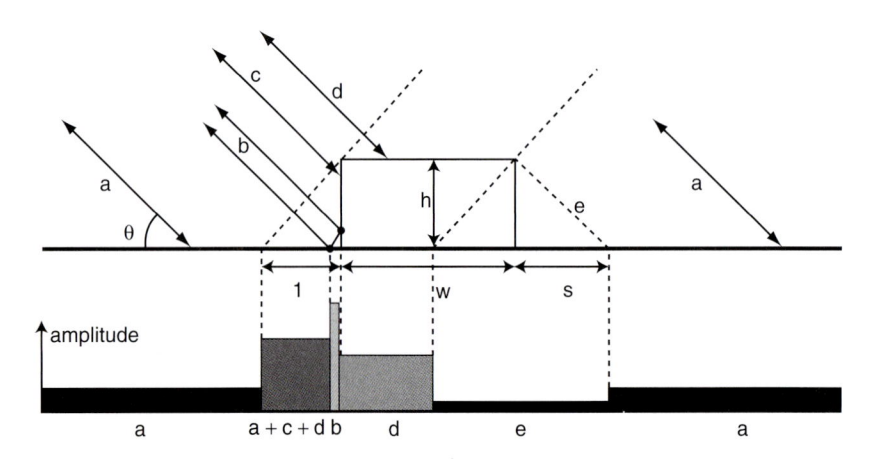

FIGURE 5.1 Along-range appearance of a very simple building in a SAR image.

TABLE 5.1 Enhancement in time of spatial resolution of spaceborne SAR systems.

Sensor class	Sensor/satellite name	Country/organization	Band	Launch	Finest available spatial resolution (m)
Medium resolution	ERS-1	ESA	C	1991	30
Medium resolution	ERS-2	ESA	C	1995	30
Medium resolution	ASAR	ESA	C	2002	30
Medium resolution	JERS	Japan	L	1992	18
High resolution	ALOS	Japan	L	2006	7
High resolution	Radarsat - 2	Canada	C	2007	3
High resolution	COSMO/SkyMed	Italy	X	2007	1 (civilian use)
High resolution	TerraSAR-X	Germany	X	2007	1

1990). Taket, Howarth and Burge (1991) developed a model to simulate SAR images of urban built targets, and Dong, Forster and Ticehurst (1997) tried to connect building features with coarse-scale characteristics of satellite SAR data.

For several years the main application of satellite SAR has been for small-scale mapping of urban areas due to the spatial resolution constraints. However, at least at a conceptual level the systematic use of such sensors for urban monitoring was proposed by Henderson and Xia (1997).

One further advantage of radar over optical data is related to the phase information, which is provided by the former and is instead unavailable on the latter. Exploitation of this additional information allows better understanding the backscattering properties of urban materials and extracting three-dimensional information using interferometric SAR data (Rosen et al., 2000). This type of data can be used for different urban applications, such as subsidence monitoring (Ferretti, Prati and Rocca, 2000) just to mention one. Further exploitation of the phase information relies upon the computation of coherence, i.e. the complex correlation between two radar scenes in a small window around each pixel. This measure has been widely considered in urban areas that generally contain many strong and long-lasting backscatterers. This translates into typical coherence levels much higher in urban areas than in agricultural fields and forestry (Usai and Klees, 1999), except where significant changes take place between the dates of the two radar images. Coherence has been used to map human settlements (Dammert, Askne and Kuhlmann, 1999) and assess damage after earthquakes (Yonezawa and Takeuchi, 2001) and other types of changes (Dierking and Skriver, 2002). Finally, although still somehow impaired by the limited availability of data, polarimetry is becoming increasingly important (Sauer et al., 2006, 2008) because multiple polarizations allow discriminating among the various backscattering mechanisms in urban areas. With such data at hand, recognition of different building structures is easier (Dong, Forster and Ticehurst, 1997). Polarimetric interferometry can also provide 3D information from a single polarimetric sensor, further widening the ability to reconstruct building information using radar data. As shown in Fig. 5.2, where the result of Pauli decomposition on a polarimetric SAR image over an urban area in Brazil is displayed, different types and structures of buildings generate different relative backscatter levels and thus different colours in the false colour representation of the data.

The next, intermediate step towards fine-resolution was represented by satellites such as Radarsat-1 (Weydahl, Bretar and Bjerke, 2003) in its *fine* mode and ALOS PALSAR (Rosenqvist et al., 2004) descending below the 10 m limit in their spatial resolutions.

Moving from ERS-like spatial resolutions (30 m) to below 10 m, a number of issues arise. Human settlements reveal their

FIGURE 5.2 Pauli decomposition on a Radarsat-2 image over an area around São Paulo, Brasil. Red = even bounce, Green = tilted even bounce, Blue = odd bounce.

non-homogeneity to a wider extent, and even different parts of the same objects tend to appear as separate scatterers. This adds to the traditional unreliability of pixel contents in a coherent radar image, and as a result pixel-by-pixel analysis becomes less efficient when handling this variety of possible responses within the same land cover class. Segmentation approaches become a more reasonable choice (Lombardo et al., 2003) and can be based either on statistical (e.g., Macrì Pellizzeri et al., 2003) or spatial (e.g., Dell'Acqua and Gamba, 2003) analysis. The latter has a more immediate meaning and is directly related to the spatial structure of the settlement, which is one of the basic indicators of its usefulness. It has been shown that formal and informal settlements differ in the "order-vs.-disorder" appearance in remotely sensed images (Niebergall, Loew and Mauser, 2007), and different land use classes can be discriminated even using coarser data (Gamba, Dell'Acqua and Trianni, 2006), with a feasible scale-adaptive approach. This latter paper exploits a priori knowledge about road direction distribution in urban areas: an adaptive filtering procedure captures the principal directions of these roads and this information allows enhancing the extraction results. After road element extraction, a special perceptual grouping algorithm is devised, exploiting co-linearity as well as proximity concepts to reduce redundant segments and fill in gaps. Finally, the road network topology is considered, checking for road intersections and regularizing the overall patterns using these focal points.

Another example of urban area delineation can be found in (Stasolla and Gamba, 2008; Gamba et al., 2009) where a map was extracted from an ASAR frame over the urban area of Beijing using a combination of morphological and textural methods, recently developed.

5.3 High-resolution SAR

Despite the limited availability of satellite HR SAR data until recently, techniques that were developed for HR airborne SAR are also useful for spaceborne HR data. For example: a technique was initially developed for road extraction from airborne SAR data (Gamba, Dell'Acqua and Lisini, 2006), combining directional filtering, perceptual grouping, and topological concepts; this technique was later applied to spaceborne data with a similar resolution (Hedman et al., 2010). In other cases, the change in ground resolution means a significant discontinuity in processing techniques. Features of spaceborne and airborne radar can indeed be very different.

Although spatial resolution is independent of altitude, viewing geometry and swath coverage can be greatly affected by altitude variations. At aircraft operating altitudes, airborne radars must image over a wide range of incidence angles (up to 70 degrees), and this has a significant effect on the backscatter from surface features and on their appearance on an image. Foreshortening, layover, and shadowing will be subject to wide variations, across the image. Spaceborne radars operate at much higher altitudes than airborne radars and are thus able to avoid some of these imaging geometry problems, with a typical range of incidence angles from 20 to 50 degrees, also resulting in more uniform illumination.

More differences are connected with the need for airborne data to correct artifacts due to the random platform motion caused by air turbulence, absent in spaceborne platforms, while on the other hand spaceborne data has to be explicitly corrected for curvature and rotation of the Earth.

Regardless of the issues connected with a switch in platform, however, a scaling down of the spatial resolution by more than one order of magnitude means very different features to emerge, such as:

- foreshortening issues appear at a much smaller scale; while in ERS-like images single buildings were not distinguishable, in meter-resolution images the different floors of the same buildings are imaged separately and end up in different locations in the image;
- the number of scatterers within the same pixel becomes smaller and the interference mechanisms change, so do image statistics;
- double-bounce phenomena become visible as such, rather than simply pushing up the average reflectivity sensed on an urban area.

Some aspects connected to meter-resolution data can be better explained by referring to representative applications. A selection of these is presented and discussed in the next subchapters.

5.3.1 Extraction of single buildings

A typical application of high-resolution radar data is three-dimensional building recognition (Bolter and Leberl, 2000; Quartulli and Datcu, 2004; Tison and Tupin, 2004; Thiele et al., 2007).

SAR and InSAR data have been widely exploited, for example, in city cores with high-rise buildings (Gamba, Houshmand and Saccani, 2000), rural areas, and industrial plants (Simonetto et al., 2005). These types of data still show limits, especially when compared with the more expensive yet more accurate LIDAR data (Stilla, Soergel and Thoennessen, 2003). LIDAR supplies 3D data with a ground sampling rate of a few points per square meter and an RMS error on elevation on the order of tenths of a meter, but it costs between 60–135€ per square kilometre. On the other hand, a spaceborne SAR scene with 8 m resolution (e.g. RADARSAT-1) is distributed at less than 1€ per square kilometre, but the resolution and height accuracy (on the order of meters) achievable are definitely lower for spaceborne SAR when used to produce a DEM in place of LIDAR.

In some cases, those two types of data may be merged to combine the pros of both; purchasing a large SAR scene but only a small parcel of LIDAR data on the same site and comparing them to derive distortion information which will drive correction of the rest of the SAR scene may be a good approach to fusion (Gamba, Dell'Acqua and Houshmand, 2003).

A first reason why SAR is difficult to use is obviously its side-looking nature, which results in phenomena such as foreshortening, layover, occlusion, total reflection, and multi-bounce scattering. All these phenomena are also found in natural scenes, but they show up much more frequently in urban areas. This happens because the urban environment is rich in smooth planar patches, often at right angles with each other (resulting in corner-cube-like reflection) or parallel to each-other (resulting in several bounces of the incident electromagnetic wave).

FIGURE 5.3 A cluster of skyscrapers in the Shanghai, China, Business District as seen in a TerraSAR-X image.

Depending on the viewing direction, acquisition of a large urban scene can be plagued with corner-cube or multi-bounce phenomena (Dong, Forster and Ticehurst, 1997); regions of radar shadow (e.g., cast behind buildings) coincide with noisy InSAR elevation data. One may overcome such obstacles by acquiring SAR data from different vantage points and fusing interpretation results (Michaelsen, Soergel and Thoennessen 2005; Soergel et al., 2005) or fusing InSAR with other 3D data such as LIDAR (Gamba, Dell'Acqua and Houshmand, 2003; Stilla, Soergel and Thoennessen, 2003). In any case, the appearance of buildings in radar images is complex enough to make perceptual approaches a good option (Michaelsen, Soergel and Thoennessen, 2006), as well as gestalt perception rules (Michaelsen, Middelmann and Sörgel, 2007). Figure 5.3 shows how a cluster of skyscrapers from the centre of Shanghai is represented in a TerraSAR-X spotlight image, which gives an idea of how necessary it is to consider approaches capable of some degree of modelling and interpretation. Just to highlight a few of the problems one encounters, the reader is invited to not the following:

In optical images a closer object occludes and thus cancels completely another object behind it; in radar images the location of backscatterers is basically determined by the optical path length, and thus structures in different locations may appear overlapping because their line-of-sight distances from the sensor are similar. So do the two skyscrapers at the top of Fig. 5.3.

For the same reason, double or triple bounces make the optical path longer that would be for line-of-sight acquisition, and thus objects illuminated by mirror-reflected rays from a building appear mislocated. Such phenomenon is not clearly visible in Fig. 5.3.

In optical images, bright spots may cause saturation of the sensor and thus results in an underestimation of local reflectivity; in radar images, due to the impulse response of the SAR system, strong backscatterers produce an overestimation of radar response over a number of pixels along the same row and column where the backscatterer is located. This phenomenon shows as "bright crosses" and it is visible in Fig. 5.3 at the right end of one skyscraper in the middle of the image.

When the complexity of the problem itself is the main issue, the adoption of techniques developed for airborne radar on spaceborne data can be a natural choice. Also, papers specifically devoted to spaceborne data (Franceschetti et al., 2008) can now be found, presenting new applications unsuitable for airborne radar (Suchandt et al., 2008). Extraction of electromagnetics instead of geometric properties of the buildings seems to be at the research forefront (Guida et al., 2008).

For building reconstruction, the greatest benefits of spaceborne VHR seem to derive from data availability (all-weather operation, inherent repeatability, large-scale acquisition) rather than from data characteristics. Although the ground resolution (1 m) would allow discriminating similarly-sized details in the scene, the actual situation shows images that do not allow easy extraction of the buildings, as shown in Fig. 5.4. Here, the historical centre of L'Aquila, Italy, is represented, and it is clear how difficult it is, not only to exactly locate the backscatter contributions presented in Fig. 5.1, but even in some cases to separate neighboring buildings.

5.3.2 Damage assessment with VHR SAR data

The use of radar data for damage assessment was proposed long before high-resolution satellite SAR data became available. Imhoff et al. (1987) used SRTM data to determine flooded areas in Bangladesh and to inventory affected assets based on a land cover classification derived from Landsat data. Though, more difficult goals as earthquake damage assessment took longer to

FIGURE 5.4 A portion of the historical centre of L'Aquila as seen by the COSMO/SkyMed SAR in spotlight mode. The original radar data was generated by the Italian Space Agency, and provided by the Italian Civil Protection Department.

be addressed. Generally speaking, damage assessment is a change detection problem; in radar images, the changes in the observed objects manifest themselves in intensity changes (a feature shared with optical images) and decorrelation of scatterers (peculiar to radar images).

Decorrelation due to the change in orientation of strong scatterers was considered by Usai and Klees (1999), and Matsuoka and Yamazaki (2000). Change in amplitude characteristics generated a different investigation stream (Aoki, Matsuoka and Yamazaki, 1998; Matsuoka and Yamazaki, 2004). A comparison between the two approaches (orientation of strong scatterers vs. change in amplitude) was attempted by Yonezawa and Takeuchi (1999). Apparently similar behaviours were reported, although the actual correlation between the damage level and the data features was not outstanding.

Multitemporal stacks of images were proposed as a way to improve accuracy of results (Trianni and Gamba, 2008), as well as the use of ancillary information (Gamba, Dell'Acqua and Trianni, 2007). Further improvement of the results is expected from the much larger amount of information conveyed by VHR SAR images, also because of the significant increase in the number of samples in the statistics.

For example, one may use statistical knowledge to develop a different approach for damage assessment, that is investigation of post-event statistics only, rather than classical change detection. This approach makes sense given the young age of spaceborne VHR SAR systems, for which the absence of any pre-event image at a given location is not an unlikely event.

Building on the idea developed by Gamba, Dell'Acqua and Trianni (2007) of exploiting ancillary data, this approach was tested on a real case, namely, the 6 April 2009 Italy earthquake in L'Aquila, Italy.

Visual inspection of the Google Earth image allowed producing a GIS layer with 58 city blocks, as shown in Fig. 5.5; the average block size is 0.1176 km^2, where the smallest polygon is 0.0146 km^2 and the biggest one is 0.698 km^2. Every block in the city centre contains 100 to 150 buildings. A comparison of each block with a layer containing footprints of severely damaged buildings was performed. These latter buildings were visually extracted from post-event aerial images acquired by the Italian Air Force and kindly provided by the Italian Civil Protection Department. Each block was marked with the percentage of its area covered by damaged buildings (DAR = damaged area ratio), a number between 0 and 46.4%, with an average value of roughly 4%. Of the total 58 blocks, 37 reported DAR = 0. A number of different texture measures were computed on a post-event, amplitude spotlight COSMO/SkyMed image over the same area, geocoded terrain corrected, provided by the Italian Space Agency through the Italian Civil Protection Department. Every single texture map was averaged over each polygon in the series, creating a series of 58 average texture values per each selected texture.

Pearson correlation coefficients were then computed between the series of DAR values and each of the texture average values. Most correlations, namely between DAR and data range, mean and entropy, were found to be negligible (less than 0.1); yet, correlation with variance was significantly higher, namely 0.275 with a window size of 3 × 3 pixels, decreasing to 0.246 at a window size of 19 × 19 pixels. Similar levels of correlation were found in analyzing COSMO/SkyMed data on Guan Xian, China (Dell'Acqua, Lisini and Gamba, 2009), after the Sichuan

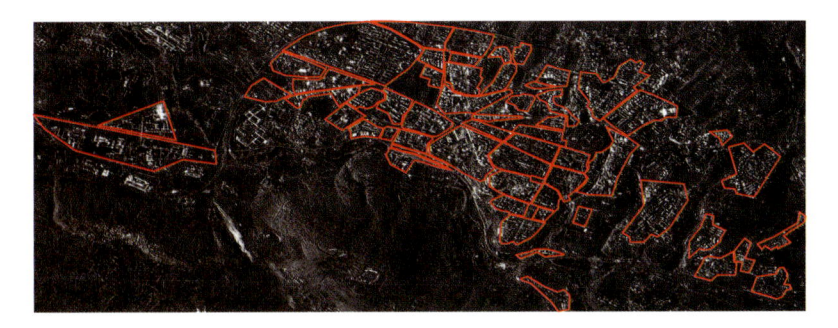

FIGURE 5.5 Partition of the urban area of L'Aquila, Italy, overlaid on a COSMO/SkyMed spotlight image. GIS polygons are outlined in red. The original radar data was generated by the Italian Space Agency and provided by the Italian Civil Protection Department.

earthquake in May 2008. In that case more complex texture measures were used, i.e. homogeneity on a 51×51 pixel window and displacements $dx = 21$, $dy = 21$, yet this appears to be a confirmation of the fact that post-event images alone may provide statistical clues on the damage level. Suppressing the elements with $DAR = 0$ in the series, the correlation reaches 0.338. This means that if one had a criterion to select blocks reporting actual damage, a more reliable estimate of the damage level could be done.

The apparent correlation with variance may be tentatively explained with a stronger speckle connected to the wider presence of small reflectors due to the randomly shaped debris stacks. It is reasonable to hypothesize that the stability of the texture statistics with varying windows sizes comes from the large number of sample at one's disposal, reckoning around 120 000 pixels on average for every single block in the case at hand, to be compared with less than one hundred for a corresponding ERS-like image.

The research is still in progress; latest results, not published yet, indicate that correlations as high as 0.7 may be reached by suitable selection of texture measures, and fusion with information from optical data allows classifying damage level into three classes with an overall accuracy of around 84%.

5.3.3 Vulnerability mapping with VHR SAR data

As mentioned above, the concept of using remote sensing to assess the effect of disasters is not new, as low-resolution systems could be used e.g. to map flooding extent (Tholey, Clandillon and De Fraipont, 1998) or to assess seismic damage (Trianni and Gamba, 2008), forest fires (Kasischke, Bourgeau-Chavez and French, 1994), and so on. The new generation of spaceborne SAR, however, opens more possibilities to look also at the opposite side of the disaster management cycle, and address preparedness and vulnerability. It is known that low resolution SAR could be successfully used to assess hazards, such as landslides (Metternicht, Hurni and Gogu, 2005), because this kind of threat depends on large scale features such as terrain slope and material; less frequently it was used to assess actual vulnerability (Sanyal and Lu, 2005), but again only where the considered threat depended on large-scale features visible at ERS-like resolutions.

The ability of acquiring small details in the scene lets scientists access new departments in vulnerability assessment. A representative example is that of seismic vulnerability.

Evaluating the vulnerability of existing building stock has a long history of method proposed along the years (Calvi et al., 2006), based either on empirical, analytical or even hybrid approaches. The various methods proposed need a considerable amount of information to be collected either on past events (for empirical methods) or on the physical features and characteristics of the considered buildings (for analytical approaches), or both (for hybrid methods). This generally constrains the estimation procedure to small areas because the *in-situ* collection of data is too expensive and time-consuming to be practical beyond some limits. It would be desirable to trade some accuracy for a wider geographic scope; enabling such change is the goal of a research activity (Polli, Dell'Acqua and Gamba, 2009) in progress in the framework of GEO (GEO, 2005).

Recently, new algorithms have been developed for seismic vulnerability assessment, requiring fewer data than their predecessors. There is one named SP-BELA (Simplified Pushover-Based Earthquake Loss Assessment) (Borzi, Crowley and Pinho 2008) capable of providing a sensible output for comparison purposes with a smallest set of inputs including footprint of the building and number of storeys, which can be extracted from remotely sensed data.

In the scientific literature it is possible to find lots of building height extraction methods, both for optical and SAR imagery. Existing methodologies are either based on shadow analysis (Hill, Mote and Blacknell, 2008) or interferometric data (Bennet and Blacknell, 2003). The interferogram calculation however fails if all of the roof backscattering is sensed before the double bounce area and therefore superimposes with the ground scattering in the layover region, which is usually the case for high buildings. To tackle the problem of signal mixture from different altitudes methods founded on interferometric or polarimetric data or stereoscopic SAR are proposed (Simonetto, Oirio and Garello, 2005; Cellier and Colin, 2006).

Generally speaking, as testified by the amount of relevant literature, the problem of extracting a building 3D shape is quite a complex one. For the purposes of seismic vulnerability estimation, however, such problem can be split into two sub-problems, namely footprint extraction and determination of the number of storeys. This is where VHR SAR data may help with its side-looking geometry. Some, still unpublished, experiments have indeed shown that analyzing the rows of reflectors found in the building façade with image processing methods one can actually infer the number of storeys in the building without a need for precise determination of its height, which is less relevant when seismic vulnerability is concerned. The final goal of the

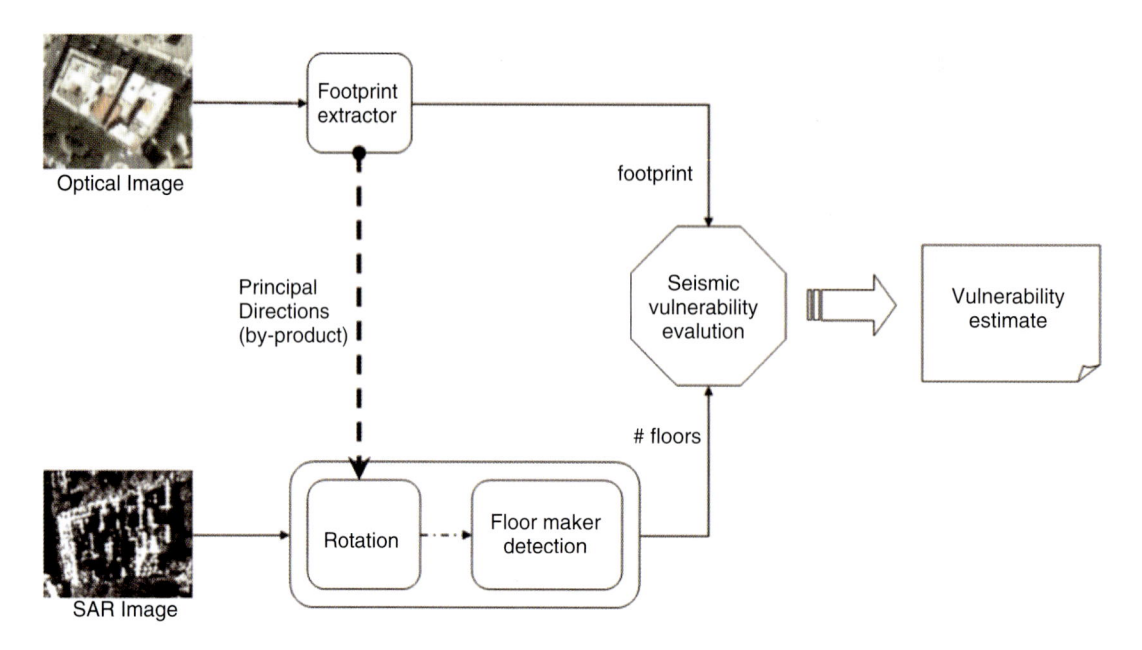

FIGURE 5.6 An EO-based scheme for geographically extensive seismic vulnerability estimation.

work is to build a processing chain like the one outlined in Fig. 5.6, where a fusion process is shown: a VHR optical and a VHR SAR images are acquired over the same area and registered. On every single building, a footprint extraction is performed on the optical image; the building orientation is passed on to the SAR analysis algorithm and used to seek lines of scatterers indicative of floors. The floor count is then passed, together with more relevant information extracted from EO data, to the seismic vulnerability model. Ideally, this method should enable production of seismic vulnerability maps useful for risk scenario analysis.

This research highlighted various difficulties inherent in fusing optical and radar VHR images. Among the most important findings were:

- In optical and radar images the effect of elevation generate foreshortening effects that are opposite in nature; elevated objects bend towards the sensor in radar images, whereas they bend away in optical images. Since most objects in urban areas are elevated, this phenomenon results in extensive mismatch between object images.

- Geocoding and orthorectification of images is frequently not sufficient to match objects even in the absence of foreshortening effects (e.g., a horizontal shape at zero elevation). Different acquisition geometries, different resolution of the images, different resolutions of the DEMs used, residual errors, all sum up to a displacement between images of the same object in the two types of data. The displacement may reach as high as several pixels and needs to be corrected manually *a posteriori*.

Conclusions

In this chapter it has been shown that EO applied to urban areas benefited in many ways from the recent refinement in spatial resolution of spaceborne radar sensors. Although some algorithms and methods exploiting meter and submeter resolution radar data had previously been developed in the airborne domain and could in principle have been reused on spaceborne domain, a large gap still remains probably due to the following reasons. First, despite the significant progress in SAR systems development, the sheer difference of two orders of magnitude in the instrument-target distance unavoidably introduces a certain degree of dissimilarity between the two types of data, and negatively biases the quality of spaceborne data. On the other hand, pointwise acquisitions of airborne data did not push towards the development of suitable algorithms capable of exploiting long time sequences, which become available with the new generation of spaceborne sensors. Nevertheless, research is in progress and some results are slowly coming up. The advent of satellite constellations like COSMO/SkyMed that are capable of delivering large amounts of data in a very short time will further thrust applications in the domains of security and disaster management, especially in terms of preparedness and vulnerability assessment, which is at least as important as damage assessment in a global perspective.

Acknowledgments

The authors wish to thank Massimiliano Aldrighi for the data processing resulting in images displayed in Fig. 5.2 and Fig. 5.3. They also wish to acknowledge the support of the Italian Space Agency and the Italian Civil Protection Department in providing COSMO/SkyMed data and ancillary information.

References

Aoki, H., Matsuoka, M. and Yamazaki, F. (1998) Characteristics of satellite SAR images in the damaged areas due to the

Hyogoken-Nanbu earthquake, in *Proceedings of the 19th Asian Conference of Remote Sensing*, pp. C7/1–6.

Bender, O., Gilg, W. and Ottl, H. (1990) X-band synthetic aperture radar for earth observation satellites (X-EOS), in *Proceedings of the Geoscience and Remote Sensing Symposium, 1990. IGARSS '90.*, pp. 2289–2294, 20–24 May 1990.

Bennett, A and Blacknell, D. (2003) Infrastructure analysis from high resolution SAR and INSAR imagery, *2nd GRSS/ISPRS Joint Workshop on Remote Sensing and Data Fusion over Urban Areas*, Berlin, Germany, 2003.

Bolter, R. and Leberl F. (2000) Detection and Reconstruction of Human Scale Features from High Resolution Interferometric SAR Data. *Proceedings of the International Conference on Pattern Recognition 2000 (ICPR2000)*, pp. 291–294.

Borzi B., Crowley H. and Pinho R. (2008) Simplified pushover-based earthquake loss assessment (SP-BELA) method for masonry buildings. *International Journal of Architectural Heritage*, 2(4), 353–376.

Calvi G. M., Pinho, R., Bommer, J. J., Restrepo-Vélez, L. F. and Crowley, H. (2006) Development of seismic vulnerability assessment methodologies over the past 30 years. *ISET Journal of Earthquake Technology*, 43(3), 75–104.

Cellier F. and Colin E. (2006) Building height estimation using fine analysis of altimetric mixtures in layover areas on polarimentric interferometric X-band SAR images. *International Geoscience and Remote Sensing Symposium 2006 (IGARSS 2006)*, Denver, CO, USA, 2006.

Dammert, P. B. G., Askne, J. I. H. and Kuhlmann. S. (1999) Unsupervised segmentation of multitemporal interferometric SAR images. *IEEE Transactions on Geoscience and Remote Sensing*, 37(5), 2259–2271.

Dell'Acqua F. and Gamba, P. (2003) Texture-based characterization of urban environments on satellite SAR images. *IEEE Transactions on Geoscience and Remote Sensing*, 41(1), 153–159.

Dell'Acqua F., Lisini, G. and Gamba, P. (2009) Experiences in optical and SAR imagery analysis for damage assessment in the Wuhan, May 2008 earthquake, in *Proceedings of IGARSS 2009*, Cape Town, South Africa, 13–17 July 2009.

Dierking, W. and Skriver, H. (2002) Change detection for thematic mapping by means of airborne multitemporal polarimetric SAR Imagery. *IEEE Transactions on Geoscience and Remote Sensing*, 40(3), 618–635.

Dong Y., Forster, B. and Ticehurst C. (1997) Radar backscatter analysis for urban environments. *Intrnational Journal of Remote Sensing*, 18 (6), 1351–1364.

Ferretti, A., Prati, C. and Rocca, F. (2000) Permanent scatterers in SAR interferometry. *IEEE Transactions in Geoscience and Remote Sensing*, 38, 2202–2212.

Franceschetti, G., Guida, R., Iodice, A. and Riccio, D. (2008) Electromagnetic modelling for information extraction from high resolution SAR images of urban areas, in *Geoscience and Remote Sensing Symposium, 2008. IGARSS 2008. IEEE International*, vol. 1, no., pp. I-78–I-81, 7–11 July 2008.

Gamba P., Houshmand, B. and Saccani M. (2000) Detection and extraction of buildings from interferometric SAR data. *IEEE Transactions on Geoscience and Remote Sensing*, 38 (1), 611–617.

Gamba P., Dell'Acqua, F. and Houshmand B. (2003) Comparison and fusion of LIDAR and InSAR digital elevation models over urban areas. *International Journal of Remote Sensing*, 24(22), 4289–4300.

Gamba, P., Dell'Acqua, F. and Lisini, G. (2006) Improving urban road extraction in high-resolution images exploiting directional filtering, perceptual grouping, and simple topological concepts. *IEEE Geoscience and Remote Sensing Letters*, 3(3), 387–391.

Gamba P., Dell'Acqua, F. and Trianni, G. (2006) Semi-automatic choice of scale-dependent features for satellite SAR image classification. *Pattern Recognition Letters*, 27(4), 244–251.

Gamba, P., Dell'Acqua, F. and Trianni, G. (2007) Rapid damage detection in the Bam area using multitemporal SAR and exploiting ancillary data. *IEEE Transactions on Geoscience and Remote Sensing*, 45(6), 1582–1589.

Gamba, P., Aldrighi, M., Stasolla, M. and Sirtori, E. (2009) A detailed comparison between two fast approaches to urban extent extraction in VHR SAR images. *2009 Joint Urban Remote Sensing Event*, pp. 1–6, 20–22 May 2009.

GEO (2005). The Group on Earth Observation. [online] www.earthobservations.org (accessed 22 November 2010).

Guida, R., Iodice, A., Riccio, D. and Stilla, U. (2008) Model-Based interpretation of high-resolution SAR images of buildings. *IEEE Journal of Selected Topics in Applied Earth Observations and Remote Sensing*, 1(2), 107–119.

Hedman K., Stilla, U., Lisini, G. and Gamba, P. (2010) Road network extraction in VHR SAR images of urban and suburban areas by means of class-aided feature-level fusion. *IEEE Transactions on Geoscience and Remote Sensing*, 48 (3), 1294–1296.

Henderson F. M. and Xia, Z.-G. (1997) SAR applications in human settlement detection, population estimation and urban land use pattern analysis: a status report. *IEEE Transactions on Geoscience and Remote Sensing*, 35 (1), 79–85.

Hill R., Moate, C. and Blacknell, D. (2008) Estimating building dimensions from synthetic aperture radar image sequences. *IET Radar, Sonar and Navigation*, 2 (3), 189–199.

Imhoff, M., Vermillion, C., Story, M. H., Choudhury, A. M. and Gafoor, A. (1987) Monsoon flood boundary delineation and damage assessment using space borne imaging radar and Landsat data. *Photogrammetric Engineering and Remote Sensing*, 53, 405–413.

Impagnatiello F., Bertoni, R. and Caltagirone, F. (1998) The SkyMed/COSMO system: SAR payload characteristics, in *Proceedings of IGARSS'98*, vol. 2, pp. 689–691, 6–10 July 1998, Seattle (WA).

Kasischke, E. S., Bourgeau-Chavez, L. L. and French, N. H. F. (1994) Observations of variations in ERS-1 SAR image intensity associated with forest fires in Alaska. *IEEE Transactions on Geoscience and Remote Sensing*, 32(1), 206–210.

Lombardo P., Sciotti, M., Macrì Pellizzeri, T. and Meloni, M. (2003) Optimum model-based segmentation techniques for multifrequency polarimetria SAR images of urban areas. *IEEE Transactions on Geoscience and Remote Sensing*, 41(9, Part 1), 1959–1975.

Macrì Pellizzeri T., Gamba, P., Lombardo, P., and Dell'Acqua, F. (2003) Multitemporal/multiband SAR classification of urban areas using spatial analysis: statistical versus neural kernel-based approach. *IEEE Transactions on Geoscience and Remote Sensing*, **41**(10, Part 1), 2338–2353.

Matsuoka M. and Yamazaki, F. (2000) Use of interferometric satellite SAR for earthquake damage detection, in *Proceedings of 6th International Conference on Seismic Zonation*, EERI, 2000, pp. 103–108.

Matsuoka M. and Yamazaki, F. (2004) Use of satellite SAR Intensity imagery for detecting building areas damaged due to earthquakes. *Earthquake Spectra*, **20**, 975, doi:10.1193/1.1774182.

Metternicht G., Hurni L. and Gogu R. (2005) Remote sensing of landslides: An analysis of the potential contribution to geo-spatial systems for hazard assessment in mountainous environments. *Remote Sensing of Environment*, **98**(2–3) 284–303, doi: 10.1016/j.rse.2005.08.004.

Michaelsen, E., Soergel, U. and Thoennessen U. (2005) Potential of building extraction from multi-aspect high-resolution amplitude SAR data, in Joint Workshop of ISPRS and the German Association for Pattern Recognition Object Extraction for 3D City Models, Road Databases and Traffic Monitoring Concepts, Algorithms, and Evaluation. Wien (eds F. Rottensteiner and U. Stilla), *International Archives of Photogrammetry, Remote Sensing and the Spatial Information Sciences*, **XXXVI** Part 3/ W24.

Michaelsen, E., Soergel, U. and Thoennessen, U. (2006) Perceptual grouping for automatic detection of man-made structures in high-resolution SAR data. *Pattern Recognition Letters*, **27**(4), 218–225. doi: http://dx.doi.org/10.1016/j.patrec.2005.08.002

Michaelsen E., Middelmann W. and Sörgel U. (2007) Cognitive vision and perceptual grouping by production systems with blackboard control – an example for high-resolution SAR-images. *Advances in Computer Graphics and Computer Vision, Communications in Computer and Information Science*, **4** (8), 293–304.

Morena L. C., James, K. V. and Beck, J. (2004) An introduction to the RADARSAT-2 mission. *Canadian Journal of Remote Sensing*, **30**(3), J221–234.

Niebergall S., Loew, A. and Mauser, W. (2007). Object-oriented analysis of very-high-resolution Quick Bird data for mega city research in Delhi/India, in *Proceedings of 2007 Urban Remote Sensing Joint Event*, Unformatted CD-ROM, 11–13 April 2007, Paris (France).

Polli D., Dell'Acqua F. and Gamba P. (2009). First steps towards a framework for earth observation (EO)-based seismic vulnerability evaluation. *Environmental Semeiotics*, **2**(1), 16–30.

Quartulli, M. and Datcu, M., (2004) Stochastic geometrical modeling for built-up area understanding from a single SAR intensity image with meter resolution. *IEEE Transactions on Geoscience and Remote Sensing*, **42**, 1996–2003.

Rosen, P. A., Hensley, S., Joughin, I. R. et al. (2000) Synthetic aperture radar interferometry (invited paper). *Proceedings of the IEEE*, **88**(3), 333–382.

Rosenqvist A., Shimada, M., Watanabe M., Tadono T., and Yamauchi, K. (2004) Implementation of systematic data observation strategies for ALOS PALSAR, PRISM and AVNIR-2, in *Proceedings of IGARSS'04*, vol. 7, pp. 4527–4530, Sept. 2004, Anchorage, AK.

Roth A. (2003). TerraSAR-X: a new perspective for scientific use of high resolution spaceborne SAR data, in *Proceedings of the 2nd GRSS/ISPRS Joint Workshop on Remote Sensing and Data Fusion over Urban Areas*, pp. 4–7, 22–23 May 2003, Berlin (Germany).

Sanyal, J. and Lu, X. X. (2005) Remote sensing and GIS-based flood vulnerability assessment of human settlements: a case study of Gangetic West Bengal, India. *Hydrological Processes*, **19** (18), 3699–3716.

Sauer S., Ferro-Famil, L., Reigber, A. and Pottier E. (2006) Analysing urban areas using multiple track POL-InSAR Data at L-Band, in *Proceedings of EUSAR'06*, May 2006.

Sauer, S., Ferro-Famil, L., Reigber, A. and Pottier, E. (2008) Multi-aspect Pol-InSAR 3D Urban Scene Reconstruction at L-Band, in *Proceedings of the European Conference on Synthetic Aperture Radar (EUSAR)*, p. 4. VDE. European Conference on Synthetic Aperture Radar (EUSAR), 2008-06-02 – 2008-06-05, Friedrichshafen, Germany.

Simonetto E., Oriot, H. and Garello R. (2005) Rectangular building extraction from stereoscopic airborne radar images. *IEEE Transactions on Geoscience and Remote Sensing*, **43**(10), 2386–2395.

Soergel U., Schulz, K., Thoennessen, U. and Stilla, U. (2005) Integration of 3D data in SAR mission planning and image interpretation in urban areas. *Information Fusion*, **6**(4), 301–310.

Stasolla, M. and Gamba, P. (2008) Spatial indexes for the extraction of formal and informal human settlements from high-resolution SAR images. *IEEE Journal of Selected Topics in Applied Earth Observations and Remote Sensing*, **1**(2), 98–106.

Stilla U., Soergel, U. and Thoennessen, U. (2003) Potential and limits of InSAR data for building reconstruction in built-up areas. *ISPRS Journal of Photogrammetry and Remote Sensing*, **58** (1–2), 113–123. Algorithms and Techniques for Multi-Source Data Fusion in Urban Areas, doi: 10.1016/S0924–2716(03)00021-2.

Suchandt, S., Runge, H., Breit, H., Kotenkov, A., Weihing, D. and Hinz, S. (2008) Traffic measurement with TerraSAR-X: processing system overview and first results, in *Proceedings of EUSAR 2008*, pp. 55–58. VDE Verlag GmbH. EUSAR 2008, Friedrichshafen, Germany.

Taket, N. D., Howarth, S. M. and Burge, R. E. (1991) A model for imaging of urban areas by synthetic aperture radar. *IEEE Transactions on Geoscience and Remote Sensing*, **29**, 432–443.

Thiele A., Cadario E., Schulz K., Thönnessen, U. and Soergel, U. (2007) Building recognition from multi-aspect high-resolution InSAR data in urban areas. *IEEE Transactions on Geoscience and Remote Sensing*, **45**(11), 3583–3593.

Tison, C. and Tupin, F. (2004) Retrieval of building shapes from shadows in high resolution SAR interferometric images, in *Proceedings of IEEE International Geoscience and Remote Sensing Symposium (IGARSS)*, **3**, 1788–1791.

Tholey N., Clandillon, S. and De Fraipont, P. (1998) The contribution of spaceborne SAR and optical data in monitoring flood events: examples in northern and southern France. *Hydrological Processes*, **11** (10), 1409–1413.

Trianni G. and Gamba P. (2008) Damage detection from SAR imagery: application to the 2003 Algeria and 2007 Peru Earthquakes. *International Journal of Navigation and Observation*, Volume 2008, Article ID 762378, 8 pages. doi:10.1155/2008/762378

Usai S. and Klees R. (1999) SAR interferometry on a very long time scale: A study of the interferometric characteristics of man-made features. *IEEE Transactions on Geoscience and Remote Sensing*, **37**(4), 2118–2123.

Weydahl D. J., Bretar F. and Bjerke P. (2003) Comparing RADARSAT-1 and IKONOS satellite images for urban features detection, in *Proceedings of the 2ndGRSS/ISPRSJoint Workshop on Remote Sensing and Data Fusion over Urban Areas*, pp. 305–308, 22–23 May 2003, Berlin (Germany).

Yonezawa, C. and Takeuchi S. (1999) Detection of urban damage using interferometric SAR decorrelation. *Proceedings of IEEE International Geoscience and Remote Sensing Symposium*, 1999, Hamburg, Germany.

Yonezawa C. and Takeuchi, S. (2001) Decorrelation of SAR data by urban damages caused by the 1995 Hyogoken-Nambu earthquake. *International Journal of Remote Sensing*, **22** (8), 1585–1600.

6

3D building reconstruction from airborne lidar point clouds fused with aerial imagery

Jonathan Li and Haiyan Guan

Lidar mapping technology can be used to provide building data that is three-dimensional (3D), accurate, timely, and increasingly affordable in complex urban areas. However, the automated creation of reliable and accurate 3D city models is still challenging. Commercial software tools for building modeling generally require a high degree of human interaction and most automated approaches described in literature stress steps of such a workflow individually. This chapter first briefly reviews the existing techniques for building extraction using airborne lidar range data. Then, we present a comprehensive approach for automated creation of 3D building models from airborne lidar point cloud data fused with color aerial imagery. Our two-step building extraction strategy, building detection followed by building reconstruction, is then detailed and implemented using a set of lidar data covering a Toronto study area collected by Optech ALTM 3100 system. Finally, the results obtained are presented and discussed.

Urban Remote Sensing: Monitoring, Synthesis and Modeling in the Urban Environment, First Edition. Edited by Xiaojun Yang.
© 2011 John Wiley & Sons, Ltd. Published 2011 by John Wiley & Sons, Ltd.

6.1 Lidar-drived building models: related work

This section first provides an overview of lidar data processing towards 3D builing reconstruction. The output of a lidar mapping system is a cloud of irregularly spaced 3D points which include not only the bare ground but also all kinds of objects (buildings, trees, cars, etc.). Therefore, the generation of reliable and accurate building models from lidar data requires a number of processes, including building detection, outline extraction, roof shape reconstruction, model generation and regularization, and finally, model quality analysis (Dorninger and Pfeifer, 2008). The majority of available literature concentrates on individual aspects only. For example, methods on building region detection in rasterized lidar data were described in the literature (e.g., Hug and Wehr, 1997; Maas, 1999; Morgan and Tempfli, 2000; Nardinocchi and Forlani, 2001; Alharthy and Bethel, 2002; Matikainen, Hyyppä and Hyyppä, 2003; Tóvari and Vögtle, 2004; Gross, Thoennessen and Hansen, 2005; Li, Li and Chapman, 2010). Techniques on roof reconstruction in lidar point clouds with known building boundaries were presented in the literature (e.g., Vosselman and Dijkman, 2001; Hofmann, Maas and Streilein, 2003). Approaches considering detection and reconstruction were presented in the literature (e.g., Rottensteiner and Briese, 2002; Lafarge *et al.*, 2008). The reconstructed models presented in these two references are, however, restricted. In both cases digital surface model (DSM) data of relatively low density is processed. This does not allow for exact positioning of building outlines and prevents the reconstruction of small roof features. Furthermore, in the latter reference the complexity of building models is restricted to a composition of predefined building parts. A general and up-to-date overview on the topic of lidar mapping technology and data processing, in particular, information extraction can be found in Shan and Toth (2008). This chapter does not aim at covering a complete bibliography, but it gives a brief summary of the existing building extraction methods developed thus far mainly in Europe and North America.

To increase the reliability of building reconstruction, additional knowledge on buildings has to be incorporated into the modeling process. Typical assumptions are to define walls as being vertical and roofs as being a composition of planar faces. This leads to an idealization of the buildings. The transition zone of two neighboring roof faces, for example, becomes a straight line defined by the intersection of two roof planes (Dorninger and Pfeifer, 2008).

Many methods have been developed for semiautomated or fully automated extraction of buildings using lidar data in the past 15 years. Recognizing that fully automation is not attainable yet, we aim to reduce the complexity of the building reconstruction task by including a user and automated processes in the system to work sequentially. For example, the user supplies the automated processes with inputs and cues. Then the automated processes produce a scene model based on these inputs. Finally, the user corrects mistakes in that scene model. In this chapter, the building extraction task is addressed by a two-step strategy: building detection followed by building reconstruction (Li, 1999; Hu, 2003).

6.1.1 Building detection

Building detection is often performed on resampled (i.e., interpolated) grid data, thus simplifying the 3D content of lidar data to 2.5D. Roughness measures, i.e. local height variations, are often used to identify vegetation. Open areas and buildings can be differentiated by first computing a digital terrain model (DTM) with so-called filtering methods (Kraus and Pfeifer, 1998; Sithole and Vosselman, 2004). Thereafter, a normalized digital surface model (nDSM) is computed by subtraction of the DTM from the DSM (Weidner and Forstner, 1995; Haala and Brenner, 1999; Gamba and Houshmand, 2002; Rottensteiner and Briese, 2002; Hu, Tao and Collins 2003), hence representing local object heights (Hug and Wehr, 1997; Maas and Vosselman, 1999; Alharthy and Bethel, 2002; Tóvari and Vögtle, 2004; Gross, Thoennessen and Hansen, 2005). High objects with low roughness correspond to building areas. Other approaches identify blobs in the DSM, based on height jumps, high curvature, etc. (Morgan and Tempfli, 2000; Nardinocchi and Forlani, 2001; Matikainen, Hyyppä and Hyyppä, 2003; Rutzinger *et al.*, 2006).

Building reconstruction may include two parts: building footprint detection and roof reconstruction. For those studies focusing on geometric reconstruction of upper roof lines instead of building footprints, the reliable reconstruction of complex building roof boundaries is a key step. Most algorithms work well only under specific assumptions, which limit roofs to simple shapes such as rectangles or low quality polygons (Weidner and Forstner, 1995; Vosselman, 1999; Wang and Schenk, 2000). Other algorithms, which do not make such assumptions, often got distorted boundaries expressed by edges detected from lidar DSMs (Baltsavias, 1999; Weidner, 1995; Yoon *et al.*, 1999; Wang and Schenk, 2000; Rottensteiner and Briese, 2002). These boundaries need to be refined using a set of geometric regularity constraints (Vestri and Devernay, 2001).

To distinguish buildings from vegetated regions, the classification is often based on shape measures assuming some geometric regularity constraints (Wang and Schenk, 2000) or the roughness of the point clouds. These measures limit the detectable buildings to a narrower spectrum, and also are not very reliable for complex scenes such as densely forested areas. The shape measures often make use of 2D properties such as area and perimeter; while complex building roofs may present close values when calculating the roughness measures. The use of the lidar multiple return data can benefit the separation of buildings and vegetation since building roofs must be solid surface (Zhan, Molenaar and Tempfli, 2002; Hu, Tao and Collins, 2003). Lidar cannot penetrate solid surfaces and will get a single return only for them. That is, the first and last returns are same in elevation at solid surfaces but are different at vegetated regions. However, lidar gets the similar effect at building boundaries as that at vegetated areas.

6.1.2 Building reconstruction

Building reconstruction recovers the geometrical parameters of the roofs and walls of a located building (Weidner and Forstner,

1995). Building boundaries, the intersection of the buildings with its surroundings (e.g., ground), need to be derived from the classified building points. Typically, the building boundary generation is initiated by detecting a coarse approximation of the outline, followed by a generalization and a regularization (Sampath and Shan, 2007; Jwa et al., 2008). Two fundamentally different approaches for building reconstruction can be distinguished: model-driven and data-driven approaches.

In the model-driven methods a predefined catalog of roof forms is prescribed (e.g., flat roof, gable roof, etc.). The shape and position parameters are determined by fitting models to lidar point clouds (Weidner and Forstner, 1995; Haala, Brenner and Anders, 1998; Maas and Vosselman, 1999). These methods will lead to a reliable reconstruction if all the constraints in building models are well satisfied and produce pretty results. For instance, an algorithm may assume that there exists a main orientation of the building and all edges are either parallel or perpendicular to that orientation. This is especially appropriate for low point densities. An advantage is that the final roof shape is always topologically correct. A disadvantage is, however, that complex roof shapes cannot be reconstructed, because they are not included in the catalog.

The data-driven methods do not assume specific building shapes for a scene, only making a few realistic assumptions such as the vertical wall constraint. A building can be expressed by bounding surfaces that may be described as planar surfaces or triangles. Although some algorithms can produce good building shapes, they often use complex plane detection techniques such as clustering of triangles based on triangulated irregular networks (TINs), 3D Hough transformation and clustering of 3D points, which often result in a heavy computational burden (Maas and Vosselman, 1999; Wang and Schenk, 2000; Vosselman and Dijkman, 2001). Some algorithms produced polyhedral models with low quality in shape (Gamba and Houshmand, 2002). In the data-driven methods the roof is "reassembled" from roof parts found by segmentation algorithms. The results of the segmentation process are sets of points, each one ideally describing exactly one roof face. Some roof elements (e.g., small dormers, chimneys, etc.) may not be represented. The challenge is to identify neighboring segments and the starting and ending point of their intersection.

Building reconstruction is based on the assumption that individual buildings can be modeled properly by a composition of a set of planar surfaces. Hence, it is based on a reliable 3D segmentation algorithm, detecting planar faces in a point cloud. This segmentation is of crucial importance for the outline detection and for the modeling approach (Dorninger and Pfeifer, 2008). Segmentation allows for a decomposition of a building in a lidar point cloud into planar surfaces and other objects. This requires the definition of a homogeneity criterion according to which similar items (e.g., points) are grouped. As homogeneity criteria, approximate height similarity or/and approximate normal vector similarity are commonly used. Determination of planar faces for roof modeling from point clouds acquired from airborne platforms is studied in the literature (Maas and Vosselman, 1999; Lee and Schenk, 2002; Rottensteiner et al., 2005). To reduce the complexity of the problem and to increase the performance of the implementation, again, 2.5D grid representations are commonly

used instead of the original points. This requires the definition of a reference direction (e.g., the vertical z-axis) to resample the given points to a regular grid defined as a scalar function over the horizontal xy-plane. Thus, only one distinct height value can be assigned to an arbitrary pair of xy-coordinates. Advantages of 2.5D approaches are a possible reduction of the amount of input data and the implicitly defined neighborhood by means of the grid representation. By contrast, for processing original point clouds, such a neighborhood (e.g., for the estimation of normal vectors) has to be defined explicitly (e.g., Filin and Pfeifer, 2006). Unfortunately, the grid resampling process introduces smoothing effects especially at sharp surface structures. Segmentation approaches based on range images suffer from these restrictions as well (e.g., Hug and Wehr, 1997; Maas and Vosselman, 1999).

6.2 Our building reconstruction method

In recent years, lidar data has been widely applied in urban 3D building extraction. A variety of methods have been proposed for this purpose. However, difficulties in building detection and reconstruction exist when lidar point cloud data is used alone due to its irregular distribution, lacks of spectral, texture and shape information. Also, edges from lidar point cloud data are jaggy due to the relatively low sample rate. Aerial images provide detailed texture and color information in high-resolution, making them necessary for texture data and appealing for extracting detailed model features. However, reconstruction from stereo aerial images only leads to sparse points, which makes them unsuitable for reconstruction of complex surfaces, such as curved surfaces and roofs with slopes. As such the building extraction based on either single image data or single lidar data cannot reach a satisfying result. To this end, this chapter presents a combination of lidar point cloud data and color aerial image data for 3D building extraction.

6.2.1 Our strategy using fused data

The proposed building extraction strategy is based on the sequential determination of individual building models from both lidar point cloud data and aerial image data. Figure 6.1 shows the workflow of our strategy.

Both an unstructured lidar point cloud with the first and last returns and a color aerial image covering the same area are used as the input. Spatial registration of lidar point cloud data and optical image is performed as data preprocessing.

At the building detection step, filtering is first used to categorize the lidar data into two classes: on-terrain and off-terrain points. The filtered lidar points are then classified into three feature classes, named buildings, trees, and ground (or on-terrain points). The line features extracted from the aerial image, geometric information, such as height information, discrete measurement, and the height difference between the first and last

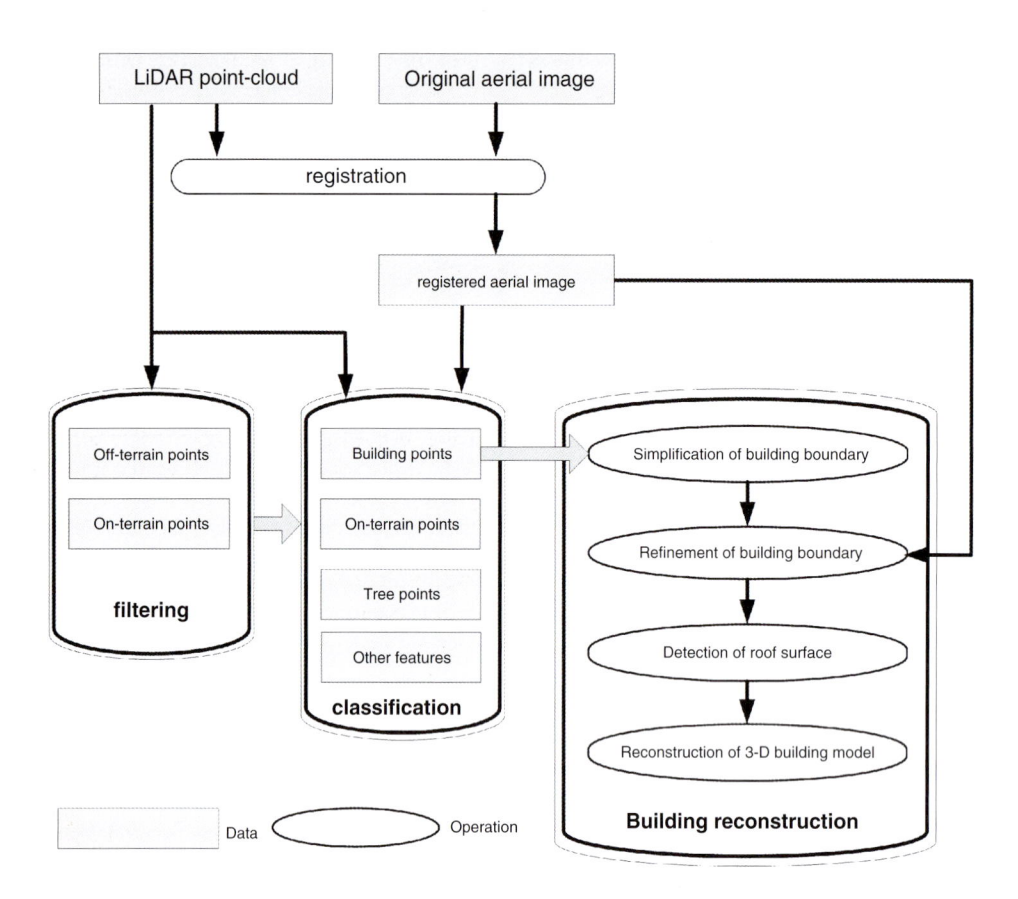

FIGURE 6.1 The workflow of our strategy from building detection to building reconstruction.

returns, provided by lidar point cloud data, are all used as the clues for classification. The homogeneous regions of the aerial image can be attained by the watersheds transform segmentation algorithm and boundary tracking which can be treated as the object of classification. In each homogeneous region, both spectral and geometric information can be comprehensively used for the classification of lidar data in accordance with the experiments to determine the conditions and rules of classification.

At the building reconstruction step, first, the coarse building roof boundary (a polygon) is determined based on the classified building regions. Boundary reconstruction is formulated as a regularization or simplification problem. Because lidar points are randomly collected, the traced boundary cannot be directly used as the final building boundary due to its irregular shape and possible artifacts introduced in the previous steps. Second, the well-known Douglas–Peucker algorithm (Douglas and Peucker, 1973; its improved algorithm by Hershberger and Snoeyink, 1992) is employed for simplification of building region polygons. Third, the least-squares template matching (Ackermann, 1983) with and the right-angle constraint are used for extraction of simple right-angle shaped buildings. Due to disadvantages of lidar data mentioned earlier, some bias exists in the building boundaries. Therefore, the simplified building boundaries are projected onto the registered aerial image for further refinement. Fourth, the random sample consensus (RANSAC) algorithm (Fischler and Bolles, 1981) is used to detect the roof surfaces and then the ridge points of a gable building can be reconstructed through building the neighbor correlation of the roof surfaces, from which the ridge lines of the roof can be attained. The most challenging process is the modeling of the individual roof faces representing the roof. We define a roof surface as the closed, polygonal boundary of a roof segment.

6.2.2 Building detection

Separation of individual buildings from lidar point cloud data is the key to an accurate reconstruction. Like buildings, trees are also one of the dominant features in urban areas. Thus, a new object-oriented supervised classification method is proposed to detect individual buildings and differentiate trees from lidar point clouds at the same time.

6.2.2.1 Region-based segmentation

Segmentation, a process of partitioning an image space into some non-overlapping meaningful homogeneous regions (polygons), is crucial to the classification result. Segmentation of color aerial image contributes to the quality of classification because they can provide more additional information than gray level images. On the other hand, abundant spectrum information of color image may lead to increase of the segmentation difficulty to some extent which can be discriminated based on the height and textural information from lidar data as additional channels. The color aerial image segmentation includes two issues: (1) the choices of color space, and (2) the choice of segmentation methods. The RGB color space is suitable for color expression, but not good for color image segmentation and analysis because of the high correlation among the red, green, and blue bands. The

CIE color space can control color and intensity information more independently and simply than the three RGB primary colors. It is especially efficient in the measurement of small color difference. The CIELAB color space, which is a color-opponent space with dimension L for lightness and a and b for the color-opponent dimensions, based on non-linearly compressed CIE XYZ color space coordinates, is adopted because it is more consistent with human perception.

In general, the segmentation techniques for monochrome images can be extended to segment color images by using R, G and B or their linear or non-linear transformations. Gonzalez and Woods (2008) grouped the existing segmentation methods into three categories: cluster-based approach, edge-based approach, and region-based approach. The cluster-based approach sometimes cannot handle the imagery with high speckle noise due to the assumption that pixels are spatially independent in image space (Li, Li and Chapman, 2010). Moreover, one of the drawbacks of feature space clustering is that the cluster analysis does not utilize any spatial information (Haralick and Shapiro, 1985). Due to image noise, the most three common problems of the edge-based approach are false detections (edge presence in a location where there is no border), missed detections (no edge presence where a real border exists), and selecting an appropriate threshold for the gradient image (this threshold separates significant from non-significant edge information) (Sonka, Hlavac and Boyle, 2002).

The region-based approach, comparing with edge-based segmentation, works generally better on noisy images, where borders are extremely difficult to detect. Watersheds transform is a powerful tool, referring to region-based algorithm, for image segmentation. The concept of watershed and catchment basin are well known in topography. Image data can be interpreted as a topographic surface where the gradient image gray-levels represent altitudes (Sonka, Hlavac and Boyle, 2002). Thus, the expression watershed can be used to denote a labeling of the image, such that all points of a given catchment basin have the same unique label, and a special label, distinct from all the labels of the catchment basins, is assigned to all points of the watersheds (Roerdink and Meijster, 2001). In this chapter, watershed segmentation is used to generate closed contours for each region in the original image since it can effectively divide individual catchment basins in a gradient image.

6.2.2.2 Lidar data preprocessing by filtering

Before the lidar data can be used further, it has to be preprocessed. Filtering is commonly used to separate the on-terrain points and the off-terrain points (Kraus and Pfeifer, 1998). Commonly used filtering algorithms are the morphological filtering, the progressive TIN densification, or the robust filtering. An extensive overview of different filtering approaches can be found in (Sithole and Vosselman, 2004). In this study, the progressive TIN densification that was first proposed by Axelsson (1999) is used through assigning different thresholds to various land cover types. The algorithm works particular well when modeling surfaces with discontinuities, which is common in urban areas.

The filtering method assumes that objects on the ground, such as trees, cars and buildings, etc. are usually higher than those of on-terrain points. In other word, the lowest point in lidar data must belong to on-terrain points. But there are points, generated from both multipath errors and system errors from the laser scanner, do not exist in the landscape. If these points, also called outliers, are misclassified the lowest points, erosion of points in the neighborhood of outliers will be occurred. It is necessary to remove outliers from the dataset before filtering. Because outliers generally are a single point or a small number of points, KD-tree (known as multidimensional binary search tree) structure provide quick query of region neighbors to remove outliers from data, because its performance far surpass the best currently known algorithms for query time and effectiveness (Bentley, 1985). If the number of points within a region is below the given threshold, these points are referred to as outliers.

First, the lowest points in lidar data, as seed points, within a user-defined grid of a size greater than the largest type of features are selected to join an on-terrain dataset. In order to minimize grid size sensitivity to the selection of seed points, a parametric plane is derived. Seed points will be removed if their perpendicular distances to the fitting plane do not satisfy with the median value estimated from the histograms. Then the rest of seed points generate a sparse TIN as initial digital elevation model (DEM). For each iteration, points in each TIN facet are added to the on-terrain data set if they meet the criteria based on the calculated parameters, distances to the TIN facets, and angles to the nodes. At the end of each iteration, the TIN and the thresholds are recomputed. Iterative process continues until no more points meet the threshold values.

6.2.2.3 Extraction of height difference from first and last returns

Difference of the first and last returns usually differentiates tree features from lidar data. As described in Fig. 6.2(a), one of the lidar system's characteristics is the ability of a laser beam to penetrate the tree's canopy through small openings. The number of echoes counts on the object within the travel path of the laser pulse. Many commercial lidar systems can measure multiple returns. Although the laser beam can penetrate the tree's canopy to the ground, it is unreliable to distinguish the tree from the lidar data with only the difference between the first- and the last-echo. This is because, first, shown in Fig. 6.2(a), the laser beam hits on the edge of building also generates two returns. Second, if the density of trees is high, the small-footprint lidar cannot penetrate the tree's canopy.

6.2.2.4 Extraction of spatial features by eigen-analysis

Discrete measurement of lidar data is defined as the spatial feature of each point by calculating dispersion matrix of its neighbors. It is another indicator for distinguishing tree points from the other features. Eigen-values of a point's dispersion matrix can reflect this point spatial information.

For each point v_j under consideration, its neighborhood points can be found by the TIN. A 3×3 dispersion matrix S_j of point v_j is given by

$$S_j = \sum_{i=0}^{n} (P_i - M)^T (P_i - M) \quad \forall (j = 0, 1, 2 \ldots N-1) \quad (6.1)$$

where, S_j is the dispersion matrix of point v_j, n is the number of neighborhood points of point v_j, P_i is the coordinate (x_j, y_j, z_j)

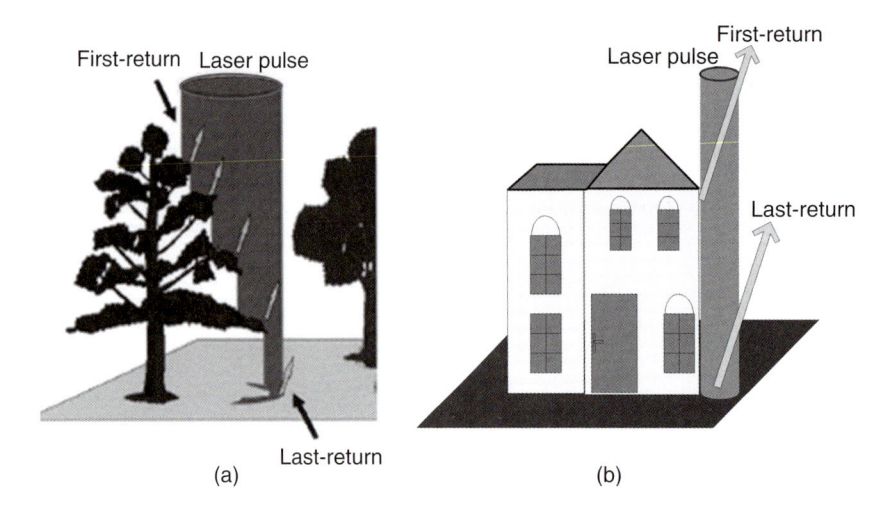

FIGURE 6.2 Height difference of first and last returns.

of point v_j, N is the number of all lidar points, M is the mean matrix of its neighborhood points.

In this 3×3 dispersion matrix, each point v_j has three eigenvalues. An eigenvalue represents the spatial information of a lidar point because it is a scalar of association with the eigenvector that reflects the spatial distribution of a lidar point. Three possibilities are considered. (1) If one of the three eigenvalues is much larger than the others, a lidar point is labelled as an "edge" point. (2) If two of the three eigenvalues are much larger than the other, the lidar point is labelled as a "plane" point. (3) If the three eigenvalues are all larger than a given threshold, the lidar point is labelled as a "discrete" point. Figure 6.3(a) shows the result of dispersion matrices of lidar points in Fig. 6.3(b), in which the white, black, and gray points represent "discrete", "edge," and "plane" points, respectively. As illustrated in Fig. 6.3, trees show the divergence property, while bare ground and most of buildings exhibit the local planarity.

6.2.2.5 Extraction of linear features from aerial imagery

In general, a building consists of regular geometric primitives (lines, corners, etc.). Given the fact that the lines (or edge) of building boundaries appear clearer that those in lidar point clouds, we apply the Canny edge operator and 2D Hough transform (Sonka, Hlavac and Boyle, 2002) to extract the line segments and use them as one of the clues in further classification.

6.2.2.6 Object-oriented supervised classification

Comparing with an unsupervised approach, a supervised approach is preferred by most researchers because it generally gives more accurate class definitions and higher classification accuracy. As a rule, a statistical supervised classification can be carried out by the following three steps (Tso and Mather, 2001):

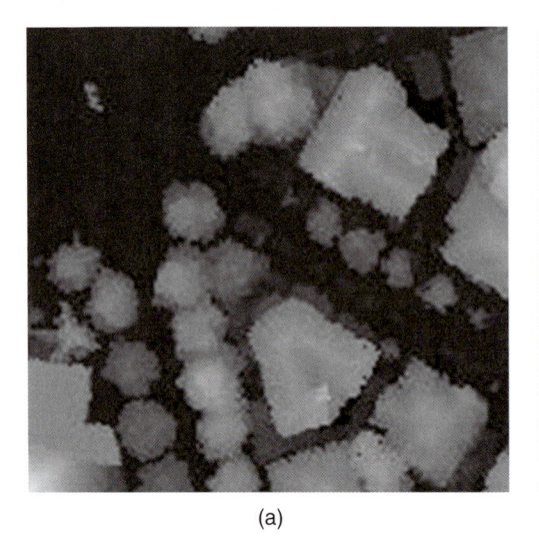

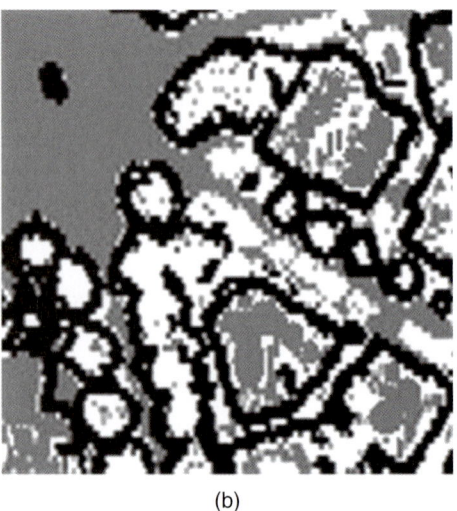

FIGURE 6.3 Discrete measurements by eigen-analysis.

(1) to define the number and nature of the information classes, and collect sufficient and representative training data for each class; (2) to estimate the required statistical parameters from the training data; and (3) to apply an appropriated decision rule.

Given the class $w_i (i = 1, 2, \cdots n)$, where n is the number of the classes. Features mentioned above yield a D dimension feature space F. So, the probability of a region represented by its feature vector $X(X \in F)$ belongs to class w_i, is defined by Bayes' rule:

$$P(w_i/X) = \frac{P(X/w_i) \bullet P(w_i)}{P(X)} \quad (6.2)$$

where $P(w_i)$ is the prior probability of class w_i, $P(X/w_i)$ is the conditional probability of class w_i has data X, $P(w_i/X)$ is the posterior probability of data X belonging to class w_i, $P(X)$ can be considered as constant value for class w_i. Therefore, Equation 6.2 can be reduced to

$$d_i(X) = P(X/w_i) \bullet P(w_i) \quad (6.3)$$

where, $I = 1, 2,$ and 3, defining three distinctive classes: buildings w_1, bare ground w_2, and trees w_3. The objects that do not belong to these three classes are labelled as unclassified ones w_4. The prior probabilities of buildings $P(w_1)$, bare ground $P(w_2)$, trees, $P(w_3)$ and unclassified objects $P(w_4)$ are obtained according to specific training data set that can represents the typical features of urban areas, and meet

$$P(w_1) + P(w_2) + P(w_3) + P(w_4) = 1 \quad (6.4)$$

Therefore, the next step is to quantify features before determining conditional probability of each class.

1. The spatial information of the lidar point using eigen-analysis $X_1(f)$ is the ratio of the number of points belonging to "scatter" and "edge" points to the amount points in the homogenous region. $1 - X_1(f)$ means the percentage of "plane" points in the region.
2. The filtering result $X_2(f)$ is the ratio of the number of points labelling as on-terrain points to the amount points in the homogenous region. $1 - X_2(f)$ implies the percent of off-terrain points in region.
3. The height difference of first-echo and last-echo $X_3(f)$ is the ratio of the number of points that height differences are over the given threshold, to the amount points in the homogenous region.
4. The linear features of the aerial image $X_4(f)$ are the ration of the length of line segments near to the region to the length of region boundary.

The conditional probability of the class w_i is ascertained by the choices of the weights (m_1, m_2, m_3, m_4) to features $(X_1(f), X_2(f), X_3(f), X_4(f))$ in

$$P(X/w_i) = m_1 \times X_1(f) + m_2 \times X_2(f) + m_3 \times X_3(f)$$
$$+ m_4 \times X_4(f) \quad (6.5)$$

For each region, we can get three class values $\{d_1(X), d_2(X), d_3(X)\}$ according to Equation 6.3. The maximum of three results $d_i(X) = \max\{d_1(X), d_2(X), d_3(X)\}$ labels the region as to which of classes it belongs. If $d_1(X) \approx d_2(X) \approx d_3(X)$, the region is temporary to be labeled as the unclassified class.

6.2.3 Building reconstruction

After having successfully detected the isolated building regions from lidar data, the next step is building reconstruction.

6.2.3.1 Simplification of building boundaries

The majority of buildings in real world are the vertical walls which form the boundaries of buildings. As shown in Fig. 6.4(a), the extracted building boundary at first is recorded as an order sequence of feature points which form a set of ragged small line segments. Thus, the boundaries of buildings are very noisy. The Douglas–Peucker algorithm is first employed to generalize boundary segments so that redundant points can be removed. Figure 6.4(b) illustrates the simplification of building regions by the Douglas–Peuker algorithm.

However, the Douglas–Peuker algorithm can only remove redundant points from those small line segments that are not perpendicular to each other. Therefore, least-squares template matching with right-angle constraint is further applied to accurately determine the building boundaries with a rectangular shape.

6.2.3.2 Least-squares template matching with right-angle constraint

The least-squares matching proposed by Ackermann (1983) uses plenty of information in image to construct an adjustment model, by which the matching precision can reach 0.1 pixels, even 0.01 pixels. Gruen (1985) extended the least-squares matching to the least-squares template matching. Assuming two image regions

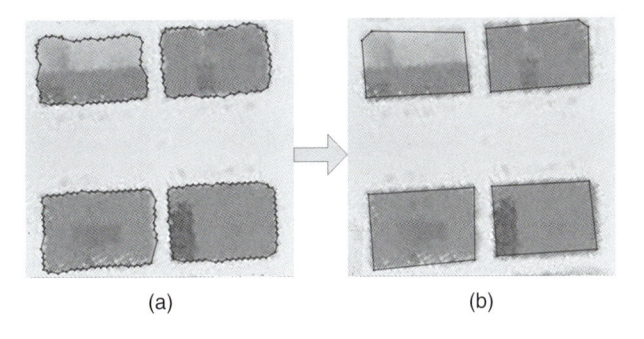

FIGURE 6.4 An example of boundaries simplified by Douglas–Peuker algorithm.

are given as discrete 2D functions $g_1(x,y)$ and $g_2(x,y)$, which might have been derived from analog images. $g_1(x,y)$ and $g_2(x,y)$ can be defined as conjugate regions of a stereo pair in the 'left' and the 'right' image respectively. $g_1(x,y)$ is interpreted in the following as the 'template', $g_2(x,y)$ as the 'picture'. Correlation is (ideally) established if

$$g_1(x,y) = g_2(x,y) \quad (6.6)$$

However, because of the random noise n, Equation 6.6 is not consistent. In other word, there is error v between the template and the picture.

$$v = g_1(x,y) - g_2(x,y) \quad (6.7)$$

Equation 6.7 can be considered as a non-linear observation equation, from which the distances between the gray levels in the template and the picture can be estimated by the least-squares adjustment.

Depending on the edge shape, the edge template can be designed in different forms. Figure 6.5 shows the designed edge template profile and the generated image patch template with the size of 7×5 pixels, respectively.

If the edge template is defined as $g_1(x,y)$ and the aerial image is defined as $g_2(x,y)$, then the image matching from Equation 6.7, can be given by

$$\begin{cases} v(x,y) = \dfrac{\partial g}{\partial x} dx + \dfrac{\partial g}{\partial y} dy - \Delta g \\ \Delta g = g_1(x,y) - g_2(x,y) \end{cases} \quad (6.8)$$

where $\partial g/\partial x$ is the derivative in the x direction, $\partial g/\partial y$ is the derivative in the y direction. They can be approximated by the first-order difference.

As mentioned above, most of the buildings in the area have rectangular or near rectangular shape. Therefore, we could assume that the line segments as shown in Fig. 6.5 (right), $L_{(i,i+1)}$ and $L_{(i-1,i)}$ are perpendicular each other. Each line segment can be represented by its two end points.

$$L_{(i,i+1)} :: \quad (y - y_i) = \dfrac{y_{i+1} - y_i}{x_{i+1} - x_i}(x - x_i) \quad (6.9)$$

$$L_{(i-1,i)} :: \quad (y - y_i) = \dfrac{y_{i-1} - y_i}{x_{i-1} - x_i}(x - x_i) \quad (6.10)$$

where, $L_{(i,i+1)}$ is perpendicular to $L_{(i-1,i)}$, which can be represented as

$$(x_{i+1} - x_i)(x_{i-1} - x_i) + (y_{i+1} - y_i)(y_{i-1} - y_i) - l_i = 0 \quad (6.11)$$

By linearizing Equation 6.11, we can write

$$\begin{aligned}(x_{i+1} - x_i) \cdot dx_{i-1} - (x_{i+1} + x_{i-1} - 2x_i) \cdot dx_i \\ + (x_{i-1} - x_i) \cdot dx_{i+1} + (y_{i+1} - y_i) \cdot dy_{i-1} \\ + (y_{i+1} + y_{i-1} - 2y_i) \cdot dy_i + (y_{i-1} - y_i) \cdot dy_i \end{aligned} \quad (6.12)$$

A combination of Equations 6.8 and 6.12 represents the indirect adjustment with geometric right-angle constraint by which the corners coordinate of building will be attained accurately in range image. This iterative matching processing will go on until the correction value of the two endpoints of a line meet the criteria or arrive the given maximal iteration times.

6.2.3.3 Roof surfaces detection

After regularization of the building boundary, the roof points of the lidar data in the boundary are available to reconstruct 3D building models. The automated detection of planes is an essential operation because it can eventually determine the roof shapes. Currently, there are mainly three proposed methods to carry out this task such as region growing, Hough transform and random sample consensus (RANSAC) (Fischler and Bolles, 1981). In a recent work, Tarsha-Kurdi, Landes and Grussenmeyer (2007) compared the Hough transform method with the RANSAC method for the automated detection of building roof planes and concluded that the RANSAC algorithm was more reliable in detecting the roof planes. In this study, we use the RANSAC algorithm to search the best plane from 3D points. The RANSAC algorithm is implemented by the following four steps:

1 To randomly select three points and calculate the parameters of the corresponding plane P, because the smallest number of three points is sufficient to determine the parameters of 3D plane model.

2 To detect every single building point and determine whether or not belonging to the calculated plane P with a predefined distance threshold τ.

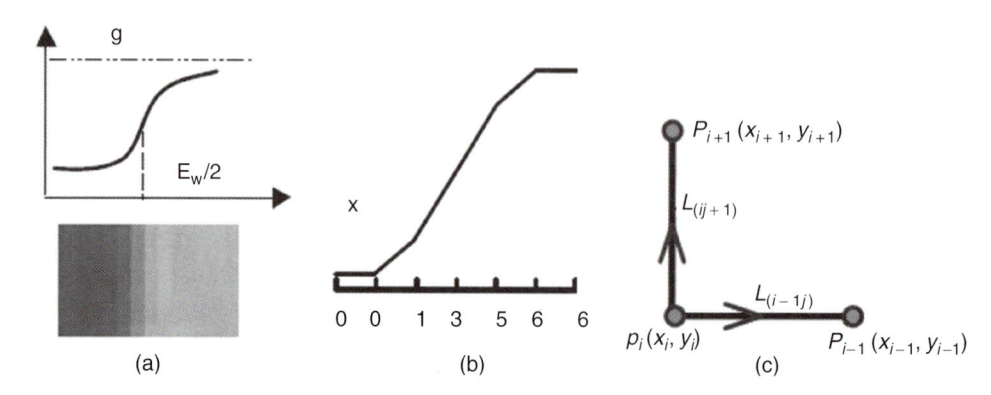

FIGURE 6.5 Least-squares template matching with the right-angle constraint: (a) edge profile template, (b) discrete template, (c) perpendicular condition.

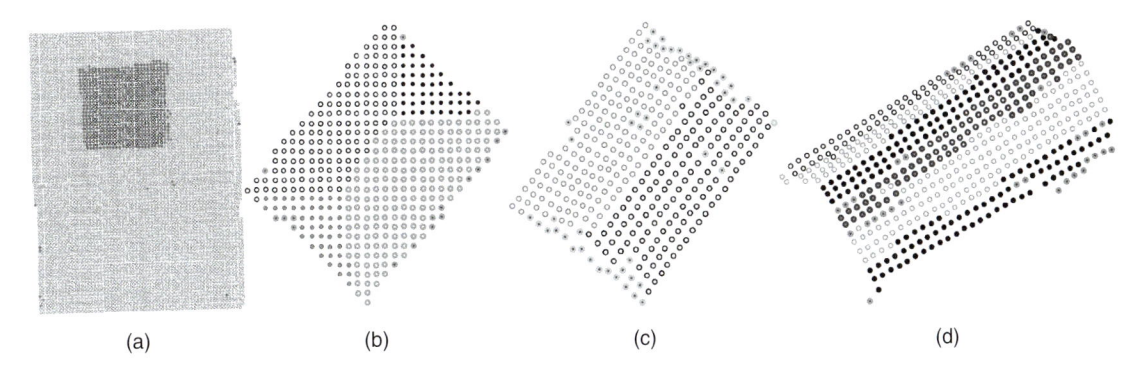

FIGURE 6.6 Results of roof detection by RANSAC.

3. To repeat Steps (1) and (2) M times. In each one, it compares the obtained result with the last one. If the new result is better, then it replaces the saved result with the new one.
4. To exclude the points belonging to the best plane from the building data when the best plane P is found.

In order to detect all roof surfaces within an individual building, the above process has to be repeated several times until the number of points is smaller than the given threshold δ (in this study, $\delta = 20$). Namely, if the number of points within a plane is below 20, these points are treated as noisy points and ready to be rejected. Defining a minimum number of points per surface helps to remove small surfaces that are not of further interest, e.g., chimneys. Beside the distance threshold τ and the minimum number of points δ, the RANSAC algorithm also uses the following three variables to control the plane model estimation process:

- ε, the probability that any selected data agrees with the plane model,
- α, ranging from 0.90 to 0.99, the minimum probability of finding at least one good set of observations in M trials, and
- M, the number of iterations, M, which can be defined as:

$$1 - (1 - (1 - \varepsilon)^m)^M = \alpha \quad (6.13)$$

Taking the logarithm of both sides of Equation 6.13, we have

$$M = \log(1 - \alpha) / \log(1 - (1 - \varepsilon)^m) \quad (6.14)$$

The RANSAC algorithm uses a pure mathematical principle to detect the best planes from a 3D point cloud. That means it looks for the maximum number of points which represent statistically the best planes without considering whether those building points are consecutive in 3D space or not. In other words, it detects a set of points which represents several roof planes or belongs to several planes. Therefore, we use a simple region-growing algorithm to find consecutive roof surfaces. It is necessary to build a TIN structure to search neighborhood for points within each individual building. Randomly starting from a point, a roof surface grows until it cannot find any neighbor point. Points being part of this roof surface are removed from the available points and the algorithm continues until all potential points are used. As shown in Fig. 6.6, several rooftop structures are successfully extracted from building points using the RANSAC algorithm.

6.2.3.4 Building boundary refinement by aerial imagery

By comparison with its vertical accuracy, the planimetric accuracy of lidar data is relatively lower due to the errors in the global positioning system (GPS), inertial measurement unit (IMU), mirror angles, and range measurements. However, lidar data cannot provide the linear edges of buildings because it is a set of sample points about the Earth's landscape. The aerial image compensates the weakness of lidar data. Locations of linear features of roofs within building boundaries are not affected by the distribution of points because they are estimated from intersection of two extracted roof planes. Thus, aerial image can be utilized to improve the quality of building boundary refinement.

The procedures of boundary refinement can be listed as follows:

1. To extract edge line segments from aerial image.
2. To project the building boundaries extracted from lidar data into the aerial image; see Fig. 6.7 for an example of a 2D projected boundary $\{B_1, B_2, B_3, B_4\}$.
3. To determine a buffer zone for each line segment of the boundaries- such as buffer of the segment B_1B_2 in Fig. 6.7. The buffer zone of the line segment of the building boundary is formed by expanding to its perpendicular direction.
4. To search for edge line segments extracted in a buffer zone and construct the new boundary by replacing the old line

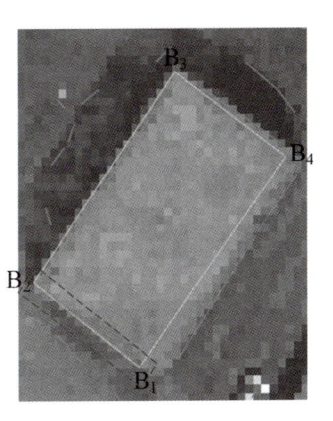

FIGURE 6.7 Refinement of building boundary.

segments of the boundary with the new correspondent ones from aerial image if it exists. If there are several edge line segments in a buffer zone, their slope and length will be analyzed to determine the new ones. If there is not one in the buffer zone, the old one is retained.

5 To refine the new boundary by least-squares template matching with orthogonal constraints.

6.2.3.5 Computation of building models

The adjacency relationship of roof surfaces will be set up to the reconstruction of 3D building model after detecting the roof surfaces by an adjacent matrix graph. Under the adjacent matrix relationship graph, the small adjacent roof surfaces were merged if its normal vectors meet the predefined threshold. Then line segments of boundary are treated as vertical walls to add into adjacent matrix graph, from which the ridge points will be decided by the intersection of adjacent surfaces and the building corners will be calculated. Finally, the 3D building models can be reconstructed.

Assume a set of roof surfaces $P_i (i = 0, 1 \ldots n-1)$ are detected, the adjacency matrix can be represented as shown in Fig. 6.8(a), where, n is the number of the roof surfaces, $R_{i,j}(i=0,1,\ldots n-1; \; j=0,1,\ldots n-1)$ is the adjacency relationship between P_i and P_j, whose elements will be labelled to 1 if the roof surfaces P_i and P_j are adjacent to each other. Otherwise, it will be labelled to 0. In Fig. 6.8(b), the diagonal elements of the adjacent relationship matrix are set to 0 because it is symmetric. It can be reconstructed with the help of the TIN data structure. For example, given a point V_m within the surface P_i, find out all of its adjacent points $\{v_0, v_1, v_2, v_3, v_4, v_5\}$. As shown in Fig. 6.9, points v_0 and v_1 are located within surfaces P_k and P_j, respectively, and points v_2 and v_3 are both within surface P_m, so the value of $R_{i,j}, R_{i,k}$ and $R_{i,m}$ in the matrix equals to 1, where, i, j and k are the indices of the roof surfaces.

With the aid of adjacent matrix, the small adjacent roof surfaces are merged according to their normal vectors. In general, the threshold of merging surfaces is set to 5°. In other words, if the angle of the normal vectors of the two adjacent surfaces is smaller than this threshold, these two surfaces are merged. For reconstruction of a 3D building model, normally building corners are reconstructed first because they are the primitives of a computer-aided drafting (CAD) model. Thus, the new relationship between roofs and vertical walls is remodeled by adding the vertical walls into the adjacent matrix of the roof surfaces.

For each single line segment of a building boundary, we first find all roof surfaces, then expand along its perpendicular direction to form a buffer zone. According to the building points within the buffer zone belonging to which one of the roof surfaces, the adjacency of the vertical walls and roof surfaces can be detected. Figure 6.10(a) shows an example of a building with

$$R_{n,n} = \begin{bmatrix} R_{0,0} & R_{0,1} & \cdots & R_{0,j} & \cdots & R_{0,n-1} \\ R_{1,0} & R_{1,1} & \cdots & R_{1,j} & \cdots & R_{1,n-1} \\ \vdots & \vdots & & \vdots & & \vdots \\ R_{i,0} & R_{i,1} & \cdots & R_{i,j} & \cdots & R_{i,n-1} \\ \vdots & \vdots & & \vdots & & \vdots \\ R_{n-1,0} & R_{n-1,1} & \cdots & R_{n-1,j} & \cdots & R_{n-1,n-1} \end{bmatrix} \Rightarrow R_{n,n} = \begin{bmatrix} 0 & R_{0,1} & \cdots & R_{0,j} & \cdots & R_{0,n-1} \\ 0 & 0 & \cdots & R_{1,j} & \cdots & R_{1,n-1} \\ \vdots & \vdots & & \vdots & & \vdots \\ 0 & 0 & \cdots & 0 & \cdots & R_{i,n-1} \\ \vdots & \vdots & & \vdots & & \vdots \\ 0 & 0 & \cdots & 0 & \cdots & 0 \end{bmatrix}$$

(a)   (b)

FIGURE 6.8 Adjacent matrix graph of roof surfaces.

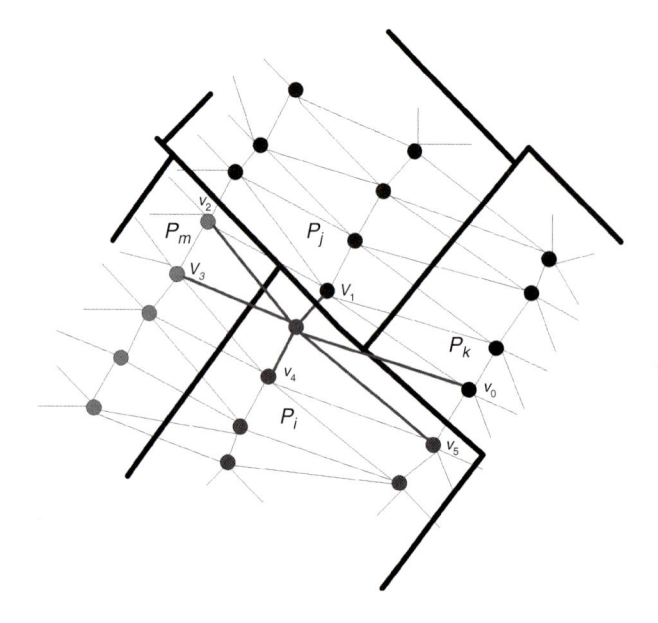

FIGURE 6.9 An example of searching neighborhood.

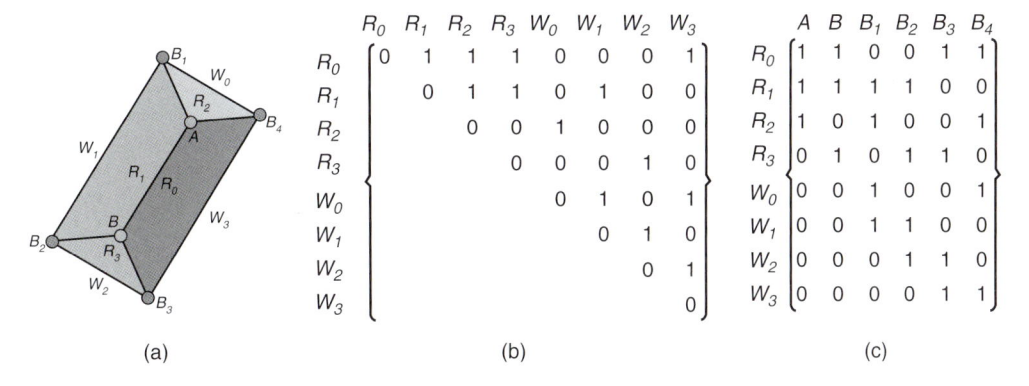

FIGURE 6.10 A hipped roof model.

a hipped roof. Figure 6.10(b) presents the adjacent matrix of the roof surfaces and vertical walls in Fig. 6.10(a), where $R_0 \sim R_4$ are the roof surfaces, $W_0 \sim W_4$ are the vertical walls. Under the adjacent matrix, the corner-surface matrix can be formed as shown in Fig. 6.10(c), whose purpose is to get the coordinates of the building corners. Typically, the coordinate of a spatial point needs at least three surfaces to solve. Thus, in the corner-surface matrix, each column represents a corner of the 3D building model. Corners located within the boundary can be calculated by the intersection of the roof surfaces. For example, corner A can be calculated by the intersection of the adjacent roofs $R_0, R_1,$ and R_2. Corner B can be calculated by the roofs $R_0, R_1,$ and R_3. In reality, the ridge line is horizontal, so the height value of the corner A is equal to that of the corner B. Similarly, corners $B_1 \sim B_4$ can also be calculated by the intersection of roofs and vertical walls. For instance, four adjacent surfaces (two roof surfaces R_1 and R_2, two vertical walls W_0 and W_1) are used to get the corner B_1. More attention needs to be paid to the heights of the corners $B_1 \sim B_4$. They should be the same value because they generally are horizontal in the real world. For a building with a hipped roof (see Fig. 6.10a), the coordinates of six corners with 14 unknown parameters are retrieved by the least-squares adjustment.

6.3 Results and discussion

This section first describes the lidar point cloud data and color aerial image data used in this chapter and then presents the results of building extraction.

6.3.1 Datasets

The lidar datasets used in this study were collected over 1000 m above ground level by an Optech ALTM 3100 system. The system also carried a 4k × 4k digital camera. The datasets include the first and the last returns of the laser beam and the true color aerial imagery. Figure 6.11 shows a residential scene, which is a subset of a typical urban area in the City of Toronto, Ontario, Canada. As shown in Fig. 6.11(a) presents the raster-based DSM, containing a total of 105,298 points, which was interpolated with both the first and last pulse returns by the bilinear interpolation method. The width and height of grid equals to the ground sample distance (GSD) of the aerial image (0.5 m). The elevation of the study area ranges from 150.00 m to 178.11 m. Besides buildings, there are several clusters of trees. Figure 6.11(b) shows the true color aerial image that was resampled to 0.5 m ground pixel. The majority of buildings appeared in the color image are with gable roofs or hipped roofs. Figure 6.11(c) and 6.11(b) illustrate the lidar range images of the first and last returns, respectively.

6.3.2 Results

Figure 6.12(a) shows the filtered results overlaid on the color aerial image. The results demonstrate that all the on-terrain points have been removed quite well. The parameter of the iterative distance in the filtering algorithm which is usually below

FIGURE 6.11 Lidar range data and colour aerial image of the study area: (a) DSM, (b) true colour aerial image, (c) first return range image, and (d) last return range image.

the height of cars, and it gradually decreases with the increase of iterative times.

The classification accuracy relies heavily on the quality of segmentation. The abundant spectral information from color aerial image is beneficial to the classification. On the other hand, it increases the difficulty in segmentation due to spectral confusion between-class and spectral variation within-class. For example, the roof material of a building along a road sometimes is similar to that of the road. Figure 6.12(b) shows the materials of the building and ground are similar, so they will always be partitioned into a homogenous region in Fig. 6.12(c). However, the quality of segmentation is not guaranteed when the range image (see Fig. 6.12d) is used alone. For instance, if a tree is close to a building, it is very difficult to separate one from another, as shown in Fig. 6.12(e). Therefore, height information from lidar data as an additional channel can improve the quality of color image segmentation. Figure 6.13(a) shows the result of segmentation by fusing lidar data and the aerial image. Figure 6.13(b) presents the spatial discrete measurement result of lidar data by eigen-analysis, in which most of tree regions are described as "discrete" and "edges," while majority of building regions are presented as "planes."

Some parameters for the supervised object-oriented classification need to be predetermined. First, the parameters of the prior probability can be derived from the typical training set. The most important parameters in this study are the weights of features. The

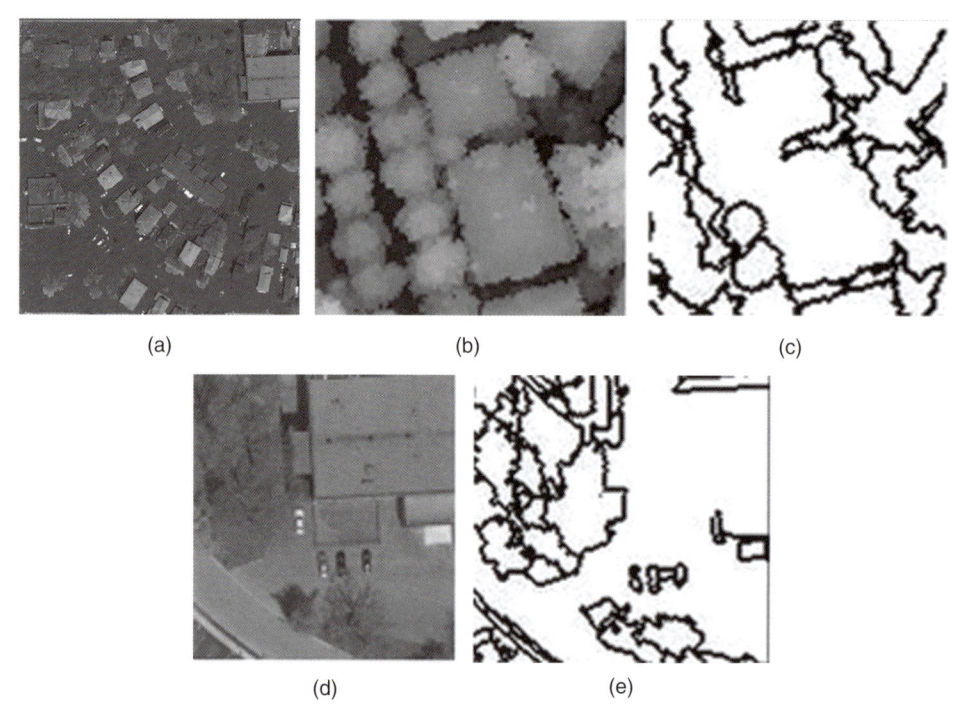

FIGURE 6.12 (a) Filtered result overlaid on the aerial image, (b) aerial image and (c) its corresponding segmentation results, (d) lidar range image and (e) its corresponding segmentation results.

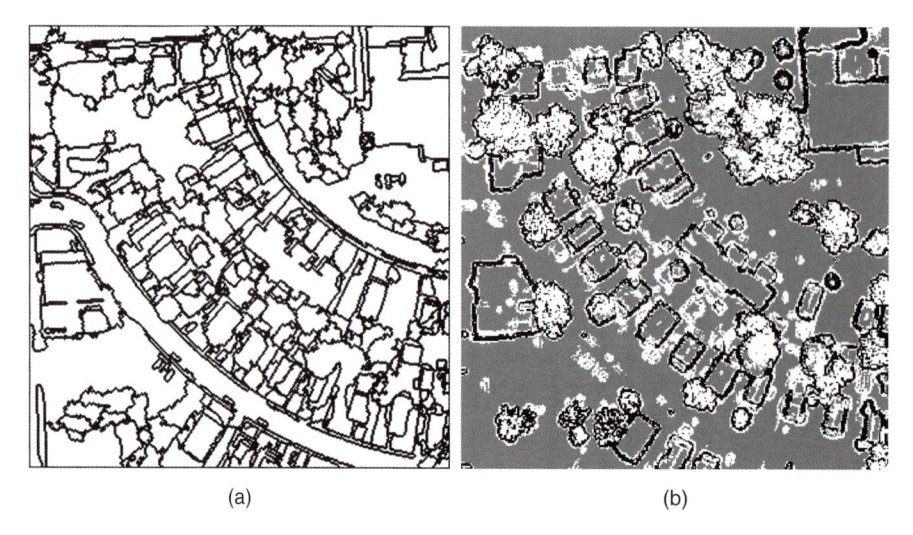

FIGURE 6.13 (a) The segmentation result of the aerial image fused with the height and texture information from lidar data; (b) The spatial distribution result of lidar data by eigen-analysis.

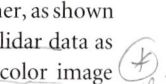

way to choose weights is self-adaptive to features of each homogeneous region. Because features $(X_1(f), X_2(f), X_3(f), X_4(f))$ have already normalized, the weights used for each feature is its value

$$m_i = X_i(f)/(X_1(f) + X_2(f) + X_3(f) + X_4(f)) \quad i = 1, 2, 3, 4 \tag{6.15}$$

This method matches the observation that a larger percentage of a feature is more important than a smaller percentage of the feature. For each homogeneous region, the weights vary with different features in the region. For example, if the values of four features are {0.8, 0.8, 0.2, 0.1} in a region, the correspondent weights are {0.42, 0.42, 0.1, 0.06}, respectively. Figure 6.14(a) shows the result of the object-oriented classification. By visual inspection of the aerial image, a few buildings were not separated from the adjoining trees, due to large crowns of those trees covering the major part of the buildings. Some small areas, which most likely belong to trees, were misclassified as buildings. These areas are so small that some of them can be eliminated by an area threshold. Then, the building blobs can be separated (see Fig. 14b and c).

In order to reconstruct the 3D building models, the building boundaries should be extracted as its vertical walls from the lidar range image. In the proposed algorithm, the boundary of a building is assumed to be a rectangular-shaped object. That is, the line segments of the boundary are either parallel or perpendicular from one to another. We first use the Douglas–Peuker algorithm to simplify these ragged building boundaries (see Fig. 6.14d). Then the simplified irregular boundaries can be regularized to rectangular-shaped ones using the least-squares template matching with orthogonal constraints. Figure 6.15 shows the extracted building boundaries by simplification using the Douglas–Peuker algorithm (labelled as B in a circle) followed by regularization using the least-squares template matching with geometric right-angle constraints.

Figures 6.16 and 6.17 show the building reconstruction process, in which "A" represents roof boundaries extracted from lidar data, "B" represents roof boundaries and ridge lines only using the least-squares template matching, "C" represents roof boundaries and ridge lines by the least-squares template matching with orthogonal constraints, and "D" represents line segments extracted from aerial image. Figure 6.16 shows one building with a gable roof (having two planar surfaces), in which (a) shows the points of the planer surfaces detected from the classified building points using the RANSAC algorithm, (b) shows the roof boundary (the "A" labels) projected onto the aerial image from the lidar-derived polygon, (c) shows the initial ridge line (the "B" labels) projected onto the aerial image detected from the intersection edges of neighboring surfaces together with final ridge line (the "C" labels) rectified using image information, and (d) shows the reconstructed 3D model of a single building with a gable roof. Figure 6.17 shows the reconstruction process of a building with a rooftop having four planar surfaces.

Figure 6.18(a) shows the building boundaries extracted from range image, which are then projected onto the color aerial image to improve the planimeric accuracy. The boundaries before and after refinement are presented by light-gray and darker lines, respectively. As shown in Fig. 6.18(a), the three individual detached houses are misclassified as a single house in the previous processes. Given the fact that these individual

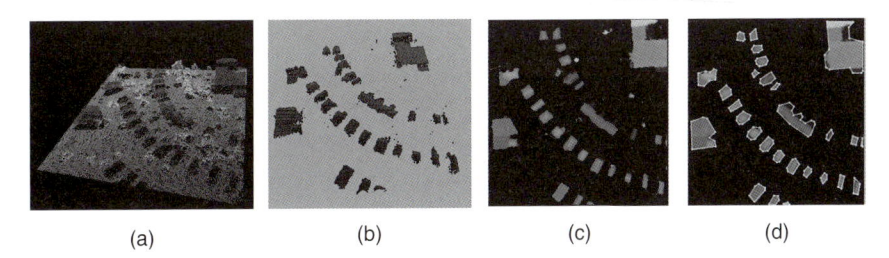

(a) (b) (c) (d)

FIGURE 6.14 (a) The result of object-oriented classification; (b) The classified buildings' lidar points; (c) The classified buildings' range image. (d) The extracted building roof boundaries.

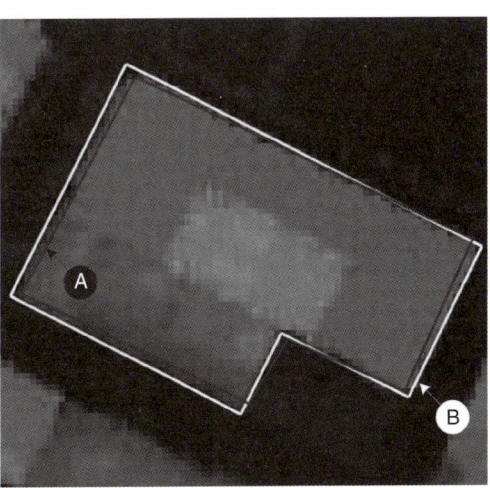

FIGURE 6.15 Extracted roof outlines. The "A" labels represent the result of Douglas–Peuker simplification. The "B" labels show the result of regularization using least-squares template matching with geometric constraint.

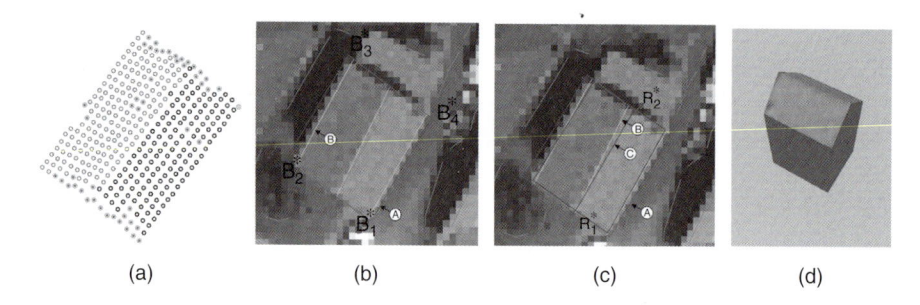

88 PART II REMOTE SENSING SYSTEMS FOR URBAN AREAS

FIGURE 6.16 The reconstructed building with a gable roof.

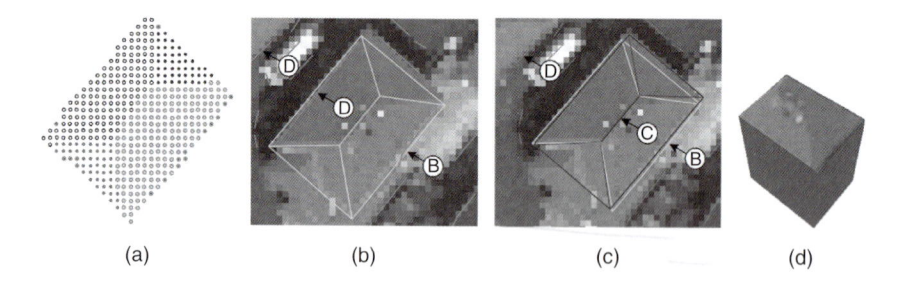

FIGURE 6.17 The reconstructed building with a complex roof having four planar surfaces.

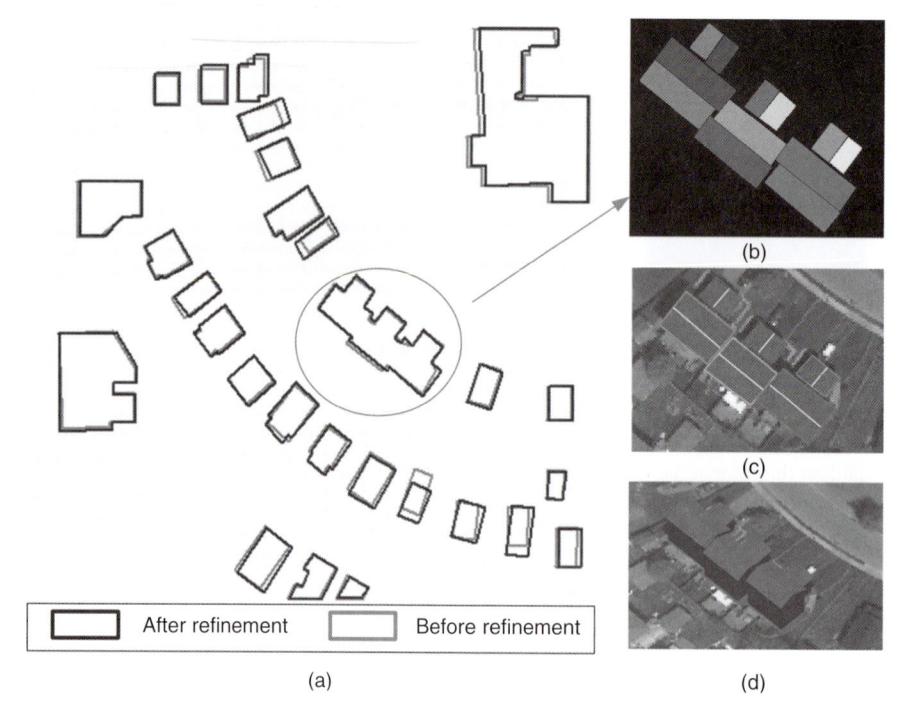

FIGURE 6.18 (a) Refined building roof boundaries, (b) Detected planer surfaces of roofs, (c) accurately located rood boundaries, (d) 3D reconstructed buildings overlaid on the aerial image.

CHAPTER 6 3D BUILDING RECONSTRUCTION FROM AIRBORNE LIDAR POINT CLOUDS FUSED WITH AERIAL IMAGERY 89

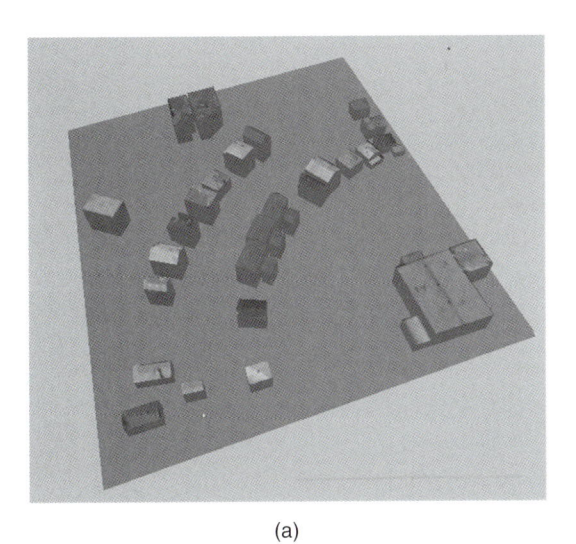

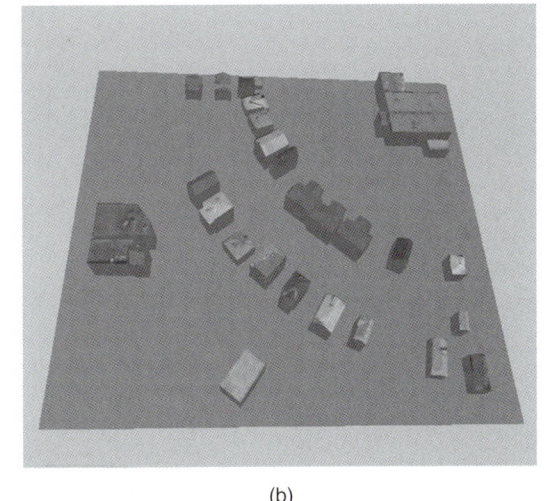

(a) (b)

FIGURE 6.19 Perspective views of the 3D building models with textured roofs reviewed from two different viewpoints.

FIGURE 6.20 The created 3D city model with textured building models, trees and terrain.

buildings are different in height, lidar point clouds are used to separate them with a total of six planar surfaces detected using the RANSAC algorithm (see Fig. 6.18b). Then the outlines of these surfaces are generated, simplified and regularized. Finally, the individual building with a gable roof can be determined by the roof ridge line from the intersection of the two planar surfaces (see Fig. 6.18c). Figure 6.18(d) shows the three reconstructed building models overlaid on the color aerial image.

The color aerial image (orthoimagery) was also used as texture and draped onto both building rooftops and terrain surface. From a data-structure point view, orthoimagery is expressed as 2D raster data, which can be stored or manipulated as a special layer in a 3D geospatial system. This work focuses on a 3D scene with only three object classes: buildings, trees and terrain. The textured building roofs are required for creation of a photorealistic 3D city model towards fly- or walk-throughs and simulation (particularly for the planning and visualization of building models). Figure 6.19 shows the perspective reviews of the study area from two different viewpoints. Figure 6.20 shows the resulting 3D city model with buildings, trees and textured terrain. It presents useful opportunities for site visualization, which enhances community participation in decision-making.

Concluding remarks

In this chapter we have presented a comprehensive approach for the determination of 3D building models from lidar point cloud data fused with color aerial imagery. A two-step extraction strategy, building detection followed by building reconstruction, has been developed and implemented. Building detection is first done by filtering to separate on-terrain and off-terrain points, and further, to integrate information from color aerial imagery into an object-oriented classification process to catalog three classes (buildings, trees, and terrain).

To ensure completeness, it is advisable to initialize the very first step, namely the coarse selection of building regions, interactively. All subsequent steps (i.e., outline extraction and regularization, planar surface detection, and building model generation) are applied automatically. Possibly erroneous buildings can be improved by interactive post-processing (i.e., manual editing).

Compared to lidar point cloud data, the advantage of color aerial imagery is its higher accuracy in sharp linear features, which can be utilized to refine building boundaries. The rectangular-shaped buildings can be delineated well by using the least-squares template matching with orthogonal constraints.

Since our method is a data-driven approach, we attempt the reconstructed building models to approximate the given point clouds the best. Compared to the model-driven methods, the data-driven approach is more flexible. While the building model reconstructed using the model-driven method is restricted by a predefined catalog of buildings or building parts, a data-driven approach requires more flexible assumptions. A typical assumption in this study is that buildings can be modeled by a composition of planar surfaces. However, this might be a restriction, considering curved roof structures.

In our case study, due to the complexity and diversity of natural and man-made objects in urban areas, manual input is required in the classification stage, which reduces the automation level. Another disadvantage is that those buildings partially covered by trees cannot be reconstructed completely and correctly due to the failure of the roof detection, even these building areas had been classified correctly. Finally, this study assumes a building as a rectangular-shaped object only. Further investigations on methods for dealing with non-rectangular-shaped buildings are needed. Nevertheless, the results met the requirement of being accurate and reliable. Subsequent processing like texture mapping is possible. The resulting models are well structured and topologically correct and are, therefore, directly applicable for 3D urban visualization.

In addition, for those applications requiring realistic 3D city models, reconstructing photorealistic building facade models should be applied. This can be done from terrestrial laser point clouds and close-range images. Considering quick creation of such building facade models in a large urban area, a vehicle-borne mobile lidar mapping system would be an effective means for fast collection of terrestrial lidar point clouds along with color digital camera images of the roadside buildings. Moreover, accurate fusion of the point clouds acquired by both airborne and terrestrial mobile lidar mapping systems would open a new avenue to reconstruction of realistic 3D city models.

Acknowledgments

The work presented in this chapter was supported by a NSERC discovery grant and a Wiser Foundation research grant. The datasets used in this chapter were provided by Optech, Inc., Toronto, Canada. The authors would like to thank the anonymous reviewers for their valuable comments.

References

Ackermann, F. (1983) High precision digital image correlation. In *Proceedings of 39th Photogrammetric Week*, University of Stuttgart, pp. 231–243.

Alharthy, A. and Bethel, J. (2002) Heuristic filtering and 3D feature extraction from lidar data. *International Archives of the Photogrammetry, Remote Sensing and Spatial Information Sciences*, **34**(3A+B), 6.

Axelsson, P. (1999) DEM generation from laser scanner data using adaptive TIN models. *International Archives of the Photogrammetry, Remote Sensing and Spatial Information Sciences*, **33**(B4/1), 110–117.

Baltsavias, E.P. (1999) Airborne laser scanning – relations and formulas. *ISPRS Journal of Photogrammetry and Remote Sensing*, **54**, 199–214.

Bentley J.L. (1985) Multidimensional binary search trees used for associative searching. *Communications of the Association for Computer Machinery (ACM)*, **18**, 509–517.

Dorninger, P. and Pfeifer, N. (2008) A Comprehensive automated 3D approach for building extraction, reconstruction, and regularization from airborne laser scanning point clouds. *Sensors*, **8**(11), 7323–7343.

Douglas, D. and Peucker, T. (1973) Algorithms for the reduction of the number of points required to represent a digitized line or its caricature. *The Canadian Cartographer*, **10**(2), 112–122.

Filin, S. and Pfeifer, N. (2006) Segmentation of airborne laser scanning data using a slope adaptive neighborhood. *ISPRS Journal of Photogrammetry and Remote Sensing*, **60**(2), 71–80.

Fischler, M.A. and Bolles, R.C. (1981) Random sample consensus: a paradigm for model fitting with applications to image analysis and automated cartography. *Communications of the ACM*, **24**, 381–395.

Gamba, P. and Houshmand, B. (2002) Joint analysis of SAR, lidar and aerial imagery for simultaneous extraction of land cover, DTM and 3D shape of buildings. *International Journal of Remote Sensing*, **23**(20), 4439–4450.

Gonzalez, R.C. and Woods, R.E. (2008) *Digital Image Processing*, 3rd edition, Prentice Hall, Upper Saddle River, NJ.

Gross, H., Thoennessen, U. and Hansen, W. (2005) 3D modeling of urban structures. *International Archives of Photogrammetry, Remote Sensing and Spatial Information Sciences*, **36**(3-W24).

Gruen A. (1985) Adaptive least squares correlation: a powerful image matching technique. *South African Journal of Photogrammetry and Remote Sensing*, **14** (3), 175–187.

Haala, N., Brenner, C. and Anders, K.-H. (1998) 3D urban GIS from laser altimeter and 2D map data. In *International Archives of Photogrammetry, Remote Sensing and Spatial Information Sciences*, **32**(3/ W1), 321–330.

Haala, N. and Brenner, C. (1999) Extraction of buildings and trees in urban environments. *ISPRS Journal of Photogrammetry and Remote Sensing*, **54**, 130–137.

Haralick, R.M. and Shapiro, L.G. (1985) Image segmentation techniques. *Computer Vision Graphics Image Process*, **29**, 100–132.

Hershberger, J. and Snoeyink, J. (1992) Speeding up the Douglas–Peucker line-simplification algorithm. *Proceedings of the 5th Symposium on Data Handling*, 134–143.

Hofmann, A.D., Maas, H.-G., and Streilein, A. (2003) Derivation of roof types by cluster analysis in parameter spaces of airborne laserscanner point clouds, in *Proceedings of the ISPRS WG III/3 Workshop on 3-D Reconstruction from Airborne Laserscanner and InSAR Data*, Dresden, October 8–10.

Hu, Y. (2003) Automated Extraction of Digital Terrain Models, Roads and Building Using Airborne Lidar Data, PhD Thesis, University of Calgary.

Hu, Y., Tao, V. and Collins, M. (2003). Automatic extraction of buildings and generation of 3D city models from airborne lidar data. *ASPRS Annual Conference*, 3–9 May, Anchorage, AK, 12 p.

Hug, C. and Wehr, A. (1997) Detecting and identifying topographic objects in imaging laser altimeter data. *International Archives of Photogrammetry, Remote Sensing, and Spatial Information Sciences*, **32**, 19–26.

Jwa, Y., Sohn, G., Tao, V. et al. (2008) An implicit geometric regularization of 3D building shape using airborne lidar data. *International Archives of Photogrammetry, Remote Sensing, and Spatial Information Sciences*, **36**(5), 69–76.

Kraus, K. and Pfeifer, N. (1998) Determination of terrain models in wooded areas with airborne laser scanner data. ISPRS Journal of Photogrammetry and Remote Sensing. 53, 193–203.

Lafarge, F., Descombes, X., Zerubia, J. et al. (2008) Automatic building extraction from DEMs using an object approach and application to the 3D-city modeling. *ISPRS Journal of Photogrammetry and Remote Sensing*, 63, 365–381.

Lee, I. and Schenk, T. (2002) Perceptual organization of 3D surface points. *International Archives of Photogrammetry, Remote Sensing and Spatial Information Sciences*, **34**(3), 6.

Li, J. (1999) *Informal Settlement Modeling using Digital Small-format Aerial Imagery*, PhD Thesis, University of Cape Town.

Li, Y., Li, J. and Chapman, M.A. (2010) Segmentation of SAR intensity imagery with a Voronoi tessellation, Bayesian inference and reversible jump Markov chain Monte Carlo algorithm. *IEEE Transactions on Geoscience and Remote Sensing*, **48**(4), 1872–1881.

Maas, H.-G. (1999) Potential of height texture measurement for the segmentation of airborne laser scanner data. *ISPRS Journal of Photogrammetry & Remote Sensing*, **54**(2/3), 245–261.

Maas, H.-G. and Vosselman, G. (1999) Two algorithms for extracting building models from raw laser altimetry data. *ISPRS Journal of Photogrammetry and Remote Sensing*, **54**(2/3), 153–163.

Matikainen, L., Hyyppä, J. and Hyyppä, H. (2003) Automatic detection of buildings from laser scanner data for map updating. *International Archives of Photogrammetry, Remote Sensing and Spatial Information Sciences*, **34**(3/ W13).

Morgan, M. and Tempfli, K. (2000) **Automatic building extraction from airborne laser scanning data**. *International Archives of Photogrammetry, Remote Sensing and Spatial Information Sciences*, **33**(B4): 616–622.

Nardinocchi, C. and Forlani, G. (2001) Building detection and roof extraction in laser scanning data. In *Proceedings of 3rd International Workshop on Automatic Extraction of Man-Made Objects from Aerial and Space Images*, Ascona, Switzerland.

Roerdink, J.B.T.M. and Meijster, A. (2001) The watershed transform: definitions, algorithms and parallelization strategies. *Fundamenta Informaticae*, **41**(1–2), 187–228.

Rottensteiner, D.F. and Briese, C. (2002) A new method for building extraction in urban areas from high-resolution lidar data. *International Archives of Photogrammetry, Remote Sensing and Spatial Information Sciences*, **34** (3A/B), 295–301.

Rottensteiner, F., Trinder, J., Clode, S. et al. (2005) Automated delineation of roof planes from lidar data. *International Archives of Photogrammetry, Remote Sensing and Spatial Information Sciences*, 36(3/ W19).

Rutzinger, M.B. Höfle, B., Pfeifer, N. et al. (2006) Object based analysis of airborne laser scanning data for natural hazard purposes using open source components. In *the First International Conference on Object-based Image Analysis*, Salzburg, Austria.

Sampath, A. and Shan, J. (2007) Building boundary tracing and regularization from airborne lidar point clouds. *Photogrammetric Engineering & Remote Sensing*, **73**(7), 805–812.

Shan, J. and Sampath, A. (2006) Urban terrain and building extraction from airborne lidar data. In *Urban Remote Sensing*, (eds Q. Weng and D. Quattrochi), CRC/Taylor & Francis, Boca Raton, pp. 21–45.

Shan, J. and Toth, C. (2008) *Topographic Laser Ranging and Scanning: Principles and Processing*. CRC Press, London, UK.

Sithole, G. and Vosselman, G. (2004) Experimental comparison of filtering algorithms for bare-earth extraction from airborne laser scanning point clouds. *ISPRS Journal of Photogrammetry and Remote Sensing*, **59**(1–2), 85–101.

Tarsha-Kurdi, F., Landes, T. and Grussenmeyer, P. (2007) Hough-transform and extended RANSAC algorithms for automatic detection of 3D building roof planes from lidar data. *International Archives of Photogrammetry and Remote Sensing*, **36**(3/ W52), 407–412.

Tóvari, D. and Vögtle, T. (2004) Object classification in laser-scanning data. In *International Archives of Photogrammetry, Remote Sensing and Spatial Information Sciences*, **36**(8/ W2), 45–49.

Tso, B. and Mather, P.M. (2001) *Classification Methods for Remotely Sensed Data*. Taylor & Francis, London, p. 75.

Vestri, C. and Devernay, F. (2001) Using robust methods for automatic extraction of buildings, in *2001 IEEE Computer Society Conference on Computer Vision and Pattern Recognition (CVPR'01)*, Vol. 1, pp. 133.

Vosselman, G. (1999) Building reconstruction using planar faces in very high-density data. *International Archives of Photogrammetry and Remote Sensing and Spatial Information Sciences*, **32**(3–2W5).

Vosselman, G. and Dijkman, S. (2001) 3D building model reconstruction from point clouds and ground plans, In *International Archives of Photogrammetry and Remote Sensing*, **34**(3/ W4), 37–44.

Wang, Z. and Schenk, T. (2000) Building extraction and reconstruction from lidar data. *International Archives of Photogrammetry, Remote Sensing and Spatial Information Sciences*, **33**(B3), 958–964.

Weidner, U. and Forstner, W. (1995) Towards automatic building extraction from high resolution digital elevation models. *ISPRS Journal of Photogrammetry and Remote Sensing*, **50**(4), 38–49.

Yoon, T. et al. (1999) Building segmentation using an active contour model. In *ISPRS Conference on Sensors and Mapping from Space*, Hannover, 27–30 September.

Zhan, Q., Molenaar, M. and Tempfli, K. (2002) Building extraction from laser data by reasoning on image segments in elevation slices. *International Archives of Photogrammetry, Remote Sensing and Spatial Information Sciences*, **34**(3A+B), 305–308.

iii

ALGORITHMS AND TECHNIQUES FOR URBAN ATTRIBUTE EXTRACTION

Developing improved image processing algorithms and techniques for working with different types of remote sensor data is an important research area in urban remote sensing. Part III (Chs 7–11) reviews some of the most exciting developments in this aspect. The first three chapters within Part III are dedicated to a set of image processing techniques that can be used to improve urban mapping performance at per-pixel, subpixel, or object level. Chapter 7 discusses some algorithmic parameters affecting the performance of artificial neural networks in image classification at per-pixel level. Chapter 8 reviews the spectral mixture analysis (SMA) technique that allows the decomposition of each pixel into independent endmembers or pure materials to map urban sub-pixel composition. Chapter 9 provides an overview on the principles of object-based image analysis (OBIA) and demonstrates how the OBIA can be applied to achieve satisfactory urban mapping accuracy. The last two chapters included in Part III deal with two important aspects for urban mapping: image fusion technique (Ch. 10) and temporal lag between urban structure and function (Ch. 11).

7

Parameterizing neural network models to improve land classification performance

Xiaojun Yang and Libin Zhou

Neural networks are an attractive machine intelligence technique increasingly being used for pattern recognition in remote sensing. However, the performance of a neural network is contingent upon various algorithmic and non-algorithmic parameters. Despite significant processes over the past two decades, there is no consistent guidance on the use of neural networks for image classification. In this chapter, we review and evaluate a set of algorithmic parameters affecting the performance of neural networks in land classification from remote sensor data. We begin with an introduction to the basic structure of neural networks emphasing upon the multi-layer perceptron networks due to their robustness and popularity. Then, we discuss two focused studies we recently conducted with a satellite imagery covering an urban area to assess the sensitivity of image classification by neural networks in relation to various internal parameter settings and the performance of several training algorithms in image classification. Based on literature review and our own experiments, we further propose a framework that can guide the use of neural networks in remote sensing. Finally, we identify several areas for future research. Overall, this work can help design better neural network models for improved performance, which can further promote the use of neural networks as a routine tool in image classification.

7.1 Introduction

A neural network is a massively parallel distributed processor comprised of simple processing units, attempting to simulate the powerful capabilities for knowledge acquisition, synthesis, and problem solving of the human brain (Haykin, 1999). It originated from the concept of artificial neuron introduced by McCulloch and Pitts in 1943. Over the past several decades, neural networks have evolved from the preliminary development of artificial neuron, through the rediscovery and popularization of the back-propagation training algorithm, to the implementation of neural networks using dedicated hardware (Dawson and Wilby, 2001). Because of the distributed structure and adaptive learning process, neural networks are capable of handling non-linear, complex phenomena; they can also be effective to process incomplete, noisy and ambiguous data (Bishop, 1995). These advantages make neural networks an attractive pattern classifier (Duda, Hart and Stork, 2001).

The use of neural networks in remote sensing began in late 1980s (Atkinson and Tatnall, 1997; Kanellopoulos and Wilkinson, 1997). Over the past two decades, numerous studies have demonstrated that neural networks can produce identical or improved classification accuracies when compared to the outcome from conventional classifiers (e.g., Benediktsson, Swain and Ersoy, 1990; Bischof, Schneider and Pinz, 1992; Civco,1993; Paola and Schowengerdt,1995a; Gopal and Woodcock, 1996; Serpico, Bruzzone and Roli, 1996; Mannan, Roy and Ray, 1998; Ji, 2000; Seto and Liu, 2003; Del Frate *et al.*, 2007; Petropoulos *et al.*, 2010). Nevertheless, the performance of neural networks is contingent upon a wide range of algorithmic and non-algorithmic parameters, such as input data dimensionality, training data, network structure, and learning process (Paola and Schowengerdt, 1995b; Foody and Arora, 1997; Kavzoglu and Mather, 2003; Mas and Flores, 2008). With the incorporation of neural networks as a standard classifier in some popular image processing software packages (see Mas and Flores, 2008), handling these diverse parameters presents a challenge to beginners and even some experienced users as an inappropriate treatment can lead to suboptimal or unacceptable classification performance (Zhou and Yang, 2010).

Investigating the sensitivity of neural networks with respect to various parameter settings has been the subject in an increasing number of studies since this knowledge is critical to the design of efficient neural network models for improved performance (e.g., Korczak and Hammadimesmoudi, 1994; Yoshida and Omatu, 1994; Jarvis and Stuart, 1996; Kanellopoulos and Wilkinson, 1997; Ozkan and Erbek, 2002; Stathakis, 2009). These studies have been conducted by using a trial-and-error approach (e.g., Paola and Schowengerdt, 1997; Kavzoglu and Mather, 2003) or some advanced methods such as generic algorithm and pruning algorithms (e.g., Kavzoglu and Mather, 2003; Benediktsson and Sveinsson, 2003). As a result, some practical guidelines have been proposed to deal with various non-algorithmic issues including input data dimensionality and training sample quantity and quality (e.g., Zhuang *et al.*, 1994; Foody, McCulloch, and Yates, 1995; Kanellopoulos and Wilkinson, 1997; Mas and Flores, 2008). Nevertheless, there is no consistent guidance to help configure neural networks. Specific to the multi-layer-perceptron (MLP) neural networks, for example, there is no consensus on the number of hidden layers, type of activation functions, or training parameters that should be used to achieve optimal performance (Jain, Mao and Mohiuddin, 1996; Paola and Schowengerdt, 1997; Kavzoglu and Mather, 2003; Mas and Flores, 2008). These inconsistencies justify further research on the algorithmic issues in order to promote the routine use of neural networks in image classification.

In this chapter, we review and assess a set of algorithmic parameters affecting the performance of neural networks in image classification. The chapter comprises several major components. First, we introduce and review some fundamental aspects of neural networks including network architectures and knowledge representation. Second, we discuss two focused studies we recently conducted with an urban–suburban area as the test site to assess the sensitivity of neural networks with respect to various internal parameter settings and the performance of several training algorithms in image classification by neural networks. Third, based on literature review and our own focused studies, we propose a framework to guide the use of neural networks in image classification, considering data acquisition and preprocessing, network model design, training process, and validation in a sequential mode. Finally, we identify several areas for future research in order to improve the success of neural networks in image classification.

7.2 Fundamentals of neural networks

This section will discuss neural network architectures with the emphasis upon the MLP networks, along with neural training methods.

7.2.1 Neural network types

There are two fundamentally different types of neural network architectures: feed-forward networks and recurrent networks. The former includes single-layer networks comprising an input layer that projects onto an output layer as well as multilayer networks having at least one hidden layer that allows the networks to extract high-order statistics. A recurrent network distinguishes itself from feed-forward networks by having at least one feedback loop whose presence can greatly affect the training capability and performance (Haykin, 1999).

Considering neural network structures and training paradigms, we can find a large number of different types of neural networks, and some of the most commonly used ones are listed in Table 7.1. Each type has advantages and disadvantages depending upon specific applications. Detailed discussions about these neural network types are given elsewhere (e.g., Bishop, 1995; Jain, Mao and Mohiuddin, 1996; Rojas, 1996; Haykin, 1999; Principe, Euliano and Lefebvre, 2000). Here, we focused on the MLP feed-forward networks due to their technological robustness and overwhelming popularity.

The MLP neural networks are relatively easy to understand and implement. As the workhorse of neural networks, they have been increasingly used in remote sensing (cf. Mas and Flores, 2008). They comprise distributed neurons and weighted links (Fig 7.1). Arranged in an input-hidden-output layered structure, each neuron contains a simple processing function (i.e., activation

TABLE 7.1 List of some commonly-used neural network types in machine learning.

No	Network types	Brief description[a]
1	Multilayer feed-forward perceptron networks	They comprise an input-hidden-output layered structure, and the input signal propagates through the network in a forward direction. They are the most widely used networks.
2	Radial basis function networks	They use radial basis functions to replace the sigmoidal hidden layer transfer function in multilayer perceptrons, thus transferring the design of a neural network as a curve-fitting problem in a high-dimensional space.
3	Self-organizing networks	They use a supervised or an unsupervised learning method to transform an input signal pattern of arbitrary dimension into a lower dimensional (usually one or two dimensional) space with topological information preserved as much as possible.
4	Adaptive resonance theory (ART) networks	They provide the functionality for creating and using a supervised or un unsupervised neural network based on the Adaptive Resonance Theory.
5	Recurrent network (e.g., Hopfield network)	Contrary to feed-forward networks, recurrent neural networks use bi-directional data flow and propagate data from later processing stages to earlier stages.
6	Modular neural networks (e.g., Committee of machine)	They use several small networks that cooperate or compete to solve problems.
7	Stochastic neural networks (e.g., Boltzmann machine)	This type of networks introduces random variations, often viewed as a form of statistical sampling, into the networks.
8	Dynamic neural networks	They not only deal with non-linear multivariate behavior, but also include learning of time-dependent behavior.
9	Neuro-fuzzy networks	They are a fuzzy inference system in the body which introduces the processes such as fuzzification, inference, aggregation and defuzzification into a neural network.

[a] Detailed discussions on these neural network types are given elsewhere (e.g., Bishop,1995; Haykin, 1999; Duda, Hart and Stork, 2001).

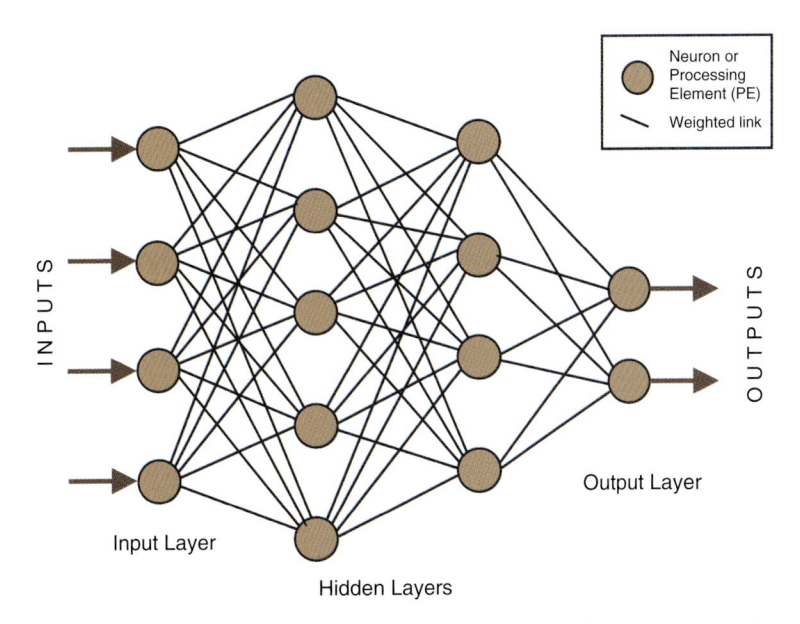

FIGURE 7.1 A fully-connected multilayer perceptron (MLP) neutral network with a 4 × 5 × 4 × 2 structure. This is a feed-forward architecture. Data flow starts from the neurons in the input layer and moves along weighted links to neurons in the hidden layers for processing. Each hidden or output neuron contains a linear discriminant function that combines information from all neurons in the preceding layers. The output layer is a complex function of inputs and internal network transformations.

function) that individually handles pieces of complex problems; the weighted links between neurons determine the direction of data flow and the contribution of the "from" neuron to the "to" neuron. These weights can be determined through an iterative training process that learns from known samples and adjusts the weights between neurons until the minimum error of the performance function is achieved. While the MLP structure and the concept of back-propagation algorithm are relatively simple, network topology and training parameter settings can complicate the overall performance of neural networks for image classification (see Foody and Arora, 1997; Kavzoglu and Mather, 2003; Mas and Flores, 2008).

7.2.2 Network topology

The topology of a neural network is critical for neural computing to solve problems with reasonable training time and satisfactory performance. For the MLP networks, the topology is determined by the number of hidden layers, neurons and connections, and the type of activation function.

7.2.2.1 Number of hidden layers, neurons and connections

The complexity of neural networks is largely defined by the total number of input features, the number of output classes, and the number of hidden layers and neurons (Duda, Hart and Stork, 2001). While the first two factors are less flexible for a particular classification problem, we can adjust the last two to find an appropriate size of neural networks. A trade-off is needed to balance the processing purpose of the hidden layers and the training time needed. Large neural networks with more hidden layers may be more effective to represent non-linear complex relationships, but usually demand more training samples and longer training time. And larger neural networks are more likely to get caught in undesirable local minima or overly fit the training data. On the other hand, compact neural networks may overly simplify the phenomena and lead to unsatisfactory results although they may be easier to train. Several prior studies investigated the issue of the optimal number of hidden layers, but their recommendations are not consistent. For example, Shupe and Marsh (2004) recommended single-hidden-layer networks while Civco (1993) preferred two-hidden-layer networks. Kanellopoulos and Wilkinson (1997) suggested that single-hidden-layer networks are suitable for most classification problems but two-hidden-layer networks may be more appropriate for those applications with more than 20 output classes. Some other researchers concluded that the number of hidden layers may not have a significant influence upon classification accuracies (e.g., Foody and Arora, 1997; Kavzoglu and Mather, 2003).

The number of neurons for the input, hidden, and output layers determines the number of weighted links, particularly for a fully connected network. Since the weight for each link is determined through neural training, more links tend to increase the training time. Thus, every effort should be made to minimize unnecessary neurons in order to improve the efficiency in neural training. Various feature extraction techniques, such as principal component analysis and discriminant analysis, have been used to reduce the data dimensionality and hence the number of neurons in the input layer (e.g., Benediktsson and Sveinsson, 1997; Liu and Lathrop, 2002). While the number of output neurons can be defined according to the research objective in a specific application, a challenging issue is to choose the number of neurons in hidden layers. If there are too few neurons in hidden layers, the network may be unable to approximate very complex functions because of insufficient degrees of freedom. On the other hand, if there are too many neurons, the network tends to have a large number of degrees of freedom which may lead to overtraining and hence poor performance in generalization (Rojas, 1996). Thus, it is crucial to find the 'optimum' number of neurons in hidden layers that adequately capture the relationship in the training data. This optimization can be achieved by using trial and error or several systematic approaches such as pruning and constructive algorithms (Reed, 1993).

7.2.2.2 Activation function

Activation function is an algorithm that transforms the weighted sum of inputs and produces outputs of a neuron. Duda, Hart and Stork (2001) suggested a non-linear activation function should be used to deal with non-linear relationships; otherwise, neural networks would provide no computational power above linear classifiers. The commonly-used non-linear activation functions include the logistic sigmoid (log-sig) function (Equation 7.1) and tangent sigmoid (tan-sig) function (Equation 7.2).

$$f_1(x) = \frac{1}{1 + e^{-(1-a)x}} \quad (7.1)$$

$$f_2(x) = \frac{e^{(1-a)x} - e^{-(1-a)x}}{e^{(1-a)x} + e^{-(1-a)x}} \quad (7.2)$$

where $f_1(x)$ refers to the log-sig activation function; $f_2(x)$ is the tan-sig activation function; x refers to the input data; and a is the user-defined training threshold.

Several researchers examined different activation functions and concluded that the tan-sig function performed better (e.g., Ozkan and Erbek, 2002; Shupe and Marsh, 2004). However, little research has been conducted to investigate how the training threshold could affect the performance of different activation functions. Based on Equations 7.1 and 7.2, the training threshold determines the size of the contribution of the input data to the output of the neuron. That is, it defines the slope of activation functions in their midrange (Fig. 7.2). Therefore, the same type of activation function with various training threshold values may perform differently in image classification.

7.2.3 Neural training

Training is a learning process by which the connection weights are adjusted until the network is deemed to be optimal. This involves the use of training samples, an error measure, and a learning algorithm. Training samples are presented to the network with input and output data over many iterations; they should not only be large in size but also be representative to ensure sufficient generalization ability. There are several different error measures, such as the mean squared error (MSE), the mean squared relative error (MSRE), the coefficient of efficiency (CE), and the coefficient of determination (r^2). The MSE has been

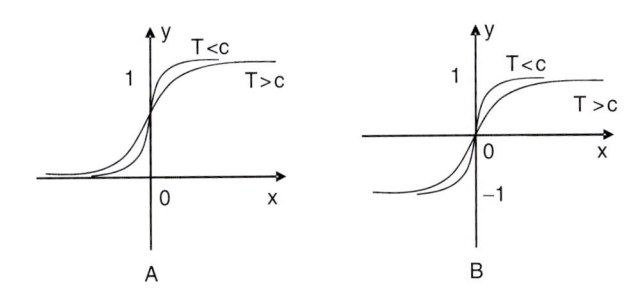

FIGURE 7.2 Curves of activation functions as related to different threshold (T) values, given that c is any constant between 0 and 1. (A) log-sig function and (B) tan-sig function.

most commonly used (Dawson and Wilby, 2001). There is a rich pool of training algorithms varying by their origins, local or global perspective, or learning mode. They are rooted in various techniques, such as optimum filtering, neurobiology, or statistical mechanics. Local learning algorithms adjust weights by using localized input signals and localized derivative of the error function, while global algorithms consider all input signals. Based on the learning mode, training algorithms can be grouped into either a supervised, unsupervised, or hybrid paradigm. Reviews on these algorithms are given elsewhere (e.g., Bishop, 1995; Jain, Mao and Mohiuddin, 1996; Haykin, 1999; Principe, Euliano and Lefebvre, 2000). Here, we focus on three groups of supervised training methods that have been commonly used to train the MLP neural networks: back-propagation, conjugate gradient, and quasi-Newton methods; they differ in the ways by which the direction and magnitude of weight adjustments are calculated (Table 7.2).

7.2.3.1 Back-propagation method

Due to its transparency and effectiveness, the back-propagation method has been widely used in neural network training (Duda, Hart and Stork, 2001). It adjusts the weights of networks through iterations based upon training samples and their desired output. On each iteration, the derivatives of the classification error as functions of the weights are computed to determine the direction and magnitude of the weight update. Through these iterations, the weights of networks are gradually optimized. Several commonly used back-propagation training algorithms include the steepest gradient descent (SGD), the gradient descent with momentum (GDM), and the resilient propagation (RP) algorithms.

The SGD algorithm is the simplest back-propagation algorithm, which determines the direction and magnitude of the weights by the derivatives of the classification error with respect to any weight. It incorporates a user-defined learning rate. It is difficult to choose an appropriate learning rate. If the learning rate is set too large, the algorithm may end up with oscillations around the minimum error; otherwise, the training may be inefficient. In addition, a fixed learning rate causes small changes in the weights even though the weights are far from their optimal values. Consequently, this algorithm usually takes more iterations to converge.

The GDM algorithm is an advanced gradient descent approach adding a momentum item in the weight adjustment formula of the SGD algorithm to improve training efficiency. Momentum is a variable used to reduce the training time and avoid oscillations around the minimum. While the SGD algorithm adjusts the weights along the steepest gradient descent of the performance function, the GDM algorithm concerns both the current gradient descent and the recent changes of the weights. Therefore, the GDM algorithm can use a larger learning rate to speed up the training process with lower risk of oscillations.

The SGD and GDM algorithms consider both the magnitude and the sign of the gradient of the performance function. These two algorithms usually take a long time to converge. The resilient propagation (RP) algorithm proposed by Riedmiller and Braun (1993) attempts to speed up the training process by eliminating the negative impacts of the magnitudes of the partial derivatives. It uses the sign of the derivatives of the classification error to determine the direction of the weight update. The magnitude of the weight update is defined by an individual variable which increases when the derivatives for two successive iterations have the same sign, otherwise decreases. Generally, the RP algorithm is much faster than other back-propagation training algorithms.

7.2.3.2 Conjugate gradient method

The basic back-propagation method adjusts the weights along the steepest descent direction. However, this can be quite time-consuming due to the fixed learning rate. Rather than using the magnitude of the gradient descent and a user-defined learning rate, the conjugate gradient method employs a series of line searches along conjugate directions to determine the optimal step size of the weight update, thus allowing fast convergence. Specifically, the conjugate gradient training comprises two steps. The first step is to search the local minimum in error along a certain search direction; the weights will be adjusted to the local minimum point. The second step is to compute the conjugate of the previous search direction as the new search direction; the global minimum error is approached through iterations. This group of training method includes several commonly used algorithms, such as Fletcher–Reeves (CGF), Polak–Ribiere (CGP), Powell–Beale (CGB), and scaled conjugate gradient (SCG) algorithms (see Table 7.2).

The CGF algorithm calculates the mutually conjugate directions of search with respect to the Hessian matrix directly from the function evaluation and the gradient evaluation, but without the direct evaluation of the Hessian of the function. It computes the coefficient (beta) as a ratio of the norm squared of the current gradient to the norm squared of the previous gradient. The CGP algorithm is similar to the CGF algorithm, differing only in the coefficient (beta) computation. It computes beta as the inner product of the previous change in the gradient with the current gradient divided by the norm squared of the previous gradient. Using the same learning model used by the CGF algorithm to compute the conjugate direction, the CGB algorithm, however, resets the search direction to the negative of the gradient if the error function is non-quadratic. Finally, the SCG algorithm is a variation of the conjugate gradient method with a scaled step size. It combines the model-trust region approach with the conjugate gradient approach to scale the step size.

TABLE 7.2 List of some training algorithms that are particularly tied with a feedforward network.

Training methods	Training algorithms	Acronym	Brief description
Back-propagation	Steepest gradient descent	SGD	It determines the direction and magnitude of the weights by the derivatives of the classification error.
	Gradient descent with momentum	GDM	This is an advanced gradient descent approach by adding a momentum item in the SGD weight adjustments formula.
	Resilient propagation	RP	It attempts to speed up the training using an adoptive variable to define the magnitude of the weight update while the direction of the weight update is defined by the sign of the derivatives of the classification error.
Conjugate gradient	Fletcher–Reeves	CGF	It calculates the conjugate of the previous search direction using the Fletcher–Reeves algorithm and then employs a line search to determine the optimal size of the weight update.
	Polak–Ribiere	CGP	It is similar to the CGF method, differing only in the conjugate direction computation. It updates the conjugate direction using the Polak-Ribiere algorithm.
	Powell–Beale	CGB	Using the same learning model used by the CGF algorithm to compute the conjugate direction, it, however, resets the search direction to the negative of the gradient.
	Scaled conjugate gradient	SCG	It is a variation of the conjugate gradient method with a scaled step size. It combines the model-trust region approach with the conjugate gradient approach to scale the step size.
Quasi-Newton	BFGS (Broyden, Fletcher, Goldfarb, and Shanno)	BFG	This algorithm approximates the Hessian matrix by a function of the gradient to reduce the computational and storage requirements.
	Levenberg–Marquardt	LM	This method is considered as a combination of gradient descent and the Gauss-Newton method. It locates the minimum of a function that is expressed as the sum of squares of nonlinear functions.

7.2.3.3 Quasi-Newton method

The quasi-Newton method is an improved optimization method developed from Newton's method that uses standard numerical optimization techniques to train neural networks (Demuth, Beale and Hagan, 2008). Although Newton's method can usually converge faster than the conjugate gradient method, there are two major drawbacks. First, it requires computing, storing, and inverting the Hessian matrix, which rapidly increases the computation complexity and requires a large memory space. On the other hand, Newton's method may not converge in non-quadratic error surfaces (Duda, Hart and Stork, 2001). The quasi-Newton method updates an approximate Hessian matrix on each iteration, which is a function of the gradient, instead of the second-order derivatives matrix used by Newton's method. This update substantially reduces the computational complexity. When comparing to other methods, however, the quasi-Newton method is more suitable for small networks with limited number of weights.

The most successful quasi-Newton method is the Broyden, Fletcher, Goldfarb, and Shanno (BFGS) algorithm that approximates the Hessian matrix by a function of the gradient to reduce the computational and storage requirements. Additional, the Levenberg–Marquardt (LM) algorithm is considered as a quasi-Newton method since it also seeks for second-order training speed by using an approximate Hessian matrix. A combination of steepest descent and the Gauss–Newton method, the LM algorithm locates the minimum of a function that is expressed as the sum of squares of non-linear functions.

7.3 Internal parameters and classification accuracy

7.3.1 Experimental design

In this focused study, we assessed a set of topological and training parameters affecting image classification accuracy by the MLP neural networks. Three of the four parameters controlling the MLP network topology were considered, including number of hidden layers, type of activation function, and training threshold; the number of neurons was excluded since prior studies have found that image classification accuracies were less sensitive to this factor (e.g., Gong, Pu and Chen, 1996; Foody and Arora, 1997; Paola and Schowengerdt, 1997; Shupe and Marsh, 2004). The GDM algorithm was used to train the MLP networks, and three related training parameters were considered, namely, learning rate, momentum, and number of iterations. These six internal parameters considered are summarized in Table 7.3.

TABLE 7.3 List of the internal network parameters tested[a].

Parameters	Abbreviation	Description	Value
Number of hidden layers	HL	A key factor controlling the topology of neural networks	$0 \leq HL \leq 4$
Activation function	AF	A linear or non-linear function for processing input data of neurons.	Log-sigmoid or Tan-sigmoid
Training threshold	TT	A user-defined threshold determining the contribution of the input data to the outcome.	$0 \leq TT < 1$
Learning rate	LR	A user-defined parameter defining the step size of the weight update	$0.001 \leq LR \leq 0.3$
Momentum	MO	A user-defined parameter controlling the influence of previous weight update on current weight update	$0 \leq MO < 1$
Number of iterations	IT	A parameter specifying how many times the training algorithm may iterate toward the targeted training goal	$400 \leq IT \leq 2800$

[a]Gradient descent with momentum (GDM) algorithm was used for network training in this focused study.

The entire experiment comprised several major components. First, we carefully constructed and trained a set of MLP neural network models with different topologies and training parameters. Then, we used these models to classify a satellite image into several major land cover categories, and we evaluated the accuracy of each classified map. Based on the classification accuracies, we further analyzed the sensitivity of these algorithm factors. Second, we compared the classification accuracies achieved by using the best neural network model and the Gaussian maximum likelihood (GML) classifier. Finally, we summarized our major findings and recommended several practical guidelines when parameterizing the MLP neural networks for image classification.

7.3.2 Remotely sensed data and land classification scheme

The remote sensor data used was a Landsat Enhanced Thematic Mapper Plus (ETM+) image dated on 9 September 1999, which was georeferenced to the UTM map projection. The image covers the northern Atlanta metropolitan area, Georgia, USA. The landscape in this area is characterized by a mosaic of urban use, agricultural use, and natural lands, making it an excellent site to test the effectiveness of different neural network configurations in image classification. In addition to the ETM+ image, we collected ancillary data through GPS-guided field observations and the use of high-resolution images available from Google Earth.

Based on the remote sensor image and the ancillary data, we designed a mixed Anderson Level I/II land use/cover classification scheme (Anderson et al., 1976) with the following six major classes:

1 High-density urban use: mostly large commercial and industrial buildings, large transportation facilities, and high-density residential areas in the city cores;
2 Low-density urban use: mostly single/multiple family houses, apartment complexes, yards, local roads, and small open spaces;
3 Exposed land: mainly non-impervious areas with sparse vegetation, such as clear-cuts, quarries, barren rock or sand along river/stream beaches;
4 Cropland/grassland: crop fields and pasture as well as cultured grasses (such as golf courses, lawns, city parks);
5 Forest: deciduous, coniferous, and mixed forest land; and
6 Water: streams, rivers, lakes, and reservoirs.

7.3.3 Network configuration and training

We carefully configured 53 MLP neural network models with different internal parameters combinations. For each neural network model, the input neurons comprised seven ETM+ image bands (excluding the thermal band due to the coarse spatial resolution) and the output neurons were six major land use/cover classes. The general rule is that for the six internal parameters, only one parameter is allowed to alter at one time while holding the other unchanged. In this way, the sensitivity of neural classification performance with respect to a specific internal parameter can be assessed. Specifically, to investigate the impact of hidden layer number, five neural network models were constructed with the number of hidden layers ranging from 0 to 4 (Table 7.4, Nos 1–5). Then, twenty neural network models were constructed to address the issues of activation function and training threshold (Table 7.4, Nos 6–25). Both log-sig function and tan-sig function were considered, and each was combined with a set of training threshold values ranging from 0 to 0.9. Finally, 28 neural network models were configured to assess the three training parameters, and the range and step of each training parameter are listed in Table 7.4.

Each of the 53 neural network models was trained with an identical training sample set that contains 250 pixels for each land cover class. The training performance was measured by the root mean square error (RMS). The training goal was set to 0.1 in RMS. The training process stopped when either the maximum number of iterations or the training goal was reached. Most of the neural network models successfully converged except the model with four hidden layers (Table 7.4, No. 5).

TABLE 7.4 List of the 53 neural network models configured with various internal parameter settings and classification accuracies obtained.

No.	HL	AF	TT	LR	MO	IT	Classification Accuracy (%)
1	0	log-sig	0.2	0.01	0.6	1000	79.33
2	1	log-sig	0.2	0.01	0.6	1000	80.00
3	2	log-sig	0.2	0.01	0.6	1000	77.67
4	3	log-sig	0.2	0.01	0.6	1000	66.67
5	4	log-sig	0.2	0.01	0.6	1000	N/A
6	1	log-sig	0	0.01	0.6	1000	82.00
7	1	log-sig	0.1	0.01	0.6	1000	79.67
8	1	log-sig	0.2	0.01	0.6	1000	80.00
9	1	log-sig	0.3	0.01	0.6	1000	80.33
10	1	log-sig	0.4	0.01	0.6	1000	80.00
11	1	log-sig	0.5	0.01	0.6	1000	79.67
12	1	log-sig	0.6	0.01	0.6	1000	79.00
13	1	log-sig	0.7	0.01	0.6	1000	78.33
14	1	log-sig	0.8	0.01	0.6	1000	79.00
15	1	log-sig	0.9	0.01	0.6	1000	72.00
16	1	tan-sig	0	0.01	0.6	1000	34.67
17	1	tan-sig	0.1	0.01	0.6	1000	34.00
18	1	tan-sig	0.2	0.01	0.6	1000	26.67
19	1	tan-sig	0.3	0.01	0.6	1000	44.67
20	1	tan-sig	0.4	0.01	0.6	1000	44.67
21	1	tan-sig	0.5	0.01	0.6	1000	44.00
22	1	tan-sig	0.6	0.01	0.6	1000	40.67
23	1	tan-sig	0.7	0.01	0.6	1000	38.67
24	1	tan-sig	0.8	0.01	0.6	1000	40.33
25	1	tan-sig	0.9	0.01	0.6	1000	35.00
26	1	log-sig	0	0.3	0.6	1000	53.67
27	1	log-sig	0	0.25	0.6	1000	60.00
28	1	log-sig	0	0.2	0.6	1000	75.00
27	1	log-sig	0	0.15	0.6	1000	64.67
30	1	log-sig	0	0.10	0.6	1000	72.33
31	1	log-sig	0	0.05	0.6	1000	75.33
32	1	log-sig	0	0.01	0.6	1000	82.00
33	1	log-sig	0	0.005	0.6	1000	81.33
34	1	log-sig	0	0.001	0.6	1000	75.00
35	1	log-sig	0	0.01	0	1000	81.33
36	1	log-sig	0	0.01	0.1	1000	81.33
37	1	log-sig	0	0.01	0.2	1000	81.33
38	1	log-sig	0	0.01	0.3	1000	82.67
39	1	log-sig	0	0.01	0.4	1000	82.00
40	1	log-sig	0	0.01	0.5	1000	81.33
41	1	log-sig	0	0.01	0.6	1000	82.00

TABLE 7.4 (*continued*).

No.	HL	AF	TT	LR	MO	IT	Classification Accuracy (%)
42	1	log-sig	0	0.01	0.7	1000	83.33
43	1	log-sig	0	0.01	0.8	1000	84.00
44	1	log-sig	0	0.01	0.9	1000	82.33
45	1	log-sig	0	0.01	0.8	400	83.00
46	1	log-sig	0	0.01	0.8	700	83.33
47	1	log-sig	0	0.01	0.8	1000	84.00
48	1	log-sig	0	0.01	0.8	1300	84.67
49	1	log-sig	0	0.01	0.8	1600	84.00
50	1	log-sig	0	0.01	0.8	1900	83.33
51	1	log-sig	0	0.01	0.8	2200	81.67
52	1	log-sig	0	0.01	0.8	2500	81.00
53	1	log-sig	0	0.01	0.8	2800	80.33

Note that model 5 failed to converge during the training phase.
HL, number of hidden layers; AF, activation function; TT, training threshold; LR, learning rate; MO, momentum; and IT, number of iterations.

7.3.4 Image classification and accuracy assessment

Each trained neural network model was used to classify the ETM+ image into the six land cover classes and hence a total of 52 land cover maps were produced with the exception of Model 5 that the training process failed to converge (Table 7.4, No. 5). The classification accuracy of each map was assessed by using the confusion matrix method that is based on the use of a reference dataset (Congalton, 1991). It computes the overall accuracy, user's accuracies, producer's accuracies, and the Kappa statistic through the comparison of the predicted values and the actual values of the reference samples. To correctly perform the accuracy assessment, a reference dataset was collected by using the stratified random sampling scheme with approximately 50 samples for each land cover class. For each of the 52 land use/cover maps, a confusion matrix was created, and the overall classification accuracies were used to evaluate the performance of each neural network model for land cover classification.

For comparison purposes, the GML classifier was also trained with the identical training samples and then used to produce a land use/cover map from the ETM+ image. The classification accuracy was assessed with the same reference samples. We further compared the classification accuracies achieved by using the best neural network model (Table 7.4, No. 48) and the GML classifier.

7.3.5 Interpretation and analysis

7.3.5.1 Hidden layer number and classification accuracy

Based on Table 7.4 (Nos 1–5) and Fig. 7.3(A), it is found that the performance of neural networks was quite sensitive to the hidden layer number, as indicated by the relatively high standard deviation. Among these models, the one with single hidden layer produced the best overall classification accuracy, followed by the model with zero hidden layer, with two hidden layers, and with three hidden layers. This finding concurs with those from Shupe and Marsh (2004) and Kanellopoulos and Wilkinson (1997). Theoretically, neural network models with more hidden layers can deal with more complex problems but require a large sample size to train. When input and output neurons are limited in number and training size is moderate or relatively small, neural network models equipped with more hidden layers can become less effective or even fail to converge in the training phase as they may end with local minima or overly fit the training data. Thus, the selection of an appropriate hidden layer number should consider the complexity of input and output neurons as well as the training sample size.

7.3.5.2 Activation function, training threshold, and classification accuracy

Table 7.4 (Nos 6–25) and Fig. 7.3(B) suggest that the activation function type can greatly affect the performance of neural network models. Clearly, the models equipped with the log-sig function substantially outperformed the ones with the tan-sig function, as indicated by the average overall accuracies. When incorporating training threshold in the comparison, we found that the models with the log-sig function were less sensitive to the training threshold values used when comparing to the ones equipped with the tan-sig function, as indicated by their standard deviations (2.65 vs. 5.79). For the former group of models, the best classification accuracy was obtained when the training threshold value was set as 0. Although the variation of classification accuracies by this group of models was relatively small, a decline trend emerged when the training threshold value was raised to 0.9. This suggests that a smaller training threshold value should be used for this group of models equipped with the log-sig function. For the models with the tan-sig function, the variation of their classification accuracies was larger, and relatively higher accuracies were obtained with moderate training threshold values (0.3–0.5).

7.3.5.3 Training parameters and classification accuracy

Table 7.4 (Nos 26–53) and Fig. 7.3(C, D, and E) show the overall classification accuracies in relation to the three training parameters, namely, learning rate (Fig. 7.3C), momentum (Fig. 7.3D), and number of iterations (Fig. 7.3E). Overall, neural network models were highly sensitive to the learning rate used, as indicated by the large standard deviation. The models with the learning rate ranging from 0.005 to 0.01 yielded higher classification accuracies. As the learning rate increased, the classification accuracy plunged by more than 20%, although another peak did occur when the learning rate increased to 0.2 (Fig. 7.3C). Overall, increasing the momentum value helped boost the classification accuracy (Fig. 7.3D), especially when this value was larger than 0.6; by adjusting the value of momentum, the classification

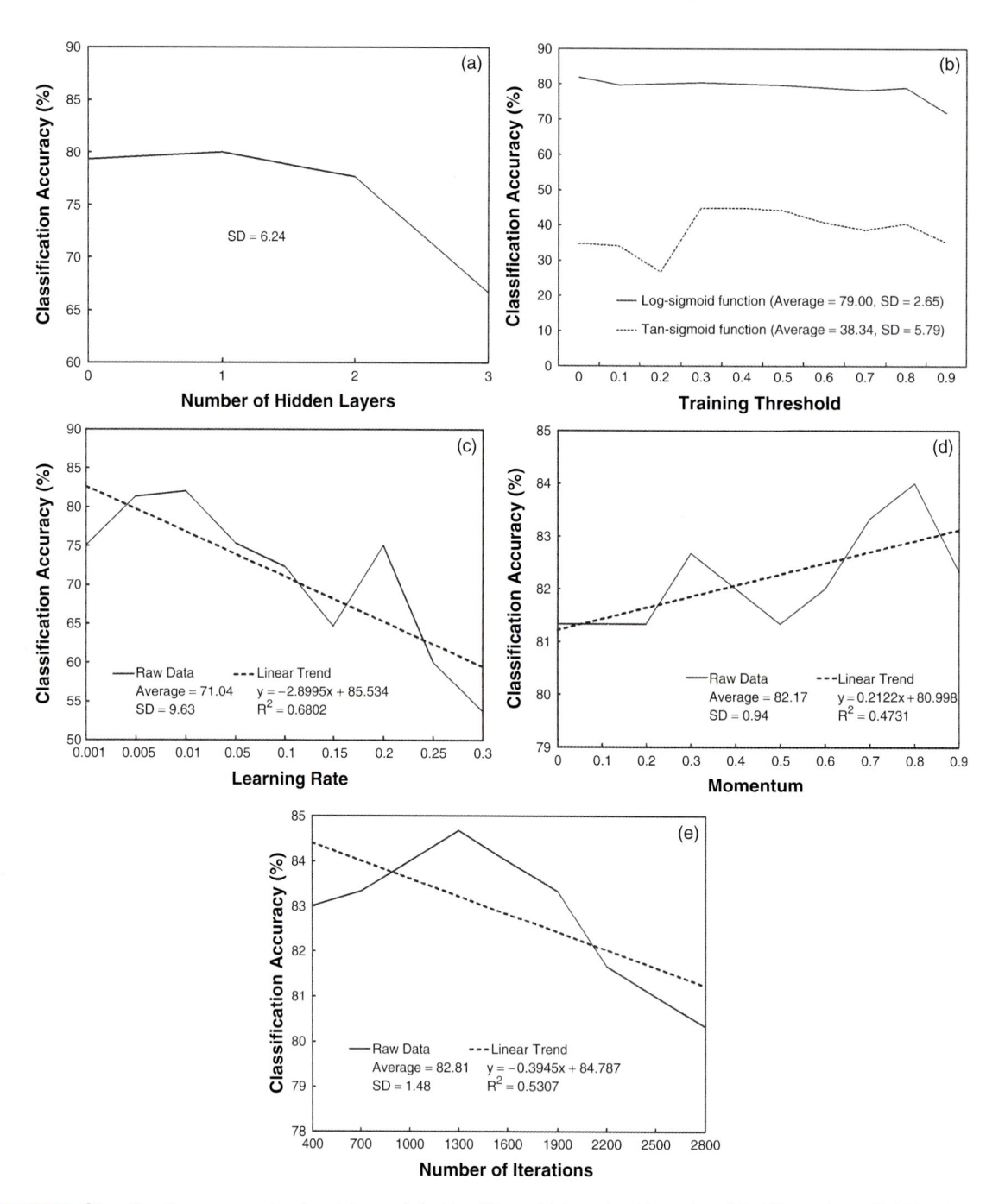

FIGURE 7.3 Classification accuracies (y-axis) as related to different internal settings (x-axis): (A) number of hidden layers, (B) activation function and training threshold, (C) learning rate, (D) momentum, and (E) number of iterations. For each figure, the average accuracy and the standard deviation (SD) are provided, and for C–E, both raw data (solid lines) and linear trends (dash lines) with linear regression equations and R square values are shown.

TABLE 7.5 Comparison of classification accuracies by neural networks (ANNs, Model 48 of Table 7.4) and Gaussian Maximum Likelihood (GML) classifier.

Classifier	Conditional Kappa Coefficients						Overall Kappa Coefficients	Overall accuracy (%)
	High-density urban use	Low-density urban use	Exposed land	Cropland/ grassland	Forest	Water		
ANNs	0.74	0.67	0.80	0.87	0.85	1.00	0.82	84.67
GML	0.61	0.52	0.92	0.92	0.93	1.00	0.77	81.00

accuracy increased from 81.33% to 84% (Table 7.4, Nos 35–44). Nevertheless, the impact of momentum on classification accuracy was quite marginal, as indicated by the relatively small standard deviation. The number of iterations had a moderate impact upon the classification accuracy, as shown by the standard deviation. The overall classification accuracy increased as the number of iterations increased to 1300; after that the accuracy began to decline (Fig. 7.3E).

7.3.5.4 MLP neural networks and GML classifier

Table 7.5 summarizes the classification accuracies by the best neural network model (Table 7.4, No. 48) and the GML classifier. Clearly, the neural network model showed a moderate improvement in the overall classification accuracy and the overall Kappa coefficient when comparing to the outcome by the GML classifier. When looking at the conditional Kappa coefficients for different land use/cover types, however, the neural network model performed much better when classifying the two spectrally complex urban classes, which further confirms the robustness of the neural network technique in dealing with non-linear, complex phenomena. On the other hand, the GML classifier performed better in classifying several spectrally homogenous land use/cover classes, such as forest, cropland, or exposed land, confirming the applicability of this popular parametric classifier.

7.3.6 Summary

In this focused study, we investigated the sensitivity of neural networks to six topological and training parameters. We found that the performance of neural networks was highly sensitive to number of hidden layers, type of activation function, and training rate. And the three other parameters, i.e., training threshold, momentum, and number of iterations, had a marginal impact upon the classification accuracy. A careful neural network configuration can lead to a moderate overall accuracy improvement and a substantial improvement for the two urban classes when comparing to the outcome by the GML classifier. These observations suggest the importance of internal parameter settings when using neural networks for image classification.

On the other hand, several practical guidelines emerged from this study, which can be useful when parameterizing the MLP neural network architecture for image classification. Specifically, a small number of hidden layers should be used when the training sample size is moderate or the number of input and output neurons is small. Using a large number of hidden layers will increase the computational complexity and lead to a suboptimal or unsatisfactory performance when the training sample size or the number of input and output neurons is not large enough. The log-sig function should be used as it can help yield much better classification accuracies and is relatively less sensitive to training threshold values. For better classification accuracies, a small learning rate, large momentum, and moderate number of iterations should be used.

7.4 Training algorithm performance

7.4.1 Experimental design

In this focused study, we evaluated the performance of several popular training algorithms in image classification by the MLP networks. They include SDE, GDM, RP, CGF, CGP, CGB, SCG, BFGS, and LM algorithms (Table 7.2). We used each algorithm to train the MLP networks multiple times using identical training samples, and then applied each of the resultant network models to derive land cover information from the ETM+ image described earlier. The training algorithms were further evaluated according to their training efficiency, capability of convergence, classification accuracy, and stability of the classification accuracy.

7.4.2 Network training and image classification

We constructed a MLP network with seven input neurons, twenty neurons in a single hidden layer, and ten output neurons. The activation functions for the hidden and output neurons were hyperbolic tangent function and logistic sigmoid function, respectively. The input data were the seven ETM+ image bands excluding the thermal band because of its coarse spatial resolution, and the output layer consisted of 10 land use/cover classes or sub-classes. While the land classification scheme used here was the same as the one described earlier, the high-density urban comprised three subclasses (i.e., open space, large roof building, and small roof building in the city core), cropland/grassland included two subclasses (i.e., well-vegetated grassland and less-vegetated land), and forest consisted of two subclasses (i.e., coniferous/mixed forest and deciduous forest). This is why the output layer comprised 10 neurons. For each subclass/class,

TABLE 7.6 Summary of the training algorithm performance.

Training algorithms[a]	Mean iterations	Convergence rate (%)[b]	Classification Accuracy[c] Mean (%)	Classification Accuracy[c] Standard deviation
SGD	65460	100	77.8	2.3
GDM	65432	100	76.1	3.3
RP	805	100	76.4	2.3
CGF	1452	30	61.7	16.6
CGP	865	100	75.1	4.4
CGB	627	50	64.5	17.8
SCG	1486	100	76.1	4.5
BFG	389	0	36.8	17.8
LM	15	100	77.7	2.7

[a] Full names of the training algorithms are given in Table 7.2.
[b] The convergence rate is defined as the ratio between the number of the converged experiments and the total number of the experiments for each training algorithm.
[c] The results are based on 10 experiments.

250 pixels were collected as the training data, and the training performance was measured by the MSE.

Several network training parameters were initiated before actually training. Specifically, the training goal was set at 0.03 in terms of MSE, the training time and iterations were infinite, and both minimum gradient and minimum step were defined as 1×10^{-6}. Note that the learning rate was set as 0.01 for the SGD algorithm or 0.02 for the GDM algorithm. The momentum factor for the GDM algorithm was defined as 0.6. The training process was stopped when the MSE error reached the training goal, indicating that the training successfully converged, or when the minimum gradient or the minimum step was met, showing that the training failed to converge. To minimize the impacts of the initial weights, we used each of the nine training algorithms to train the network ten times with the above training parameters settings. As a result, 90 network models were created, which were further used to classify the ETM+ scene into 10 land cover classes or subclasses which were finally merged into the six major land use/cover classes. In total, 90 land use/cover maps were produced.

7.4.3 Performance evaluation

The performance of each training algorithm was evaluated by using the four criteria, namely, training efficiency, capability of convergence, classification accuracy, and stability of the classification accuracy.

7.4.3.1 Training efficiency

Different training algorithms vary in their computational intensity and the time to reach the training goal. An efficient training algorithm can considerably reduce the time cost for image classification by neural networks. Therefore, the training efficiency has been considered as a critical criterion for examining the usefulness of training algorithms (Skinner and Broughton, 1995). The number of iterations used for training is a good indicator of the efficiency of training algorithms (Kanellopoulos and Wilkinson, 1997; Kisi, 2007). Here, we used the average number of iterations during the ten experiments to quantify the training efficiency (Table 7.6).

From Table 7.6, it is clear that the training efficiency of different algorithms varied greatly. The LM algorithm was the most efficient. Several algorithms, such as RP, CGF, CGP, CGB, SCG, and BFG, showed a moderate training efficiency. Both the SGD and GDM algorithms were extremely poor in terms of the training efficiency.

7.4.3.2 Capability of convergence

The capability of convergence provides the information on how often a training algorithm can reach the training goal. Failure to converge usually leads to a poor classification performance. Therefore, the capability of convergence has been considered as an important criterion for measuring the performance of training algorithms (Skinner and Broughton, 1995; Kanellopoulos and Wilkinson, 1997). Here, the capability of convergence was defined as the rate of convergence, which is actually the ratio between the number of the converged experiments and the total number of the experiments by a specific algorithm (Table 7.6).

All the three back-propagation algorithms (SGD, GDM, and RP) successfully converged in every experiment. Two of the conjugate gradient algorithms, namely, CGP and SCG, and the LM algorithm were also quite good in this regard. However, the other two conjugate gradient algorithms, namely, CGF and CGB, were quite poor in terms of their capability of convergence. The BFG algorithm failed to converge in all experiments, which may be due to the emergence of non-quadratic error surfaces.

7.4.3.3 Classification accuracy

The performance of a pattern classifier is usually assessed by estimating its classification accuracy. Training algorithms resulting in a poor classification accuracy are less useful in practice. Here, we used the confusion matrix method described earlier to measure the classification accuracy. A test dataset with approximately 50 samples randomly selected for each land use/cover subclass/class was used for this purpose. The overall classification accuracies were compared to evaluate the performance of each training algorithm.

Based on Table 7.6, it is clear that the classification accuracy varies by training algorithms. If using the average classification accuracy as the reference, the SGD and LM algorithms clearly performed the best, followed by the GMD, RP, CGP, and SCG algorithms. The BFG, CGB, and CGF algorithms generated relatively lower average classification accuracies, which were mostly caused by their failure to converge in one or more experiments.

7.4.3.4 Stability of the classification accuracy

The stability of the classification accuracy was another criterion that we used to evaluate the performance of a training algorithm. The training methods which result in relatively stable classification accuracies are generally preferred. To evaluate the stability of the classification accuracy, the standard deviation of the classification accuracies from the ten experiments for each algorithm was computed. A smaller standard deviation suggests a more stable performance. Based on Table 7.6, it is found that the three back-propagation algorithms, namely, SGD, GDM, and RP, and the LM algorithm were quite stable, the CGP and SCG algorithms were less stable, and the CGF, CGB and BFG algorithms were least stable. The poor performance of the last group of training algorithms was clearly attributed to their poor capability of convergence.

7.4.3.5 Overall evaluation

When combining the four categories of comparison, it is found that the LM algorithm performed the best. It was the most efficient algorithm with a strong capability of convergence, providing the most accurate and stable land use/cover classification accuracies. However, this algorithm may become less efficient when a large memory space is needed to accommodate the high complexity of computation due to increasing network size. The RP algorithm performed almost identical to the LM algorithm except that the former used more iterations to converge. Since the RP algorithm does not require computing the matrix of the second-order derivatives of its performance function, the networks size increase will have a limited impact upon the performance. With a much lower computational cost and less memory usage, the RP algorithm can outperform the LM algorithm when training large networks with many parameters. The CGP and SCG algorithms also performed well although they showed moderate training efficiency and the resultant classification accuracies were not as stable as those from the LM and RP algorithms. The SGD and GDM algorithms were not competitive when compared to the LM, RP, CGP, and SCG algorithms due to their extremely low training efficiency. Finally, the CGF, CGB, and BFG algorithms were not recommended as they showed a poor capability of convergence which led to poor classification accuracies. The above observations suggest the importance of selecting an appropriate training algorithm when using neural networks for land use/cover classification.

7.5 Toward a systematic approach to image classification by neural networks

While the prospect of neural networks for image classification has been quite promising, the capability of neural networks tends to be oversold as an all-inclusive 'panacea' that is capable to outperform other classifiers regardless of network architecture, training algorithms, or data quality. Consequently, this field has been characterized by inconsistent research designs and immature operational practices. Nevertheless, recent studies suggested the need to adopt a systematic approach to pattern recognition by neural networks considering data acquisition and pre-processing, network configuration, training algorithms, and validation in a sequential mode (e.g., Maier and Dandy, 2000; Principe, Euliano and Lefebvre, 2000; Mas and Flores, 2008; Yang, 2009a, b).

Here, we propose a systematic approach that can guide the use of neural networks for image classification from remote sensor data (Fig. 7.4). It comprises several core technical components, beginning with data collection and acquisition that have been considered as a critical component in any remote sensing-based land use/cover mapping projects (Yang and Lo, 2002). At this stage, both remote sensor data and ancillary data should be collected to help prepare a land use/cover classification scheme, the training and validation data sets, and the primary data actually used in image classification. Every effort should be made to acquire sufficient data that represent the conditions that the neural network may encounter later. The information contents and data quality should be emphasized during the phase of data acquisition (Yang, 2009a).

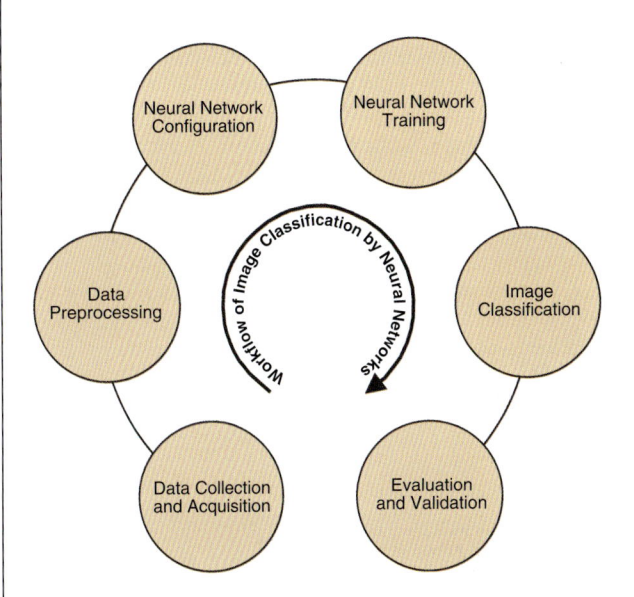

FIGURE 7.4 A systematic approach to image classification by neural networks. While data acquisition and pre-processing are normally quite time consuming, neural network configuration is most challenging. Nevertheless, neural network training can be quite computationally intensive.

Data pre-processing is necessary as it can have a significant effect on neural network performance (Foody, McCulloch and Yates, 1995; Foody and Arora, 1997; Mas and Flores, 2008). There are two major tasks during this phase. First, it is necessary to determine which image bands should be actually included as the input data. If large image scenes or many image bands are being considered, a data dimensionality reduction technique like principal component analysis (e.g., Liu and Lathrop, 2002) should be used to extract salient features prior to the actual classification. This procedure can greatly reduce the computational burden. The second task is to identify the training, test, and validation datasets by using reference data in combination with some image interpretation procedures. The training dataset is for network training, the test dataset is used to assess the performance of the network at the training stage for cross-validation purposes, and the validation dataset is used to evaluate the performance of a network against independent data. Both the test and validation datasets should be much smaller in size when comparing to the training set but each dataset should be representative of the same population. If the available dataset is limited, the division of data may be difficult, and some other methods, such as bootstrapping (Kohavi, 1995) or the hold-out method (Masters, 1995), could be attempted to maximize utilization of the available data.

Prior to training, it is important to define an appropriate neutral network architecture and training parameters. Begin with a multilayer perceptron neural network and a back-propagation learning algorithm as the benchmark to evaluate any other network types and learning methods. Specify an appropriate number of hidden layers and nodes unless a pruning algorithm or cascade correlation is used. Begin with one hidden layer as a starting point. Choose either logistic sigmoid or hyperbolic tangent function as the activation function. Also choose appropriate values for learning parameters. As demonstrated in the two focused studies, a number of trial-and-error experiments may need in order to optimize the network architecture. The initial weights should be randomly chosen. Use the training data set in the training, and the test set for cross-validation in order to determine when to terminate the training process. Neural networks training may adopt a strategy to avoid overtraining by calculating the classification error of a test dataset on each iteration, and once the error goes up, then stop training. In our focused studies, the conditions of stopping training were defined by the training goal and other parameters, such as minimum gradient size. Once the training is completed, save the weights and architecture of the neural model, and classify the image to produce a land use/cover map. The classification performance is assessed by using the independent validation data through the error matrix method described earlier.

Future research directions

There are a few areas where further research is needed. Firstly, there are many arbitrary decisions involved in the construction of a neural network model, and therefore, there is a need to develop guidance that helps identify the circumstances under which particular architectures should be adopted and how to optimize the parameters that control them. For this purpose, more empirical, intermodel comparisons, and rigorous assessment of neural network performance with different inputs, architectures, and internal parameters are needed. Secondly, data preprocessing is an area where more guidance is needed. There are many theoretical assumptions that have not been confirmed by empirical trials. It is not clear how different pre-processing methods could affect the classification outcome. Future investigation is needed to explore the impact of data quality and different methods in data division, data standardization, or data reduction upon the land classification. Lastly, continuing research is needed to develop effective strategies and probing tools for mining the knowledge contained in the connection weights of trained neural network models in image classification. This can help uncover the 'black-box' construction of the neural network, which can help improve the success of neural network applications to image classification.

References

Anderson, J. R., Hardy, E. E., Roach, J. T. and Witmer, R. E. (1976) *A Land Use and Land Cover Classification System for Use with Remote Sensor Data*, USGS Professional Paper 964, Sioux Falls, SD, USA.

Atkinson, P. M. and Tatnall, A. R. L. (1997) Introduction: neural networks in remote sensing. *International Journal of Remote Sensing*, **18**, 699–709.

Benediktsson, J. A. and Sveinsson, J. R. (1997) Feature extraction for multisource data classification with artificial neural networks. *International Journal of Remote Sensing*, **18**, 727–740.

Benediktsson, J. A. and Sveinsson, J. R. (2003) Multisource remote sensing data classification based on consensus and pruning. *IEEE Transactions on Geoscience and Remote Sensing*, **41**, 932–936.

Benediktsson, J. A., Swain, P. H. and Ersoy, O. K. (1990) Neural network approaches versus statistical methods in classification of multisource remote sensing data. *IEEE Transactions on Geosciences and Remote Sensing*, **28**, 540–551.

Bishop, C. M. (1995) *Neural Networks for Pattern Recognition*, Oxford University Press, Oxford, UK.

Bischof, H., Schneider, W. and Pinz, A. J. (1992) Multispectral classification of Landsat-images using neural networks. *IEEE Transactions on Geoscience and Remote Sensing*, **30**, 482–490.

Civco, D. L. (1993) Artificial neural networks for land-cover classification and mapping. *International Journal of Geographical Information Systems*, **7**, 173–186.

Congalton, R. G. (1991) A review of assessing the accuracy of classification of remotely sensed data. *Remote Sensing of Environment*, **37**, 35–46.

Dawson, C. W. and Wilby, R. L. (2001). Hydrological modelling using artificial neural networks. *Progress in Physical Geography*, **25**, 80–108.

Demuth, H., Beale, M. and Hagan, M. (2008) *Neural Network Toolbox^{TM} User's Guide*, The MathWorks, Inc., Natick, MA.

Duda, R. O., Hart, P. E. and Stork, D. G. (2001) *Pattern Classification*, John Wiley & Sons, Inc, New York.

Del Frate, F., Pacifici, F., Schiavon, G. and Solimini, C. (2007) Use of neural networks for automatic classification from high-resolution images. *IEEE Transactions on Geoscience and Remote Sensing*, **45**, 800–809.

Foody, G. M. and Arora, M. K. (1997) An evaluation of some factors affecting the accuracy of classification by an artificial neural network. *International Journal of Remote Sensing*, **18**, 799–810.

Foody, G. M., McCulloch, M. B. and Yates, W. B. (1995) The effect of training set size and composition on artificial neural-network classification. *International Journal of Remote Sensing*, **16**, 1707–1723.

Gong, P., Pu, R. and Chen, J. (1996) Mapping ecological land systems and classification uncertainties from digital elevation and forest-cover data using neural networks. *Photogrammetric Engineering and Remote Sensing*, **62**, 1249–1260.

Gopal, S. and Woodcock, C. (1996) Remote sensing of forest change using artificial neural networks. *IEEE Transactions on Geoscience and Remote Sensing*, **34**, 398–404.

Haykin, S. (1999) *Neural Networks: A Comprehensive Foundations*, 2nd edition, Prentice Hall, Upper Saddle River, NJ.

Jain, A. K., Mao, J. and Mohiuddin, K. M. (1996) Artificial neural networks: a tutorial. *Computer*, **29**, 31–44.

Jarvis, C. H. and Stuart, N. (1996) The sensitivity of a neural network for classifying remotely sensed imagery. *Computers & Geosciences*, **22**, 959–967.

Ji, C. Y. (2000) Land-use classification of remotely sensed data using Kohonen Self-Organizing Feature Map neural networks. *Photogrammetric Engineering and Remote Sensing*, **66**, 1451–1460.

Kanellopoulos, I. and Wilkinson, G. G. (1997) Strategies and best practice for neural network image classification. *International Journal of Remote Sensing*, **18**, 711–725.

Kavzoglu, T. and Mather, P. M. (2003) The use of backpropagating artificial neural networks in land cover classification. *International Journal of Remote Sensing*, **24**, 4907–4938.

Kisi, Ö. (2007) Streamflow forecasting using different artificial neural network algorithms. *Journal of Hydrologic Engineering*, **12**, 532–539.

Kohavi, R. (1995) A study of cross-validation and bootstrap for accuracy estimation and model selection, in *1995 International Joint Conference on Artificial Intelligence*. Available http://robotics.stanford.edu/%7Eronnyk/accEst.pdf accessed 15 June 2010.

Korczak, J. and Hammadimesmoudi, F. (1994) A way to improve an architecture of neural-network classifier for remote-sensing applications. *Neural Processing Letters*, **1**, 13–16.

Liu, X. and Lathrop, R. G. (2002) Urban change detection based on an artificial neural network. *International Journal of Remote Sensing*, **23**, 2513–2518.

Maier, H. R. and Dandy, G. C. (2000) Neural networks for the prediction and forecasting of water resources variables: a review of modeling issues and applications. *Environmental Modelling & Software*, **15**, 101–124.

Mannan, B., Roy, J. and Ray, A. K. (1998) Fuzzy ARTMAP supervised classification of multi-spectral remotely-sensed images. *International Journal of Remote Sensing*, **19**, 767–774.

Mas, J. F. and Flores, J. J. (2008) The application of artificial neural networks to the analysis of remotely sensed data. *International Journal of Remote Sensing*, **29**, 617–663.

Masters, T. (1995) *Advanced Algorithms for Neural Networks: A C++ Sourcebook*, John Wiley & Sons, Inc., New York.

McCulloch, W. S. and Pitts, W. (1943) A logical calculus of the ideas immanent in nervous activity. *Bulletin of Mathematical Biophysics*, **5**, 115–133.

Ozkan, C. and Erbek, F. S. (2002) The comparison of activation functions for multispectral Landsat TM image classification. *Photogrammetric Engineering and Remote Sensing*, **69**, 1225–1234.

Paola, J. D. and Schowengerdt, R. A. (1995a) A detailed comparison of the backpropagation neural network and maximum-likelihood classifiers for urban land use classification. *IEEE Transactions on Geosciences and Remote Sensing*, **33**, 981–996.

Paola, J. D. and Schowengerdt, R. A. (1995b) A review and analysis of backpropagation neural networks for classification of remotely-sensed multi-spectral imagery. *International Journal of Remote Sensing*, **16**, 3033–3058.

Paola, J. D. and Schowengerdt, R. A. (1997) The effect of neural-network structure on a multispectral land-use/land-cover classification. *Photogrammetric Engineering and Remote Sensing*, **63**, 535–544.

Petropoulos, G. P., Vadrevu, K. P., Xanthopoulos, G., Karantounias, G. and Scholze, M. (2010) A comparison of spectral angle mapper and artificial neural network classifiers combined with Landsat TM imagery analysis for obtaining burnt area mapping. *Sensors*, **10**, 1967–1985.

Principe, J. C., Euliano, N. R. and Lefebvre, W. C. (2000) *Neural and Adaptive Systems: Fundamentals Through Simulations*, John Wiley & Sons, Inc., New York.

Reed, R. (1993) Pruning algorithms – a survey, *IEEE Transactions on Neural Networks*, **4**, 740–747.

Riedmiller, M. and Braun, H. (1993) A direct adaptive method for faster backpropagation learning: the RPROP algorithm, in *Proceedings of the IEEE International Conference on Neural Networks*, 28 March–1 April 1993, San Francisco, CA, pp. 586–591.

Rojas, R. (1996) *Neural Networks: A Systematic Introduction*, Springer-Verlag, Berlin.

Serpico, S. B., Bruzzone, L. and Roli, F. (1996) An experimental comparison of neural and statistical non-parametric algorithms for supervised classification of remote sensing images. *Pattern Recognition Letters*, **17**, 1331–1341.

Seto, K. C. and Liu, W. G. (2003) Comparing ARTMAP neural network with the maximum-likelihood classifier for detecting urban change. *Photogrammetric Engineering and Remote Sensing*, **69**, 981–990.

Shupe, S. M. and Marsh, S. E. (2004) Cover- and density-based vegetation classifications of the Sonoran desert using Landsat TM and ERS-1 SAR imagery. *Remote Sensing of Environment*, **93**, 131–149.

Skinner, A. J. and Broughton, J. Q. (1995) Neural networks in computational material science: training algorithms. *Modelling and Simulation in Materials Science and Engineering*, **3**, 371–390.

Stathakis, D. (2009) How many hidden layers and nodes, *International Journal of Remote Sensing*, **30**, 2133–2147.

Yang, X. (2009a) Artificial neural networks. In *Handbook of Research on Geoinformatics* (ed H. Karimi), IGI Global, Philadelphia, pp. 129–136.

Yang, X. (2009b) Artificial neural networks for urban modeling. In *Manual of Geographic Information Systems* (ed. M. Madden), American Society for Photogrammetry and Remote Sensing, MD, pp. 647–658.

Yang, X. and Lo, C. P. (2002) Using a time series of normalized satellite imagery to detect land use/cover change in the Atlanta, Georgia metropolitan area. *International Journal of Remote Sensing*, **23**, 1775–1798.

Yoshida, T. and Omatu, S. (1994) Neural-network approach to land-cover mapping. *IEEE Transactions on Geoscience and Remote Sensing*, **32**, 1103–1109.

Zhou, L. and Yang, X. (2010) Training algorithm performance for image classification by neural networks. *Photogrammetric Engineering and Remote Sensing*, **76**, 945–951.

Zhuang, X., Engel, B. A., Lozanogarcia, D. F., Fernandez, R. N. and Johannsen, C. J. (1994) Optimization of training data required for neuro-classification. *International Journal of Remote Sensing*, **15**, 3271–3277.

8

Characterizing urban subpixel composition using spectral mixture analysis

Rebecca Powell

Deriving accurate measures of urban land cover from remote sensing data is fundamentally a challenging endeavor due to the high spectral and spatial heterogeneity of urban areas. One strategy to address spatial variability is mapping sub-pixel components of land cover using spectral mixture analysis (SMA), which models each pixel as the linear sum of spectrally "pure" endmembers. Spectral variability can be addressed by multiple endmember spectral analysis (MESMA), which allows the number and type of endmembers to vary on a per-pixel basis. The high spectral diversity of urban materials can be generalized in terms of three fundamental components – vegetation (V), impervious surfaces (I), and bare soil (S) – facilitating comparisons of urban landscapes across regions and through time. This chapter presents an overview of the SMA technique to characterize urban landscapes, followed by two case studies that highlight the flexibility of MESMA to map V-I-S components. The first study characterizes the morphology of settlements along the "arc of deforestation" in the Brazilian Amazon, leveraging assumptions about the unidirectional trajectory of urbanizing land cover to improve accuracy of fraction estimates. The second study quantifies urban tree and grass cover in a Western US city, integrating phenological information to discriminate dominant vegetation types.

8.1 Introduction

Deriving meaningful, quantitative measures to characterize urban land cover remains a challenge in remote sensing applications because urban areas are complex landscapes of built structures and human-modified land cover (Forster, 1983; Zipperer, et al., 2000). Globally, urban land cover varies in terms of density of habitation, structure of buildings, type of construction materials, abundance of vegetation, and size of open spaces, among other factors. As a result, there is no globally consistent spectral signature for urban land cover, and there is no straightforward means of comparing urban environments from different regions (Ridd, 1995; Small, 2005). Even within a single urban area, image analysis is complicated because of the high spectral and spatial variability of built-up materials which results in a large number of "mixed pixels" across a range of spatial scales and limits the utility of traditional classification approaches (Forster, 1983; Ridd, 1995; Small, 2001; Franke et al., 2009; Myint and Okin, 2009; Wu, 2009).

In order to compare urban systems, whether between countries, between cities within the same country, or between the same city at different time periods, standardized units of measurement and descriptive parameters are needed. One strategy is to focus on characterizing the bio-physical composition of urban land cover, because measures based on land cover do not depend on human interpretation, nor on the economic, historic, or cultural development of the city (Whyte, 1985; Ridd, 1995). Additionally, the problem of mixed pixels can be addressed by characterizing the landscape in terms of continuous variables rather than assigning discrete, mutually exclusive classes (DeFries et al., 1999; Ji and Jensen, 1999; Clapham Jr., 2003; Small, 2005; Weng and Lu, 2009). Modeling each pixel as the percent cover of basic urban materials preserves the heterogeneity of urban land cover (Clapham Jr., 2003; Hansen, et al., 2002) and captures more detail than the minimum resolution imposed by the pixel.

One conceptual framework that addresses these issues is the V-I-S model of urban ecosystem analysis (Ridd, 1995), which decomposes the urban landscape as a combination of three fundamental components, in addition to water: vegetation (V), impervious surfaces (I), and bare soil (S). Spectral mixture analysis (SMA) is a technique to derive the sub-pixel abundance of each land-cover component, thereby characterizing the urban landscape as continuous surfaces of V-I-S components. This strategy has several advantages. First, the V-I-S model does not depend on land-use classes that may be subjective or regionally and temporally specific (e.g., Anderson et al., 1976; Ridd, 1995; Small, 2005). Fractions derived from SMA represent a physical model of the landscape (Adams and Gillespie, 2006), and such continuous descriptions of urban land cover are more readily compared with other datasets, incorporated in environmental models, and scaled-up in a physically meaningful manner (Jensen, 1983; Carlson and Sanchez-Azofeifa, 1999; DeFries et al., 1999; Hansen et al., 2002; Clapham Jr., 2003; Lepers, et al., 2005). Finally, urban extent can be arbitrarily specified by assigning a threshold value for the impervious component without loss of information (Ridd, 1995).

This chapter presents an overview of the SMA technique to characterize the urban physical environment in terms of V-I-S components. The analysis involves compiling a regionally specific library of the spectra that best represent the diversity of urban materials. The spectral response of each pixel in the urban landscape is "unmixed" to determine the fraction of each material component present. The number and type of materials can vary on a per-pixel basis to accommodate the high spectral and spatial variability of urban land cover. Sub-pixel fractions of specific materials are grouped into generalized V-I-S classes, and continuous maps of each component are generated. Accuracy is assessed in terms of agreement between modeled fractions and reference fractions measured in the field or from finer spatial resolution imagery.

The overview of SMA methodology is followed by a summary of two case studies that demonstrate the flexibility of MESMA in mapping V-I-S components in very different environments. In both studies, SMA is applied to map the fundamental spectral components of the urban environment generalized in terms of Ridd's V-I-S model. Additionally, each of these studies leverages temporal information relevant to the mapping objectives to guide model development and implementation. The first case study analyzes the evolution of urban morphology in settlements located along the "Arc of Deforestation" in the Brazilian Amazon (Fig. 8.1a). Assumptions about the unidirectional trajectory of urban land cover as a function of time informs spectral library construction and model selection. Additionally, this study illustrates for the first time that the same spectral library and model selection rules can be applied across a region and through time. The second case study quantifies the abundance of the two dominant urban vegetation types (i.e., trees and lawn) in the western US city of Denver, Colorado, (Fig. 8.1b) and demonstrates that phonological information (i.e., differences in the timing of "green-up") can complement spectral information in order to discriminate broad vegetation types. Together, these case studies highlight a major strength of the SMA approach: that model implementation can be customized to optimally capture the spatial and spectral variations of land cover specific to a particular region and that temporal information can be integrated in the analysis to reduce spectral ambiguities.

8.2 Overview of SMA implementation

8.2.1 SMA background

SMA is a method that explicitly accounts for mixed pixels and derives information at the subpixel scale (Adams et al., 1986; Small, 2001; Powell et al., 2007; Franke et al., 2009). SMA assumes that (a) the landscape can be modeled as mixtures of a few basic spectral components, known as "endmembers," and (b) the measured spectrum of each pixel can be modeled as a linear combination of endmember spectra, weighted by the fraction of each endmember within the pixel's instantaneous field of view (IFOV) (Adams et al., 1986; Roberts et al., 1998a; Song, 2005; Powell et al., 2007). The objective of SMA is to model per-pixel abundance of the "pure" components of the scene.

In virtually any urban environment, the signal recorded by a sensor will include reflectance from multiple land-cover components. For example, 30-m Landsat pixels overlaid on high-resolution imagery from an urbanizing landscape in Rondônia, Brazil, consist of mixtures of green vegetation, a variety of impervious surface materials, and soil (Fig. 8.2a). The response

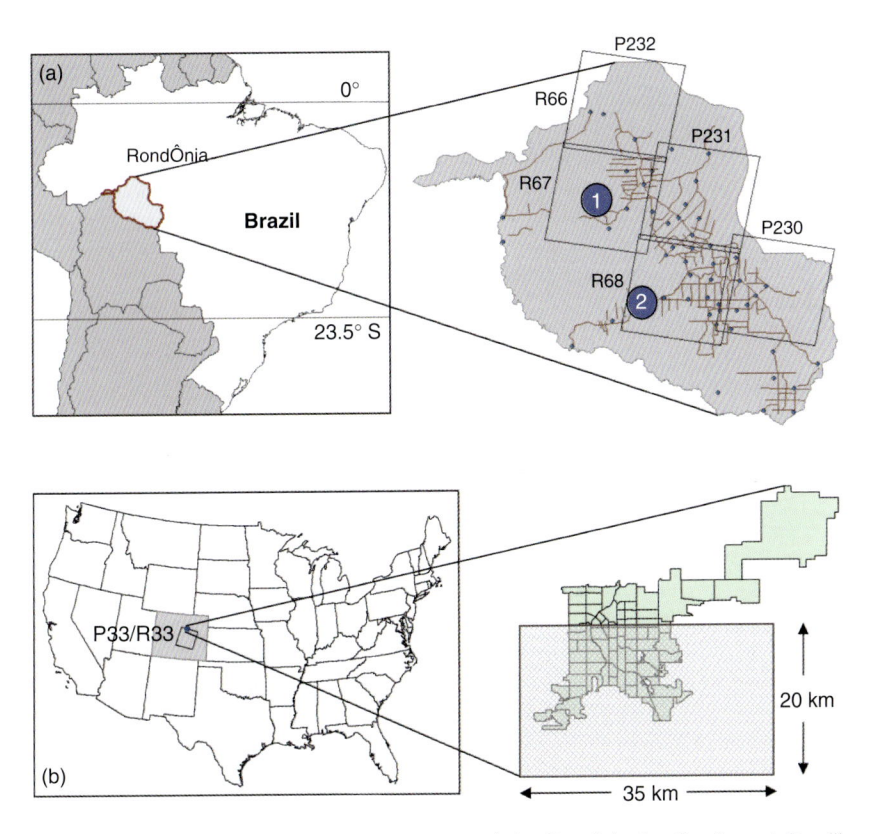

CHAPTER 8 CHARACTERIZING URBAN SUBPIXEL COMPOSITION USING SPECTRAL MIXTURE ANALYSIS 113

FIGURE 8.1 Location of two case studies. (a) Settlements in Rondônia, Brazil, in the Southwest Brazilian Amazon (adapted from Powell and Roberts, 2008), two are highlighted in this chapter; and (b) the southern portion of Denver, Colorado.

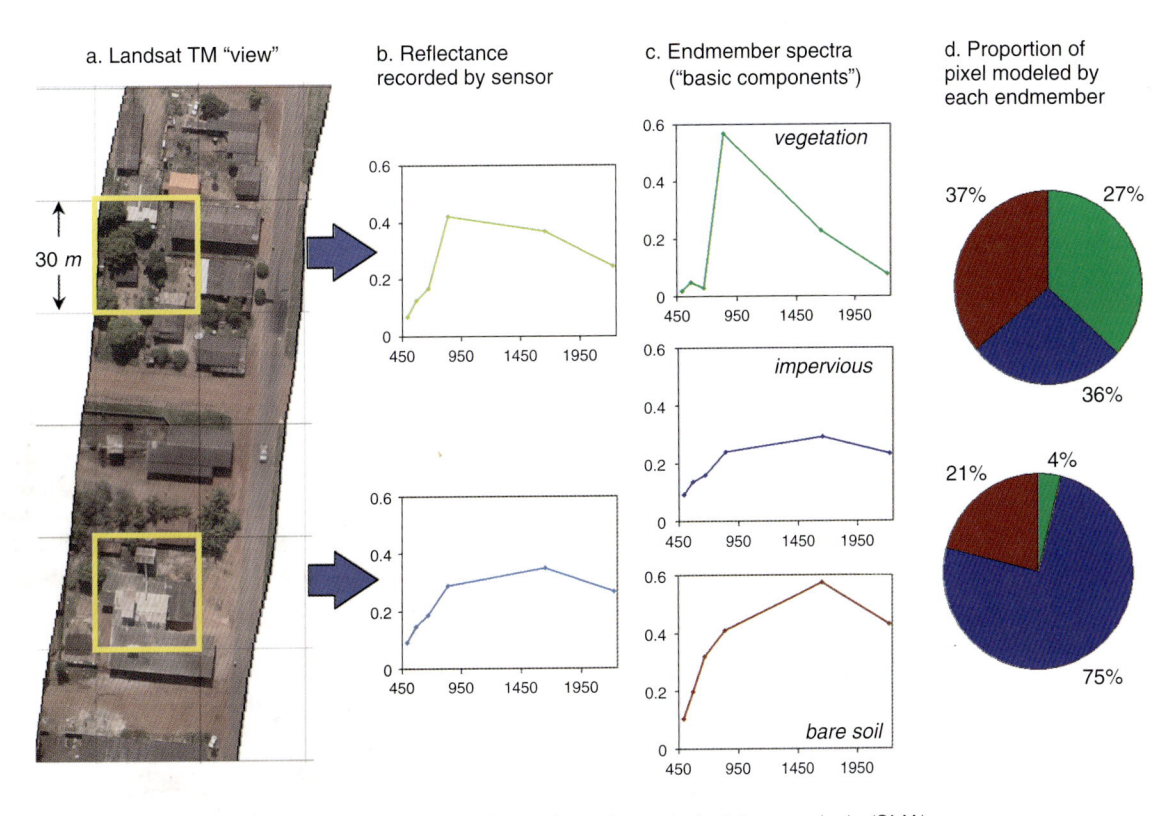

FIGURE 8.2 Conceptual overview of spectral mixture analysis (SMA).

recorded by the sensor for each pixel (Fig. 8.2b) is the weighted sum of the pure spectra of each material in the pixel's IFOV (Fig. 8.2c). The goal is SMA is to "unmix" the relative proportion of endmembers that best model the measured spectrum; in general, these are assumed to be correlated with the area covered by each material present (Fig. 8.2d). The output of SMA is a set of images representing the fractional cover of each endmember, with digital numbers (DN) values typically scaled between zero and 1 (i.e., zero representing "not present" and 1 representing 100% cover), in addition to an image that summarizes the difference between the modeled and measured spectrum (Adams, et al., 1986), usually in terms of the root mean square (RMS) error.

For a given pixel, SMA can be described mathematically as follows (Adams *et al.*, 1986; Smith *et al.*, 1990; Roberts *et al.*, 1998a):

$$DN_i = \sum_{j=1}^{K} F_j \cdot DN_{i,j} + e_i \qquad (8.1)$$

where DN_i is the measured value of a pixel in band i in DNs recorded by the sensor or in units of radiance or reflectance, F_j is the fraction of endmember j present in the pixel's IFOV, DN_{ij} is the value of the endmember j in band i, and e_i is the residual, or the difference between observed and modeled DNs for band i. There are N bands in the dataset, and K endmembers in the mixture model. In addition, the mixing equation is subject to the following constraint:

$$\sum_{j=1}^{K} F_j = 1, \qquad (8.2)$$

i.e., the fraction of endmembers for each pixel must sum to 1 (or 100% cover). Per-pixel RMS error is effectively the mean residual across all bands, given by:

$$RMSE = \sqrt{\frac{\sum_{i=1}^{N} e_i^2}{N}}. \qquad (8.3)$$

SMA models consist of a set of endmembers that represent the basic spectral components of the landscape, in addition to an endmember that represents shade to account for variations in brightness due to illumination, topography, or other sources of surface variability (Adams *et al.*, 1986; Smith *et al.*, 1990; Dennison *et al.*, 2004). If pixel values in an image are recorded in units of reflectance, the shade endmember will have values of zero in all bands; for an image recorded in units of radiance or raw DNs, the shade endmember will consist of non-zero values related to atmospheric scattering.

Given an image that consists of N bands, the goal is to simultaneously solve a total of $N + 1$ linear equations (Equations 8.1 and 8.2 above) for each pixel, thereby estimating the fractions of each endmember present (F_j), while minimizing the residuals (e_i) on a per-pixel basis. For a given application, the validity of the SMA model is assessed based on (a) whether the estimated fractions are physically realistic, i.e., fraction values are between 0 and 1, (b) the overall goodness of fit, usually measured by RMS error, and (c) the spatial distribution of model error throughout the scene. Based on such assessments of model fit, model parameters may be adjusted until the results are satisfactory (Roberts *et al.*, 1998a).

Despite the strengths of SMA to estimate sub-pixel components, the implementation of SMA is based on several simplifications that limit its applicability in urban environments. First, all pixels in the scene are modeled with an invariant set of endmembers, usually between three and five (Powell *et al.*, 2007; Franke *et al.* 2009; Myint and Okin, 2009). As a result, the selected endmembers may not effectively model all elements in the scene or a pixel may be modeled with endmembers that are not present in its IFOV. Either situation results in decreased accuracy of estimated fractions (Sabol *et al.*, 1992; Roberts *et al.*, 1998b). Second, simple SMA assumes that the spectral properties of each endmember are constant across the scene, while only the fractions themselves vary; therefore, the procedure cannot effectively account for endmember variability (Song, 2005). This can be especially problematic in urban landscapes, which are generally characterized by a high degree of within-class variability, especially for impervious materials (Kressler and Steinnocher, 2001; Herold *et al.*, 2004; Franke *et al.* 2009; Myint and Okin, 2009; Weng and Lu, 2009).

An extension of simple SMA that addresses these challenges is multiple endmember spectral mixture analysis (MESMA), which allows both the number and type of endmembers to vary on a per-pixel basis (Roberts *et al.*, 1998b). Implementation of MESMA involves building a large, comprehensive spectral library that accounts for the spectral variability of the scene and applying a series of simple SMA models, based on different combinations of library endmembers, to every pixel in the image. The best model is selected for each pixel based on several measures of fit, which can include the allowed range of endmember fractions, maximum residuals allowed for each wavelength, and the maximum allowed RMS error (Roberts *et al.*, 1998b). Once the best model for each pixel is selected, the per-pixel fractions are often generalized into land-cover components of interest (e.g., vegetation, impervious surface, soil), and an image of fractional cover is generated for each component. Fractions generated from MESMA are usually compared to fractions derived from high spatial resolution imagery (e.g., aerial photographs), and agreement between modeled and reference fraction cover is used to refine endmember selection and/or model constraints until a satisfactory level of accuracy is achieved (Rashed *et al.*, 2003; Powell *et al.*, 2007; Powell and Roberts, 2008; Rashed, 2008).

MESMA has been successfully applied to moderate-resolution imagery to map subpixel fractional V-I-S components in several environments, including Los Angeles (Rashed *et al.*, 2003; Rashed, 2008), Phoenix (Myint and Okin, 2009), and several cities in the Amazon region of Brazil (Powell *et al.*, 2007; Powell and Roberts, 2008). Additionally, Franke *et al.* (2009) applied MESMA to hyperspectral imagery to generate a hierarchical map of urban land cover, ranging from pervious/impervious layers at the coarsest level to maps of specific urban materials and vegetation species at the finest level. Steps to implement MESMA for mapping V-I-S components in urban environments are summarized in Fig. 8.3 and discussed in detail below.

8.2.2 Endmember selection

Many researchers have noted that the most important step in applying SMA is the selection of appropriate endmembers (e.g., Tompkins, *et al.*, 1997) (Fig. 8.3, Step 1). Endmembers used in SMA may be collected in a laboratory or field setting using

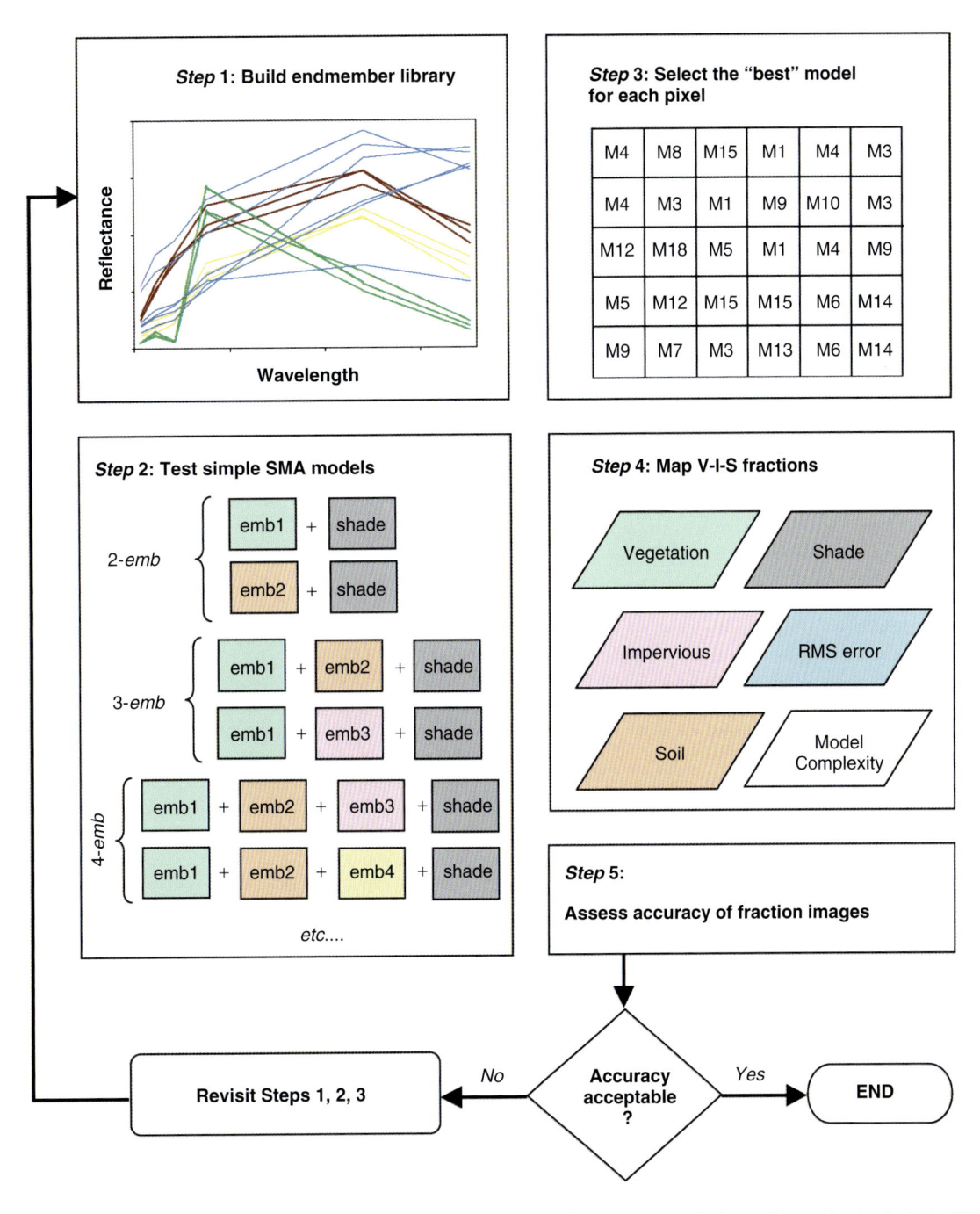

FIGURE 8.3 Overview of the MESMA technique to map urban V-I-S components (adapted from Rashed et al., 2003).

an imaging spectrometer; these are known as "reference endmembers." Additionally, "image endmembers" may be derived from the image itself or from other images. Reference endmembers have the advantage that they are collected under controlled conditions and that each spectrum is associated with a specific, known material. Because they are collected in reflectance mode, they are easily transported between image platforms and across dates (Adams et al., 1995; Roberts et al., 1998a). In contrast, image endmembers have the advantage that no additional data or field campaigns are required and that spectra are collected at the same scale as the data (Roberts et al., 1998a; Rashed et al., 2003). While image endmembers are not as readily portable across imaging platforms, they can be compared across dates and across adjacent scenes if they are radiometrically intercalibrated (e.g., Roberts et al., 1998a). A spectral library may include both reference and image endmembers, as long as all endmembers in the library have been radiometrically intercalibrated (e.g., Powell et al., 2007; Powell and Roberts, 2008).

Endmembers selected for a spectral library should be spectrally "pure," though the definition of purity will depend on the scale of analysis (Smith et al., 1990). An endmember should also be representative of other spectra in its own material class (Song, 2005). Ideally, an endmember will also be spectrally distinct from spectra in other material classes, though this may be more or less true, as the primary goal of SMA is to model materials of interest, not to identify spectra that are mathematically orthogonal in spectral space (Adams and Gillespie, 2006). These may represent conflicting goals; for example, in urban environments bright impervious spectra such as concrete may be spectrally similar to bare soil, even when hyperspectral data are analyzed (Herold et al., 2004). Additionally, not all spectral components in the scene can be represented in SMA models because the technique greatly simplifies reality (Song, 2005), even when MESMA is implemented. The goal in building a spectral library, therefore, is to capture the dominant spectral components of the scene that are physically meaningful and specific to the goals of analysis. Spectral libraries should be constructed on a case-by-case basis and may utilize a number of strategies simultaneously (Smith et al., 1990; Franke et al., 2009).

One method for endmember selection is to extract spectra of pixels from homogeneous areas of known materials; often finer spatial resolution imagery, such as an aerial photograph, is used to guide this process (e.g., Rashed et al., 2003; Wu and Murry, 2003; Myint and Okin, 2009). While this methodology is straightforward, it may not capture the most representative endmembers for a given material type. A second, very common, approach for selecting endmembers is to generate a series of two-dimensional plots; the bounding envelope of the plotted pixels can be thought of as the "mixing space" for that image. The vertices of the mixing space are considered the "purest" pixels in the scene and serve as candidate endmembers (e.g., Adams, et al., 1986; Roberts, et al., 1998a; Small, 2005). These two strategies may also be applied in tandem, as each provides complementary information (e.g., Weng, Hu and Lu, 2008; Weng and Lu, 2009).

One commonly used, semiautomated approach to identify pixels that represent the extremes of mixing space for a particular image is the pixel purity index (PPI). The PPI algorithm projects the spectrum of each pixel onto multiple unit vectors randomly oriented in N-dimensional space (where N is the number of image bands) and records the number of times the pixel is found to land on vertices of the mixing space (Boardman et al., 1995). Pixels with high PPI counts are candidate image endmembers, though they should be visually inspected to verify that they can be assigned to a specific material class, because spectrally extreme pixels are not always physically meaningful (Rashed et al., 2003; Powell et al., 2007; Rashed, 2008; Franke et al., 2009). For example, the PPI algorithm will often identify as extreme pixels that are saturated in one or more bands or pixels located on sharply defined borders, such as between land surface and water (Powell et al., 2007).

More recently, several methods have been proposed for endmember selection, particularly when building a spectral library that contains multiple spectra per material class. These approaches include count-based endmember selection (CoB), endmember average root mean square error (EAR), and minimum average spectral angle (MASA). In contrast to PPI and N-dimensional visualization, which depend on the mixing space as defined by a particular image for endmember selection, each of these newer approaches starts with a large library of candidate endmembers and applies the two-endmember case of MESMA to identify representative spectra (Roberts et al., 1998b; Dennison, et al., 2004). In other words, every spectrum in the candidate library (plus shade) is used to unmix every other spectrum in the library. The goal is to rank endmembers that best represent their material class (e.g., vegetation, impervious surface, soil) and/or are distinctly different from spectra belonging to other classes.

CoB identifies the endmembers within a library that model the greatest number of endmembers within their class and the fewest endmembers outside of their class (Roberts et al., 2003). Candidate endmembers are used to model all other spectra in the library, subject to fraction, RMS, and/or residual constraints. For each potential endmember, a count of the total number of spectra that are modeled within the same class (in_CoB) and the number of spectra modeled outside the class (out_CoB) are recorded. The optimal endmember for each class is that which has the highest in_CoB and lowest out_CoB values, where an ideal endmember would model all of the spectra within its class and none of the spectra in other classes (Roberts et al., 2003; Franke et al., 2009).

The goal of EAR is to identify the most representative spectrum for a material class in a library based on RMS error (Dennison and Roberts, 2003). The procedure starts with a library of potential endmember spectra, grouped by class. In this case, the two-endmember model is applied without constraints. The average RMS error (EAR) for a given endmember modeling all other spectra in its class is calculated and assumed to measure how representative that endmember is of other spectra in its class. The optimal endmember for each class is the endmember with the lowest EAR (Dennison and Roberts, 2003; Franke et al., 2009).

The goal of MASA is similar to EAR, except that the measure of fit is based on a spectral angle metric instead of RMS error (Dennison et al., 2004). Spectral angle mapping (SAM) measures similarity between two spectra by calculating the spectral angle between the two spectral vectors (Kruse et al., 1993). The angle itself is analogous to RMS error; if the angle falls below a user-defined threshold, the spectra are identified as belonging to the same class. MASA is the mean spectral angle between each potential endmember modeling the other spectra in its own class. The endmember with the lowest MASA is considered most representative of its class (Dennison et al., 2004; Franke et al., 2009). For a given endmember library, the optimal endmembers selected by MASA and EAR will be similar, but may not be identical. RMS error calculated from MESMA tends to better identify optimal endmembers for classes with high-albedo spectra, while the spectral angle calculated by SAM better identifies endmembers for classes with dark-albedo spectra (Dennison et al., 2004).

Because each method of selecting endmembers is sensitive to different criteria, the analyst may choose to simultaneously implement multiple methods and experiment with how well each set of "optimal" endmembers models the data being analyzed. Ultimately, performance of potential endmembers should be assessed based on how well the spectra model the image being analyzed (Franke et al., 2009).

8.2.3 SMA models

After a preliminary endmember library has been constructed, the next step is to determine the rules for selecting spectral mixture models. MESMA tests multiple models and identifies the one "best-fit" model for each pixel based on selection criteria and a measure of goodness-of-fit (Franke et al., 2009). This process involves three steps: (a) specifying which combinations of endmembers are allowed, (b) identifying model constraints so that candidate models more accurately represent reality, and (c) determining the criteria to select the overall "best-fit" model for each pixel (Fig. 8.3, Step 2).

The maximum number of endmembers that can be modeled per pixel is equal to the number of spectral bands in the image, if a measure of fit is desired (i.e., residuals for each band) (Adams *et al.*, 1986). However, even given data with very fine spectral resolution (i.e., a large number of bands), there is a practical limit to the number of endmembers that can be modeled, because a large enough number of endmembers can model virtually any measured spectrum, though the resulting combination may be physically meaningless. Thus, accuracy of fraction estimates tends to decrease as the number of endmembers increases, and as a result, the maximum number of allowed endmembers is usually no more than four or five, regardless of the spectral resolution of the data (Sabol *et al.*, 1992; Song 2005). The most rigorous and exhaustive implementation of MESMA is to test all possible combinations of two-, three-, and four- (e.g., Rashed *et al.*, 2003; Powell *et al.*, 2007), and in some environments, five-endmember models (e.g., Myint and Okin, 2009). However, limiting allowed combinations of endmembers reduces computational time and can improve fraction results. Specification of allowed model combinations, therefore, should be supported by close inspection of high resolution reference data to identify the most common combinations of materials in a specific urban area.

An endmember representing shade is commonly included in every model in order to account for variations in surface brightness (Adams *et al.*, 1986; Dennison and Roberts, 2003; Rashed *et al.*, 2003; Powell *et al.*, 2007). To increase computational efficiency and spectral separability, models are often limited to permutations of different classes of materials (e.g., Rashed *et al.*, 2003; Powell and Roberts, 2008; Rashed 2008; Myint and Okin, 2009). However, because impervious materials exhibit high sub-pixel spatial and spectral heterogeneity, it may be valuable to test models that include more than one impervious surface endmember (e.g., Powell *et al.*, 2007).

Once allowed endmember combinations are identified, every model is tested for every pixel in the image. Constraints are defined to select candidate models, most commonly based on the following: (a) Bright fractions are physically realistic, i.e., ideally bright fractions are constrained between zero and 1, though some applications relax the constraint to account for sensor noise and the imprecision of per-pixel analysis introduced by the modular transfer function (MTF) of the sensor (Roberts *et al.*, 1998b; Townshend *et al.*, 2000). (b) The maximum shade fraction is limited (most commonly between 0.50 and 0.80), because dark pixels can often be modeled by a high shade fraction and a small fraction of virtually any bright endmember. (c) A RMS error constraint is also applied, most commonly 2.5% reflectance, roughly equivalent to 6.4 DN for Landsat data, so that all candidate models have a minimal goodness-of-fit (Roberts *et al.*, 1998b; Okin *et al.*, 1999; Dennison and Roberts, 2003; Powell *et al.*, 2007).

For each pixel, the best model at each level of complexity (i.e., two-, three-, four-, or five-endmember) is identified as the model which meets all constraints and has the minimum RMS error, as this is assumed to be the best fit (e.g., Painter *et al.*, 1998). If no model meets all of the constraints, the pixel remains unmodeled. At this stage, each pixel could be associated with several candidate models. Selection of the overall best-fit model depends on the goals of the analysis, but several factors concerning model complexity should be considered (Fig. 8.3, Step 3). First, there is a negative correlation between increasing model complexity and estimated endmember fraction accuracy (Sabol *et al.*, 1992). There is also a negative correlation between model complexity and RMS error; i.e., RMS error tends to decrease as model complexity increases because additional degrees of freedom reduce overall error (Okin *et al.*, 1999; Dennison and Roberts, 2003; Powell and Roberts, 2008). Finally, there is a positive correlation between model complexity and computational expense (Roberts, *et al.*, 1998b). Several methods of comparing models of different degrees have been proposed, including ranking models by RMS error, which almost always results in selection of the most complex model (Painter *et al.*, 1998); ranking by degree of complexity, where the simplest model is favored (Roberts *et al.*, 1998b; Powell *et al.*, 2007); comparing the relative magnitude of fractions and eliminating models with very small fractions (Okin *et al.*, 1999); and considering more complex models only when they result in a significant reduction of RMS error (Roberts *et al.*, 2003; Powell and Roberts, 2008).

Endmember combinations, model constraints, and model selection rules are specific to the application. Each of these factors can be adjusted to better fit the conditions of the scene and goals of the analyst. For example, a commonly recognized source of confusion in urban environments is the spectral similarity between bright impervious spectra and dry soil spectra (e.g., Ridd, 1995; Small, 2001; Kressler and Steinnocher, 2001). One strategy to reduce confusion between impervious and soil spectra is to segment the dataset into "urban" and "non-urban" land cover and apply different models to each portion of the landscape (e.g., Myint and Okin, 2009). Another strategy is to include impervious endmembers only in the most complex models (e.g., Powell and Roberts, 2008), as urban environments are characterized by high spectral complexity relative to other land-cover types. In a given study, therefore, the specific models tested and constraints employed may be adjusted following preliminary accuracy assessment of fractions.

8.2.4 Mapping fraction images

The output of MESMA is a set of images: (a) an image indicating the "winning" model for each pixel; (b) fractional abundance images, indicating the estimated fraction of each endmember; and (c) an image of RMS error. Often, the fraction images are combined to create abundance images of each generalized class of materials (i.e., % vegetation, % impervious surfaces, % soil). For each generalized material, the pixels values are scaled to represent the fractional abundance of that component (Fig. 8.3, Step 4). However, information on the winning model can be retained, providing multiple "scales" of class information, particularly when hyperspectral data are analyzed (Franke *et al.*, 2009).

While shade is an integral component of an urban landscape, it is often a function of topographic effects, shadowing due to surface roughness, and solar zenith angle. Shade is therefore usually considered a variant on endmember brightness rather than a land-cover component itself (Adams *et al.*, 1986; Smith *et al.*, 1990; Rashed *et al.*, 2003). To more accurately represent the physical abundance of materials in the scene, the fractions of each pixel may be shade-normalized; i.e., each non-shade fraction is divided by the sum of all non-shade fractions used to model the pixel (Adams *et al.*, 1986). For example, if a pixel is modeled by a three-endmember combination of shade, vegetation, and impervious, the shade-normalized impervious fraction (Imp_{sh_norm}) would be calculated as follows:

$$Imp_{sh\_norm} = \frac{Imp}{Imp + Veg}. \qquad (8.4)$$

For two-endmember models, the resulting shade-normalized fraction is 1.0. Shade normalization allows comparison of modeled fraction estimates with measures derived from other data sources or from other dates (Powell et al., 2007; Myint and Okin, 2009).

8.2.5 Model complexity

An additional product that can be generated from MESMA is a map of per-pixel model complexity by generating a thematic layer that represents the number of endmembers included in the overall best-fit model for each pixel. The number of endmembers required to adequately model a pixel is a measure of the spectral complexity of the land cover in the pixel's IFOV. Small (2005) noted that spectral heterogeneity at scales of 10–20 m may be "the most characteristic feature" of urban land cover globally. Characterizing the spectral complexity of a landscape, therefore, may facilitate the delineation of built-up areas from undeveloped categories of land cover (Small, 2005). In a study of a rapidly urbanizing region of the Brazilian Amazon, Powell and Roberts (2008) found that model complexity is correlated with the degree of human impact on a landscape: four-endmember models corresponded to built-up (urban) land cover, three-endmember models corresponded to disturbed land cover, and two-endmember models corresponded to natural land cover. Thus, maps of model complexity can serve as a summary measure of human impact.

8.2.6 Accuracy assessment

One measure of model fit is generated by the SMA algorithm itself – RMS error. However, a limitation of SMA is that RMS error only measures model fit, but provides no information on fraction accuracy. It is possible, therefore, that a model could fit the data very well but use an incorrect combination of endmembers (Sabol et al., 1992; Roberts et al., 1998b). Therefore, assessing accuracy of SMA results cannot depend on RMS error alone, and an independent dataset is required for validation. Accuracy assessment is not only a quantitative measure of the quality of the final product, but also serves as an important intermediate step to assess the appropriateness of endmembers and guide refinement of allowed model combinations, model constraints, and selection rules (Fig. 8.3, Step 5).

Some applications of SMA use fractional abundance images as input to traditional classifiers (e.g., maximum likelihood) and generate maps of land-use/land-cover classes. In this case, accuracy assessment is usually applied to the final classified image using a standard error matrix and accuracy measures such as overall, User's, and Producer's accuracy (e.g., Roberts et al., 2002; Weng and Lu, 2009). However, if the final product of SMA is the set of fractional abundance images themselves, the most common form of accuracy assessment involves comparing modeled fractions with reference fractions, usually derived from finer spatial resolution imagery, e.g., aerial photography (Rashed et al., 2003; Wu and Murray, 2003; Rashed, 2008), aerial videography (Powell et al., 2007; Powell and Roberts, 2008), or QuickBird imagery (Small and Lu, 2006; Myint and Okin, 2009). *In situ* data have also been used to evaluate continuous vegetation fractions estimated from multispectral data (Smith et al., 1990; Elmore et al., 2000), though in urban environments physical access to many areas may be restricted and thereby limit the feasibility of collecting field data.

Commonly, modeled and reference fractions are compared using correlation analysis; a scatter plot of modeled vs. reference fractions is generated for a given material class, and the "goodness" of the correlation is assessed in terms of the slope, intercept, and coefficient of determination (R^2) or correlation coefficient (Pearson's r) of the relationship. If the fractions were perfectly modeled, the best-fit regression line would have a slope equal to one and an intercept equal to zero. Other measures used to evaluate the accuracy of fraction estimates include RMS error, mean absolute error (MAE), and bias. RMS error is calculated between the modeled and reference fractions for a given material class (e.g., Wu and Murray, 2003); MAE is the average absolute residual between modeled and reference fractions; and bias is the average residual, indicating trends of over- or underestimation (Schwarz and Zimmermann, 2005; Powell et al., 2007; Weng et al., 2008).

To compare modeled and reference fractions, most studies average fraction values over a region, either a window of pixels (e.g., 3 × 3) or an area that corresponds to the reference sampling unit. Pixel-to-pixel comparison of urban landscapes is problematic for several reasons. First, because of the high spectral and spatial variability, a small error in georegistration can result in significantly different subpixel compositions (Clapham Jr., 2003; Small and Lu, 2006; Powell et al., 2007). Second, the signal recorded at the sensor for a given pixel is influenced by the spectral properties of the surrounding pixels due to the MTF of the sensor (Forster, 1983; Townshend et al., 2000; Small, 2001). Third, the process of averaging fractions over larger areas generally reduces variance that may be caused by signal noise or geolocation error (Woodcock and Strahler, 1987; Powell et al., 2007). Therefore, averaging endmember fractions over zones can generate more robust estimates of land-cover composition. One of the strengths of SMA is that sub-pixel fractions are ratio data, and aggregating fraction values generates statistically valid results (Clapham Jr., 2003).

8.3 Two case studies

This section presents an overview of several MESMA products and applications in the context of two case studies: the first characterizes the evolution of urban land cover on the Brazilian "arc of deforestation" and the second quantifies vegetation cover in a temperate Western US city. Rather than recount every detail of each study, these examples are meant to highlight the flexibility of MESMA in characterizing the fundamental components of urban landscapes. These examples also illustrate the difficulty of fully automating or generically applying MESMA; rather, determination of the most appropriate strategy for endmember selection and the most effective model parameters requires iterative assessment by the analyst (Powell and Roberts, 2008).

8.3.1 Evolution of urban morphology on a tropical forest frontier

Urban morphology (i.e., the structure and composition of a city) is the physical manifestation of complex social and economic forces, in the context of a specific natural environment (Moudon, 1997; Rashed et al., 2003, 2005). A first step in understanding connections between urban landscape change and socio-economic drivers is to characterize urban land cover in terms of the composition and distribution of its bio-physical components (Rashed et al., 2005; Powell and Roberts, 2008). The objective of this case study was to characterize trajectories of intraurban and periurban landscapes on a tropical forest frontier in terms of V-I-S components. This study demonstrates for the first time the possibility of applying the same endmember library and same model rules across multiple scenes and through time, implying that modeled fractions are directly comparable from one date to another. Additionally, this case study illustrates how assumptions concerning the unidirectional trajectory of urban change through time can inform spectral library construction and model selection.

Data: The full analysis involved ten settlements in the state of Rondônia, in the Southwest Brazilian Amazon, a location of government-directed settlement programs initiated in the early 1970s (Fig. 8.1a). The region is a unique environment to study urbanizing landscapes because the history of urban development essentially coincides with the Landsat data archive. The dominant land-cover types in this region are primary forest (dark green regions in the false color display of the Landsat sub-image, Fig. 8.4a), pasture (pink areas), and second-growth forest (brighter green areas). The region of analysis was covered by five Landsat TM/ETM+ scenes. Four dates were analyzed for each scene, corresponding as closely as possible to the years 1985, 1990, 1995, and 2000. Two settlements are highlighted in this chapter: Buritis (Settlement #1 in Fig. 8.1a) and Seringueiras (Settlement #2 in Fig. 8.1a).

A single date for each scene was converted to surface reflectance using a radiative transfer model, and all other dates were intercalibrated to the reflectance images using relative radiometric calibration (Roberts et al., 1998a; Furby and Campbell, 2001). Subsequent data processing occurred in reflectance space. An area of approximately 20 × 20 km centered on each settlement in the study was clipped for analysis. A water-burn mask was generated for each sub-scene and each date by applying a threshold to band 7, and those pixels were excluded from further analysis. Validation data were derived from high-resolution aerial videography collected in 1999, which transected three settlements in the study area (Hess et al., 2002; Powell and Roberts, 2008).

Endmember library: Endmembers were selected to represent each V-I-S category. Vegetation was subdivided into green vegetation (GV) and non-photosynthetic (i.e., senesced) vegetation (NPV), because these seasonal states of vegetation have distinctly different spectral properties (e.g., Roberts et al., 1998a). Candidate endmembers were collected from a combination of reference and image endmembers; the former were selected from a spectral library that included laboratory- and field-collected spectra. Candidate image endmembers were identified by applying the PPI algorithm to all ten samples from the year 2000 and selecting spectra that could be visually classified with high confidence as belonging to one of the four materials of interest.

TABLE 8.1 Allowed models by generalized material class.

2-emb (15)	3-emb (27)	4-emb (108)
NPV + shade	NPV + GV + shade	NPV + Soil + Imp + shade
GV + shade	NPV + Soil + shade	GV + Soil + Imp + shade
Soil + shade	GV + Soil + shade	
Imp + shade		

Numbers in parentheses indicate the total number of models generated for all permutations of endmembers included in the final MESMA library.
NPV = non-photosynthetic vegetation, *GV* = green vegetation, *Imp* = impervious.

The final MESMA library was selected in a two-step process. First, candidate endmembers were ranked in terms of how representative they were of other endmembers in the library based on EAR. Second, low-EAR candidate endmembers were ranked in terms of how well they represented materials in the image by sequentially applying low-EAR endmembers to four-endmember models and assessing model fraction accuracy using reference data. Similarly, the appropriate number of endmembers for each category was determined by iteratively adding low-EAR endmembers to the MESMA library and assessing resulting model accuracy based on the visual inspection of fraction and RMS error images, as well as quantitative comparison of model results and high-resolution reference data, following the procedure summarized in Powell and Roberts (2008). The final MESMA library consisted of 15 endmembers, in addition to photometric shade: three GV, three NPV, six impervious surfaces, and three soil (Fig. 8.3, Step 1).

Allowed models: After experimentation with different endmember combinations concurrent with inspection of the high spatial-resolution reference data, allowed models were specified as those combinations presented in Table 8.1. All permutations of two-, three-, and four-endmember models were tested for each pixel. Inspection of reference data indicated that four-endmember models were sufficient to capture the spectral complexity of this landscape; therefore, five-endmember models were not tested. Impervious material spectra were excluded from three-endmember models, because inspection of the high spatial-resolution reference data revealed that 30-m samples of built-up land cover in this region consisted almost exclusively of three different material types, in addition to shade (e.g., see Fig. 8.2a); the only exceptions were major roads, which were sometimes modeled as a two-endmember model of impervious and shade. The tightness of model constraints minimized confusion between two-endmember models. When impervious materials were included in 3-endmember models, bright impervious surfaces were confused with soil and/or NPV. By only including impervious surface endmembers in the highest complexity models, impervious fractions were restricted to urban land cover; non-urban land-cover types were spectrally less complex and therefore could be better modeled with simpler models (i.e., 2- or 3-endmember models).

Model constraints: Candidate models were selected based on the following criteria: (a) only models with RMS error below 2.5%

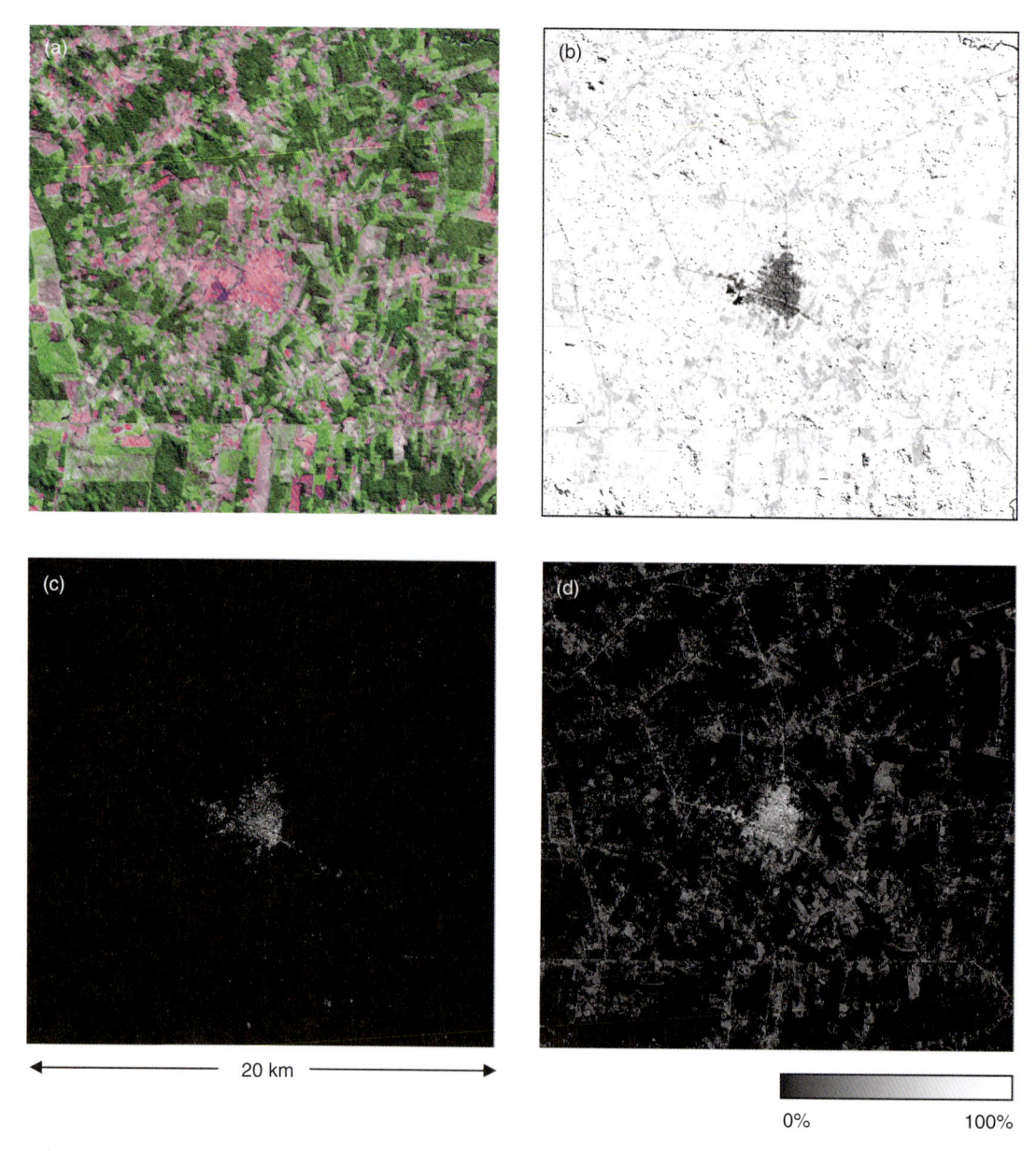

FIGURE 8.4 Shade-normalized V-I-S fractions for the settlement of Buritis, Rondônia, located in the center of the subscene. (a) False color composite of 2001 Landsat ETM+ sub-scene bands 543; (b) Vegetation fraction (GV + NPV); (c) Impervious fraction; (d) Soil fraction.

reflectance were considered, (b) the maximum shade fraction was constrained to 0.55, and (c) the minimum bright (i.e., non-shade) fractions allowed was 0.10. Model constraints were determined by experimentation and visual inspection of resulting fraction and RMS error images. In particular, assessment was facilitated by the *a priori* assumption that impervious fractions should be spatially constrained in the region of built-up land cover, and conversely, impervious surfaces should be a near-zero component of other land-cover types. Constraints were therefore chosen that minimized impervious fraction "speckle" in non-built-up land-cover types (e.g., pasture, primary forest).

Selection rules: The model with the lowest RMS error was selected for each level of complexity. To determine the overall best-fit model, lower complexity models were favored, and higher complexity models were selected only if their RMS error was significantly lower. A threshold of 0.25% was empirically determined to be a "significant" difference (Roberts *et al.*, 2003; Powell and Roberts, 2008).

Results: The final products of this analysis were shade-normalized fraction images for each generalized material type (V-I-S). GV and NPV fractions were combined in the vegetation layer, as their relative composition is related to the strength of the wet season that year, as well as the length of the dry season relative to the date of image collection. Shade-normalized fractions for the settlement of Buritis in 2001 are presented in Fig. 8.4. High fractions appear bright, and low (or zero-value) fractions appear dark. The vegetation component (Fig. 8.4b) is quite low in the urban center, higher in pasture areas (which are a mixture of GV and NPV), and highest in forested areas (which are dominated by GV). Impervious fractions (Fig. 8.4c) are concentrated in the urban land cover in the center of the image, and are near-zero in the rest of the image. Soil fractions (Fig. 8.4d)

TABLE 8.2 Summary measures for accuracy assessment for three validation sites in Rondônia. A total of 41 samples were included; average sample size was 2 × 2 pixels.

	Vegetation	Impervious	Soil
Slope	0.787	0.814	0.704
Intercept	10.75	5.99	5.99
Pearson's r	0.93	0.87	0.84
MAE	7.71	5.55	8.59
Bias	1.71	2.99	−6.20

"Pearson's r" refers to the Pearson correlation coefficient between modeled and measured fractions; "MAE" refers to mean absolute error; "Bias" refers to mean error.

are also an important component of the built-up landscape and dominate roadways in this scene, indicating that the roads are either unpaved or dust-covered. The soil component is also high in pasture areas, but remains low in forested land cover.

Correlation between reference fractions and modeled fractions was assessed for the three 2000 sub-scenes for which reference data were available (Table 8.2). Correlation between reference and modeled fractions was highest for impervious fractions, with a slope of 0.814 and a Pearson's correlation coefficient (r) of 0.87, and lowest for soil fractions, with slope equal to 0.704 and Pearson's r equal to 0.84. The vegetation fraction, representing the sum of GV and NPV fractions, had a relatively high slope (0.787) and the highest correlation ($r = 0.93$). For all classes, the intercept was approximately 10% or lower, and the MAE was also less than 10%, within the uncertainty expected due to the MTF of Landsat TM (Townshend et al., 2000).

The biases (i.e., mean residuals) for all classes were quite low (Table 8.2). The impervious and vegetation fractions had very small, positive biases (below 3%), indicating a slight overestimation of the modeled fractions relative to the reference fractions. The soil fractions had a slightly larger negative bias (approximately 6%), indicating a general trend of underestimation of the soil fractions. The overestimation of impervious and vegetation fractions was approximately equal to the underestimation of the soil fractions, a result of the SMA constraint that requires the fractions of any pixel to sum to 1.0. Much of the uncertainty of fraction estimation is due to spectral ambiguity between soil, NPV, and/or bright impervious spectra. Unfortunately, comparing reference and modeled fractions does not allow the specific nature of spectral confusion to be quantified.

However, at every time step, impervious materials were modeled with the highest density in the vicinity of the urban core, and in general were not modeled in periurban areas of the landscape, indicating that errors of commission were relatively low for the impervious fractions. Additionally, the fact that overall measures of area covered by impervious materials increased or remained constant through time for all of the urban areas in the sample increased confidence in the accuracy of impervious fractions. Still lacking is extensive data to assess errors of omission for impervious surface mapping.

Maps of model complexity (i.e., the number of endmembers in each "winning" model, assumed to be the number of endmembers needed to adequately model per-pixel spectral variability) illustrate the relationship between spectral complexity and the degree of human impact on the landscape. Model complexity for the settlement of Seringueiras is presented for four dates: 1986 (Fig. 8.5a), 1990 (Fig. 8.5b), 1995 (Fig. 8.5c), and 2000 (Fig. 8.5d). Four-endmember models correspond to urban land cover, i.e., spectrally the most complex land-cover type, while three-endmember models correspond to cleared land and regenerating vegetation. Two-endmember models correspond to primary forest, spectrally the least complex land cover in this landscape at this spatial resolution. The population of Seringueiras remained low, but increased almost 15-fold during this period. Population was estimated as approximately 260 inhabitants in 1986 and reported as approximately 3800 inhabitants in the 2000 Brazilian Census (IBGE, 2000). The increase in population was accompanied by an increase in both extent and density of built-up land cover. Areas along roads and adjacent to the settlement center were cleared first (i.e., two-endmember models are replaced with three-endmember models, indicating human alteration of the landscape), and through time, the areas of disturbance expanded outward from the settlement and the roads. Because maps of model complexity clearly separate built-up from non-built-up land cover, this methodology could be further explored to facilitate semi-automated mapping of urban extent at regional to global scales (Powell and Roberts, 2008).

Fraction images through time not only capture change in urban extent (i.e., land-cover transitions), but also quantify change in urban composition (i.e., land-cover modifications) (Rashed et al., 2005). To compare urban composition through time, a fixed urban boundary was defined based on urban extent in the year 2000. Mean fraction values within each urban boundary were calculated, corresponding to percent area covered by each material type. In a tropical frontier environment, an idealized trajectory of urban land-cover change might be imagined as follows: (i) settlers move into a tropical forest (sub-pixel fractions dominated by vegetation); (ii) initial clearing occurs (sub-pixel fractions are dominated by soil, with small fractions of GV or NPV); and (iii) a settlement emerges (soil fraction diminishes and impervious fraction increases). The trajectories of the settlements of Seringueiras (Fig. 8.6a) and Buritis (Fig. 8.6b), illustrate this pattern, though change occurs at different rates in each case. In Seringueiras, conversion to urban land cover followed a gradual, consistent trend, as indicated by the steady increase of impervious and soil fractions and corresponding decrease of vegetation fractions (Fig. 8.6a). In contrast, the location where Buritis emerged was dominated by natural land cover (100% vegetation) until 1995, and then experienced explosive conversion to urban land cover, as indicated by the steep slopes associated with changes in impervious and soil fractions (Fig. 8.6b), and concurrent loss of vegetation fractions.

The physical composition and structure of urban areas represent a complex interaction between physical and social processes (Rashed et al., 2005; Sánchez-Rodriguez et al., 2005). A logical extension of the research presented here, therefore, is to compare urban land-cover change with socio-economic characteristics of settlements through time, both at inter-urban and intra-urban scales (Boucek and Moran, 2004; Rashed et al., 2005). Linking physical descriptions of development trajectories to socio-economic drivers of urban expansion in the Amazon could contribute to assessment of the role of urbanization in regional land-cover change and global environmental change (Lambin et al., 2003).

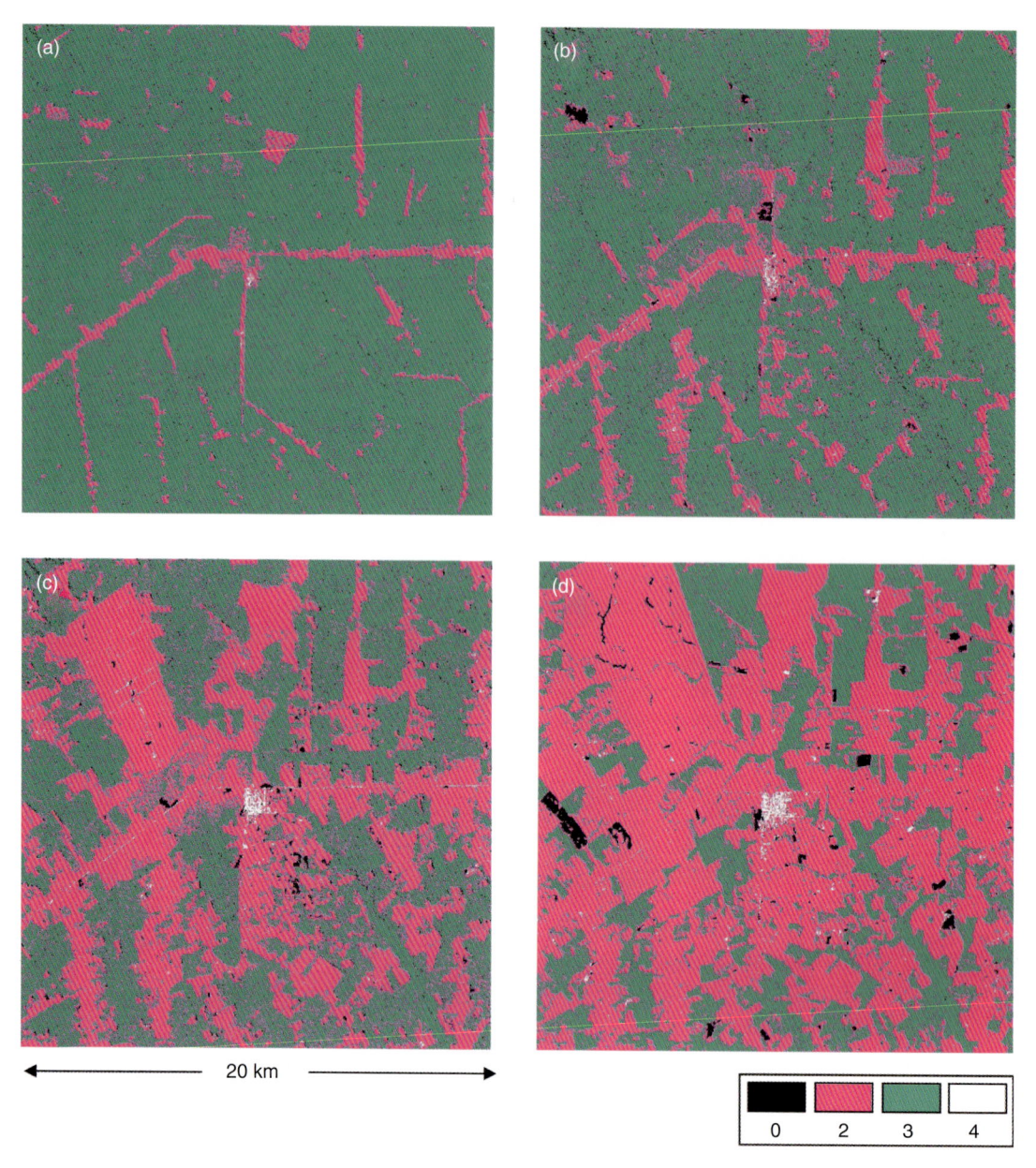

122 PART III ALGORITHMS AND TECHNIQUES FOR URBAN ATTRIBUTE EXTRACTION

FIGURE 8.5 Maps of model complexity for the settlement of Seringueiras, Rondônia, for the years (a) 1986, (b) 1990, (c) 1995, (d) 2000. Water, burned, and unmodeled pixels are black.

8.3.2 Discriminating urban tree and lawn cover in a western US city

Knowledge of the distribution and characteristics of urban vegetation is necessary to quantify and model urban environmental conditions at various spatial scales (Small and Lu, 2006; Franke *et al.*, 2009). The distribution, type, and quality of urban vegetation strongly influence urban environmental processes, such as the cycling of water, nutrients, energy, and carbon (Small, 2001; Lee and Lathrop, 2006; Petaki *et al.*, 2006), as well as substantially impact the social and economic characteristics of a neighborhood (Nowak *et al.*, 2001). In most US cities, the dominant vegetation types are trees and turf grass (i.e., lawn); both contribute to moderating the urban climate, reducing storm water runoff, and providing habitat for wildlife. Trees, in particular, contribute to urban environmental quality by mitigating air quality, sequestering carbon, and lowering building energy use through shading, evapotranspiration, and wind speed reduction (Small and Lu, 2006). While an additional benefit of turf grass is enhanced carbon uptake by soil, lawns are also associated with a number of negative effects, including non-point source pollution from over-use of chemicals and fertilizer, as well as increased summer water use, especially in Western cities (Milesi, *et al.*, 2005). For example, fifty-four percent of annual household water use in Denver, Colorado, is dedicated to landscaping (Denver Water, 2009).

The objective of this study was to inventory urban lawn and tree cover in Denver to provide city resource managers with a rapid assessment of the distribution and total area covered by each vegetation type (Fig. 8.1b). The challenge of such an assessment is the spectral similarity between lawn and tree vegetation; one

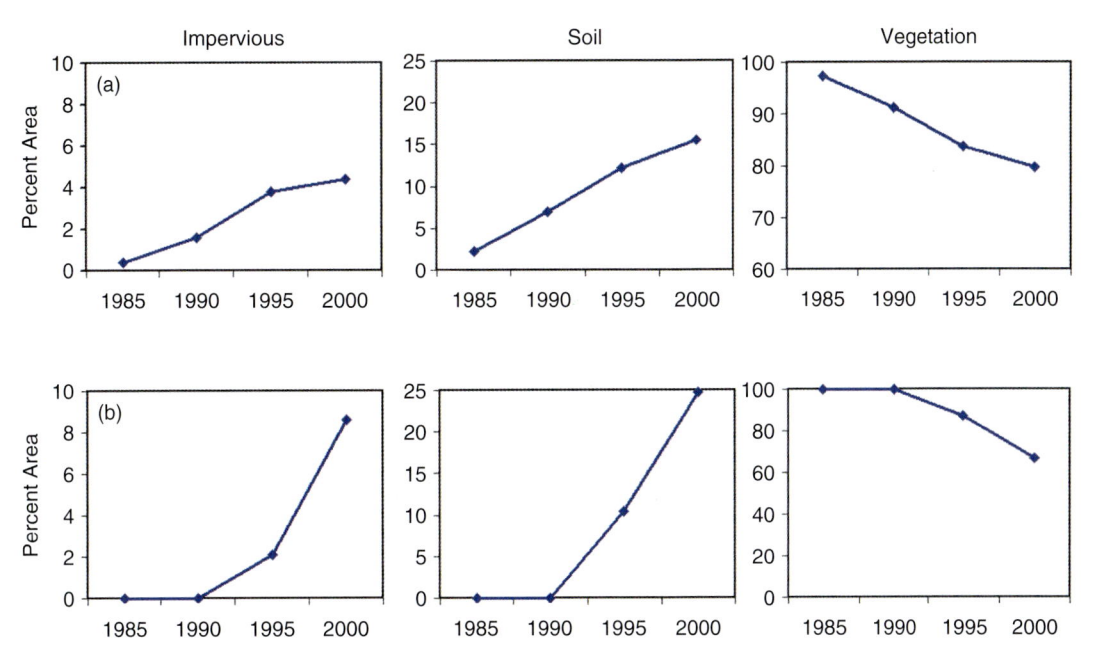

FIGURE 8.6 Generalized urban growth trajectories represented by percent area of 2000 urban extent covered by each V-I-S component for (a) Seringueiras and (b) Buritis.

strategy to separate materials that are spectrally similar is to integrate temporal information in the analysis. This strategy has been most commonly applied to traditional land-cover classification; for example, Yuan et al. (2005) distinguished between forests and agriculture in a peri-urban environment using two images collected at different times in the crop cycle – the first in late spring before crop green-up and the second in late summer shortly before harvest. Kuemmerle, Röder and Hill, (2006) separated percent herbaceous and percent woody cover in a Mediterranean shrubland by comparing SMA results applied to two dates – the first image corresponded to the end of the rainy season when all vegetation was green, and the second image corresponded to the end of the dry season when grasses were senesced while shrubs remained green.

A similar methodology was tested here to separate the lawn and tree components of urban vegetation. MESMA was applied to two dates in the growing season to separate vegetation fractional cover from other materials in the urban environment. Vegetation types were separated based on two simple assumptions concerning the phenology (i.e., seasonal cycle) of urban vegetation in this climate: (a) grass cover can be quantified in early spring, leaf-off conditions, because deciduous trees and shrubs do not leaf out until approximately 6–8 weeks after lawns green up, and (b) tree canopy can be estimated by comparing green vegetation measured in late summer (i.e., full-canopy conditions) to green vegetation measured in early spring.

Data: Two Landsat ETM+ images (P33/R33) were acquired to compare different seasonal states of urban vegetation – the first from early spring, leaf-off conditions (16 April 2003), and the second from late summer, full-canopy conditions (18 August 2002). Only the southern half of Denver was included for this pilot study because the city is located at the boundary of two Landsat scenes. Aerial photographs collected in 2005 by the National Agriculture Imagery Program (NAIP) were used to guide endmember selection, SMA model specification, and SMA model constraints and selection rules. The reference dataset for accuracy assessment was based on a fine-resolution inventory of tree-canopy cover for the city and county of Denver, generated in 2006 using object-oriented analysis of pan-sharpened QuickBird imagery (approximately 61-cm spatial resolution); however, the publically available estimates of tree-canopy cover were aggregated to the neighborhood level (NCDC, 2006).

Endmember library: Image endmembers were selected from the 2003 Landsat image (leaf-off conditions); candidate endmembers were identified by applying PPI to the image subset and verifying the material composition of selected pixels using the aerial photographs. To determine which candidate endmembers were most representative of materials in the image, two-endmember models (i.e., bright endmember + shade) were tested using each candidate endmember to unmix the scene, and only those endmembers that modeled at least a portion of the scene well were selected for the final MESMA library. Library spectra were radiometrically calibrated to the 2002 image using relative radiometric calibration (Roberts et al., 1998a; Furby and Cambell, 2001). The same endmember library was applied to both dates, and all MESMA processing occurred in DN-space. The final spectral library consisted of nine endmembers, in addition to non-zero shade: one green vegetation endmember, two non-photosynthetic vegetation, two soil, two dark impervious endmembers, and two bright impervious endmembers. A proxy for shade was selected from a deep water body. The selection of a single GV endmember for the MESMA library may seem inconsistent, as the goal of this study was to discriminate between two different types of vegetation. However, while the mixture of basic spectral components (e.g., the proportion of GV, NPV, Shade) was expected to vary between grass cover and tree canopy cover, the basic spectral component representing green vegetation was assumed to be the same for both vegetation types, especially given the coarse spectral resolution of Landsat data (Smith et al., 1990).

Allowed models and model constraints: All models included the green vegetation endmember; any pixel that did not include green vegetation was presumed to remain unmodeled given the constraints imposed below. Only three- and four-endmember models were tested, because visual inspection of the high spatial resolution aerial photographs indicated that very few Landsat pixels in the area of interest had fewer than three basic spectral components (i.e., every 30-m pixel was expected to contain at least two different material types and shade). Non-shade endmember fractions were constrained between −0.05 and 1.05, a standard choice in the literature to account for sensor noise and measurement uncertainty (e.g., Dennison and Roberts, 2003). The maximum shade fraction allowed was 0.50 (water bodies therefore remained unmodeled). Only models with an RMS error less than 6.24 DNs (equivalent to approximately 2.5% reflectance) were considered.

Model selection rules: For each level of complexity (three- or four-endmember), the model that met all the constraints and had the lowest RMS error was selected. Three-endmember models were selected only when no four-endmember models met the criteria; i.e., more complex models were favored over simpler models. This decision was based on evaluation of the average spectral complexity of this urban landscape by visually inspecting the aerial photographs overlaid with a 30-m grid, as well as testing different model selection rules and evaluating the resulting fraction and RMS images.

Results: Shade normalized fractions for each date are presented as RGB (NPV, GV, Soil) composites in Fig. 8.7(a) (April, leaf-off image) and Fig. 8.7(b) (August, leaf-on image). Because impervious fractions are not included in the RGB display of the image, variations in brightness are related to the presence or absence of impervious surfaces (i.e., dark areas have high impervious fractions, and bright areas have low impervious fractions). It should be noted that the relative abundance of soil in the April image most likely indicates confusion between soil and bright impervious spectra. Because the purpose of this analysis was solely to assess abundance of vegetation, no attempt was made to further refine the soil and impervious endmembers.

Another way to display fraction images that facilitates visual interpretation, especially to assess change between two images, is to bin the fraction values into meaningful categories. For example, green vegetation fractions have been binned into categories in Fig. 8.7c (April, leaf-off image) and Fig. 8.7d (August, leaf-on image), where values of the bins were selected to highlight change between the two dates and to enable direct comparison of vegetation fractions.

Estimation of percent tree cover and percent grass cover was based on the simple assumption that the green vegetation measured in April predominantly corresponded to grass cover, and the green vegetation that measured in August was a combination of tree and grass cover. Therefore, the crudest estimate of tree cover would be to subtract the estimated April GV fraction from the estimated August GV fraction. However, pixel-to-pixel comparisons can be problematic for several reasons, as discussed in Section 8.2.6. In addition, while the reference dataset was derived from very fine resolution imagery, it was only reported at the neighborhood level; as a result, preliminary accuracy assessment of the MESMA-generated estimates of tree cover was conducted at the neighborhood scale of analysis. To estimate percent tree cover at the neighborhood level, the mean values of shade-normalized, green vegetation cover were calculated for each neighborhood for each date. Percent tree cover was calculated as percent green vegetation cover estimated from the August image, minus percent grass cover estimated from the April image.

The estimates of zonal averages of percent tree were compared to those derived from the Quickbird imagery for 47 neighborhoods in South Denver. Modeled estimates of tree cover agreed with reference measures within ±5% for 35 neighborhoods, and within ±10% for 44 neighborhoods. Only three neighborhood estimates disagreed by 10% or more. A scatterplot of modeled versus reference tree-cover fractions by neighborhood is presented in Fig. 8.8a, with the 1:1 line plotted for reference. The Pearson's correlation coefficient between the two datasets is 0.78, while the slope of the best-fit line is 0.641. While ideally the slope of the best-fit line would be closer to one, the MAE was 3.4 and the bias was −0.7, indicating a high average agreement between modeled and reference fraction when aggregated to the neighborhood level. Additional information is provided by plotting the residuals (modeled − observed fractions) as a function of the reference fractions; the best-fit line is superimposed to indicate general trends of over- and under-estimation (Fig. 8.8b). This plot indicates that tree fractions tend to be over-estimated by MESMA for very small fractions (positive residuals) and under-estimated for larger fractions (negative residuals). While some of the disagreement is certainly due to oversimplified assumptions used by the analysis as discussed below, a portion of the disagreement could also be due to the discrepancy between the dates of reference data collection (2005 and 2006) and image collection (2002 and 2003).

This study involved several obvious oversimplifications of urban vegetation, ignoring potentially confounding factors, such as grass cover that was obscured by tree canopy in summer imagery and the presence of coniferous tree canopy in the early spring imagery. In addition, accuracy assessment to date has only included tree canopy cover and neglected assessment of the grass cover estimates. Despite these limitations, estimates of tree cover aggregated to a neighborhood level were within ∼ 10% agreement of estimates derived from very high spatial resolution imagery, indicating the potential of MESMA to provide a relatively quick and effective method for discriminating the two plant functional types in urban environments in the temperate West of North America. Because this methodology is conceptually simple and can be applied to data that are regularly and freely available, assessments of urban vegetation cover can be applied on a regular basis with very low cost. Such analysis can support urban resource managers, as well as researchers investigating relationships between urban vegetation and local and regional environmental processes. The "synoptic view" offered by moderate resolution remote sensing data can complement analyses of finer spatial resolution imagery and *in situ* measurements that cannot be supported as frequently because of high cost (Small, 2001).

Conclusions

This chapter has summarized the implementation of MESMA to quantify the V-I-S components of urban land cover. The analysis involves compiling a regional library of the spectra that best represent the diversity of urban materials. The spectral response of each pixel in the urban landscape is "unmixed" to determine the fraction of each material component present. The number and type of materials can vary on a per-pixel basis to accommodate the high spectral and spatial variability of urban land cover. The sub-pixel fractions of specific materials are

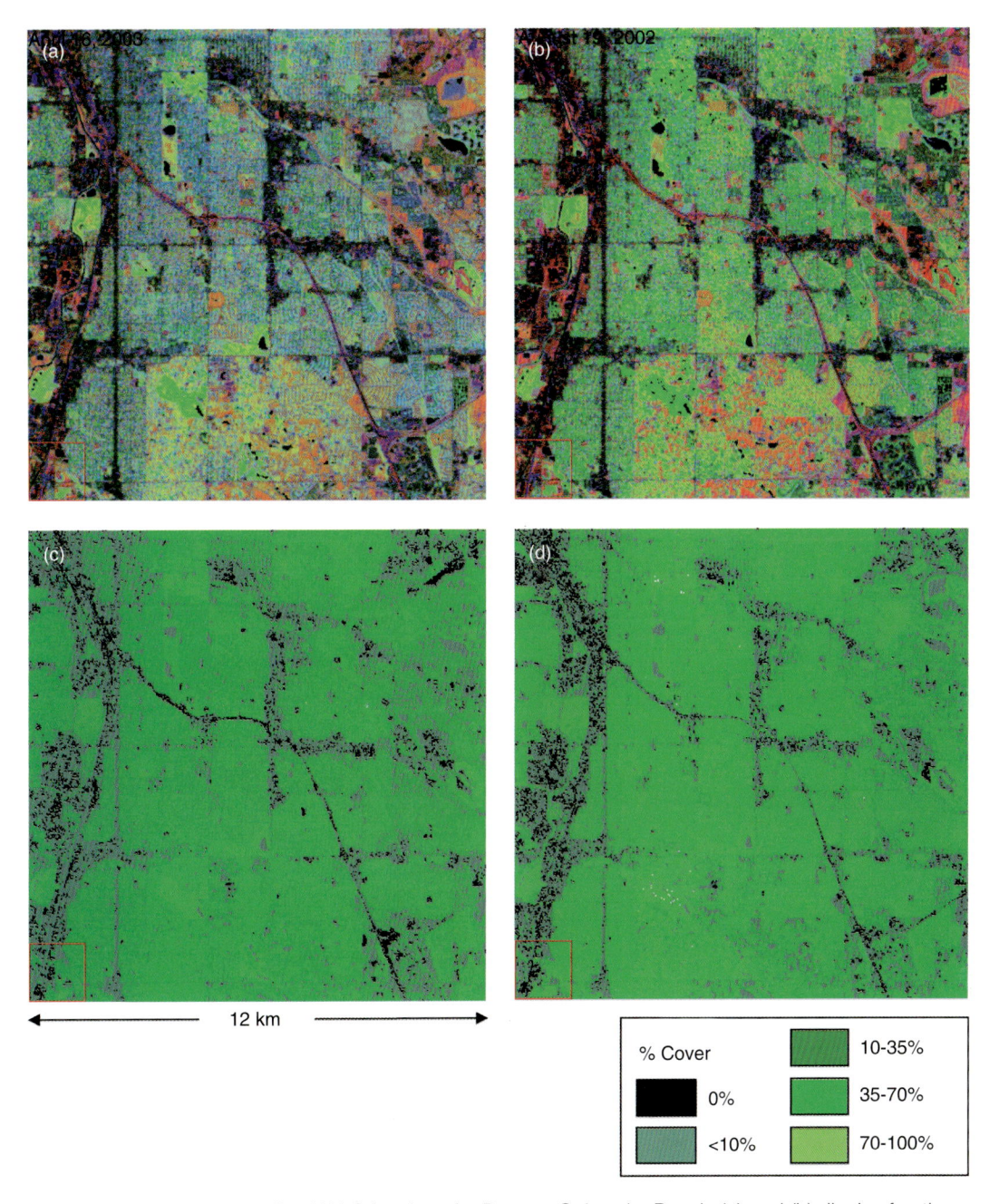

FIGURE 8.7 Subset of shade-normalized V-I-S fractions for Denver, Colorado. Panels (a) and (b) display fractions results as RGB composites: non-photosynthetic vegetation (R), green vegetation (G), and soil (B). Panels (c) and (d) display the green vegetation fractions results, binned into classes that highlight change in vegetation cover.

grouped into generalized V-I-S classes, and continuous maps of each component are generated. Accuracy is assessed in terms of agreement between modeled fractions and reference fractions as measured from finer spatial resolution imagery or from the field.

A potential disadvantage of this approach is that the process will never be fully automated, as iterative revisions and contextual meaning must be supplied by an analyst in the context of the goals of each project; however, the examples presented here highlight a major strength of this approach – that model implementation can easily be customized to best capture local spatial and spectral variations. The output can be generalized within the V-I-S framework, facilitating the comparison of urban landscapes across regions and through time. Maps that summarize model (i.e., spectral) complexity demonstrate the potential to delineate urban extent at regional and global scales. Because SMA fractions model the landscape in terms of continuous variables, landscape change can be quantified in terms of land-cover conversion (i.e., urban expansion) and land-cover modification (i.e., internal changes within the urban fabric itself) (Rashed et al., 2005).

While the examples presented in this chapter analyzed multispectral datasets, future applications of MESMA will increasingly be applied to hyperspectral datasets, especially when such systems become operational on satellite platforms. The primary advantage of hyperspectral data is the increased spectral information content

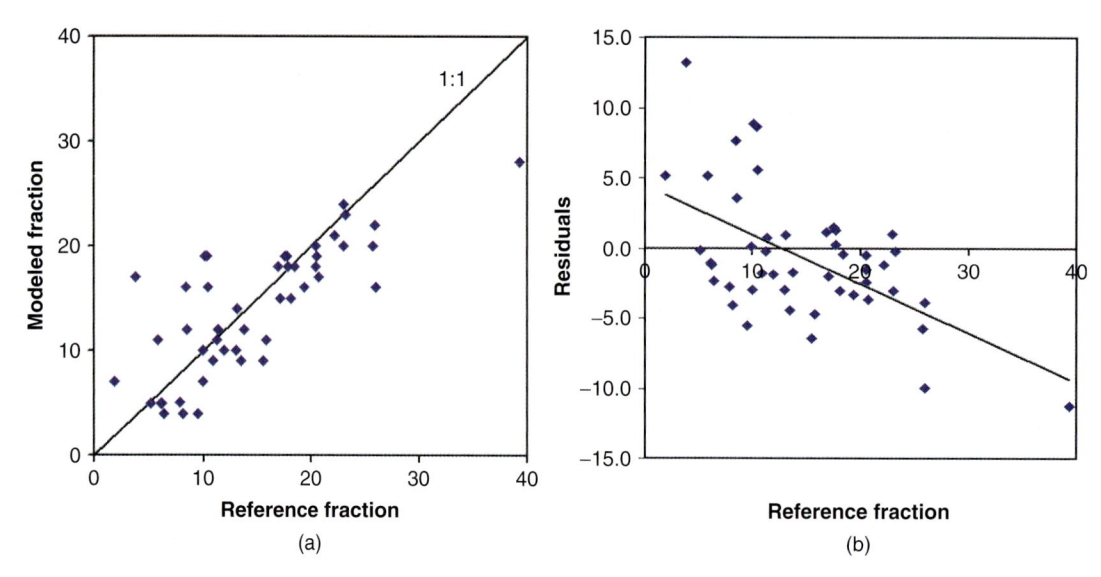

FIGURE 8.8 (a) Scatterplot comparing reference and modeled tree fractions; each point represents the mean tree cover fraction for a neighborhood in South Denver ($n = 47$). The 1:1 line represents perfect agreement. (b) Plot of residuals (*modeled - reference*) as a function of the reference fractions. The best-fit line indicates general trends of over- and under-estimation.

that facilitates spectral discrimination (e.g., Herold *et al.*, 2004). However, even with increased spectral resolution, some surfaces are likely to remain confusing, such as dark, featureless surfaces (e.g., asphalt roads and tar-shingled roofs), and highly reflective surfaces (e.g., dry, sandy soil and concrete) (Herold *et al.*, 2004; Franke *et al.*, 2009). One promising approach that addresses the issues of spectral confusion and within-class variability in urban landscapes is hierarchical (multilayer) MESMA. This methodology applies MESMA to multiple levels of class complexity, from general categories (e.g., pervious or impervious) to specific urban material and vegetation species types. Hierarchical MESMA incorporates the spatial dimension of imagery by using the coarsest levels of analysis, associated with the highest degree of confidence, to constrain the more complex levels (Franke *et al.*, 2009). Use of MESMA to characterize urban land cover across different levels of information will advance the development of quantitative, generalized measures needed to characterize urban landscapes globally, while maintaining information on the types and distributions of locally specific materials to support urban planning and ecosystem management.

Acknowledgments

The author is grateful to Dar Roberts, Philip Dennison and Kerry Halligan, who have successfully transformed multiple MESMA codes into a user friendly, open-source extension of ENVI, known as "Viper Tools" and available for download at http://www.vipertools.org. The author also acknowledges Sharolyn Anderson for helpful comments on drafts of this manuscript. Research in Rondônia was funded in part by a NASA Earth System Science Graduate Student Fellowship, as well as NASA Grant NCC5−282 as part of LBA-Ecology. Digital videography was acquired as part of LBA-Ecology investigation LC-07. The Landsat ETM+ imagery was acquired from the Tropical Rain Forest Information Center (TRFIC). The author also thanks two anonymous reviewers for helpful comments.

References

Adams, J.B., Smith, M.O. and Johnson, P.E. (1986) Spectral mixture modeling: A new analysis of rock and soil at the Viking Lander 1 site. *Journal of Geophysical Research*, **91**(B8), 8098–8112.

Adams, J.B., Sabol, D.E., Kapos, V. *et al.* (1995) Classification of multispectral images based on fractions of endmembers: application to land-cover change in the Brazilian Amazon. *Remote Sensing of Environment*, **52**, 137–154.

Adams, J.B. and Gillespie, A.R. (2006) *Remote Sensing of Landscapes with Spectral Images: A Physical Modeling Approach*. Cambridge University Press, Cambridge.

Anderson, J.R., Hardy, E.E., Roach, J.T. and Witmer, R.E. (1976) *A land use and land cover classification system for use with remote sensor data*, US Geological Survey, Professional Paper 964, United States Government Printing Office, Washington, D.C..

Boardman, J.W., Kruse, F.A. and Green, R.O. (1995) Mapping target signatures via partial unmixing of AVIRIS data, in *AVIRIS Airborne Geoscience Workshop Proceedings*. JPL Publications, Pasadena.

Boucek, B. and Moran, E.F. (2004) Inferring the behavior of households from remotely sensed changes in land cover: Current methods and future directions, in *Spatially Integrated Social Science* (eds M.F. Goodchild and D.G. Janelle), Oxford University Press, Oxford, pp. 23–47.

Carlson, T.N. and Sanchez-Azofeifa, G.A. (1999) Satellite remote sensing of land use changes in and around San José, Costa Rica. *Remote Sensing of Environment*, **70**, 247–256.

Clapham Jr., W.B. (2003) Continuum-based classification of remotely sensed imagery to describe urban sprawl on a watershed scale. *Remote Sensing of Environment*, **86**, 322–340.

DeFries, R.S., Townshend, J.R.G. and Hansen, M.C. (1999) Continuous fields of vegetation characteristics at the global scale at 1-km resolution. *Journal of Geophysical Research*, **104**(D14), 16911–16923.

Dennison, P.E. and Roberts, D.A. (2003) Endmember selection for multiple endmember spectral mixture analysis using endmember average RMSE. *Remote Sensing of Environment*, **87**, 123–135.

Dennison, P.E., Halligan, K.Q. and Roberts, D.A. (2004) A comparison of error metrics and constraints for multiple endmember spectral mixture analysis and spectral angle mapper. *Remote Sensing of Environment*, **93**, 359–367.

Denver Water, 2009. *Water Use*. [Online] Available www.denverwater.org (accessed 11 November 2010)..

Elmore, A.J., Mustard, J.F., Manning, S.J. and Lobell, D.B. (2000) Quantifying vegetation change in semiarid environments: precision and accuracy of spectral mixture analysis and the normalized difference vegetation index. *Remote Sensing of Environment*, **73**, 87–102.

Forster, B. (1983) Some urban measurements from Landsat data. *Photogrammetric Engineering and Remote Sensing*, **49** (12), 1693–1707.

Franke, J., Roberts, D.A., Halligan, K. and Menz, G. (2009) Hierarchical multiple endmember spectral mixture analysis (MESMA) of hyperspectral imagery for urban environments. *Remote Sensing of Environment*, **113**, 1712–1723.

Furby, S.L. and Campbell, N.A. (2001) Calibrating images from different dates to 'like-value' digital counts. *Remote Sensing of Environment*, **77**, 186–196.

Hansen, M.C., DeFries, R.S., Townshend, J.R.G., Sohlberg, R., Dimiceli, C. and Carroll, M. (2002) Towards an operational MODIS continuous field of percent tree cover algorithm: examples using AVHRR and MODIS data. *Remote Sensing of Environment*, **83**, 303–319.

Herold, M., Roberts, D.A., Gardner, M.E. and Dennison, P.E. (2004) Spectrometry for urban area remote sensing–Development and analysis of a spectral library from 350 to 2400nm. *Remote Sensing of Environment*, 91, 304–319.

Hess, L.L., Novo, E.M.L.M., Slaymaker, D.M., *et al*. (2002) Geocoded digital videography for validation of land cover mapping in the Amazon basin. *International Journal of Remote Sensing*, **23**(7), 1527–1556.

IBGE, Instituto Brasileiro de Geografia e Estatstica (2000) *Censo Demográfico*. [Online] Available at http://www.ibge.gov.br (accessed 11 November 2010).

Jensen, J.R. (1983) Biophysical remote sensing. *Annals of the Association of American Geographers*, **73**(1), 111–132.

Ji, M. and Jensen, J.R. (1999) Effectiveness of subpixel analysis in detecting and quantifying urban imperviousness from Landsat Thematic Mapper Imagery. *Geocarto International*, **14**(4), 31–39.

Kressler, F.P. and Steinnocher, K.T. (2001) Monitoring urban development using satellite images, in *Remote Sensing of Urban Areas* (ed. C. Jürgens), Regensburger Geographische Schriften, Regensburg, pp. 140–147.

Kruse, F.A., Lefkoff, A.B., Boardman, J.W., *et al*. (1993) The spectral image processing system (SIPS – interactive visualization and analysis of imagining spectrometer data. *Remote Sensing of Environment*, **44**, 145–163.

Kuemmerle, T., Röder, A., and Hill, J. (2006) Separating grassland and shrub vegetation by multidate pixel-adaptive spectral mixture analysis. *International Journal of Remote Sensing*, **27**(15), 3251–3271.

Lambin, E.F., Geist, H.J., and Lepers, E. (2003) Dynamics of land-use and land-cover change in tropical regions. *Annual Review of Environment and Resources*, **28**, 205–241.

Lee, S. and Lathrop, R.G. (2006) Subpixel analysis of Landsat ETM+ using self-organizing map (SOM) neural networks for urban land cover characterization. *IEEE Transactions on Geoscience and Remote Sensing*, **44**(6), 1642–1654.

Lepers, E., Lambin, E.F., Janetos, A.C. *et al*. (2005) A synthesis of information on rapid land-cover change for the period 1981–2000. *BioScience*, **55**(2), 115–124.

Milesi, C., Running, S.W., Elvidge, C.D., Dietz, J.B., Tuttle, B.T., and Nemani, R.R. (2005) Mapping and modeling the biogeochemical cycling of turf grasses in the United States. *Environmental Management*, **36**(3), 426–438.

Moudon, A.V. (1997) Urban morphology as an emerging interdisciplinary field. *Urban Morphology*, **1**, 3–10.

Myint, S.W. and Okin, G.S. (2009) Modelling land-cover types using multiple endmember spectral mixture analysis in a desert city. *International Journal of Remote Sensing*, **30**(9), 2237–2257.

NCDC Native Communities Development Corporation, Imaging and Mapping Division (2006) *Urban Tree Canopy Inventory: City and County of Denver, Colorado*. NCDC Contract #29-114-1031. submitted to The City and County of Denver, CO, 11 pp.

Nowak, D.J., Noble, M.H., Sisinni, S.M., and Dwyer, J.F. (2001) People and trees. *Journal of Forestry*, **99**(3), 37–42.

Okin, W.J., Okin, G.S., Roberts, D.A. and Murray, B. (1999) Multiple endmember spectral mixture analysis: endmember choice in an arid shrubland, in *AVIRIS Airborne Geoscience Workshop Proceedings*, JPL Publications, Pasadena.

Painter, T.H., Roberts, D.A., Green, R.O. and Dozier, J. (1998) The effect of grain size on spectral mixture analysis of snow-covered area from AVIRIS data. *Remote Sensing of Environment*, **65**, 320–332.

Petaki, D.E., Ehleringer, J.R., Flanagan, L.B., *et al*. (2006) Urban ecosystems and the North American carbon cycle. *Global Change Biology*, **12**, 2092–2102.

Powell, R.L. and Roberts, D.A. (2008) Characterizing variability of the urban physical environment for a suite of cities in Rondônia, Brazil. *Earth Interactions*, **12**, Paper No. 13

Powell, R.L., Roberts, D.A., Dennison, P.E. and Hess, L.L. (2007) Sub-pixel mapping of urban land cover using multiple endmember spectral mixture analysis: Manaus, Brazil. *Remote Sensing of Environment*, **106**, 253–267.

Rashed, T. (2008) Remote sensing of within-class change in urban neighborhood structures. *Computers, Environment and Urban Systems*, **32** (5), 343–354.

Rashed, T., Weeks, J.R., Roberts, D., Rogan, J., and Powell, R. (2003) Measuring the physical composition of urban morphology using multiple endmember spectral mixture models. *Photogrammetric Engineering and Remote Sensing*, **69** (9), 1011–1020.

Rashed, T., Weeks, J.R., Stow, D. and Fugate, D. (2005) Measuring temporal compositions of urban morphology through spectral mixture analysis: towards a soft approach to change analysis in crowded cities. *International Journal of Remote Sensing*, **26** (4), 699–718.

Ridd, M.K. (1995) Exploring a V-I-S (vegetation-impervious surface-soil) model for urban ecosystem analysis through remote sensing: comparative anatomy for cities. *International Journal of Remote Sensing*, **16** (12), 2165–2185.

Roberts, D.A., Batista, G.T., Pereira, J.L.G., Waller, E.K., and Nelson, B.W. (1998a) Change identification using multitemporal spectral mixture analysis: applications in Eastern Amazonia, in: *Remote Sensing Change Detection: Environmental Monitoring Methods and Applications* (eds R.S. Lunetta and C.D. Elvidge), Ann Arbor Press, Chelsea, MI, pp. 137–161.

Roberts, D.A., Gardner, M., Church, R., Ustin, S., Scheer, G., and Green, R.O. (1998b) Mapping chaparral in the Santa Monica Mountains using multiple endmember spectral mixture models. *Remote Sensing of Environment*, **65**, 267–279.

Roberts, D.A. Numata, I., Holmes, K., *et al.* (2002) Large area mapping of land-cover change in Rondônia using multitemporal spectral mixture analysis and decision tree classifiers. *Journal of Geophysical Research*, **107** (D20), 8073, doi:10.1029/2001JD000374.

Roberts, D.A., Dennison, P.E., Gardner, M.E., Hetzel, Y., Ustin, S.L., and Lee, C.T. (2003) Evaluation of the potential of Hyperion for fire danger assessment by comparison to the Airborne Visible/Infrared Imaging Spectrometer. *IEEE Transactions on Geoscience and Remote Sensing*, **41** (6), 1297–1310.

Sabol, D.E., Adams, J.B. and Smith, M.O. (1992) Quantitative subpixel spectral detection of targets in multispectral images. *Journal of Geophysical Research*, **97** (E2), 2659–2672.

Sánchez-Rodriguez, R. Seto, K.C., Simon, D., Solecki, W.D., Kraas, F., and Laumann, G. (2005) *Science Plan: Urbanization and Global Environmental Change*, IHDP Report No. 15. International Human Dimensions Programme on Global Environmental Change: Bonn, Germany.

Schwarz, M. and Zimmermann, N.E. (2005) A new GLM-based method for mapping tree cover continuous fields using regional MODIS reflectance data. *Remote Sensing of Environment*, **95**, 428–443.

Small, C. (2001) Estimation of urban vegetation abundance by spectral mixture analysis. *International Journal of Remote Sensing*, **22** (7), 1305–1334.

Small, C. (2005) A global analysis of urban reflectance. *International Journal of Remote Sensing*, **26** (4), 661–681.

Small, C. and Lu, J.W.T. (2006) Estimation and vicarious validation of urban vegetation abundance by spectral mixture analysis. *Remote Sensing of Environment*, **100**, 441–456.

Smith, M.O., Ustin, S.L., Adams, J.B., and Gillespie, A.R. (1990) Vetetation in deserts: I. A regional measure of abundance from multispectral images. *Remote Sensing of Environment*, **31**, 1–26.

Song, C. (2005) Spectral mixture analysis for subpixel vegetation fractions in the urban environment: how to incorporate endmember variability. *Remote Sensing of Environment*, **95**, 248–263.

Tompkins, S., Mustard, J.F., Pieters, C.M. and Forsyth, D.W. (1997) Optimization of endmembers for spectral mixture analysis. *Remote Sensing of Environment*, **59**, 472–489.

Townshend, J.R.G., Huang, C., Kalluri, S.N.V., DeFries, R.S., Liang, S., and Yang, K. (2000) Beware of per-pixel characterization of land cover. *International Journal of Remote Sensing*, **21** (4), 839–843.

Weng, Q., Hu, X. and Lu, D. (2008) Extracting impervious surfaces from medium spatial resolution multispectral and hyperspectral imagery: a comparison. *International Journal of Remote Sensing*, **29** (11), 3209–3232.

Weng, Q. and Lu, D. (2009) Landscape as a continuum: an examination of the urban landscape structures and dynamics of Indianapolis City, 1991–2000, by using satellite images. *International Journal of Remote Sensing*, **30** (10), 2547–2577.

Whyte, A. (1985) Ecological approaches to urban systems: retrospect and prospect. *Nature and Resources*, **21** (1), 13–20.

Woodcock, C.E. and Strahler, A.H. (1987) The factor of scale in remote sensing. *Remote Sensing of Environment*, **21**, 311–332.

Wu, C. (2009) Quantifying high-resolution impervious surfaces using spectral mixture analysis. *International Journal of Remote Sensing*, **30**(11), 2915–2932.

Wu, C. and Murray, A.T. (2003) Estimating impervious surface distribution by spectral mixture analysis. *Remote Sensing of Environment*, **84**, 493–505.

Yuan, F., Sawaya, K.E., Loeffelholz, B.C., and Bauer, M.E. (2005) Land cover classification and change analysis of the Twin Cities (Minnesota) Metropolitan Area by multitemporal Landsat remote sensing. *Remote Sensing of Environment*, **98**, 317–328.

Zipperer, W.C., Wu, J., Pouyat, R.V. and Pickett, S.T.A. (2000) The application of ecological principles to urban and urbanizing landscapes. *Ecological Applications*, **10**(3), 685–688.

9

An object-oriented pattern recognition approach for urban classification

Soe W. Myint and Douglas Stow

In contrast to subpixel and per-pixel image classification approaches, object-based image analysis (OBIA) attempts to exploit spatially and spectrally similar groups of pixels in order to identify objects within an imaged scene. Object-based approaches are most applicable to high spatial resolution data, where objects of interest are larger than the ground resolution element. Such objects in urban scenes could be related to natural features of urban landscapes (e.g., trees and lakes) or man-made features (e.g., buildings or roads). This chapter introduces readers to the principles of OBIA and demonstrates how it can be applied to achieve satisfactory accuracy in urban mapping. We employed two case studies with two example subsets extracted from Quickbird multispectral satellite data and demonstrated two object-based analysis procedures, namely decision rule (i.e., membership function) and nearest neighbor classifiers. The object-oriented classification employed to identify urban classes in this chapter is specifically based on Definiens software and the various routines it contains to achieve OBIA. However, an overview of the object-oriented approach, parameters generally available for object analysis, image segmentation procedure, rule set approaches, nearest neighbor classifier using training samples, and limitations/uncertainties associated with object-based techniques reported in this chapter are applicable to urban mapping using any OBIA software.

Urban Remote Sensing: Monitoring, Synthesis and Modeling in the Urban Environment, First Edition. Edited by Xiaojun Yang.
© 2011 John Wiley & Sons, Ltd. Published 2011 by John Wiley & Sons, Ltd.

9.1 Introduction

Remotely sensed image data have been used extensively for many years for the identification and mapping of land-use and land-cover (LULC) classes for urban environments. Image analysis techniques such as object-based image analysis (OBIA) have been developed in the recent past to facilitate semiautomated mapping of LULC classes. For instance, the Definiens commercial OBIA software utilizes image segmentation and object-based classification to delineate and classify objects within an imaged scene. Such software automates processes and incorporates expert knowledge to enable object-based classification. A further benefit of partially automating the OBIA process is that once a processing sequence has been created, a model incorporating this sequence can be distributed to and used by others, and only site-specific calibrations are normally necessary to achieve comparable and consistent results.

The human mind is extraordinarily adept at pattern recognition, as when extracting relevant image-derived information such as urban LULC classes from remotely sensed data. A human interpreter can recognize and identify a large number of urban objects such as roads, buildings, or LULC polygons using tone/color, texture, contextual, size, pattern, orientation, height, and shape information contained within fine spatial resolution satellite data. These are traditionally known as basic elements of image interpretation (Lillesand, Kiefer and Chipman, 2008). While our eyes and brains can distinguish the difference among different types of LULC sharing similar spectral responses (e.g., cement roads, sidewalks, and driveways), these tasks have been challenging for traditional single or per-pixel classifiers that attempt to identify LULC classes in a more automated fashion.

Attempts to perform more automated and effective image analysis and information extraction from digital image data have been pervasive for a few decades. Yet, fundamental advances in automated or semi-automated digital image analysis remain a challenge, especially when classifying detailed LULC classes from fine spatial resolution data. Particularly challenging is the generation of computer algorithms that perform in the same manner that the human brain functions to extract image information.

Object-based approaches to semi-automated LULC mapping have been a major focus area of remote sensing and image processing research in the past decade. Relative to single- or per-pixel approaches, OBIA attempts to exploit spatial relationships of groups of pixels in order to delineate and identify objects within an imaged scene (Benz et al., 2004). Object-based approaches are most applicable to the high or H-resolution remote sensing scene model (Strahler, Woodcock and Smith, 1986), where objects of interest are larger than the ground resolution element associated with a pixel. Such objects may be related to natural features of urban landscapes (e.g., trees and lakes) or human-made features (e.g., buildings or roads). Object-based image processing techniques have been developed to support environmental remote sensing for over 30 years (Ketting and Landgrebe, 1976). However, a greater research emphasis on such techniques has occurred in the past several years due to: (1) the greater availability of digital remote sensing image data having fine spatial resolution that are generally not amenable to achieving highly accurate mapping and monitoring results when generated with per-pixel image classification routines, and (2) the greater availability and affordability of high performance computers and object-based image processing software.

The purpose of this chapter is to introduce readers to the principles of OBIA and demonstrate how it can be applied to achieve satisfactory accuracy in urban LULC mapping. We employed two case studies with two example subset images extracted from Quickbird multispectral satellite data and demonstrated two object-based classification approaches, namely rule set (i.e., membership function) and nearest neighbor classifiers. These classifiers are supported in the Definiens Developer 7.0 OBIA software. Background is provided from the Definiens software perspective. The software package uses a region-based, local mutual segmentation routine, a type of region growing approach to generate image objects or segments, prior to performing object-based classification (Baatz and Schäpe, 2000; Benz et al., 2004; Yu et al., 2006).

9.2 Object-oriented classification

As explained earlier, the object-oriented classification employed to map urban LULC classes in this chapter focuses specifically on Definiens software known as eCognition and the various routines it contains to achieve OBIA. Here we provide background on the components of an OBIA approach to LULC classification, while describing some of the specific processing steps and parameter choices associated with the Definiens software environment. Even though this chapter is specifically centered around the use of Definiens software package, an overview of object-oriented classification approach, consideration of many different features and parameters, image segmentation procedure in relation to scale levels, overview and general procedure to perform a rule set approach, nearest neighbor classifier using training samples, and limitations and uncertainties associated with object-based classification technique are useful and applicable to any image processing software that contains object-based image analysis functions.

9.2.1 Image segmentation

The first step in the process of an object-oriented analysis is to segment an image, representing a scene or the ground area covered by the image extent, into image objects (Lee and Warner, 2006; Stow et al. 2007; Im, Jensen, and Tullis 2008). An image object is generally defined as a group of pixels sharing similar spectral and/or textural properties (Navulur, 2007). When we interpret a particular subset of the image, our brain tends to focus-in on small areas or patches consisting of homogeneous tone/color or texture that may have characteristic shapes or sizes. Hence, visually such portions of the image are recognized as objects. For example, we see and can delineate a green rectangular area and then identify it first as landscaped grassland, and then in its urban context, as the lawn of a park or an institutional complex. Similarly, a house, a tree, a group of trees, an open cement parking lot, a street segment, or a swimming pool can be considered as image objects. Thus, an image object is a group of connected pixels having similar or characteristic image properties. Each image object represents a specific, small region in an image, and an urban image contains many different objects. Image segmentation is a primary function of OBIA that splits an image into separated regions or objects (Myint et al., 2008). Lizarazo and Barros (2010) proposed a fuzzy-based image segmentation procedure that combines properties of fuzzy and crisp image regions for urban land-cover classes. Walker and Blaschke (2008) employed a two-tier segmentation procedure that originally segmented the entire scene with the same parameterization for all objects and re-segmented based on spectral heterogeneity of neighboring urban objects. Even though there are many different segmentation algorithms exist (Pal and Pal 1993, Blaschke, Burnett and Pekkarinen, 2004), there are only very few operational commercial software package available that support image segmentation and object-based classification. Currently, the majority of applications are built within the software environment found in Definiens (Benz et al. 2004; Walker and Blaschke, 2008). This software utilizes multiple segmentations based on a global heterogeneity criterion (Baatz and Schäpe, 2000). The segmentation function in Definiens software (Baatz and Schäpe, 1999 and 2000) is based on three parameters, namely shape (S_{sh}), compactness (S_{cm}), and scale (S_{sc}) parameters.

9.2.1.1 Compactness and shape factors

Users are required to specify values ranging from 0 to 1 for the shape and compactness factors to determine objects at different level of scales of image objects. These two parameters control the homogeneity of different objects. The shape factor adjusts spectral homogeneity vs. shape of objects. A lower value of the shape parameter leads to lesser influence of color on the segmentation. The compactness factor, balancing compactness and smoothness, determines the object shape between smooth boundaries and compact edges. The more compact image objects can be achieved by specifying a higher compactness value. The criteria for optimizing objects in relation to compactness is used when delineating image objects that are compact, and are separated from non-compact objects, but contain relatively weak spectral contrast (Definiens, 2007).

9.2.1.2 Object scale levels

Segmented objects are organized into image object levels also known as scale levels. An image object level serves as an internal working area for the object-based image analysis. The scale parameter that controls the object size that matches the user's required level of detail is considered the most crucial parameter of image segmentation. Different levels of object sizes can be determined by applying different numbers in the scale function. The higher number of scale (e.g., 50) generates larger homogeneous objects (smaller scale – lower level of detail), whereas the smaller number of scale (e.g., 10) will lead to smaller objects (larger scale). A smaller number used in the scale parameter is considered higher level in the segmentation procedure. The decision on the appropriate level of scale depends on the size of objects required to achieve the mapping goal(s). The software also allows users to assign different level of weights to different bands in the selected image during image segmentation. This type of scale is a spatially aggregated scale (few pixels vs. more pixels in objects). Each image object level consists of a layer of segmented objects. It is important to note that image layers are the data that exist in the original image when it is imported. However, object levels store image objects that represent different types of image data. Figure 9.1 demonstrates how hierarchical image segmentation delineates image objects at different levels.

An entire image can be segmented based on the three key parameters (shape, compactness, and scale), with parameter selection depending on the size and shape characteristics of objects associated with the classes of interest. At this stage, we do not know which objects belong to which class, but still need to assign predetermined classes to segmented objects.

9.2.2 Features

In Definiens (or eCognition) software, a feature is an attribute that represents certain information concerning segmented objects in

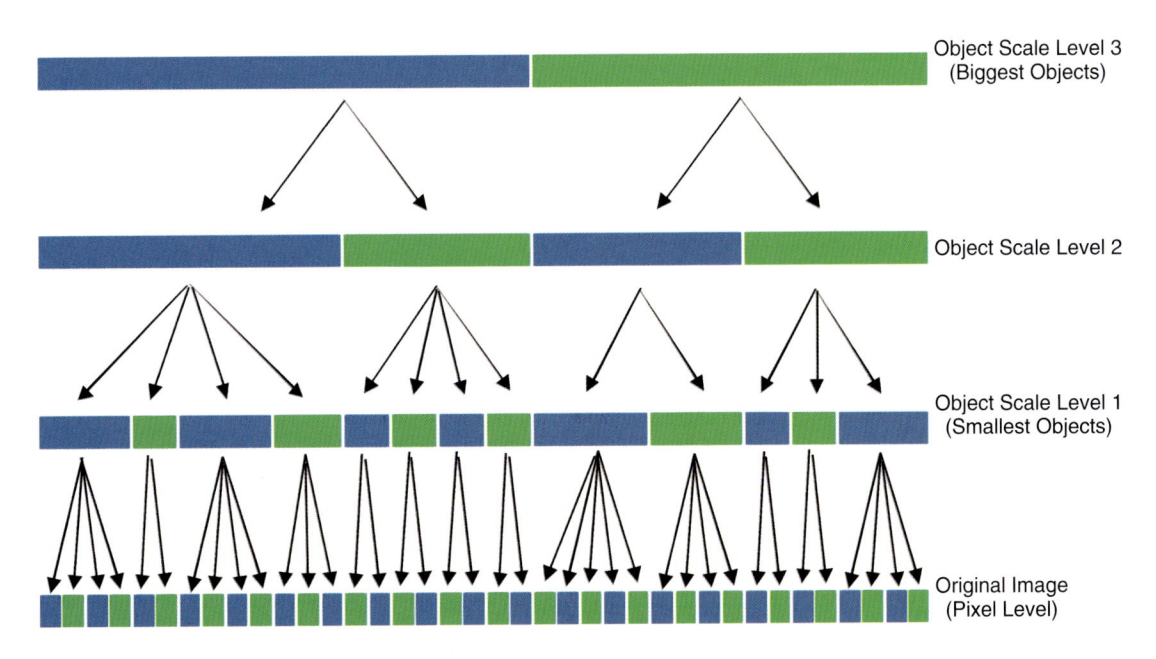

FIGURE 9.1 Image objects at each image object level.

an image. For example, measurements, attached data, or values can be part of features. There are several types of features available in the software namely Object Features, Class-Related Features, Scene Features, and Process-Related Features. Each type of feature includes different categories, and each category can have many different sub-categories. There may also be several features under each subcategory. Thus, features may be considered available bands or information channels containing different types of information.

For example, Object Features type contains several categories such as Customized, Layer Values, Shape, Texture, Variables, and Hierarchy. We have listed available sub-categories under each category below. They all are available under the type called Object Features.

1. Customized – Create New "Arithmetic Feature"; Create New "Relational Feature".
2. Layer Values – Mean, Standard Deviation; Pixel-based; To Neighbors; To Super-object; To Scene, Hue-Saturation-Intensity.
3. Shape – Generic, Position; To Super-object, Based on Polygons; Based on Skeletons.
4. Texture – Layer Value Texture Based on Sub-objects; Shape Texture Based on Sub-Objects; Texture After Haralick.
5. Variables – Create new 'Object Variable'.
6. Hierarchy.

As mentioned earlier, each subcategory may contain several different features depending on which category we consider. Some subcategories contain features of the original bands in the image (e.g., mean band 1, mean band 2, mean band 3, mean band 4). If we have added three principal component bands to the original Quickbird multispectral image data (four bands), the total bands in this case would be seven bands. In addition to the original bands, mean subcategory also generates two new features called Brightness and Max. Diff. For example, Brightness band represents an overall intensity measurement, which is the mean value of selected image layers. To obtain Max Diff. band, the minimum mean value belonging to an object is subtracted from its maximum value. By comparing the means of all layers belonging to an object are compared with each other, the maximum and minimum values can be obtained. Subsequently the result is divided by the brightness value (Definiens, 2007). In the above example, there will be seven features (seven bands) available under Mean or Standard Deviation subcategories. Mean and standard deviation features are the mean and standard deviation values of each feature type derived for all pixels within each segmented object in each band.

However, this is not the same situation for other sub-categories such as "Shape." For example, if we select Generic under Shape subcategory, there are 21 features that can be derived that are independent of the original bands. In other words, there are no Generic (under "Shape") features of the original bands. This is because a particular shape of a segmented object will be exactly the same regardless of the number of original bands. Features under this sub-category are provided in Table 9.1.

Other feature categories are organized in a different manner. For example, there are three sub-categories under "Texture." There are 24 different divisions under one of the subcategories called Texture After Haralick. If we select one of the 24 divisions called GLCM Homogeneity, we can have five more sub-divisions pertaining to the direction of texture features (i.e., All Directions, Direction 0°, Direction 45°, Direction 90°, Direction 135°). Under each sub-division, there are texture transformed bands (features) of the original spectral bands and GLCM Homogeneity (all direction) band. We can also create our own features such as normalized difference vegetation index (NDVI) under the type called "Customized."

9.2.3 Classifiers

Definiens software provides two general types of classifiers for assigning class labels to segmented objects, rule set approach and the nearest neighbor classifier. Before initiating the classification procedure either with the rule set approach or the nearest neighbor classifier, it is important to select a small subset of the data that represents fairly reasonably the entire study area and contains a class or a number of classes to be identified. To achieve this, we need to evaluate the whole data set and then choose a small area to start with. This procedure will not only save development time for developing classification rules, but also enables better understanding of the nature of the data and effectiveness of different rule sets and parameters. The processing times for smaller subsets can be a lot lower than testing every step on the whole data set. However, test results may not provide a reliable rule set or discriminant function for the entire dataset. We may still need to adjust the preliminary decision rules to achieve satisfactory results for the whole dataset.

9.2.3.1 Rule set approach

By using a user's expert knowledge we can define rules and constraints in the membership function to control the classification procedure. The rule set approach also known as membership function describes intervals or range of feature characteristics that determine whether the objects belong to a particular class or not. A rule set using only one feature (band) may not be able to identify a class with a satisfactory accuracy. In this case, we may need to use one or more additional rule sets to exclude some other covers that are not part of the selected class. However, in some cases, a threshold value in a particular feature (band) may still be reasonably effective in identifying some classes. For example, mean values of the original brightness value above a certain value identifies a particular class (same type of objects). The rule set approach or membership function is apparently a non-parametric rule. Hence, this approach is independent of the assumption that data values follow a normal distribution. This part of the analysis largely involves the development of strategy for transforming the way the human brain functions to recognize objects into a decision support system.

9.2.3.2 Nearest neighbor classifier

To classify image objects using the nearest neighbor classifier, we need to select features (e.g., original bands, transformed bands, indices), define training samples (objects), classify, review the outputs, and optimize the classification. The nearest neighbor classification procedure uses a set of training samples in a supervised manner to represent different classes in order to assign class values to segmented objects. The procedure consists of two steps:

TABLE 9.1 Available features belong to Generic sub-category under Shape category.

Feature Type	Category	Sub-category	Features
1. Object features	1.1 Customized		
	1.2 Layer Values		
	1.3 Shape	1.3.1 Generic	1.3.1.1 Area
			1.3.1.2 Asymmetry
			1.3.1.3 Border Index
			1.3.1.4 Border Length
			1.3.1.5 Compactness
			1.3.1.6 Density
			1.3.1.7 Elliptic Fit
			1.3.1.8 Elliptic Fit (legacy feature)
			1.3.1.9 Length
			1.3.2.0 Length/Width
			1.3.2.1 Main Direction
			1.3.2.2 Radius of Largest Enclosed Ellipse
			1.3.2.3 Radius of Largest Enclosed Ellipse (legacy feature)
			1.3.2.4 Radius of Smallest Enclosed Ellipse
			1.3.2.5 Radius of Smallest Enclosed Ellipse (legacy feature)
			1.3.2.6 Rectangular Fit
			1.3.2.7 Rectangular Fit (legacy feature)
			1.3.2.8 Roundness
			1.3.2.9 Roundness (legacy feature)
			1.3.3.0 Shape Index
			1.3.3.1 Width
		1.3.2 Position	
		1.3.3 To super object	
		1.3.4 Based on Polygons	
		1.3.5 Based on Skeletons	
	1.4 Texture		
	1.5 Variables		
	1.6 Hierarchy		

(1) identify certain image objects as samples, and (2) classify image objects due to their nearest sample neighbors in feature space (Definiens, 2008). The nearest neighbor classifier is also based on a non-parametric rule and is therefore independent of normally distributed input data. The nearest neighbor approach enables transportability of a classification system to other areas, requiring only the selection or modification of new objects (training samples) until a satisfactory result is obtained (Ivits and Koch, 2002). Application of the nearest neighbor method is also advantageous when using spectrally similar classes that are not well separated using a few features or just one feature (Definiens, 2008).

There are two options available with the nearest neighbor function in Definiens software, namely (1) Standard Nearest Neighbor, and (2) Nearest Neighbor. The Standard Nearest Neighbor option automatically select mean values of objects for all the original bands in the selected image, whereas the second option requires users to specify selected features (e.g., shape, texture, hierarchy).

9.3 Data and study area

We used two subsets of a Quickbird image data over the city of Phoenix acquired on 22 August 2007 to demonstrate how decision rules and nearest neighbor classifiers can be used to identify land cover classes using an object-oriented approach. Figure 9.2 presents the first subset used for different decision rules to extract swimming pools and Figure 9.3 shows the second subset used

FIGURE 9.2 A false color infrared Quickbird subset covering swimming pools and surrounding land covers.

FIGURE 9.3 A subset covering a commercial and a residential area.

for nearest neighbor classifier to identify multiple classes using different composite bands. The dataset has 2.4 m spatial resolution with four wavebands: blue – B1 (0.45–0.52μm), green – B2 (0.52–0.60μm), red – B3 (0.63–0.69μm), and near infrared – B4 (0.76–0.90μm). The radiometric resolution of the dataset is 11 bits. The subsets cover urban segments (commercial and residential), grassland, unmanaged soil, desert landscape, and pool, giving a general coverage of urban land-use and land-cover classes. The selected land-cover classes that we identified for the study include buildings, shadows, other impervious surfaces (e.g., roads and parking lots), exposed soil, trees and shrubs, grass, and swimming pool. These particular land-cover classes are important to the analysis of the urban energy budget using a model that requires them (Grimmond and Oke, 2002). In addition to the original bands, principal component analysis (PCA) bands stretched to 16 bit were used in the analysis.

9.4 Methodology

To demonstrate the applicability and effectiveness of two different classification approaches within object-oriented image analysis paradigm, we used two subsets of a Quickbird multispectral data to identify and map urban land-cover types. In the first application example, we indentified two swimming pools in a park using different decision and expert system rules. We employed the nearest neighbor classifier to identify different urban land-cover classes using two different combinations of multispectral bands in the second application example.

We tested four different scale levels with the same set of segmented parameter values for both application examples. The default values of shape parameter $(S_{sh})0.1$ was used to give less weight on shape and give more attention on spectrally more homogeneous pixels for image segmentation. We also used default value of compactness parameter $(S_{cm})0.5$ to balance compactness and smoothness of objects equally. We employed four different scale levels (S_{sc}) to segment objects: 10, 25, 50, and 100. Figure 9.4 shows segmented images of a subset at object scale level 1, scale level 2, scale level 3, and scale level 4 (scale parameters 10, 25, 50, and 100) using shape $(S_{sh})0.1$ and compactness $(S_{cm})0.5$.

9.4.1 Rule-based detection of swimming pools

One's expert knowledge is the most important tool for creating a rule set for extracting a particular LULC class. In this case, we need to use our expert knowledge to discriminate swimming pools from other land covers. We attempted to delineate and identify two swimming pools within a subset of the full study area. First, we visually analyzed many different image feature types (i.e., wavebands and transforms) by displaying and interpreting them on the screen. This was to see if swimming pools stand out as unique objects of particular tones or colors in the various features. We determined that mean values of the original Quickbird band 4, PCA band 2, and brightness band were effective in identifying pools. The brightness band is a surrogate for surface albedo and represents an overall intensity measurement, which is the mean value of selected image layers. Figure 9.5 (a–c) represent mean values of image segments for the original band 4 (near infrared band), PCA band 3, and brightness band features, respectively. The pool objects observed in these three features/bands appear to be darker than surrounding areas. There is high potential for discriminating pools from other areas based on these three features, with some adjustment of decision rules.

We used the Definiens "update range" function for each of the three features to determine threshold values for delineating and identifying pools. Four different scale levels were used to obtain the threshold values, based on visual analysis and iteration. We found the mean band 4 feature values to be less than 10 000, PCA 2 feature value less than 22 000, and brightness feature value less than 22 000 at object scale level 4 lead to Pool class. The output maps derived using the above rule sets are presented in Figs 9.6, 9.7, and 9.8. The pool areas extracted with the rules used for the mean value of the original band 4 and brightness features also include shadow areas cast by trees and buildings in the image. Hence, shadow areas were removed using a different feature and rule set.

After evaluating different combinations of scale parameters, input features, and decision rules, we determined the threshold value of the original band 5 at the same scale level that was effective for identifying shadows in the image. Figure 9.9 shows the output map of shadow areas in the scene using the above rule set. We used the shadow map to exclude shadow objects that had

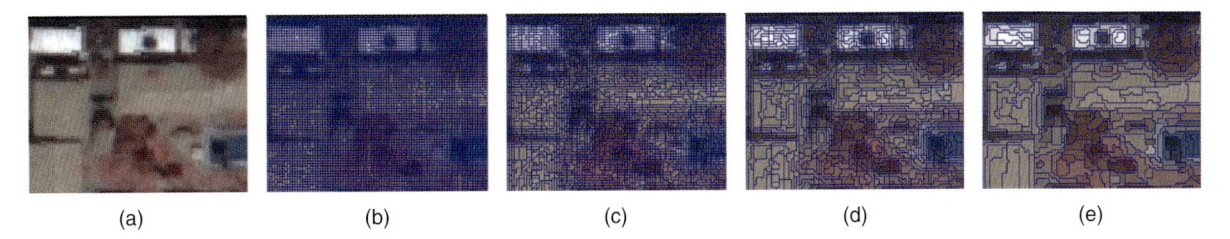

CHAPTER 9 AN OBJECT-ORIENTED PATTERN RECOGNITION APPROACH FOR URBAN CLASSIFICATION 135

FIGURE 9.4 A subset image and segmented images at different object scales using shape (S_{sh})0.1 and compactness (S_{cm})0.5. (a) Original subset; (b) level 1 (scale parameter 10); (c) level 2 (scale parameter 25), (d) level 3 (scale parameters 50), (e) level 4 (scale parameter 100).

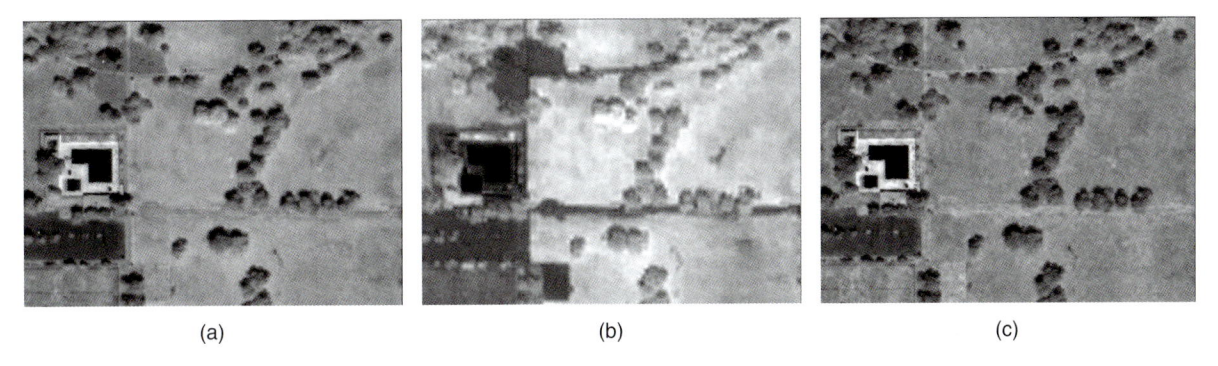

FIGURE 9.5 (a) Mean value of the original band 4 feature (Dark tone in pool areas). (b) Mean value of PCA 2 feature (Dark tone in pool areas), (c) Mean value of brightness feature (Dark tone in pool areas).

FIGURE 9.6 Mean value of the original band 4 feature (a threshold value of 10 000 to extract pools – blue areas). Note: Some other land-cover areas were mistakenly identified as pools.

FIGURE 9.7 Mean value of PCA 2 feature (a threshold value of 22 000 to extract pools – blue areas). Note: No mistakenly identified areas.

been mistakenly identified as pools, when generated by the mean values of the original band 4 and brightness features.

Figures 9.10 and 9.11 show the output maps after masking the shadows. It was found that the rule set with PCA 2 feature did not produce shadows in the area and hence, the above procedure was not necessary. However, we provided the output of Rule set 4 (PCA 2 feature value less than 22 000 at scale level 1) in Fig 9.12. To demonstrate the effect of using different image object levels we also identified the pool using the same approach with PCA 2 feature at level 1. By visual inspection, the output at level 1 provided more realistic representations of pool edges than the one produced at level 4. However, we found three small objects identified as pools adjacent to the actual pools, though they are not very noticeable. It is uncertain which one of these four outputs has the highest accuracy since delineated features are similar among the four outputs. Thus, an accuracy assessment was conducted for all four classified outputs. We provided the rule sets developed to extract pools in Table 9.2.

We produced error matrices in order to analyze and evaluate each approach. These error matrices show the contingency of the class to which each pixel truly belongs (columns) on the map unit to which it is allocated by the selected analysis (rows). From the error matrix, overall accuracy, producer's accuracy, user's accuracy, and kappa coefficient were generated.

To evaluate different rule sets applied to extract swimming pool more effectively we digitized the swimming pool in the first subset using a visual interpretation approach via heads-up digitizing to produce a reference map that contains two classes (i.e., pool and non-pool class). We used each pixel in the whole

FIGURE 9.8 Mean value of brightness feature (a threshold value of 22 000 to extract pools – blue areas). Note: Some other land-cover areas were mistakenly identified as pools.

FIGURE 9.10 Output generated by rule set 1 (Mean band 4 feature < 10 000 and not mean PCA 1 < 19 000 at scale level 4). Note: A few other land-cover areas were mistakenly identified as pools.

FIGURE 9.9 Mean value of the original band 5 feature (a threshold value of 19 000 to extract shadows – yellow areas).

FIGURE 9.11 Output generated by rule set 2 (Mean brightness feature < 22 000 and not mean PCA 1 < 19 000 at scale level 4). Note: A few other land-cover areas were mistakenly identified as pools.

TABLE 9.2 Selected rule sets to identify swimming pools.

	Decision Rule
Rule 1	Means band 4 feature < 10,000 and not mean PCA 1 < 19,000 at scale level 4
Rule 2	Means brightness feature < 22,000 and not mean PCA < 19,000 at scale level 4
Rule 3	PCA 2 feature value less than 22,000 at scale level 4
Rule 4	PCA 2 feature value less than 22,000 at scale level 1

(Congalton, 1991). To be consistent and for precise comparison purposes, we applied the same sample points generated for the output generated by the first data set (mean values of the original bands and PCA bands 1, 2, and 3) to the output produced by the second dataset (mean values of the original bands, brightness band, and maximum difference band).

9.4.2 Nearest neighbor classifier to extract urban land covers

We attempted to identify seven urban classes including building, grass, impervious, swimming pool, shadow, soil, and tree. There are two features or data sets used for the analysis: (1) mean values of the original bands and PCA bands 1, 2, and 3; and (2) mean values of the original bands, brightness band, and maximum difference band. We observed that tree land-cover class stands out as bright objects and unmanaged soil cover turned out to be one of the darkest object types for the maximum difference band

subset to perform the accuracy assessment instead of randomly sampling pixels from the image. The reference map can be assumed to contain a negligible amount of error. There are total of 175 062 pixels in the image, and only 1973 pixels belong to the pool area.

We selected 200 samples points that led to approximately 30 points per class (seven total classes) for the accuracy assessment. A minimum of 15 points per class was selected to generate 200 test points using a stratified random sampling approach

FIGURE 9.12 Output generated by the rule set 4 (PCA 2 feature value less than 22 000 at scale level 1). Note: A few pixels mistakenly identified as pool.

FIGURE 9.13 Output map produced by the nearest neighbor classifier using mean value of all the original bands and the first three PCA bands. Note: soil = yellow; building = white; impervious = gray; trees = dark green; pool = cyan, shadows = black; grass = light green.

and, buildings appeared to be brighter than other land-cover objects in the brightness band.

Substantial confusion in spectral signatures between unmanaged soil and some impervious surface types was also evident in the image subset. For example, unmanaged soil in a false color composite display (near infrared, red visible, and green visible bands in red, green, and blue) appears as light brown or dark colors, as do some rooftop materials and roads. After examining different levels of object scales, we felt that an object scale level of 1 (scale parameter 10) was appropriate for this dataset. Hence, we decided to use scale level 1 for the nearest neighbor classifier.

The object-oriented approach allows additional selection or modification of new objects (training samples) each time after performing a nearest neighbor classification quickly until the satisfactory result is obtained (Myint, *et al.*, 2008). Hence, we attempted to use many different training samples (objects) per class as a trial and error approach. Figures 9.13 and 9.14 present output maps showing swimming pool and other urban land-cover classes generated by the nearest neighbor classifier with the above approach with the above two sets of features. There were 29 soil object samples, 15 building samples, 28 impervious samples, six tree samples, two pool samples, nine shadow samples, and two grass samples used to identify the selected classes using mean values of the original bands and PCA bands 1, 2, and 3. Hence, there were a total of 91 training samples for the first combination of features. We used 15 soil object samples, 16 building samples, 17 impervious samples, 14 tree samples, two pool samples, six shadow samples, and two grass samples to identify the same classes using mean values of the original bands, brightness, and maximum difference (Max. Diff.). There were 72 total training samples for this combination.

9.5 Results and discussion

9.5.1 Decision rule set to extract pool

Classification accuracies are not significantly different among the four outputs generated by different decision rules or object scale levels. The overall accuracies achieved by the rule sets 1, 2, 3, and 4 for the pool detection are 99.59, 99.82, 99.69,

FIGURE 9.14 Output map produced by the nearest neighbor classifier using mean value of all the original bands, Brightness band, and max. Dif. Band. Note: soil = yellow; building = white; impervious = gray; trees = dark green; pool = cyan, shadows = black; grass = light green.

and 99.65% respectively (Tables 9.3, 9.4, 9.5, 9.6). However, we should not be satisfied with these overall accuracies since only 1973 pixels belong to the pool area and the entire image contains 175 062 pixels. Hence, the overall accuracy in this case does not reflect appropriately the effectiveness of a particular rule set in identifying the swimming pool.

Producer's accuracies produced by the rule sets 1, 2, 3, and 4 are 82.82, 88.60, 95.39, and 94.48% respectively. While rule set 3 yielded the highest producer's accuracy, it had a slightly lower user's accuracy (80.77%). This implies that even though 95.93% of the pool area has been correctly identified as pool, only 80.77% of the area identified as pool in the output map is truly of that category. The highest user's accuracy resulted from rule set 2 (95.26%). The same rule s*et als*o produced relatively high producer's accuracy (88.60%). In fact this was the rule set that yielded the highest overall accuracy. However, we cannot make a conclusive decision on which one of the two rule sets is the best to extract the pool. In practice, one could apply both

TABLE 9.3 Overall accuracy, producer's accuracy, user's accuracy, and Kappa coefficient produced by the rule set 1.

	Referece Data Non-Pool	Referece Data Pool	Total Pixels	Producer's Accuracy	User's Accuracy
Non-Pool	172705	339	173044	99.78%	99.80%
Pool	384	1634	2018	82.82%	80.97%
Total Pixels	173089	1973	175062		

Overall Accuracy = 99.59%
Kappa Coeficient = 0.82%

TABLE 9.4 Overall accuracy, producer's accuracy, user's accuracy, and Kappa coefficient produced by the rule set 2.

	Referece Data Non-Pool	Referece Data Pool	Total Pixels	Producer's Accuracy	User's Accuracy
Non-Pool	173002	225	173227	99.95%	99.87%
Pool	87	1748	1835	88.60%	95.26%
Total Pixels	173089	1973	175062		

Overall Accuracy = 99.82%
Kappa Coeficient = 0.92%

approaches to the entire dataset and evaluate the outputs visually and quantitatively.

9.5.2 Nearest neighbor classifier to extract urban land covers

The second application example and corresponding data set (mean values of the original bands, brightness band, and maximum difference band) yielded an overall accuracy of 73.50% (Table 9.7). The two lowest producer's accuracies were associated with Grass (47.83%) and Trees (50.00%). The two lowest user's accuracies were associated with Impervious (57.38%) and Trees (66.67%). This implies that there is some signature confusion among grass, trees, and impervious surface.

The higher overall accuracy (87.50%) resulted with the first set of data (mean values of the original bands and PCA bands 1, 2, and 3) (Table 9.8). This is a very high accuracy for an urban LULC map derived semi-automatically from fine spatial resolution data. The two lowest producer's accuracies were associated with shrubs (75.86%) and impervious (80.00%). The two lowest user's accuracies were associated with pool (68.18%) and soil (81.25%). This implies that greater signature confusion occurs among shrubs, impervious, pool, and soil classes. Low producer's and user's accuracies were not consistent between the two different

TABLE 9.5 Overall accuracy, producer's accuracy, user's accuracy, and Kappa coefficient produced by the rule set 3.

	Referece Data Non-Pool	Referece Data Pool	Total Pixels	Producer's Accuracy	User's Accuracy
Non-Pool	172641	91	172732	99.74%	99.95%
Pool	448	1882	2330	95.39%	80.77%
Total Pixels	173089	1973	175062		

Overall Accuracy = 99.69%
Kappa Coeficient = 0.87%

TABLE 9.6 Overall accuracy, producer's accuracy, user's accuracy, and Kappa coefficient produced by the rule set 4.

	Referece Data Non-Pool	Referece Data Pool	Total Pixels	Producer's Accuracy	User's Accuracy
Non-Pool	172585	109	172694	99.71%	99.94%
Pool	504	1864	2368	94.48%	78.72%
Total Pixels	173089	1973	175062		

Overall Accuracy = 99.65%
Kappa Coeficient = 0.86%

TABLE 9.7 Overall accuracy, producer's accuracy, user's accuracy, and Kappa coefficient produced by the nearest neighbor classifier with mean value of the original bands and three PCA bands.

	\multicolumn{7}{c}{Referece Data}	Total	Producer's	User's
	B	I	T	G	P	S	Sh	Pixels	Accuracy	Accuracy
B	31	0	0	0	0	3	0	34	93.94%	91.18%
I	1	32	0	1	0	0	2	36	80.00%	88.89%
T	0	0	27	1	0	0	0	28	93.10%	96.43%
G	0	0	−2	22	0	0	0	24	91.67%	91.67%
P	0	5	0	0	15	0	2	22	93.75%	68.18%
S	1	2	0	0	0	26	3	32	89.66%	81.25%
Sh	0	1	0	0	1	0	22	24	75.86%	91.67%
Total Pixels	33	40	29	24	16	29	29	200		

Overall Accuracy = 87.50%
Kappa Coeficient = 0.85%

TABLE 9.8 Overall accuracy, producer's accuracy, user's accuracy, and Kappa coefficient produced by the nearest neighbor classifier with mean value of the original bands, brightness band, and maximum difference band.

	\multicolumn{7}{c}{Referece Data}	Total	Producer's	User's
	B	I	T	G	P	S	Sh	Pixels	Accuracy	Accuracy
B	33	2	1	4	1	0	0	41	91.67%	80.49%
I	2	35	9	1	0	8	6	61	85.37%	57.38%
T	0	1	12	5	0	0	0	18	50.00%	66.67%
G	0	0	0	11	0	0	1	12	47.83%	91.67%
P	0	0	0	0	14	0	2	116	93.33%	87.50%
S	1	1	2	2	0	25	2	33	76.76%	75.76%
Sh	0	2	0	0	0	0	0	17	60.71%	89.47%
Total Pixes	36	41	24	23	15	33	28	200		

Overall Accuracy = 73.50%
Kappa Coeficient = 0.68%

sets of features. Differences in accuracies may partly be due to the fact that different training samples were taken for both sets of data. The same training samples should not be used for different sets of band combinations as is normally done with supervised per-pixel classifiers. This is because objects derived from one set of features (different combination of bands) are different from those generated by another set of features, even though segmentation parameters (compactness, shape) and scale parameter may be the same. The overall classification accuracy could be increased by more effectively identifying swimming pools using the rule set approach developed in this study and overlaying the output pool map with the land-use and land-cover map generated by nearest neighbor approach.

Conclusion

The rule set classifier is generally more time consuming and difficult to implement than identifying classes using a nearest neighbor classifier. We recommend that different rule sets be developed and that the effectiveness of potentially strong decision rules be tested with a small subset. As demonstrated in this chapter, it is a good idea to subset a small part of the data that represents reasonably the entire study area and all classes of interest before performing the classification for the whole data set. If one feels that a small subset may not represent most of the study area, then several different small subsets could be used to develop the segmentation and classification parameters and procedures. However, selected decision rules will still need to be tested for an entire dataset upon developing and testing different decision rules. The threshold values and data ranges used for different features may need to be modified for the entire dataset, especially when implementing different decision rules for purposes of separating overlapping classes.

A general procedure to perform the image classification phase based on a rule set approach is:

1 Understand the overview of the classification specificity, nature of classes and data, and potential signature confusion among selected classes.
2 Select the image subset.
3 View image (e.g., waveband) features and develop a classification strategy.
4 Transform the strategy into a rule set.
5 Classify the subset image.
6 Qualitatively evaluate the results.

7. Refine the rule set, if necessary.
8. Reclassify (if not unsatisfactory, develop a new strategy).
9. Apply the best rule set to the entire dataset.
10. Classify the entire image.
11. Export results.

The same strategy should also be considered for the classification with nearest neighbor classifier. However, the results obtained from testing data (small subsets) should be considered to provide general guidance for implementing the nearest neighbor classifier. One still needs to select object training samples when using the entire dataset. The classification accuracy is greatly influenced by the selection of training samples. It is important that one selects as many spectrally distinct training objects as possible when classifying an entire study area.

Our experience with Definiens/eCognition software package is that some objects that appear visually similar and belong to two different classes may still be identified separately and accurately by selecting many different types of training samples from within problem areas. This may stem for the Definiens classifiers being based on non-parametric rules that are independent of a normal distribution. The object-oriented approach allows additional selection or modification of new training samples (training objects) after multiple iterations of a nearest neighbor classification, until a satisfactory result is obtained. There are many possible combinations of functions, parameters, object scale levels, features, and variables available with the object-oriented approach. The successful use of nearest neighbor classifier in the object-based paradigm largely relies on repeatedly modifying training objects as a trial-and-error approach. Nonetheless, the object-based classification approach seems to be very effective at classifying urban LULC categories, especially when based on fine spatial resolution multispectral imagery.

References

Baatz, M. and A. Schäpe. (1999) Object-Oriented and Multi-Scale Image Analysis in Semantic Networks. In *Proceedings of the 2nd International Symposium on Operationalization of Remote Sensing, 16–20 August 1999*. Enschede, ITC.

Baatz, M. and Schäpe, A. (2000) Multiresolution segmentation-an optimization approach for high quality multi-scale image segmentation, in *Angewandte Geographische Informationsverarbeitung* (eds XJ. Strobl, T. Blaschke and G. Griesebner), Wichmann-Verlag, Heidelberg, pp. 12–23.

Benz, U.C., Hofmann, P., Willhauck, G., Lingenfelder, I. and Heynen, M. (2004) Multi-resolution, object-oriented fuzzy analysis of remote sensing data for GIS-ready information. *ISPRS Journal of Photogrammetry and Remote Sensing*, **58**, 239–258.

Blaschke, T., Burnett, C. and Pekkarinen, A. (2004) New contextual approaches using image segmentation for object-based classification, in *Remote Sensing Image Analysis: Including the Spatial Domain* (eds F. de Meer and S. de Jong), Kluwer Academic, Dordrecht, pp. 211–236.

Congalton, R.G. (1991) A review of assessing the accuracy of classifications of remotely sensed data. *Remote Sensing of Environment*, **37**, 35–46.

Definiens. 2007. *Definiens Developer 7.0, Reference Book*, München, Germany, p. 195.

Definiens. 2008. *Definiens Developer 7.0, User Guide*, München, Germany, p. 506.

Grimmond, C.S.B. and Oke, T.R. (2002) Turbulent heat fluxes in urban areas: Observations and local-scale urban meteorological parameterization scheme (LUMPS). *Journal of Applied Meteorology*, **41**, 792–810.

Im, J., Jensen, J.R., and Tullis, J.A. (2008) Object-based change detection using correlation image analysis and image segmentation techniques. *International Journal of Remote Sensing*, **29**, 399–423.

Ivits, E. and Koch, B. (2002) Object-oriented remote sensing tools for biodiversity assessment: a european approach, in *Proceedings of the 22nd EARSeL Symposium, Prague, Czech Republic, 4–6 June 2002*, Millpress Science Publishers, Rotterdam, Netherlands.

Ketting R.L. and Landgrebe, D.A. (1976) Classification of multispectral image data by extraction and classification of homogeneous objects. *IEEE Transactions on Geoscience Electronics GE*, **14**, 19–26.

Lee, J.Y. and Warner, T.A. (2006) Segment based image classification. *International Journal of Remote Sensing*, **27**, 3403–3412.

Lillesand, T.M., Kiefer, R.W. and Chipman, J.W., (2008) *Remote Sensing and Image Interpretation*, 6th edition, John Wiley & Sons, New York.

Lizarazo, I. and Barros, J. (2010) Fuzzy image segmentation for urban land-cover classification. *Photogrammetric Engineering & Remote Sensing*, **76**, 151–162.

Myint, S.W., Giri, C.P., Wang, L., Zhu, Z. and Gillette, S. (2008). Identifying mangrove species and their surrounding land use and land cover classes using an object oriented approach with a lacunarity spatial measure. *GIScience and Remote Sensing*, **45**, 188–208.

Navulur, K. (2007) *Multispectral image analysis using the object-oriented paradigm*. CRC Press, Taylor & Francis Group, Boca Raton, FL.

Pal, R. and Pal, K. (1993) A review on image segmentation techniques. *Pattern Recognition*, **26**, 1277–1294.

Stow, D., Lopez, A., Lippitt, C., Hinton, S. and Weeks, J. (2007) Object-based classification of residential land use within Accra, Ghana based on QuickBird satellite data. *International Journal of Remote Sensing*, **28**, 5167–5173.

Strahler A.H., Woodcock C.E. and Smith, J.A. (1986) On the nature of models in remote sensing. *Remote Sensing of Environment*, **20**, 121–139.

Walker, J.S. and Blaschke, T. (2008) Object-based land-cover classification for the Phoenix metropolitan area: optimization vs. transportability. *International Journal of Remote Sensing*, **29**, 2021–2040.

Yu, Q., Gong, P., Clinton, N., Biging, G., Kelly, M. and Shirokauer, M. (2006) Object-based detailed vegetation classification with airborne fine spatial resolution remote sensing imagery. *Photogrammetric Engineering & Remote Sensing*, **72**, 799–811.

10

Spatial enhancement of multispectral images on urban areas

Bruno Aiazzi, Stefano Baronti, Luca Capobianco, Andrea Garzelli and Massimo Selva

Spatial enhancement, usually denoted as Pan-sharpening, consists in increasing the spatial resolution of a multispectral (MS) image by means of a panchromatic (Pan) observation of the same scene, acquired with a higher spatial resolution, by preserving or enhancing (Thomas et al., 2008) the radiometric quality of the original MS image. Many techniques have been proposed so far for Pan-sharpening. Multiresolution analysis (MRA) and component substitution (CS) are the two basic frameworks to which image fusion algorithms can be reported. State-of-the-art algorithms add the spatial details extracted from the Pan image into the MS data set by considering different injection strategies. The capability of efficiently modeling the relationships between MS and Pan images is crucial for the quality of fusion results and particularly for a correct recovery of local features. Context-adaptive (CA) injection models have been proposed in the MRA and more recently in the CS frameworks. In this chapter some of the most recent state-of-the-art Pan-sharpening algorithms are reported. Their performances are discussed in terms of objective and visual quality, taking into account the specific objective of spatial accuracy that is crucial for the analysis of urban areas.

Urban Remote Sensing: Monitoring, Synthesis and Modeling in the Urban Environment, First Edition. Edited by Xiaojun Yang.
© 2011 John Wiley & Sons, Ltd. Published 2011 by John Wiley & Sons, Ltd.

10.1 Introduction

The spatial resolution increase of recent satellite imagers is impressive. IKONOS, QuickBird, Orbview, GeoEye-1, World-View, KOMPSAT-2 are recording panchromatic (Pan) Earth images with a spatial resolution that ranges from 0.41 m to 1 m. Such other satellite imagers as Pleiades and GeoEye-2 are to be launched in next future with a spatial resolution of 0.5 m and 0.25 m, respectively. All these sensors are therefore particularly interesting for the analysis of urban areas exhibiting also small spatial details. Most of these imagers also collect multispectral (MS) data at a resolution that is four times lower than the Pan image. In order to increase the resolution of MS to the resolution of Pan data, remote-sensing image fusion techniques aims at integrating the information conveyed by data acquired with different spatial and spectral resolutions. The most straightforward goal is photoanalysis, but also automated tasks such as features extraction and segmentation/classification in high spatial detail areas have been found to benefit from fusion (Wahlen 2002; Zhang and Wang 2004; Bruzzone et al. 2006; Colditz et al 2006).

An extensive number of image fusion methods have been proposed in the literature, starting from the second half of the 1980s (Chavez, 1986). Most of them are based on a general protocol in which high-frequency spatial information is extracted from the Pan image and injected into the resampled MS bands by exploiting different models. In general, the image fusion methods described by this protocol can be divided into two main families: the techniques based on a linear spectral transformation followed by a component substitution (CS) and the algorithms that exploit a spatial frequency decomposition usually performed by means of multiresolution analysis (MRA). Under general and likely assumptions, CS methods are insensitive to aliasing and little sensitive to mis-registrations of moderate extent, unlike MRA methods are. Conversely, MRA methods are less sensitive to temporal shifts, i.e. MS and P acquired at different times that may introduce spectral distortions, or color changes (Ehlers, Klonus and Astrand, 2008).

10.1.1 Component substitution fusion methods

Basically, CS techniques linearly transform the MS data set into a more uncorrelated vector space. Then, one of the transformed bands, usually the smooth intensity I, is *replaced* by the Pan image P, histogram-matched to the I component itself, before the inverse transformation is applied. This procedure is equivalent to *inject*, i.e., add, the difference between P and I into the resampled MS data set as shown in (Tu et al., 2001). One of the first CS techniques that was applied is the intensity-hue-saturation (IHS) method (Carper, Lillesand and Kiefer, 1990) that is appropriate when exactly three MS bands are to be fused since the IHS transform is defined for three components only. When more than three bands are available, a generalized IHS (GIHS) transform can be defined by including the response of the near-infrared band into the intensity component I (Tu et al., 2004). In this case, I is obtained by weighting the MS bands with a set of coefficients whose choice can be related to the spectral responses of the Pan and MS bands (Tu et al., 2004; Gonzáles-Audícana et al., 2006). Principal component analysis (PCA) is an alternative to the IHS technique and works for an arbitrary number of MS bands. PCA is analogous to the IHS scheme since the Pan image is substituted to the first principal component (PC1). Histogram matching of Pan to PC1 is mandatory before substitution because the mean and variance of PC1 are generally far greater than those of Pan. It is well established that PCA performances are better than those of IHS (Chavez, Sides and Anderson, 1991) since PCA is data dependent and thus capable to fit data statistics. Consequently, the spectral distortion in the fused bands is usually less noticeable than IHS, even if it cannot completely be avoided. Generally speaking, if the spectral responses of the MS bands are not perfectly overlapped with the bandwidth of Pan, as it happens with the most advanced very high resolution imagers, IHS- and PCA-based methods may yield poor results in terms of spectral fidelity (Zhang, 2004). Another CS technique reported in the literature is Gram–Schmidt (GS) spectral sharpening, which was invented by Laben and Brower in 1998 and patented by Eastman Kodak (Laben and Brower, 2000). The GS method has been implemented in the Environment for Visualizing Images (ENVI) software, is widely used and produces *good* fused images. Its efficacy is mainly due the injection gain that is proportional, for each band, to the covariance of the synthesized intensity and the expanded MS band as reported in (Aiazzi, Baronti and Selva, 2007). As a matter of fact, since the sharp P and the smooth I have generally a different local radiometry, spectral distortions can arise in the fusion results. A mitigation of the consequent color changes can be obtained if I matches as much as possible the spectral response of Pan. This result is achieved by designing I as a linear combination of the MS bands by means of a set of coefficients according to the overlaps existing among the spectral responses of MS and Pan images (Tu et al., 2004). Such coefficients can be further optimized by minimizing the distance between P and I, for example in the minimum mean square error (MMSE) sense (Aiazzi, Baronti and Selva, 2007), utilizing a genetic algorithm (Garzelli and Nencini, 2006) that optimizes the Q4 score parameter defined in (Alparone et al., 2004b) or imposing MMSE constraints on the multispectral images (Garzelli, Nencini and Capobianco, 2008).

10.1.2 Multiresolution analysis fusion methods

Although the spectral quality of CS fusion results may be sufficient for most applications and users, methods based on injecting zero-mean high-pass spatial details, extracted from the Pan image without resorting to any transformation, have been extensively studied to overcome the inconvenience of spectral distortion. In fact, since the pioneering high-pass filtering (HPF) technique (Chavez, Sides and Anderson, 1991), fusion methods based on injecting high-frequency components into resampled versions of the MS data have demonstrated a superior spectral fidelity (Wald, Ranchin and Mangolini, 1997; Alparone et al., 2007). HPF basically consists of an addition of spatial details, taken from a high-resolution Pan observation, into a bicubically resampled version of the low resolution MS image. Such details are obtained by taking the difference between the Pan image and its low-pass version achieved through a simple local pixel averaging, i.e., a box filtering. Later improvements have been obtained with the introduction of multiresolution analysis (MRA), by employing several decomposition schemes, specially based on the discrete wavelet transform (DWT) (Garguet-Duport et al., 1996; Yocky, 1996), uniform rational filter banks (Aiazzi et al., 2000), curvelet transform (Nencini et al., 2007) and Laplacian pyramids (LP) (Wilson, Rogers and Kabrisky, 1997; Alparone et al., 1998). The rationale of high-pass detail injection as a problem of spatial frequency spectrum substitution from one signal to another, was formally developed in a multiresolution framework as an application of filter banks theory (Argenti and Alparone, 2000). The DWT has been extensively employed for remote sensing data fusion (Zhou, Civco, and Silander, 1998; Ranchin and Wald, 2000; Scheunders and De Backer, 2001). According to the basic DWT fusion scheme (Li, Manjunath and Mitra, 1995), pairs of subbands of corresponding frequency content are merged together. Afterwards, the fused image is synthesized by taking the inverse transform. Fusion schemes based on the "*à trous*" wavelet algorithm and Laplacian pyramids were successively proposed (Núñez et al., 1999; Chibani and Houacine, 2000; Garzelli et al., 2000). The "*à trous*" wavelet and the LP are oversampled, unlike the DWT, which is critically subsampled. The missing out of the decimation step allows an image to be decomposed into nearly disjoined band-pass channels in the spatial frequency domain, without losing the spatial connectivity (translation invariance property) of its high-pass details, e.g., edges and textures. This property is fundamental because, for critically subsampled schemes, spatial distortions, typically ringing or aliasing effects, may be present in the fused products and originate shifts or blur of contours and textures. As a simple outcome of multirate signal processing theory (Vaidyanathan, 1992), the LP can be easily generalized (GLP) to deal with scales whose ratios are whatsoever integer or even fractional numbers (Aiazzi et al., 1997, 1999).

10.1.3 Injection model of details

The definition of an efficient, physically consistent, and computationally practical Pan-sharpening method lies on three main issues. The first regards how to extract the spatial-detail information from the Pan image (using MRA or CS frameworks for example). The second concerns how the spatial details are injected into the resampled MS data and involves two aspects: (a) the estimation of a mathematical relationship to transform the spatial details extracted from the Pan image into those suitable for the MS representations, this step being defined within the *Amélioration de la Résolution Spatiale par Injection de Structures* (ARSIS) concept as interband structure model (IBSM) (Ranchin and Wald, 2000; Ranchin et al., 2003); (b) the adaptation of this mathematical relationship to the high resolution representation in order to transform the high-resolution details of Pan by defining a high-resolution IBSM (HRIBSM). The third issue is related to the quality assessment of the spatially enhanced MS images and to possibly drive the fusion process.

Regardless of how the spatial details are extracted from the Pan image, data fusion methods require the definition of a model that establishes how the missing high-pass information is injected into the resampled MS bands (Ranchin and Wald, 2000). In other words, the model, referred to as IBSM, deals with the radiometric transformation (gain and offset) of spatial structures (edges and textures) when passing from the Pan to MS images. The model is generally calculated at the coarser resolution and extrapolated to the finest resolution. This condition has been proved to be satisfactory for the MS and Pan data whose scale ratio is equal to four by investigating a Kalman-based fusion method which performs a prediction of fusion parameters across scales (Garzelli and Nencini, 2007). It should also be advisable to compute an HRIBSM by considering additional information on the MS-imaging system. The goal is to make the fused bands the most similar to what the narrow-band MS sensor would capture if it had the same spatial resolution as the broadband Pan. Notable examples of injection models are additive combination of "*à-trous*" wavelet frames, as in the additive wavelet to the luminance component (AWL) technique (Núñez et al., 1999), the injection of wavelet details after applying intensity–hue–saturation (IHS) transformation or principal component analysis (González-Audícana et al., 2004), the spectral distortion minimization (SDM) with respect to the resampled MS data (Alparone et al., 2003), or the spatially adaptive injection, as in the context-based-decision (CBD) algorithm (Aiazzi et al., 2002) and in the RWM method (Ranchin, et al., 2003). More efficient schemes can be obtained by taking into account the modulation transfer functions (MTFs) of the MS scanner and of the Pan sensor in order to design the MRA reduction filters or the decimation filters generating the MS and Pan data at degraded scales. In this way, it is possible to avoid a poor enhancement that sometimes occurs when MTFs are assumed to be ideal filters (Aiazzi et al., 2006). Theoretical considerations on injection models and experimental comparisons among MRA-based Pan-sharpening methods can be found in (Garzelli and Nencini, 2005). A further question concerns the adoption of global or local (context adaptive – CA) injection models. Computational cost is lower for global models but results are superior in general for local ones (Aiazzi et al., 2009) even if some caution should be adopted for local models since due to their nature they can bring local improvements but also cause possible local distortions or impairments. Since urban areas are characterized by the presence of important spatial details, investigations on injection models for such areas should be considered a relevant topic for future research.

10.1.4 Quality assessment

The evaluation of fusion results underlies on both qualitative and quantitative assessment. Qualitative assessment is usually demanded to visual analysis of skilled experts and depends on the application targets. Quantitative results of data fusion are provided thanks to the availability of reference originals obtained either by simulating the target sensor by means of high-resolution data from an airborne platform (Laporterie-Déjean et al., 2005), or by degrading all available data to a coarser resolution and carrying out fusion from such data. In practical cases, the first strategy is not feasible. Concerning the second strategy, however, the underlying assumption is that fusion performances are invariant to scale changes (Wald, Ranchin and Mangolini, 1997; Wald, 1999). Hence, algorithms optimized to yield the best results at coarser scales, i.e., on spatially degraded data, should still be optimal when the data are considered at the finest scales. This assumption may be reasonable in general, but may not hold for very high-resolution data, especially in a highly detailed urban environment, unless the spatial degradation is performed by using low-pass filters whose frequency responses match the shape of the MTFs of the sensors (Aiazzi et al., 2006). Generally, the original MS images are used as reference data for pan-sharpened images

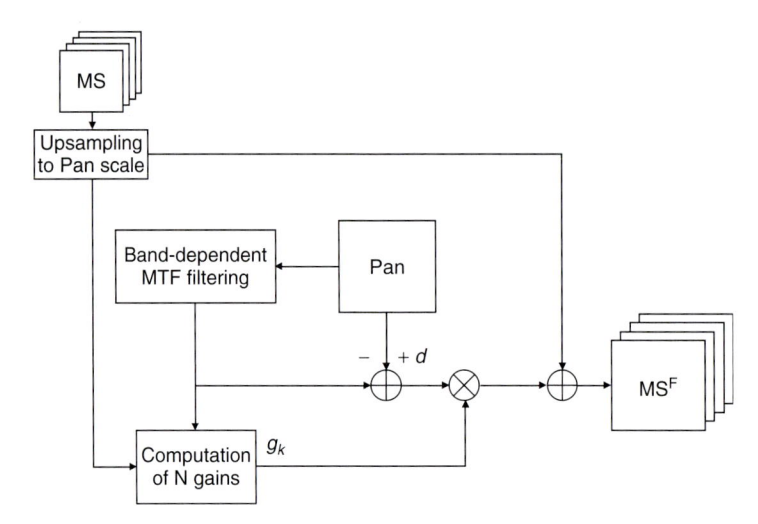

FIGURE 10.1 Multiresolution fusion scheme: spatial details d are extracted from the Pan image and injected in the expanded MS images. Images g_k rule the injection gain.

obtained from spatially degraded MS and Pan images. The degradation factor equals the spatial resolution ratio between original MS and Pan data. This evaluation protocol is used, among others, by the ERGAS (Ranchin and Wald, 2000) and the Q4 indexes (Alparone et al., 2004b). More recently, it has been shown that quality may be assessed without a reference image, i.e., directly at the spatial resolution of Pan, by evaluating the QNR index (Alparone et al., 2008). QNR is based on the invariance of the Q index defined by (Wang and Bovik 2002) and measures the quality of the fusion process by merging two factors, denoted as D_λ and D_s, which quantify the spectral and spatial distortions of the fused products, respectively. QNR, D_λ and D_s are the indices that are adopted in this work to assess the quality of the fused products.

10.2 Multiresolution fusion scheme

One of the most powerful and efficient framework for MRA algorithms is represented by the generalized Laplacian pyramid (GLP) decomposition. Its performances are practically the same as those of the "à trous" wavelet transform (Aiazzi et al., 2002). The method can take into account the MTF of each MS channel, thus fitting the detail extraction from the Pan image to each MS band (Aiazzi et al., 2006). In addition, local CA models can be adopted in the injection process, with the aim to preserve and sometimes increase the spectral information of the fused products, by unmixing the coarse MS pixels through the sharp Pan image. A noticeable example is given by the GLP with context-based decision (GLP-CBD) algorithm (Aiazzi et al., 2002) that employs a CA model aimed at inserting or not the spatial details for each pixel. Although the GLP-CBD model is very efficient when computing evaluation scores, its fused images can sometimes suffer from a poor contrast in some localized areas. In order to prevent this effect, the decision rule can be modified as described in (Aiazzi et al., 2009) by avoiding the on-off decision in the CA model.

A simplified scheme for MRA fusion that holds once the MS images have been registered with the PAN image is reported in Fig. 10.1. The scheme evidences the characteristic by which an algorithm is classified as MRA: details are extracted by subtracting a low-pass filtered version of the Pan image to the Pan image itself.

According to this scheme, and by simplifying the notation here and in the following by avoiding to explicitly indicate the spatial indices for all the images, the MRA fusion algorithm is expressed by

$$MS_k^F = (P - P_k^L)g_k + MS_k^\uparrow \quad k = 1, \cdots, N \quad (10.1)$$

Where MS_k^F is the fused multispectral image, k denotes the k-th band and N indicates the number of MS bands; the low-pass filtered Pan version P_k^L is obtained by convolving P by means of the MTF of each MS band and $MS_k^\uparrow$ represents the k-th MS component expanded to the scale of the Pan image. The injection model is accomplished by the weight image g_k that modulates, through an element-by-element multiplication, the image of the high spatial details. If g_k is constant for each band, the injection model is defined as *global*; in this case g_k is usually derived by computing global statistical parameters on the whole MS and P images. Otherwise, g_k is related on the local context measured on a sliding window of the current pixel, can be computed by taking local statistics of the MS and P images and the model is denoted as *local*. The choice adopted in (Aiazzi et al., 2009) is to take g_k as the regression coefficients $\beta(P_k^L, MS_k^\uparrow)$, relating the MS bands $MS_k^\uparrow$ and the Pan MTF filtered version P_k^L. $\beta(P_k^L, MS_k^\uparrow)$ is given by the covariance of P_k^L and $MS_k^\uparrow$, normalized by the variance of P_k^L. The scheme is denoted with GLP when the *global* model is adopted. Conversely, GLP-CA denotes the scheme when the *local* model is adopted.

10.3 Component substitution fusion scheme

Let us consider the general CS fusion scheme in Fig. 10.2. An algorithm is classified as CS when details are extracted by subtracting a synthetic intensity image to the Pan image and are thus produced without filtering the Pan image. Filtering can occur for estimating weights w_k but not for producing details. In the scheme, the original MS bands are preliminarily expanded to the same spatial scale of the full-resolution Pan image P to obtain the data set $MS_k^\uparrow$.

Weights w_k are fixed or can be computed as a function of P and $MS_k^\uparrow$, as in Fig. 10.2, to generate the synthetic intensity I component:

$$I = \sum_{k=1}^{N} w_k MS_k^\uparrow, \qquad (10.2)$$

Afterward, I is subtracted from P to produce a detail image d, i.e., $d = P - I$. A second set of coefficients g_k is adopted to modulate the detail image d before its addition to the expanded data set $MS_k^\uparrow$ in order to yield the final fused multispectral images MS_k^F. A histogram matching is usually performed on P to match its mean and standard deviation to those of the synthesized I component. This procedure has the objective of reducing the spectral distortions that are caused by the spectral mismatch between I and P images. The scheme described in Fig. 10.2 is general and agrees with the derivation presented by (Tu et al., 2004) in which it is demonstrated that the substitution of I with P and the successive inverse transform is equivalent to add $d = P - I$ to the expanded $MS_k^\uparrow$ data set. Depending on the values of the two sets of weights w_k and g_k, the scheme is suitable for describing any CS-based fusion algorithm (Aiazzi, Baronti and Selva 2007). In particular, the GS algorithm for which results are reported in the following is characterized by input weights w_k all equal to 1/4 and output weights g_k given by the regression coefficients $\beta(I, MS_k^\uparrow)$, expressed by the covariance of I and $MS_k^\uparrow$, normalized by the variance of I. The efficient CS algorithm we briefly review in this section is the Gram–Schmidt (GS) adaptive (GSA) algorithm reported in (Aiazzi, Baronti and Selva 2007). The algorithm can be also applied adaptively as reported in (Aiazzi et al., 2009) originating the context adaptive version (GSA-CA). As observed in (Otazu et al., 2005), the radiometric values of vegetated areas are likely to be much smaller in the synthetic intensity I image than in the true Pan. This effect causes the injection in the fused MS bands of a radiance offset, which may give rise to color distortion. In order to avoid this drawback, the idea reported in (Alparone et al., 2004a) is to generate the synthetic intensity in such a way that the spectral response of the sensor is considered, by differently weighting the contributions coming from the MS spectral channels. Conventional methods, however, only consider the nominal spectral responses when available (Tu et al., 2004). Actually, the influence of other phenomena, such as on-orbit working conditions, variability of the observed scene, and post processing effects, can significantly modify the nominal spectral response. In particular, atmospheric influence depends on the viewing angle since the scattering effect is related to the wavelength and on atmospheric conditions. The solution proposed in (Aiazzi, Baronti and Selva 2007) is to perform a linear regression between the low-pass (MTF) filtered Pan and the MS bands by computing a synthetic intensity I that has a minimum mean-square error (MMSE) with respect to the Pan. This solution allows the histogram matching to be skipped since the similarity of I and P is guaranteed by the MMSE criterion. The steps of the procedure when working at the PAN scale are as follows.

1 Obtain the $MS_k^\uparrow$ images, by expanding the MS image to the scale of P image and reduce the resolution of P image at the same spatial scale by means of a low-pass filter that should ideally represent an MTF capable to produce a P^L image with spatial frequency components similar to those of the MS image.

2 Assume that the I component is given as in Equation 10.2 and estimate the set of coefficients w_k by means of a linear regression algorithm (Ross 2004) in order to minimize the MSE between P^L and I.

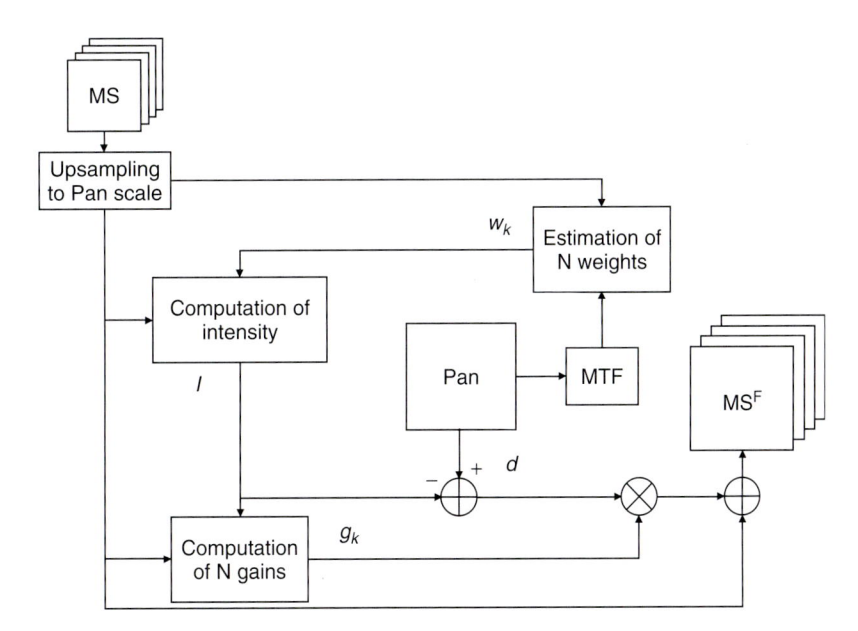

FIGURE 10.2 Component substitution fusion scheme: spatial details d are computed by subtracting the intensity component I from the Pan image and injected in the expanded MS images. Images g_k rule the injection gain. Weights w_k are estimated in order to minimize the difference between Pan and I in order to improve the quality of the fused images.

3 Calculate intensity I at the scale of P by utilizing the set of coefficients w_k determined at step (2) by means of Equation 10.2.

The implicit assumption that is done in step (3) is that the regression coefficients that are computed at the resolution of the MS image are practically the same as those that would be computed at the resolution of the original Pan image, if MS observations were available at full spatial resolution. This is equivalent to assume that the spectral responses of the dataset are practically unaffected by the change of the spatial resolution. The fused image is then given by:

$$MS_k^F = (P - I)g_k + MS_k^\uparrow \qquad (10.3)$$

Also in this case the injection model is accomplished by the weight image g_k that can be constant or varying pixel by pixel, the injection model being global or local, respectively. The choice adopted in (Aiazzi, et al., 2009) is to take g_k as the regression coefficients $\beta(I, MS_k^\uparrow)$. The scheme is denoted with GSA when a *global* model is adopted, while GSA-CA denotes the scheme with the *local* model.

10.4 Hybrid MRA – component substitution method

A Pan-sharpening algorithm for very-high resolution MS images has been proposed by (Garzelli, Nencini and Capobianco, 2008), which is optimal in the minimum mean squared error sense and computationally practical, even when local optimization is performed. This solution adopts an injection model in which a detail image extracted from the panchromatic band is calculated for each MS band by evaluating a band-dependent generalized intensity I_k from the N multispectral bands. The block diagram of the GMMSE (Global MMSE) fusion method is reported in Fig. 10.3.

From one side the MS image is up-sampled to the Pan scale ($MS_k^\uparrow$) with a quasi-ideal interpolator to contribute to the estimation of parameters and to generate the synthetic intensities I_k. From the other hand, each MS band is filtered with its MTF and up-sampled to the Pan scale to be fed to the parameter estimation procedure together with the low-pass MTF filtered P image. The estimation procedure is thus performed at reduced resolution (MTF filtered MS and Pan) and the parameters obtained are then utilized at the finest resolution. For each MTF a different I_k is generated. Details d_k are obtained for each band as $P - I_k$. Fusion is afterwards obtained by adding d_k, multiplied by the injection gains g_k, to each expanded $MS_k^\uparrow$ band. The fusion equation, which accounts for $N \times (N + 1)$ optimized parameters, i.e., N gains g_k and N^2 weights $w_{k,j}$ can be written as follows:

$$MS_k^F = MS_k^\uparrow + g_k \left(P - \sum_{j=1}^{N} w_{k,j} MS_j^\uparrow \right), \quad k = 1, \ldots, N. \qquad (10.4)$$

By separating MS and P factors, the equation may be reformulated as:

$$MS_k^F = MS_k^\uparrow + \sum_{j=1}^{N} \gamma_{k,j} MS_k^\uparrow + \gamma_{k,N+1} P, \, k = 1, \ldots, N \qquad (10.5)$$

Where the complete set of $N \times (N + 1)$ parameters is represented by

$$\gamma_{k,j} = \begin{cases} g_k & j = N + 1 \\ -g_k w_{k,j} & j = 1, \ldots, N \end{cases} \qquad (10.6)$$

and finally, in the compact form using lexicographically ordered column-wise images,

$$MS_k^F = MS_k^\uparrow + H\gamma_k \quad k = 1, 2, \ldots, N \qquad (10.7)$$

Where $\gamma_k = [\gamma_{k,1}, \gamma_{k,2}, \cdots, \gamma_{k,N+1}]^T \, k = 1, 2, \ldots N$, and $H = [MS_1^\uparrow, MS_2^\uparrow, \ldots, MS_N^\uparrow, P]$, represents the observation matrix of

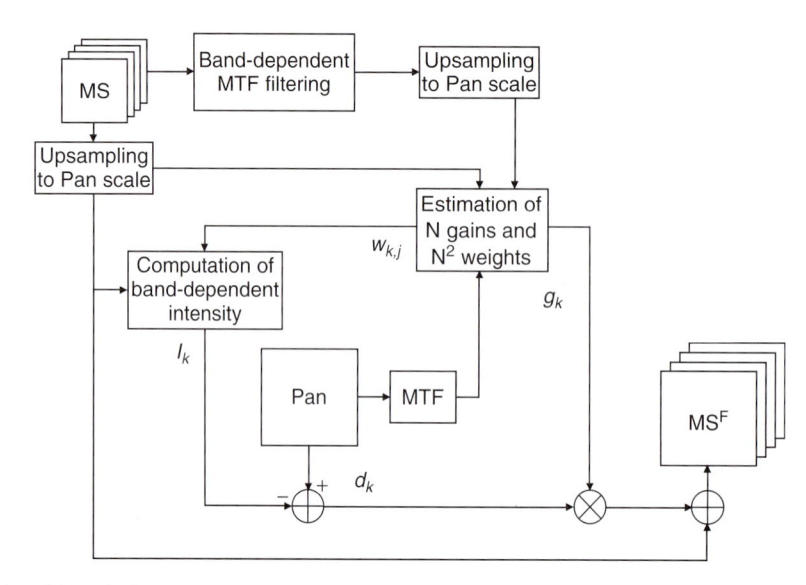

FIGURE 10.3 Hybrid multiresolution-component substitution fusion scheme: spatial details d_k are extracted from the Pan image and injected in the expanded MS images. Weights $w_{k,j}$ and gains g_k are jointly estimated at reduced resolution.

the linear model, having $N_r \times N_c$ rows and $(N+1)$ columns, N_r and N_c being the Pan dimensions.

The $N \times (N+1)$ parameters are jointly optimally estimated in the MMSE sense at degraded resolution (both P and MS are filtered by the corresponding MTF filters), as shown by the following least squares (LS) solution:

$$\gamma_k = (H_L^T H_L)^{-1} H_L^T (MS_k^{\uparrow} - MS_k^{L\uparrow}) \quad k=1,2,\ldots,N \quad (10.8)$$

Where H_L is the observation matrix formed by image data with resolution reduced by means of proper MTF filtering. The MMSE solution is stable because of the high dimensionality of H_L. Constraints can be imposed on the MMSE solution in order to obtain positive weights, but such a choice usually causes a little degradation in performances.

By observing Eq. 10.4, one may conclude that the method follows a component substitution approach since the Pan image is not filtered. However, it is evident from Equation 10.8 and from Fig. 10.3, that the MMSE fusion method adopts MRA for parameter estimation. As a conclusion, the MMSE Pan-sharpening algorithm may be classified as a hybrid component-substitution/MRA fusion method.

10.5 Results

Four Pan-sharpening algorithms have been selected for comparison purposes. The GLP-CA scheme has been selected as representative of MRA advanced techniques, the GSA-CA algorithm has been chosen among the CS schemes, while the GMMSE scheme with global gain is the representative of hybrid MRA/CS schemes. A fourth algorithm, the GS method as implemented in ENVI, has been reported in order to provide a reference with a well-established and rather efficient scheme. Also the plain resampled MS image (EXP) is reported as reference.

Results of the algorithms are reported for two different test images. The first is an 11-bit QuickBird satellite scene consisting of an MS image of Trento area in Northern Italy, with spatial sampling interval (SSI) of 2.8 m and of a Pan image with an SSI of 0.7 m. The whole scene has been fused by means of the four selected algorithms and quantitative results have been reported for a fused image of 1520×1520 pixels where urban feature are more concentrate. The second data set consists of an 11-bit IKONOS satellite MS image, referring to Umbria Region in Central Italy, with an SSI of 4 m and of a Pan image with an SSI of 1 m. Also in this case the whole scene has been fused and quantitative results refer to a 2048×2048 subimage where urban features appear.

Quantitative results have been judged by the novel metric quality with no reference (QNR) that is computed at full scale even if the reference is not available as occurs in this case (Alparone et al., 2008). QNR is based on the invariance of the Q index defined by (Wang and Bovik 2002) and measures the quality of the fusion process by merging two factors, denoted as D_λ and D_s, which quantify the spectral and spatial distortions of the fused products, respectively. D_λ and D_s should be ideally 0 in the best case and 1 in the worst case. Once D_λ and D_s have been defined (Alparone et al., 2008), QNR is given by $(1-D_\lambda)^\alpha \ldots (1-D_s)^\beta$. QNR should be ideally 1 and 0 in the best and worst case, respectively. α and β are factors chosen to give a different weight to spatial or spectral distortion. In the reported case study $\alpha = \beta = 1$, i.e., spatial and spectral distortions are given the same weight.

Table 10.1 reports D_λ, D_s and QNR values for the selected algorithms measured on the QuickBird image. Concerning the EXP image, as expected from its definition, D_λ is very near to 0 while D_s is moderate. Concerning GS algorithm, QNR is comparable with the expanded MS image but lower than the QNR values of the other methods. As a general consideration, all other algorithms perform very well and the indices exhibit similar values. GSA-CA obtains the highest QNR score and is also the best concerning spectral distortion D_λ. GLP-CA obtains the best result concerning spatial distortion. GMMSE exhibits a low spectral distortion, practically equal to GSA-CA; this determines a good score that is intermediate between GSA-CA and GLP-CA.

Table 10.2 reports D_λ, D_s and QNR values for the selected algorithms measured on the IKONOS image. Apart from the expanded MS image that is comparable, all the algorithms obtain scores that are a little bit lower than QuickBird. This occurs because of some residual local subpixel mis-registrations between the Pan and the MS image. The presence of misregistration artifacts explains why GLP-CA scores on IKONOS are less competitive than the scores of the other algorithms. In fact GLP-CA is based on MRA and suffers from misregistration more than the other algorithms that are CS based, as already recognized in the research community. GSA-CA and GMMSE obtain the best QNR with nearly the same score. GSA-CA exhibits the best D_s while the best D_λ is obtained by GMMSE. As in the case of QuickBird, GMMSE exhibits a very stable behaviour.

Qualitative analysis has been performed by visual inspection on the test areas. From these areas two 512×512 details have been extracted from the Pan image, the expanded MS original and the four selected Pan-sharpening algorithms. These images are displayed in Figs 10.4 and 10.6, respectively.

The 512×512 details have been selected in such a way to contain some buildings with sharp edges and vegetated areas as well, in which color distortion may appear.

TABLE 10.1 Spatial distortion D_s, spectral distortion D_λ and quality with no reference (QNR) index are reported for the selected algorithms for the QuickBird test image. The best results are shown in bold.

Metric	EXP	GS	GSA-CA	GLP-CA	GMMSE
D_s	0.1474	0.1031	0.0441	**0.0334**	0.0532
D_λ	0.0011	0.0721	**0.0623**	0.0875	0.0630
QNR	0.8431	0.8322	**0.8963**	0.8820	0.8872

TABLE 10.2 Spatial distortion D_s, spectral distortion D_λ and quality with no reference (QNR) index are reported for the selected algorithms for the IKONOS test image. The best results are shown in bold.

Metric	EXP	GS	GSA-CA	GLP-CA	GMMSE
D_s	0.1454	0.1146	**0.0685**	0.0782	0.0723
D_λ	0.0222	0.0704	0.0684	0.0754	**0.0657**
QNR	0.8356	0.8231	**0.8678**	0.8523	0.8668

The Pan detail relative to the QuickBird image is reported in Fig. 10.4(a) and constitutes the visual reference for spatial quality. The expanded MS is shown in Fig. 10.4(b) and represents the reference for spectral, i.e., color, quality. Figure 10.4(c) portrays the GS pan-sharpened image obtained by ENVI. Spatial enhancement is apparent when comparing this result with Fig. 10.4(b); some spectral distortions can be noticed on roofs, whose red color is too bright, and on vegetated areas that appear too light when compared to the correspondent MS areas. Indeed, notwithstanding that visualization parameters have been set in the same way for all the images, the GS fused image exhibits an apparent change in colors with respect to all the other images. Figure 10.4(d) shows the GLP-CA fused image. Both spectral and spatial quality is high; details appear sharp and accurate.

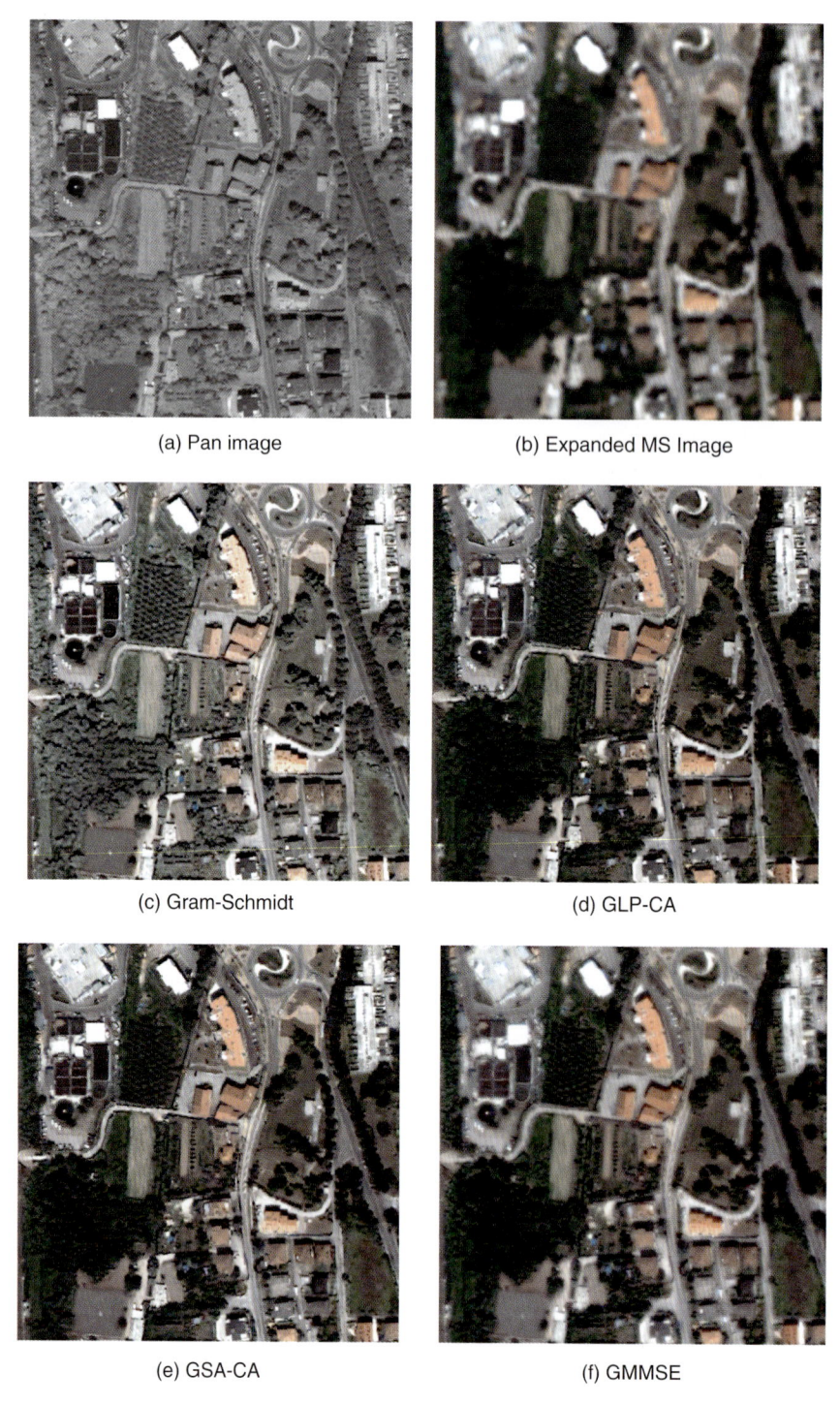

(a) Pan image

(b) Expanded MS Image

(c) Gram-Schmidt

(d) GLP-CA

(e) GSA-CA

(f) GMMSE

FIGURE 10.4 Detail of the QuickBird test area (512 × 512) processed by the selected Pan-sharpening algorithms. The original Pan (a) and expanded MS (b) images are reported in order to give a reference. Spatial enhancement is noticeable for all the fused images. Spectral distortion of GS algorithm (c) is apparent while all other algorithms show colors very similar to the expanded MS.

CHAPTER 10 SPATIAL ENHANCEMENT OF MULTISPECTRAL IMAGES ON URBAN AREAS   149

Only some minor color impairments can be revealed at pixel level by properly zooming some details. The result of GSA-CA algorithm is shown in Fig. 10.4(e). Only a careful analysis on the display can reveal some small differences with respect to the GLP-CA result of Fig. 10.4(d), the GSA-CA result being a little bit sharper than GLP-CA; on the overall the quality of the two algorithms is very high and their behavior is nearly the same since they feature a similar CA model. Figure 10.4(f) reports the image fused with the GMMSE algorithm. Also in this case quality is noteworthy. Spectral features are preserved and spatial details are correctly injected. The comparison of this method with GLP-CA and GSA-CA reveals that GMMSE has a stable behavior probably due to the multiresolution estimation of its model parameters. On the other hand, GLP-CA and GSA-CA fused images show a higher contrast on sharp edges.

In order to add a further element to evaluate the selected algorithms, Fig. 10.5 reports the maps of the details that each method injects into the expanded multispectral images. Details are represented in full color for the red, green and blue bands. On each map the contours have been superimposed of those blocks of size 32 × 32, on which the spatial distortion D_s exceeds a threshold value of 0.3. Figure 10.5(a), (b), (c) and (d) report the maps of the GS, GLP-CA, GSA-CA and GMMSE methods, respectively. Among the evaluated methods, GMMSE obtains the best results since only very few blocks (7) are flagged. Most of these blocks occur on vegetated areas and it is likely that errors in D_s are mainly due to texture rather than an incorrect injection of spatial details on buildings. GSA-CA and GLP-CA present a greater number of blocks exceeding the threshold, GLP-CA (18 blocks) being a little bit better than GSA-CA (22). Concerning GS, the number of errors of the algorithm is significant: about

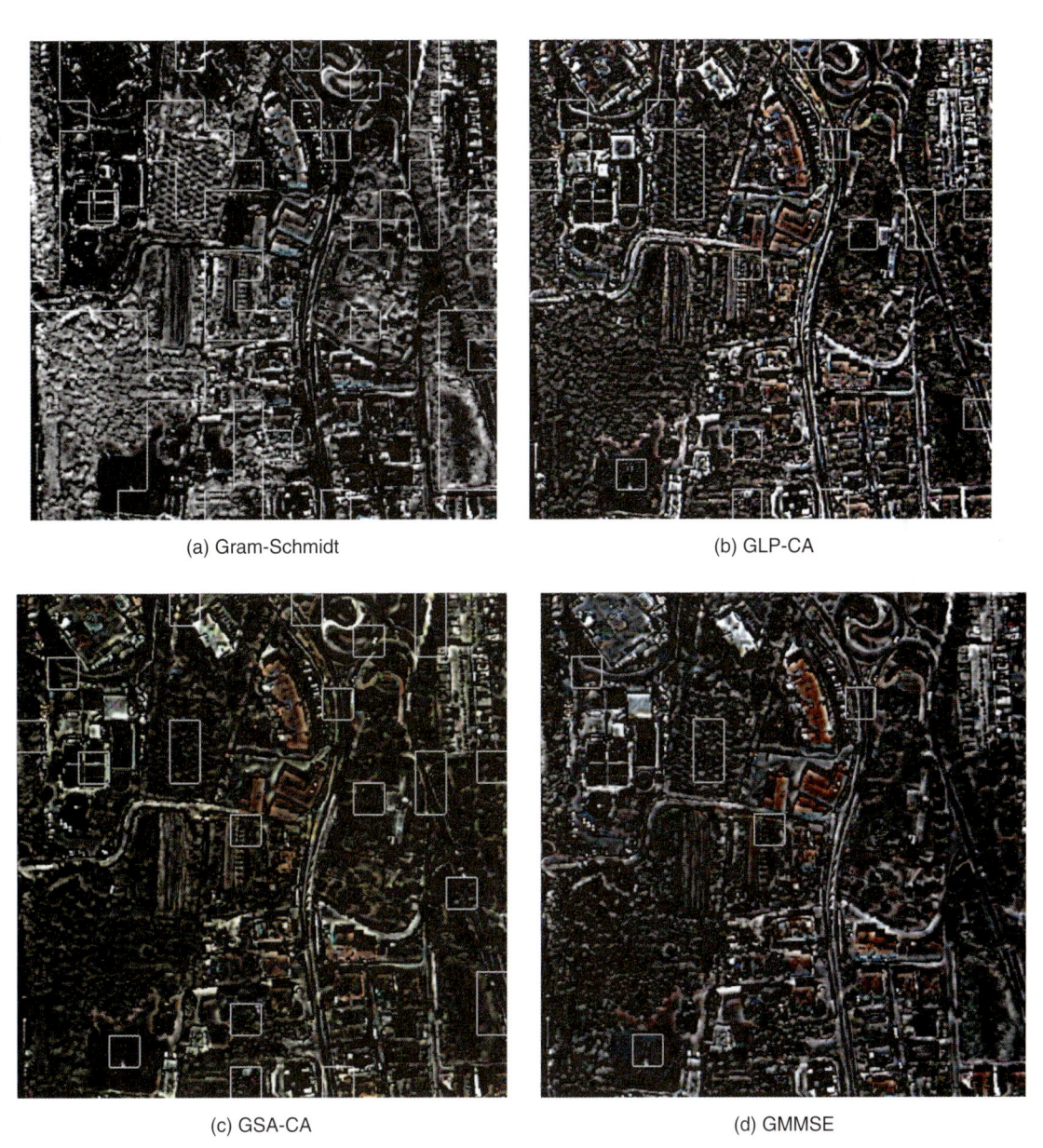

(a) Gram-Schmidt

(b) GLP-CA

(c) GSA-CA

(d) GMMSE

FIGURE 10.5 Maps of the details injected by each method into the expanded MS bands of the QuickBird scene. Details are represented in full color for the red, green and blue bands. The blocks of size 32 × 32, on which the spatial distortion D_s exceeds a threshold value of 0.3, have been superimposed.

(a) Pan image

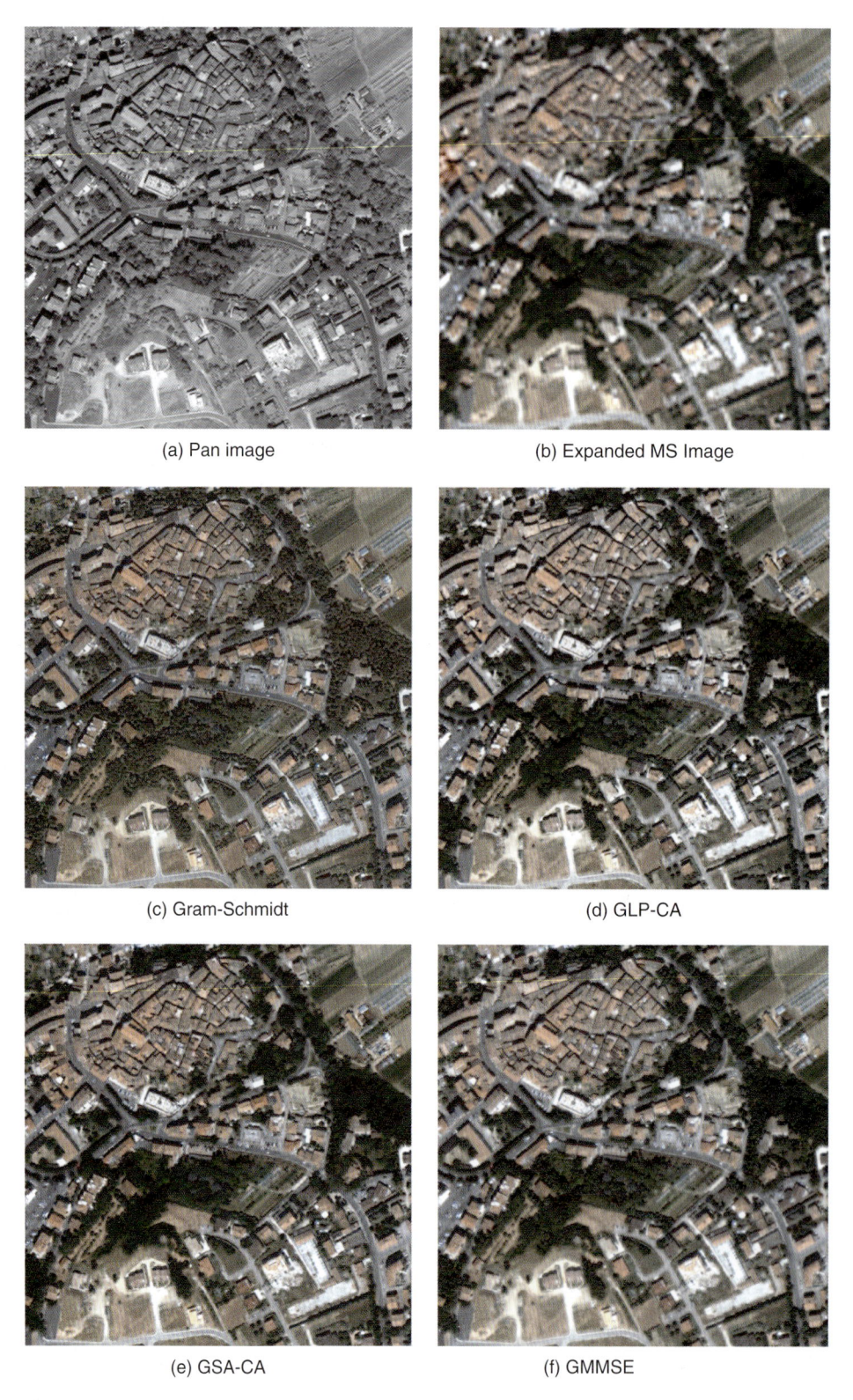

(b) Expanded MS Image

(c) Gram-Schmidt

(d) GLP-CA

(e) GSA-CA

(f) GMMSE

FIGURE 10.6 Detail of the IKONOS test area (512 × 512) processed by the selected Pan-sharpening algorithms same as Fig. 10.4. Spatial enhancement all the algorithms is apparent. Only GLP-CA (d) suffers from a residual spatial misregistration between the Pan and the expanded MS image, which causes some blur on contours. Concerning colors, only GS (c) shows some mismatch with the expanded MS image on vegetated areas that appear again too bright and textured.

one third (88) of the blocks is flagged. It is interesting to link the results of Fig. 10.5 with Table 10.1. Since in Table 10.1 GLP-CA is the best algorithm in terms of average D_s errors, followed by GSA-CA and GMMSE, it is evident that the distribution of D_s errors is different for the three algorithms. In particular, the distribution of errors for GMMSE is flatter than that of the other methods. This fact may indicate that the algorithm exhibits a behavior that is more stable with respect to landscape variations, which may occur frequently in an urban scenario.

Results relative to IKONOS images are reported in Fig. 10.6. The order by which the images are presented is the same as in Fig. 10.4. GS method appears more sharpen than the others on vegetated areas because of a heavy and excessive injection of Pan details in the blue and green bands. GLP-CA and GSA-CA colors are quite similar to the expanded MS on vegetated areas due to the local injection model. GMMSE colors are very similar to the EXP image. On urban areas all the algorithms properly inject spatial details. Only a careful analysis reveals a little blur on GLP-CA and GMMSE contours because of the misregistration problem outlined above.

Figure 10.7 reports the maps of the details that each method injects into the expanded multispectral images with the same order of Fig. 10.5. The maps are obtained with the same procedure described for Fig. 10.5. GMMSE and GSA-CA obtain the best results. Only one block is flagged. Also GLP-CA is efficient and only two blocks are flagged. Conversely a higher number of blocks is flagged by the GS algorithm (15).

By analyzing these results with the scores reported in Table 10.2 and in Table 10.1, it appears that for the IKONOS image the distribution of D_s errors is extremely flat and stable.

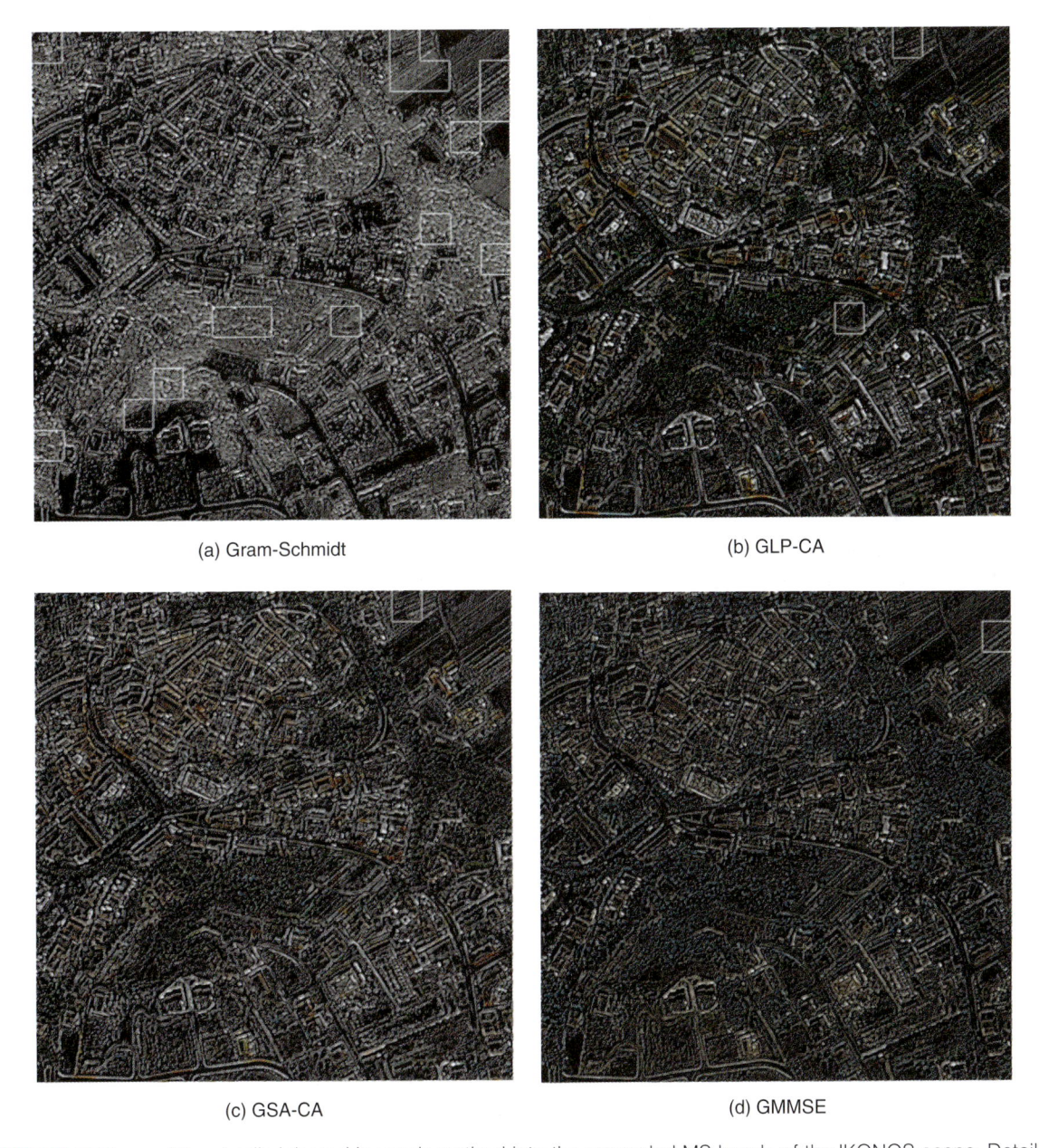

(a) Gram-Schmidt

(b) GLP-CA

(c) GSA-CA

(d) GMMSE

FIGURE 10.7 Maps of the details injected by each method into the expanded MS bands of the IKONOS scene. Details are represented in full color for the red, green and blue bands. The blocks of size 32 × 32, on which the spatial distortion D_s exceeds a threshold value of 0.3, have been superimposed.

TABLE 10.3 Main characteristics of GS, GSA-CA, GLP-CA and GMMSE algorithms. $\beta(\cdot,\cdot)$ denotes the covariance of the two arguments normalized by the variance of the first argument.

Algorithm	Type	Input weights w_k	Output weights g_k	Performance
GS	CS	Fixed Global for n bands $1/n$	Adaptive global $\beta(I, MS_k^\uparrow)$	Good spatial quality with some color distortions especially on vegetated areas
GSA-CA	CS	Adaptive global Minimization of MSE between P^L and I	Adaptive on local window $\beta(I, MS_k^\uparrow)$	High spatial quality with good color preservation.
GLP-CA	MRA	Not Applicable: details obtained by MTF filtering of Pan	Adaptive on local window $\beta(P_k^L, MS_k^\uparrow)$	High spatial quality with good color preservation. Sensitive to spatial mis-registrations.
GMMSE	HYBRID	Adaptive global See Equation 10.8	Adaptive global See Equation 10.8	High spectral quality and good spatial quality.

Notwithstanding D_s errors are higher in Table 10.2 than in Table 10.1, a lower number of blocks is flagged for the IKONOS test and all the algorithms exhibit a stable behavior, especially on urban areas.

Finally, Table 10.3 summarizes the main characteristics of the selected algorithms. Each algorithm is classified according to its type (CS, MRA and Hybrid) and the input and output weights are reported for a quick reference. A synthetic judge is also given as an indication for use.

Conclusions

In this chapter the main issues concerning Pan-sharpening methods are introduced and discussed. Some advanced methods are described in the general frameworks of component substitution and multiresolution analysis and they are compared with the well established Gram-Schmidt Pan-sharpening algorithm. Quantitative results evaluated in terms of the quality with no reference (QNR) index and qualitative results show the efficacy of the selected algorithms both in terms of spatial and spectral quality and further improvements appear hard to be obtained. Among the reviewed algorithms, the context adaptive component substitution method (GSA-CA) obtains the best scores concerning QNR, the multiresolution context adaptive algorithm (GLP-CA) appears the best regarding spatial sharpness in the absence of misregistration. The hybrid scheme (GMMSE) represents an efficient trade-off since it guarantees good performances with the advantage of a stable behavior when spatial details are injected.

References

Aiazzi, B., Alparone, L., Baronti, S., Cappellini, V., Carlà, R. and Mortelli, L. (1997) A Laplacian pyramid with rational scale factor for multisensor image data fusion, in *Proceedings of the International Conference on Sampling Theory and Application*, pp. 55–60.

Aiazzi, B., Alparone, L., Argenti, F. and Baronti, S. (1999) Wavelet and pyramid techniques for multisensor data fusion: a performance comparison varying with scale ratios, in *Proceedings of SPIE Image Signal Processing and Remote Sensing V* (ed. S.B. Serpico), 3871, 251–262.

Aiazzi, B., Alparone, L., Argenti, F., Baronti, S. and Pippi, I. (2000) Multisensor image fusion by frequency spectrum substitution: subband and multirate approaches for a 3:5 scale ratio case, in *Proceedings of the IEEE International Geoscience and Remote Sensing Symposium*, pp. 2629–2631.

Aiazzi, B., Alparone, L., Baronti, S. and Garzelli A. (2002) Context-driven fusion of high spatial and spectral resolution data based on oversampled multiresolution analysis. *IEEE Transactions on Geoscience and Remote Sensing*, **40**(10), 2300–2312.

Aiazzi, B., Alparone, L., Baronti, S., Garzelli, A. and Selva M. (2006) MTF-tailored multiscale fusion of high resolution MS and Pan imagery. *Photogrammetric Engineering & Remote Sensing*, **72**(5), 591–596.

Aiazzi, B., Baronti, S. and Selva M. (2007) Improving component substitution pansharpening through multivariate regression of MS+Pan data. *IEEE Transactions on Geoscience and Remote Sensing*, **45** (10), 3230–3239.

Aiazzi, B., Baronti, S., Lotti, F. and Selva M. (2009) A Comparison Between Global and Context Adaptive Pansharpening of Multispectral Images. *IEEE Geoscience and Remote Sensing Letters*, **6** (2), 302–306.

Alparone, L., Cappellini, V., Mortelli, L., Aiazzi, B., Baronti, S. and Carlà, R. (1998) A pyramid-based approach to multisensor image data fusion with preservation of spectral signatures, in *Future Trends in Remote Sensing* (ed. P. Gudmandsen) Balkema, Rotterdam, pp. 418–426.

Alparone, L., Aiazzi, B., Baronti, S. and Garzelli, A. (2003) Sharpening of very high resolution images with spectral distortion minimization, in *Proceedings of the IEEE International Geoscience and Remote Sensing Symposium*, pp. 21–25.

Alparone, L., Baronti, S., Garzelli, A. and Nencini, F. (2004a) Landsat ETM+ and SAR image fusion based on generalized intensity modulation. *IEEE Transactions on Geoscience and Remote Sensing*, **42** (12), 2832–2839.

Alparone, A., Baronti, S., Garzelli A. and Nencini, F. (2004b) A global quality measurement of Pan-sharpened multispectral imagery. *IEEE Geoscience and Remote Sensing Letters*, **1** (4), 313–317.

Alparone, L., Wald, L., Chanussot, J., Thomas, C., Gamba, P. and Bruce, L.M. (2007) Comparison of Pansharpening algorithms: Outcome of the 2006 GRS-S data fusion contest. *IEEE Transactions on Geoscience and Remote Sensing*, **45** (10), 3012–3021.

Alparone, L., Aiazzi, B., Baronti, S., Garzelli, A., Nencini, F. and Selva, M. (2008) Multispectral and panchromatic data fusion assessment without reference. *Photogrammetric Engineering & Remote Sensing*, **74**(2), 193–200.

Argenti, F. and Alparone, L. (2000) Filterbanks design for multisensor data fusion. *IEEE Signal Processing Letters*, **7** (5), 100–103.

Bruzzone, L., Carlin, L., Alparone, L., Baronti, S., Garzelli, A. and Nencini, F. (2006) Can multiresolution fusion techniques improve classification accuracy?, in *Proceedings of SPIE*, 6365.

Carper, W., Lillesand, T. and Kiefer, R. (1990) The use of intensity-hue-saturation transformations for merging SPOT panchromatic and multispectral image data. *Photogrammetric Engineering and. Remote Sensing*, **56** (4), 459–467.

Chavez, Jr., P.S. (1986) Digital merging of Landsat TM and digitised NHAP data for 1: 24 000 scale image mapping. *Photogrammetric Engineering and Remote Sensing*, **52** (10), 1637–1646.

Chavez, Jr., P.S., Sides, S.C. and Anderson J.A. (1991) Comparison of three different methods to merge multiresolution and multispectral data: Landsat TM and SPOT panchromatic. *Photogrammetric Engineering and Remote Sensing*, **57** (3), 295–303.

Chibani, Y. and Houacine, A. (2000). Model for multispectral and panchromatic image fusion, in *Proceedings of SPIE*, 4170, 238–244.

Colditz, R. Wehrmann, T., Bachmann, M. *et al.* (2006) Influence of image fusion approaches on classification accuracy: A case study. *International Journal of Remote Sensing*, **27** (15), 3311–3335.

Ehlers, M., Klonus, S. and Astrand, P. (2008). Quality assessment for multisensor multi-date image fusion, in *The International Archives of the Photogrammetry, Remote Sensing and Spatial Information Sciences*, **XXXVII** Part B4, 499–506.

Garguet-Duport, B., Girel, J., Chassery, J.-M. and Pautou, G. (1996) The use of multiresolution analysis and wavelet transform for merging SPOT Panchromatic and multispectral image data. *Photogrammetric Engineering and Remote Sensing*, **62** (9), 1057–1066.

Garzelli, A., Benelli, G., Barni, M. and Magini, C. (2000) Improving wavelet-based merging of panchromatic and multispectral images by contextual information, in *Proceedings of SPIE* 4170, 82–91.

Garzelli, A. and Nencini, F. (2005) Interband structure modeling for Pansharpening of very high-resolution multispectral images. *Information Fusion*, **6** (3), 213–224.

Garzelli, A. and Nencini, F. (2006) PAN-sharpening of very high resolution multispectral images using genetic algorithms. *International Journal of Remote Sensing*, **27** (15), 3273–3292.

Garzelli, A. and Nencini, F. (2007) Panchromatic sharpening of remote sensing images using a multiscale Kalman filter. *Pattern Recognition*, **40** (12), 3568–3577.

Garzelli, A., Nencini, F. and Capobianco, F. (2008) Optimal MMSE Pan sharpening of very high resolution multispectral images. *IEEE Transactions on Geoscience and Remote Sensing*, **46** (1), 228–236.

González-Audícana, M., Saleta, J.L., Catalan, R.G. and Garcia, R. (2004) Fusion of multispectral and panchromatic images using improved IHS and PCA mergers based on wavelet decomposition. *IEEE Transactions on Geoscience and Remote Sensing*, **42** (6), 1291–1299.

Gonzáles-Audícana, M., Otazu, X., Fors, O. and Alvarez-Mozos, J. A. (2006) A low computational-cost method to fuse IKONOS images using the spectral response function of its sensors. *IEEE Transactions on Geoscience and Remote Sensing*, **44** (6), 1683–1691.

Laben, C.A. and Brower, B.V. (2000) Process for enhancing the spatial resolution of multispectral imagery using Pansharpening. US Patent 6 011 875, Technical Report, Eastman Kodak Company.

Laporterie-Déjean, F., de Boissezon, H., Flouzat, G. and Lefévre-Fonollosa, M.-J. (2005) Thematic and statistical evaluations of five panchromatic/multispectral fusion methods on simulated PLEIADES-HR images. *Information Fusion*, **6** (3), 193–212.

Li, H., Manjunath, B. S. and Mitra, S. K. (1995) Multisensor image fusion using the wavelet transform. *Graphical Models and Image Processing*, **57** (3), 235–245.

Nencini, F., Garzelli, A., Baronti, S. and Alparone, L. (2007) Remote sensing image fusion using the curvelet transform. *Information Fusion*, **8** (2), 143–156.

Núñez, J., Otazu, X., Fors, O., Prades, A., Palà, V. and Arbiol, R. (1999) Multiresolution-based image fusion with additive wavelet decomposition. *IEEE Transactions on Geoscience and Remote Sensing*, **37** (3), 1204–1211.

Otazu, X., Gonzáles-Audícana, M., Fors, O. and Núñez, J. (2005) Introduction of sensor spectral response into image fusion methods. Application to wavelet-based methods. *IEEE Transactions on Geoscience and Remote Sensing*, **43** (10), 2376–2385.

Ranchin, T. and Wald, L. (2000) Fusion of high spatial and spectral resolution images: The ARSIS concept and its implementation. *Photogrammetric Engineering and Remote Sensing*, **66** (1), 49–61.

Ranchin, T., Aiazzi, B., Alparone, L., Baronti, S. and Wald, L. (2003) Image fusion - the ARSIS concept and some successful implementation schemes. *ISPRS Journal of Photogrammetry and Remote Sensing*, **58** (1/2), 4–18.

Ross, S. M. (2004) *Introduction to Probability and Statistics for Engineers and Scientists*, Elsevier Academic, Burlington, MA.

Scheunders, P. and De Backer, S. (2001) Fusion and merging of multispectral images with use of multiscale fundamental forms. *Journal of the Optical Society of America A, Optical Image Science*, **18** (10), 2468–2477.

Thomas, C., Ranchin, T., Wald, L. and Chanussot, J. (2008) Synthesis of multispectral images to high spatial resolution: a critical review of fusion methods based on remote sensing physics. *IEEE Transactions on Geoscience and Remote Sensing*, **46** (5), 1301–1312.

Tu, T.-M., Su, S.-C., Shyu, H.-C. and Huang, P.S. (2001) A new look at IHS-like image fusion methods. *Information Fusion*, **2** (3), 177–186.

Tu, T.-M., Huang, P.S., Hung, C.-L. and Chang, C.-P. (2004) A fast intensity–hue–saturation fusion technique with spectral adjustment for IKONOS imagery. *IEEE Geoscience and Remote Sensing Letters*, **1** (4), 309–312.

Vaidyanathan, P.P. (1992) *Multirate Systems and Filter Banks*, Prentice Hall, Englewood Cliffs, NJ.

Wahlen, J. (2002) Comparison of standard and image-filter fusion techniques, in *Data Mining III* (eds A. Zanasi, C.A. Brebbia, N.F.F. Ebecken, and P. Melli), WIT Press, Southampton, UK.

Wald, L. (1999) Some terms of reference in data fusion. *IEEE Transactions on Geoscience and Remote Sensing*, **37** (3), 1190–1193.

Wald, L., Ranchin, T. and Mangolini, M. (1997) Fusion of satellite images of different spatial resolutions: Assessing the quality of resulting images. *Photogrammetric Engineering and Remote Sensing*, **63** (6), 691–699.

Wang, Z. and Bovik, A.C. (2002) A universal image quality index. *IEEE Signal Processing Letters*, **9** (3), 81–84.

Wilson, T.A., Rogers, S.K. and Kabrisky, M. (1997) Perceptual-based image fusion for hyperspectral data. *IEEE Transactions on Geoscience and Remote Sensing*, **35** (4), 1007–1017.

Yocky, D. A. (1996) Multiresolution wavelet decomposition image merger of Landsat Thematic Mapper and SPOT panchromatic data. *Photogrammetric Engineering and Remote Sensing*, **62** (9), 1067–1074.

Zhang, Y. (2004) Understanding image fusion. *Photogrammetric Engineering and Remote Sensing*, **70** (6), 657–661.

Zhang, Y. and Wang, R. (2004) Multi-resolution and multispectral image fusion for urban object extraction, in *Proceedings of the 20th ISPRS Congress*, pp. 960–966.

Zhou, J., Civco, D.L. and Silander, J.A. (1998) A wavelet transform method to merge Landsat TM and SPOT panchromatic data. *International Journal of Remote Sensing*, **19** (4), 743–757.

11
Exploring the temporal lag between the structure and function of urban areas

Victor Mesev

Urban areas are complex assemblages of tangible physical structures and human behavioral functionality. Remote sensing is routinely employed to measure land cover types (structures) from increasingly finer spatial resolutions, while census and planning data infer land use (function). Traditionally, the relationship is assumed linear, but classic urban theory suggests that city dynamics contain a temporal lag between structure and function; in other words, a gap of time between when decisions are made to change a city by its population to when those changes actually physically materialize. This chapter explores this so-called temporal lag from a conceptual standpoint, especially in relation to urban theory, as well as outlining structural-functional links that may empirically determine the scope and rate of the lag. It also deliberates the search for an appropriate scale of analysis for using remote sensing in urban studies; discussing the continuum between micro and macro urban remote sensing. The chapter ends with calls for developing a research agenda that aims to measure temporal lags more precisely, along with a need to build urban remote sensing methodologies that appreciate the dynamic nature of cities by carefully linking process to both structure and function simultaneously.

11.1 Introduction

Multitemporal analysis is one of the strengths of remote sensing. By using sensor data across two or more points in time consistent and frequent changes in land cover and land use can be measured routinely and cost-effectively. When applying multitemporal analysis to spectrally complex urban areas, further emphasis is placed on establishing consistent temporal relationships between remotely sensed data and ancillary information – normally used to improve classification accuracy as well as to improve the thematic description of urban classes. Much of the rationale for this strict temporal consistency, where remotely sensed data are used from a time period that is very similar to the date that ancillary information is collected, is based on the notion that urban areas are static surfaces and that its buildings and citizens are intertwined and evolve at the same rate of change. This is an assumption that seems to be grounded more on convenience rather than on theory. This chapter will challenge this assumption of static cities and open the debate on whether there is instead a time-induced relationship between the physical structures of urban areas and their human functionality. A time-induced relationship implies the investigation of a so-called *temporal lag*; in other words a difference in time between *when* urban structures appear and *when* decisions are made by their inhabitants to implement those structural changes. To this end the chapter calls for the inclusion of urban theory when using remote sensing to measure multitemporal changes of land cover and land use in urban areas. It investigates whether urban theory – which embraces a more process-led dynamic city – can be can be applied and tested on empirical datasets representing urban structure from high spatial resolution IKONOS sensor data, as well as urban functionality from point-based mailing addresses and rasterized census surfaces. Preliminary evidence is given of temporal differences between the three data sets, highlighting possible temporal lags between physical structure and socioeconomic function. A discussion follows on whether these temporal differences can be related to urban processes as predicted by theory or whether they are unrelated random differences. The chapter concludes with calls for further testing and a research agenda on how urban theory can be incorporated directly into urban multitemporal analysis, specifically on how temporal lags between urban structure and urban function can predict urban change. First, it is important to establish the appropriate scale of analysis for the investigation of temporal lags; in other words, at what scale remote sensing should be applied to urban theory.

11.2 Micro and macro urban remote sensing

Research focused on the remote sensing of urban areas using satellite sensor data is at an intersection (*inter alia* Mesev, 2003; Gamba, Dell'Acqua and Dasarathy, 2005; Weng and Quattrochi, 2007; Xian, 2010). In one direction lie opportunities for developing methodologies for precision mapping of urban structural configuration from very high spatial resolution imagery – with a focus on pragmatic applications; and in the other direction lie challenges for exploring the more ontological questions surrounding the fusion of structural and functional representations – focusing on more holistic views of urban growth and urban economic and social sustainability. The former includes examples of *micro urban remote sensing* and is a domain commonly visited by photogrammetists and scientists involved in civil engineering and planning applications, while the latter constitutes *macro urban remote sensing* and is far more in line with the construction of deductive and reductionist urban geographic models.

The distinction between micro and macro remote sensing is based on the scale of analysis and not necessarily on the multidimensionality of the remotely sensed data, in particular its spatial resolution. Indeed, urban remote sensing was expected to benefit in both micro and macro scales of analysis at the advent and subsequent prevalence of higher spatial resolution satellite sensor data (IKONOS, QuickBird, WorldView, etc.). Much of the potential for new areas of research was to revolve around precision mapping where space-borne imagery were anticipated to aid the delineation of buildings and transport structures – very much in the same manner as air-borne photography was traditionally used to update topographic maps (Couloigner and Ranchin, 2000). However, to date, the level of expectation for these high spatial resolution satellite sensor datasets seems to have far exceeded the number of practical urban applications. Despite the perceived advances in clarity and detail stemming from pixels representing smaller instantaneous fields of view, most of the criticism, in direct contrast, has been linked with the increased spectral heterogeneity resulting from the finer scaled spatial resolution. It means that urban classifications remain highly tenuous and any reliable micro remote sensing, usually in the form of precision mapping, is extracted directly from the spatial orientation of pixels – in the similar vein to conventional interpretation of aerial photography, but with slightly lower clarity and with limited stereoscopic capabilities. However, the spectral heterogeneity problem is less of a restriction for macro remote sensing, which instead of measuring individual objects such as buildings, roads and even side-walks, is more concerned with a generalized view of an urban area such as neighborhoods, zones or even the whole city. Classification accuracy is less important, with the emphasis more on interpreting generalized land cover/land use, measuring overall building density, and understanding urban processes such as growth, congestion/pollution, and poverty. Arguably, it is this understanding of urban processes that many researchers consider as the more important benefits of remote sensing when applied to urban areas. However, to fully appreciate the scale of dynamic urban changes remotely sensed data need to be embellished with ancillary information measuring socioeconomic characteristics, housing descriptors, and zoning restrictions. But even the remote sensing-ancillary data combination only provides an essentially empirically derived model of a static city. What is needed is a theoretical basis from which to interpret and understand urban land cover and land use change; a theoretical basis built on the concept of a temporal lag between what an urban society demands and what urban physical consequences materialize.

11.3 The temporal lag challenge

Urban areas are routinely represented by mixtures of land cover and land use, where land cover types are interpreted from remotely sensed data and land use characteristics from ground-based surveys such as population censuses, housing, and planning maps. The unique spatial configuration of urban land cover types such as vegetation, soil, water and impermeable surfaces determines the physical structure of an urban area, while human occupation and the functioning of social and economic activities are all indications of urban land use. When shaping spatial models of any given urban area through the combination of land cover types and land use characteristics, the conventional view is to select remotely sensed data and surveys that have been taken at similar points in time. This is the intuitive approach and the basis for many remote sensing and general geospatial integrative methodologies and applications.

However, this assumption of strict time-dependence may be theoretically fallible. When consulting the established literature on theoretical urban geography, the time relationship between physical structure and societal functioning is generally regarded as anything but instantaneous; and that urban structures are an eventual consequence of functional manifestations taken and decided upon years or sometimes decades earlier (see Whitehand, 1977; Brenner, 2000; Longley, 2002). In other words, there is an inherent temporal lag between the reasons why society decides to build (urban function) and when it is built (urban structure). This lag is the basis to much of the theory on urban life cycles, urban sociological interactions, and even municipal engineering (see Herbert and Thomas, 1982; Carter, 1985; Clark, 2008).

There is also evidence to suggest that, because cities are closed systems, temporal causality is bidirectional; that structure sometimes affects function. For instance, when the building of dense apartment complexes in one point in time sometimes leads to higher levels of poverty and even crime rates in subsequent years, or alternatively when a mixture of condominiums and green space promotes gentrification. Conventional urban theory would further suggest that, if measured accurately, these so-called temporal lags would be appropriate indicators for measuring the pace of change in urban processes. In particular, changes in actual physical growth resulting from construction work, as well as the possibility to examine less obvious indications of change such as social and economic deprivation levels, severity of cultural segregation, exacerbation of housing overcrowding, and traffic congestion. Any insights, in either direction, of how one or more of these urban processes fluctuate would be considered essential for understanding the rapid dynamics of urban morphologies, and in turn provide valuable information for the pursuit of urban planning, legislative zoning and environmental sustainability policies, both for today's cities and for their long-term future.

Before investigating evidence for a temporal lag, it's important to establish the relationship between structure and function at the static scale of analysis; in particular how structural configuration can be linked with functional characteristics.

11.4 Structural–functional links

Research in urban remote sensing has been generally confined to measuring impermeable surfaces by classifying image pixels into urban built land cover. Notable breakthroughs have augmented remotely sensed with socioeconomic information to generate models of urban function as well as structure (Chen, 2002; Harvey, 2002; Barnsley, Steel and Barr, 2003). Figure 11.1 represents three types of datasets that are frequently used for structural–functional models; high spatial resolution sensor images to measure structure, and point-based mailing addresses and rasterized area-based census surfaces to tessellate socioeconomic characteristics of urban areas. Each of the three types represents the study site of the city of Belfast, Northern Ireland. The high spatial resolution image is from the IKONOS sensor (Space Imaging) at 4 m, pan-sharpened and taken in July 2001, the point-based mailing addresses are from the COMPAS database from the Ordnance Survey of Northern Ireland, and the surface is of the 2001 Census Population is rasterized at a 200 m grid (see Martin, Langford and Tate, 2000 for calculations).

One approach to combining structure and function, as represented by Fig. 11.1, is to perform a direct spatial comparison or formulate a linear statistical relationship between the IKONOS image and point-based addresses and area-based census surfaces at variant time slices. Commendable research includes the estimation of population and housing units by Lo (2003) and by Harvey (2002) by limiting census attributes within the spatial boundaries of built land cover extracted from remotely sensed data. In terms of point-based address data, work by Aubrecht et al. (2009) and Mesev (2005, 2007) attempted to replicate the spatial configuration of urban neighborhoods. The assumption is that urban neighborhoods exhibit distinctive spatial expressions in terms of their architectural, structural, and morphological composition – the complex assemblage of different land covers (bare soil, concrete, tarmac, grass, water etc.). By employing spatial metrics to quantify these attributes it is possible to demonstrate how individual urban neighborhoods may be distinguished and delineated from second order imagery (Pasaresi and Bianchin, 2001; Herold, Scepan and Clarke, 2002; Barnsley, Steel and Barr, 2003).

On-going research is exploring an agenda for building disaggregated urban models that infer spatial urban syntactic structural and functional configurations within vector-determined spectral limitations using high spatial resolution IKONOS imagery. Disaggregated models can be built either from point-based address data extracted from COMPAS in Northern Ireland and the United Kingdom (postal records), or from area-based parcel data in the United States. Knowing the spatial distribution of these point data introduces a number of key indicators that measure parameters such as density (compactness versus sparseness) and arrangement (linearity versus randomness) (Mesev, 2007). Commercial neighborhoods exhibit different levels of complexity and irregularity to residential neighborhoods, so too does high density residential from low density residential. This can be measured by even the most elementary metrics, such as area, density, and percent land cover. Fractal geometry is well suited to measuring the structural irregularity of the morphology of

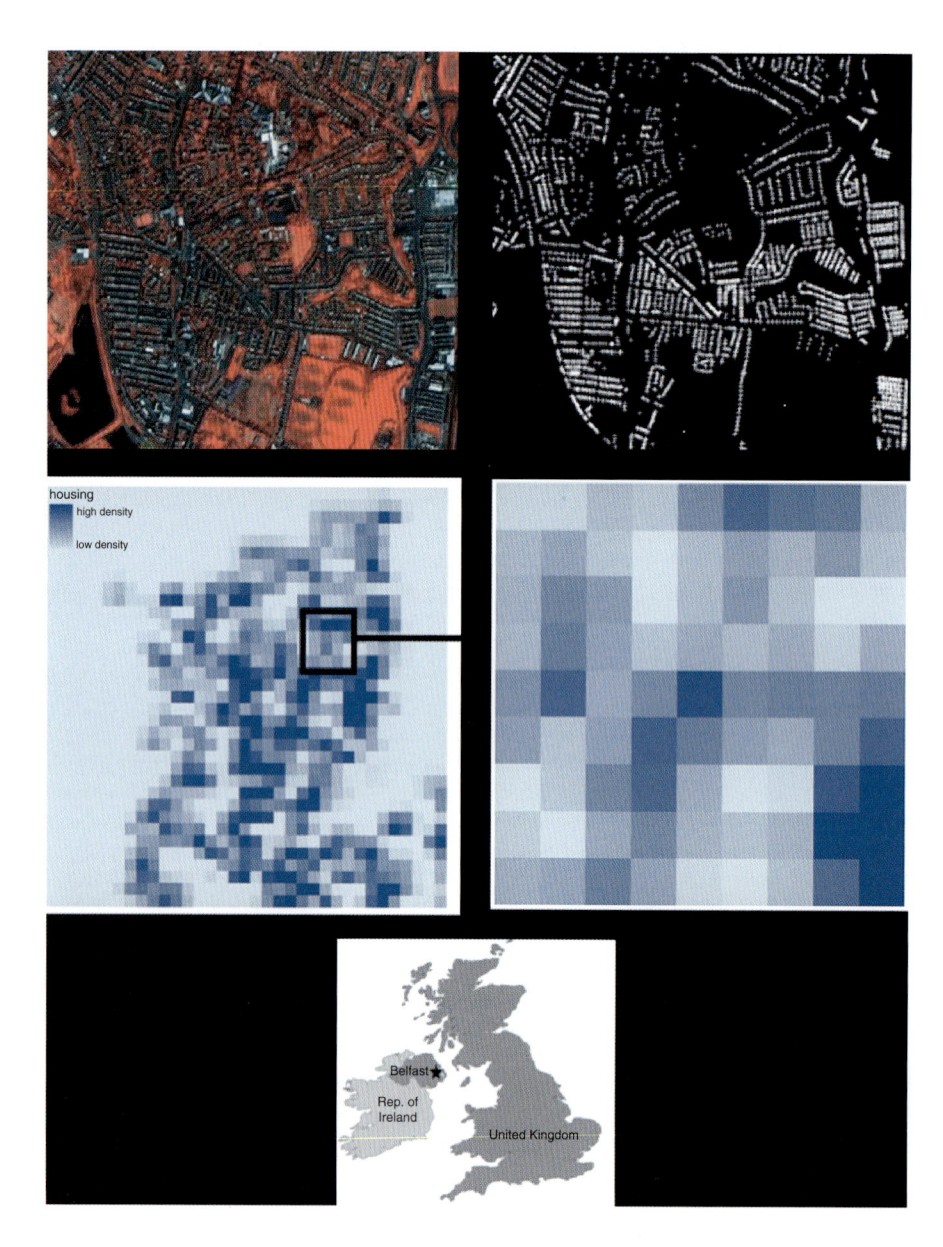

FIGURE 11.1 Structural and functional representations of the Belfast study area: IKONOS image (top left), postal addresses (top right), census housing surface (bottom left, for whole of NE Belfast; bottom right, the same spatial dimensions as IKONOS image and postal addresses).

urban neighborhoods where increasing irregularity is reflected in less space-filling of urban structure and summarized by a greater fractal dimension (D). A useful complement to D is the contagion index, which measures the degree of fragmentation within the neighborhood, and a reciprocal to D is the lacunarity measurement of the spatial distribution of gaps and holes in the overall urban structural fabric (for full explanation see Mesev, 2005; 2007, Myint, Mesev and Lam, 2006).

Another example is provided by Fig. 11.2 which illustrates the coupling of structural representation from unclassified IKONOS imagery with the COMPAS postal data representing buildings (solid boxes). The goal of structural–functional links is to use the location and the 'area' attributes from COMPAS to help classify built land cover from the image by, for example in Bayesian modifications of maximum likelihood, constraining *a priori* spectral space. The dashed boxes represent the desired classification of individual residential buildings that have similar areal dimensions to those of the measured building areas as determined from the postal data. Incidentally, the areal dimensions of buildings are derived from total floorspace area and represented as graduated squares. Encouraging results are documented in Mesev (2007) from preliminary empirical testing on IKONOS imagery using aerial photography at 15 cm spatial resolution. An iterative computational procedure is being operationalized which links the spatial delineation of buildings from high resolution sensor data with the functional characteristics from postal records. Using the software Definiens, a spectra-spatial classification using IKONOS and COMPAS address points and implementing a nearest neighbor contextual rule reported accuracies of 92.8% compared to 86.6% from a multispectral-only classification.

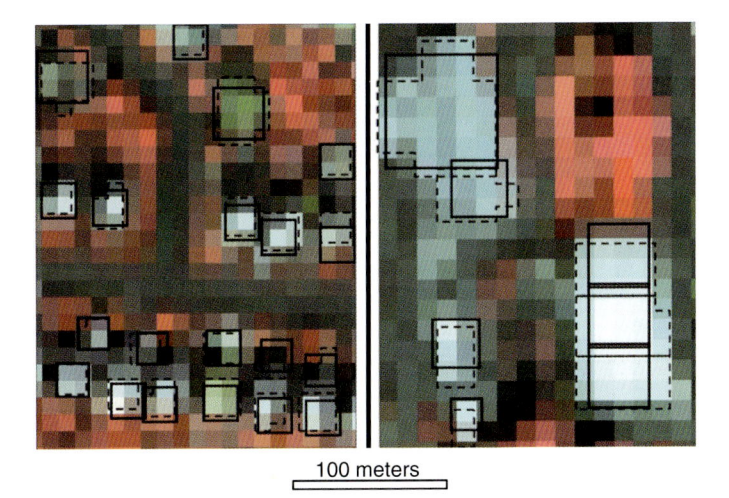

FIGURE 11.2 Structural–functional links: IKONOS and area-determined postal addresses (residential on the left, and commercial on the right) for the Belfast study site.

11.5 Temporal–structural–functional links

A more recent perspective on research into urban structural–functional models is the pursuit of time-dependence; understanding how temporal lags affect the causal links between societal and political functional demands and physical ramifications. Thus far integrative remote sensor models have assumed temporal equality. This is where the same time period is assumed for both when the image is taken and when functional attributes are collected. Instead, Fig. 11.3 illustrates how a temporal integrative model at two time periods (T_1 and T_2) can be formulated by combining urban structural patterns (derived from classified remote sensor data) post T_1 as T_{1+1} and post T_2 as T_{2+1} and urban functional demands and decisions (derived predominantly from population censuses and urban plans) pre T_1 as T_{1-1} and pre T_2 as T_{2-1} respectively.

The relationship states that decisions and trends in urban functions at T_{1-1} determine the type and density of urban structure at T_{2-1}. Precisely how urban functions *determine* urban structure (and maybe even how structure determines functions) is reflective of theories of urban process; for instance, demand for new housing type and housing density, suburbanization, decentralization of businesses, segregation levels, deprivation and congestion and pollution. Changes in urban population, including changes in demographic profiles (family, ethnic minorities and affluent levels), demand for housing (both size and value), and local government plans are the main drivers behind urban processes that link function and structure.

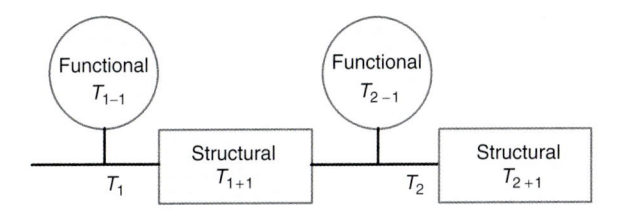

FIGURE 11.3 Temporal–structural–functional relationship.

11.6 Empirical measurement of temporal lags

There are a variety of conceivable methods to determine and even measure temporal lags between structure and function. These can be data-driven, process-driven, or theory-driven. The review by Lo (2007) investigated the development of geospatial technology and its use and potential use in urban morphology. The fields of remote sensing, photogrammetry and GIS when combined are particularly important and suitable for use in urban morphological research. Among the many data-driven applications based on these geospatial technologies, Taubenbock *et al.* (2009), in particular, explored the benefits of multitemporal remote sensing for analyzing long-term changes in temporal and spatial urban sprawl, redensification and urban development for large cities in India. Most other studies also develop analytical models to simulate and evaluate temporal changes. Benguigui, Czamanski and Marinov (2004) investigated the temporal lag of towns in the Tel Aviv, Israel area by comparing an analytic model with a computer simulation to predict population growth. In the dynamic analytic model they used time in two phase; in the first, the derivative was an increasing function – a town was very attractive and there was a short delay between decision to build and complete realization of the process – and there was no shortage of land. However in the other time phase the delay began to increase and there was a lack of available land, leading to a decreased the rate of the population variation until saturation.

Cheng and Masser (2004) acknowledged the inherent spatial and temporal complexity of urban growth by developing a process-oriented cellular automata methodology at both the local spatial and the global dynamic scales based planning and decision-making processes. They linked spatial and temporal patterns using an innovative nonlinear function of land development and dynamic weighting. The model approach was more recently developed by Chen (2009) who investigated spatial interactions between cities based on a time-lag parameter and time functions, and developed a Newtonian-type model that integrated a temporal dimension into the spatial processes of city distributions.

In terms of theory-driven work, Dietzel *et al.* (2005) attempted to develop theories in spatio-temporal dynamics urban geography by using remotely sensed data to determine the historical extent of urban areas and spatial metrics patterns of urban growth over a hundred year period. Changes in these metrics produced a general temporal oscillation between phases of diffusion and coalescence in urban growth. In more abstract terms, Latham and McCormack (2004) postulated that any increased attention to the 'material' requires a more expansive engagement with the "immaterial" when attempting to understand the complex spatialities of the urban. And finally, Aubrecht *et al.* (2009) developed an integrative model of three-dimensional urban structure and function centered on airborne laser scanning, geocoded address point data and raster population surface producing accurate functional classifications.

In summarizing the literature on urban temporal lags, it seems research is either on-going or in theoretical development. Any models dealing in any all or some combination of urban temporality, structuralism, and functionality have or are currently being designed to create more complete representations of urban morphology and socioeconomic characteristics, where temporal lags are non-linear indicators of land cover/land use changes. Current work outlined by the author in this chapter is at a similar prototype stage and sensitivity analyses are presently being developed to determine the concept of the optimum temporal lag, i.e. whether it should be decennial to coincide with most population censuses or interdecennial. Research is underway to establish whether such integrative and dynamic models have also the ability to predict urban growth based on the relationship between land cover/land use (as measured by remotely sensed data) and population and demographic demands (as measured by the census and other socioeconomic and housing sources). It is hoped in the fullness of time that such models of temporality will become vital components in the monitoring of city-wide variations of social deprivation, housing density, traffic congestion, heat island effects, non-point source pollution and others issues of urban sustainability.

In the meantime, temporal lags may be observed without the need for an analytical model. Figure 11.4 illustrates a simple spatial comparison between point-based POINTER (successor to COMPAS) data from 2007 with an IKONOS image taken in May

FIGURE 11.4 Temporal lag between IKONOS image (colored pixels representing structure) and POINTER postal addresses (black points representing function).

2009. Even with a two-year difference, there is evidence (in the top middle section of the figure) of a temporal lag. This is where the pixels from the IKONOS image clearly show spectral colors associated with new built land cover yet the POINTER dataset has yet to be updated. This is obviously a situation of temporal discrepancy and the real question is whether such new structural developments in urban areas can be linked by functional demand by the population and if they can be linked whether they are predictive of future urban development. These are issues that need further research.

Conclusions

This chapter is a review of urban remote sensing research and the challenges for developing more complete representations of urban areas from remotely sensed data and socioeconomic data sources. It is an exploration into the feasibility of addressing a temporal lag between the structural configuration of a city (as measured by satellite imagery) and the city's functional characteristics (as measured by social surveys, such as population censuses). Such research can be developed to predict urban growth based on previous demands and policies for new residential and commercial developments. The chapter is also a call for scientists engaged in urban remote sensing research to advocate models and methodologies that are more pragmatic and prescriptive; with the distinct objective of informing policy makers of possible demands for residential development and commercial expansion. And finally, this chapter is an attempt to stimulate conceptual thinking and develop research agendas on how temporal lags help define prescriptive methodologies that help target possible changes in urban structure with careful reference to population censuses that were taken well before these structural changes materialized. Possible items on a research agenda may include a more precise measurement of a temporal lag between function and form; should it be decennial to coincide with population censuses or linked to specific urban building regimes; do temporal lags vary with city size and urbanization rates; and are temporal lags uniform across an entire city or variable within neighborhoods.

Urban remote sensing is gaining in prominence on the world stage yet has far to go before being able to foster rigorous and reliable models of the urban hierarchy – the most spatially diffuse and functionally dynamic landscapes on the earth's surface. The distinction between micro and macro remote sensing equates to a distinction between precision urban structural (syntactic) configuration and city-wide functional representation using integrative models that link spectral information from high spatial resolution sensor data with spatial and temporal indicators from auxiliary sources. In each the focus is on integrative models that explore metrics and maximization procedures in an attempt to summarize the cartographic and geocomputation potential of the burgeoning urban remote sensing technology. The author is currently testing sensitivity analyses to determine optimum lags and it is hoped that the resulting models of multi-temporality will become vital components in the monitoring of city-wide variations of social poverty, housing density, traffic congestion, heat island effects, non-point source pollution and others issues of urban sustainability.

References

Aubrecht, C., Steinnocher, K., Hollaus, M. and Wagner, W. (2009) Integrating earth observation and GIScience for high resolution spatial and functional modeling of urban land use. *Computers, Environment and Urban Systems*, **33**, 15–25.

Barnsley M.J., Steel, A. and Barr, S. (2003) Determining urban land use through an analysis of the spatial composition of buildings identified in LIDAR and multispectral image data. *Remotely Sensed Cities*, (ed. V. Mesev), London, Taylor & Francis, pp. 83–108.

Benguigui, L., Czamanski, D. and Marinov, M. (2004) Scaling and urban growth. *International Journal of Modern Physics C*, **15** (7), 989–996.

Brenner, N. (2000) The urban question as a scale question: reflections on Henri Lefebvre, urban theory and politics of scale. *International Journal of Urban and Regional Research*, **24** (2), 361–378.

Carter, H. (1985) *The Study of Urban Geography*, Arnold, Maryland, USA.

Chen, K. (2002) An approach to linking remotely sensed data and areal census data. *International Journal of Remote Sensing*, **23**, 37–48.

Chen, Y.G. (2009) Urban gravity model based on cross-correlation function and Fourier analyses of spatio-temporal process. *Chaos Solutions and Fractals*, **41** (2), 603–614.

Cheng, J. and Masser, I. (2004) Understanding spatial and temporal processes of urban growth: cellular automata modelling. *Environment and Planning B*, **31**, 167–194.

Clark, W.A.V. (2008) Geography, space, science: perspectives from studies of migration and geographical sorting. *Geographical Analysis*, **40**, 258–275.

Couloigner, I. and Ranchin, T. (2000) Mapping of urban areas: A multiresolution modeling approach for semi-automatic extraction of streets. *Photogrammetric Engineering and Remote Sensing*, **66**, 867–874.

Dietzel, C., Herold, M., Hemphill, J.J. and Clarke, K.C. (2005) Spatio-temporal dynamics in California's central valley: empirical links to urban theory. *International Journal of Geographical Information Science*, **19** (2), 175–195.

Gamba, P., Dell'Acqua, F. and Dasarathy, B.V. (2005) Urban remote sensing using multiple data sets: past, present, and future. Information Fusion, **6**, 319–326.

Harvey, J.T. (2002) Estimating census district populations from satellite imagery: some approaches and limitations. *International Journal of Remote Sensing*, **23**, 2071–2095.

Herbert, D.T. and Thomas, C.J. 1982. *Urban Geography: A First Approach*. John Wiley & Sons, Inc., New York.

Herold, M., Scepan, J. and Clarke, K.C. (2002) The use of remote sensing and landscape metrics to describe structures and changes in urban land uses. *Environment and Planning A*, **34**, 1443–1458.

Latham, A. and McCormack, D.P. (2004) Moving cities: rethinking the materialities of urban geography, *Progress in Human Geography*, **28** (6), 701–724.

Lo, C.P. (2003) Zone-based estimation of population and housing units from satellite-generated land use/land cover maps, in *Remotely Sensed Cities* (ed. V. Mesev), Taylor & Francis: London, pp. 157–180.

Lo, C.P. (2007) The application of geospatial technology to urban morphological research. *Urban Morphology*, **11**(2), 81–90.

Longley, P.A. (2002) Geographical information systems: will developments in urban remote sensing and GIS lead to 'better' urban geography? *Progress in Human Geography*, **26**, 231–239.

Martin, D.J., Langford, M. and Tate, N.J. (2000) Refining population surface models: experiments with Northern Ireland census data. *Transactions in GIS*, **4**, 343–360.

Mesev, V. (2003) *Remotely Sensed Cities*. Taylor & Francis, London.

Mesev, V. (2005) Identification and characterisation of urban building patterns using IKONOS imagery and point-based postal data. *Computers, Environment and Urban Systems*, **29**, 541–557.

Mesev, V. (2007) Fusion of point-based urban data with IKONOS imagery for locating urban neighborhood features and patterns. *Information Fusion*, **8**, 157–167.

Myint, S., Mesev, V. and Lam, N. (2006) Urban textural analysis from remote sensor data: Lacunarity measurements based on the differential box counting method. *Geographical Analysis*, **38**, 371–390.

Pasaresi, M. and Bianchin, A. (2001) Recognizing settlement structure using mathematical morphology and image texture, in *Remote Sensing and Urban Analysis* (eds J.P. Donnay, M.J. Barnsley and P.A. Longley), Taylor & Francis, London, pp. 55–67.

Taubenbock, H., Wegmann, M., Roth, A., Mehl, H. and Dech, S. (2009) Urbanization in India – spatiotemporal analysis using remote sensing data. *Computers, Environment and Urban Systems*, **33**(3), 179–188.

Weng, Q. and Quattrochi, D.A. (2007) *Urban Remote Sensing*, CRC Press, Boca Raton.

Whitehand, J.W.R. (1977) The basis for an historico-geographical theory of urban form. *Transactions of the Institute of British Geographers*, **2**(3): 400–416.

Xian, G. (2010) *Remote Sensing Applications for the Urban Environment*, CRC Press, Boca Raton.

iv

URBAN SOCIOECONOMIC ANALYSES

Applying remote sensing to urban socioeconomic analyses has been an expanding research area in urban remote sensing. Part IV (Chs 12–16) examines some latest developments in the synergistic use of remote sensing and other types of geospatial information for developing urban socioeconomic indicators. It begins with a chapter (Ch. 12) discussing a pluralistic approach to defining and measuring urban sprawl. This topic is included as part of urban socioeconomic analyses because defining urban sprawl involves not only urban spatial characteristics but also socioeconomic conditions such as population density and transportation. The remaining chapters in Part IV examine several exciting areas of urban socioeconomic analyses. Chapter 13 details a method for small area population estimation by combined use of high-resolution imagery with lidar data. Chapter 14 reviews various areal interpolation techniques emphasizing dasymetric mapping, followed by an example in which population estimates and sociodemographic data are derived for different spatial units by using dasymetric mapping methods. Chapter 15 discusses a method that has been developed to estimate the global percent population having electric power access based on the presence of satellite detected nighttime lighting. Finally, Chapter 16 examines the roles of remote sensing and GIS for urban environmental justice research.

12

A pluralistic approach to defining and measuring urban sprawl

Amnon Frenkel and Daniel Orenstein

The term "urban sprawl" is often used as a synonym for undesired low-density or otherwise unplanned urban spatial development. However, the precise definition and its desirability are debated. Remote sensing practitioners can contribute to our understanding of urban spatial development by measuring its spatial characteristics and dynamics and providing the data to planners and policy makers. By extension, such data can assist in defining sprawl and assessing its presence and intensity in a given metropolitan area. In this chapter, we review the extensive literature and controversial debate around the definition of urban sprawl, emphasizing common themes in definitions and those quantifiable spatial characteristics that would be of specific interest to remote sensing practitioners. The chapter shows that sprawl can be described by multiple quantitative measures, but that different sprawl measures may yield conflicting results. As a complex and multi-faceted phenomenon, we suggest that sprawl is best defined for a given case study, and quantified using a range of indicators specially selected to suit the researcher's definition of sprawl, spatial scale of analysis and specific characteristics of the study site.

12.1 Introduction

> ... I know it when I see it
>
> **Justice Potter Stewart, 1964**[1]

Justice Stewart's frequently quoted statement was not a reference to urban sprawl, but considering the widespread debate about its very definition, it is particularly appropriate and widely used in this context. Urban sprawl is indeed something that many people seem to recognize and have an opinion about, but when it comes to quantifying its dimensions, we become less certain regarding what we are measuring on the ground.

The term "urban sprawl" was first coined by Buttenheim and Cornick (1938), and its use became common throughout the latter half of the 20th century. Sprawl has been used as the descriptive, yet generic, term of choice to describe a variety of urban development forms that shared low density of buildings and population as a unifying trait. These types of urban spatial development played a predominant role in modern urban form in North America and Europe (Glaeser and Kahn, 2004) and a contentious debate regarding their desirability erupted and continues through the present (Ewing, 1997, 2008; Gordon and Richardson, 1997, 2000).

Despite broad interest that developed around the issue of urban sprawl, establishing a clear and unambiguous definition has proven to be an elusive task (Chin, 2002; Hasse and Lathrop, 2003a; Wolman *et al.*, 2005; Hasse, 2007). Commentators on sprawl refer to a broad array of defining characteristics (Hess *et al.*, 2001; Johnson, 2001; Ewing Pendall and Chen, 2002; Wolman *et al.*, 2005; Cutsinger and Galster, 2006; Hasse, 2007). Galster and colleagues (2001) write that the term "urban sprawl" became a metaphor used alternatively to describe (or imply) the patterns, processes, causes and/or consequences of particular urban spatial development patterns. A concise definition has been further muddled because the term is ultimately a cultural construct (Bruegmann, 2005). Therefore, cultural milieu, ideology, and personal experience are intimately linked to how people define sprawl. The lack of a single definition has logically led to difficulty in establishing a unified methodology for measuring the phenomenon; after all, how can we measure what we don't know we're measuring (Burchell *et al.*, 1998; Malpezzi, 1999; Torrens and Alberti, 2000; Galster *et al.*, 2001; Johnson, 2001; Ewing, Pendall and Chen, 2002)?

Most scholars and practitioners agree that a first step towards defining sprawl is to quantify various characteristics of urban spatial development and the dynamics guiding them. Once this is done, scholars, policy-makers and others can then debate the desirability of such phenomena and discuss, if needed, policies to address them. Therefore recent research efforts have focused on establishing and measuring quantifiable variables that capture various characteristics of urban spatial development.

We begin this chapter by integrating several definitions of sprawl derived from a comprehensive survey of the academic and professional literature in order to extract quantifiable spatial characteristics recurring throughout the literature. It is our belief that despite the constant refrain that there is no consensus on sprawl, there is enough agreement to move forward in quantifying relevant forms of urban spatial development. To this we add two caveats. First, sprawl researchers must be explicit in their qualitative definition of sprawl and use quantitative variables that complement their definition. Second, since different variables may yield different results, a pluralistic approach should be adopted which allows for the possibility that sprawl is a multifaceted phenomenon that appears differently on the landscape depending on how, where and when it is measured. We allow the researcher and/or end-user to determine which variables are relevant to their location-specific research and their own sprawl definitions. We conclude our overview of sprawl with a short historical narrative of urban spatial development that was/is considered sprawl.

Next, we provide an extended list of spatial variables for measuring the state of sprawl and associated processes and explore how these variables have been applied empirically. We taxonomize the variables and rank them according to criteria for what constitutes a good measure and suggest when and where the application of each variable would be recommended. We conclude by comparing results of four macro-studies of sprawl in US metropolitan regions to elucidate how the use of different measures produces similar or different results.

While we direct our narrative to remote sensing experts, we emphasize that "sprawl" is often considered as much a socio-economic phenomenon as a physical one. As such, the remote sensing literature is somewhat limited with regard to sprawl discourse, primarily measuring certain physical manifestations of urban development, like building density, time series of urban growth, and geometric parameters of urban form (Sutton, 2003; Hasse, 2007; Irwin and Bockstael, 2008; Bhatta, Saraswati and Bandyopadhyay, 2010). We note that all of these, when combined with geographically-specific socioeconomic and demographic data (e.g., Martinuzzi, Gould and Ramos Gonzalez, 2007), greatly expand our options for measuring sprawl. We assume here that professionals employing remote sensing would benefit by knowing what variables would be useful for them to quantify, and after doing so, provide the results to urban planners, the policy-making community and other stakeholders.

12.2 The diversity of definitions of sprawl

Several syntheses of sprawl definitions exist in the literature (Burchell *et al.*, 1998; Galster *et al.*, 2001; Hess *et al.*, 2001; Malpezzi and Guo, 2001; Chin, 2002; Ewing, 2008; Frenkel and Ashkenazi, 2008a, 2008b; Torrens, 2008). According to these sources, urban sprawl has been defined primarily in three ways: (1) definitions relating to describing a physical and spatial phenomenon of urban spatial development (qualitatively and quantitatively); (2) definitions that focus on the purported social, economic and/or ecological consequences of the phenomenon (described in various ways), and by extension, by normative desires to avoid perceived undesirable urban spatial development patterns, and; (3) definitions focusing on particular socio-economic trends that lead to particular urban spatial

---

[1] Jacobellis v. Ohio (378 US 184; 1964); available from http://caselaw.lp.findlaw.com/scripts/getcase.pl?court=USandvol=378andinvol=184 (accessed 15 November 2010).

development patterns. Examples for each of the three definitions are provided below.

12.2.1 Definitions describing an urban spatial development phenomenon

Sprawl is most often considered a particular spatial pattern of urban development characterized by low density residential and commercial development. Low density could be considered in terms of building density or population density. This development may be adjacent to existing development, as with suburbs, or scattered and discontinuous development physically separated from the central city, as with leapfrog development (Harvey and Clark, 1965; Downs, 1994; Ewing, 1997; Burchell et al., 1998; Hess et al., 2001; Chin, 2002; Ewing, Pendall and Chen, 2002; Glaeser and Kahn, 2004; Tsai, 2005; Torrens, 2008). As Chin points out, both forms of development are classified as sprawl, although "the forms and resulting impact are vastly different" (Chin, 2002). This is at least partly understood when considering that sprawl can have different definitions at different spatial scales (Tsai, 2005) – for instance at the scale of a single urban settlement or at the scale of a region of multiple settlements (see below). Sprawl is also defined as developed land highly segregated into single uses (Ewing, 1997). The presence of large blocks of exclusively residential land or commercial strip development, for example, is considered sprawl (Chin, 2002).

While there are multiple, measurable characteristics of urban sprawl (about which we expand upon in a following section), we note that there are no settled values, or quantitative thresholds, that define sprawl in absolute terms. Proposed absolute values or thresholds that separate 'good' spatial development from 'bad' are subject to debate, as is the question of what residential density constitutes sprawl (Chin, 2002). Superlatives are common throughout the sprawl literature, describing the phenomenon as "excessive" (Bruekner, 2000), "wasteful" (Torrens and Alberti, 2000) and "inefficient" (Fulton et al., 2001; Peiser, 2001; Frenkel and Ashkenazi, 2008a; Thompson and Prokopy, 2009). Others describe the kind of urban growth considered to be sprawl as "dysfunctional" (Ewing, Pendall and Chen, 2002). But sprawl is clearly a relative, rather than absolute, phenomenon (Frenkel and Ashkenazi, 2008b; Bhatta, Saraswati and Bandyopadhyay, 2010). This is explicitly recognized in the work of Sutton (2003), for example, for whom sprawl is relative to an average relationship between population size and developed area across US metropolitan regions.

We suggest that one way of working towards consensus on the matter is to define sprawl as a directional process (Harvey and Clark, 1965; Hess et al., 2001), rather than an absolute state of being. Accepting this, the dynamic temporal and spatial patterns of urban spatial growth become crucial to measure and monitor. Noting how these patterns change over time and space change the debate from one about sprawl (a state) into one about sprawling (a process). In other words, while we may not be able to agree that a given density constitutes sprawl, we can call a process of declining density, for example, as sprawling.

12.2.2 Definitions based on consequences of sprawl; sprawl is as sprawl does

"Ultimately," write Ewing and colleagues, "sprawl must be judged by its consequences" (Ewing, Pendall and Chen, 2002). Thus, the definition of sprawl becomes the socio-economic or ecological consequences of a particular kind of urban spatial development. Consequences might include (1) lack of accessibility between regions in the urban area (Ewing, Pendall and Chen, 2002); (2) high rates of driving and vehicle ownership (Burchell et al., 1998; Ewing, Pendall and Chen, 2002); (3) increased air pollution (Ewing, Pendall and Chen, 2002); (4) undesirable ecological impacts, such as impact on ecosystem cycles or species composition (Perry and Dmi'el, 1995; Cam et al., 2000; Kreuter et al., 2001; McKinney, 2002; Hasse and Lathrop, 2003b; Robinson, Newell and Marzluff, 2005); (5) consumption of exurban open space and agricultural land (Burchell et al., 1998; Hasse and Lathrop, 2003b; Frenkel, 2004; Czamanski et al., 2008; Koomen, Dekkers and van Dijk, 2008; Thompson and Prokopy, 2009), and/or (6) catalyzing socio-economic and racial segregation (Squires and Kubrin, 2005). Some of these variables can be measured directly, particularly loss and fragmentation of open space, while others depend on proxy measures and non-remotely sensed data.

Most, if not all, of these claims are contested. For instance, Glaeser and Kahn (2004) note that while sprawl and associated increases in private automobile use may have increased air pollution, technological improvements in fuel efficiency and emissions control have led to an overall reduction in most air pollutant emissions in the United States. The claim of sprawl leading to socio-economic and racial segregation is also challenged (Glaeser and Kahn, 2004; Wheeler, 2008). Further research attests to the potential benefits of sprawl in terms of maximizing consumer preference, efficient distribution of business and residential areas, low cost relative to high-rise or high concentration settlement (Gordon and Richardson, 1997), and increasing species and ecological habitat diversity (Czamanski et al., 2008).

The great interest that urban planners, policy makers, scholars and activists share regarding sprawl is, to a large degree, derived from opinions regarding how a city should develop spatially, and what the role (if any) the planner and policy maker should serve in promoting or preventing sprawl. Researchers have noted that the debate around sprawl is often the result of its ideological framing (Burchell et al., 1998; Chin, 2002; Hasse, 2004, 2007). Thus, some researchers and activists define sprawl in a pejorative way in order to advocate or oppose a particular policy or plan. One's description of sprawl characteristics can thus be seen as a subjective extension of values-laden planning goals; that is, sprawl is in the eye of the beholder. Opponents of sprawl define it in terms of what it is not: highly centralized, compact cities with mixed land uses, whose transportation systems de-emphasize the role of the private automobile in lieu of public and/or non-motorized transportation (e.g., Duany, Plater-Zyberk and Speck, 2000). Advocates of more laissez faire policy approach define it in a more positive light: benign at worst and the desired expression of people's residential preferences at best (Gordon and Richardson, 1997). Simultaneously, these latter scholars provide research results that challenge the claims of the former group.

Consider two examples: the advocacy organizations Smart Growth America and The Cato Institute. Smart Growth America is an advocacy organization self-described as "a nationwide coalition promoting a better way to grow: one that protects farmland and open space, revitalizes neighborhoods, keeps housing affordable, and provides more transportation choices" (Smart Growth America, 2009). The organization also commissions research on sprawl. Their assessments of sprawl (e.g., Ewing, Pendall and Chen, 2002) are based on a positive vision of what constitutes good urban development. On the other end of the spectrum, the Cato Institute, a free market advocacy organization that seeks to "increase the understanding of public policies based on the principles of limited government, free markets, individual liberty, and peace" (Cato Institute, 2009). This institute also selects its own characteristics of what constitutes good urban development based on their ideological world view (Gordon and Richardson, 2000). As can be assumed, the reports produced by each organization, produced by reputable scholars, advocate two opposing views on sprawl, how it should be measured, its impact and its policy implications.

12.2.3 Definitions according to the social and/or economic processes that give rise to particular urban spatial development patterns

Research suggests that socioeconomic trends may lead to the aforementioned characteristics of spatial development, and thus these trends are included in the definition of sprawl. Some research defines sprawl processes as characterized by the flight of stronger income classes away from the urban center and towards the urban fringe (Ewing, Pendall and Chen, 2002), and the decline of city centers (van den Berg *et al.*, 1982, Mills and Hamilton, 1994; Golledge and Stimson, 1997). The flight of economically strong populations and retail businesses that leave for fringe areas in search of more lax building regulations and/or preferable tax remission lead to a severe decline in the municipal tax base of the region from where they came (Hadly, 2000). Squires and Kubrin (2005), consider urban sprawl to operate simultaneously with concentration of poverty and racial segregation, where sprawl is catalyzed by and catalyzes socioeconomic and racial segregation. Other researchers discuss sprawl as a result of lack of integrated land-use planning (Burchell *et al.*, 1998).

On the other hand, social and economic processes leading to sprawl are sometimes couched in positive terms, as when decentralization of employment and population is considered a desirable process (Glaeser and Kahn, 2004). Sprawl has been also described as the inevitable result of increased mobility due to an automobile-based transportation system (Glaeser and Kahn, 2004), the logical response of markets to consumer demand (Gordon and Richardson, 1997), or possibly as an expression of efficiency maximization among multiple economic players (Batty and Longley, 1994).

Again, the processes emphasized by the various researchers and/or advocates often reflect their ideological disposition. Either way, these definitions are less relevant to a volume on urban remote sensing because they depend on data other than remotely sensed data to measure them.[2]

12.2.4 Sprawl redux: focusing on the concerns of remote sensing experts

Of these definitions, we believe that the characteristics around which we can extract the most objective information, spatial characteristics of urban growth, are of particular concern to remote sensing experts. Therefore, for the remainder of this chapter, we focus on those definitions of sprawl that are physical–spatial in nature, e.g., low density building along the edges of an urban center, tracts of single land use types (e.g. separation of residential, employment and commercial centers), or development not contiguous to existing built-up areas. The distribution of such development should be measured at the neighborhood, metropolitan and regional level, as definitions of sprawl vary depending on spatial scale. These are characteristics that can be readily measured by remote sensing experts (Hasse, 2007, Martinuzzi, Gould and Ramos Gonzalez, 2007; Bhatta, Saraswati and Bandyopadhyay, 2010), and their quantification is of utmost importance in tracking sprawl over time.

The questions of whether or not these spatial characteristics are good or bad, whether they are caused by particular processes and whether they lead to particular desired or undesired environmental, economic or social processes are left aside at this point. The spatial measurements described later in the text provide a crucial foundation of data on which to build further analyses, and they are characteristics that can be derived through remotely sensed data and quantified.

12.3 Historic forms of ``urban sprawl''

To understand the origins of the particular form of urban spatial development described as sprawl, we consider two major points in the history of modern urban development when profound demographic, urban and spatial changes were taking place. The first period was the industrial revolution of the 19th century. This period was marked by the massive migration from rural areas to industrial cities and their transformation into centers of activities, primarily in Europe. The period was also characterized by the massive immigration from Europe to the core cities in the United States, leading to an out-migration of the middle and upper classes out of the cities to the urban fringe (Paddison, 2001). Following the industrialization of cities and rapid rise in population densities, the quality of life in cities fell and people romanticized for life in the adjoining open spaces. The squalid conditions that developed in these major industrial cities gave rise to zoning reforms in cities and to suburban development outside of them (Gillham, 2002). From the mid-1800s in the

---

[2] It is possible to measure some of these socio-economic phenomena spatially through proxies (e.g. the use of "night lights" as proxies for GDP, Henderson, Storeygard and Weil, 2009), but our focus here is the measurement of physical–spatial characteristics of urban development.

United States, homes and neighborhoods began to appear in the countryside, soon to be connected with railroads and streetcars that would catalyze additional demographic movement from city to suburb (Gillham, 2002).

The process was greatly expedited in the post World War II mid 20th century, a period that was characterized by spatial diffusion of residents and activities to the outskirts of urban centers (Mills and Hamilton, 1994). In the United States, and to a lesser extent in Europe, the process of suburbanization (commonly associated with sprawl, but see below) started in earnest following World War II, with a combination of high population growth and an inability of city centers to absorb this growth.[3] A rapidly growing post-war economy, improvements in technology and a rise in standard of living all contributed to increasing demand for large-lot, single-family homes on the outskirts of cities. Concurrently, city centers were in decline (Batty, Xie and Zhanly, 1999, Golledge and Stimson, 1997). The rise of the automobile as a predominant form of transportation facilitated and expanded this process (Glaeser and Kahn, 2004) and set in motion a positive feedback mechanism: the more car-dependent society became, the more suburbs held appeal; the more suburbs proliferated, the more dependent society became on the automobile.

The actual use of the term "sprawl" began in the United States in the 1950s, and became widely used from the 1960s (Belser, 1960, Harvey and Clark, 1965, Gans, 1967, Real Estate Research Corporation, 1974).[4] From the 1970s, the term "sprawl" was often accompanied by "suburbanization," although the two are conceptually unique from one another. Suburbanization refers to the migration of urban residents to the peripheries or outside of cities in a metropolitan area in order to establish new residential neighborhoods (Angotti, 1993). Fishman (1987) differentiates between English/American suburbanization and that of continental Europe in that the former was characterized by middle and upper class residents leaving the cities for green, low-density homes in the urban periphery, while the latter was led by industry leaving the cities, followed by the working class.

Sprawl is a broader concept as defined in our introduction that includes social, demographic and economic characteristics and a broader diversity of urban spatial development characteristics of which suburbanization is just one. Other characteristic development forms include edge cities and exurban development and also included is the demographic and socioeconomic decline of urban centers.

While sprawl may be considered a global phenomenon, the history of sprawl seems largely to have been written in the United States and to a lesser degree in Europe. As early as the 1920s, planners in the United States began noting an acceleration of the rate of loss of open and agricultural land in favor of development (Burchell et al., 1998). The rise of zoning regulations, which provided the legal foundation for separating land uses, is considered a major contributor to later sprawl patterns (Gillham, 2002). Later, in the United States, the strong belief in individual property rights and free markets, along with a distrust of strong, central government is posited to have had a significant impact on the shape of sprawling land development patterns. As Gordon and Richardson suggest (2000), the history of American movement from cities to suburbs might reasonably be viewed as people realizing their residential preferences.

In the United States, sprawl, as defined by loss of farms and open space, was noted by planners as early as 1929 in New York (Burchell et al., 1998). Sprawl critics point to Federal zoning policies from 1922 onward, that gave rise to segregated land use, which in turn laid the foundation for an automobile-centered transportation network (Burchell et al., 1998). In the 1950s and 1960s in the United States, sprawl terminology began entering the planning literature, once again emphasizing low density development and the predominance of automobiles. Leapfrog development, complemented by the rise of federal highway system, fed the critique of spatial growth patterns. By 1972, McKee and Smith (cited in Burchell et al., 1998) would distill the definition of sprawl into four forms: (1) very low density development; (2) ribbon-variety development extending along access routes; (3) leapfrog development; and (4) a "haphazard intermingling of developed and vacant land."

In 1991, Garreau introduced the concept of "edge cities" (Garreau, 1991) as the evolution of non-residential urban cluster development along junctions of beltways and interstate roads. Edge cities introduced a new dimension to sprawl, in that it was not low-density residential development around a single urban core, but entirely new urban cores developing as satellites to main cities. Unlike suburbs, edge cities serve all the functions of the urban core with an emphasis on employment centers. The European analogy to edge cities have been called Functional Urban Areas (van den Berg et al., 1982), and in this case, they are considered a collection of urban communities that together include residential, employment and recreational centers, developed on former agricultural land, and within functional proximity of a major urban center.

The development of edge cities added a new dimension to thinking about sprawl – the dimension of spatial scale. Now rather than envisioning only the urban core and sprawled development in connection to it, a broader scale of analysis was needed to consider regional development patterns. Whereas suburbs emphasized sprawl at a municipal level with a particular emphasis on a decline in building density, at the regional scale, terms such as satellite towns, edge cities, exurbs, and megalopolis become relevant to describe spatial broader phenomenon for which density is only one of many relevant spatial characteristics. Consequently, while density remains the most intuitive and popular spatial variable for measuring sprawl, the list of variables becomes longer when considering the multi-scalar dimensions of sprawl.

12.4 Qualitative dimensions of sprawl and quantitative variables for measuring them

In this section, we present quantifiable variables that have been suggested in the literature and employed empirically to measure sprawl. We first present several criteria – our own and drawn from the literature – that a variable measuring sprawl should

---

[3] Although Jackson (1985) suggests that suburbanization, which is defined as a situation when peripheral areas develop at a faster pace than central urban areas, appears as early as 1815 in the US and Britain.

[4] For a compilation of early references to sprawl from the early to mid-20th century, see Hess and colleagues (2001).

meet. Next, we present variables that have been used to measure sprawl in empirical studies. We conclude the section by ranking the variables according to the criteria we set at the outset of the section.

12.4.1 Criteria for a good sprawl measurement variable

In order to minimize discord between various studies on sprawl, spatial variables used to measure sprawl should be held up to certain criteria. These criteria include:

1. *Objectivity*. The variable must be quantifiable and reproducible (Ewing, Pendall and Chen, 2002; Lopez and Hynes, 2003; Torrens, 2008). Since sprawl is a subjective term, researchers should provide all measured values and the values at which they consider sprawl to be occurring, thereby allowing users to decide for themselves if the values suggest sprawl or not (Wilson *et al.*, 2003).

2. *Applicability to a large number of places*. The variable must be generalizable to a wide range of study sites and times and not be specific only to the study site of the current examination (Lopez and Hynes, 2003; Wilson *et al.*, 2003; Irwin and Bockstael, 2008; Torrens, 2008). If it is applicable in only particular situations, the researcher should be explicit regarding the limitations of the variable's application.

3. *Appropriateness for multiple spatial scales of investigation*. Sprawl may occur at a variety of spatial scales (e.g. housing unit, neighborhood, town. region, metropolis, state or country). A good variable is robust enough to apply to multiple scales of investigation, while others may be appropriate to only a certain scale.

4. *Meaningfulness, usefulness, and simplicity*. The variable must capture one of the descriptive elements of sprawl (Ewing, Pendall and Chen, 2002; Lopez and Hynes, 2003; Wilson *et al.*, 2003; Torrens, 2008). The data emerging from sprawl studies must be relevant to stakeholders, and therefore it is crucial that the variables are easily explained, understood and relevant to them (Lopez and Hynes, 2003; Wilson *et al.*, 2003).

5. *Ease of application*. An additional quality of a good sprawl indicator is one that is not overly dependent on complex calculations, software that requires a highly specialized skill set, or inaccessible data such that other researchers or practitioners would not be able to employ the measures in their research. Some variables are good in theory, but the data may be inaccessible or not available at the scale of resolution or historical period desired for research. On the other hand, some methodologies for data preparation demand a high level of computational or spatial analysis skills, an advanced understanding of spatial metrics, or access to computer hardware and software that may make the variables less desirable for the intended end user.

12.4.2 What shall we measure?

Prior to the calculation of sprawl measures, total built area must be measured. Urban spatial growth and its rate of change over time are not, on their own, sprawl measures. They are, however, the most important variables to measure because most sprawl measures that follow are dependent on them. Urban land cover is referred to analogously as built space or impermeable surface cover, although each has slightly different implications for how much land will ultimately be quantified as urban (Orenstein *et al.*, 2010).

Estimations of values for urban land cover and changes in land cover over time are also the most important contributions of remote sensing experts to studying sprawl processes. The sheer amount of published literature on urban remote sensing (this book included) testifies to its importance as well as to the rapidly advancing state of the art (Ward, Phin and Murray, 2000; Stefanov, Ramsey and Christensen, 2001; Zhang *et al.*, 2002; Sutton, 2003; Rogan and Chen, 2004; Xian and Crane, 2005; Martinuzzi, Gould and Ramos Gonzalez 2007; Jat, Garg and Khare, 2008; Pu *et al.*, 2008; Bhatta, Saraswati and Bandyopadhyay, 2010). Aside from estimating generic urban land cover, rapid improvements in the quality of data and interpretive methodologies make it possible to differentiate between types of urban land cover (Foresman, Pickett and Zipperer, 1997; McCauley and Goetz, 2004). Differentiating growth in residential area (as contrasted with industrial, business and commercial areas) and in low-density residential area is particularly important, as they are two sub-variables commonly used for characterizing sprawl (McCauley and Goetz, 2004; Irwin and Bockstael, 2008). Computing the amount of and change in availability of developable land, assessed in conjunction with ancillary data like statutory land use plans, also provides important data for sprawl characterization.

Sprawl measures suggested in the literature can be divided into five major groups (Table 12.1):

- density (building and population);
- relative population growth rates;
- spatial geometry of built and open space;
- accessibility between residential, commercial and business areas;
- aesthetic measures.

Due to the nature of the current volume with its emphasis on remote sensing, we focus on those variables whose values can be derived through remote sensing data and analysis. We briefly mention other variables as well, but those are generally quantified using other, non-remote sensed data, such as census and survey data. As such, aesthetic measures as a category are not included in Table 12.1, but see below).

12.4.2.1 Density

There are various types of densities, as well as many ways and scales at which to measure them (Churchman, 1999; Burton, 2000; Chin, 2002; Tsai, 2005). Density can be defined as the ratio between the amount of a certain urban activity and the area on which it exists, for instance population size (Lopez and Hynes, 2003) or housing units per unit area (Razin and Rosentraub, 2000). Population density is considered a key theme in sprawl literature (Galster *et al.*, 2001) and while some argue that it is the most important measure (Fulton *et al.*, 2001; Maret, 2002; Lopez and Hynes, 2003), they are careful to specify that, while important, it is not the only measure of sprawl. As a sprawl measure, population density fails to take into account

TABLE 12.1 Sprawl measurements assessment.

Group of sprawl measurements		Measurement	Criteria				
			Objectivity	Applicable to a large number of places	Appropriate for the spatial scale	Meaningful, useful, and simple to understand	Ease of application
Urban land cover and spatial growth[a]		Change in total amount of urban land cover	+++	+++	A	+++	+++
		Growth in residential area	+++	+++	B	++	++
		Growth in low-density residential area	++	++	B	++	++
		Amount of and change in availability of developable land	++	+	B	+++	++
Density		Gross population density	+++	+++	A	+++	+++
		Net population density	+++	+++	B	++	++
		Current and expected population divided by developed and developable land	++	+	B	++	++
		Density as a function of accessibility to the C.B.D.	++	++	A	++	++
		Density gradients	+++	++	A	++	+
		Amount of population living in low-density	++	++	B	++	++
Relative population growth rates		Sprawl Quotient	+++	+++	A	++	+++
		Suburbs versus central city	++	+	C	++	++
Spatial Geometry[b]	Composition (Degree of homogeneity/heterogeneity in land use)	Percentage contribution of each patch type	+++	++	C	+++	++
		Mean patch size	+++	++	C	++	++
	Configuration	Edge shape (e.g. circularity)	+++	++	A	++	++
		Area to circumference ratio	+++	++	B	++	++
		Continuity indices (applied primarily to open space)	++	++	C	++	++
		Leapfrog indices (applied primarily to built space). Equal to % built in urban core/% built outside urban core.	++	++	A	+++	++
		Fractal dimension	+++	++	A	+	+
Accessibility measures		Road length/area	+++	+++	A	+++	++
		Household traveling time	+++	+++	B	+++	++
		Mean Proximity Index	+++	+++	A	+++	++
		Gravity/logit models	+++	+++	A	++	++

Key: +++ Meets criteria well   A Applicable at multiple spatial scales
     ++ Meets criteria with some exceptions   B Applicable at some spatial scales
     +   Does not meet criteria   C Applicable at limited spatial scale

[a] Note that growth rate of urban land cover variables are not, by themselves, indicators of sprawl. Rather, these variables are crucial for calculating the values of many of the sprawl variables that follow.
[b] The user must be cautious in applying spatial indicators to measure sprawl, as such indicators may be influenced by municipal borders. For instance, in comparing a long, narrow locality with a circular locality, one may reach the conclusion that the former locality is more sprawled due to circularity.

any aspect of spatial geometry, ecological impact, or land use composition – all of which are significant elements of sprawl by all conventional definitions discussed here and elsewhere (Frenkel and Ashkenazi, 2008b).

Sprawl is generally defined as a condition in which one or more types of density is relatively low or decreases over a certain time period. But what constitutes low density? Burchell and colleagues, among others, make it clear that density a relative value:

> **Sprawl is not simply development at less-than-maximum density; rather, it refers to development that, given a national and regional framework (i.e. suburbs in various locations of the United States), is at a low relative density, and one that may be too costly to maintain**
>
> **(Burchell et al., 1998).**

Population density can be calculated in several ways, depending on the extent of data and knowledge of the urban landscape. These include gross population density (total population/built area) (Fulton et al., 2001; Sutton, 2003), net population density (total population/built residential area), and population plus expected population divided by developed plus developable land. Density gradient analyses consider density as a function of distance from urban centers or central business districts, where population per unit area declines with distance from urban centers (Batty and Longley, 1994; Alperovich, 1995, Jordan, Ross and Usowski, 1998). Researchers point out that during the past few decades density gradients have been falling (i.e. sprawl is increasing) in developed as well as developing countries (Ingram, 1998). This, they suggest, emphasizes the universality of urban sprawl.

12.4.2.2 Relative population growth rates

If it is possible to differentiate between built land use types and if municipal scale population data is available, a "Sprawl Index" (SI) or "Sprawl Quotient" can be estimated. These are defined as the ratio between the growth rate of built-up areas and the population growth rate in that area. A quotient higher than one implies urban sprawl (Weitz, 1999; Hadly, 2000).

Another example applying density measures is the use of the relative amount of population living in low-density as compared to high-density census tracts in US metropolitan areas (Lopez and Hynes, 2003). Similarly, sprawl has also been defined as a condition in which population growth rates in the suburbs are higher than inside the central city (Jackson, 1985).

12.4.2.3 Spatial-geometry of built and open space

Spatial geometry constitutes the largest group of sprawl measures. These are numerous geometric measures, many of which have been adopted from ecological research (Irwin and Bockstael, 2008) or from fractal geometry (Torrens and Alberti, 2000; Herold and Menz, 2001). They have particular relevance to remote sensing experts and others seeking to quantify spatial measures of sprawl.

As in the discipline of landscape ecology, the landscape is considered to be composed of spatially distinct "patches" with distinct ecological qualities and parameters including area, circumference, edge shape, area/circumference ratio and others.

The distribution of patches across the landscape are characterized and quantified with measures including relative abundance, connectivity and degree of separation between like patches. The individual patch geometry and the aggregate distribution patterns of patches is posited to affect ecological function at the landscape scale (Turner, 1989; Gustafson, 1998). When patch theory is transferred to the domain of urban spatial analysis, patches are defined as land use types (e.g. residential, industrial, commercial, open-natural space, open-agricultural space), and the metrics transfer as well. Here, the patch geometry and distribution of urban patches are suggested to have wide ranging implications for environmental quality, economics, social relations and other social variables.

Geometric-ecological measures can be grouped into two types for urban landscape analysis: composition and configuration (Torrens and Alberti, 2000). Composition refers to how heterogeneous an area is with regard to its mix of patch types and provides information regarding the relationship between and among patches in a matrix. Configuration refers to the geometry of individual land use patches, or how regular or irregular their shape. Patch circumference, and various descriptors of shape of the patch and its edge, like circularity (Gibbs, 1961) and area/edge ratio (McGarigal et al., 2002), are common measures for configuration.[5]

Fractal dimensions provide a second approach to measuring sprawl, where fractal measures replace Euclidean geometry (Batty and Longley, 1994). Fractals are defined as "objects of any kind whose spatial form is nowhere smooth, hence termed 'irregular', and whose irregularity repeats itself geometrically across many scales" (Batty and Longley, 1994). Although the measures are related to configuration, fractals arise from a conceptually different way of looking at the spatial development of cities. Research on fractal dimensions has contributed to our understanding of urban spatial development and our understanding of the forces that may shape a city's form (Benguigui et al., 2000, Benguigui, Blumenfeld-Lieberthal and Czamanski, 2006, Thomas, Frankhauser and Biernacki, 2008). Torrens (2008) and Frenkel and Ashkenazi (2008b) integrate fractal variables into a list of geometric variables with which they characterize sprawl.

The degree of homogeneity/heterogeneity in built land uses (e.g. residential, commercial, industrial) is measured by composition variables (Fulton, 1996; Ewing, Pendall and Chen, 2002). Urban sprawl has been defined as a homogeneous development pattern, characterized by the absence of mixed land use (in particular, residential areas separated from trade and services) at the neighborhood and city scale (Fulton, 1996). Built-up areas with a high rate of mixed land uses are regarded as compact and sustainable (Jenks, Burton and Williams, 1996, Burton, 2000), whereas a high percentage of residential land use is considered homogenous and non-mixed, and thus, sprawling. Another way of looking at this aspect is the balance that exists between the amount of population and the number of jobs (Ewing, Pendall and Chen, 2002). A non-balanced situation where population is large relative to jobs in a single geographic unit is considered a component of sprawl.

---

[5] For readers interested in the mathematical equations for each of these indicators derived from landscape ecology, the easily accessible Fragstats Users Guide provides a comprehensive and detailed overview of each landscape indicator, its equation and its strengths and weaknesses (McGarigal et al., 2002); see: http://www.umass.edu/landeco/research/fragstats/documents/fragstats_documents.html (accessed 15 November 2010).

Many of the variables used to measure homogeneity/heterogeneity are again drawn from the discipline of landscape ecology (McGarigal et al., 2002). Variables such as leapfrog and connectivity indices describe the mix of urban "patches" within a matrix of open space and measure the proximity of similar patch types from one another (Galster et al., 2001). These measures quantify the level of scatter and fragmentation of the urban landscape. When built areas are separated from one another by open space then the landscape is considered fragmented, which is another sprawl characteristic (Torrens and Alberti, 2000).

One of the most intuitive measures derived from landscape metrics is the number of patches of a certain land use type. The larger this number, the more heterogeneous mix of land use patches at the landscape scale. Mean patch size takes the average size of all of the patches of a given land use, and the smaller the average patch size, the more heterogeneous or fragmented the landscape and thus the more sprawled (Torrens and Alberti, 2000; Herold and Menz, 2001).

Three additional sample sprawl measures derived from landscape metrics that measure patch composition are contagion, connectance, and proximity. Contagion is the tendency of patch types to be aggregated (McGarigal et al., 2002); high contagion value at the municipal scale could suggest large tracts of homogeneous land use or sprawl. However, at the regional scale high contagion value might suggest a low amount of fragmentation of the landscape, with built patches clustered and not fragmenting the open space 'matrix'. Connectance and proximity both compute the functional closeness of patches of similar type, and their values are interpreted such that greater dispersal of patches (i.e. built patches in an open space matrix) represents greater sprawl.

Finally, several variables measure the degree of irregularity of the patch including circularity and edge to area ratio. In terms of sprawl, irregularity is considered sprawling, with a perfectly circular patch synonymous to compact development, as opposed to linear or irregular development (Gibbs, 1961).

12.4.2.4 Accessibility between residential, commercial and business areas

Sprawl is defined as a condition of poor accessibility, followed by the massive use of private vehicles (Ewing, 1994, 1997, Ewing, Pendall and Chen, 2002), or as Al Gore put it, "A gallon of gas can be used up just driving to get a gallon of milk."[6] Accessibility can be quantified by measuring road length, road areas, and the traveling times of households (Hadly, 2000).

Landscape ecology metrics can also be used to analyze accessibility. For example, the size and distribution of residential "patches" relative to other land uses may provide a proxy measure for accessibility between these patches and commercial and industrial "patches." Accessibility can also be assessed by calculating the fractal dimensions of road networks (Benguigui, 1998). Further, some ecological measures are useful to measure accessibility, such as "mean proximity index" (MPI) (Gustafson, 1998; Torrens and Alberti, 2000). Another group of accessibility measures is used in transportation models, including: the isochrones measurements through which one counts the number of possible trip destinations in a given area; gravity indices based on the classical gravitation model used in urban planning – the movements of goods, people and information between different spatial locations, often referred to as origins and destinations, and; utility function index gathered from discrete choice models customary to transportation planning discipline (Torrens and Alberti, 2000). Degree of dependency on private automobiles for transport is also considered to be a proxy for sprawl. Where accessibility is lower, there is a higher reliance on private automobiles to connect between the residential and other land uses (Ewing, 1997, 1994; Ewing, Pendall and Chen, 2002).

12.4.2.5 Aesthetic measures

Sprawl is often considered a boring, homogeneous form of development (Fulton, 1996; Gordon and Richardson, 1997). Being subjective by definition, it is difficult to measure and quantify the aesthetics of sprawl unless by consumer preference or survey data. Several recent studies have attempted to define archetypes of urban development or sprawl, such as residential sprawl or strip-mall sprawl, and to compare various landscapes to those archetypes. It seems that much work is still needed in this area (Torrens and Alberti, 2000), and, as noted above, these measures are less relevant to remote sensing experts.

12.4.3 Choosing among the sprawl measures

The advantages and disadvantages of each of these sprawl measures can be considered on the basis of the five criteria outlined at the beginning of this section. We selected a representative sample from the span of possibilities and ranked them according to how well they comply with the five criteria. Our ranking is based on a comprehensive literature review (and thus the experiences of other researchers), as well as our own experiences measuring sprawl and conveying concepts and empirical findings to colleagues, students, professionals and stakeholders. Rankings are on a scale of one to three, with three being the highest ranking variables for the given criteria. Applicability for different spatial scales receives its own ranking system; from A to C, with A being applicable to multiple spatial scales and C to only one scale. Results are summarized in Table 12.1.

We offer three caveats to our ranking of the sprawl variables. First, a measurement that receives high marks across all categories does not necessarily make it the best measurement for all case studies or for measuring all aspects of sprawl. The sprawl quotient, for example, has disadvantages (noted below) that are only picked up in one of the five criteria and ranks highly in the other four. Further, there are measures that received low marks, but still provide important information regarding sprawl – sometimes only at particular spatial scales or for particular places, but useful nonetheless.

Second, while the user can choose from among these variables in a way that suits their specific research question, data sources may also determine the variables selected. The relevant considerations here are (1) availability of historical data and (2) the scale of resolution. With regard to the first consideration, as we have emphasized, sprawl is both a state and a process. As a process, directionality of trends is important. Therefore, it is crucial that

---

[6] Quote from a speech by Al Gore during his campaign for the US presidency, January 1999. Available from http://www.greenclips.com/00issues/139.htm (accessed 15 November 2010).

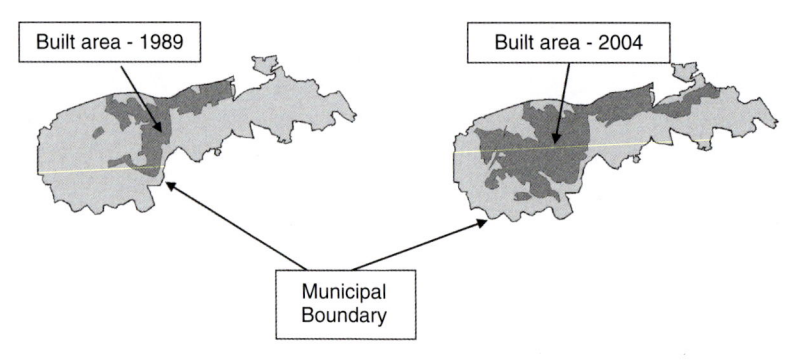

Sprawl Indexes	1989	2004	% change	Equation
Gross density (residents per sq)	6,487	5,286	−19%	$D_{gi} = \frac{P_i}{UA_i}$
Shape index (SH)	3.0	3.3	12%	$SH_i = \frac{L_i}{2\sqrt{\pi A_i}}$
Leap frog index (LFI)	2%	1%	−50%	$LFI_{gi} = \frac{Aout_i}{UA_i}$
Mean Patch Size (Ha) (MPS)	36.4	73.0	101%	$MPS_{ij} = \sum_{j=1}^{n} \frac{a_{ij}}{n_{ij}}$

Where:
P_i = number of residents in urban settlement i
A_i = centralbuilt-up area of urban settlement i
UA_i = urban built-up area of settlement i
RA_i = Residential area of settlement i (land-use no. 1)
L_i = Perimeter of central built-up area of settlement i
$Aout_i$ = leapfrog areas in settlement i
a_{ij} = area of land-use j in urban settlement i (j = 1...n)
n_{ij} = number of polygons of land-use j in urban settlement i (j = 1...n)

FIGURE 12.1 A temporal comparison of urban spatial growth for the Israeli city of Carmiel, 1989 and 2004. The built space area estimated using maps and verified with aerial photographs and ground survey. Four sprawl measures are provided for the two dates to compare temporal trends.

spatial data be available for two or more points in time, so that temporal changes in spatial variables can be measured (see Fig. 12.1). With regard to the second consideration, the user must reconcile the tradeoff between low resolution needed to capture large areas and for comparative research between regions, and high resolution needed to capture fine-grained processes that would be lost when resolution is too low (Irwin and Bockstael, 2008). An example of this tradeoff is the utility of Landsat data: Landsat provides excellent data for large areas with high frequency of data capture, but it lacks the resolution to capture very low density development (Orenstein et al., 2010). Low density development is of utmost importance when considering the extent of sprawl (Irwin and Bockstael, 2008).

Third, by ranking individual sprawl measures, we do not imply that a single measure will suffice in capturing this multifaceted phenomenon. On the contrary, since sprawl has so many dimensions, simultaneous application of multiple indicators is not only recommended, but required (Torrens and Alberti, 2000; Galster et al., 2001; Ewing, Pendall and Chen, 2002; Hasse and Lathrop, 2003a; Sutton, 2003; Cutsinger and Galster, 2006). It is apparent that quantitative indicators for measuring sprawl yield ambiguous and often contrary results. Urban areas could be considered sprawled using some measures, yet compact using others (Hasse, 2004; Frenkel and Ashkenazi, 2008b; Torrens, 2008). This fact is exemplified through the use of four urban areas in Israel (Fig. 12.2). In the figure, the "Type A" urban area ranks compact using four sample sprawl measures. The "type D" urban area, on the other hand, ranks sprawled using these measures. Types B and C both rank sprawled on two of the four measures, but they are different measures in both cases. This point is further emphasized in the previous figure (Fig. 12.1), where over time at one location, population density and shape index both suggest more sprawl, while leapfrog index and mean patch size suggests less sprawl. Clearly the use of one or even two measures misses the complexity of sprawl characterization.

In response to this challenge, researchers are measuring multiple sprawl characteristics simultaneously (Ewing, Pendall and Chen, 2002; Hasse and Lathrop, 2003a; Hasse, 2004; Irwin and Bockstael, 2008; Frenkel and Ashkenazi, 2008b; Torrens, 2008), or integrating multiple measures into a single index after narrowing down the range of variables using reduction techniques (Ewing, Pendall and Chen, 2002; Frenkel and Ashkenazi, 2008b). Cutsinger and Galster (2006) argue that since metropolitan areas may be considered sprawled according to some indicators while simultaneously considered not sprawled in other dimensions, a

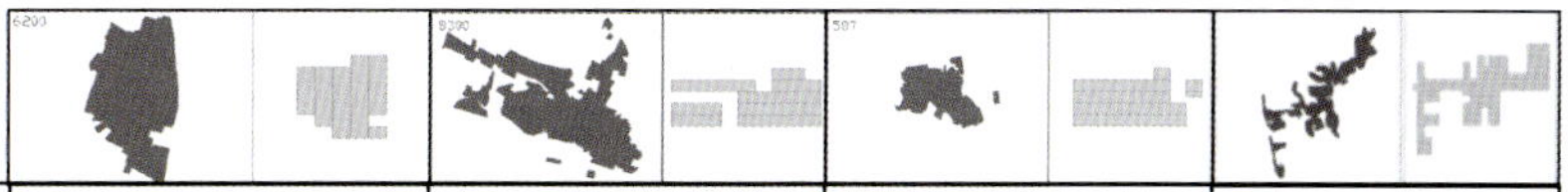

Sprawl Indices		Type A		Type B		Type C		Type D	
Sprawl measurement	Gross density	22,356	Compact	9,213	Compact	1,304	Sprawl	2,470	Sprawl
	Density growth rate (1985-2004)	37%	Compact	52%	Compact	1%	Sprawl	−8%	Sprawl
	Shape index	1.68	Compact	3.15	Sprawl	1.95	Compact	4.63	Sprawl
	Leap frog index	0%	Compact	2%	Sprawl	0%	Compact	9%	Sprawl
Population size 2004		138,900		211,600		2,500		13,300	

FIGURE 12.2 A comparison of four urban areas and their ranking according to four selected sprawl measures. Each urban area is illustrated with an estimate of built area derived from maps and verified with aerial photographs (left; for which the sprawl measures were calculated) and with a visually abstracted illustration (right) to simplify its geometry (right).

new typology for urban land use patterns is needed in place of a misleading 'more or less sprawled' dichotomy.

Most of the sprawl measures score high in *objectivity*; that is, they can be quantified and the methods by which they are obtained can be replicated. While the value obtained for any measure is subject to user interpretation (e.g., what density value constitutes high or low density), some measures have an added layer of subjectivity in that they require user decisions prior to calculating the value of the measure. For instance, some measures require deciding *a priori* what constitutes a low-density neighborhood, a suburb, or a central business district so that their area or distances between them can be quantified. This is straightforward in theory, but can be challenging in practice and subject to much deliberation. Other examples are those measures that depend on user-input for determining thresholds by which the measure will be calculated. Continuity index received a lower ranking for this reason because its calculation depends on the designation of threshold distances by which to calculate whether a patch is continuous (adjoining) with a similar, nearby patch. This decision is not trivial, as the choice of threshold may have significant effect on the outcome of the calculation of the measure. The other variables that depend on measuring patch types are objective as long as there is general *a priori* agreement regarding what characterizes a patch and what differentiates it from other patches.

With regard to the criteria of applicability to a large number of *diverse research sites*, we found that some of the measurements were formulated to fit uniquely to urban development patterns in the specific country being researched (primarily for the United States or Europe). Suburbs and low-density residential areas, for example, are fairly distinct to the United States context, and may have a different meaning or even irrelevance in other country case studies. Quantifying developable land is also very specific to particular countries – some countries may have statutory plans that define what is developable, while in some countries, topography and ecological conditions or indigenous land tenure rules may dictate what is developable. Thus, this variable would be difficult to use in international comparative work. Similarly, using measures that depend on a central business district (CBD) is becoming increasingly difficult. In recent years there has been great change in the evolution of big cities and metropolitan regions from the classic spatial monocentric pattern into polycentric pattern (Gar-on Yeh and Wu, 1997; Parr, 2004).

Spatial–geometric measures, on the other hand, are considered to be suitable in most places, especially when investigating landscapes partitioned into dichotomous built and open space. Their use appears most frequently in the interdisciplinary literature focusing on ecology and urban/regional planning (e.g., Leitao and Ahern, 2002; Taylor, Brown and Larsen, 2007). A caveat to this is that the relevant type of patches to measure may differ greatly from site to site. Because of this, comparative studies between sites using these measures maybe more difficult and therefore we ranked them in the middle range for this criterion.

Regarding our *multiple-scale applicability* criterion, some measures are excellent for a particular spatial scale, but difficult to apply or not applicable at another scale. Our ranking system for this criterion is based on whether the given measure is appropriate for many spatial scales (A) or only a very specific scale (C). We suggest that the researcher must carefully select an appropriate measure for the spatial scale under investigation and make no assumptions with regard to the application to other spatial scales. For example, at the scale of a single neighborhood within a city, measuring growth in residential area is a relatively straightforward task (as distinct from industrial or commercial areas).

However, scaling up to the level of an entire metropolitan area, residential area cannot always be reliably differentiated from other forms of development using standard data sources (satellite imagery, aerial photographs) unless researchers have access to detailed ancillary data sets (Vogelmann *et al.*, 1998; Yang and Lo, 2002). This is even more relevant at broader spatial scales like regions and countries. From our research in Israel, we find that using suburban development as an indicator does not work at the scale of an individual city because suburbs, defined as low density, residential neighborhoods, generally occur outside of the urban locality jurisdiction in satellite "bedroom" communities. So in our case, using suburban versus urban population growth rates is relevant mostly at the metropolitan or regional scale.

Using patch measures can be useful at multiple spatial scales, though the patch types may vary depending on the scale of analysis. Some patch types become difficult to measure or irrelevant to sprawl characteristics at certain spatial scales. Contiguity between patch types, for example, is important for regional-scale analyses where ecological open space issues are important. At the local scale, this measure has less utility.

Ranking of whether the measurement is *meaningful, useful* and *simple to understand* was conducted based on our assessment of how much professional knowledge was required for a stakeholder to understand the concept behind the measure. Many of the measures are very straightforward (how much land is built, how many people live in a certain block of land), but others require more nuanced understanding, such as density gradients, and certain geometric measurements. Fractal dimensions, as useful as they may be, are difficult to explain to a diverse audience.

The sprawl quotient is a very popular measure, but we find several instances where its application is problematic and its meaning misconstrued. For example, in compact, high density cities with aging demographic profiles, a small amount of spatial growth can result in an illogically high sprawl quotient relative to sprawling, low density (but demographically young) cities (Frenkel and Ashkenazi, 2008b). Similarly, the comparison of percentage of population living in low density versus high density urban areas can be heavily influenced by the particular demographics of each area (e.g. young families in low density versus aging individuals in high density tracts). Values may change, perhaps suggesting sprawl even in the absence of urban expansion. Negative population growth has been shown to introduce complications for the use of other per capita indices to measure sprawl, as well (Hasse and Lathrop, 2003b). On the other hand, sprawl has been shown to occur where the amount of developed land grows, even while population falls (Kasanko *et al.*, 2006), thereby producing negative values of the sprawl quotient.

Finally, we consider *ease of application*. This consideration depends on how much data is required, the need for software and associated technical ability, and/or whether quantifying the measures depends on complex calculations. At the extreme, data for some measures can be extracted and estimated with a paper map and a marking pen. Others demand access to digitized maps, remotely sensed data, GIS software and survey/census data. For instance using patch type measures may demand a high degree of *a priori* knowledge about the area and ancillary data sets to complement remotely sensed data, such that the user can define each urban patch type (e.g., residential, industrial, commercial, etc.). Still others demand proficiency at applying complex computational or mathematical calculations using professional software and programming. We give high ranking to those measurements that could be used outside of a university or well-funded government research institution, and low ranking to those measures that would be difficult to collect without large budgets and high levels of technical proficiency. Some spatial metrics received lower rankings due to the challenge of clearly defining and measuring patch types. Once patch types are defined, however, the landscape metrics can be calculated using the popular and free Fragstats software (McGarigal *et al.*, 2002), assuming GIS software proficiency.

12.4.3.1 Does choice of measures matter?

In order to assess how important the choice of sprawl measures is to the characterization of sprawl, we compared four studies that estimated sprawl across metropolitan areas in the United States (Jordan, Ross and Usowski, 1998; Razin and Rosentraub, 2000; Ewing, Pendall and Chen, 2002; Lopez and Hynes, 2003).

Each of the studies employed a different sprawl measure or set of measures and in some cases, unique datasets (see Table 12.2).

The results of the comparative analysis of the studies' findings show many similarities, but also several differences (Table 12.3). In many cases the same metropolitan region appeared at the extremes (i.e., the most sprawled or the most compact) in all of the studies. These regions have characteristics of sprawl or compactness that were robust enough to manifest themselves across many measures. On the other hand, there were multiple inconsistencies in the rankings, where regions would rank highly as either sprawled or compact in one or more studies, but fall into the mid-range in other studies, being neither sprawled nor compact.

Third, there were several cases where a metropolitan region would be characterized as compact by one or more studies, but sprawled in another. In these cases, it was generally the study by Jordan and colleagues (1998) that provided a contrary result for a given region, as it did with Los Angeles, Miami, and Chicago metropolitan regions. In each of these cases, the regions were considered compact according to the measures in two or three of the other studies, while they rated sprawled in the Jordan *et al.* study. This may be due to at least three reasons. First, three studies were measuring *state* of sprawl at a given time. The fourth study measured both state and **process** between 1970 and 1990. In the case of Miami and Chicago, the areas were becoming more sprawled over time relative to their status in 1970 and 1980. Likewise, the Oklahoma City metropolitan region, rated as sprawling in one study and in the middle range in two others, was becoming more compact over time according to Jordan *et al.*[7] Second, spatial extent of metropolitan areas may have been defined differently by each author. Even though most of the researchers were working with US Census Bureau definitions, there is room for selectivity regarding which metropolitan boundaries to employ. Third, as several authors have suggested, different sprawl measures can yield disparate results regarding a single location (Frenkel and Ashkenazi, 2008b; Torrens, 2008), as also shown in Fig. 12.1. It appears that density gradients (used by Jordan, Ross and Usowski, 1998) capture elements of sprawl differently than the measures used in other studies.

This emphasizes three parallel considerations for sprawl research. First, the element of time deserves a more central role in the study of sprawl. Some scholars discuss relative values of sprawl measures, either changing in time or between places, where the difference between sprawl and compact is not an absolute, but rather, a relative change along a continuum (Pendall, 1999; Johnson, 2001). They investigate temporal changes in urban spatial development, such as increases or decreases in residential density (Hasse and Lathrop, 2003a; Frenkel and Ashkenazi, 2008b), or changes in density gradients from CBDs (Jordan, Ross and Usowski, 1998) to determine dynamic patterns of sprawl. Relative sprawl and the direction of sprawl indicators over time are thus key considerations in sprawl studies (Torrens and Alberti, 2000; Galster *et al.*, 2001; Malpezzi and

---

[7] A fifth study which classified level of sprawl across US metropolitan areas, and which makes use of remotely sensed data (nighttime satellite imagery) is Sutton (2003). Sutton's results support, for the most part, the classifications in Table 12.3. Little Rock, Knoxville, Greenville and Atlanta ranked as sprawled (support for the consensus), as did Oklahoma City. Lincoln, Chicago and Miami were ranked as relatively compact, although New York and Phoenix rank neither compact nor sprawled, but near the national average. Syracuse ranked as relatively compact, as did Los Angeles, adding to the ambiguity about that city.

TABLE 12.2 Five studies ranking degree of sprawl in United States metropolitan regions. Relevant results from four of the studies are summarized in Table 12.3 The fifth study, Sutton, is not included in Table 12.3, but is discussed in the text (See footnote 7).

Study	Unit of measure	Sprawl measure
Jordan, et al. (1998)	79 Metropolitan Statistical Areas (MSAs) and Primary Metropolitan Statistical Areas (PMSAs) as defined by the US Statistical Bureau.	Population density gradient from Central Business District (CBD) and change in density gradient over time.
Razin and Rosentraub (2000)	PMSAs and/or consolidated MSAs. Article included results for only 20 cities – the 10 most sprawled and the 10 most compact.	Index of three measures: • the percentage of dwellings in single-unit detached houses • population per square kilometer • housing units per square kilometer
Ewing, et al. (2002)	Every metropolitan area in the United States for which they could access all the necessary data (83 areas in total)	Index including: 7 variables representing aspects of population and residential density 6 variables representing land use heterogeneity 6 variables representing population distribution relative to city centers (which they term variables measuring "strength of metropolitan centers." 3 variables representing accessibility of street networks (related to size of city blocks)
Lopez and Hynes (2003)	330 US metropolitan areas with a population of over 50 000	Sprawl index (proportion of the metropolitan area population living in high density tracks relative to that living in low density tracks)
Sutton (2003)	244 urban clusters with populations over 50 000	Relationship of each urban cluster relative to a regression of all urban clustersIn urban area/urban population (area derived from remotely sensed nighttime data of light intensity)

TABLE 12.3 Similar and contrary results from the comparison of four studies (Jordan, Ross and Usowski, 1998; Razin and Rosentraub, 2000; Ewing, Pendall and Chen, 2002; Lopez and Hynes, 2003) ranking metropolitan region on scale from sprawl to compact in the United States.

Consensus – sprawled[a]	Consensus – compact[b]	General agreement – sprawled[c]	General agreement – compact[d]	Ambivalent results[e]
Little Rock, AR (4/4)	New York, NY (4/4)	Oklahoma City, OK (3/4)	Colorado Springs (2/3)	Los Angeles and Honolulu – ranked among the most compact in two studies, in the mid-range in a third study and most sprawled in a fourth study
Knoxville, TN (4/4)	San Francisco, CA (4/4)	Syracuse, NY (2/3)	Fort Lauderdale, FL (2/3)	Chicago – ranked among the most compact in three studies and among the most sprawled in the fourth study
Greenville, SC (3/3)	Jersey City, New Jersey (3/3)		Chicago, III (3/4)	Miami – ranked by 3 studies as among the most compact, and one as among the most sprawled.
Atlanta, GA (3/3)	Lincoln, NE (2/2)			Phoenix – one study places it among the most compact and another among most sprawled

[a] All studies that included this metropolitan region placed it near the top of their list of sprawled metropolises.
[b] All studies that included this metropolitan region placed it near the top of their lists for compact metropolises.
[c] The majority of studies (3 of 4, or 2 of 3) that included this metropolitan region placed it near the top of their list of sprawled metropolises.
[d] The majority of studies (3 of 4, or 2 of 3) that included this metropolitan region placed it near the top of their list of compact metropolises.
[e] Some studies list these metropolitan regions as sprawled while others list them as compact.

Guo, 2001). Second, in comparative studies, the precise area under investigation (Wolman *et al.*, 2005) and the definition of built area (Orenstein *et al.*, 2010) must be consistent. Third, sprawl should be conceptualized as a multidimensional phenomenon that requires a different measure or set of measures for each dimension (Torrens and Alberti, 2000; Galster *et al.*, 2001; Ewing, Pendall and Chen, 2002; Cutsinger and Galster, 2006).

Conclusion

In this chapter, we presented multiple definitions of sprawl, the historic development of the term and associated urban development patterns, and the measures employed to quantify these patterns. We find an increasingly nuanced discussion regarding definition of and measures for quantifying sprawl. This is a productive result of a longstanding debate around all aspects of the subject.

Today, sprawl is understood to be both a pattern at a given time and a process of change over time. Sprawl is generally accepted as a relative state, warranting cross-site comparisons and multi-temporal analyses. It is defined by multiple quantitative, spatial characteristics whose values do not necessarily lead to similar conclusions; different sprawl measures may yield conflicting results. As such, the state of the art in measuring sprawl involves the application of multiple variables either in the form of an integrated index or considered in parallel and the acceptance that a defined area may display sprawl-like characteristics in some, but not all, measures.

We therefore note the importance of comparative studies such as the US metropolitan regions analyses mentioned above. These analyses each provide a comparison of multiple sites, and in the case of the study by Jordan and colleagues (1998), an analysis of change over two decades. However three of these studies (representative of much of the sprawl literature), employed only one or a few variables culled from relatively easily accessible statistical data to measure sprawl. This may be understandable considering the amount of data that would need to have been collected and processed for such a broad comparison. The result, however, is that important characteristics of sprawl, such as spatial geometry as in the cases above, were not assessed. A remotely sensed meta-analysis of US metropolitan regions would be a welcome contribution to this discussion.

The debate around the desirability of sprawl is a values-driven discussion. So, it is imperative that researchers are forthright and explicit in their chosen definition and their objectives, such that their readers, critics and end-users can assess their research in the proper context. The research community can contribute a broad range of quantitative measures that can be used to elucidate the processes and allow stakeholders to assess where we have been and where we are going. It will then be up to all of the stakeholders (researchers included) to decide whether or not they are observing sprawl, and if so, whether it is desirable process or not.

References

Alperovich, G. (1995) The effectiveness of spline urban density functions: an empirical investigation. *Urban Studies*, **32**, 1537–1548.

Angotti, T. (1993) *Metropolis 2000: Planning, Poverty and Politics*, Routledge, New York.

Batty, M. and Longley, P. (1994) *Fractal Cities: A Geometry of Form and Function*, Academic Press, London.

Batty, M., Xie, Y. and Zhanli, S. (1999) The dynamics of urban sprawl. *Working Paper Series*. London, Centre for Advanced Spatial Analysis, University College.

Belser, K. (1960) Urban dispersal in perspective. In Engelbert, E.A. (Ed.) *The Nature and Control of Urban Dispersal*, California Chapter of the American Institute of Planners, Berkeley, CA.

Benguigui, L. (1998) A fractal analysis of the public transportation system of Paris. *Environment and Planning A*, **27**, 1147–1161.

Benguigui, L., Blumenfeld-Lieberthal, E. and Czamanski, D. (2006) The dynamics of the Tel Aviv morphology. *Environment and Planning B: Planning and Design*, **33**, 269–284.

Benguigui, L., Czamanski, D., Marinov, M. and Portugali, Y. (2000) When and where is the city fractal? *Environment and Planning B*, **27**, 507–519.

Bhatta, B., Saraswati, S. and Bandyopadhyay, D. (2010) Urban sprawl measurement from remote sensing data. *Applied Geography*, **30**, 731–740.

Bruegmann, R. (2005) *Sprawl, A Compact History*, The University of Chicago Press, Chicago.

Brueckner, J.K. (2000) Urban sprawl: diagnosis and remedies. *International Regional Science Review*, **23**, 160–171.

Burchell, R.W., Shad, N.A., Listokin, D. *et al.* (1998) The Costs of Sprawl – Revisited. Transit Cooperative Research Program (TCRP) Report 39, Transportation Research Board, Washington.

Burton, E. (2000) The compact city: just or just compact? A preliminary analysis. *Urban Studies*, **37**, 1969–2006.

Buttenheim, H.S. and Cronick, P.H. (1938) Land reserves for American cities. *The Journal of Land and Public Utility Economics*, **14**, 254–265.

Cam, E., Nichols, J.D., Sauer, J.R., Hines, J.E. and Flather, C.H. (2000) Relative species richness and community completeness: birds and urbanization in the Mid-Atlantic states. *Ecological Applications*, **10**, 1196–1210.

Cato Institute (2009) About Cato. Cato Institute, Washington DC.

Chin, N. (2002) Unearthing the roots of urban sprawl: a critical analysis of form, function and methodology. *CASA Working Paper 47*. London, Centre for Advanced Spatial Analysis, University College London.

Churchman, A. (1999) Disentangling the concept of density. *Journal of Planning Literature*, **13**, 389–411.

Cutsinger, J. and Galster, G. (2006) There is no sprawl syndrome: A new typology of metropolitan land use patterns. *Urban Geography*, **27**, 228–252.

Czamanski, D., Benenson, I., Malkinson, D., Marinov, M., Roth, R. and Wittenberg, L. (2008) Urban sprawl and ecosystems – can nature survive? *International Review of Environmental and Resource Economics*, **2**, 321–366.

Downs, A. (1994) *New Visions for Metropolitan America*, Brookings Institution Press, Washington DC.

Duany, A., Plater-Zyberk, E. and Speck, J. (2000) *Suburban Nation: The Rise of Sprawl and the Decline of the American Dream*, North Point Press, New York, NY.

Ewing, R. (1994) Characteristics, causes, and effects of sprawl: a literature review. *Environmental and Urban Issues*, **64**, 1–15.

Ewing, R. (1997) Is Los Angeles-style sprawl desirable? *Journal of the American Planning Association*, **63**, 107–126.

Ewing, R. (2008) Characteristics, causes, and effects of sprawl: a literature review. *Environmental and Urban Issues*, **21**, 1–15.

Ewing, R., Pendall, R. and Chen, D. (2002) Measuring sprawl and its impact, Smart Growth America, Washington DC.

Fishman, R. (1987) Bourgeois utopias: visions of suburbia. *Bourgeois Utopias: The Rise and Fall of Suburbia*. Basic Books, New York.

Foresman, T.W., Pickett, S.T.A. and Zipperer, W.C. (1997) Methods for spatial and temporal land use and land cover assessment for urban ecosystems and application in the greater Baltimore-Chesapeake region. *Urban Ecosystems*, **1**, 201–216.

Frenkel, A. (2004) The potential effect of national growth-management policy on urban sprawl and the depletion of open spaces and farmland. *Land Use Policy*, **21**, 357–369.

Frenkel, A. and Ashkenazi, M. (2008a) The integrated sprawl index: measuring the urban landscape in Israel. *The Annals of Regional Science*, **35**, 56–79.

Frenkel, A. and Ashkenazi, M. (2008b) Measuring urban sprawl: how can we deal with it? *Environment and Planning B*, **35**, 56–79.

Fulton, W. (1996) *The New Urbanism*, Lincoln Institute of Land Policy, Cambridge, MA.

Fulton, W., Pendall, R., Nguyen, M. and Harrison, A. (2001) Who Sprawls Most? How Growth Patterns Differ Across the US. Brookings Institute, Washington DC.

Galster, G., Hanson, R., Ratcliffe, M.R., Wolman, H., Coleman, S. and Freihage, J. (2001) Wrestling sprawl to the ground: defining and measuring an elusive concept. *Housing Policy Debate*, **12**, 681–717.

Gans, H. J. (1967) *The Levittowners: Ways of Life and Politics in a New Suburban Community*, Pantheon Books, New York.

Gar-on Yeh, A. and Wu, F. (1997) Changing spatial distribution and determinants of land development in Chinese cities in the transition from a centrally planned economy to a socialist market economy: a case study of Guangzhou. *Urban Studies*, **34**, 1851–1879.

Garreau, J. (1991) *Edge City: Life on the New Frontier*, Doubleday, New York.

Gibbs, J. P. (1961) A method for comparing the spatial shapes of urban units. In Gibbs, J. P. (Ed.) *Urban Research Methods*. D Van Nostrand, Princeton, NJ.

Gillham, O. (2002) *The Limitless City*, Island Press, Washington DC.

Glaeser, E.L. and Kahn, M.E. (2004) Sprawl and urban growth. In Henderson, V. and Thisse, J.F. (Eds.) *Handbook of Regional and Urban Economics*. Elsevier, Amsterdam.

Golledge, R.G. and Stimson, R.J. (1997) Urban patterns and trends. In Golledge, R.G. (Ed.) *Spatial Behavior: A Geographic Perspective*. The Guilford Press, New York.

Gordon, P. and Richardson, H.W. (1997) Are compact cities a desirable planning goal. *Journal of the American Planning Association*, **63**, 95–106.

Gordon, P. and Richardson, H. W. (2000) Critiquing sprawl's critics. The Cato Institute, Washington DC.

Gustafson, E.J. (1998) Quantifying landscape spatial pattern: what is the state of the art? *Ecosystems*, **1**, 143–156.

Hadly, C.C. (2000) Urban sprawl: indicators, causes and solutions. Bloomington Environmental Commission.

Harvey, R.O. and Clark, W.A.V. (1965) The nature and economics of urban sprawl. *Land Economics*, **41**, 1–9.

Hasse, J. (2004) A geospatial approach to measuring new development tracts for characteristics of sprawl. *Landscape Journal*, **23**, 52–67.

Hasse, J. E. (2007) Using remote sensing and GIS integration to identify spatial characteristics of sprawl at the building-unit level. In Mesev, V. (Ed.) *Integration of GIS and Remote Sensing*. John Wiley & Sons Ltd., Chichester.

Hasse, J. and Lathrop, R.G. (2003a) A housing-unit-level approach to characterizing residential sprawl. *Photogrammetric Engineering and Remote Sensing*, **69**, 1021–1030.

Hasse, J.E. and Lathrop, R.G. (2003b) Land resource impact indicators of urban sprawl. *Applied Geography*, **23**, 159–175.

Henderson, V., Storeygard, A. and Weil, D.N. (2009) Measuring economic growth from outer space. *NBER Working Paper Series*. Cambridge, MA, National Bureau of Economic Research.

Herold, M. and Menz, G. (2001) Landscape metric signatures (LMS) to improve urban land use information derived from remotely sensed data. In Buchroithner, M. F. (Ed.) *A Decade of Trans-European Remote Sensing Cooperation*. A A Balkema, Rotterdam.

Hess, G., Daley, S.S., Dennison, B.K. et al. (2001) Just what is sprawl, anyway? *Carolina Planning*, **26**, 11–26.

Ingram, G. K. (1998) Patterns of metropolitan development: what have we learned? *Urban Studies*, **35**, 1019–1035.

Irwin, E. G. and Bockstael, N.E. (2008) The evolution of urban sprawl: Evidence of spatial heterogeneity and increasing land fragmentation. *Proceedings of the National Academy of Sciences*, **104**, 20672–20677.

Jackson, K. T. (1985) *Crabgrass Frontier: The Suburbanization of the United States*, Oxford University Press, New York.

Jat, M.K., Garg, P.K. and Khare, D. (2008) Monitoring and modeling of urban sprawl using remote sensing and GIS techniques. *International Journal of Applied Earth Observation and Geoinformation*, **10**, 26–43.

Jenks, M., Burton, E. and Williams, K. (Eds.) (1996) *The Compact City – A Sustainable Urban Form?* E and FN Spon, New York, NY.

Johnson, M.P. (2001) Environmental impacts of urban sprawl: a survey of the literature and proposed research agenda. *Environment and Planning A*, **33**, 717–735.

Jordan, S., Ross, J.P. and Usowski, K.G. (1998) US suburbanization in the 1980s. *Regional Science and Urban Economics*, **28**, 611–627.

Kasanko, M., Barredo, J.I., Lavalle, C. et al. (2006) Are European cities becoming dispersed?: A comparative analysis of

15 European urban areas. *Landscape and Urban Planning*, **77**, 111–130.

Koomen, E., Dekkers, J. and van Dijk, T. (2008) Open-space preservation in the Netherlands: Planning, practice and prospects. *Land Use Policy*, **25**, 361–377.

Kreuter, U. P., Harris, H. G., Matlock, M. D. and Lacey, R. E. (2001) Change in ecosystem service values in the San Antonio area, Texas. *Ecological Economics*, **39**, 333–346.

Leitao, A.B. and Ahern, J. (2002) Applying landscape ecological concepts and metrics in sustainable landscape planning. *Landscape and Urban Planning*, **59**, 65–93.

Lopez, R. and Hynes, H.P. (2003) Sprawl in the 1990s: measurement, distribution, and trends. *Urban Affairs Review*, **38**, 325–355.

Malpezzi, S. (1999) Estimates of the measurements and determinants of urban sprawl in US metropolitan areas. *CULER WP 99–06*. Madison, WI, Center for Urban Land Economics Research, University of Wisconsin.

Malpezzi, S. and Guo, W.-K. (2001) Measuring "sprawl": alternative measures of urban form in US metropolitan areas. Madison, WI, Center for Urban Land Economics Research, University of Wisconsin.

Maret, I. (2002) Understanding the diversity of sprawl and the need for targeted policies. *Paper presented at the 5th Congress on Environment and Planning*, 23–27 September, Oxford.

Martinuzzi, S., Gould, W. A. and Ramos Gonzalez, O. M. (2007) Land development, land use, and urban sprawl in Puerto Rico integrating remote sensing and population census data. *Landscape and Urban Planning*, **79**, 288–297.

McCauley, S. and Goetz, S.J. (2004) Mapping residential density patterns using multi-temporal Landsat data and a decision-tree classifier. *International Journal of Remote Sensing*, **25**, 1077–1094.

McGarigal, K., Cushman, S.A., Neel, M.C. and Ene, E. (2002) FRAGSTATS: Spatial Pattern Analysis Program for Categorical Maps. Produced by the authors at the University of Massachusetts, Amherst.

McKinney, M.L. (2002) Urbanization, biodiversity, and conservation. *Bioscience*, **52**, 883–890.

Mills, E. S. and Hamilton, B.W. (1994) *Urban Economics*, Harper Collins, New York.

Orenstein, D., Bradley, B., Albert, J., Mustard, J. and Hamburg, S.P. (2010) How much is built? Quantifying and interpreting patterns of built space from different data sources. *International Journal of Remote Sensing*, in press.

Paddison, R. (Ed.) (2001) *Handbook of Urban Studies*, London, Sage Publications.

Parr, J. (2004) The polycentric urban region: a closer inspection. *Regional Studies*, **38**, 231–240.

Peiser, R. (2001) Decomposing urban sprawl. *Town Planning Review*, **72**, 275–298.

Pendall, R. (1999) Do land-use controls cause sprawl? *Environment and Planning B*, **26**, 555–571.

Perry, G. and Dmi'el, R. (1995) Urbanization and sand dunes in Israel: Direct and indirect effects. *Israel Journal of Zoology*, **41**, 33–41.

Pu, R., Gong, P., Michishita, R. and Sasagawa, T. (2008) Spectral mixture analysis for mapping abundance of urban surface components from the Terra/ASTER data. *Remote Sensing of Environment*, **112**, 939–954.

Razin, E. and Rosentraub, M. (2000) are fragmentation and sprawl interlinked? *Urban Affairs Review*, **35**, 821–836.

Real Estate Research Corporation (1974) The costs of sprawl: detailed cost analysis. Government Printing Office, Washington, DC.

Robinson, L., Newell, J.P. and Marzluff, J.M. (2005) Twenty-five years of sprawl in the Seattle region: growth management responses and implications for conservation. *Landscape and Urban Planning*, **71**, 51–72.

Rogan, J. and Chen, D. (2004) Remote sensing technology for mapping and monitoring land-cover and land-use change. *Progress in Planning*, **61**, 301–325.

Smart Growth America (2009) Our Mission. [Online] Available http://www.smartgrowthamerica.org/mission.html (accessed 2 December 2010).

Squires, G.D. and Kubrin, C.E. (2005) Privileged places: race, uneven development and the geography of opportunity in urban America. *Urban Studies*, **42**, 47–68.

Stefanov, W.L., Ramsey, M.S. and Christensen, P.R. (2001) Monitoring urban land cover change: An expert system approach to land cover classification of semiarid to arid urban centers. *Remote Sensing of Environment*, **77**, 173–185.

Sutton, P.C. (2003) A scale-adjusted measure of "urban sprawl" using nighttime satellite imagery. *Remote Sensing of Environment*, **86**, 353–369.

Taylor, J. J., Brown, D. G. and Larsen, L. (2007) Preserving natural features: A GIS-based evaluation of a local open-space ordinance. *Landscape and Urban Planning*, **82**, 1–16.

Thomas, I., Frankhauser, P. and Biernacki, C. (2008) The morphology of built-up landscapes in Wallonia (Belgium): A classification using fractal indices. *Landscape and Urban Planning*, **84**, 99–115.

Thompson, A.W. and Prokopy, L.S. (2009) Tracking urban sprawl: Using spatial data to inform farmland preservation policy. *Land Use Policy*, **26**, 194–202.

Torrens, P. M. (2008) A Toolkit for Measuring Sprawl. *Applied Spatial Analysis*, **1**, 5–36.

Torrens, P. M. and Alberti, M. (2000) Measuring sprawl. *WP 27*. Centre for Advanced Spatial Analysis, University College London.

Tsai, Y.H. (2005) Quantifying urban form: compactness versus 'sprawl'. *Urban Studies*, **42**, 141–161.

Turner, M. G. (1989) Landscape ecology: the effect of pattern on process. *Annual Review of Ecology and Systematics*, **20**, 171–197.

van den Berg, L., Drewett, R., Klaasen, L.H., Rossi, A. and Vijverberg, C.H.T. (1982) *Urban Europe: A Study of Growth and Decline*, Pergamon Press, Oxford.

Vogelmann, J.E., Sohl, T.L., Cambell, P.V. and Shaw, D.M. (1998) Regional land cover characterization using Landsat Thematic Mapper data and ancillary data sources. *Environmental Monitoring and Assessment*, **51**, 415–428.

Ward, D., Phinn, S.R. and Murray, A.T. (2000) Monitoring growth in rapidly urbanizing areas using remotely sensed data. *Professional Geographer*, **52**, 371–386.

Weitz, J. (1999) From quiet revolution to smart growth: state growth management programs, 1960–1999. *Journal of Planning Literature*, **14**, 266–337.

Wheeler, C. H. (2008) Urban decentralization and income inequality: is sprawl associated with rising income segregation across neighborhoods? *Regional Economic Development Journal of the Federal Reserve Bank of St. Louis*, **4**, 41–57.

Wilson, E.H., Hurd, J.D., Civco, D.L., Prisloe, M.P. and Arnold, C. (2003) Development of a geospatial model to quantify, describe and map urban growth. *Remote Sensing of Environment*, **86**, 275–285.

Wolman, H., Galster, G., Hanson, R., Ratcliffe, M., Furdell, K. and Sarzynski, A. (2005) The fundamental challenge in measuring sprawl: which land should be considered? *The Professional Geographer*, **57**, 94–105.

Xian, G. and Crane, M. (2005) Assessments of urban growth in the Tampa Bay watershed using remote sensing data. *Remote Sensing of Environment*, **97**, 203–215.

Yang, X. and Lo, C.P. (2002) Using a time series of satellite imagery to detect land use and land cover changes in the Atlanta, Georgia metropolitan area. *International Journal of Remote Sensing*, **23**, 1775–1798.

Zhang, Q., Wang, J., Peng, X., Gong, P. and Shi, P. (2002) Urban built-up land change detection with road density and spectral information from multi-temporal Landsat TM data. *International Journal of Remote Sensing*, **23**, 3057–3078.

13

Small area population estimation with high-resolution remote sensing and lidar

Le Wang and Jose-Silvan Cardenas

Timely small-area population estimates are critical for both public and private sector decision-making. State and local governments must allocate resources, and private businesses need to identify and delineate customer profiles, market areas and site locations. Currently, such population data is only available for one date per decade through the national census. This study focuses on developing methods for intercensal small-area population estimation from integrative use of airborne light detection and ranging (lidar) and high spatial resolution aerial photographs. Particularly, it addresses the following question: What level of information extracted from lidar and high-spatial resolution imagery can be effectively infused in deriving small-area population estimation? This question was addressed through a comparative study of seven linear models parameterized in terms of building count, building area and/or building volume, at two different land use levels: single-family dwelling, multifamily dwelling and other types, versus residential and other types. Results showed that while building volume is more relevant to population counts at census block level; it also represents the most challenging parameter to measure by automated analyses of lidar and high resolution remote sensing imagery. Because of that, a simple model that primarily utilizes residential building counts resulted in more reliable population estimation.

Urban Remote Sensing: Monitoring, Synthesis and Modeling in the Urban Environment, First Edition. Edited by Xiaojun Yang.
© 2011 John Wiley & Sons, Ltd. Published 2011 by John Wiley & Sons, Ltd.

13.1 Introduction

Small-area population estimates are essential for understanding and responding to many social, political, economic, and environmental problems (Liu, 2003). The size and distribution of the population often are key determinants for resource allocation for state and local governments (Smith, Nogle and Cody, 2002). Population estimates are critical in decisions about when and where to build public facilities such as schools, libraries, sewage treatment plants, hospitals, and transportation infrastructure. For example, in public transit route design, population density is considered a primary indicator of the number of potential daily trips originating in an area (Benn, 1995), with population densities below approximately 4000 persons per square mile found to generate low demand for public transit (Downs, 1992; Transportation Research Board, 1996, 1997). Population estimates are also often used by the private sector for customer profile analysis, market area delineation, and site location identification (Martin and Williams, 1992; Plane and Rogerson, 1994). In addition, population estimates are also extensively utilized as denominators in generating many diagnostic indicators for studies of environmental and socioeconomic conditions and trends. Based on these indicators, such as unemployment rates, mortality and morbidity rates, etc., billions of dollars in public funds are allocated every year. Lastly, population information is an important input in many urban and regional models, such as land use and transportation interaction models, urban sprawl analysis, environment equity studies, and policy impact analysis (Rees, Norman and Brown, 2004). In short, the generation of accurate and timely population estimates is crucial (Smith, Nogle and Cody, 2002). Although small-area population estimates are of great significance and have many applications, detailed and accurate population and socioeconomic information is only available for one date per decade through the national census. Therefore, the need for frequent intercensal updates, particularly in rapid growth areas, is critically apparent for public and private sector planning.

Remote sensing has long been used to estimate population, particularly for large areas. The earliest application of remote sensing for population estimates involved manually counting the number of houses using aerial photos (Lo, 1986a, b). A survey is then conducted to estimate the average number of persons per house. The product of the number of houses and average household size then produces the total population of the study area. Although this method is relatively accurate, it involves manual interpretation of aerial photos. This is quite time consuming and labor intensive, which prohibits its application in large urban areas. This method therefore is rarely used by state and local agencies for small area population estimates. Moreover, accurate "persons per household" information, which is required for a variety of dwelling types, is difficult to obtain (Lo, 1989, Watkins and Morrow-Jones, 1985).

Automatic approaches with satellite remote sensing imagery, therefore, have been proposed for estimating population density (Lo, 1995). These approaches can be classified into: (1) implicit estimation, in which spectral radiance/reflectance information and their transformations are utilized for population estimation, and (2) explicit estimation, which utilizes urban physical parameters extracted from remote sensing imagery for population generation (Cowen and Jensen, 1998, Jensen and Cowen, 1999). For implicit estimation, Lo (1995) utilized the radiances of bands 1, 2, and 3 of SPOT HRV imagery to obtain population density information in Hong Kong, China. He discovered that strong negative correlations exist between population density and the radiances of band 3 (0.79–89 μm) and band 1 (0.500.59 μm), and a positive correlation exists between population and the radiances of band 2 (0.6–0.68 μm). In Lo's study, although urban biophysical parameters were not utilized, it is clear that the radiances of bands 1 and 3 in SPOT HRV imagery are closely associated with concentration of vegetation, which has a negative relationship with population density. Moreover, the radiance of band 2 is highly related to urban built-up areas; therefore, it has a positive correlation with population density. In addition to the radiances in individual bands, Harvey (2002) utilized radiance transformations, including radiance squares, cross-product of radiances in different bands, radiance ratio from different bands, difference-sum-ratio, and other transformations. In another study, Sutton *et al.* (1997) utilized light energy extracted from the Defense Meteorological Satellite Program Operational Linescan System (DMSP-OLS) imagery for population estimation. The light energy represents the intensity of urban land uses, with higher energy existing in commercial and residential areas, and lower energy in agricultural areas. In summary, in these implicit estimations, although urban physical parameters are not directly utilized, specific urban physical environments are represented by radiances and their transformations. The second category of regression models utilizes urban biophysical and land use information extracted from remote sensing imagery for population estimation. For example, Lo (1995, 2003) utilized high and low urban land use areas to estimate zonal population counts. Chen (2002) used three levels of residential density for projecting population density in Sydney, Australia. Li and Weng (2005) also applied land use types as independent variables for developing separate regression models for different residential regions in Indianapolis, Indiana. A detailed review of existing methods based on remote sensing can be found in Wu, Qiu and Wang, (2005).

Although very high-spatial resolution satellites (IKONOS with 1 m and QuickBird with 0.65 m in their respective panchromatic bands) have been available for around a decade now (IKONOS launched in 1999 and QuickBird launched in 2001), studies taking advantages of this fine spatial resolution for small-area population estimations are still scarce and generally limited to the application of traditional visual interpretation for housing units count (e.g., Yagoub, 2006). Likewise, airborne light detection and ranging (lidar) devices have allowed rapid access to vertical information of urban structures, but the integration of this new level of information for population estimation has not been fully investigated. Recently, it was suggested that the volume information provided by lidar could serve to best improve small-area population estimations (Wu, Wang and Qiu, 2008). Whether or not lidar measurements coupled with automated techniques for building extraction and land use classification can lead to improved small-area population estimation is still an unresolved matter.

This study sought to address the following questions: Can lidar and high-spatial resolution imagery be employed for automating and refining small-area population estimation? If so, what types and levels of information extracted from these sensors are mostly effective? These questions were addressed through a comparative study of seven linear models that drew upon inputs from building count, building area and/or building volume, at two different land use levels. At the finer land use level, buildings were classified as single-family dwelling, multifamily dwelling, or other non-residential land use types. While at the coarser level, buildings were classified as either residential or other non-residential land use types. In order to investigate the aforementioned questions, a number of automated building extraction and land use classification methods were developed. In addition, the impact of the errors arisen from the remote sensing analysis to the final population estimation was quantitatively assessed.

13.2 Study sites and data

The study area is located in the city of Austin, Texas. Austin is currently the third fastest growing large city in the United States with a population of 750 000. According to the US Censuses of 1990 and 2000, the city grew an impressive 41% (from 465 622 in 1990 to 656 562 in 2000), with an average annual growth rate of 3.5%. The city's population has been projected to top 800 000 by 2010 (City of Austin, 2009a). The selected area for this study covers approximately 4.8 × 6.4 km (nearly 12 square miles), representing 4.5% of the entire city area. A major interstate highway, IH-35, runs south–north through the city and splits it in two sides. The west side is dominated by civic, commercial, as well as some residential land uses located in the northern and southern ends. In total of 1153 census blocks fall within this area with over 15 000 residential buildings (representing 85% of the total). In this area, residential buildings correspond primarily to single-family detached and two-family attached (94%), and secondarily to multifamily three/fourplex and apartment/condo (6%).

In order to perform detailed analysis of building detection and land use classification methods, we selected four small study areas within the larger study area. The inset boundaries are shown in Fig. 13.1. These sites were carefully selected to represent the wide spectrum of living environments found in the study area:

- Inset 1 contains multifamily dwelling units located in a sparsely vegetated area,
- Inset 2 contains single-family dwelling units located in a densely vegetated area,

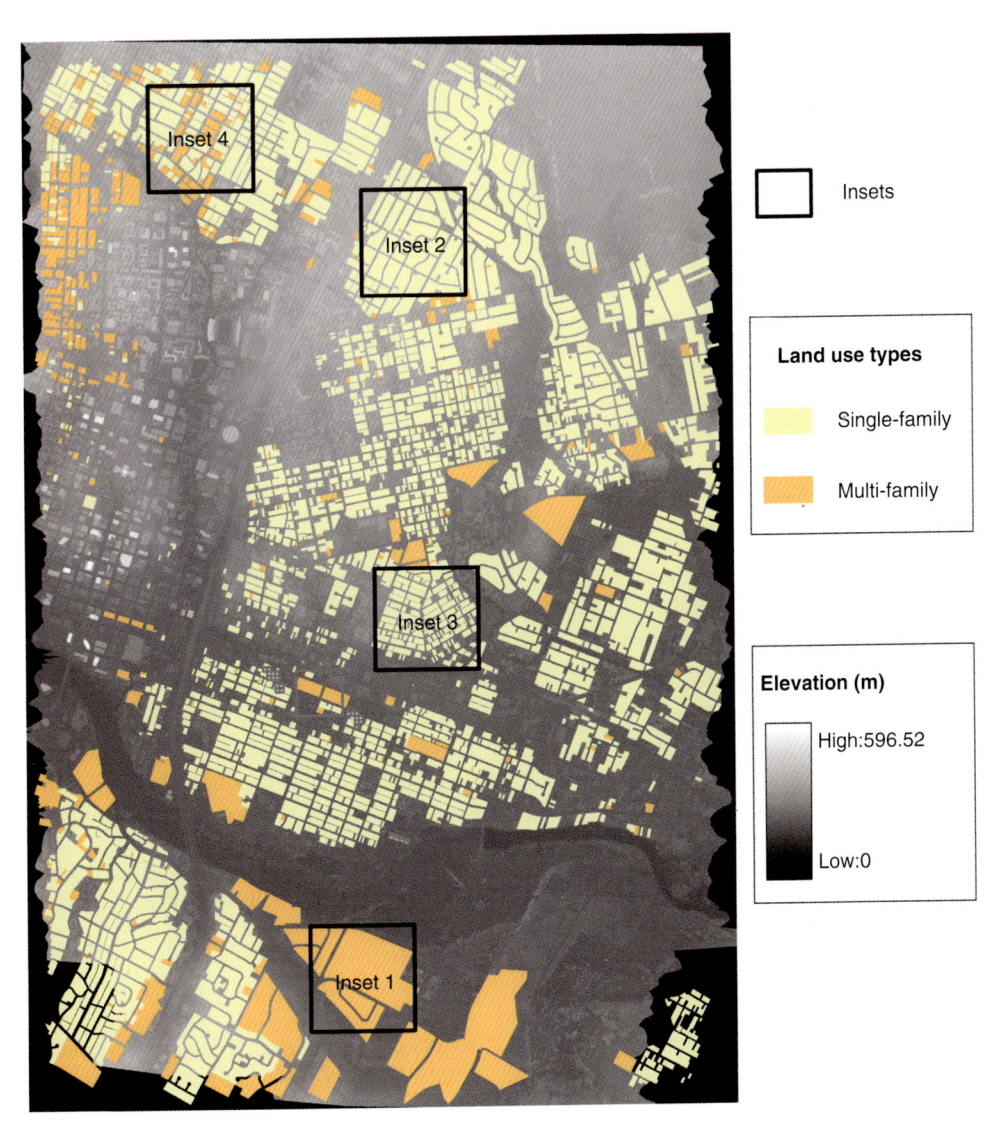

FIGURE 13.1 Location of the four insets in the study site. Background are lidar data second return and land use types.

- Inset 3 contains single-family dwelling units located in sparsely vegetated area,
- Inset 4 contains both single-family and multifamily dwelling units located in a sparsely vegetated area.

Datasets acquired for the study area includes lidar altimetry measurements, demographic and geographic census data, building footprints and land use layers, high resolution aerial photography and a Landsat TM image. All datasets were contemporarily acquired around year 2000. The lidar data was acquired in 2000 using an Optech Inc. Airborne Laser Terrain Mapper (ALTM) 1225 instrument mounted on a single-engine craft. The ALTM instrument delivers a cloud of three-dimensional (3-d) points for the first and last return of a laser pulse. There were around 40 million points in the study area, for which the backscatter intensity was also available from each return. Demographic and geographic data were acquired at census block level through the US Census Bureau's American FactFinder (US Census Bureau, 2009a) and TIGER/Line shapefiles (US Census Bureau, 2009b) web sites, respectively. Building footprint and land use layers, together with a high-resolution (60 cm, 2 ft) color-infrared (CIR) aerial photography, were acquired through the City of Austin Neighborhood Planning and Zoning Department (NPZD, City of Austin, 2009b).

13.3 Methodology

This study involved development and application of a number of methods, including: (1) data preprocessing, (2) building extraction, (3) land use classification, (4) population estimation and (5) accuracy assessment. Due to space limit, methods 1–3 are briefly introduced below while methods 4–5 are presented in more detail.

13.3.1 Data preprocessing

For the purpose of building extraction and land use classification, several raster layers were derived from the lidar point cloud and the CIR photograph. The layers derived from the CIR photograph included the following masks: vegetation, bare ground, impervious surface, and pervious, non-bare ground. The datasets derived from lidar consisted of the feature height, ground mask, intensity difference, and the gray level co-occurrence matrix (GLCM) angular second moment of a digital surface model (DSM). The DSM was produced at a spatial resolution of 1 m by applying a point to raster conversion tool to the point cloud elevation values associated with the last return of the lidar data. Then a ground mask was produced using the multi-resolution ground filtering approach developed by Silvan-Cardenas and Wang (2006). A bare earth digital terrain model (DTM) was generated by utilizing the ground mask. The recovered DTM was then subtracted from the DSM to produce a feature height layer.

A reference building footprint layer was derived by editing the NPZD's building footprint layer, which was produced by manual digitization using aerial photos and lidar datasets collected in 2003 (City of Austin, 2009b). A reference land use layer was obtained from the NPZD at city of Austin.

13.3.2 Building extraction

Automatic extraction of buildings from remote sensing data has become more common over the years, particularly with the availability of high spatial resolution remote sensing images and lidar. In order to select an appropriate building detection method for population estimation, four building extraction methods were tested in this study. These methods were based on region-growing segmentation (Zhang, Yan and Chen, 2006), the Hermite transform (Silvan-Cardenas and Escalante-Ram'irez, 2006), the Dempster–Shafer theory of evidence (Shafer, 1976; Lu, Trinder and Kubik, 2006), and Definiens' eCognition segmentation method. Details can be found in Silvan-Cardenas et al. (2009). As a result, building footprints as well as building volumes were generated.

13.3.3 Land use classification

Population counts have a close relationship with land use type. Given the premise that population is only counted in residential areas, an immediate need will be to single out residential land use type from other land use types. Within the category of residential land use, it is also necessary to separate different land use types such as single family, multi-family, etc., since they are associated with different population densities.

Conventional per-pixel classification may not be adequate for the urban land use classification (Jensen and Cowen, 1999). This is particularly true when high spatial resolution data, such as the CIR photography and lidar, is employed in our analyses. Therefore, the "per-field" approach appears more promising (Pedley and Curran, 1991; Aplin, Atkinson and Curran, 1999; Erol and Akdeniz, 2005). This method classifies land use/cover by predetermined field boundaries with the premise that each field pertains to a single, homogeneous class. In this study, we adopted tax parcels as the basic mapping unit. A Meta classifier was chosen for conducting the land use classification. The Meta classifier transforms a multiclass problem into several binary problems (Ichino, 1979). The Meta method was applied to classify tax parcels into nine land use classes: single family (SF), multi-family (MF), commercial, office, industrial, civic, open space, transportation, and undeveloped. Later, the nine classes were subsequently aggregated to three land use classes: SF, MF, and non-residential land use. The parcel attributes were derived from parcel boundaries (area, perimeter and shape), neighboring parcels (distance to nearest parcel, similarity of area between parcel and the nearest parcel, similarity of perimeter between parcel and the nearest parcel, and similarity of shape between parcel and the nearest parcel, where similarity was defined as a normalized absolute difference), NDVI zonal statistics (average and standard deviation), vegetation mask (percent of vegetation cover in parcel), land cover proportions (percent of impervious surface, percent of bare ground, percent of other land cover type), building footprint layer (number of buildings in parcel, fraction of building cover in parcel, average, minimum, and maximum statistics of area, perimeter, shape, volume and height of buildings within the parcel), neighboring buildings (average, minimum and maximum of distance to nearest neighbors) and building density. This method is hereafter referred to as the MultiClass-tax parcel method.

13.3.4 Population estimation models

Seven linear models of population estimations at census blocks level were tested in this study, with alternating explanatory variables that incorporated different information of building structure and land use types.

Table 13.1 summarizes the building statistics and land use information incorporated by each model. The first six models were constructed by combining three building statistics at block level (building count, footprint area and total volume) with two levels of land use information (Residential vs. SF and MF). For example, Model 1 uses the per-block counts of residential buildings (N) regardless of whether it is SF or MF, whereas Model 2 uses the split of the count of residential building into the count of SF buildings (N1) and the count of MF buildings (N2). These two models are inspired in the previous housing unit method (Watkins and Morrow, 1985; Smith and Cody, 2004), where the number of buildings replaces the number of housing units. Likewise, Model 3 and Model 4 are inspired in the broadly used area-based methods, but with a linear form. We used the linear form of the area–population relationship because a preliminary exploratory analysis confirmed that the linear form fitted the data better than the allometric form. Models 5 and 6 were proposed under the premise that building volume can better describe the living space, and thus may allow for more accurate population estimates. Regarding Model 7, an optimal linear model was constructed by selecting a few explanatory variables out of 16 variables. The set of initial variables included building count, area, volume, perimeter, shape and height, from both SF and MF buildings. The variable selection strategy sought to include the smallest number of explanatory variables while maintaining a high correlation with the dependent variable.

The coefficients for each of the seven models were estimated through ordinary least squares procedure (Selvin, 1995). We also calculated normalized coefficients or path coefficients, which correspond to the regression coefficients multiplied by the standard deviation of the explanatory variable divided by the standard deviation of the dependent variable. The normalized forms are useful for inter-comparisons as they represent the sensitivity of the dependent variable to the variation of the independent variable (Selvin, 1995).

13.3.5 Accuracy assessment

Accuracy for building detections was assessed at both pixel and object level. At the pixel level, the overall accuracy (percent of correctly classified pixels) and the kappa statistics were derived from the standardized confusion matrix (Congalton and Green, 1999). At the object level, we calculated the detection rate, i.e. the percentage of correctly detected buildings to total number of reference buildings, and the commission error, i.e. percentage of false detections to total number of detected buildings.

To assess land use classifications accuracy, both the reference and the classified land use labels were attached to each building footprint. Then a confusion matrix was built by cross-comparison of the reference and extracted land use labels. Finally, we calculated the overall and per-class accuracy ratios and the kappa coefficient of agreement from the confusion matrix.

The goodness of fit of population estimation models was assessed through the coefficient of determination (R^2), whereas the model validation was based on statistics derived from the absolute difference between census and estimated populations, i.e. absolute error. The statistics included the mean, standard deviation, median (50-percentile), maximum (100-percentile), lower quartile (25-percentile), upper quartile (75-percentile), and interquartile range (the difference upper and lower quartiles) of the absolute error. Among these error measures, the median absolute error (MAE) and the interquartile range (IQR) were more extensively considered as they are most common in population estimation studies. Relative errors were not considered due to sensitivity issues in areas of low population density, or even indetermination in non-populated areas.

13.4 Results

13.4.1 Building detection results

The results from each building detection method are illustrated in Fig. 13.2. These error maps were built through comparing the detection mask from each method with the reference building footprint layer in raster format. Errors of omission and commission are colored in blue and red for easy identification. In addition, the overall per-pixel accuracy, the kappa statistics (Congalton, 1991), the detection rate, and the commission error were

TABLE 13.1 Seven Population Estimation Model.

Models	Equation	Land use type	Adopted variables
1	$P_1 = \alpha_1 N + \varepsilon$	Residential	Building counts
2	$P_2 = \alpha_1 N_1 + \alpha_2 N_2 + \varepsilon$	SF and MF	Building counts
3	$P_3 = \alpha_1 A + \varepsilon$	Residential	Footprint area
4	$P_4 = \alpha_1 A_1 + \alpha_2 A_2 + \varepsilon$	SF and MF	Footprint area
5	$P_5 = \alpha_1 V + \varepsilon$	Residential	Building total volume
6	$P_6 = \alpha_1 V_1 + \alpha_2 V + \varepsilon_2$	SF and MF	Building total volume
7	$P_7 = \alpha_1 A_1 + \alpha_2 N_1 + \alpha_3 V_1 + \varepsilon$	SF and MF	Building counts, footprint area, Building volume

P stands for the population counts at Census block, N stands for building counts within Census blocks, N1 stands for single family building counts, N2 stands for multi-family building counts, V stands for total building volumes, alpha stands for coefficients to be estimated from the models, epsilon stands for residual of the model.

188 PART IV URBAN SOCIOECONOMIC ANALYSES

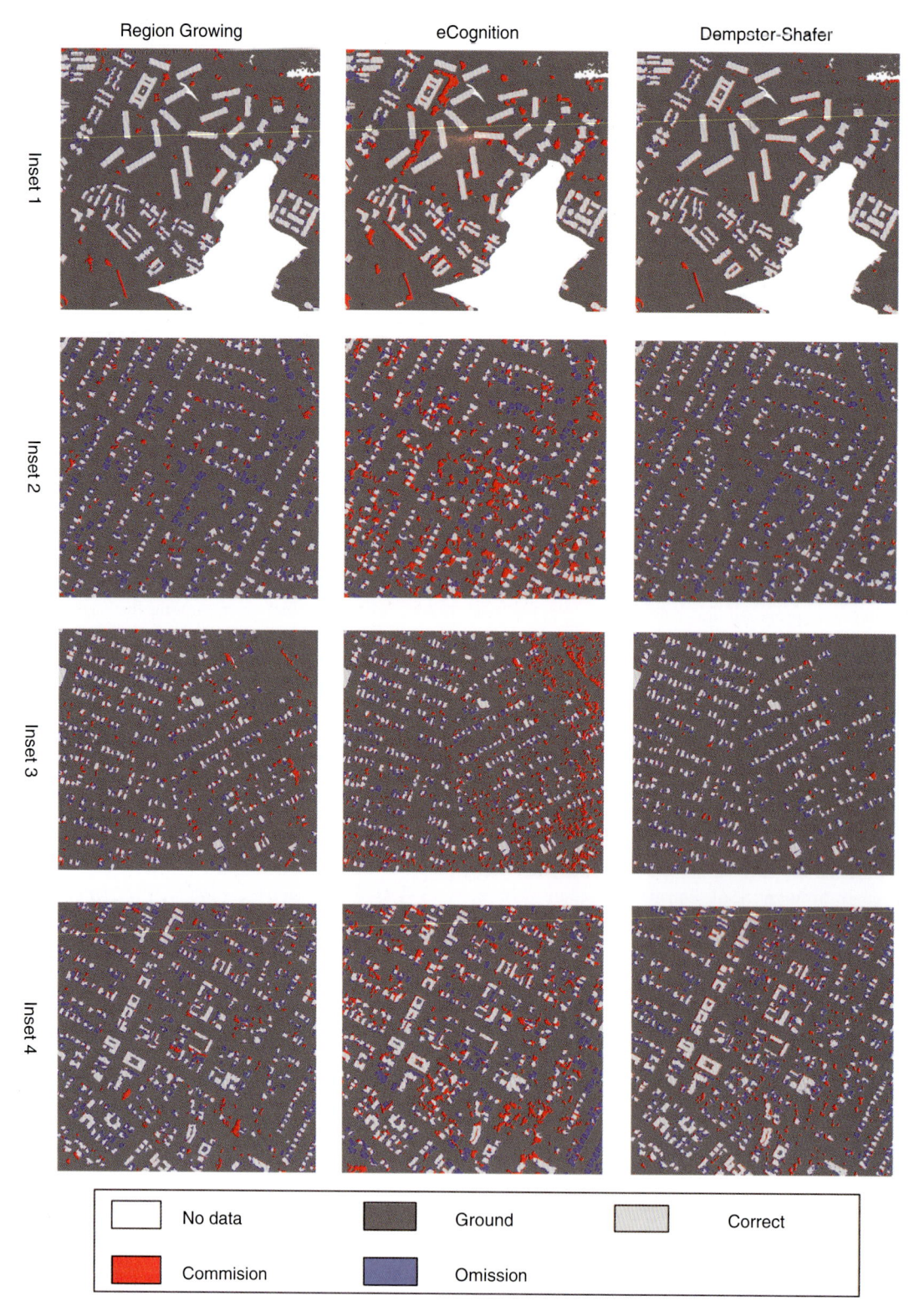

FIGURE 13.2 Error maps from each building detections method and inset. Rows from top to bottom correspond to Dempster-Shafer, Region Growing, Hermite Transform and eCognition, whereas columns from left to right correspond to subsets from inset 1 through inset 4, respectively.

TABLE 13.2 Accuracy assessment of the four building detection methods at both pixel and object levels.

Method	% Pixels Overall accuracy	% Pixels Kappa	% Objects Commission error	% Objects Detection rate
Region Growing	91	64	19	83
eCognition	87	55	43	84
Dempster–Shafer	92	66	30	90

calculated for each method and inset. The average statistics over all insets are provided in Table 13.2. It was observed that although most methods had acceptable performance, with average overall accuracy ranging from 87 to 92% and detection rate ranging from 84 and 90%, there was a considerable variability across insets. Specifically, inset 1 was the most accurately classified area, which had a per-pixel accuracy ranging from 92 to 95% by all the methods. This was due to the relatively high and large structure of MF buildings. On the other hand, inset 2 represented the most challenging area due to the relatively small size of SF buildings and the high chance of occlusions by trees. In this case, per-pixel accuracies ranged from 83 to 89%. Accuracy with the inset 3 and inset 4 stays in between the two former extremes.

Based on the accuracy levels and its consistency over the various insets, the Dempster–Shafer method was selected for detecting buildings over the entire study area. The accuracy of the detected buildings for the entire study area was not assessed, but it is reasonable to assume a detection rate of 90%, as this was the average performance from selected insets (Table 13.2). The building mask for the entire study area was imported into ArcGIS and converted to vector format (yielding 15 211 building polygons) for further analysis.

13.4.2 Land use classification results

The land-use classification results with the MultiClass-tax parcel method are given in Table 13.3.

13.4.3 Population estimation results

13.4.3.1 Construction of Model 7

The construction of Model 7 followed the procedure outlined in the previous section using a threshold of 0.1 for the part correlation. We selected 16 block-level statistics from buildings as the initial set of explanatory variables. These explanatory variables were building count, total building footprint area, total building volume, maximum building volume, average building height, maximum building height, average building footprint perimeter and average building footprint shape, for both SF and MF dwelling types. This procedure yielded a model of the form:

$$P_7 = \alpha_1 A_1 + \alpha_2 C_2 + \alpha_3 V_2 + \varepsilon \quad (13.1)$$

where A_1 is the area of SF dwellings, C_2 and V_2 are count and volume of MF dwellings, and α's are model parameters. Interestingly, the building perimeter, building shape and building height did not provide further explanation power to the variability of block population. Equation 13.1 also disclosed that the volume information of SF buildings is not as important as its area.

13.4.3.2 Calibration of models

The estimated coefficients for all the models are provided in Table 13.4 together with the normalized form ($\tilde{\alpha}$s). The values for the normalized coefficients suggest that population estimates are more sensitive to MF than to SF building characteristics.

The bar chart of Fig. 13.3 compares the goodness of fit in terms of R^2 values obtained from each model. As it turns out, building volume information, at both land use levels, fits better than building area and building counts. It is also noticeable that all models that distinguish between SF and MF dwelling types yielded superior goodness of fit with respect to their counterpart that did not made this distinction. This is largely due to the fact that persons per dwelling and persons per unit area of footprint are different for SF and MF. Interestingly, the improvement achieved when making the distinction of dwelling types was most notable for building counts (from $R^2 = 0.34$ to $R^2 = 0.59$), and building area (from $R^2 = 0.71$ to $R^2 = 0.81$) than was for building volume (from $R^2 = 0.81$ to $R^2 = 0.82$). This result suggests that building volume already accounts for the dwelling type characteristic. The observation is important because, unlike area-based or count-based models, a volume-based model would not rely on detailed land use information (at least not to the level of distinguishing between dwelling types), which is generally difficult to extract by automated means.

TABLE 13.3 The accuracy assessment of the MultiClass-tax classification method.

Method	% Correct SF	MF	Res	Overall accuracy	Kappa
MultiClass-tax parcel	91	55	85	83	71

TABLE 13.4 Estimated coefficients for each model defined in Table 13.1.

Model	Regression coefficients α_1	α_2	α_3	Path coefficients $\tilde{\alpha}_1$	$\tilde{\alpha}_2$	$\tilde{\alpha}_3$
1	3.4335	–	–	0.65	–	–
2	1.8129	11.8573	–	0.35	0.75	–
3	0.0303	–	–	0.76	–	–
4	0.0153	0.0416	–	0.32	0.87	–
5	0.0036	–	–	0.81	–	–
6	0.0025	0.0039	–	0.04	0.89	–
7	0.0153	4.7673	0.0031	0.32	0.30	0.72

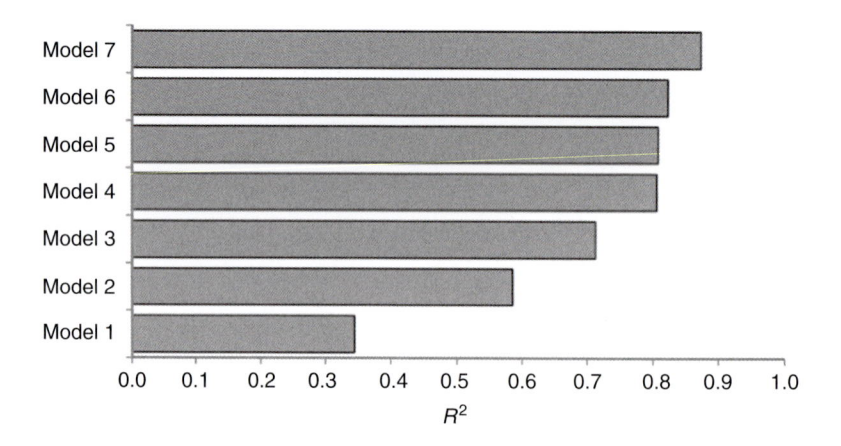

FIGURE 13.3 Goodness of fit of seven population estimation models.

13.4.3.3 Validation of models

In order to confirm the fitting trends observed during the calibration stage, an independent set of samples was used to validate each model. The R^2 and median absolute error with interquartile ranges are plotted in Fig. 13.4 for a visual comparison. Contrary to expectation, Models 2–4 performed comparably to, or even better than, models that incorporate volume information (Models 5–7). This may be due to the fact that the estimated building height was influenced by trees partially (or even totally) occluding the building, especially for SF residential buildings. The inaccurate height information negatively impacted the population estimations from building volumes. Presumably, this also explains why Model 7 was no longer the best performing model. Nevertheless, the differential improvement by the incorporation of the finest land use information seemed to hold in the validation sample (compare Model 1 with Model 2, Model 3 with Model 4, and Model 5 with Model 6).

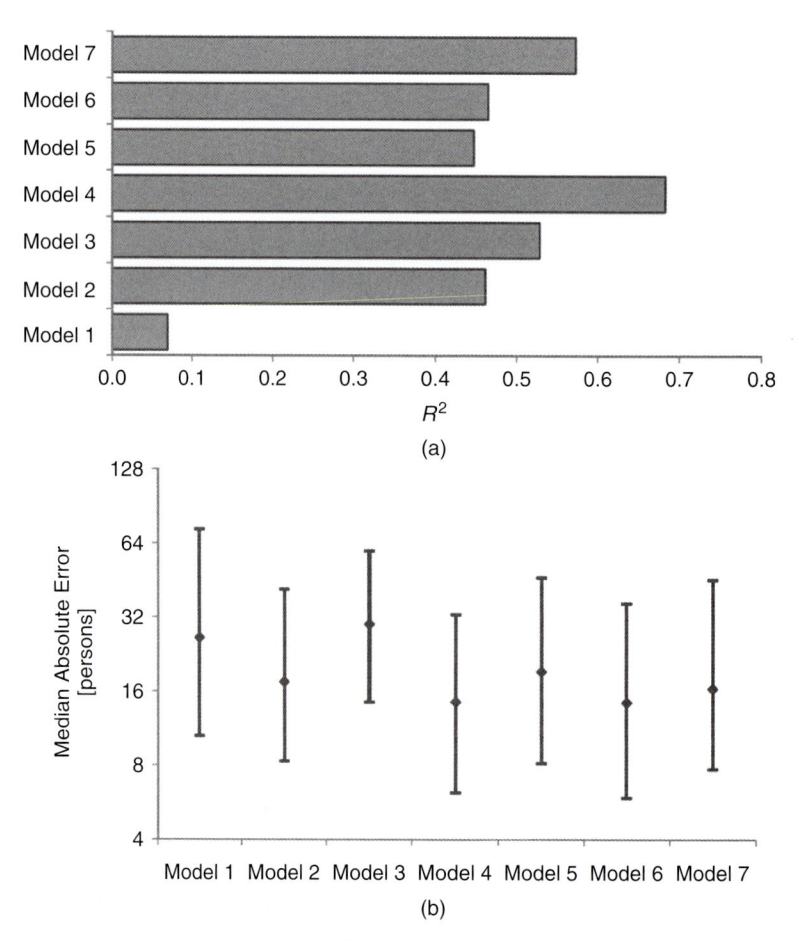

FIGURE 13.4 R^2 statistics (a) and median absolute error (b) with interquartile rage (vertical bars) based on validation samples. Note the vertical axis is in a logarithmic scale in (b).

13.4.3.4 Population estimation uncertainty analysis

The calibrated models were applied to the entire study area using the extracted buildings layer and the classified land-use layers. The goal here was to examine the effect of errors from building detection and land-use classification to population estimation. The reference building footprint layer was first replaced by the automatically detected buildings layer, while the reference land use layer was still used for assigning the type of dwelling. Geometric attributes were calculated from the extracted buildings layer and aggregated at census block level. Figure 13.5 shows the scatter plot of estimated versus measured building attributes at census block level. These scatter plots indicated that three detected buildings attributes (building count, area, and volume) generally yielded lower values than those derived from the reference building footprint layer. This trend was most obvious for blocks dominated by SF dwelling units. Therefore, population counts were also underestimated from all the models. The sensitivity of each model to the underestimation errors of explanatory variables was then assessed through the MAE of population estimates.

Figure 13.6 compares the MAE yielded by each model when using the reference building layer and when using the extracted buildings layer. The two curves are negatively correlated ($R^2 = 0.86, n = 7, p < 0.05$), showing that the trend observed in the model fitting stage (Fig. 13.3) is inverted due to building detection errors alone. Although building volume fits block population better than building area, and building area fits better

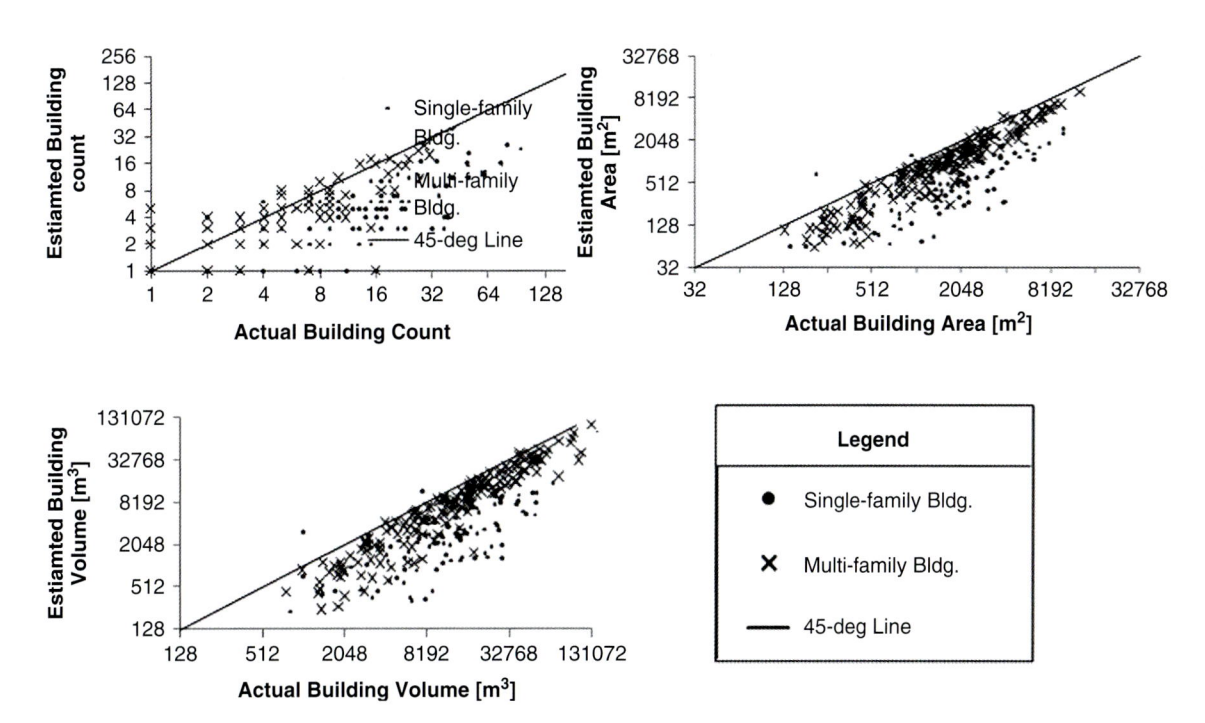

FIGURE 13.5 Scatter plots of estimated versus reference building count, building area and building volume at census block level. Axes are in a logarithmic scale.

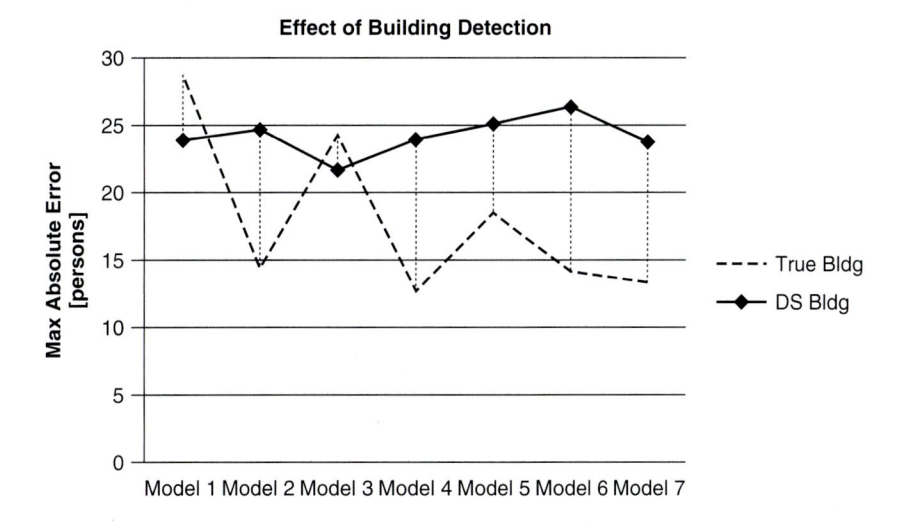

FIGURE 13.6 Effect of building detection accuracy on population estimation.

than building count, it appears that building count is less sensitive to building detection errors than is building area, and building area is less sensitive than building volume. Since both building area and building volume measurements heavily depend on lidar measurements, they are more severely affected by detection and measurement errors. This is the case especially for SF residential buildings located in densely vegetated areas. Moreover, the models that distinguish between dwelling types (Models 2, 4, and 6) tend to have larger errors than their counterpart that consider residential buildings altogether (Models 1, 3, and 5). This result suggests that the best strategy is to use the minimal amount of information: residential building counts.

Discussion and conclusions

Small-area population estimation is an important task that has received considerable attention by the remote sensing community in the past four decades. The wealth of related studies reveals that the notion of living space had been considered a key linkage between population and remote sensing measurements. Unfortunately, a formal definition for this important variable has proved difficult due in part to the relatively coarse spatial resolution of the remote sensing data used for population estimation. The advent of fine spatial resolution satellite images (1 to 5 m) coupled with lidar measurements opened new opportunities for considering the 3-d nature of living space in urban environments and for improving small-area population estimations. In the study reported here, we tested the potential of high spatial resolution lidar measurements coupled with automated and semiautomated techniques for building extraction and land use classification. We compared seven linear models for small-area population estimations, each of which is parameterized in terms of one, two or three explanatory variables representing building statistics on a per-block basis (count, area, and volume) at one of two land use classification levels (residential or SF/MF). These explanatory variables were meant to more closely represent the living space because the great majority of population lives inside buildings. Interestingly, when considering other geometric characteristics of building, such as perimeter, shape and height, their contribution to the regression was not significant (Model 7).

At the model fitting stage, it was observed that the incorporation of either the fine-level land use information or the volume information led to higher correlation coefficients. At a validation stage, however, the differential improvement achieved by including building volume appeared not as important as that of using fine land use information. Presumably, this was due to errors introduced by the method used for calculating the height information from lidar data. At the estimation stage, we first replaced the reference building layer by the detected buildings and tested the effect on the population estimation errors. The original trend observed during the fitting state was totally inverted; suggesting that the incorporation of finer land use or building volume (or even building area) did not improve the population estimations. The sensitivity of each model to the errors in the land use information further favored the simplest model based on counts of residential buildings. While the reason for this inversion appeared to be the violation of the fundamental assumption that the high-quality calibration sample was representative of the remote sensing-derived parameters, it was also apparent that the most important parameters for population estimation, namely residential subtype land uses and building volume, were also the most difficult to accurately extract from remote sensing. Based on results reported herein, future lidar-based population estimation should focus on improving building detection methods first, particularly in reconstructing the 3-d building shape more accurately, and then on improving land use classification methods.

Acknowledgments

This study was supported by grants to Le Wang from the National Science Foundation (BCS-0822489, DEB-0810933), and from National Key Basic Research and Development Program, China (2006CB701304). The authors are thankful to three Co-PIs of the NSF project (BCS-0822489) Dr. Changshan Wu at University of Wisconsin at Milwaukee, Dr. Peter Rogerson at SUNY-Buffalo, and Dr. Frederick Day at Texas State University-San Marcos. Ms. Tiantian Feng and Mr. Benjamin D. Kamphaus are thanked for their help on developing building detection and land use classification.

References

Aplin, P., Atkinson, P. and Curran, P. (1999) Fine spatial resolution simulated satellite sensor imagery for land cover mapping in the United Kingdom. *Remote Sensing of Environment*, **68**(3), 206–216.

Benn, H. (1995) TCRP synthesis of transit practice 10: Bus route evaluation standards. *Transportation Research Board*, National Research Council, Washington, DC.

Chen, K. (2002) An approach to linking remotely sensed data and areal census data. *International Journal of Remote Sensing* **23**, 37–48.

City of Austin (2009a) City of Austin demographics. Online. Available http://www.ci.austin.tx.us/demographics/ (accessed 16 November 2010).

City of Austin (2009b) City of Austin GIS data sets. Online. Available ftp://ftp.ci.austin.tx.us/GIS-Data/Regional/coa_gis.html. (Last accessed Jan. 16, 2011)

Congalton, R.G., 1991. A review of assessing the accuracy of classifications of remotely sensed data, Remote Sensing of Environment, 3735–3746.

Congalton, R., and K. Green, 1999. Assessing the Accuracy of Remotely Sensed Data: Principles and Practices, CRC/Lewis Press, Boca Raton, Florida, 137 p.

Cowen, D.J. and Jensen J.R. (1998) Extraction and modeling of urban attributes using remote sensing technology. In *People and Pixels: Linking Remote Sensing and Social Science* (ed. D. Liverman, E.F. Moran, R.R. Rindfuss, and P.C. Stern), National Academy Press, Washington DC.

Downs, A. (1992) Stuck in traffic: coping with peak-hour traffic congestion. Brookings Institution, Washington, DC, Lincoln Institute of Land Policy, Cambridge, MA.

Erol, H., and Akdeniz F. (2005) A per-field classification method based on mixture distribution models and an application to Landsat Thematic Mapper data. *International Journal of Remote Sensing*, **26** (6), 1229–244.

Harvey, J. (2002) Estimating census district populations from satellite imagery: some approaches and limitations. *International Journal of Remote Sensing*, **23**(10), 2071–2095.

Ichino, M. (1979) A nonparametric multiclass pattern classifier. *IEEE Transactions on Systems, Man and Cybernetics*, **9**(6), 345–352.

Jensen, J.R. and Cowen. D.J. (1999) Remote sensing of urban/suburban infrastructure and socio-economic attributes. *Photogrammetric Engineering & Remote Sensing*, **65** (5), 611–622.

Li, G. and Weng, Q. (2005) Using Landsat ETM imagery to measure population density in Indianapolis, Indiana, USA. *Photogrammetric Engineering & Remote Sensing*, **71**(8), 947–958.

Liu, X. (2003). Estimation of the spatial distribution of urban population using high spatial resolution satellite imagery. University of California, Santa Barbara.

Lo, C.P. (1986a). Accuracy of population estimation from medium-scale aerial photography. *Photogrammetric Engineering and Remote Sensing*, **52**: 1859–1869.

Lo, C.P. (1986b) *Applied Remote Sensing*. Longman, London.

Lo, C.P. (1989). A raster approach to population estimation using high-altitude aerial and space photographs. *Remote Sensing of Environment*, **27**, 59–71.

Lo, C.P. (1995). Automated population and dwelling unit estimation from high-resolution satellite images: A GIS approach. *International Journal of Remote Sensing*, **16**(1), 17–34.

Lo, C.P. (2003). Remotely sensed cities. In *Zone-Based Estimation of Population and Housing Units from Satellite-Generated Land Use/Land Cover Maps* (ed. Victor Mesev), CRC Press, Boca Raton, p. 157.

Lu, Y., Trinder, J., and Kubik, K. (2006) Automatic building detection using the Dempster-Shafer algorithm. *Photogrammetric engineering and remote sensing*, **72**(4), 395–403.

Martin, D. and Williams, H. (1992) Market-area analysis and accessibility to primary health-care centres. *Environment and Planning A*, **24**(7), 1009–1019.

Plane, D.A., and Rogerson, P.A. (1994) The geographical analysis of population with applications to business and planning. New York: John Wiley.

Pedley, M, and Curran, P. (1991) Per-field classification: an example using SPOT HRV imagery. *International Journal of Remote Sensing* **12**(11): 2181–2192.

Rees, P., Norman, P. and Brown, D. (2004) A framework for progressively improving small area population estimates. *Journal of the Royal Statistical Society Series A*, **167**, Part 1: 5–36.

Selvin, S. (1995) *Practical Biostatistical Methods*. Duxbury Press, Belmont, CA. USA.

Shafer, G. (1976) *A Mathematical Theory of Evidence*. Princeton University Press, Princeton, NJ.

Silvan-Cardenas, J. and Escalante-Ram'irez, B. (2006) The multiscale Hermite transform for local orientation analysis. *IEEE Transactions on Image Processing*, **15** (5), 1236–1253.

Silvan-Cardenas, J. and Wang, L. (2006). A multi-resolution approach for filtering LiDAR altimetry data. *ISPRS Journal of Photogrammetry and Remote Sensing*, **61**(1), 11–22.

Silvan-Cardenas, J., Wang, L., Rogerson, P., Wu, C., Feng, T. and Kamphaus, B. (2009) Assessing high spatial resolution remote sensing for small-area population estimation, *International Journal of Remote Sensing*, in press.

Smith, S. and Cody, S. (2004) An evaluation of population estimates in Florida: April 1, 2000. *Population Research and Policy Review*, **23**(1), 1–24.

Smith, S., Nogle, J. and Cody, S. (2002) A regression approach to estimating the average number of persons per household. *Demography*, **39** (4) 697–712.

Sutton, P., Roberts, D., Elvidge, C., and Meij, H. (1997) A comparison of nighttime satellite imagery and population density for the continental United States. *Photogrammetric Engineering and Remote Sensing*, **63** (11), 1303–1313.

Transportation Research Board. (1996) Transit And Urban Form. Transit cooperative research program. *TCRP report 16*, vol. 1 and 2. National Research Council, Washington, DC.

Transportation Research Board. (1997) Towards A Sustainable Future. Special Report 251. National Academy Press, Washington, DC.

US Census Bureau (2009a) American FactFinder. Online. Available http://factfinder.census.gov (accessed 16 November 2010).

US Census Bureau (2009b) TIGER, TIGER/Line and TIGER-related products. Online. Available http://www.census.gov/geo/www/tiger (accessed 16 November 2010).

Watkins, J.F. and Morrow-Jones, H.A. (1985) Small area population estimates using aerial photography. Photogrammetric Engineering and Remote Sensing **51** (12): 1933–1935.

Wu, S., Qiu, X. and Wang, L. (2005) Population estimation methods in GIS and remote sensing: a review. *GIScience & Remote Sensing*, **42** (1), 80–96.

Wu, S., Wang, L. and Qiu, X. (2008) Incorporating GIS building data and census housing statistics for sub-block-level population estimation. *The Professional Geographer*, **60** (1), 121–135.

Yagoub, M. (2006) Application of remote sensing and geographic information systems (GIS) to population studies in the gulf: A case of Al Ain City (UAE). *Journal of the Indian Society of Remote Sensing*, **34**(1), 7–21.

Zhang, K., Yan, J. and Chen, S. (2006) Automatic construction of building footprints from airborne LIDAR data. *IEEE Transactions on Geoscience and Remote Sensing*, **44** (9), 2523–2533.

14

Dasymetric mapping for population and sociodemographic data redistribution

James B. Holt and Hua Lu

The analysis of geographically referenced sociodemographic data often requires the use of data collected at different spatial scales and/or different temporal resolutions. In addition, population data that are collected at one spatial resolution may be unsuitable for a particular research project, and these data must be redistributed to different spatial units of analysis. Researchers must overcome these challenges by employing a variety of areal interpolation techniques to make the data spatially compatible. Dasymetric mapping techniques have been demonstrated to be one means through which this can be achieved successfully. We provide an overview of areal interpolation techniques with an emphasis on dasymetric mapping. We illustrate an example in which population estimates and sociodemographic data are derived for different spatial units by employing dasymetric mapping methods that rely upon ancillary data from a variety of sources, including remotely sensed satellite imagery.

14.1 Introduction

Statistical data are commonly required for social science research and practice. These data originate from a variety of sources, such as government agencies, non-governmental organizations, and the private sector. Because of variations in data collection purposes and jurisdictions, statistical datasets are compiled for different administrative units. Quite often, these administrative units differ in spatial extent. Furthermore, concerns over individuals' rights to privacy have given rise to data release restrictions aimed at preserving confidentiality; this often results in the aggregation of individual data to arbitrary administrative units prior to data dissemination. For survey data, limitations in sample sizes can lead the data collection agency to aggregate responses to geographic levels that do not meaningfully coincide with spatial variations in the data, but rather are chosen for the statistical stability of the resulting data estimates. The choices of areal unit to which statistical data are aggregated vary by agency and purpose of the dataset. For example, some health-related datasets aggregate to the level of respondents' ZIP code. Other datasets are aggregated to the county-level. Census data are available at various aggregations that correspond to a complex hierarchy of US Census geography. These constraints in data collection, aggregation, and reporting create great difficulties and pose significant methodological challenges to data users.

An additional complication that is related to the issue of data spatial incompatibility is the possibility of temporal changes in administrative boundaries. Data from a particular agency may be reported consistently for the same administrative boundaries, yet these boundaries may be revised over time. For example, changes in census geography such as block groups and census tracts (Howenstine, 1993), ZIP codes (or their spatial corollaries, ZIP code Tabulation Areas), US Congressional Districts, or census enumeration and dissemination areas in other countries (e.g., Canada) may require the use of methods designed to account for temporally-induced spatial mismatches (Martin, Dorling and Mitchell, 2002; Gregory, 2002; Schuurman et al., 2006; Schroeder, 2007).

Gotway and Young (2002) present a detailed discussion of the problems associated with combining data from multiple spatially incompatible datasets. They describe the Change of Support Problem, or COSP, in which the area or volume associated with the data, as well as the shape and orientation of the spatial units for which the data are collected, can affect statistical associations based on those data. The Modifiable Areal Unit Problem or MAUP (Openshaw, 1984) is a special case of the COSP, in which varying scales and spatial aggregations of units may result in a change in statistical analysis results for areal data.

Where it is necessary to use data that are reported in incompatible spatial units, a commonly used approach is to employ areal interpolation techniques. Areal interpolation is the process of transforming a dataset that is collected and/or reported in one set of areal units to another set of areal units (Goodchild and Lam, 1980; Lam, 1983; Flowerdew and Green, 1994). For example, data reported at the ZIP code level can be apportioned to census tracts through a variety of areal interpolation techniques. These techniques, which will be described further below, vary in terms of methodological approach, complexity, and accuracy. The focus of this chapter is to describe in detail one of these areal interpolation approaches: dasymetric mapping. We will also discuss the specific role of remotely sensed satellite imagery as a source of ancillary data for dasymetric mapping, with emphasis on both population mapping and sociodemographic data redistribution. We illustrate some of these dasymetric mapping approaches with examples from Gwinnett County, Georgia, which is part of the Atlanta metropolitan area. While areal interpolation, along with other statistical techniques, is often used for population estimation, population estimation *per se* is not the focus of this chapter. Readers interested in that particular application are referred to the chapter in this text by Wang.

14.2 Dasymetric maps, dasymetric mapping, and areal interpolation

There exists the potential for confusion between the terms and concepts of dasymetric maps (as a product), dasymetric mapping (as a process), and areal interpolation. A dasymetric map is an area-based cartographic tool that enables representation of a statistical phenomenon, such as population density (Langford, 2003). Dasymetric maps are meant to convey both the magnitude of a statistical surface as well as the spatial extent of the phenomenon being mapped. Slocum et al. (2009) describe dasymetric maps as an alternative to choropleth maps:

> Like the choropleth map, a dasymetric map displays standardized data using areal symbols, but the bounds of the symbols do not necessarily match the bounds of enumeration units (e.g., a single enumeration unit might have a full range of gray tones representing differing levels of population density)

> (Slocum et al., 2009, p. 271).

In contrast, the term "dasymetric mapping" represents the process of transforming data that are aggregated arbitrarily (usually due to the constraints and requirements of data collection) into a dasymetric map that is a more accurate depiction of the statistical distribution of the data. Mennis (2009) convincingly argues that the dasymetric mapping process usually does not yield true dasymetric maps, because the dasymetric mapping process results in mapped distributions that are *estimated* from a disaggregation of choropleth maps units.

To better illustrate this principle, consider a choropleth map, in which the entire land surface is divided into a space-filling tessellation and each area is assigned a value of the phenomenon and color-shaded accordingly. It is implicitly assumed in choroplethic mapping that the phenomenon of interest is homogeneously distributed throughout each particular mapping unit and that the distribution of the phenomenon takes places over the entire mapped area. It is also assumed that sharp breaks in data values occur at the boundaries of mapping units. In the population mapping example, this would translate to the assumption there are people residing in all areas and conversely there are no unpopulated areas. Crampton (2004) provides a critical historical overview of choropleth and dasymetric maps and argues in favor of the use of dasymetric maps on both a conceptual/theoretical basis as well for reasons of geostatistical soundness.

A dasymetric map can be considered a variation of the choropleth map in the sense that the mapping units are spatially refined to include only those areas that contain the phenomenon of interest (e.g., populated areas). This spatial refinement is achieved through the use of ancillary data, which represent a separate phenomenon that is considered to be related to the distribution of the mapped phenomenon itself (Flowerdew, Green and Kehris, 1991). The result is not a continuously varying surface, nor a space-filling surface, but instead a discretized set of spatially refined areas (either polygons or pixels, depending upon the geospatial data model used) onto which the cartographic representations for the phenomenon are assigned. The set of spatially-refined areas are estimates of the spatial distribution of the phenomenon, which in fact may be unknown.

Finally, the term "areal interpolation" refers to the process of transforming data from one set of source zones (e.g., ZIP codes) to a set of spatially incompatible target zones (e.g., census tracts) (Goodchild and Lam, 1980). Therefore, the spatial redistribution of population and other sociodemographic data, as we will focus our discussion in this chapter, involves two transformation processes: dasymetric mapping, which involves data disaggregation, to in turn enable areal interpolation, which involves data reaggregation to a set of desired areal units.

While we focus on methodological aspects of areal interpolation, there are numerous examples of its use in applied settings, for example: examining the economic impact of water usage in California (Goodchild, Anselin and Deichmann, 1993), analyzing spatially-aggregated crime data (Poulsen and Kennedy, 2004), revealing urban sprawl (Lo, 2004), understanding spatial dynamics of asthma (Maantay, Maroko and Herrmann, 2007; Maantay, Maroko and Porter-Morgan, 2008), mapping urban risks for flood hazards (Maantay and Maroko, 2009), environmental justice/equity analysis (Mennis, 2002; Boone, 2008); generating burden of disease estimates (Hay et al., 2005), refining spatial accessibility measures for healthcare (Langford and Higgs, 2006), catastrophic loss estimation (Chen et al., 2004), and developing a National Agriculture Land Use Dataset (Comber, Proctor and Anthony, 2008).

Various methods for areal interpretation have been developed. These can be distinguished by whether or not the method involves the use of ancillary data to estimate the relationship between population and a related variable (e.g., land use). Areal interpolation methods that *do not* rely upon ancillary data can be further distinguished as either area-based methods or point-based methods (Wu, Qiu and Wang, 2005). Area-based methods include polygon overlay (Lam, 1983), also known as areal weighting (Goodchild and Lam, 1980; Howenstine, 1993; Sadahiro, 2000) and pychnophylactic interpolation (Tobler, 1979; Rase, 2001). Point-based methods include kriging (Kyriakidis, 2004) and kernel-based methods (Bracken and Martin, 1989; Martin, 1989, 1996).

14.2.1 Ancillary data

Ancillary data are a necessary component for areal interpolation through dasymetric mapping as they are the basis on which the unknown statistical surface of the phenomenon of interest is modeled. The key assumption in the use of ancillary data is that their distribution is related to the variation in the statistical surface (Eicher and Brewer, 2001). The use of ancillary data, regardless of the type or source of the data, is an attempt to add 'intelligence' to the process of understanding this relationship (Flowerdew, Green and Kehris, 1991; Flowerdew and Green, 1992, 1994; Eicher and Brewer, 2001).

Several types of ancillary data exist and have been evaluated for their utility and accuracy in areal interpolation, including: expert knowledge of the study area (Wright, 1936; Mennis and Hultgren, 2006), street network data (Xie, 1995; Mrozinski and Cromley, 1999; Chen et al., 2004; Reibel and Bufalino, 2005), zoning/parcel/cadastral data (Maantay, Maroko and Herrmann, 2007; Maantay, Maroko and Porter-Morgan, 2008; Maantay and Maroko, 2009), scanned raster maps (Langford, 2007), and remotely sensed data (Langford, Maguire and Unwin, 1991; Langford and Unwin, 1994, Eicher and Brewer, 2001; Harvey, 2002; Holt, Lo and Hodler, 2004; Liu, Kyriakidis and Goodchild, 2008; Bhaduri et al. 2007; Briggs et al., 2007; Kressler and Steinnocher, 2008).

Further distinctions can be made based on how the relationship between the ancillary data and the variable of interest is estimated. These can be either *a priori* or empirically specified; the former is based on expert knowledge and/or assumptions (which can be arbitrary) and the latter is determined through statistical techniques. Examples of *a priori* specification include Wright (1936), Langford and Unwin (1994), Eicher and Brewer (2001) and Holt, Lo and Hodler (2004). Statistical methods include regression techniques (Flowerdew and Green, 1989, 1992; Flowerdew, Green and Kehris, 1991; Langford, Maguire and Unwin, 1991; Goodchild, Anselin and Deichmann, 1993; Harvey, 2002; Yuan, Smith and Limp, 1997; Langford, 2006; Lo, 2008). The latter three papers accounted for the possibility of spatial non-stationarity in the relationships between ancillary data and population, and Lo (2008) addressed this through the application of geographically weighted regression (Fotheringham, Brusndon and Charlton, 2002) to a case study of data for metropolitan Atlanta, Georgia.

14.2.2 Dasymetric mapping

The focus of this chapter is the use of dasymetric mapping for areal interpolation. As previously defined, dasymetric mapping is the process of transforming data from one spatial aggregation (typically a choropleth map) into a map that is a more accurate depiction of the magnitude and spatial extent of the data. This process involves the use of additional information about the distribution of the data, so as to make the resulting redistribution more meaningful. We first discuss the origins of dasymetric mapping, which has been somewhat misunderstood until recently, followed by methodological variations to the dasymetric approach, including binary dasymetric and three-variable dasymetric mapping. We then discuss different approaches to incorporating intelligence into the process, through different forms of ancillary data.

14.2.3 Origins

The seminal article on dasymetric mapping for most English-speaking audiences is J.K. Wright's (1936) description of a dasymetric method for redistributing population densities for Cape Cod, Massachusetts. Wright relied on information on population distributions from a combination of an almost (at the time) half-century-old US Geological Survey (USGS) topographic sheet and, more importantly, his personal knowledge of Cape Cod in order to make inferences about which portions of Cape Cod were populated and to what degree of density. He assigned population densities, *a priori*, to sparsely populated township subdivisions and then computed the population densities for the remaining more densely-populated subdivisions in magnitudes that would ensure that the overall population densities for each township remained the same as before dasymetric recalculation. Wright's method can be described more formally as:

$$D_n = \frac{D}{1-a_m} - \frac{D_m a_m}{1-a_m}$$

where D is the mean population density per area (in Wright's case, the township), D_m is the estimated population density for subarea m (in Wright's case, the larger and less densely-population township subdivision, D_n is the estimated population density for subarea n (in Wright's example, the smaller and more densely-population township subdivision), and a_m is subarea m's proportion of the total area. Knowing D and a_m, and treating D_m as a known quantity, Wright was able to solve for D_n, for each township. He then created a dasymetric map of the resulting township subdivision population densities. Wright admits that his methods of assigning population densities are not exact, but represent a good approximation, based on "controlled guesswork...perhaps the best that could be accomplished without intensive field work" (Wright, 1936, p. 104).

Most follow-on dasymetric mapping researchers refer to this as the first-published use of dasymetric mapping, although some authors did acknowledge, including Wright, the Russian origin of at least the word *dasymetric*: "It is a map to which the Russians have applied the term 'dasymetric' (density measuring)" (Wright, 1936, p. 104). In a footnote, Wright also refers to a similar type map that was printed in an English language publication three years earlier by the American Geographical Society. There are at least two additional examples of the use and description of dasymetric mapping, *per se*, in the English-language literature prior to Wright in 1936 (De Geer, 1926; and Semenov-Tian-Shansky, 1928).

Fortunately, Petrov (2008) demystifies the origins of the term "dasymetric map" and discusses early uses by Russians, particularly Benjamin Semenov-Tian-Shansky. Petrov and others (MacEachren, 1979; McCleary, 1969; Robinson, 1982), as cited in Petrov, 2008; and Maantay, Maroko and Herrmann, 2007) describe perhaps the earliest map that appears to use dasymetric mapping principles: George Julius Poulett Scrope's world population density map of 1833 (Scrope, 1833). Four years later, Henry Drury Harness created a population-density map for Ireland (Robinson, 1955). Both Scrope's and Harness' maps distinguished between areas based on population density; however, as Petrov (2008) points out, neither described their process for constructing the maps. The first description of the dasymetric methodology in the English-speaking literature was by De Geer (1926) who reviewed the series of population density maps for Europe that were compiled by Semenov-Tian-Shansky. Semenov-Tian-Shansky subsequently described his dasymetric maps two years later in the same journal (Semenov-Tian-Shansky, 1928).

Further development and use of dasymetric mapping stagnated for several decades, perhaps largely due to the analytic complexity of the methods along with the absence of automated computational devices and the dearth of suitable ancillary data. Beginning in the late 1980s and early 1990s, interest was renewed in dasymetric mapping, as several researchers in the United States and the United Kingdom investigated various methods of areal interpolation (Goodchild and Lam, 1980; Lam, 1983; Goodchild, Anselin and Deichmann, 1993; Flowerdew and Green, 1989; Flowerdew, Green and Kehris, 1991; Langford, Maguire and Unwin, 1991; Langford and Unwin, 1994). The development and proliferation of geographic information systems (GIS) in the 1990s along with the advent of systematically collected satellite remote sensor data presented new opportunities for the practical application of dasymetric mapping techniques.

14.2.4 Dasymetric mapping variations

14.2.4.1 Binary dasymetric

The most basic technique for adding intelligence to the process of creating a dasymetric map is to differentiate between populated and unpopulated areas. Population or other sociodemographic data are reapportioned to those areas that are considered to be populated. The mapping units themselves are spatially refined, and the resulting population densities are changed due to the change in areal support; in a binary approach, this results in higher population densities than in the original population map.

It is necessary to use ancillary data to make the distinction between populated and unpopulated areas. For the binary approach, it is assumed that population can only occur, and for dasymetric mapping should only be depicted, in residential areas. Thus, the ancillary data or information should accurately estimate those areas that can be considered residential in nature. These may include land-use/landcover categories (LULC) such as Urban or Built-up Land, which may involve subcategories such as Residential, Mixed Urban or Build-up Land, Other Urban or Built-up Land (Anderson *et al.*, 1976), as well as rural areas in which people may reside, such as forested and agricultural land (e.g., Anderson's Level I categories 2 and 4). The choice depends upon the assumptions of the researchers, the scale at which LULC data are available, and the particular characteristics of the study area. In cases where multiple LULC classes are used, these classes are collapsed into one overall class that represents populated areas; by default the remainder of the study area is considered unpopulated; thus, the approach is considered "binary". This methodology for dasymetric mapping was described in detail by Langford and Unwin (1994) and subsequently referred to as the "binary dasymetric method" by Eicher and Brewer (2001). This technique also has been referred to at least once as "filtered areal weighting" (Maantay, Maroko and Herrmann, 2007). Many examples of its use exist (Langford and Unwin, 1994; Holt, Lo snd Hodler, 2004; Lo, 2004; Poulsen and Kennedy, 2004; Chen *et al.*, 2004; Langford and Higgs, 2006; Langford, 2007) and Eicher and Brewer (2001) have compared this method to other dasymetric mapping techniques. We provide a current example of binary dasymetric mapping and describe the methodological steps in further detail later.

Fisher and Langford found that binary dasymetric mapping was very robust to classification error in LULC derivation from satellite imagery. Classification errors up to 40% could be experienced before the resulting binary dasymetric map accuracy degraded to the level of the next-best dasymetric method (Fisher and Langford, 1996). Binary dasymetric mapping also can be more accurate than regression-based models of areal interpolation; Langford (2003) attributes this to the fact that binary dasymetric maps are calibrated to each areal unit of the source data, whereas regression models are typically fitted at the global level. An additional property of binary dasymetric maps (and of three-class maps as well) is the preservation of volume for each areal unit; this property was noted by Lam (1983) to hold true for area-based areal interpolation methods in general. In other words, the sum of population counts (dasymetrically derived) within areal units remains the same as for the original source zones prior to dasymetric mapping. This property, also known as the pycnophylactic property (Tobler, 1979), is important in areal interpolation tasks because of the desire to avoid statistically creating or eliminating population, which would introduce error in the resulting population or sociodemographic data estimates. Despite its outward simplicity, binary dasymetric mapping is equally applicable at a variety of spatial scales (Langford, 2003), which makes it a flexible approach, adaptable to a variety of application settings.

14.2.4.2 Three-class (or N-class) dasymetric

Several researchers have noted the utility of binary dasymetric mapping yet have suggested avenues for improving on the method. Flowerdew, Green and Kehris (1991) may have been the first to suggest that the binary method is crude and somewhat limited. They suggest that additional information about the distribution of the variable of interest may be helpful (they used clay versus limestone soil belts and fitted the data using an expectation-maximization (EM) algorithm). Such an extension to the binary method is an approach that apportions population to specific LULC categories, such as residential, urban, forested, and agricultural (Holloway, Schumacher and Redmond, 1996; Eicher and Brewer, 2001) or high-density urban, low-density urban, and nonurban (Mennis, 2003); this has been termed the "three-class method" (Eicher and Brewer, 2001, p. 129). Conceptually, there is no limitation to the number of classes to which population may be apportioned; therefore a more-appropriate name for this method in general may be the "N-class dasymetric" method. In Eicher and Brewer's example for Southwestern Pennsylvania, they utilized landcover data provided by the US Geological Survey, and assigned 70% of each county's population to urban land-use polygons, 20% to agricultural and woodland polygons, and the remaining 10% to forest polygons.

The three-class (or N-class) method is made feasible by improvements in classification accuracy and calibration of population densities for multiple residential classes. The challenge in this method is determining the appropriate assignment of density values per class (Langford, 2003); a difficulty that is avoided in binary dasymetric mapping because 100% of population is assigned to just one class and 0% to the other class. A second weakness, identified by Eicher and Brewer (2001) as the major weakness of their approach, is that "the three-class method does not account for the area of each particular LULC within each county" (Eicher and Brewer, 2001, p. 130).

Recent research has addressed both of these methodological concerns. Mennis (2003) introduces a technique for estimating the population density fraction of each LULC type, in which he sampled the population of all block groups that are completely contained within their respective LULC classes, on a county-by-county basis to account for spatial heterogeneity of densities across his study area of metropolitan Philadelphia, Pennsylvania. He used the population density fractions to assign the percentage of each block group's population to each of three LULC classes in each county. These percentages were further adjusted by an area-weighted fraction to account for the differential areas of each LULC within each block group. An extension to Mennis' 2003 work is the "Intelligent Dasymetric Method", or "IDM" (Mennis and Hultgren, 2006). They combine analysts' expert judgment along with a flexible empirical sampling approach in order to quantitatively derive population densities for individual LULC classes. They apportion population to target zones based on a ratio of these empirically-derived densities. In their tests, they found the IDM method to be more accurate than a simple areal weighting approach and at least as accurate as a binary dasymetric approach.

14.2.4.3 Dasymetric mapping with vector ancillary data

In contrast to the previously described dasymetric methods that rely upon ancillary data from remote sensors, other researchers have used either vector data or a combination of vector, raster, and census data to delineate dasymetric mapping units. In two examples of using vector data for dasymetric mapping, Xie (1995) and Reibel and Bufalino (2005) use street network data derived from TIGER/Line files produced by the US Census Bureau. Xie illustrates three techniques to apportion population from source zones to the street network segments, including segment length, street segment class, and the number of houses presumed to be contained within the address ranges of each street segment. This set of methodologies is similar to areal weighting with the main difference being that population is assumed to be evenly distributed along a one-dimensional line as opposed to a two-dimensional area. Xie's research indicated that using a weighted approach – namely the use of street segment classes – provided the most accurate results for population distribution. Reibel and Bufalino (2005) replicated the street-weight count approach originally proposed by Xie (1995) and conducted an error analysis, in which they obtained smaller errors than from using an area-weighting technique. Reibel and Bufalino chose their vector-only methodology as a way to promote the use of areal interpolation in applied situations. They state that "dasymetric weighting using remote sensing has been almost completely restricted to computational experiments by geographers and allied spatial scientists" (Reibel and Bufalino, 2005, p. 129), and it is their contention that using remote sensing data requires specialized raster data processing skills and raster GIS software.

Maantay, Maroko and Herrmann (2007), use cadastral data for New York City, and incorporate an expert system (the Cadastral-based Expert Dasymetric Systems, or CEDS) to determine the cadastral variable (either number of residential units or the adjusted residential areas for each tax lot) that fits the data best. They critique the use of remote sensing data for ancillary data on four points: first, they suggest that remote sensing data do not adequately capture the precision needed to adequately model population density in highly urbanized areas, due to limitations to satellite imagery spatial resolution and intrapixel heterogeneity (citing a report by Forster from 1985, prior to the advent of very high resolution imagery and recent image processing techniques); second, they suggest that using impervious surface extents or other morphological variables as proxies for the degree of urbanized area can result in misclassification errors; third, they note that building heights are not revealed from imagery (although they subsequently admit that lidar may provide a feasible method for estimating building heights); and fourth, they express concerns over expense, data storage and processing burdens, and difficulties in differentiating between residential and nonresidential areas. Their fourth criticism is based on a quote taken from Moon and Farmer (2001, p. 42), which Moon and Farmer attributed to Mesev, Longley and Batty (1996). With respect to this particular concern, which may have been valid fifteen years ago, developments in computing capabilities and image processing techniques have made this problem less insurmountable. Maantay, Maroko and Herrmann (2007) also criticize the use of street network data as ancillary data, suggesting that this type of data is not accurate enough for densely settled urban core areas.

Xie (2006) illustrates a related approach to using parcel data. Rather than beginning with a cadastral database, Xie used high resolution imagery (digital orthophotos and lidar) to extract building-related pixels, which were then differentiated as residential and non-residential through the overlay of parcel data and tax roll information. The key emphasis of Xie's approach is the use of individual housing units as the most basic and meaningful unit from which to construct a population surface. Xie's approach represents a very innovative combination of population-related data that are derived from various sources, which are collected for different purposes, in order to produce a data layer with enhanced informational value.

Wu, Wang and Qiu (2008) present a slightly different approach to obtaining housing-unit-level population estimates. They used building footprint data, in vector polygon format, obtained from a metropolitan planning and zoning department, along with census-derived housing measures. By dividing total building volumes (the sum of building footprint area) by average space per housing unit, they derived an estimate of the total number of housing units. They multiplied that estimate by the housing unit occupancy rate and the average household size to obtain an estimate of the population at the block level.

14.3 Application example: metropolitan Atlanta, Georgia

To demonstrate the practical utility of dasymetric mapping for population and sociodemographic data redistribution, we applied some basic techniques that can be performed in a standard, commercial geographic information system (GIS) for Gwinnett County, Georgia, USA. Gwinnett County is part of the metropolitan Atlanta area (Fig. 14.1a), a region that has experienced substantial population growth in the past three decades and has been the site for numerous urban geography research studies (Yang, 2002; Yang and Lo, 2003; Lo and Yang, 2002; Lo and Quattrochi, 2003; Lo, 2004, 2008; Holt, Lo and Hodler, 2004; Holt and Lo, 2008). Gwinnett County (Fig. 14.1b) represents a portion of the Atlanta metropolitan area that contains a diverse

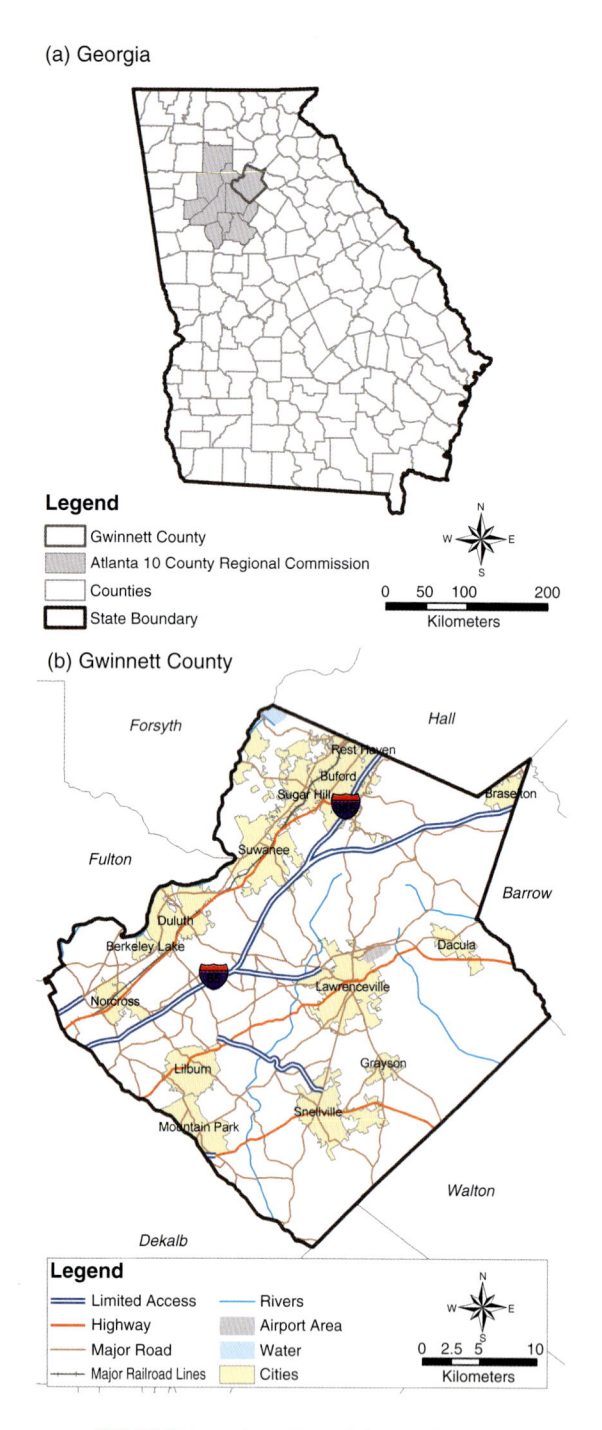

FIGURE 14.1 Location of the study area.

mix of LULC classes and associated population densities. This diversity in LULC classes and the clarity of dasymetric population density maps were observed in previous studies of the Atlanta area (Holt, Lo and Hodler, 2004; Lo, 2004, 2008).

14.3.1 Data

We obtained population and sociodemographic data for the year 2000 from the US 2000 decennial census (US Census Bureau,

(a) Census 2000 total population by blocks

Total Persons
- 0
- 1 – 100
- 101 – 350
- 351 – 850
- 851 – 1800
- 1801 – 5735
- County boundary

(b) Census 2000 population density by blocks

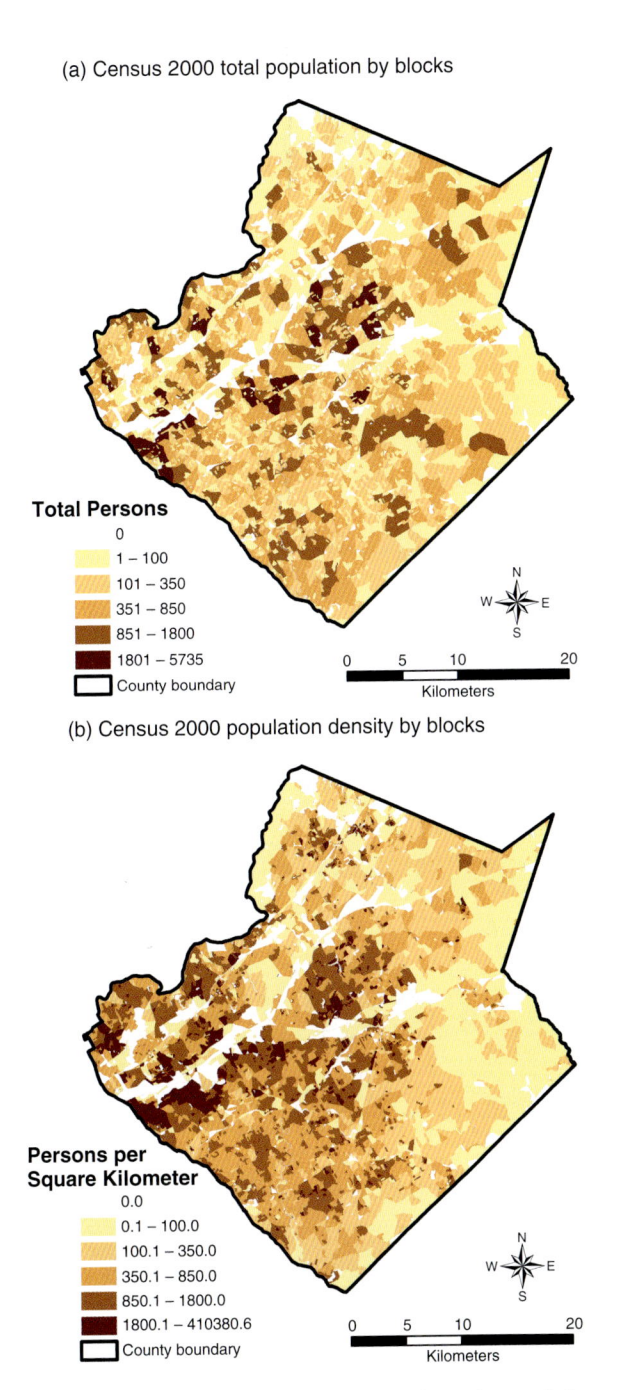

Persons per Square Kilometer
- 0.0
- 0.1 – 100.0
- 100.1 – 350.0
- 350.1 – 850.0
- 850.1 – 1800.0
- 1800.1 – 410380.6
- County boundary

FIGURE 14.2 Block-level population (a) and population density (b) at the block level for Gwinnett County.

2000) at the block and block group levels for Gwinnett County, Georgia (FIPS = 13135). Figure 14.2a and b depict the 2000 population and population density at the block level for Gwinnett County. We obtained three types of ancillary data in order to evaluate their impact on dasymetric mapping accuracy. LULC data for 2001 (the most proximate data available for comparison to the 2000 sociodemographic data) were obtained from the National Land Cover Database (NLCD) produced by the Multi-Resolution Land Characteristics Consortium (MRLC, 2009), a group of nine US federal government agencies whose objective is acquiring Landsat 5 and Landsat 7 imagery and generating a national LULC database, at 30 meters spatial resolution, for the 50 states and Puerto Rico (MRLC, 2009). LULC data (LandPro) for 2001 were also obtained from the Atlanta Regional Commission, the regional planning organization for metropolitan Atlanta; these data were derived from manual interpretation and GIS digitizing of true-color aerial photographs with 1 m spatial resolution obtained in 2001. Parcel data for 2001 were obtained from the Gwinnett County GIS Department. We used these three sources of ancillary data to test for differences in dasymetric population estimates based on different spatial resolutions and postulated LULC accuracies. We also obtained data on the locations of parks, airports, interstate and other primary highways, railroads, and landmarks (ESRI Data and Maps, 2008, ESRI, Redlands, CA); these data were used to create spatial masks for areas that were assumed, *a priori*, to have no residential population. Boundary data for counties, Census block groups, Census blocks, and voting districts were downloaded from US Census 2008 TIGER/Line files for the 2000 Census boundaries (US Census Bureau, 2008).

14.3.2 Dasymetric maps

We performed all the analysis using ArcGIS (ArcInfo) version 9.3.1 (Environmental Systems Research Institute, Redlands, CA) with the Spatial Analyst Extension. We used four processes to estimate the population to the pixel (e.g., grid) level: three implementations of the binary dasymetric method and one implementation of the N-class dasymetric method. First, we employed the binary dasymetric method (Langford and Unwin, 1994) using the NLCD 2001 LULC as the ancillary data layer. We refined the spatial extent of the populated area by extracting parks, airports, landmarks (excluding D10 military installations and reservations; D31 hospitals, urgent care facilities, and clinics; D37 federal penitentiaries, state prisons, or prison farms; portion of D43 educational institutions – colleges and universities for which the Census Bureau collects institutional population counts), interstates, highways and railroads as a binary mask after recoding a clipped raster layer as 1 = residential (NLCD classes: Developed Low Intensity [22] and Developed Medium Intensity [23], Developed Open Space [21] and Developed High Intensity [24] may be appropriate additions to residential area, based on the particular study area.) and 0 = non-residential (all other NLCD classes). We used the block group population as our population source layer. We used the Tabulate Area tool in ArcGIS for block groups and the reclassified NLCD 2001 layer to obtain residential population areas for each block group. We joined the table back to the block group feature class and calculated population densities for all block groups ($n = 208$). We converted block groups to a raster layer of population density. Finally, we used map algebra, multiplying the block group population density raster by Reclassified NLCD 2001 (binary raster layer), to derive the residential area population density (Fig. 14.3a).

For the second and third implementations we used binary dasymetric mapping with LandPro 2001 and the Gwinnett County 2001 parcel data as the ancillary data layers. These two ancillary layers provided increased spatial resolution and/or attribute specificity for delineating residential areas. The GIS processing steps for both implementations were identical to those of the first implementation, which we described in the preceding paragraph. The only exception was that we did not

202 PART IV URBAN SOCIOECONOMIC ANALYSES

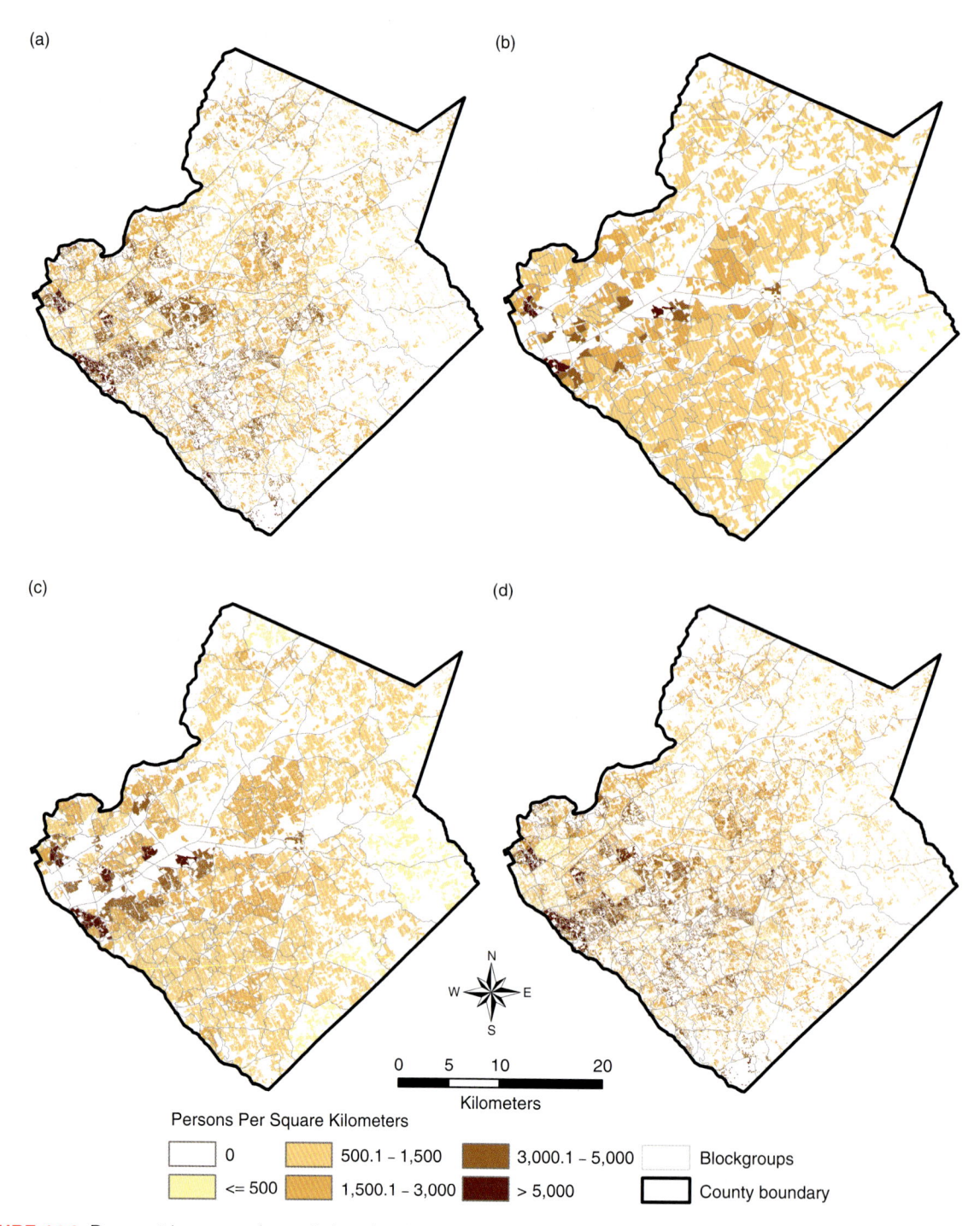

FIGURE 14.3 Dasymetric maps of population density by block group for Gwinnett County: (a) binary method using NLCD 2001; (b) binary method using LandPro 2001; (c) binary method using 2001 parcel data; (d) N-Class method using Dasymetric Mapping Extension and NLCD 2001.

need to apply the mask of highways, parks, and other similar features (as was necessary for the NLCD layer). Our definition of residential area, using the LandPro ancillary data was: 1 = residential area (LandPro codes: Low Density Single Family Residential [111], Medium Density Single Family Residential [112], High Density Residential [113], Multifamily Residential [117], and Mobile Home Parks [119]), and 0 = non-residential area (all other LandPro codes). Our definition of residential area, using the Gwinnett County parcel data was: 1 = residential area (parcel codes: Residential SFR [101], Residential Duplex [102], Residential Triplex [103], Residential Quadplex [104], Mixed Residential/Commercial [105], Condominium – Common Element [106], Condominium – Fee Simple [107], Mobile Home [108], Apartments – Three Story and Under [211], and Mobile Home Park [213]), and 0 = non-residential area (all other parcel codes). Figure 14.3b and c depict the block-group population densities using the LandPro and parcel data, respectively.

For the fourth process, we used the Dasymetric Mapping Extension (Sleeter and Gould, 2007) developed by the USGS for ArcGIS, which is based on the three-class dasymetric mapping process. We again used the NLCD 2001 as our ancillary layer, but for this implementation we reclassified the data into three classes: 2 = low-density residential area (NLCD class: Developed Low Intensity [22]), 1 = high-density residential area (NLCD class: Developed Medium Intensity [23]), (Developed Open Space [21] and Developed High Intensity [24] may be appropriate to include as residential area based on the particular study area) and 0 = non-residential area (all other NLCD classes). We used the same block group layer as before for the 2000 population. We initiated and ran the Dasymetric Mapping Extension, using the default of 80% for the minimum threshold value of the source unit that must be covered by an ancillary class. The population density estimates that we derived using this tool are depicted in Fig. 14.3d.

We compared our dasymetrically-derived population estimates to the 2000 population data for Gwinnett County Census blocks ($n = 5048$). Because we used population data from the block-group level for the dasymetric process, and because census blocks nest completely within block groups, we can use the block-level data to validate the block-group level estimates. We used the Zonal Statistic tool to calculate block level population from the four dasymetric population raster maps. Table 14.1 provides a comparison of the root mean square error (RMSE) – both standard and normalized – from the four dasymetric implementations. The lowest RMSE was obtained by using the binary method in conjunction with the LandPro 2001 ancillary data, closely followed by the binary method using the parcel data. The RMSE for the two methods (binary and N-class) that used the NLCD 2001 data were close to one another and both higher than the error associated with the LandPro and parcel data implementations. The normalized RMSE (essentially the RMSE divided by the range of the observed population counts at the block level) provides more of a relative measure of error (express in percentages). We could not compute mean percentage errors because 1209 census blocks contained no population, and thus division by zero would have been required in the calculation.

Figure 14.4 provides a map for the difference of census 2000 population and area interpolation estimates at block group level.

TABLE 14.1 Comparison of root mean squared (RMS) errors for four dasymetric mapping implementations for the Gwinnett County study area.

Dasymetric method	Ancillary data	Root means square error (population)	Normalized root mean square error (%)
Binary	NLCD 2001	115.3	2.01
Binary	LandPro 2001	102.2	1.78
Binary	Parcel 2001	105.5	1.84
N-Class with D-M Extension	NLCD 2001	117.3	2.04

14.3.3 Areal interpolation

To demonstrate the redistribution of sociodemographic data, we use the example of total population, and the population of individuals living at or below the poverty level. We began by mapping these variables using the binary dasymetric method with LandPro 2001 as the ancillary data. Figure 14.5(a, b) depicts the dasymetric estimates for total population density, and the density of the population living in poverty overlaid with voting districts, respectively. For areal interpolation, we used block groups as the source zone and Gwinnett County voting districts ($n = 132$) as the target zone. We selected voting districts because they are completely contained within the Gwinnett County boundary, and Census blocks do not nest within the voting district boundaries in all cases, therefore offering an applied example of how areal interpolation can be useful. The areal reaggregation of the data into the target zones was relatively straightforward: we used the Zonal Statistic tool to compute data values (total population, and individuals in poverty) at the voting district levels. The areally-interpolated Gwinnett County voting district estimates for the total population, and the population living in poverty are depicted in Fig. 14.6(a, b). We also selected ZIP codes as our target zone, because they are not completely contained within the county boundary. By using areal interpolation we obtained estimates for the portion of the ZIP codes that were contained within Gwinnett County. The results are depicted in Fig. 14.6(c, d) for total population and population living in poverty.

For dasymetric mapping and areal interpolation using these four implementations, we observed that accuracies depended on the classification of our ancillary layers. When using NLCD 2001, it is very difficult to differentiate high-density office buildings, high-rise residential buildings, two-story office buildings, small strip malls, residential apartments, and townhouses. When using the LandPro 2001 data, institutional populations (e.g., university housing and correctional facilities) may get eliminated if these areas are classified as nonresidential areas. It may be necessary to obtain the location of such institutions and to manually reclassify as residential. We also observed that the resulting accuracies of the four implementations did not vary greatly. We obtained slightly better RMS errors with a binary dasymetric approach, using custom-written geoprocessing algorithms written by one of the authors. These algorithms (Fig. 14.7), although straightforward, required time to conceptualize and to implement in ArcGIS Model Builder. However, the geoprocessing algorithms can be saved and reused for subsequent applications. Therefore, researchers who anticipate the need for repeated dasymetric mapping may wish to consider such an approach. This approach should be weighed against the ease of use of a custom dasymetric mapping tool such as the Dasymetric Mapping Extension, produced and distributed (free of charge) by the USGS.

We chose not to use more-complex dasymetric mapping strategies, such as the one outlined by Mennis and Hultgren (2006), simply to illustrate the ease of implementation of methods that have been shown to be accurate and produce useful results. Indeed, Langford and Higgs (2006) note that despite recent efforts to refine dasymetric mapping with three-tier density classifications, "there is little evidence to date of any clear benefit over the simpler two-tier, binary dasymetric method" (Langford and Higgs, pp. 297–8). We also support Langford's (2007) argument that "there is little evidence to suggest widespread

204   PART IV   URBAN SOCIOECONOMIC ANALYSES

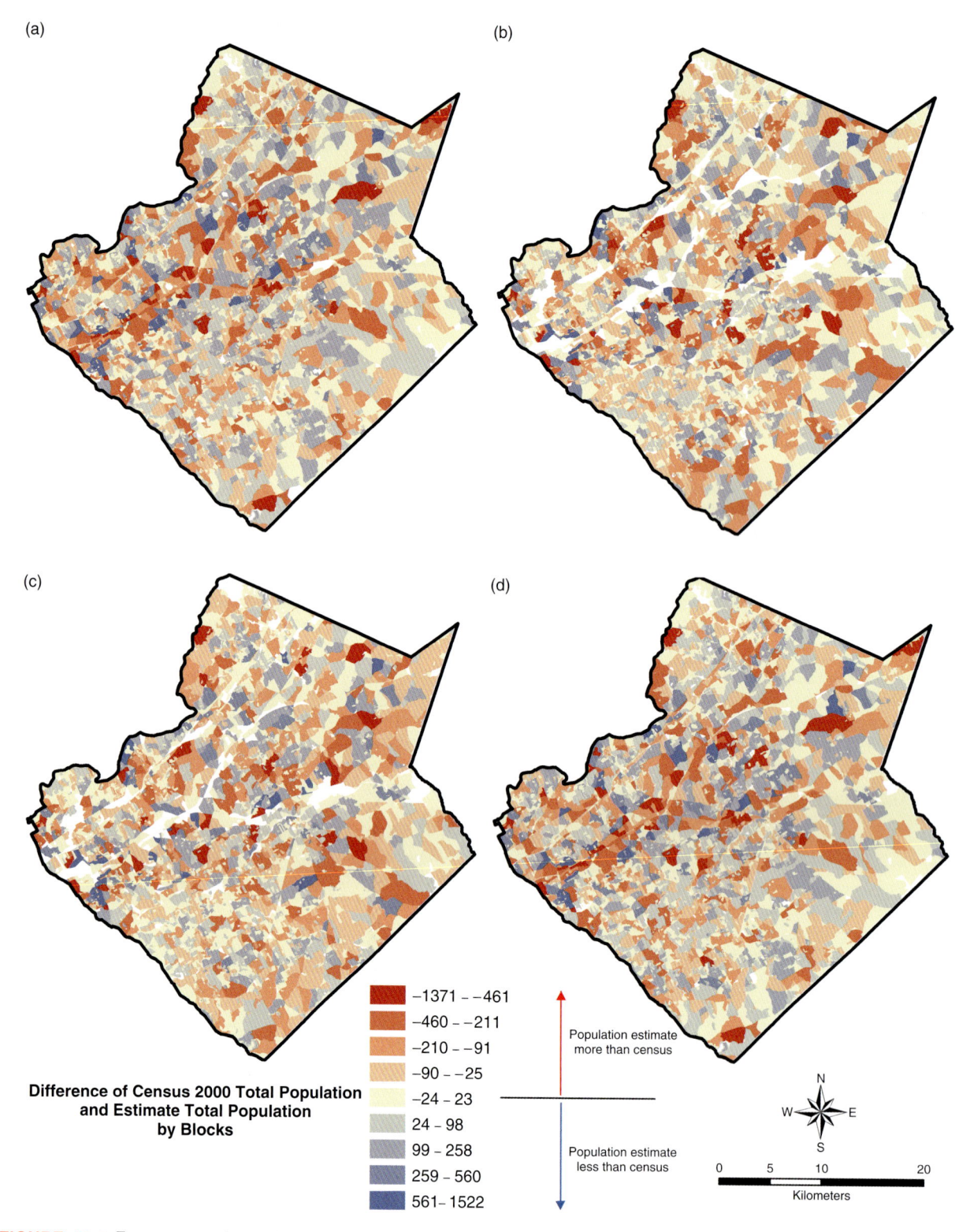

FIGURE 14.4 Error maps of population estimates by blocks for Gwinnett County: (a) binary method using NLCD 2001; (b) binary method using LandPro 2001; (c) binary method using 2001 parcel data; (d) N-Class method using Dasymetric Mapping Extension and NLCD 2001.

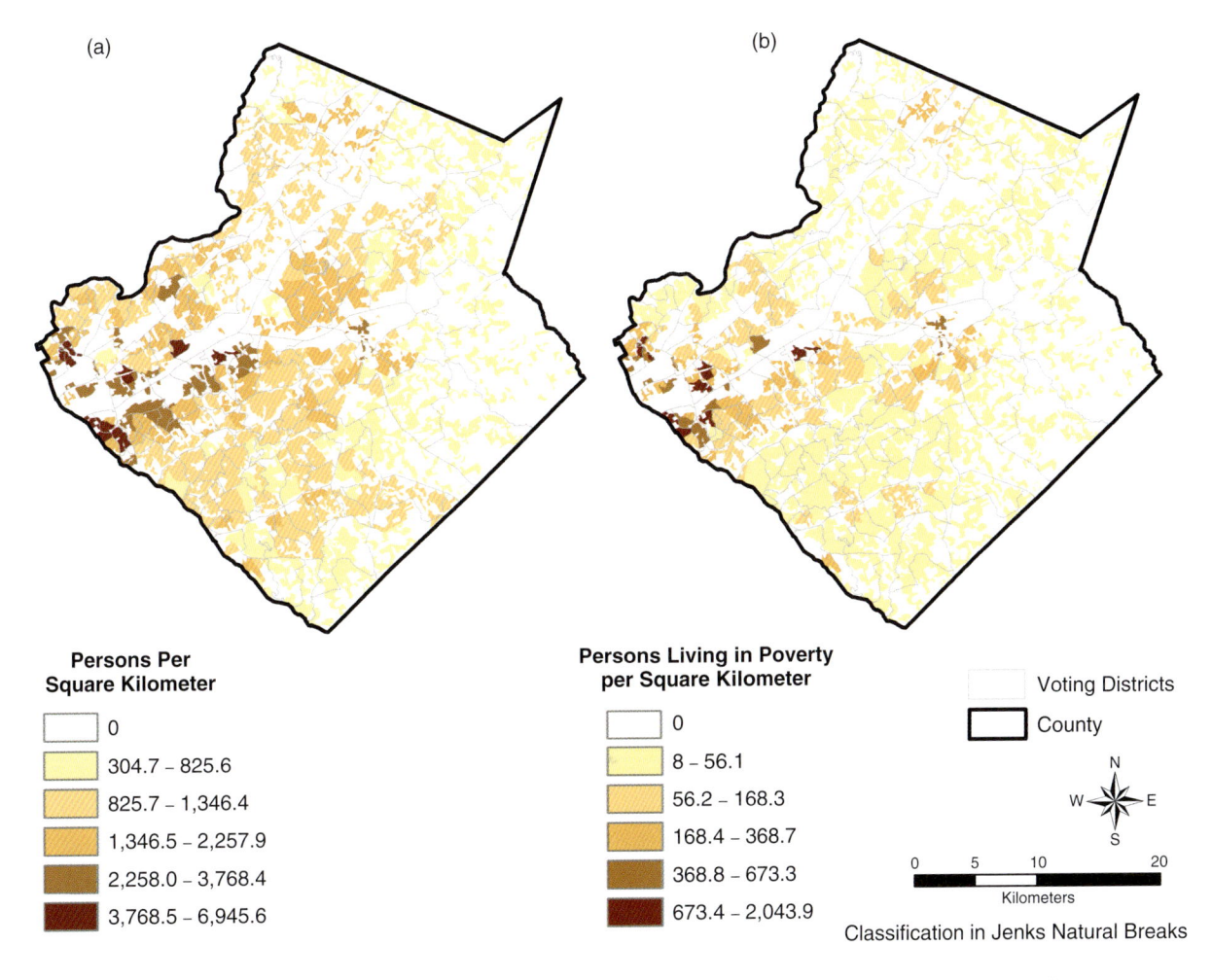

FIGURE 14.5 Dasymetric maps for Gwinnett County: (a) population density; (b) poverty density.

usage amongst the GIS user community. It is argued that to encourage greater uptake such methods must offer simplicity and convenience" (Langford, p. 19).

It is our contention that the binary dasymetric mapping method, both as a cartographic tool as well as a means for performing areal interpolation, provides accurate results and can be easily implemented using current GIS software applications with custom geoprocessing scripts or off-the-shelf extensions, such as that provided by USGS for ArcGIS. Furthermore, the availability of ancillary data such as the NLCD 2001 and other remote sensing derivative products now indicates that GIS users no longer need to be remote sensing experts in order to conduct dasymetric mapping. For example, Reibel and Agrawal (2007) performed dasymetric mapping using NLCD data and they found that this easily-obtainable dataset, which does not require image processing by the end-user, performed quite well for areal interpolation. The challenge for the remote sensing community may be to continue developing image processing techniques that will foster the proliferation of LULC data and increase the accuracy of those data.

Conclusions

Dasymetric mapping results in a more accurate representation of the magnitude and spatial extent of a phenomenon. It facilitates, and is the most accurate method for performing, areal interpolation. It preserves the pychnophylactic property, and is applicable across spatial scales. Recent research has focused on new approaches to classifying LULC, such as object-based analyses, texture measurements, and fuzzy classification, which may result in increased accuracy and precision in modeling populated land areas and may also overcome difficulties in using very high resolution satellite imagery. Other research has focused on improvements in measuring and estimating statistical relationships between various LULC classes and population, thus adding more intelligence to ancillary data. In this regard, incorporation of additional ancillary data, such as cadastral data, and the refinement of regression modeling approaches that address the existence of spatial nonstationarity (Langford, 2006; Lo, 2008) have been demonstrated to be useful. In addition, recent developments in GIS technology and computing capacity, along

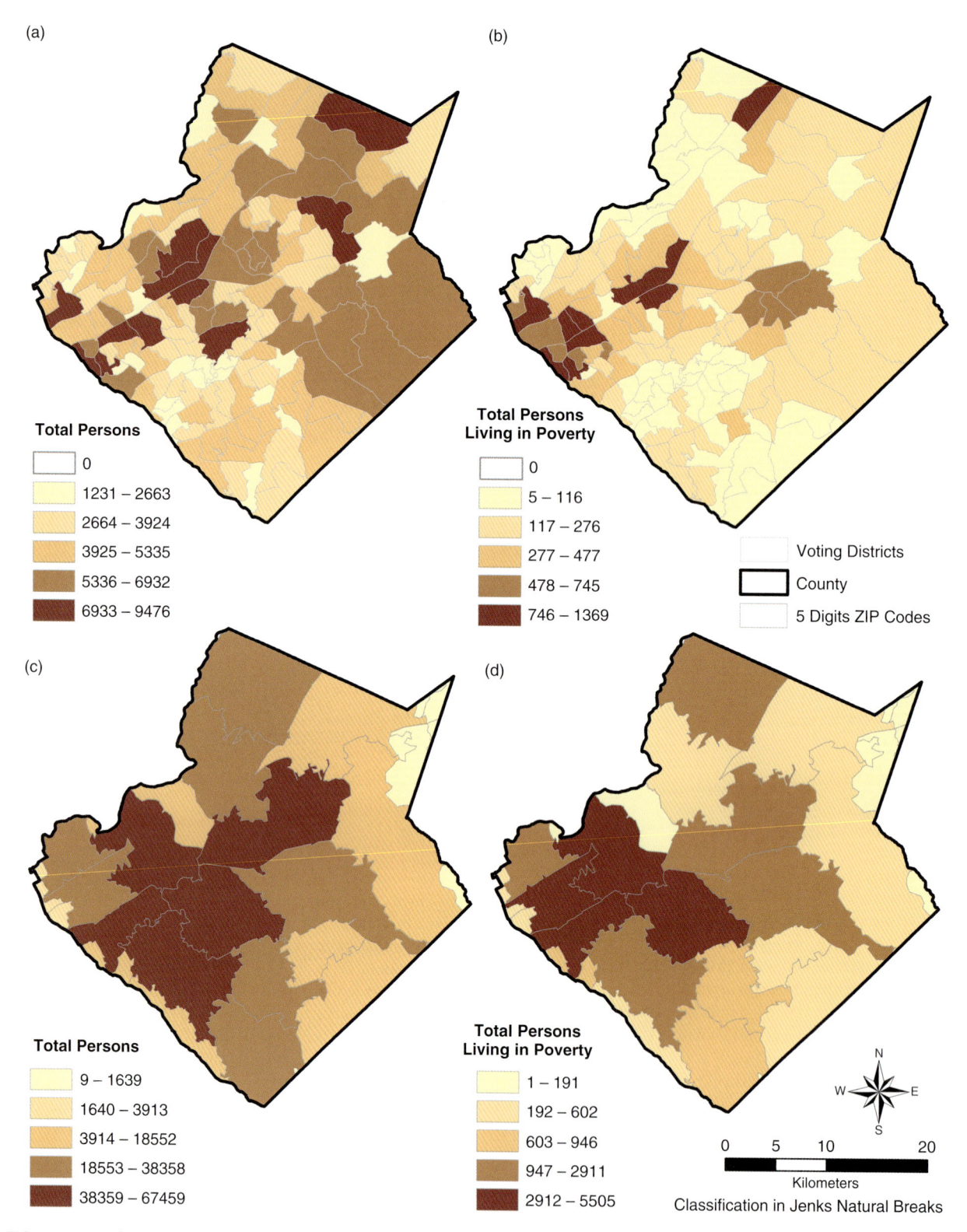

FIGURE 14.6 Areal interpolation results, Gwinnett County (a) total population by voting districts; (b) total number of individuals in poverty by voting districts; (c) total population by ZIP code inside Gwinnett County; (d) total population living in poverty by ZIP code inside Gwinnett County.

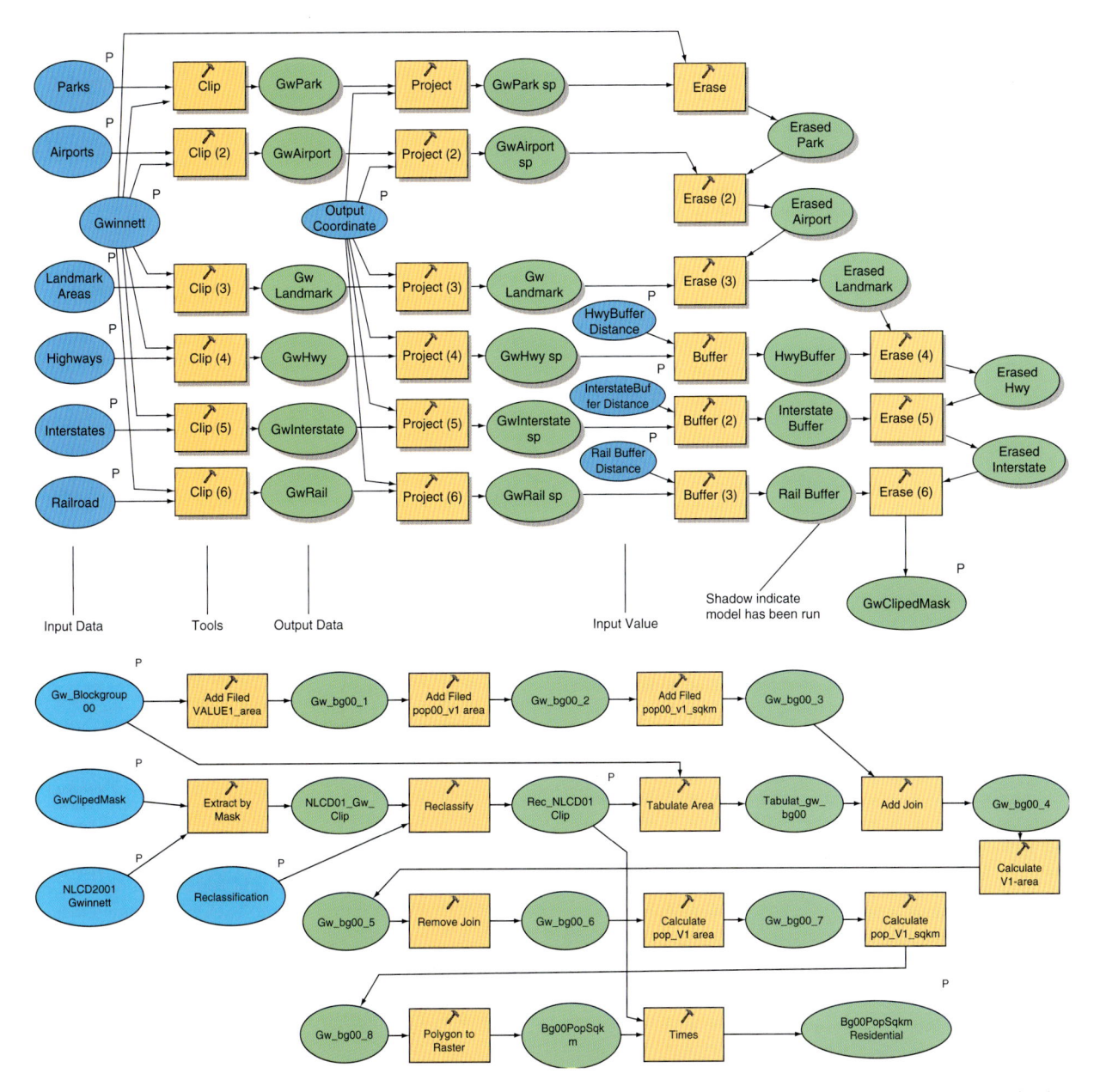

FIGURE 14.7 GIS processing flows for dasymetric mapping: (a) creation of spatial mask; (b) binary dasymetric method as implemented.

with the provision of LULC data at low-to-no cost, make dasymetric mapping a very viable approach for areal interpolation tasks.

Wright acknowledges that his dasymetric method could be used for either population "or of other phenomena, within the limits of townships or other territorial units for whose subdivisions no statistical data are available", and "...it might well be applied in mapping various phenomena for which statistics are available by counties but not by minor civil divisions" (Wright, 1936, p. 110). Eicher and Brewer (2001) demonstrate this utility by implementing three different methods of dasymetric mapping for population and data on housing values for Western Pennsylvania.

Clearly, Wright's method, along with any of the other dasymetric methods we described, could be used to map the distribution of virtually any sociodemographic data. This might include not only census-derived data, such as housing, economic, and demographic variables, but also public health surveillance data. The limitations in using this approach may result more from the origin of the data themselves as opposed to limitations in dasymetric methodology. Specifically, many public health data are collected by surveys and the resulting sample sizes may impose limitations due to concerns over small numbers and variance instability. In addition, public health agencies that collect data on rare or sensitive conditions may be required to suppress data at more granular levels due to rightful concerns over the preservation of privacy. Nonetheless, with adequate source data, researchers and data analysts can apply dasymetric methods to more accurately map the distributions of sociodemographic and health-related data.

Acknowledgments

The authors wish to thank Mike Alexander, Research Division Chief, Wei Wang, Senior Principal Planner, and Ryan Barrett, GIS Analyst, from the Atlanta Regional Commission, Atlanta, GA, for furnishing the LandPro 2001 LULC data; and Sharon Stevenson, GISP, LIS Manager, and Sam A. Rape, Jr., IT GIS Administrator, from the Gwinnett County GIS Department, Lawrenceville, GA, for furnishing the Gwinnett County parcel data. The findings and conclusions in this report are those of the authors and do not necessarily represent the official position of the Centers for Disease Control and Prevention.

References

Anderson, J.R., Hardy, E.E., Roach, J.T. and Witmer, R.E. (1976) A land use and land cover classification system for use with remote sensor data. United States Government Printing Office, Washington, DC.

Bhaduri, B., Bright, E., Coleman, P. and Urban, M. (2007) LandScan USA: a high-resolution geospatial and temporal modeling approach for population distribution and dynamics. *GeoJournal*, **69** (1), 103–117.

Boone, C.G. (2008) Improving resolution of census data in metropolitan areas using a dasymetric approach: applications for the Baltimore Ecosystem Study. *Cities and the Environment*, **1** (1), 25.

Bracken, I., and Martin, D. (1989) The generation of spatial population distributions from census centroid data. *Environment and Planning A*, **21**, 537–543.

Briggs, D. J., Gulliver, J., Fecht, D. and Vienneau, D. M. (2007) Dasymetric modelling of small-area population distribution using land cover and light emissions data. *Remote Sensing of Environment*, **108** (4), 451–466.

Chen, K. P., McAneney, J., Blong, R., Leigh, R., Hunter, L. and Magill, C. (2004) Defining area at risk and its effect in catastrophe loss estimation: a dasymetric mapping approach (vol 24, pg 97, 2004). *Applied Geography*, **24** (3), 259–259.

Comber, A., Proctor, C. and Anthony, S. (2008) The creation of a national agricultural land use dataset: combining pycnophylactic interpolation with dasymetric mapping techniques. *Transactions in GIS*, **12** (6), 775–791.

Crampton, J. (2004) GIS and geographic governance: reconstructing the choropleth map. *Cartographica: The International Journal for Geographic Information and Geovisualization*, **39** (1), 41–53.

De Geer, S. (1926) Review: a population density map of European Russia (Dazimetricheskaya Karta Evropeiskoi Rossii) (Carte dasymetrique de la Russie d'Europe) by Benjamin Semenov-Tian-Shansky. *The Geographical Review*, **16**, 341–343.

Eicher, C. L., and Brewer, C.A. (2001) Dasymetric Mapping and Areal Interpolation: Implementation and Evaluation. *Cartography and Geographic Information Science*, **28**, 125–138.

Fisher, P. F., and Langford, M. (1996) Modeling sensitivity to accuracy in classified imagery: A study of areal interpolation by dasymetric mapping. *Professional Geographer*, **48** (3), 299–309.

Flowerdew, R. and Green, M. (1989) Statistical methods for inference between incompatible zonal systems, in *Accuracy of Spatial Databases* (eds M.F. Goodchild and S. Gopal), Taylor & Francis, London, pp. 239–248.

Flowerdew, R., and Green, M. (1992). Developments in areal interpolation methods and GIS. *Annals of Regional Science*, **26** (1), 67–78.

Flowerdew, R., and Green, M. (1994) Areal interpolation and types of data, in *Spatial Analysis and GIS* (eds S. Fotheringham and R. Rogerson), Taylor & Francis, London.

Flowerdew, R., Green, M, and Kehris, E. (1991) Using areal interpolation methods in geographic information systems. *Papers in Regional Science: The Journal of the RSAI*, **70** (3), 303–315.

Fotheringham, A.S., Brunsdon, C., and Charlton, M. (2002) *Geographically weighted regression: the analysis of spatially varying relationships*, John Wiley & Sons, Chichester, UK.

Goodchild, M.F., Anselin, L., and Deichmann, U. (1993) A framework for the areal interpolation of socioeconomic data. *Environment and Planning A*, **25** (3), 383–397.

Goodchild, M.F., and Lam, N.S-N. (1980) Areal interpolation: a variant of the traditional spatial problem. *Geo-Processing*, **1**, 297–312.

Gotway, C.A., and Young, L.J. (2002) Combining incompatible spatial data. *Journal of the American Statistical Association*, **97** (458), 632–648.

Gregory, I.N. (2002) The accuracy of areal interpolation techiques: standardising 19th and 20th century census data to all long-term comparisons. *Computers, Environment and Urban Systems*, **26**, 293–314.

Harvey, J.T. (2002) Estimating census district populations from satellite imagery: some approaches and limitations. *International Journal of Remote Sensing*, **23** (10), 2071–2095.

Hay, S.I., Noor, A.M., Nelson, A., and Tatem, A.J. (2005) The accuracy of human population maps for public health application. *Tropical Medicine and International Health*, **10** (10), 1073–1086.

Holloway, S.R., Schumacher, J., and Redmond, R. (1996) People and place: Dasymetric mapping using Arc/Info. Missoula: Wildlife Spatial Analysis Lab, University of Montana.

Holt, J.B., and Lo, C.P. (2008) The geography of mortality in the Atlanta metropolitan area. *Computers, Environment and Urban Systems*, **32**, 149–164.

Holt, J.B., Lo, C.P., and Hodler, T.W. (2004) Dasymetric estimation of population density and areal interpolation of census data. *Cartography and Geographic Information Science*, **31** (2), 103–121.

Howenstine, E. (1993) Measuring demographic change: the split tract problem. *The Professional Geographer*, **45** (4), 425–430.

Kressler, F., and Steinnocher, K. (2008) Object-oriented analysis of image and LiDAR data and its potential for a dasymetric mapping application, in *Object-Based Image Analysis: Spatial Concepts for Knowledge-Driven Remote Sensing Applications* (eds T. Blaschke, S. Lang and G. J. Hay), Springer-Verlag, Berlin, pp. 611–624.

Kyriakidis, P.C. (2004) A geostatistical framework for area-to-point spatial interpolation. *Geographical Analysis*, **36** (3), 259–289.

Lam, N. S-N. (1983) Spatial interpolation methods: a review. *The American Cartographer*, **10** (2), 129–149.

Langford, M. (2003) Refining methods for dasymetric mapping using satellite remote sensing, in *Remotely Sensed Cities* (ed. V. Mesev), Taylor & Francis Ltd., London, pp. 137–156.

Langford, M. (2006) Obtaining population estimates in non-census reporting zones: An evaluation of the 3-class dasymetric method. *Computers Environment and Urban Systems*, **30** (2), 161–180.

Langford, M. (2007) Rapid facilitation of dasymetric-based population interpolation by means of raster pixel maps. *Computers Environment and Urban Systems*, **31** (1), 19–32.

Langford, M., and Higgs, G. (2006) Measuring potential access to primary healthcare services: the influence of alternative spatial representations of population. *The Professional Geographer*, **58** (3), 294–306.

Langford, M., and Unwin, D.J. (1994) Generating and mapping population density surfaces within a geographical information system. *Cartographic Journal*, **31** (1), 21–26.

Langford, M., Maguire, D.J. and Unwin, D.J. (1991) The areal interpolation problem: estimating population using remote sensing in a GIS framework, in *Handling Geographical Information: Methodology and Potential Applications*, (eds I. Masser and M. Blakemore), Longman, London, pp. 55–77.

Liu, X.H., Kyriakidis, P.C., Goodchild, M.F. (2008) Population-density estimation using regression and area-to-point residual kriging. *International Journal of Geographical Information Science*, **22** (4), 431–447.

Lo, C. P. (2004) Testing urban theories using remote sensing. *GIScience and Remote Sensing*, **41** (2), 95–115.

Lo, C.P. (2008) Population estimation using geographically weighted regression. *GIScience and Remote Sensing*, **45** (2), 131–148.

Lo, C.P. and Quattrochi, D.A. (2003) Land-use and land-cover change, urban heat island phenomenon, and health implications: a remote sensing approach. *Photogrammetric Engineering and Remote Sensing*, **69** (9), 1053–1063.

Lo, C. P. and Yang, X. (2002) Drivers of land-use/land-cover changes and dynamic modeling for the Atlanta, Georgia Metropolitan area. *Photogrammetric Engineering and Remote Sensing*, **68** (10), 1073–1082.

Maantay, J. and Maroko, A. (2009) Mapping urban risk: Flood hazards, race, and environmental justice in New York. *Applied Geography*, **29**, 111–124.

Maantay, J.A., Maroko, A.R. and Herrmann, C. (2007) Mapping population distribution in the urban environment: the cadastral-based expert dasymetric system (CEDS). *Cartography and Geographic Information Science*, **34** (2), 77–102.

Maantay, J., Maroko, A. and Porter-Morgan, H. (2008) Research note – a new method for mapping population and understanding the spatial dynamics of disease in urban areas: asthma in the Bronx, New York. *Urban Geography*, **29** (7), 724–738.

MacEachren, A.M. (1979) The evolution of thematic cartography. A research methodology and historical review. *The Canadian Cartographer*, **16** (1), 17–33.

Martin, D. (1989) Mapping population data from zone centroid locations. *Institute of British Geographers: Transactions*, **14** (1), 90–97.

Martin, D. (1996) An assessment of surface and zonal models of population. *International Journal of Geographical Information Science*, **10** (8), 973–989.

Martin, D., Dorling, D. and Mitchell, R. (2002) Linking censuses through time: problems and solutions. *Area*, **34** (1), 82–91.

McCleary, G. (1969) The dasymetric method in thematic cartography. PhD dissertation, University of Wisconsin-Madison.

Mennis, J. (2002) Using geographic information systems to create and analyze statistical surfaces of population and risk for environmental justice analysis. *Social Science Quarterly*, **83** (1), 281–297.

Mennis, J. (2003) Generating surface models of population using dasymetric mapping. *Professional Geographer*, **55** (1), 31–42.

Mennis, J. (2009) Dasymetric mapping for estimating population in small areas. *Geography Compass*, **3** (2), 727–745.

Mennis, J. and Hultgren, T. (2006) Intelligent dasymetric mapping and its application to areal interpolation. *Cartography and Geographic Information Science*, **33** (3), 179–194.

Mesev, V., Longley, P. and Batty, M. (1996) RS-GIS: Spatial distributions from remote imagery, in *Spatial analysis: Modelling in a GIS environment* (eds. P. Longley and M. Batty), GeoInformation International, Cambridge, pp. 123–148.

Moon, Z.K. and Farmer, F.L. (2001) Population density surface: A new approach to an old problem. *Society and Natural Resources*, **14** (1), 39–49.

Mrozinski R.D.J. and Cromley, R.G. (1999) Singly- and doubly-constrained methods of areal interpolation for vector-based GIS. *Transactions in GIS*, **3** (3), 285–301.

Multi-Resolution Land Characteristics Consortium (MRLC) (2009) About the MRLC Program. [Online] Available http:www.mrlc.gov/about.php (accessed 16 November 2010).

Openshaw, S. (1984) *The Modifiable Areal Unit Problem*, Geo Books, Norwich.

Petrov, A.N. (2008) Setting the record straight: on the Russian origins of dasymetric mapping. *Cartographica: The International Journal for Geographic Information and Geovisualization*, **43** (2), 133–136.

Poulsen, E. and Kennedy, L.W. (2004) Using dasymetric mapping for spatially aggregated crime data. *Journal of Quantitative Criminology*, **20** (3), 243–262.

Rase, W. (2001) Volume-preserving interpolation of a smooth surface from polygon-related data. *Journal of Geographical Systems*, **3** (2), 199–213.

Reibel, M. and Agrawal, A. (2007) Areal interpolation of population counts using pre-classified land cover data. *Population Research and Policy Review*, **26** (5), 619–633.

Reibel, M. and Bufalino, M.E. (2005) Street-weighted interpolation techniques for demographic count estimation in incompatible zone systems. *Environment and Planning A*, **37** (1), 127–139.

Robinson A.H. (1955) The 1837 maps of Henry Drury Harness. *Geographical Journal*, **121**, 440–550.

Robinson, A.H. (1982) *Early Thematic Mapping in the History of Cartography*, University of Chicago Press, Chicago, Illinois.

Sadahiro, Y. (2000) Accuracy of count data transferred through the areal weighting interpolation method. *International Journal of Geographical Information Science*, **14** (1), 25–50.

Schroeder, J. P. (2007) Target-density weighting interpolation and uncertainty evaluation for temporal analysis of census data. *Geographical Analysis*, **39** (3), 311–335.

Schuurman, N., Grund, D., Hayes, M. and Dragicevic, S. (2006) Spatial/temporal mismatch: a conflation protocol for Canada Census spatial files. *Canadian Geographer-Geographe Canadien*, **50** (1), 74–84.

Scrope, G.P. (1833) *Principles of Political Economy, Deduced from the Natural Laws of Social Welfare, and Applied to the Present State of Britain*, Longmans, London.

Semenov-Tian-Shansky, B. (1928) Russia: territory and population: a perspective on the 1926 census. *Geographical Review*, **18** (4), 616–640.

Sleeter, R. and Gould, M. (2007) Geographic information system software to remodel population data using dasymetric mapping methods. techniques and methods 11-C2. US Department of the Interior, US Geological Survey, Washington, DC.

Slocum, T.A., McMaster, R.B., Kessler, F.C. and Howard, H.H. (2009) *Thematic Cartography and Geovisualization*, 3rd edition, Pearson, Prentice Hall, Upper Saddle River, NJ.

Tobler, W.R. (1979) Smooth pycnophylactic interpolation for geographical regions. *Journal of the American Statistical Association*, **74** (367), 519–530.

US Census Bureau (2000) Census 2000, Summary Files 1 and 3; generated by Hua Lu; using American FactFinder. [Online] Available http://factfinder.census.gov (accessed 16 November 2010).

US Census Bureau (2008) 2008 TIGER/Line files. [Online] Available http://www2.census.gov/cgi-bin/shapefiles/national-files (accessed 16 November 2010).

Wright, J.K. (1936) A method of mapping densities of population with Cape Cod as an example. *Geographical Review*, **26**, 103–110.

Wu, S.-S., Qiu, X. and Wang, L. (2005) Population estimation methods in GIS and remote sensing: a review. *GIScience and Remote Sensing*, **42** (1), 80–96.

Wu, S-s., Wang, L. and X. Qiu (2008) Incorporating GIS building data and census housing statistics for sub-block-level population estimates. *The Professional Geographer*, **60** (1), 121–135.

Xie, Y. (1995) The overlaid network algorithms for areal interpolation problem. *Computers, Environment and Urban Systems*, **19** (4), 287–306.

Xie, Z. (2006) A framework for interpolating the population surface at the residential-housing-unit level. *GIScience and Remote Sensing*, **43** (3), 233–251.

Yang, X. and Lo, C.P. (2003) Modelling urban growth and landscape changes in the Atlanta metropolitan area. *International Journal of Geographical Information Science*, **17** (5), 463–488.

Yang, X. (2002) Satellite monitoring of urban spatial growth in the Atlanta metropolitan area. *Photogrammetric Engineering and Remote Sensing*, **68** (7), 725–734.

Yuan, Y., Smith, R.M. and Limp, W.F. (1997) Remodeling census population with spatial information from LandSat TM imagery. *Computers, Environment and Urban Systems*, **21** (3/4), 245–258.

15

Who's in the dark—satellite based estimates of electrification rates

Christopher D. Elvidge, Kimberly E. Baugh, Paul C. Sutton, Budhendra Bhaduri, Benjamin T. Tuttle, Tilotamma Ghosh, Daniel Ziskin and Edward H. Erwin

A technique has been developed to estimate the percent population having electric power access based on the presence of satellite detected night-time lighting. A global survey was conducted for the year 2006 using night-time lights collected by the US Air Force Defense Meteorological Satellite Program (DMSP) in combination with the US Department of Energy Landscan population dataset. The survey includes results for 232 countries and more than 2000 subnational units. The results are compared to reported electrification rates for 86 countries compiled from a variety of sources by the International Energy Agency. The DMSP derived estimate of number of people worldwide who lack access to electricity is 1.62 billion, only slightly larger than the 1.58 billion estimated by the International Energy Agency.

15.1 Introduction

The wide distribution of over 6 billion people across more than 200 countries has made it difficult to collect and synthesize consistent data on the human condition at anything more that broad national and sub-national units. The primary reporting is for population and economic variables such as Gross Domestic Product (GDP). There is a paucity of data on quality-of-life variables and where such data are collected variations in the methods, survey questions used and timetables make the reports difficult to assimilate into a global assessment. Satellite sensors provide one of the few globally consistent and repeatable sources of observations. Clearly it would be useful to have one or more satellite derived indices that could used to estimate socioeconomic parameters, such as the distribution of economic activity, population, and living conditions. Historically, earth observing systems that aim for global coverage have been designed to observe environment and weather, not human activities. It would be sheer luck to find data from one of these global earth observing systems that also made a direct observation of a human activity. But there are several examples that can be pointed to. Satellite sensors such as NOAA's AVHRR and NASA's MODIS detect fires, many of which are anthropogenic in origin, using a combination of thermal bands. These same sensors detect urban heat islands and paucity of green vegetation in heavily built up urban cores. But the most remarkable example of a global earth observing satellite sensor detection of human activity are the night-time lights collected by the US Air Force Defense Meteorological Satellite Program (DMSP) Operational Linescan System (OLS).

Human beings around the world use lights at night to enable the extension of activity past sundown. The brightness of lights is affected by multiple factors, such as population density, economic activity, infrastructure investment, lighting type, lighting fixtures, and even cultural preferences in lighting. Despite these complexities, a number of studies have used night-time lights to map phenomena which would be cost prohibitive to map based on ground surveys. This includes the distribution of economic activity (Doll, Muller and Elvidge, 2000; Ebener *et al.*, 2005; Ghosh *et al.*, 2009), the density of constructed surfaces (Elvidge *et al.*, 2007a), poverty levels (Elvidge *et al.*, 2009a), and resource consumption (Sutton *et al.*, 2009).

By overlaying lights and population (Fig. 15.1) it is possible to observe clear differences in the quantity of lighting per person around the world. Populations in the developed world generally have a surplus of lighting, yielding the blue-green and white areas on Fig. 15.1. Areas with high population count and modest lighting levels show up as pink. (in portions of India and China). The red colors on Fig. 15.1 indicate populations where no lighting was detected by the DMSP sensor.

In this study we develop a new application for the night-ime lights, the estimation of electrification rates. For year 2005 the International Energy Agency (IEA) World Energy Outlook (IEA 2006) estimated the global electrification rate at 75.6% with 1.58 billion people living without electricity. Lack of electric power is a poverty indicator with links to conditions that are detrimental to health and wellbeing such as lack of refrigeration for food, poor water quality, lack of sanitary facilities, and limited access to health care services. We map the spatial extent of electrification in 2006 based on the presence of DMSP detected lighting. Combining the spatial extent of lighting with population count we estimate electrification rates. We compare the DMSP estimates of electrification rates with reported rates for 86 countries published for year 2005 by the International Energy Agency (IEA, 2006). Finally we discuss possible sources of error and ideas for improvements.

15.2 Methods

15.2.1 Data sources

The two primary data sources for this study are DMSP night-time lights and gridded population count. Both the night-time lights and population grid were from year 2006. National level reference data on the extent of electrification were drawn from the International Energy Agency's World Energy Outlook (WEO) 2006.

The DMSP-OLS visible band was designed to enable the detection of moonlit clouds at night in the visible band. A photomultiplier tube is used to intensify the visible band signal by about a million fold. This enables the detection of moonlit clouds and lighting present at the Earth's surface. NGDC has developed a capability to make cloud-free composites of the night-time visible band OLS data (Elvidge *et al.*, 2001). Additional procedures are used to remove ephemeral lights (mostly fires) and background noise to produce gridded stable lights products.

There are several gridded population products available (Fig. 15.2). We have found the US Department of Energy Landscan data (Dobson *et al.*, 2000; Bhaduri *et al.*, 2002) to be the most compatible with the DMSP night-time lights. Both are produced in a geographic projection with the same 30 arcsecond grid resolution. Also, the recent Landscan products have not used night-time lights as an input, thus there is not circularity in using the two datasets. The Landscan data are spatial allocations of census reported population numbers based on models developed using three satellite derived data sources: (1) NASA MODIS land cover, (2) the topographic data from the Shuttle Radar Topography Mission (SRTM), and (3) high resolution outlines of human settlements derived from the Controlled Image Base (CIB) from the US National Geospatial Intelligence Agency (NGA). Landscan data are referred to as population count instead of population density, which is based on residence. On a population density grid commercial centers and airports have very low numbers, despite the fact that there are substantial numbers of people present during certain hours. Landscan attempts to represent the spatial distribution of population based on person hours. Thus population is distributed across residential, commercial, industrial and public areas such as airports and schools.

The IEA has been compiling and reporting on electrification rates since 2002 in a publication series titled "World Energy Outlook". They admit there is no internationally accepted definition for electric power access and no standard method for collecting such data. Their objective has been to report the percentage of the population has access to electricity in their home. Data are collected from various sources, ranging from government agencies, international development programs and energy research organizations. Where the country reported data appeared contradictory, out of date, or unreliable the IEA reports estimates based on consideration of data from similar countries, earlier surveys, data from the international organizations, and journal articles.

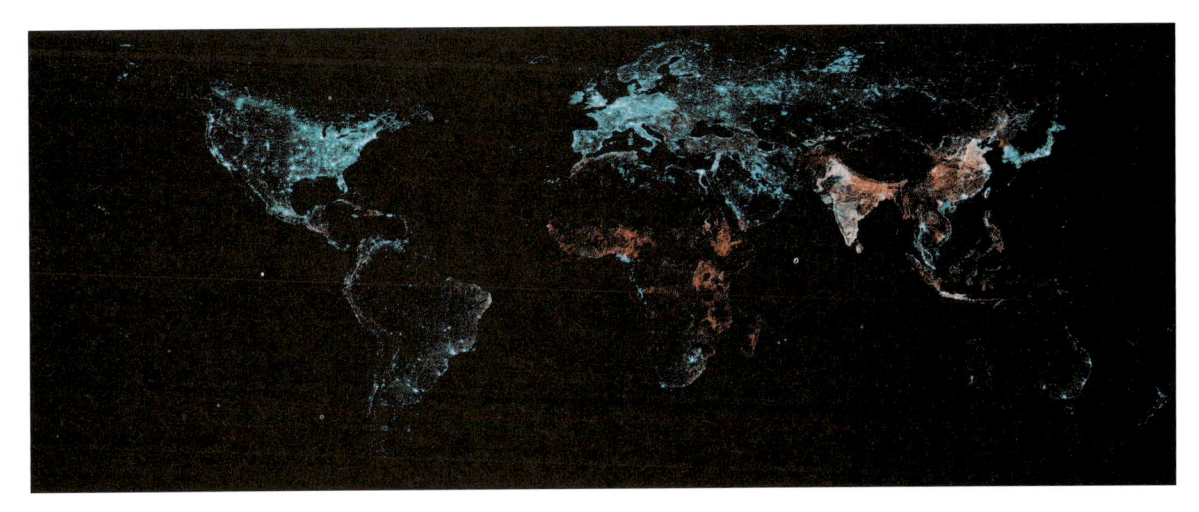

FIGURE 15.1 Color composite image made with the 2006 Landscan population grid as red and the 2006 DMSP night-time lights as green and blue. In the developed world there is an abundance of lighting yielding a cyan color. Red areas indicate dense population with no detectable DMSP lighting.

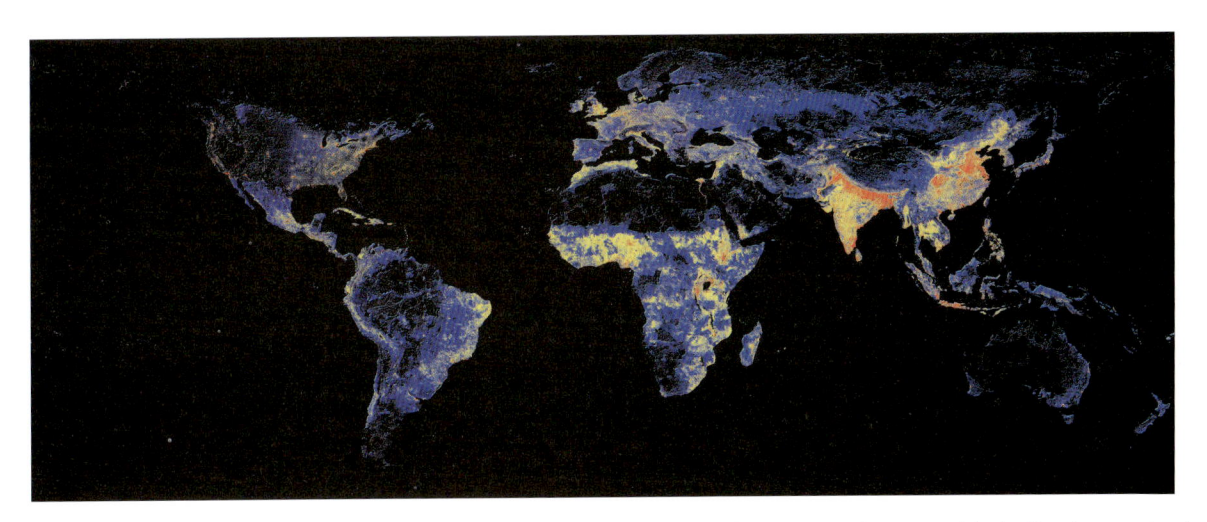

FIGURE 15.2 Landscan population count. Uninhabited areas with population counts of zero are black. Rural areas having population counts ranging from 1–10 are blue. Suburban and densely populated rural areas with population counts ranging from 11 to 99 are yellow. Red areas have population counts of 100 or more per grid cell.

15.2.2 Data processing

While the fires and background noise were removed in the production of the stable lights for 2006 – the gas flares are still present. To avoid overestimating electrification rates in countries with substantial numbers of gas flares, areas lit by gas flares were masked out and not used in the analysis. The locations of gas flares in the DMSP night-time lights had already been determined in consultation with high resolution imagery available in Google Earth (Elvidge *et al.*, 2009b). The remaining lights are all deemed to be from electric lighting. A binary mask was generated for the areas lit by the presence of gas flares. The gas flare mask was applied to the Landscan grid to zero out the population count in areas lit by gas flares. A second mask was produced for the remaining lights. This mask was used to divide the gas flare free population grid into two segments: (A) population with lighting detected (Fig. 15.3), and (B) population with no lighting detected (Fig. 15.4). The percent electrification rate is then calculated as:

$$\frac{\text{Population with DMSP lighting (A)}}{\text{Total population (A + B)}} \times 100$$

The analysis was conducted at both a national and subnational level.

15.3 Results

Using the data shown in Figs 15.2–15.4 we estimated the electrification rates for 232 countries – listed in descending population order in Table 15.1. The national level DMSP estimates are represented in map form in Fig. 15.5. The total number of people found to be without electricity is 1.62 billion, only 2.5% larger than the 1.58 billion estimated by the IEA. The IEA estimates are

FIGURE 15.3 Landscan population count in areas with DMSP detected lighting. The color coding is the same as Fig. 15.2.

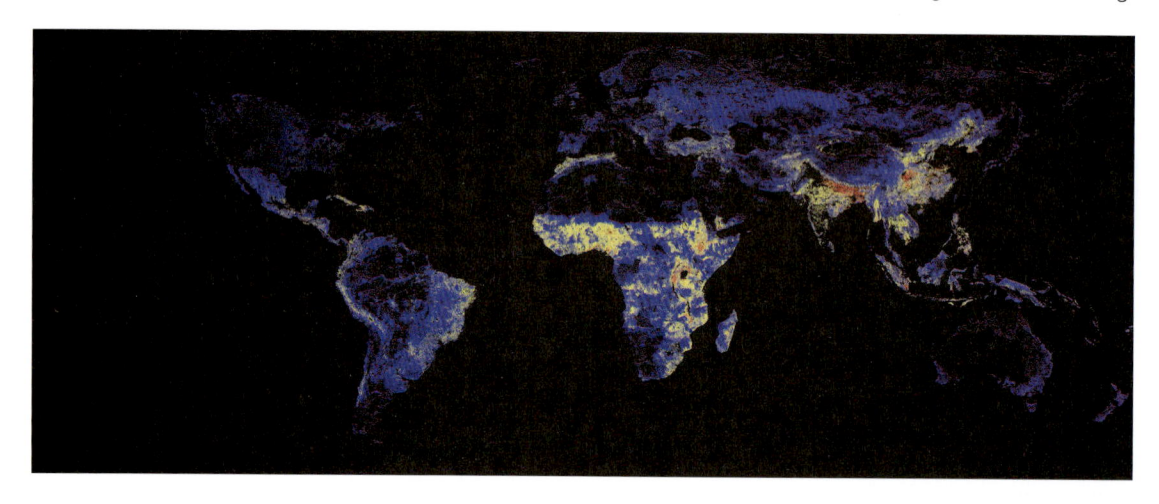

FIGURE 15.4 Landscan population count in areas with no DMSP detected lighting. The color coding is the same as Fig. 15.2.

listed in the third data column in Table 15.1 and are shown in map form in Fig. 15.6. Because the Landscan data are disaggregated it is possible to estimate electrification rates at the subnational level (Fig. 15.7) or at user defined spatial aggregations.

15.4 Discussion

Figure 15.8 compares the DMSP estimated and IEA reported electrification rates. Overall, there is general agreement between the DMSP and IEA electrification rate estimates. Developed countries with near 100% electrification rates yielded DMSP electrification rates ranging from 98 to 100% (Table 15.2). The countries having DMSP estimated electrification rates less than 20% (Table 15.3) are countries long recognized among the poorest on Earth.

However, it is possible to identify cases where the two estimates differ substantially. Table 15.4 lists the top 10 countries where the IEA reported electrification rate is higher than the DMSP estimate. Leading here are Thailand, China and Cuba, each with more than a 20% difference between the two numbers. Countries having 10–20% higher electrification rates reported by the IEA include Brazil, Philippines, Paraguay, Mongolia, Chile, Cameroon and Algeria. We do not know the source of the discrepancies.

The IEA estimates China has a 99.4% electrification rate, citing the Chinese Ministry of Science and Technology and the US Department of Energy National Renewable Energy Laboratory. In contrast, the DMSP estimated electrification rate is 75.6%. Thus, the DMSP estimate identified 320 million more people without electricity in China than the IEA had reported. That is more than the entire population of the USA! A large portion of the Chinese population identified to be without electricity are in Sichuan Province and surrounding provinces in the interior south-central China, known to be amongst the poorest regions of China. It is possible that the IEA reported electrification rate for China is valid in the wealthy coastal areas and underestimates the lack of electric power access in less wealthy the interior regions. Or it may be that the definition being used to define "access to electricity" is so broad that it encompasses 99.4% of the Chinese population. Another possibility is that the electrification rate is indeed high, but use of outdoor lighting is so sparse in some regions that the DMSP sensor is unable to detect the lighting. Similar possibilities exist for the other countries listed on Table 15.4.

Table 15.5 lists the top 11 countries where the DMSP estimates exceed the IEA reported electrification rates. Leading the list is

TABLE 15.1 Estimates of national electrification rates.

Country	Landscan 2006 population	DMSP estimated electrification rate (%)	IEA reported electrification rate (%)
China	1 308 905 728	74.9	99.4
India	1 104 764 800	75.8	55.5
United States	291 958 400	99.0	
Indonesia	223 445 600	79.5	54
Brazil	181 723 136	79.8	96.5
Pakistan	165 333 376	90.8	54
Bangladesh	146 274 784	53.0	32
Russia	137 334 752	86.5	
Nigeria	131 131 568	42.0	46
Japan	121 929 464	99.5	
Mexico	106 107 432	93.7	
Philippines	84 165 344	62.8	80.5
Vietnam	82 873 472	80.3	84.2
Germany	82 284 928	98.3	
Egypt	78 002 176	100.0	98
Ethiopia	74 580 856	12.4	15
Turkey	68 341 640	82.9	
Iran	64 260 172	94.9	97.3
Thailand	64 074 048	70.2	99
Congo, DRC	62 137 408	23.4	5.8
France	59 562 360	97.7	
United Kingdom	59 185 168	99.2	
Italy	56 513 852	99.4	
South Korea	46 776 532	100.0	
Ukraine	46 517 576	83.8	
Myanmar	46 174 136	26.2	11.3
South Africa	45 957 312	74.2	70
Colombia	43 065 580	82.0	86.1
Sudan	41 005 056	33.4	30
Argentina	39 522 200	88.1	95.4
Spain	39 451 484	97.0	
Poland	38 388 900	96.4	
Tanzania	37 158 680	17.7	11
Kenya	35 813 744	29.0	14
Canada	32 498 608	97.2	
Algeria	32 268 142	87.9	98.1
Morocco	32 187 260	77.6	85.1
Afghanistan	31 032 188	29.5	7
Uganda	29 390 972	15.2	8.9
Nepal	28 748 452	30.1	33
Peru	28 113 022	69.0	72.3

TABLE 15.1 (continued).

Country	Landscan 2006 population	DMSP estimated electrification rate (%)	IEA reported electrification rate (%)
Uzbekistan	27 253 948	94.5	
Iraq	26 810 654	93.0	15
Saudi Arabia	26 496 632	99.1	96.7
Venezuela	24 973 556	94.5	98.6
Malaysia	23 057 040	90.2	97.8
North Korea	22 510 660	36.8	22
Ghana	22 411 932	46.2	49.2
Romania	22 266 808	86.7	
Yemen	21 162 504	49.3	36.2
Mozambique	20 290 134	21.7	6.3
Sri Lanka	20 229 058	95.3	66
Australia	19 666 616	92.1	
Madagascar	18 730 568	14.5	15
Syria	18 670 416	95.5	90
Cameroon	17 516 510	36.3	47
Côte d'Ivoire	17 016 192	54.9	50
Netherlands	16 320 233	100.0	
Chile	15 528 649	87.2	98.6
Kazakhstan	15 366 266	73.0	
Burkina Faso	13 915 535	15.7	7
Cambodia	13 900 239	15.8	20.1
Malawi	13 240 124	18.4	7
Ecuador	13 057 767	83.7	90.3
Niger	12 536 623	19.8	
Guatemala	12 447 853	79.3	78.6
Zimbabwe	12 220 747	34.1	34
Senegal	12 128 555	48.1	33
Angola	11 902 221	29.8	15
Mali	11 707 643	22.5	
Zambia	11 336 282	34.7	19
Cuba	11 227 785	76.2	95.8
Serbia & Montenegro	10 787 706	90.3	
Belgium	10 431 084	100.0	
Portugal	10 328 523	98.2	
Czech Republic	10 250 290	99.5	
Greece	10 137 542	96.9	
Chad	10 059 864	18.4	
Hungary	9 981 097	95.0	
Belarus	9 762 124	76.8	
Tunisia	9 752 671	91.9	98.9
Rwanda	9 629 262	13.4	

TABLE 15.1 (*continued*).

Country	Landscan 2006 population	DMSP estimated electrification rate (%)	IEA reported electrification rate (%)
Guinea	9 362 449	20.9	
Dominican Republic	9 090 900	88.6	92.5
Bolivia	8 985 352	66.3	64.4
Somalia	8 729 994	25.1	
Sweden	8 441 214	98.1	
Burundi	8 245 397	7.7	
Austria	8 171 036	98.9	
Azerbaijan	8 046 536	81.3	
Benin	7 919 077	36.4	22
Haiti	7 808 419	28.0	36
Switzerland	7 629 250	99.6	
Bulgaria	7 319 524	92.2	
Honduras	7 182 415	71.3	61.9
Tajikistan	6 941 439	87.1	
El Salvador	6 778 252	92.8	79.5
Paraguay	6 507 533	70.2	85.8
Laos	6 389 236	22.0	
Israel	6 151 147	99.9	96.6
Sierra Leone	5 921 899	24.9	
Jordan	5 893 604	97.5	99.9
Libya	5 827 906	97.6	97
Nicaragua	5 527 336	60.5	69.3
Togo	5 522 532	32.9	17
Slovakia	5 459 714	96.1	
Papua New Guinea	5 210 907	17.9	
Denmark	5 157 683	96.3	
Kyrgyzstan	5 099 656	86.2	
Turkmenistan	5 090 828	86.3	
Finland	5 089 830	94.6	
Eritrea	4 723 351	26.9	20.2
Georgia	4 559 424	74.1	
Bosnia & Herzegovina	4 476 640	84.9	
Central African Republic	4 300 050	25.2	
Croatia	4 299 097	96.5	
Singapore	4 206 270	100.0	100
Moldova	4 176 313	77.9	
Norway	4 098 739	88.1	
Costa Rica	4 067 027	93.0	98.5
Ireland	3 937 104	94.5	
Puerto Rico	3 824 000	100.0	
New Zealand	3 801 472	87.0	

TABLE 15.1 (continued).

Country	Landscan 2006 population	DMSP estimated electrification rate (%)	IEA reported electrification rate (%)
Congo	3 752 949	60.7	19.5
Lithuania	3 603 037	75.6	
Albania	3 471 578	73.0	
Lebanon	3 458 243	99.6	99.9
Uruguay	3 424 161	90.0	95.4
Mauritania	3 197 154	36.2	
Panama	3 134 964	80.4	85.2
Armenia	2 985 608	87.5	
Oman	2 931 608	97.3	95.5
Liberia	2 923 682	30.0	
Mongolia	2 826 511	52.2	64.6
Jamaica	2 631 911	99.8	87.3
West Bank	2 517 877	100.0	
United Arab Emirates	2 419 677	99.9	91.9
Bhutan	2 327 470	33.6	
Latvia	2 204 224	75.6	
Macedonia	2 059 063	91.8	
Kuwait	2 011 370	100.0	100
Namibia	1 984 458	42.8	34
Slovenia	1 984 024	97.1	
Lesotho	1 974 952	34.8	11
Botswana	1 640 102	53.7	38.5
The Gambia	1 599 751	44.4	
Guinea-Bissau	1 407 918	28.5	
Gabon	1 358 236	69.0	47.9
Gaza Strip	1 305 448	100.0	
Estonia	1 277 632	84.7	
Mauritius	1 219 407	100.0	93.6
Swaziland	1 123 708	47.1	
Timor Leste	1 027 709	9.0	
Trinidad & Tobago	929 390	99.5	99.1
Qatar	847 314	100.0	70.5
Fiji	786 719	55.1	
Reunion	754 849	99.8	
Cyprus	753 955	99.4	
Guyana	719 444	68.6	
Comoros	600 507	33.3	
Bahrain	581 127	100.0	99
Luxembourg	483 098	100.0	
Suriname	461 789	82.8	
Equatorial Guinea	435 168	23.3	

TABLE 15.1 (*continued*).

Country	Landscan 2006 population	DMSP estimated electrification rate (%)	IEA reported electrification rate (%)
Martinique	426 392	100.0	
Guadeloupe	381 306	100.0	
Cape Verde	376 338	80.4	
Malta	368 860	100.0	
Western Sahara	331 567	92.1	
Brunei	303 062	96.6	99.2
Solomon Is.	293 641	0.5	
Belize	286 146	78.8	
The Bahamas	271 502	94.3	
Barbados	261 993	100.0	
Iceland	243 255	87.9	
Djibouti	203 931	59.4	
New Caledonia	199 394	67.1	
French Polynesia	182 166	96.9	
Sao Tome & Principe	179 070	61.7	
Netherlands Antilles	177 571	99.7	99.6
Mayotte	170 490	99.7	
St. Lucia	164 331	100.0	
Guam	158 857	100.0	
French Guiana	155 574	79.8	
Vanuatu	148 457	15.0	
Samoa	139 479	49.6	
Virgin Is.	98 686	100.0	
St. Vincent & the Grenadines	92 502	99.2	
Jersey	88 366	100.0	
Grenada	75 259	100.0	
Northern Mariana Is.	73 153	99.9	
Aruba	71 740	100.0	
Andorra	71 447	98.8	
Isle of Man	68 639	98.5	
Seychelles	68 493	100.0	
Antigua & Barbuda	64 016	99.8	
Tonga	57 901	85.4	
Dominica	57 503	90.6	
Guernsey	56 408	100.0	
American Samoa	50 729	100.0	
Faroe Is.	37 755	73.5	
Monaco	37 046	100.0	
Micronesia	35 739	79.6	
Liechtenstein	34 768	100.0	
Greenland	33 363	76.1	

TABLE 15.1 (continued).

Country	Landscan 2006 population	DMSP estimated electrification rate (%)	IEA reported electrification rate (%)
St. Kitts & Nevis	28 666	100.0	
San Marino	25 918	100.0	
Cayman Is.	25 902	100.0	
British Virgin Is.	19 578	100.0	
Wallis & Futuna	13 886	11.9	
Anguilla	13 201	100.0	
Palau	13 076	76.5	
Nauru	12 283	100.0	
Cook Is.	11 093	90.9	
Bermuda	9 391	100.0	
Turks & Caicos Is.	7 822	93.0	
Montserrat	7 638	73.8	
St. Pierre & Miquelon	6 793	98.9	
St. Helena	6 006	69.0	
Falkland Is.	2 974	76.7	
Kiribati	2 462	2.6	
Gibraltar	2 195	100.0	
Niue	2 069	61.5	
Tuvalu	1 378	1.4	
Norfolk I.	1 170	95.6	
Maldives	405	53.6	
Christmas I.	359	100.0	
Marshall Is.	347	0.0	
Vatican City	283	100.0	
Cocos Is.	233	72.5	
Pitcairn Is.	11	0.0	
Jarvis I.	2	0.0	

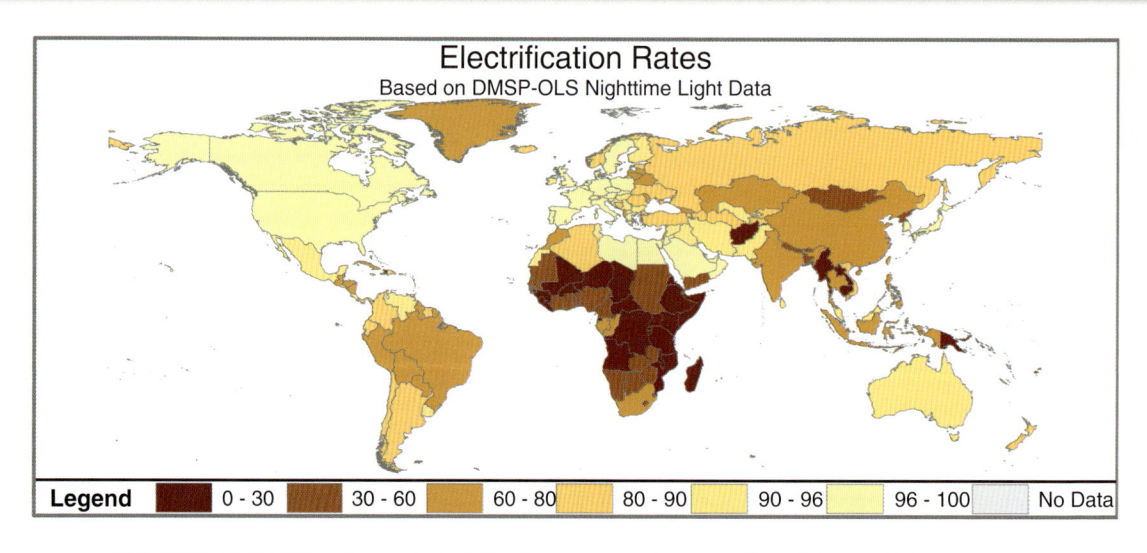

FIGURE 15.5 DMSP estimated electrification rates for the countries of the world for year 2006.

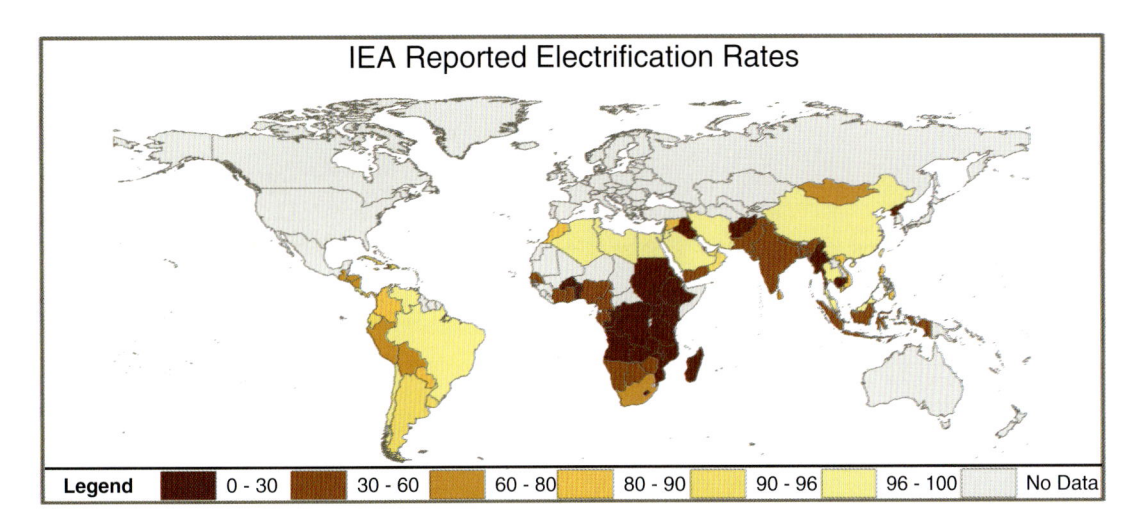

FIGURE 15.6 Electrification rates published by the IEA for year 2005.

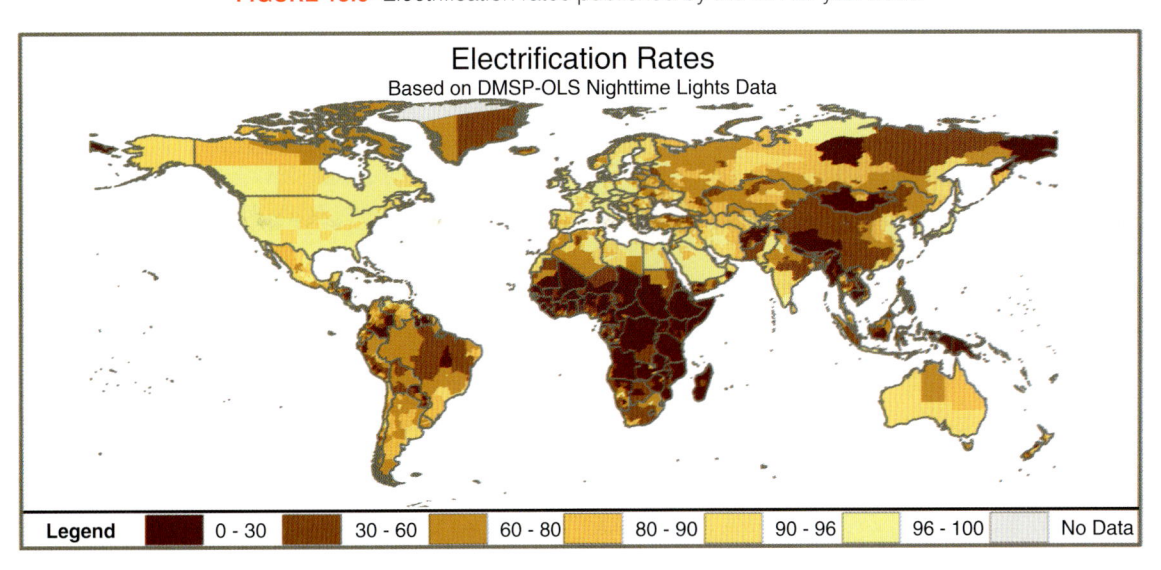

FIGURE 15.7 DMSP estimated electrification rates for primary subnational units (states and provinces) for year 2006.

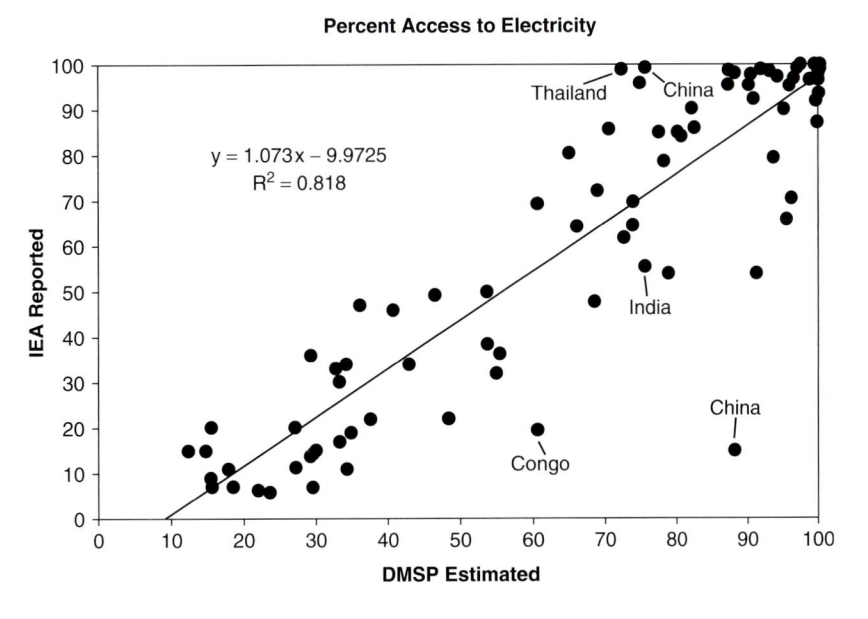

$$y = 1.073x - 9.9725$$
$$R^2 = 0.818$$

FIGURE 15.8 DMSP versus IEA estimates of national electrification rates.

TABLE 15.2 Countries where electrification rates are expected to be near 100%.

Country	DMSP Estimated (%)
Singapore	100.0
Netherlands	100.0
Belgium	100.0
South Korea	100.0
Switzerland	99.6
Czech Republic	99.5
United Arab Emirates	99.6
Japan	99.5
Italy	99.4
United Kingdom	99.2
Saudi Arabia	99.1
United States	99.0
Austria	98.9
Germany	98.3
Portugal	98.2
Sweden	98.1
France	97.7

TABLE 15.3 Countries with populations over a million and DMSP estimated electrification rates under 20%.

Country	DMSP Estimated (%)	IEA Reported %
Niger	19.8	
Malawi	18.4	7
Chad	18.4	
Papua New Guinea	17.9	
Tanzania	17.7	11
Cambodia	15.8	20.1
Burkina Faso	15.7	7
Uganda	15.2	8.9
Madagascar	14.5	15
Rwanda	13.4	
Ethiopia	12.4	15
Timor Leste	9.0	
Burundi	7.7	

TABLE 15.4 Top 10 countries where the DMSP estimated electrification rate is lower than the IEA reported rate.

Country	DMSP Estimated (%)	IEA Reported (%)	Difference
Thailand	70.2	99	−28.8
China	74.9	99.4	−24.5
Cuba	76.2	95.8	−19.6
Philippines	62.8	80.5	−17.7
Brazil	79.8	96.5	−16.7
Paraguay	70.2	85.8	−15.6
Mongolia	52.2	64.6	−12.4
Chile	87.2	98.6	−11.4
Cameroon	36.3	47	−10.7
Algeria	87.9	98.1	−10.2

TABLE 15.5 Top 11 countries where the DMSP estimated electrification rate is higher than the IEA reported rate.

Country	DMSP Estimated (%)	IEA Reported (%)	Difference
Iraq	93.0	15	78.0
Congo	60.7	19.5	41.2
Pakistan	90.8	54	36.8
Qatar	100.0	70.5	29.5
Sri Lanka	95.3	66	29.3
Indonesia	79.5	54	25.5
Lesotho	34.8	11	23.8
Afghanistan	29.5	7	22.5
Gabon	69.0	47.9	21.1
Bangladesh	53.0	32	21.0
India	75.8	55.5	20.3

Iraq, for which the IEA estimated an electrification rate of 15% and the DMSP estimate was 88.1%. In WEO 2004 (IEA, 2004) the electrification rate for Iraq was reported as 94.5% for the year 2002. The number was revised down for 2005 based on an Iraq government report (COSIT, 2005), which includes results of a household survey regarding the stability of electric power service in the months following the US invasion in 2003. The IEA summarized the COSIT surveys, concluding that only 15% of Iraqi households had reliable access to electricity. The disparity between the two electrification rate estimates for Iraq can be attributed to the DMSP's ability to detect intermittent lighting over the course of a year.

Other countries where the DMSP estimated electrification rate exceeded the IEA reported value include Congo, Pakistan, Sri Lanka, Qatar, Indonesia, Lesotho, Bangladesh, Afghanistan, Gabon, and India. The Indian estimate from DMSP is 75.5% nearly matches the DMSP estimate for China (75.6%). As with China, the core of the DMSP identified population with no lighting detected are located in a heavily populated zone known as the poorest region of the nation, in this case the Ganges River Plain stretching from Delhi to Calcutta.

There are several possible sources of error in the DMSP estimates. There may be errors of omission, or undercounting of the population with access to electricity in rural areas where the outdoor lighting is not bright enough for DMSP detection. This is the major source of error in developed countries such as the USA, France and New Zealand, which are believed to have near 100% electrification but fall 1–2% short of this in the DMSP estimates (Table 15.2). This style of error may be larger in the developing countries that have lower electric power consumption levels, such as China and India. There may errors of commission, or overcounting of the population with access to electricity in areas that have street lighting and commercial lighting, yet no electric power access in a portion of the homes in the same pixel. Another source of discrepancy arises from homes with intermittent or sporadic electric power service. In the case of Iraq the IEA only tallied population with reliable electric power service in the estimation of the electrification rate. The DMSP data were processed to detect intermittent lighting, yielding a substantially higher estimate of the electrification rate. Finally, it should be noted that the DMSP electrification rate estimates were derived from areas that are devoid of lighting from gas flares. That is to say, in areas with onshore gas flares, the electrification rate has been estimated outside of the area lit by the gas flares. This includes portions of countries listed by Elvidge et al. (2009b), including Russia, Nigeria, Iran, Iraq, Algeria, Libya, and others.

Conclusion

We derived the first systematic global assessment of electrification rates by combining DMSP night-time lights with a population density grid. In this analysis, the electrification rate was estimated by tallying the population count in areas having DMSP lighting as compared to the total population. Using this technique we have a standardized product, with reporting for 232 countries and more than 2000 sub-national units. In contrast, the only other available reporting on international electrification rates comes from the International Energy Agency (IEA), which in 2006 reported electrification rates for 86 countries.

There are several potential areas for improvement in the estimation of electrification rates based on night-time lights. It would be good to place error bars on the estimated electrification rates, to have estimates of the errors of omission and commission, and to be able to rate the stability of the electric service. In most parts of the world the DMSP is able to collect twenty to fifty cloud-free dark night coverages in a year, which may be enough repeat coverages to assess the stability of electric power service (i.e., the ratio of lights detected versus lights not detected). The most serious constraint on these improvements is the current lack of validation data on electrification rates collected at a spatial resolution compatible with DMSP and Landscan (about 1 km^2 resolution). To address the problem with gas flares obscuring lights from small towns and villages the best solution would be to collect the night-time lights data at higher spatial resolution (Elvidge et al., 2007b).

One of the applications for the full resolution grid of the population count in areas without DMSP detected lighting is to identify areas of the world that could benefit from installation of sustainable solar and wind energy systems. In many of these areas, people are burning kerosene to produce subsistence levels of heat and lighting that cannot be detected by DMSP. Mills et al. (2005) have shown that liquid fuels are extremely inefficient and costly light sources. The only thing cheap about this approach to lighting is the cost of the lanterns. Given the emphasis being placed on reducing carbon emissions, Mills (2005) developed low cost photovoltaic panels and light emitting diode (LED) fixtures that enable families to produce light using locally generated electricity without the expense of extending the electric power grid. This approach has similarity to the rapid expansion of cell phone usage in places where the land line telephone system is antiquated and decrepit.

While there are some known sources of error in the current product, the method does provide electrification rates using a standardized definition and standardized data sources, with complete global coverage. We anticipate that there will be improvements to the night-time lights approach to estimating electrification rates. We also anticipate that night-time lights will be useful for detecting changes in electric power access. This could include both expansions and contractions in access to electric power.

Acknowledgments

One of the authors of this manuscript is an employee of UT-Battelle, LLC, under contract DE-AC05–00OR22725 with the US Department of Energy. Accordingly, the United States Government retains and the publisher, by accepting the article for publication, acknowledges that the United States Government retains a non-exclusive, paid-up, irrevocable, worldwide license to publish or reproduce the published form of this manuscript, or allow others to do so, for United States Government purposes.

References

Bhaduri, B., Bright, E. Coleman, P. and Dobson, J. (2002) LandScan: Locating people is what matters. *Geoinformatics*, **5**, 34–37.

COSIT (2005) *Iraq Living Conditions Survey 2004, Volume III: Socio-economic Atlas of Iraq*, Central Organization for Statistics and Information Technology, Ministry of Planning and Development Cooperation, Baghdad, Iraq.

Dobson, J., Bright, E.A., Coleman, P.R., Durfee R.C. and Worley, B.A. (2000) LandScan: a global population database for estimating populations at risk. *Photogrammetric Engineering and Remote Sensing*, **66**, 849–857.

Doll, C.N.H., Muller, J.-P. and Elvidge, C.D. (2000) Night-time imagery as a tool for global mapping of socio-economic parameters and greenhouse gas emissions. *Ambio*, **29**, 157–162.

Ebener, S., Murray, C., Tandon A. and Elvidge, C. (2005) From wealth to health: modeling the distribution of income per capita at the sub-national level using nighttime lights imagery. *International Journal of Health Geographics*, **4**, 5–11.

Elvidge, C.D., Imhoff, M.L., Baugh, K.E. et al. (2001) Night-time lights of the world: 1994–1995. *ISPRS Journal of Photogrammetry & Remote Sensing*, **56**, 81–99.

Elvidge, C.D., Tuttle, B.T., Sutton,, P.C, et al. (2007a) Global distribution and density of constructed impervious surfaces. *Sensors*, **7**, 962–1979.

Elvidge, C.D., Cinzano, P., Pettit, D.R *et al.* (2007b) The Nightsat mission concept, *International Journal of Remote Sensing*, **28** (12), 2645–2670.

Elvidge, C.D., Sutton, P.C., Ghosh, T. *et al.* (2009a) A global poverty map derived from satellite data, *Computers and Geosciences*, **35**, 1652–1660.

Elvidge, C.D., Ziskin, D., Baugh, K.E. *et al.* (2009b) A fifteen year record of global natural gas flaring derived from satellite data. *Energies*, **2** (3), 595–622.

Ghosh, T., Anderson, S., Powell, R.L., Sutton, P.C. and Elvidge, C.D. (2009) Estimation of Mexico's informal economy and remittances using nighttime imagery, *Remote Sensing*, **1** (3), 418–444.

International Energy Agency (2004) *World Energy Outlook*, Appendix to Chapter 10: Electrification Tables. IEA, Paris.

International Energy Agency (2006) *World Energy Outlook*, Appendix B Electricity Access. IEA, Paris.

Mills, E. (2005) The specter of fuel-based lighting. *Science*, **308** (5726), 1263–1264.

Sutton, P.C., Anderson, S.J., Elvidge, C.D., Tuttle, B.T. and Ghosh, T. (2009) Paving the planet: impervious surface as proxy measure of the human ecological footprint. *Progress in Physical Geography*, **33** (4), 510–527.

16

Integrating remote sensing and GIS for environmental justice research

Jeremy Mennis

Environmental justice concerns the rights of all persons to live in a clean and safe environment and to have the ability to participate in environmental decision-making in their community. Satellite remote sensing can provide valuable information on the spatial distribution of environmental hazards and amenities, as well as population, for environmental justice research. Geographic information systems (GIS) play a key role in integrating remotely sensed and other spatial data, and in quantifying and analyzing spatial patterns of environmental hazards and amenities, as well as demographic character. A case study is presented that investigates the relationship between vegetation, as measured by Normalized Difference Vegetation Index (NDVI) data derived from Landsat imagery, and indicators of race and socioeconomic status in Philadelphia, Pennsylvania, USA. Ordinary least squares (OLS) and spatial econometric regression demonstrate that healthy green vegetation is associated with wealth and white population, though these relationships are explained in part by urban form. Future research in utilizing remote sensing in environmental justice research should incorporate advances in the spatial and spectral resolution of remotely sensed data, as well as address the use of remote sensing to analyze the distribution of environmental amenities and natural hazards.

16.1 Introduction

Environmental justice concerns the rights of all persons to live in a clean and safe environment and to have the ability to participate in environmental decision-making in their community (Bullard, 1996; Liu, 2001; EPA, 2003). The issue of environmental justice has gained increasingly widespread attention from government, industry, and academic sectors over the past two decades as civil rights groups have begun to recognize that racial inequality in the distribution of environmental risk can, indeed, be considered a civil rights issue. Analogously, political activists concerned with environmental issues have recognized that issues of race, class, and discrimination are intimately connected to environmental policy (Bullard, 1990).

The issue of environmental justice is inherently spatial in nature. One of the central questions of environmental justice research concerns describing, and understanding the causal mechanisms behind, the spatial coincidence among patterns of environmental hazards and demographic characteristics. Because satellite remote sensing provides valuable information on the spatial distribution of environmental characteristics, as well as population, remote sensing can play a valuable role in environmental justice research. Likewise, geographic information systems (GIS), as the primary software tool for handling spatial data, plays a key role in integrating remotely sensed and other spatial data, and in quantifying and analyzing spatial patterns of environmental hazards and amenities, as well as demographic character.

The objective of this chapter is to describe the role of environmental remote sensing and GIS in environmental justice research. The following section serves as an introduction to the central principles and issues in environmental justice research. The next section addresses the primary ways in which remote sensing is used in environmental justice research. The role of GIS in integrating remotely sensed and other spatial data for environmental justice analysis is described in the following section. The chapter concludes with an environmental justice case study focusing on environmental amenities in Philadelphia, Pennsylvania.

16.2 Environmental justice research

Recent academic interest in environmental justice is often traced to a series of catalyzing events in the early and middle 1980s (see McGurty, 2007). In 1982, African American residents of Warren County, North Carolina protested about the siting of a landfill built to hold polychlorinated biphenyl (PCB)-contaminated soil, claiming that the siting of the landfill in an African American community was a violation of civil rights. The protests drew the attention of the national news media as well as the National Association for the Advancement of Colored People (NAACP), who initiated legal action to prevent the development of the landfill. Although the landfill was developed, these events spurred the US General Accounting Office (GAO), the research arm of the US Congress, to investigate racial disparities in environmental hazards, which found that three out of the four largest hazardous waste treatment facilities in the Southeast United States were located in minority communities (GAO, 1983).

Members of the United Church of Christ (UCC) had been involved in the Warren County protests. In 1987, the UCC's Commission for Racial Justice (CRJ) published a national-level study of the racial composition of communities hosting hazardous waste sites which found widespread racial inequity in their distribution (CRJ, 1987). Several other academic studies soon followed (e.g. Bullard, 1990; Mohai and Bryant, 1992; Hird, 1993) providing additional, though often mixed, evidence of the relationship between race and environmental hazards. Reflecting the increasing awareness of environmental justice issues, in 1993 the US Environmental Protection Agency (EPA) established the National Environmental Justice Advisory Council (NEJAC). This was followed by Executive Order 12 898, signed by President William J. Clinton in 1994 (Clinton, 1994), which instructs federal agencies to adopt environmental policies regarding the siting of hazardous facilities and environmental decision-making that do not discriminate based on race.

Terminology concerning issues in environmental justice has been sharply contested. The term 'environmental racism' predominated following the 1982 Warren County protests, and connotes the relationship between environmental risk and the intentional racial discrimination that was the focus of the civil rights movement. However, this term was criticized for implying the presence of intentional racial discrimination when merely the association between environmental hazards and minorities may be identified. The term 'environmental equity' was coined to address this issue, where racial disparities in the burden of environmental risk are said to provide evidence of environmental inequity. The term 'environmental justice' is typically used to describe the activist movement or the academic field which concern themselves with issues of the relationship between race and environmental hazards. Remote sensing and GIS technologies can be used to analyze the spatial relationship between the spatial distributions of certain socioeconomic characteristics and environmental amenities and risks. Thus, they are key tools in environmental equity analyses, and inform the field of environmental justice.

The most common quantitative approach to analyzing environmental equity involves gathering data on the location or spatial distribution of a particular type of environmental hazard, as well as spatial data on socioeconomic characteristics, often acquired from census data. Statistical analysis, often employing a form of regression where a measure of the degree of hazard composes the dependent variable, is then used to test for associations between the degree of hazard and various socioeconomic characteristics, typically while controlling for other non-socioeconomic factors that one theorizes influence the location of the hazard.

A wide variety of research at the national and metropolitan scope in the United States has demonstrated substantial environmental inequity regarding a variety of hazards associated with industrial activity, including hazardous waste facilities (treatment storage and disposal facilities – TSDFs) (Goldman and Fitton, 1994), facilities releasing toxins to the environment that are listed in the EPAs Toxic Release Inventory (TRI) database (Mennis, 2005), and superfund sites (O'Neil, 2007). Other researchers have found racial inequity in the distribution of other types of hazards, such as ambient air pollution (Jerrett et al., 2001), those associated with transportation (Duthie, Cervenka and Walker, 2007), and industrial agriculture (Taquino, Parisi and Gill, 2002). It should be noted that a number of studies have also found no evidence, or inconclusive evidence, of widespread inequity in various types of environmental hazards (Atlas, 2002; Hird, 1993;

Anderton *et al.*, 1994; Bowen *et al.*, 1995). However, the vast majority of academic environmental justice research supports the notion that the burden of environmental risk in urban areas from pollution and waste management falls disproportionately on minorities and the poor.

A number of authors have offered critiques of environmental justice research methods (Zimmerman, 1994; Bowen and Wells, 2002; Maantay, 2002), perhaps the most fundamental of which concerns the issue of causation. Clearly, simply because there is evidence that there is an association between the spatial distribution of race and environmental hazard does not necessarily imply that the association is the direct result of acts of intentional discrimination by environmental decision-makers or others. This argument was embodied in the debate over competing explanations of "race versus class" for explaining patterns of environmental inequity (Cutter, 1995; Downey, 1998). Here, the issue concerns which socioeconomic characteristics may be the actual drivers of the relationship of socioeconomic character with environmental hazard, when collinearity among race, ethnicity, and indicators of class (e.g. poverty, income, and educational attainment) occurs. Many authors have noted that it is impossible to disentangle the effects of race and class, as these characteristics have been so closely woven together in the historical processes of industrial development and residential and labor segregation that have produced the patterns of environmental inequity that may be currently observed (Pulido, 2000). Several longitudinal quantitative studies, as well as a handful of qualitative studies employing archival research, have attempted to address such processes, with mixed evidence of intentional discrimination (Boone and Modarres, 1999).

Szasz and Meuser (1997) provide a helpful typology of explanations for environmental inequity. The typology distinguishes between a situation in which a hazardous facility location is chosen because of demographics and a situation in which a facility location is chosen for reasons other than demographics. In the former case, a location for a facility may be chosen because of the economic benefits associated with facility development, because of political disempowerment of the targeted community, or because of outright racial prejudice on the part of facility developers or policy makers. In the case where a facility is sited for reasons other than demographics, the choice of facility location may be based on the availability of inexpensive, industrial land that (coincidentally or not) coincides with socioeconomic disadvantage. Additionally, a facility may be initially located in a community that is not socioeconomically disadvantaged or in a relatively uninhabited area, and socioeconomic disadvantage proximate to the facility increases subsequent to the facility being built.

Other critiques of environmental justice research concern methodology. First, the measurement of environmental risk is subjective, in that a researcher must decide what constitutes risk and how is the magnitude of that risk calculated. Early environmental justice studies simply considered the presence versus the absence of a hazard, say, a TSDF, as a proxy for risk. This approach was criticized for being too blunt to capture magnitude of actual risk. More recently, environmental justice researchers have considered more sophisticated measures of hazard, such as the distance from a hazardous facility (Mennis, 2002), the density of hazardous facilities (Downey, 2003), the toxicity and volume of toxic chemicals released (Sadd *et al.*, 1999), measures of ambient pollution gathered from sensor networks (Jerret *et al.*, 2001), and distributions of hazards derived from computational simulations of diffusion of toxic materials through various media (Fisher, Kelly and Romm, 2005).

Another methodological challenge concerns the quantification of the relationship between environmental hazard and socioeconomic character. Since environmental data and socioeconomic data typically take different forms, with census-based socioeconomic data often only made available in aggregated form, data integration is necessary to develop case-based flat files of observations used for statistical analysis. This challenge has emerged in the debate over the appropriate "scale of analysis" for environmental justice research, which is perhaps better formulated as a debate over the appropriate analytical spatial resolution. Researchers have argued for instance, whether US zip codes, US Census tracts, or US Census block groups provide more accurate measures for capturing the demographic character of the neighborhood surrounding a hazardous facility (Williams, 1999).

The choice of scale-of-analysis is related to what is often referred to as the modifiable areal unit problem (MAUP) (Openshaw, 1984), which concerns the fact that different ways of spatially aggregating punctiform data influences statistical results. The MAUP takes two forms. The first may be considered an issue of scale, where, for instance the statistical results of analyzing population data at the unit of the US Census block group will differ from the results of an analysis of the same raw data aggregated to the US Census tract level (where many block groups spatially nest within a single tract). The second form is an issue of partitioning, where the same dataset, partitioned into different regions for the purpose of spatial aggregation, may yield different statistical results, even when each partitioning scheme uses the same number of regions (i.e. is roughly the same scale of analysis) (Fotheringham and Wong, 1991).

Other authors have criticized environmental justice research for its use of statistical methods that do not account for the spatial nature of most environmental justice data. Ordinary least squares (OLS) regression, for example, produces biased parameter estimates in the presence of spatial autocorrelation in the model residuals. Because socioeconomic and environmental characteristics tend to exhibit strong spatial dependency, models of environmental equity often violate the assumptions of OLS regression and other conventional statistics. A number of environmental justice researchers have addressed this issue by incorporating statistical methods adapted for spatial data, such as spatial econometric modeling (Buzzelli *et al.*, 2003) and geographically weighted regression (Mennis and Jordan, 2005).

16.3 Remote sensing for environmental equity analysis

The role of environmental remote sensing in environmental justice research is two-fold. First, remote sensing may be used to quantify environmental characteristics of the earth surface, including environmental hazards and environmental amenities, so that the locations of such hazards and amenities may be compared to distributions of demographic characteristics. Typically, this application of remote sensing requires medium to high resolution imagery (e.g. 1 m–30 m) that indicates land use/land cover, vegetation concentration, tree cover, or other environmental characteristics. Second, remote sensing is used to develop estimates of population over small-areas to facilitate the comparison of environmental hazard (or amenity) with demographic character. Here, remotely sensed imagery is incorporated into a dasymetric mapping or areal interpolation algorithm to derive fine-resolution population estimates where none exists, as in many developing countries. Alternatively, such an approach can be used to refine census data in many industrialized nations where spatial population data are readily available.

From the beginning of the academic interest in environmental justice, researchers have recognized that land use and land cover may be used as explanatory variables in statistical models of environmental equity. Though it is perhaps a tautology, researchers have employed independent variables that capture industrial land use/land cover to predict the locations of facilities that release air pollution and other hazards associated with industrial activity. The motivation for including such a land use variable goes to the early arguments about causation in environmental equity – is a hazardous facility located in a particular location because of purely demographic reasons, or because there is land available that suits the purpose of the facility (Been, 1994; Boer et al., 1997)? Land use/cover data products derived from remotely sensed data, such as industrial land cover data extracted from land cover maps derived from Landsat imagery, have been used in models of environmental equity (Mennis, 2005).

Another aspect of the use of remote sensing in environmental justice research concerns the measurement of environmental amenities. While much of the environmental justice literature has focused on technological hazards associated with landfills, air pollution, and toxic byproducts of industrial development, more recently researchers have begun to focus on the relationship of socioeconomic status with positive environmental characteristics (Boone, 2008). One of the primary environmental amenities upon which researchers have focused is green space, broadly defined, i.e. open space, parkland, and vegetation, particularly in urban areas, where access to non-residential, vegetated landscapes are valued for recreation, health, and general well-being (Comber, Brunsdon and Green, 2008; Geoghegan, 2002; Strife and Downey, 2009). Because remote sensing can play a key role in quantifying vegetation and identifying open space in urban environments (Small, 2001), researchers have utilized remotely sensed imagery to capture this type of environmental amenity for environmental justice studies.

Several studies linking socioeconomic status to vegetation have focused on assessment of quality of life, and have employed the Normalized Difference Vegetation Index (NDVI) as a measure of vegetation character (Tucker, 1979). The NDVI exploits the nature of healthy green vegetation to reflect relatively strongly in the near-infrared wavelengths (NIR) and weakly in the visible wavelengths (VIS), and can be considered a general measure of the amount or density of green vegetation, though it may also reflect other characteristics such as soil moisture content. It is calculated as the ratio (NIR-VIS)/(NIR+VIS), where higher values indicate higher vegetation concentration.

Lo and Faber (1997) combined NDVI derived from Landsat Thematic Mapper (TM) imagery with socioeconomic US Census data for Athens-Clarke County, Georgia. Using principle components analysis (PCA), these authors found that single dimensions of variation within the data captured associations between NDVI and socioeconomic characteristics. High NDVI was associated with high income, high educational attainment, and other indicators of socioeconomic advantage, as well as low population density. The authors concluded that NDVI could be used to capture a combined quality-of-life indicator that incorporates both biophysical and demographic characteristics.

In a similar study of the Denver, Colorado region, Mennis (2006) found analogous relationships between NDVI and US Census-derived socioeconomic characteristics in residential land. Here, higher NDVI was associated with several indicators of socioeconomic advantage, including higher educational attainment and lower population density, as well as a lower percentage of minority (i.e., non-white or Hispanic) residents. However, the relationship between vegetation and socioeconomic status was largely driven by certain types of developed land, including wealthy, older neighborhoods with mature vegetation as well as newer subdivision style developments with large lots and well-maintained lawns. The author notes a likely feedback effect between vegetation and socioeconomic status, such that concentrated vegetation raises housing values, thus prohibiting poorer segments of the population from buying into vegetated communities. At the same time, wealthier people are better able to create and maintain vegetated landscapes in a semiarid desert environment such as Denver.

NDVI is not the only indicator of vegetation used in environmental justice research. Li and Weng (2007) used spectral mixture analysis (SMA) to extract green vegetation and impervious land covers from a Landsat Enhanced Thematic Mapper Plus (ETM+) image to model quality of life in a study of Indianapolis, Indiana. Green vegetation was positively correlated with several US Census variables indicating socioeconomic advantage, including educational attainment, employment, and housing value, and was negatively correlated with poverty rate. Greater impervious surface, on the other hand, was associated with lower socioeconomic status.

Zhang, Tarrant and Green (2008) note that relationships between socioeconomic status and open space may vary among rural, suburban, and urban regions. In an analysis of Georgia, they stratify US Census block groups into rural, suburban, and urban regions and examine the relationship of a host of socioeconomic variables with proximity to federally managed open space land in rural areas. For quantifying open space in urban and suburban areas the authors use vegetation data contained within the 1992 National Land Cover Data (NLCD) data set (Vogelman et al., 2001), a digital land cover map derived from classification of Landsat imagery. Results indicate that Georgia residents living in close proximity to publicly managed land in rural areas, or in areas with high vegetation concentration in suburban and urban areas, are more likely to be wealthy, white, and have higher educational attainment.

In an analysis of Terre Haute, Indiana, Jensen et al. (2004) use Leaf Area Index (LAI) as a measure of vegetation. LAI is a ratio measure of the area of ground covered by leaves. LAI was captured in situ at points throughout the city as derived from below- and above-forest canopy measurements of photosynthetically active radiation (PAR). A continuous image of LAI for the entire study area was derived by creating an artificial neural network (ANN) model of LAI from green, red, and near-infrared reflectance values contained in an Advanced Spaceborne Thermal Emission Radiometer (ASTER) image. LAI and population density were used to model household income and housing value. These authors found that higher LAI and lower population density are associated with higher income and higher housing values.

Heynen, Perkins and Roy, (2006) used urban tree canopy data derived from aerial photography, US Census data capturing racial and other socioeconomic characteristics, and interviews with key managers of the urban forest in Milwaukee, Wisconsin. These authors find a racially inequitable distribution of tree cover, where whites are more likely to live in neighborhoods with more tree cover and Hispanics less so. However, these authors note that urban forest management has complex relationships with socioeconomic status, as the character of urban trees in socioeconomically distressed neighborhoods often differs substantially from those in more elite neighborhoods. Trees in poor neighborhoods, which often sprout up in unmanaged or, marginally managed, land may be considered a nuisance, while planted trees in managed landscapes add value to residential properties.

Another way that remote sensing may be employed for environmental justice analysis concerns the sensing of environmental hazards. The use of remote sensing for observing and planning for environmental hazards is well established, including for natural disasters such as hurricanes, volcanoes, flooding, and wildfires, among others (Hong, Adler and Verdin, 2007). While there is clear promise for using remote sensing to analyze the equity of the distribution of such environmental hazards with regards to socioeconomic character, researchers have only just begun to exploit remotely sensed data that capture hazards as a resource for environmental justice research. One particular hazard that has been addressed is extreme heat. Remote sensing researchers have analyzed the urban heat island effect for some time (Oke, 1982; Roth, Oke and Emery, 1989; Lo, Quattrochi and Luvell, 1997), and image products from sensors that record thermal emissivity can be integrated with socioeconomic data.

Both Lo and Faber (1997) and Li and Weng (2007) use temperature data derived from Landsat thermal band imagery and find that higher temperature is associated with indicators of socioeconomic disadvantage. This is not surprising, given that temperature is positively correlated with impervious surface and population density, lending supporting evidence that temperatures are higher in inner-city urban areas, which also tend to be home to the largest concentrations of the poor and minorities in the US. In an analysis of Tucson, Arizona, Harlan *et al.* (2006) found similar results, where the poor and minorities had a greater exposure to heat stress, as well as reduced access to green vegetation and open space. These populations also had fewer resources to mitigate the effects of extreme heat, such as air conditioning and swimming pools.

16.4 Integrating remotely sensed and other spatial data using GIS

One of the major methodological challenges concerning environmental justice concerns the technique used to assess the population at risk from a particular hazard. GIS is used to integrate various types of data and derive spatial relationships among environmental hazards, amenities, and population character (see Chakraborty and Armstrong, 1997). Typically the goal of the data integration is to produce a set of observations that include associated measurements of hazard and socioeconomic character that can be used in multivariate statistical analysis. The conventional approach has been to acquire socioeconomic data at a particular unit of data aggregation, typically as defined by the census organization or agency responsible for collecting the data, and aggregate (or disaggregate) other data of interest to the census units. For example, one might record the presence or absence of a hazardous facility within each US Census tract. Or, the total volume, weight, or toxicity of toxic chemicals released may be tallied for each spatial unit. Alternatively, one may define zones of degree of risk, and tally socioeconomic characteristics within each zone. These zones may be defined by, say, proximity to a hazardous facility, or by an area within a modeled plume of pollution from a point source. Such an approach typically requires disaggregation of socioeconomic data that are only available at spatial units incompatible with the spatial units that capture environmental hazard.

Approaches for disaggregation of population data are generally captured by methods of areal interpolation and dasymetric mapping (Eicher and Brewer, 2001; Mennis, 2003; Mennis and Hultgren, 2006). US Census data products, for example, typically model population and population character using a spatially exhaustive tessellation of spatial units, which assumes a homogeneous distribution of population within each unit. The simplest approach to disaggregation is simple areal weighting, where an area of overlap with the spatial unit is apportioned a percentage of the population of the unit proportional the percentage of the unit's area that is occupied by the overlap. In dasymetric mapping, an additional, ancillary, data set is used to inform the disaggregation of population data captured in choropleth map form to an improved model of population distribution. A variety of remotely sensed ancillary data sources have been used for this purpose, including pixel reflectance values and image texture, as well as classified data products (Yuan, Smith and Limp, 1997; Harvey, 2002; Wu, Qiu and Wang, 2005). In the case of classified land cover data, for example, one may disaggregate data from census units by exploiting the fact that few people live on water, and that population density is likely to be higher in urban as compared to agricultural land covers.

A handful of studies have employed dasymetric mapping to disaggregate population data for environmental justice studies. Maantay (2007) has developed the Cadastral-Based Expert Dasymetric Mapping System (CEDS) to disaggregate US Census data using parcel-level data. This approach was applied to an environmental justice analysis of flood hazard in New York City. Results indicate that simple areal weighting tended to undercount the population at risk, and particularly minorities, from flooding, suggesting that the nature of spatial data integration can substantially influence the results of environmental justice analyses (Maantay and Maroko, 2009). Higgs and Langford (2009) employ a variety of techniques to disaggregate population data to environmental risk zones in an environmental justice study of landfills in Wales, United Kingdom. These researchers experiment with a variety of dasymetric mapping techniques and find that evidence of environmental inequity is sensitive to the method by which one considers a population exposed to environmental hazard. Other comparative studies have found similar evidence (Mohai and Saha, 2006; Most and Sengupta, 2004).

Mennis (2002) used classified land cover data derived from Landsat TM imagery to disaggregate U.S. Census population data to a statistical surface representation for Philadelphia, Pennsylvania. This allowed for fine resolution modeling of the relationship between demographic character and proximity to hazardous facilities contained in EPA databases. Results suggested that minorities were clustered nearby hazardous facilities, but not in immediate proximity (i.e. within 500 m), as such facilities tended to be sited on sparsely populated industrial land. Such information would be masked by the use of spatially coarser demographic data.

Because most remotely sensed imagery used in environmental justice analyses is used to capture environmental hazards and amenities at a scale that can be matched to demographic character, fine and medium resolution image data are typically used. Thus, remotely sensed imagery used in environmental justice analyses are typically at a finer resolution than population data encoded in census spatial units. When remotely sensed imagery is used for quantifying characteristics of environmental hazards or amenity, say, for vegetation concentration, the remotely sensed image data may be aggregated to the level of the census unit to support statistical analysis. For example, the mean NDVI value of all the pixels contained within each Census unit may be calculated and attached as an attribute of that census unit.

16.5 Case study: vegetation and socioeconomic character in Philadelphia, Pennsylvania

As an example of how remote sensing is used in environmental justice research, an environmental equity study is demonstrated for the Philadelphia, Pennsylvania metropolitan area. Here, remotely sensed vegetation concentration data are integrated with data characterizing race, poverty, educational attainment, and population density to assess the relationships of socioeconomic status with the amenities associated with higher concentrations of vegetation. The study area comprises the five-county Philadelphia metropolitan area in southeast Pennsylvania, including the city/county of Philadelphia, as well as Montgomery, Bucks, Delaware, and Chester Counties.

Vegetation data are derived from an orthorectified, cloud-free, leaf-on, 30 meter resolution Landsat ETM+ image (path 14, row 32) dated 29 September 1999, acquired from the Landsat GeoCover collection as distributed by the University of Maryland's Global Land Cover Facility (GLCF) web site. A NDVI image was derived from this Landsat image. Socioeconomic data were acquired from the 2000 US Census at the tract level, including the following variables: percent of the total population self-identifying as African American (percent African American), percent of the total population self-identifying as Hispanic (percent Hispanic), percent of the total population whose income falls below the poverty line (of the population for whom poverty status has been recorded) (poverty rate), percent of the population at least 25 years of age who have a high school diploma or equivalent (percent high school), and population density (total population per square kilometer).

Both the NDVI and Census data were reprojected to the Universal Transverse Mercator projection and coordinate system for overlay. The NDVI image was then clipped using the tract data so that pixels outside the study area boundary were assigned a value of 'no data.' The NDVI data were then converted to ranks, as opposed to NDVI values, where the pixel with the very lowest NDVI value in the data set was assigned a value of "1," the next highest a value of "2," and so on. Pixels with equal NDVI values are tied, and the highest rank is equal to the number of pixels in the image (not including pixels with no data values). Thus, a low NDVI rank indicates a low NDVI value – and thus indicates a relatively low concentration of healthy green vegetation. The mean NDVI rank was then computed for each tract and assigned as an attribute of that tract. Descriptive statistics for each of the variables used in the analysis are provided in Table 16.1.

Correlation and ordinary least squares regression were employed to identify relationships among the explanatory variables representing socioeconomic status and the dependent variable representing vegetation character (mean NDVI rank). The well-known Moran's I statistic (Lloyd, 2007) was used to test for spatial autocorrelation in the residuals. Where spatial dependency in the regression residuals is found, spatial econometric modeling is employed to account for spatial effects.

The Pearson correlation between each of the explanatory variables and the mean NDVI rank is presented in Table 16.2. Clearly, higher NDVI is associated with the absence of both African Americans and Hispanics, as well as socioeconomic advantage, i.e. low poverty rate and high educational attainment. Not surprisingly, higher vegetation concentration is also associated with sparser population.

A four-class, quantile-classified choropleth map of each of the variables are presented in Fig. 16.1. The lowest values of mean NDVI rank (Fig. 16.1, top left) are clearly concentrated in the city of Philadelphia and the industrial and urban areas which lie to its south along the Delaware River. These areas are colored orange in the southeastern part of the map. The highest values of NDVI lie at the exurban and rural areas of Chester and Bucks Counties in the southwest and northwest regions of the study area, respectively. A high percentage of African Americans is strongly concentrated

TABLE 16.1 Descriptive statistics of explanatory variables.

	Mean	Standard deviation
% African American	22.68	31.86
% Hispanic	4.36	9.68
% High School	81.04	13.36
Poverty Rate	12.32	13.33
Population Density	3428	3753
Mean Rank NDVI	76 019	46 639

TABLE 16.2 Correlation of explanatory variables with mean NDVI rank.

	Pearson's r
% African American	−0.45***
% Hispanic	−0.28***
% High School	0.65***
Poverty Rate	−0.61**
Population Density	−0.65***

***$p < 0.005$.

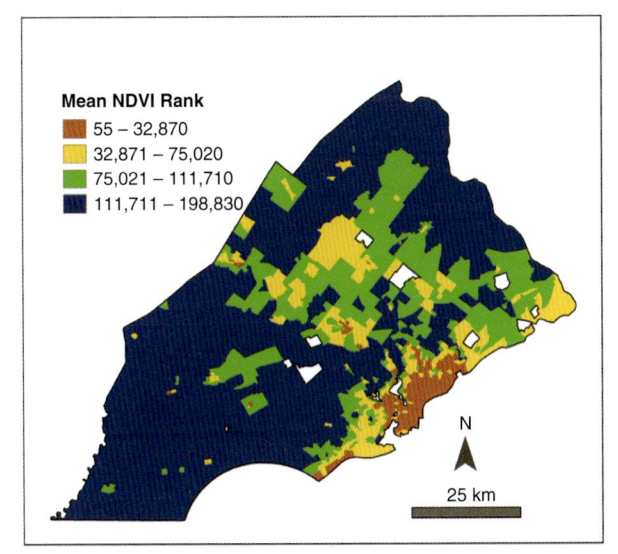

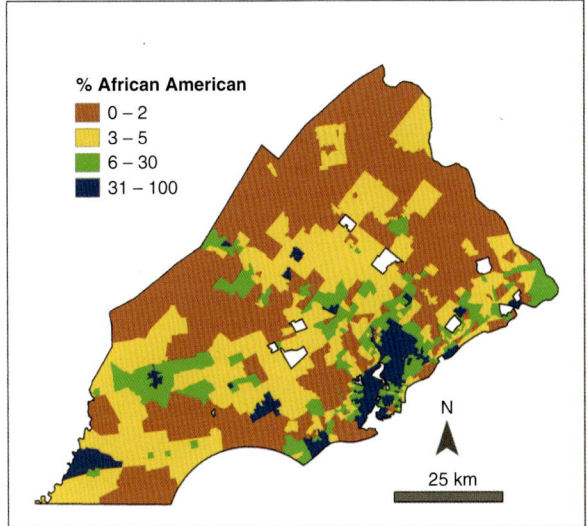

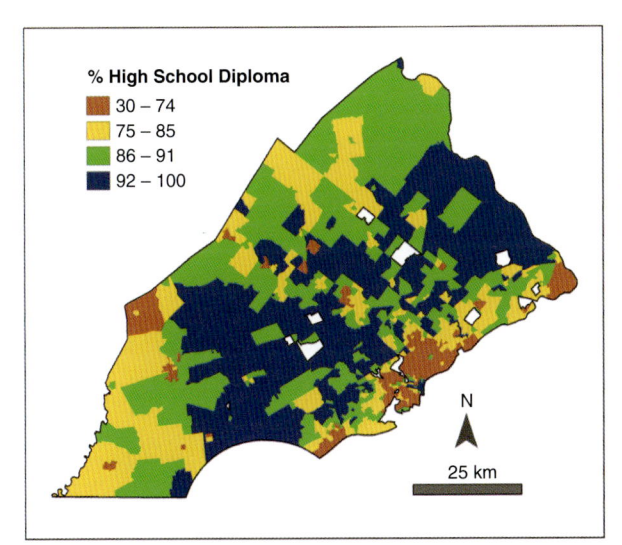

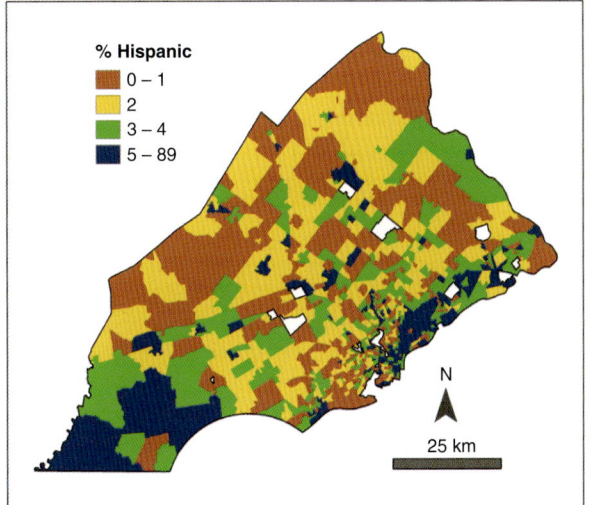

FIGURE 16.1 Quantile-classified choropleth maps used in the analysis: Mean NDVI Rank (top left), percent African American (top right), percent High School (bottom left), and percent Hispanic (bottom right).

in parts of Philadelphia (Fig. 16.1, top right), though there are several smaller cities in the suburban and exurban parts of the metropolitan area that also have high concentrations of African Americans. Hispanics are also primarily concentrated in parts of Philadelphia (Fig. 16.1, bottom right), though high concentrations also occur in Chester County in the southwest part of the study region, where demand for agricultural labor draws Hispanic farm workers. High educational attainment is concentrated primarily in the suburbs of Philadelphia (Fig. 16.1, bottom left), with lower values in the exurban and rural areas to the west and the lowest educational attainment evident in the city of Philadelphia.

Results of the OLS regression of mean NDVI rank are reported in Table 16.3. Note that poverty rate was dropped as an explanatory variable because it is highly correlated with percent high school ($r = 0.68$, significant at $p < 0.005$). The variance inflation factor (VIF) statistic indicated problematic collineary among explanatory variables when both poverty rate and percent high school were both included in a single regression equation. All regression models reported in Table 16.3 have VIF values less than 2.0. Model 1 includes only the race variables in the model and confirms the results of the Pearson correlations reported in Table 16.2 – high NDVI is associated with low concentrations of both African Americans and Hispanics. Using the race variables alone, 25% of the variation in mean NDVI rank is explained. Model 2 also includes percent high school, and indicates that high NDVI is associated with educational attainment. Notably, the influence of educational attainment on NDVI is higher than that of either of the race variables, and reduces their influence on NDVI from Model 1. When population density is added to the regression equation, as in Model 3, both race variables lose significance. The adjusted R^2 also increases to 0.57.

The Moran's I statistic indicates that all the models presented in Table 16.3 have spatial dependency in the model residuals – differences between observed and predicted values of mean NDVI rank tend to be similar for tracts nearby one another as compared to tracts farther apart. We employ spatial econometric modeling estimation to account for spatial autocorrelation in the model residuals. The spatial lag form of the spatial econometric model incorporates the spatial lag of the dependent

TABLE 16.3 Results of the OLS regression of mean NDVI rank.

Explanatory variable	Model 1	Model 2	Model 3
% African American	−0.44*** (−15.88)	−0.13*** (−4.44)	0.02 (0.66)
% Hispanic (ln)	−0.22*** (−7.88)	−0.07* (−2.55)	0.00 (0.18)
% High School		0.56*** (18.12)	0.45*** (16.01)
Population Density			−0.44*** (−16.76)
Constant	92.65*** (57.60)	−77.61*** (-8.17)	−33.01*** (−3.76)
Adjusted R^2	0.25	0.44	0.57
AIC	9911	9629	9384
Moran's I	0.684***	0.575***	0.501***
LM Lag	1329***	957***	726***
Robust LM Lag	95***	115***	108***
LM Error	1245 ***	880***	668***
Robust LM Error	10***	38***	50***

*$p < 0.05$, **$p < 0.01$, ***$p < 0.005$.
Reported values are standardized coefficients, with t-values in parentheses.

TABLE 16.4 Results of the spatial lag and spatial error regressions of mean NDVI rank.

Explanatory variable	Spatial lag model	Spatial error model
% African American	0.04 (0.03)	−0.04 (0.05)
% Hispanic (ln)	−0.19 (0.39)	0.12 (0.41)
% High School	0.48*** (0.07)	0.65*** (0.10)
Population density	−0.002*** (0.0002)	−0.003*** (0.0003)
Spatial lag term	0.75*** (0.02)	
Spatial error term		0.86*** (0.02)
Constant	−15.85** (5.81)	25.31 ** (9.77)
AIC	8705.17	8737.26
Moran's I	0.040*	0.005

*$p < 0.05$, **$p < 0.01$, ***$p < 0.005$.
Reported values are unstandardized coefficients, with standard errors in parentheses.

variable as an explanatory variable in the model, which may be expressed as

$$y = \rho W y + X\beta + \varepsilon$$

where ρ is the spatial autoregressive coefficient, and W is the spatial weights matrix. Maximum likelihood is employed to estimate the model (Anselin, 1988). Alternatively, the spatial error form of the spatial econometric model allows the error term to vary over space, such that

$$y = \rho W u + X\beta + \varepsilon$$

where $\rho W u$ represents the spatially autocorrelated component of the error term.

Generally, the spatial lag model is appropriate where one hypothesizes that a causal process of diffusion that occurs across space from one spatial unit to another is responsible for the spatial autocorrelation in the residuals. The spatial error model treats the spatial dependency as a nuisance to be accounted for so that the assumptions of regression may be met and unbiased parameter estimates may be obtained. Such spatial dependency in the model residuals may be due to model misspecification in the form of missing variables or nonlinear relationships among explanatory and dependent variables. The Lagrange multiplier (LM) statistics and their robust forms also offer some guidance as to the appropriate form of the spatial econometric model to adopt (Anselin et al., 1996). In the present study, LM statistics for the both the lag and error models are highly significant (Table 16.3), though the value of the LM lag statistic is greater than that of the LM error statistic.

As the nature and cause of the spatial dependency in the model residuals is unknown, Table 16.4 reports the results of both the spatial lag and spatial error forms of the spatial econometric model. They are similar in that for both models percent high school and population density are significant, and the race variables are not. The spatial lag and spatial error terms in each model are also significant. Spatial autocorrelation in the model residuals, as indicated by Moran's I, persists slightly in the spatial lag model but not in the spatial error model. Figure 16.2 presents the Moran scatterplot (Anselin, 1996) for the OLS (Table 16.3, model 3) and spatial error models. This figure presents the nature of the spatial autocorrelation in the model residuals, where each point represents the value of the model residual and the spatial lag of the model residual for an observation, such that the higher the slope of the line of best fit, the greater the degree of spatial autocorrelation in the model residuals. Clearly, the slope of the line of best fit is much lower for the spatial error model, and not significantly different than zero, as compared to the OLS model.

The results of the case study suggest that African Americans and Hispanics tend to live in places with less healthy green vegetation than other people in the Philadelphia, Pennsylvania metropolitan area. This relationship of race with vegetation is partly explained by class, specifically educational attainment, as highly educated people tend to live in greener areas. However, even after accounting for educational attainment, African Americans and Hispanics still tend to live in less green areas of

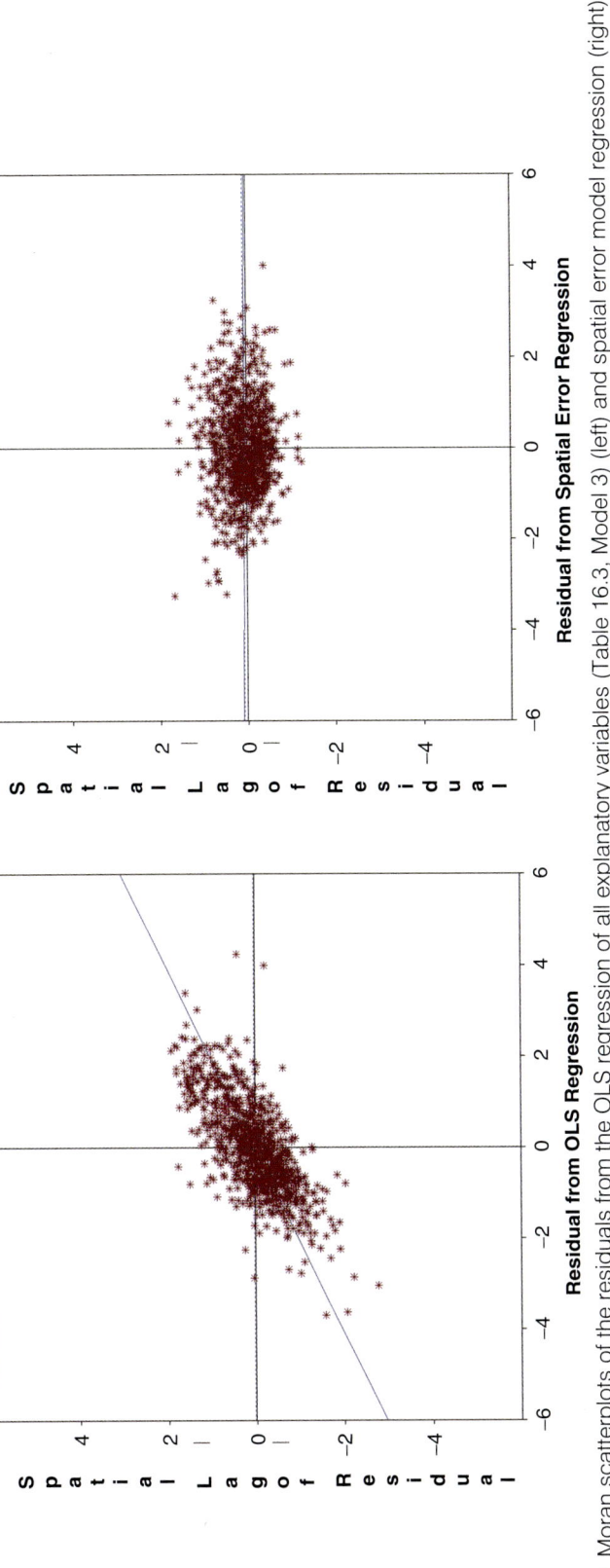

FIGURE 16.2 Moran scatterplots of the residuals from the OLS regression of all explanatory variables (Table 16.3, Model 3) (left) and spatial error model regression (right). For each scatterplot, the horizontal axis plots the observed residual value and the vertical axis plots the spatial lag of the residual value.

the region, i.e., even among the highly educated in the region, minorities are still less likely to live nearby abundantly vegetated areas. The role of population density suggests that this negative relationship between minorities and vegetation is explained in large measure by the degree of "urbanness." One narrative interpretation of these results would be that minorities tend to live in more densely populated, urban areas; and because urban areas tend to have lower vegetation concentration, minorities tend to live in less green areas than non-minorities. This interpretation suggests that racial inequity in vegetation character is a function of historical processes related to residential mobility. Racial segregation in residential housing, as well as historical processes of suburbanization and "white flight," have produced the current racialized spatial residential patterns that can be currently observed (Pulido, 2000).

Conclusion

Several prospects and challenges can be identified for the continuing integration of remote sensing and GIS for environmental justice research. One issue concerns how new remotely sensed data products can aid environmental justice research, as with recent advances in the spatial resolution of remotely sensed image products. In the case study above, for example, vegetation concentration was captured at a resolution of 30 m, potentially masking substantial spatial variation in vegetation character. Urban environments typically have high spatial heterogeneity in land cover, which can change abruptly over short distances. For instance, parkland (with high vegetation concentration) can lie directly adjacent to industrial land with effectively no vegetation present. Or, similarly, consider a main commercial street corridor running through a suburban neighborhood. The suburban area, with a single-family home style of residential development, may have relatively strong vegetation concentration due to the presence of lawn grasses and moderate tree cover. A main commercial street running through such a neighborhood, however, is likely to have far less vegetation and greater impervious surface cover. In such cases as these, remotely sensed imagery at a resolution of 30 meters is unlikely to capture the detail of such spatial variation in vegetation, as the resolution is simply too coarse to accurately capture abrupt changes in vegetation character (as with the boundary between parkland and industrial land) and narrow or other oddly shaped features in the landscape (such as commercial streets that bisect residential neighborhoods, or waterways).

Commercially available multi-spectral image products with less than five meter resolution, such as IKONOS (Space Imaging, Inc.) and Quickbird (DigitalGlobe, Inc.) imagery, allow for much finer resolution and precise spatial estimates of urban vegetation character (Nichol and Lee, 2005). Advances in spectral resolution, on the other hand, allow for the recognition of not only general measures of vegetation concentration, as with NDVI, but the recognition of specific types of vegetation and other land covers. Hyperspectral sensors carried on board airborne platforms, such as Airborne Visible Infrared Imaging Spectrometer (AVIRIS), which has 224 spectral bands, can provide imagery from which specific types of vegetation may be extracted. SMA has been applied to AVIRIS imagery to yield various types of vegetation, including woody vegetation and grasses (Wessman *et al.*, 1997), as well as to distinguish among other types of urban land covers. SMA has also been applied to multi-spectral IKONOS imagery to yield distinct categories of vegetation types, such as trees and grasses (Nichol and Wong, 2007).

Increases in spatial and spectral resolution are not only useful for measuring vegetation character as an indicator of an environmental amenity in environmental justice research, but also can serve as useful ancillary data sets for dasymetric mapping of population. For example, in the case study above, population data were attached to Census tracts and the analysis makes the assumption that population is distributed homogeneously throughout each tract, whereas in reality (nighttime) population is concentrated in residential parts of tracts. High spatial resolution imagery from IKONOS has been used in dasymetric mapping to disaggregate census population data and derive population distribution data sets at a much finer resolution than the original census data, and an even finer resolution than would be possible utilizing Landsat or other more commonly used remotely sensed ancillary data for dasymetric mapping (McKenzie, 2008). Because more accurate population data can improve the estimation of spatial relationships between population and environmental hazards and/or amenities, these advances in the spatial and spectral resolution of remotely sensed data can contribute to advances in environmental justice.

It is worth noting that remotely sensed imagery has also been used not only for dasymetric mapping of population but also for population estimation in the absence of a priori census data. For example, Mubareka *et al.* (2008) use Shuttle Radar Topography Mission (SRTM) and Landsat Thematic Mapper data to estimate settlement location and population density in northern Iraq using a small sample of settlements to calibrate the population estimation, whereas others have used high resolution satellite imagery to simply count residential dwellings and estimate population (e.g., Bjorgo, 2000). Such studies may be particularly useful in estimating population in Third World nations where census information may be very spatially coarse, of unknown accuracy, not current, or simply nonexistent. One notes, however, that these techniques may be difficult to extend to enumerating not only population, but also indicators of race, class, and other socioeconomic characteristics that form the basis of environmental justice research.

Remotely sensed data may also be used to capture other environmental amenities, besides vegetation, or hazards that may be useful in environmental justice research. As noted earlier, the majority of environmental justice studies have focused on technological hazards from industry, for which data are typically derived from environmental databases of toxic emissions and/or environmental monitoring networks. As with the use of thermal imagery for studying the environmental justice aspects of the urban heat island effect, the use of remotely sensed imagery may be used to substantially expand the scope of environmental justice research to a variety of types of environmental hazards that have not been substantially addressed in the environmental justice literature. Remotely sensed data can provide valuable data on environmental risk due to natural hazards, including volcanic eruptions, earthquakes, landslides, and flooding. High resolution elevation data in particular, as derived from light detection and ranging (LIDAR) and interferometric synthetic aperture radar (InSAR), have proven particularly useful in estimating areas at risk for a variety of types of hazards (Tralli *et al.*, 2005).

Many remote sensing technologies have now been in operation for years, and some for decades. Thus, remotely sensed data on vegetation and other environmental characteristics can be used to capture change through time, and when combined with longitudinal population data, can potentially be used to investigate the spatial and temporal aspects of the relationship between socioeconomic status and environmental risks and amenities. Certainly, such an approach has been used to investigate, and calibrate models of, urban growth and land cover change (Clarke and Gaydos, 1998). Its application to environmental justice research is still relatively novel, but may play a key role in investigating evidence for causation in the relationships between environmental risk and socioeconomic disadvantage, or at least provide information about the process by which relationships between environmental character and socioeconomic status came to occur. Such a perspective is for the most part lacking in quantitative environmental justice research due to the prevalence of cross-sectional, as opposed to longitudinal, study design. Such research designs are the result of the limited access researchers generally have to longitudinal environmental and population data.

With Landsat data going back the middle 1970s, however, and recent commercial historical US Census data products being introduced to the market, the opportunity exists for examining concordant changes in urban environmental and socioeconomic characteristics over the course of decades. Though these data are not particularly high temporal resolution (i.e., the US Census data is decadal), the longitudinal sampling provides a measure of environmental and socioeconomic trends over time that may be compared.

In sum, the use of remotely sensed imagery can play a key role in environmental justice research by providing data on environmental hazards and amenities, which may be used as the outcome variable in quantitative analyses of environmental equity. Remotely sensed data can also play a key role by providing information on land use and/or land cover, which may be used as an explanatory variable in estimating hazardous facility location, as well as for redistributing population for refined estimates of the spatial relationships between environmental characteristics and socioeconomic status. Substantial advances in environmental justice research may be had through the use of remotely sensed data for novel applications, particularly for the measurement of environmental amenities related to vegetation, as well as for environmental hazards that have been largely unexamined, such as the risk due to earthquakes, flooding, and other natural disasters.

References

Anderton, D.L., Anderson, A.B., Oakes, J.M., and Fraser, M.R. (1994) Environmental equity: the demographics of dumping. *Demography*, **31**, 229–248.

Anselin, L. (1988) *Spatial Econometrics: Methods and Models*, Kluwer Academic, Boston.

Anselin, L. (1996) The Moran scatterplot as an ESDA tool to assess local instability in spatial association, in *Spatial Analytical Perspectives on GIS* (eds M. Fischer, H. Scholten, and D. Unwin), Taylor and Francis, London, pp. 111–125.

Anselin, L., Bera, A., Florax, R.J. and Yoon, M. (1996) Simple diagnostic tests for spatial dependence. *Regional Science and Urban Economics*, **26**, 77–104.

Atlas, M. (2002) Few and far between? An environmental equity analysis of the geographic distribution of hazardous waste generation. *Social Science Quarterly*, **83**, 365–378.

Been, V. (1994) Locally undesirable land uses in minority neighborhoods: Disproportionate siting or market dynamics? *Yale Law Journal*, **103**, 1383–1422.

Bjorgo, E. (2000) Using very high spatial resolution multispectral satellite sensor imagery to monitor refugee camps. *International Journal of Remote Sensing*, **21** (3), 611–616.

Boer, J.T., Pastor Jr., M., Sadd, J.L. and Snyder, L.D. (1997) Is there environmental racism? The demographics of hazardous waste in Los Angeles County. *Social Science Quarterly*, **78**, 793–810.

Boone, C.G. and Modarres, A. (1999) Creating a toxic neighborhood in Los Angeles County: A historical examination of environmental inequity. *Urban Affairs Review*, **35**, 163–187.

Boone, C.G. (2008). Environmental justice as process and new avenues for research. *Environmental Justice*, **1** (3), 149–153.

Bowen, W.M., Salling, M.J., Haynes, K.E. and Cyran, E.J. (1995) Toward environmental justice: Spatial equity in Ohio and Cleveland. *Annals of the Association of American Geographers*, **85**, 541–663.

Bowen, W.M. and Wells, M.V. (2002). The politics and reality of environmental justice: A history and considerations for public administrators and policy makers. *Public Administration Review*, **62**, 687–698.

Bullard, R. (1990) *Dumping in Dixie: Race, Class and Environmental Quality*, Westview, Boulder, CO.

Bullard, R.D. (1996). Environmental justice: It's more than waste facility siting. *Social Science Quarterly*, **77**, 493–499.

Buzzelli, M., Jerrett, M., Burnett, R. and Finklestein, N. (2003). Spatiotemporal perspectives on air pollution and environmental justice in Hamilton, Canada, 1985–1996. *Annals of the Association of American Geographers*, **93** (3), 557–573.

Chakraborty, J. and Armstrong, M.P. (1997). Exploring the use of buffer analysis for the identification of impacted areas in environmental equity assessment. *Cartography and Geographic Information Systems*, **24**, 145–157.

Clarke, K.C. and Gaydos, L. (1998). Loose coupling a cellular automaton model and GIS: long-term growth prediction for San Francisco and Washington/Baltimore. *International Journal of Geographical Information Science*, **12** (7), 699–714.

Clinton, W.J. (1994) Federal actions to address environmental justice in minority populations and low-income populations. Executive Order 12898 of February 11, 1994. *Federal Register*, **59**, 7629–7633.

Comber, A., Brunsdon, C. and Green, E. (2008) Using a GIS-based network analysis to determine urban greenspace accessibility for different ethnic and religious groups. *Landscape and Urban Planning*, **86** (1), 103–114.

CRJ (Commission for Racial Justice) (1987) *Toxic Wastes and Race in the United States: A National Report on the Racial and Socioeconomic Characteristics of Communities with Hazardous Waste Sites*, United Church of Christ Commission for Racial Justice, New York.

Cutter, S.L. (1995) Race, class, and environmental justice. *Progress in Human Geography*, **19** (1), 107–118.

Downey, L. (1998). Environmental injustice: is race or income a better predictor? *Social Science Quarterly*, **79**, 766–778.

Downey, L. (2003) Spatial measurement, geography, and urban racial inequality. *Social Forces*, **81**, 937–952.

Duthie, J., Cervenka, K. and Waller, S.T. (2007) Environmental justice analysis: challenges for metropolitan transportation planning. *Journal of the Transportation Research Board*, **2013**, 8–12.

Eicher, C. and Brewer, C. (2001) Dasymetric mapping and areal interpolation: implementation and evaluation. *Cartography and Geographic Information Science*, **28**, 125–138.

EPA (US Environmental Protection Agency) (2003) Environmental justice fact sheet: EPA's commitment to environmental justice, US EPA Office of Environmental Justice, Washington, DC.

Fisher, J.B., Kelly, M. and Romm, J. (2005) Scales of environmental justice: combining GIS and spatial analysis for air toxics in West Oakland, California. *Health and Place*, **12**(4), 701–714.

Fotheringham, A.S. and Wong, D. (1991) The modifiable areal unit problem in multi-variant statistical analysis. *Environment and Planning A*, **23**, 1025–1044.

GAO (General Accounting Office) (1983) *Siting of Hazardous Waste Landfills and Their Correlation with Racial and Economic Status of Surrounding Communities*. US General Accounting Office, Washington, D.C.

Geoghegan, J. (2002) The value of open spaces in residential land use. *Land Use Policy*, **19**: 91–98.

Goldman, B. and Fitton, L. (1994) *Toxic Wastes and Race Revisited*, Center for Policy Alternatives, Washington, DC.

Harlan, S.L., Brazel, A.J., Proashad, L., Stefanov, W.L. and Larsen, L. (2006) Neighborhood microclimates and vulnerability to heat stress. *Social Science and Medicine*, **63**, 2847–2863.

Harvey, J.T. (2002) Estimation census district population from satellite imagery: some approaches and limitations. *International Journal of Remote Sensing*, **23**, 2071–2095.

Heynen, N., Perkins, H.A. and Roy, P. (2006) The political ecology of uneven urban green space: the impact of political economy on race and ethnicity in producing environmental inequality in Milwaukee. *Urban Affairs Review*, **42**, 3–25.

Higgs, G. and Langford, M. (2009) GIScience, environmental justice, and estimating populations at risk: the case of landfills in Wales. *Applied Geography*, **29**, 63–76.

Hird, J.A. (1993) Environmental policy and equity: The case of Superfund. *Journal of Policy Analysis and Management*, **12**, 323–343.

Hong, Y., Adler, R. and Verdin, J. (2007) Use of 21st century satellite remote sensing technology in natural hazard analysis. *Natural Hazards*, **43**(2), 165–166.

Jensen, R., Gatrell, J., Boulton, J. and Harper, B. (2004) Using remote sensing and geographic information systems to study urban quality of life and urban forest amenities. *Ecology and Society*, **9**(5), 5.

Jerrett, M., Burnett, R.T., Kanaroglou, P. et al. (2001) A GIS-environmental justice analysis of particulate air pollution in Hamilton, Canada. *Environment and Planning A*, **33**, 955–73.

Li, G. and Weng, G. (2007) Measuring the quality of life in city of Indianapolis by integration of remote sensing and census data. *International Journal of Remote Sensing*, **28**(2), 249–267.

Liu, F. (2001) *Environmental Justice Analysis: Theories, Methods, and Practice*, Lewis Publishers, New York.

Lloyd, C.D. (2007) *Local Models for Spatial Analysis*, CRC Press, New York.

Lo, C.P. and Faber, B.J. (1997) Integration of Landsat Thematic Mapper and census data for quality of life assessment. *Remote Sensing of Environment*, **62**, 143–157.

Lo, C.P., Quattrochi, D.A. and Luvall, J.C. (1997) Application of high resolution thermal infrared remote sensing data and GIS to assess the urban heat island effect. *International Journal of Remote Sensing*, **18**, 287–304.

Maantay, J. (2002) Mapping environmental injustices: pitfalls and potential of geographic information systems in assessing environmental health and equity. *Environmental Health Perspectives*, **110**(Suppl. 2), 161–171.

Maantay, J.A. (2007) Asthma and air pollution in the Bronx: methodological and data considerations in using GIS for environmental justice and health research. *Health and Place*, **13**, 32–56.

Maantay, J.A. and Maroko, A. (2009) Mapping urban risk: flood hazards, race, and environmental justice. *Applied Geography*, **29**, 111–124.

McGurty, E. (2007) *Transforming Environmentalism: Warren County, PCBs, and the Origins of Environmental Justice*, Rutgers University Press, New Brunswick, NJ.

McKenzie, S.J.P. (2008) Disaggregating job seekers allowance statistics for Belfast using IKONOS satellite imagery, **v2008**, 444–450.

Mennis, J. (2002) Using geographic information systems to create and analyze statistical surfaces of population and risk for environmental justice analysis. *Social Science Quarterly*, **83**, 281–297.

Mennis, J. (2003) Generating surface models of population using dasymetric mapping. *The Professional Geographer*, **55**(1), 31–42.

Mennis, J. (2005) The distribution and enforcement of air polluting facilities in New Jersey. *The Professional Geographer*, **57**(3), 411–422.

Mennis, J. (2006) Socioeconomic-vegetation relationships in urban, residential land: the case of Denver, Colorado. *Photogrammetric Engineering and Remote Sensing*, **72**(8), 911–921.

Mennis, J., and Hultgren, T (2006) Intelligent dasymetric mapping and its application to areal interpolation. *Cartography and Geographic Information Science*, **33**(3), 179–194.

Mennis, J. and Jordan, L. (2005) The distribution of environmental equity: exploring spatial non-stationarity in multivariate models of air toxic releases. *Annals of the Association of American Geographers*, **95**(2), 249–268.

Mohai, P. and Bryant, B. (1992) *Race and the Incidence of Environmental Hazards: A Time for Discourse*, Westview, Boulder, CO.

Mohai, P. and Saha, R. (2006) Reassessing racial and socioeconomic disparities in environmental justice research. *Demography*, **43**(2), 383–399.

Most, M.T. and Sengupta, R. (2004) Spatial scale and population assignment choices in environmental justice analyses. *The Professional Geographer*, 56(4), 574–586.

Mubareka, S., Ehrlich, D., Bonn, F. and Kayitakire, F. (2008) Settlement location and population density estimation in rugged terrain using information derived from Landsat ETM and SRTM data. *International Journal of Remote Sensing*, 29 (8), 2339–2357.

Nichol, J.E. and Lee, C.M. (2005) Urban vegetation monitoring in Hong Kong using high resolution multispectral images. *International Journal of Remote Sensing*, 26, 903–919.

Nichol, J.E. and Wong, M.S. (2007) Remote sensing of urban vegetation life form by spectral mixture analysis of high resolution Ikonos satellite images. *International Journal of Remote Sensing*, 28 (5), 985–1000.

O'Neil, S.G. (2007) Superfund: Evaluating the impact of Executive Order 12898. *Environmental Health Perspective*, 115 (7), 1087–1093.

Oke, T. (1982) The energetic basis of the urban heat island. *Quarterly Journal of the Royal Meteorological Society*, 108, 1–24.

Openshaw, S. (1984) *The Modifiable Areal Unit Problem*, Geobooks, Norwich, UK.

Pulido, L. (2000) Rethinking environmental racism: White privilege and urban development in southern California. *Annals of the Association of American Geographers*, 90, 12–40.

Roth, M., Oke, T. and Emery, W. (1989) Satellite-derived urban heat islands from three coastal cities and the utilization of such data in urban climatology. *International Journal of Remote Sensing*, 10, 1699–1720.

Sadd, J.L., Pastor Jr., M., Boer, J.T. and Snyder, L.D. (1999) "Every breath you take...": The demographics of toxic air releases in southern California. *Economic Development Quarterly*, 13, 107–23.

Small, C. (2001) Estimation of urban vegetation abundance by spectral mixture analysis. *International Journal of Remote Sensing*, 22, 1305–1334.

Szasz, A. and Meuser, M. (1997) Environmental inequalities: Literature review and proposals for new directions in research and theory. *Current Sociology*, 45, 99–120.

Strife, S. and Downey, L. (2009) Childhood development and access to nature: a new direction for environmental inequality research. *Organization and Environment*, 22(1), 99–122.

Taquino, M., Parisi, D. and Gill, D.A. (2002) Units of analysis and the environmental justice hypothesis: the case of industrial hog farms. *Social Science* Quarterly, 83, 298–316.

Tralli, D.M., Blom, R.G., Zlotnicki, V., Donnellan, A. and Evans, D.L. (2005) Satellite remote sensing of earthquake, volcano, flood, landslide, and coastal inundation hazards. *ISPRS Journal of Photogrammetry and Remote Sensing*, 59(4), 185–198.

Tucker, C.J. (1979) Red and photographic infrared linear combinations for monitoring vegetation. *Remote Sensing of the Environment*, 8, 127–150.

Vogelmann, J.E., Howard, S.M., Yang, L., Larson, C.R., Wylie, B.K. and Van Driel, N. (2001) Completion of the 1990s National Land Cover Data Set for the conterminous United States from Landsat Thematic Mapper data and ancillary data sources. *Photogrammetric Engineering and Remote Sensing*, 67, 650–662.

Wessman, C.A., Bateson, A. and Benning, T.L. (1997) Detecting fire and grazing patterns in tallgrass prairie using spectral mixture analysis, *Ecological Applications*, 7(2), 493–511.

Williams, R.W. (1999) The contested terrain of environmental justice research: Community as a unit of analysis. *The Social Science Journal*, 36, 313–28.

Wu, S.-S., Qiu, X. and Wang, L. (2005) Population estimation methods in GIS and remote sensing: a review. *GIScience and Remote Sensing*, 42(1), 80–96.

Yuan, Y., Smith, R.M. and Limp, W.F. (1997) Remodeling census population with spatial information from Landsat imagery. *Computers, Environment and Urban Systems*, 21, 245–258.

Zhang, Y., Tarrant, M.A. and Green, G.T. (2008) The importance of differentiating urban and rural phenomena in examining the unequal distribution of locally desirable land. *Journal of Environmental Management*, 88, 1314–1319.

Zimmerman, R. (1994) Issues of classification in environmental equity: How we manage is how we measure. *Fordham Urban Law Journal*, 21, 633–70.

V

URBAN ENVIRONMENTAL ANALYSES

Urban environmental analyses can help understand the status, trends, and threats in urban areas so that appropriate management actions can be planed and implemented. Part V (Chs 17–21) reviews some latest developments in the synergistic use of remote sensing and relevant geospatial techniques for urban environmental analyses. Chapter 17 details a remote sensing approach to high-resolution urban impervious surface mapping. This topic is included as part of urban environmental analyses because landscape imperviousness has recently emerged as a key indicator being used to address a variety of urban environmental issues. Chapter 18 examines the impact of different remote sensing methods for characterizing the distribution of impervious surfaces on runoff estimation and how this can affect the assessment of peak discharges in an urbanized watershed. Chapter 19 reviews the light-use efficiency models and applies them to estimate gross primary production in the eastern United States that was associated with various settlement densities. Chapter 20 discusses the utilities of remote sensing for characterizing biodiversity in urban areas, how urbanization affects biodiversity, and how remote sensing-based biodiversity research can be integrated with urban planning for biodiversity conservation. The last chapter (Ch. 21) in Part V discusses how remote sensing can be used to study the influence of urban land use/cover changes on urban meteorology, climate and air quality.

17

Remote sensing of high resolution urban impervious surfaces

Changshan Wu and Fei Yuan

Numerous approaches have been developed to quantify the distribution of impervious surfaces through remote sensing technologies in recent years. These methods can be divided into two groups: pixel-based models and object-based models. Pixel-based models utilize each individual pixel as the basic unit of analysis, while object-oriented models employ pre-identified spatial objects as the basic unit. While most of these methods have been applied to medium resolution remote sensing data (e.g., Landsat TM), high-resolution impervious surface information has also been generated. This chapter reviews major pixel-based and object-based techniques for impervious surface estimation, and compares these two groups of models through case studies in Grafton, WI and Mankato, MN, USA. Results indicate that for the pixel-based model, an integrated spectral mixture analysis and regression tree model outperforms the individual models, and the object-based model performs well if the parameters of image segmentation are defined appropriately.

17.1 Introduction

Impervious surfaces are defined as those surfaces that water cannot infiltrate. In urban areas, building rooftops, streets, highways, parking lots, and sidewalks are the typical impervious surfaces. While compacted soil or gravel, high clay content soils, and extraction areas are also classified as impervious surfaces, transportation elements have been determined to contribute the most to total impervious surface area (Schueler, 1994). Urbanization in the form of converting rural land uses to urban land uses is directly associated with the increase of urban imperviousness. Urban imperviousness directly affects water quality, the amount of runoff to streams and lakes, aquatic habitats, and the aesthetics of landscapes (Dougherty et al., 2004). The spatial structure of urban thermal patterns and urban heat balances are also associated with urban surface characteristics. Therefore, accurate measurement of impervious surface area provides an essential indicator of environmental quality and valuable input to urban planning and management activities (Schueler, 1994). Urban imperviousness has been utilized to assess adverse influences of urbanization on urban climate, air and water quality, and natural habitat (Schueler, 1994; Dougherty et al., 2004; Yuan and Bauer, 2007), and to quantify urban development and population growth (Xian and Crane, 2005; Yang and Liu, 2005; Yang, 2006; Wu and Murray, 2007; Morton and Yuan, 2009).

The most accurate methods for impervious mapping are traditional ground surveys and aerial photographic interpretation and digitizing. These methods, however, are time consuming and labor intensive. Alternatively, the decreasing costs and increasing availability of multispectral digital imagery have led to more and more successful programs of impervious surfaces mapping by automatic processing of digital remote sensing data. For regional scale studies, remote sensing of impervious surface has focused on subpixel analysis using moderate resolution Landsat and SPOT satellite data since they have the advantages of relatively large coverage, multiple spectral bands, and comparatively low cost (Adams et al., 1995; Yang et al., 2003; Wu and Murray, 2003; Wu, 2004; Lu and Weng, 2004; Bauer, Loeffelholz, and Wilson, 2007). The major subpixel level approaches to estimate percent impervious surface area include spectral mixture analysis, regression analysis, regression tree, artificial neural networks and expert systems. Detailed properties and comparison of major methods can be found in Yuan, Wu, and Bauer (2008) and Weng and Hu (2008).

On the other hand, for local studies, higher resolution satellite imagery is preferred because the 30-m resolution of the Landsat and the 20-m resolution of the SPOT data are generally not sufficient to discriminate individual features (e.g., buildings, streets, trees) within the urban mosaic (Small, 2003). Use of high resolution satellite images for impervious surface mapping has attracted more and more attention since the launches of 4-m IKONOS and 2.4-m Quickbird sensors in 1999 and 2001 respectively. In the literature, two groups of models, pixel-based and object-based approaches, have been applied to estimating both medium- and high-resolution impervious surface areas. Pixel based models consider each individual pixel as the unit of analysis, while object-based approaches employ pre-identified objects as the analysis unit.

The objective of this chapter is to review popular pixel- and object-based models for impervious surface estimation, and compare these two groups of models through case studies in Grafton WI and Mankato, MN, USA. The remainder of this chapter is as follows. Section 17.2 reviews the pixel- and object-based models for impervious surface estimation. Case studies of these two groups of models are given in Sections 17.3 and 17.4. In particular, several pixel-based models, including a spectral mixture analysis, a regression tree model, and an integrated approach, are detailed in Section 17.3, and an object-oriented model is given in Section 17.4. Further discussion and conclusions are detailed in Section 17.5.

17.2 Impervious surface estimation

17.2.1 Pixel-based models

For estimating impervious surface areas, many pixel-based models have been developed in the literature. These models, especially subpixel analysis methods, have been widely applied to medium-resolution imagery (e.g., Landsat and SPOT data). For high-resolution imagery, pixel size may be comparable to or smaller than the size of urban objects, and therefore most pixels in the imagery can be considered as pure pixels. While there are usually less mixed pixels in high-resolution satellite imagery such as 4-m IKONOS and 2.4-m Quickbird data, the mixed pixel problem still exists. In particular, Wu (2009) estimated the proportion of pure impervious surface pixels in Grafton, WI, and found that approximately 40–50% of pixels containing impervious surfaces are mixed pixels. Therefore, mixed pixel problem may still be one of the major factors that affect the accuracy of impervious surface estimation in high resolution imagery, given the heterogeneous characteristics of urban environment. Thus, this section reviews several pixel-based methods applied to both medium and high resolution remote sensing imagery. These models include regression modeling, spectral mixture analysis, regression tree, and artificial neural network (ANN).

17.2.1.1 Regression modeling

Regression modeling estimates impervious surface areas through constructing the relationship between the percentage of impervious surface areas (% ISA) and the spectral and/or spatial information of an individual pixel. Specially, greenness index, the second component of the Tasseled Cap (TC) transformation or the Normalized Difference Vegetation Index (NDVI) has been applied to both medium and high resolution remote sensing imagery (Heinert, 2002; Gillies et al., 2003; Sawaya et al. 2003; Yuan et al., 2005; Bauer, Loffelholz and Wilson, 2007). In urbanized areas with little amount of soil, the amount of impervious surface area is significantly and inversely correlated to the amount of green vegetation, represented by the TC Greenness or NDVI. With the constructed relationship, high resolution impervious surfaces have been derived with reasonable accuracy (Sawaya et al., 2003).

17.2.1.2 Spectral mixture analysis

Spectral mixture analysis (SMA) is utilized for calculating land cover fractions within a pixel and involves modeling a mixed spectrum as a combination of spectra for pure land cover types, called endmembers. SMA has been successfully applied for estimating the fractions of impervious surface for a pixel of medium resolution remote sensing imagery. For example, Phinn et al. (2002) successfully estimated impervious surface distribution using a constrained spectral mixture analysis method with endmembers chosen from aerial photos. Wu and Murray (2003) implemented a constrained linear SMA to generate impervious surface distribution in Columbus OH, and found that impervious surface fraction can be estimated by a linear model of low and high albedo endmembers. Further, Wu (2004) proposed a normalized spectral mixture analysis (NSMA) to achieve a better estimation accuracy of impervious surface distribution. For high resolution impervious surface estimation, Lu and Weng (2009) developed linear spectral mixture analysis (LSMA) and decision tree classifier (DTC) to quantify urban imperviousness. In addition, considering the spectral variations and boundary effects of IKONOS imagery, Wu (2009) developed a modified NSMA model.

17.2.1.3 Regression tree model

In addition to SMA, regression tree model is another popular method for generating impervious surface data in an urban area. Like regression analysis, a regression tree constructs the relationships between a dependent variable (e.g., impervious surface fraction) and independent variables (e.g., reflectance for a particular band). However, the regression tree model is more complicated than regression analysis. It grows a (inverted) categorical tree by repeatedly splitting the data according to specific rules depending on how the dependent variable and the independent variables interact with each other. The goal of the algorithm is to categorize the data into more homogeneous groups by uncovering the predictive structure of the problem under consideration (Breiman et al., 1984). The performance of the regression tree model was reported to be more accurate than ordinary regression models in many applications (Huang and Townshend, 2003). Yang et al. (2003) reported the results of applying regression tree models in quantifying impervious surface fraction applied to Landsat ETM+ images. This method has been adopted by United States Geological Survey (USGS) to produce 30 × 30 m national land cover dataset (NLCD). For high resolution impervious surface estimation, Goetz et al. (2003) estimated impervious surface using a classification tree approach for Montgomery County, MD, in which 11 IKONOS image tiles were acquired in six swaths to provide complete coverage of the study site. Lu and Weng (2007) used a hybrid approach based on the combination of decision tree classifier and unsupervised classification to extract impervious surface areas from IKONOS data.

17.2.1.4 Artificial neural network

In addition to the above pixel-based models, ANN is another alternative to traditional classification models and it can address the problems of nonlinearity among variables (Ji, 2000). Further, when compared with other classifiers (e.g., maximum likelihood), ANN models generated better classification results (Flanagan and Civco, 2001; Kavzoglu and Mather, 2003). ANN has been applied to estimate urban impervious surface information from both moderate resolution (Flanagan and Civco, 2001;

Pu et al., 2008) and high resolution (Mohapatra and Wu, 2007) remote sensing imagery.

17.2.2 Object-based models

Besides pixel-based models, object-based classification has also been developed for classifying impervious surface areas from high resolution remote sensing imagery (e.g., QuickBird). Compared to traditional per-pixel classification methods, object-based approaches provide unique capabilities to incorporate large-scale textural and contextual information, and numerous object-based features in the classification process. The unique characteristic of object-oriented classification is creating image objects by image segmentation and performing classification on image objects rather than image pixels. The purpose of image segmentation is to provide optimal information that simultaneously represents objects in different spatial resolutions for further classification (Gitas, Mitri and Ventura, 2004). The complexity of urban landscape in high resolution remote sensing imagery makes traditional digital image classification difficult. Object-oriented approach is considered more appropriate since it differentiates land cover classes based on both spectral and spatial information of the image data. Some studies demonstrated significantly higher accuracy for the object-oriented approach (Benz et al., 2004; Wang, Sousa, and Gong, 2004), while other investigations reported object-based method produced similar results with comparable accuracy as other methods (Willhauck, 2000; Sun, 2003).

17.3 Pixel-based models for estimating high-resolution impervious surface

17.3.1 Introduction

The objective of this section is to develop a pixel-based method for generating accurate and high-resolution impervious surface data from IKONOS imagery. In particular, this section explores whether the existing impervious surface generation methods (e.g., SMA and RT models) can be applied to high resolution remote sensing imagery. Moreover, an integrated SMA and RT model has been developed to further improve the estimation accuracy. The rest of this section is organized as follows. Section 17.3.2 describes the study area and remotely sensed data, including IKONOS imagery and aerial photographs. Two popular pixel-based impervious surface estimation methods, SMA and RT models, and the integrated SMA and RT model, are detailed in Section 17.3.3. Section 17.3.4 reports the results of accuracy assessment.

17.3.2 Study area and data

Grafton (including the village and town) in Ozaukee County, WI (see Fig. 17.1) was chosen as the study area. Being about

30 km north of the City of Milwaukee, WI, Grafton occupies a geographical area of 65.4 km², with more than 14 000 residents, and approximate 5700 housing units according to Census 2000. Land uses in this region include urban (e.g., commercial, residential, transportation) and rural (e.g., agricultural, forestry). Grafton has experienced a rapid growth in the last decades. In fact, the population of Grafton has increased approximate 58% from 1970 to 2000, and the number of housing units reached about 5800 in 2000, doubled the total housing unit number in 1970. For the study area, an IKONOS image (Fig. 17.1b) acquired on 3 September 2002 was provided by the American Geographical Society Library (AGSL) in the University of Wisconsin-Milwaukee. This IKONOS image was preprocessed and the reflectance for each pixel was calculated from its digital number (DN) values under the guide of Space Imaging technical documents (Peterson, 2001). In order to evaluate the results of automatic impervious surface estimation, a color digital aerial photograph (Fig. 17.1c) for Grafton was also obtained from the AGSL. This photograph was acquired in November 2002, with a spatial resolution about 2 ft (0.61 m). Both the IKONOS image and aerial photograph were orthorectified and re-projected to the UTM projection (zone 16, datum WGS84). Moreover, a further georeference was implemented to reduce geometry misregistration between the IKONOS image and the aerial photograph.

17.3.3 Methodology

With the preprocessed IKONOS imagery, impervious surface information was derived for the study area. In this chapter, two popularly applied impervious surface estimation methods, SMA and RT, have been examined. These two methods have been successfully applied to medium-resolution remote sensing imagery (e.g., Landsat TM and ETM+), but it is still questionable whether they are suitable with the IKONOS imagery, which has a much higher spatial resolution and less number of spectral bands

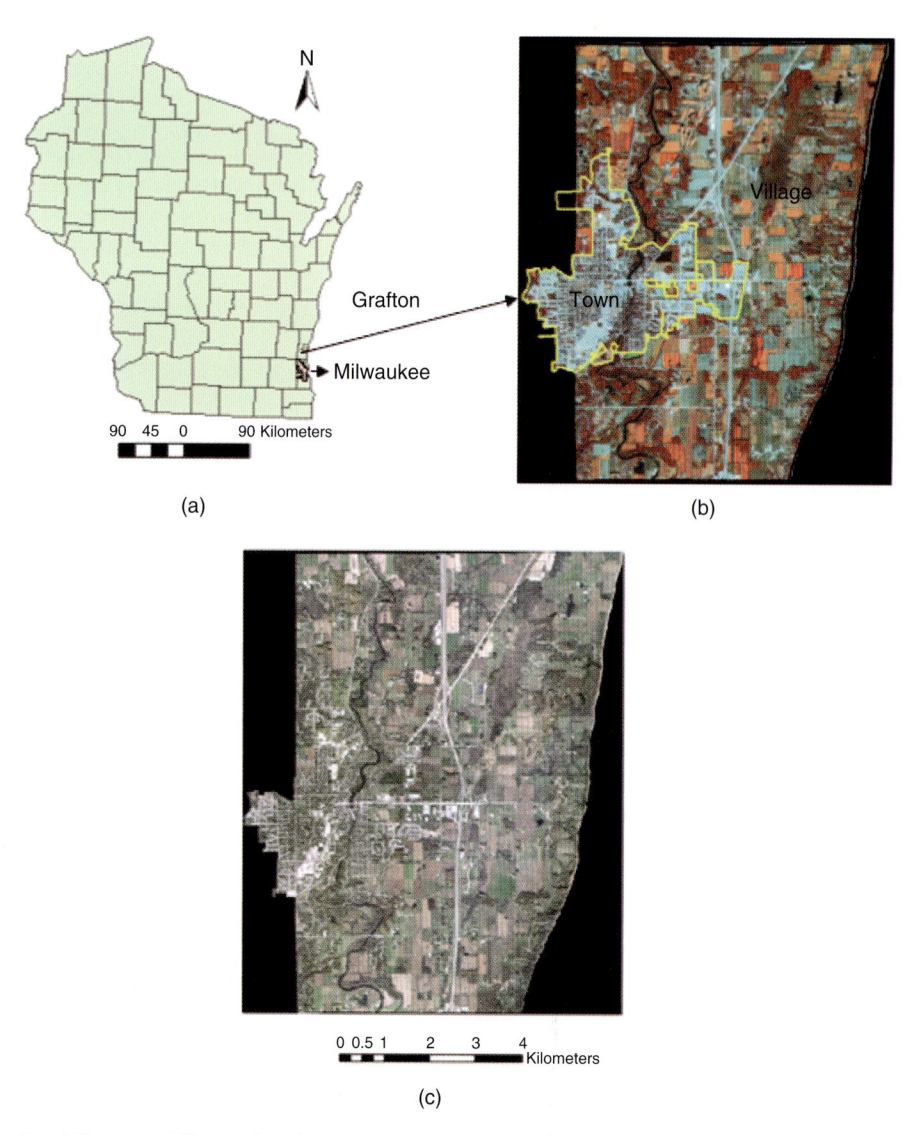

FIGURE 17.1 Grafton Village and Town, Ozaukee, Wisconsin (a) indicates the geographic location of Grafton in Wisconsin; (b) shows the IKONOS imagery for Grafton obtained on September 3, 2002, and (c) illustrates the color aerial photograph taken in November 2002.

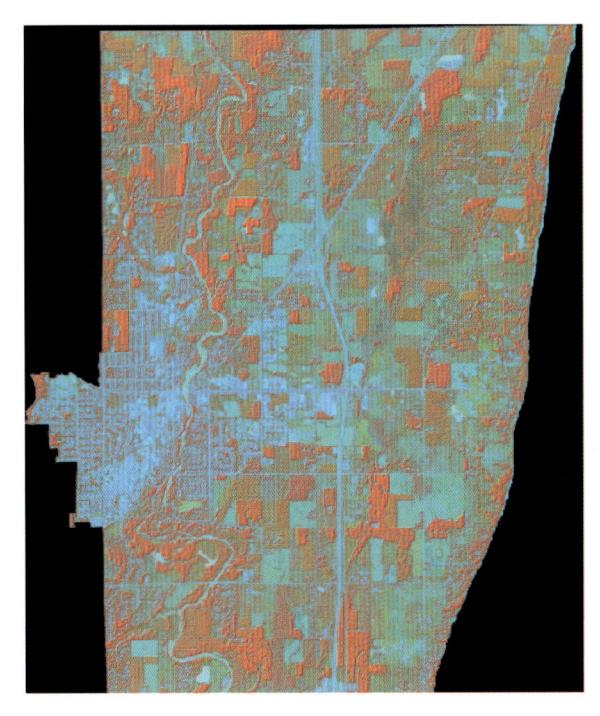

FIGURE 17.2 Normalized reflectance IKONOS imagery for the study area.

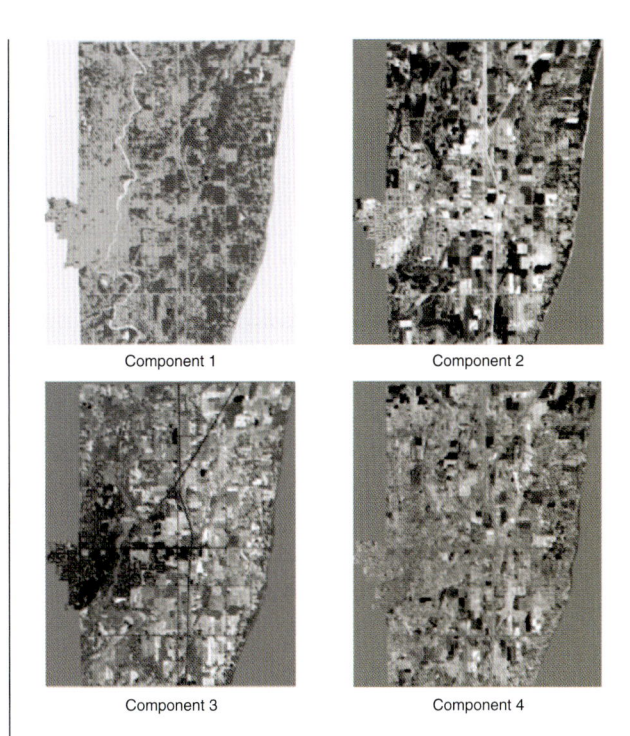

FIGURE 17.3 Maximum noise fraction (MNF) transformation for the IKONOS normalized imagery.

(e.g., four bands compared to six bands in Landsat imagery). In addition, a hybrid approach, which integrating the SMA and RT models, has been developed in this chapter.

17.3.3.1 Spectral mixture analysis model

SMA assumes that a pixel in remote sensing imagery contains a number of land covers, and the spectrum of the pixel is a combination of spectra for these pure land cover types, called endmembers. The fraction of each pure land cover type can be calculated by modeling the relation between the mixed spectrum and the spectrum of each pure land cover type. Dependent on the significance of multiple scattering of light on land cover types, spectral mixture analysis can be divided into linear and non-linear models. Linear models have been popularly applied in urban applications, and therefore, utilized in this section (Phinn et al., 2002; Wu and Murray 2003, Wu, 2004). In this study, a modified normalized SMA model has been developed for generating the fraction of impervious surfaces. This model includes three steps: (1) spectral normalization, (2) maximum noise fraction (MNF) transformation, and (3) spectral mixture analysis. Spectral normalization (see Equation 17.1) reduces spectral variations associated with absolute brightness, while maintains useful information to separate major land cover types.

$$\overline{R_b} = \frac{R_b}{m} \times 100 \quad (17.1)$$

where $m = \frac{1}{n}\sum_{b=1}^{x} R_b$.

Where $\overline{R_b}$ is the normalized reflectance for band b in a pixel; R_b is the original reflectance for band b; m is the average reflectance for that pixel; and n is the total number of bands (4 for IKONOS imagery). With the normalized image (see Fig. 17.2), the next step is to perform an MNF transformation. Unlike the principal component (PC) transformation, which places components according to their variances, the MNF transformation orders components according to their signal to noise ratios. Therefore the MNF transformation is considered superior in keeping information (not the variances) in its first several components. In this study, the MNF transformation was implemented using the minimum/maximum autocorrelation factors (MAF) procedure proposed by Green et al. (1988). The resulting MNF images (see Fig. 17.3) indicate close spatial relationships between MNF components and land cover types. In particular, for MNF component 2, high values are associated with urban areas and bare soil, and low values are related to vegetated areas. In addition, MNF component 3 is valuable for differentiating built-up areas and bare soil, with high values for bare soil, and low values for urbanized areas. This pattern can also be recognized through analyzing the feature space representation of the first three MNF components (see Fig. 17.4). It also indicates that it is feasible to model heterogeneous land uses in the study area using only three endmembers: vegetation, impervious surface, and soil. The selection of these endmembers is based on analyzing the feature space representation and visualizing the IKONOS imagery and the photographs. With the selected endmembers, an SMA model has been applied to the MNF imagery for calculating the fractions of vegetation, impervious surface, and soil in each pixel. This SMA model was formulated as follows.

$$M_b = \sum_{i=1}^{n} f_i M_{i,b} + e_b \quad (17.2)$$

Where $\sum_{i=1}^{n} f_i = 1$ and $f_i \geq 0$.

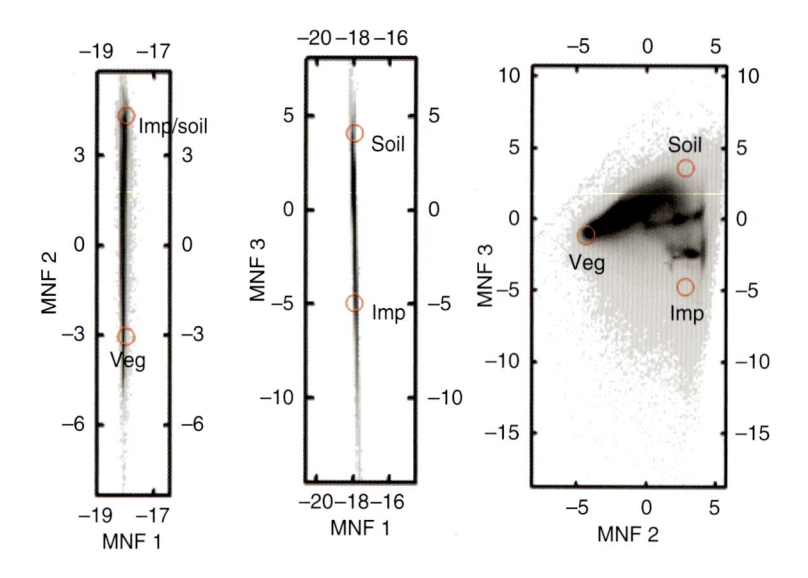

FIGURE 17.4 Feature space representation of the first three MNF components for the normalized reflectance IKONOS image.

Where M_b is MNF component b for a pixel; $M_{i,b}$ is MNF component b for endmember i; f_i is the fraction of endmember i; n' is the number of endmembers; and e_b is the residual. This model was solved using a least squares method in which the residual e_b is minimized. The outputs of this SMA model are the fraction images for vegetation, impervious surface, and soil.

17.3.3.2 Regression tree approach

In this section, a regression tree approach has been applied to the IKONOS imagery for estimating impervious surface distribution in Grafton. In particular, a regression tree program, Cubist, developed by Quinlan (1993), was utilized. Eight independent variables, including four IKONOS reflectance bands and four Tasseled Cap components, were utilized for estimating impervious surface distribution. The Tasseled Cap components were generated by following the procedure provided by Horne (2003). The selections of independent variables are consistent with the method utilized by the researchers in the United States Geological Survey (USGS) (Yang et al., 2003) for producing the 30 × 30 m national land cover dataset (NLCD). For regression tree model development and assessment, 150 training samples were generated using a stratified random methodology to ensure enough samples within urban areas. A 5 × 5 pixel sampling size was utilized to diminish the effects caused by image misregistration. With these training samples, a regression tree model was developed using the Cubist program. The rules developed by the model are listed in Table 17.1.

17.3.3.3 Integrated approach of SMA and RT

Although the impervious surface estimates from both SMA and RT models appear to be consistent with land cover types, some problems still exist. For the SMA model, a major problem may be the selection of endmember's spectra. Except the multiple-endmember spectral mixture analysis (MESMA) method (Roberts et al., 1998), most SMA models only allow a single spectrum for a particular endmember, regardless its internal variations. Although the normalized SMA applied in this paper attempted to reduce the internal spectral variations for each endmember, some variations still exist. Therefore, the choice of endmember spectra may significantly influence the

TABLE 17.1 Rule definition using the regression tree model.

Rule	Condition	Impervious surface fraction
1	tc3 > −30.25508	0
2	b1 ≤ 0.17878 tc1 > 754.3892	−2.17833 + 148 b1 − 41 b3 −0.005 tc1 + 0.003 tc2
3	tc1 ≤ 754.3892	5.45418 + 273 b1 − 72 b3 − 0.015 tc1 + 0.01 tc2 − 7 b4
4	b1 > 0.17878 tc1 > 1379.779 tc3 ≤ −30.25508 tc4 > −33.24916	22.29837 + 270 b1 − 119 b3 −0.019 tc1 − 0.059 tc3 − 0.049 tc4 − 7 b4
5	b1 > 0.17878 tc1 ≤ 1379.779 tc3 ≤ −30.25508 tc4 > −33.24916	39.21432 + 273 b1 − 119 b3 −0.019 tc1 − 0.059 tc3 − 0.049 tc4 − 7 b4
6	b4 > 0.14026 tc2 > −713.7446 tc4 ≤ −33.24916	29.32636 − 0.526 tc2 + 1439 b4 −0.352 tc1 + 79 b1 − 45 b3
7	b4 ≤ 0.14026 tc4 ≤ −33.24916	64.3753 − 0.257 tc4
8	b4 > 0.14026 tc2 ≤ −713.7446	19.65423 − 0.847 tc2 + 2273 b4 −0.558 tc1 + 79 b1 − 45 b3

b1, b2, b3, b4 indicate reflectance for band 1, 2, 3, and 4 in the IKONOS image, tc1, tc2, tc3, and tc4 indicate the Tasseled Cap component 1, 2, 3, and 4.

impervious surface estimation accuracy (Song, 2005). For the RT model, the selection of independent variables may influence the final estimates. In this paper, four reflectance bands and four Tasseled Cap components have been chosen following the studies of Yang et al. (2003). Although this selection seems reasonable, other variables may be more closely related to impervious surface distribution. As an approach to partially address these problems, an integrated approach of the SMA and RT model is proposed. In this approach, the SMA provides an 'initial' estimate, and the RT method is utilized to further calibrate the initial estimate. In particular, the estimates of vegetation, impervious surface, and soil fractions, generated by the SMA model, were inputted to the RT model as independent variables, and the dependent variable is the 'true' impervious surface fraction digitized from the aerial photograph. The same sets of training and testing samples were applied, and the Cubist program was utilized to construct the RT model. The resulting regression tree rules are shown in Table 17.2.

TABLE 17.2 Rule definition for the integrated spectral mixture analysis and regression tree model.

Rule	Condition	Impervious surface fraction
1	imp $\leq$ 15.36628	$-0.05834 + 0.78$ imp
2	imp $>$ 15.36628	$26.59405 + 0.69$ imp $- 0.51$ veg

17.3.3.4 Accuracy assessment

In order to compare the performances of these three models, an accuracy assessment has been carried out. In detail, 150 testing samples with a sampling unit of 5×5 pixels were generated using a stratified random methodology for evaluating the estimation accuracy. For each sample, a 20×20 m sampling area was identified, and the impervious surfaces were digitized using ERDAS IMAGINE™ Area of Interest (AOI) tools. The fraction

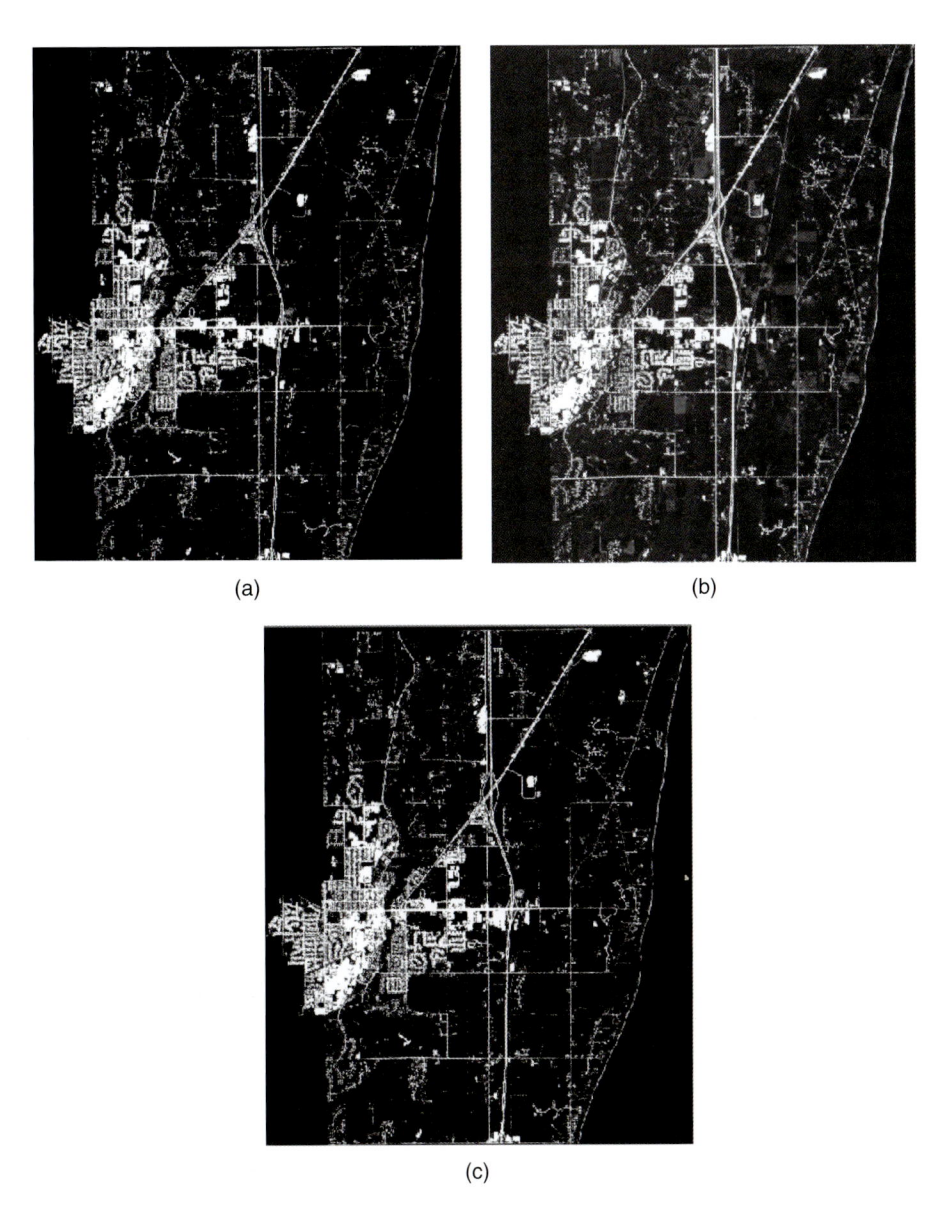

FIGURE 17.5 Impervious surface fraction imagery generated from (a) spectral mixture analysis model, (b) regression tree method, and (c) integrated model of spectral mixture analysis and regression tree.

of impervious surfaces within the sample was calculated through dividing the impervious surface area by the total sampling area. Three parameters were utilized to quantify the accuracy of impervious surface estimation. In particular, the root-mean-square-error (RMSE) (Equation 17.3), the mean average error (MAE) (Equation 17.4), and the correlation coefficient (R^2) (Equation 17.5) between estimated and measured impervious surface fraction, have been utilized.

$$RMSE = \sqrt{\frac{\sum_{i=1}^{N}(\hat{I}_i - I_i)^2}{N}} \quad (17.3)$$

$$MAE = \frac{1}{N}\sum_{i=1}^{N}|I_i - \hat{I}_i| \quad (17.4)$$

$$R^2 = \frac{\sum_{i=1}^{N}(I_i - \bar{I})^2}{\sum_{i=1}^{N}(\hat{I}_i - \bar{I})^2} \quad (17.5)$$

Where $\hat{I}_i$ is the estimated impervious surface fraction for sample i; I_i is the 'true' impervious surface fraction digitized from aerial photos; $\bar{I}$ is the mean of the samples; and N is the total number of samples. With these three measures, the RMSE and MAE measure the relative prediction errors of the model, and the correlation coefficient (R^2) measures the fitness of the model. In this study, all three measures have been calculated for comparison purposes. In addition to these percentage based accuracy assessment, traditional hard accuracy assessment was also performed. In particular, a pixel is classified as impervious surface if the percentage of impervious surface in that pixel is over 50%, otherwise this pixel is classified as non-impervious surface.

17.3.4 Results

The impervious surface fraction images derived from the SMA, RT, and the integrated SMA and RT approaches are reported in Fig. 17.5(a, b, and c), respectively. Figure 17.5(a) indicates that the SMA method successfully quantifies the general pattern of impervious surface distribution. In particular, the fraction of impervious surfaces is high in commercial areas (City of Grafton), and near zero in agricultural and forest areas. Moreover, the confusion between urban areas and bare soil has been successfully addressed for the study area. Similarly, the RT method (Fig. 17.5b) has successfully identified the general pattern of land cover in the study area, with high impervious surface fraction in commercial, transportation, and residential areas, and low impervious surface fraction in rural areas. This method, however, may slightly over-estimate impervious surface fractions in some regions with bare soils. Finally, the integrated SMA and RT method has also clearly illustrated the land cover patterns in the study area (see Fig. 17.5c).

For an objective assessment of these three methods, the estimation accuracies are calculated and reported in Fig. 17.6 and Table 17.3. Results (see Table 17.3) indicate that regression tree has the highest overall classification accuracy (90.7%), while SMA and the integrated approach have slightly lower accuracy (87.3%). In addition, through visualizing Fig. 17.6 and Table 17.3, several conclusions may also be obtained. First, the SMA and RT models, which are primarily applied to coarse-resolution remotely sensed data, can be transferred to IKONOS imagery with similar accuracy. As shown in Table 17.3, the RMSEs for both SMA and RT models are about 10–12%, and MAEs are about 6–8%, which are comparable to the reported accuracies when these models applied to coarse-resolution data (Wu and Murray 2003; Yang

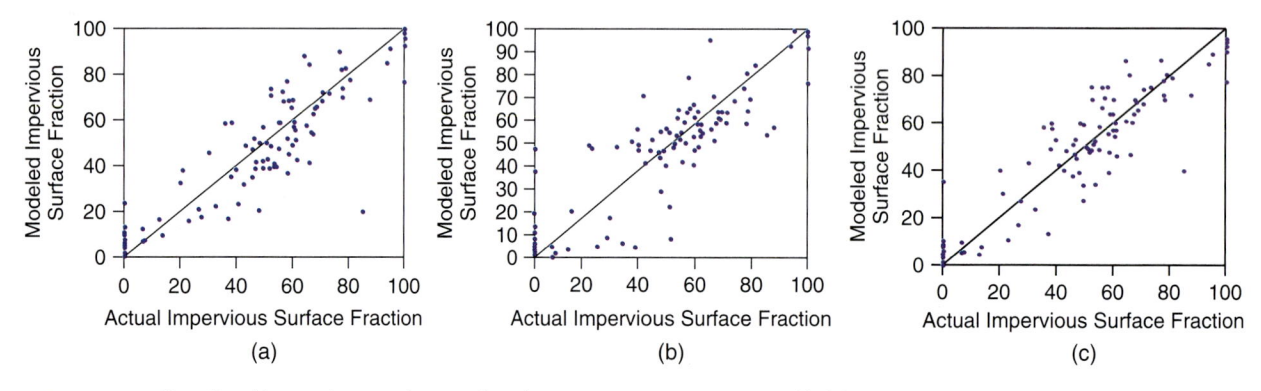

FIGURE 17.6 Results of impervious surface estimation accuracy assessment with (a) spectral mixture analysis, (b) regression tree analysis, and (c) integrated model of spectral mixture analysis and regression tree.

TABLE 17.3 Accuracy assessments of pixel-based impervious surface estimation models.

	Root-mean-square-error (RMSE)	Mean average error (MAE)	Correlation coefficient (R^2)	Overall classification accuracy
Spectral mixture analysis	10.697	6.716	0.903	87.3%
Regression tree	11.835	7.396	0.865	90.7%
Integrated spectral mixture analysis and regression tree	9.687	6.052	0.922	87.3%

et al., 2003). As a further comparison, the SMA method performs slightly better than the RT model. Comparing with the RT model, the SMA method has a slightly lower estimation error (e.g. $RMSE = 10.7\%$ and $MAE = 6.7\%$) and higher model-fit ($R^2 = 0.90$). Moreover, Fig. 17.6(b) indicates the RT method may be problematic when applied in areas with low impervious surface factions (e.g., < 40%) because of the confusions between impervious surfaces and bare soil. Finally, the integrated SMA and RT model outperforms the individual SMA and RT models, as it has the lowest $RMSE$ (9.69%) and MAE (6.05%), and the highest correlation coefficient (0.92). This may be because the integrated model partially addresses the endmember selection problem in the SMA model and the variable selection issue in the RT approach.

17.4 Object-based models for estimating high-resolution impervious surface

Unlike the pixel-based impervious surface estimation models detailed in Section 17.3, object-based models use both spectral and spatial contextual information of the image data. A bottom up region-merging approach is usually used to segment the image into objects consisting of multiple pixels. This case study demonstrates an example of impervious surface mapping using an object-oriented approach.

17.4.1 Study area and data preparation

The study site is a 2.1×1.7 km^2 area surrounding the Minnesota State University, Mankato campus, with the central coordinates of 44.15° N and 93.99° W. Given the varied land covers it includes, it provides a good test site for the purposes of the study. A Quickbird image acquired on 6 October 2003 was obtained from the Digital Globe archive collection. The data were recorded in 11 bits and were radiometrically and geometrically corrected. The image includes four multispectral bands at 2.4-m resolution and one panchromatic band at 0.6-m resolution. The sun azimuth and elevation angle were 158.1 and 38.5 degrees and the satellite azimuth and elevation view angles were 27.7 and 76.3 degrees. The primary ancillary data used was the National Agriculture Imagery Program (NAIP) color digital ortho-rectified aerial photography, acquired on 26 June 2003 by the US Department of Agriculture. This color aerial imagery was utilized as visual reference for selecting training and testing samples.

All data were projected to UTM (spheroid NAD83, Zone 15) and Datum GRS 1980 system. The training and test samples for the object-oriented classification were manually delineated in eCognition™. The locations of samples were selected randomly and distributed evenly across the study area. All samples were checked against the 1-m NAIP color aerial imagery. Compared to pixel-based classifications, the object-based approach using the nearest neighbor classifier requires fewer training samples since one sample object includes several to many typical pixel samples and their variations (Baatz et al., 2004).

17.4.2 Object-oriented classification

Three major steps were involved in the impervious surface mapping procedure. First, object-based classification was applied to the Quickbird imagery to classify the data into five general land cover classes: impervious surface, forest, water, non-forested rural, and shadow. In the next step of the classification, shadow was further classified to impervious or non-impervious area using class-related features. Lastly, final refined impervious maps are generated by combining the "shadowed" imperious areas with the classified impervious class.

Segmentation is the first and most important process of the eCognition™-based object-oriented classification. It extracts image objects at modifiable scale parameters, single layer weights, and the mixing of the homogeneity criterion concerning color and shape. In other words, the outcome of segmentation is directly related to several adjustable criteria – scale parameter, color, and shape – defined by users. In particular, the scale parameter is a measure for the maximum change in heterogeneity that may occur when merging two image objects. A larger scale parameter value leads to bigger objects and vice versa. Adjusting the shape factor will also affect the overall fusion value computed based on both spectral heterogeneity and the shape heterogeneity. Therefore, how to choose an optimal scale parameter and shape factor is critical to the quality of classification. To set the appropriate objects for use and to evaluate how classification accuracy changes when adjusting these two criteria, different scale parameters ranging from 5 to 50 and shape factors ranging from 0.1 to 0.9 were tested.

Besides the scale parameter and shape factor, the result of object-oriented classification is also dependent on the object-based metrics, including measures of texture, length, and shape as well as measures of spatial relationships to other super-, sub-, and neighboring objects, utilized in the classification process. In this study, two extra object-based metrics – "Ratio of Band 1", and "Ratio of Band 4" – were selected and tested in addition to the brightness values. The "Ratio of Band L" is the band L mean value of an image object divided by the sum of all spectral band mean values (Baatz et al., 2004). For impervious surfaces, the "Ratio of the Band 1" (blue spectral band) is comparatively high, and the "Ratio of Band 4" (near infrared spectral band) is relatively low, which may help differentiate impervious areas from the other land cover classes. To determine if the use of the metrics would improve the classification, step-wise classifications by adding these metrics were also performed.

To classify shadow into different land cover types, class-related features – relative to their adjacent forest or impervious – were added in the next level process. For example, if more than 55% of the relative border of the shadow object was impervious surfaces or forest, then the shadow object was assigned a new value of impervious or forest respectively. Otherwise, the shadow was classified as non-forest.

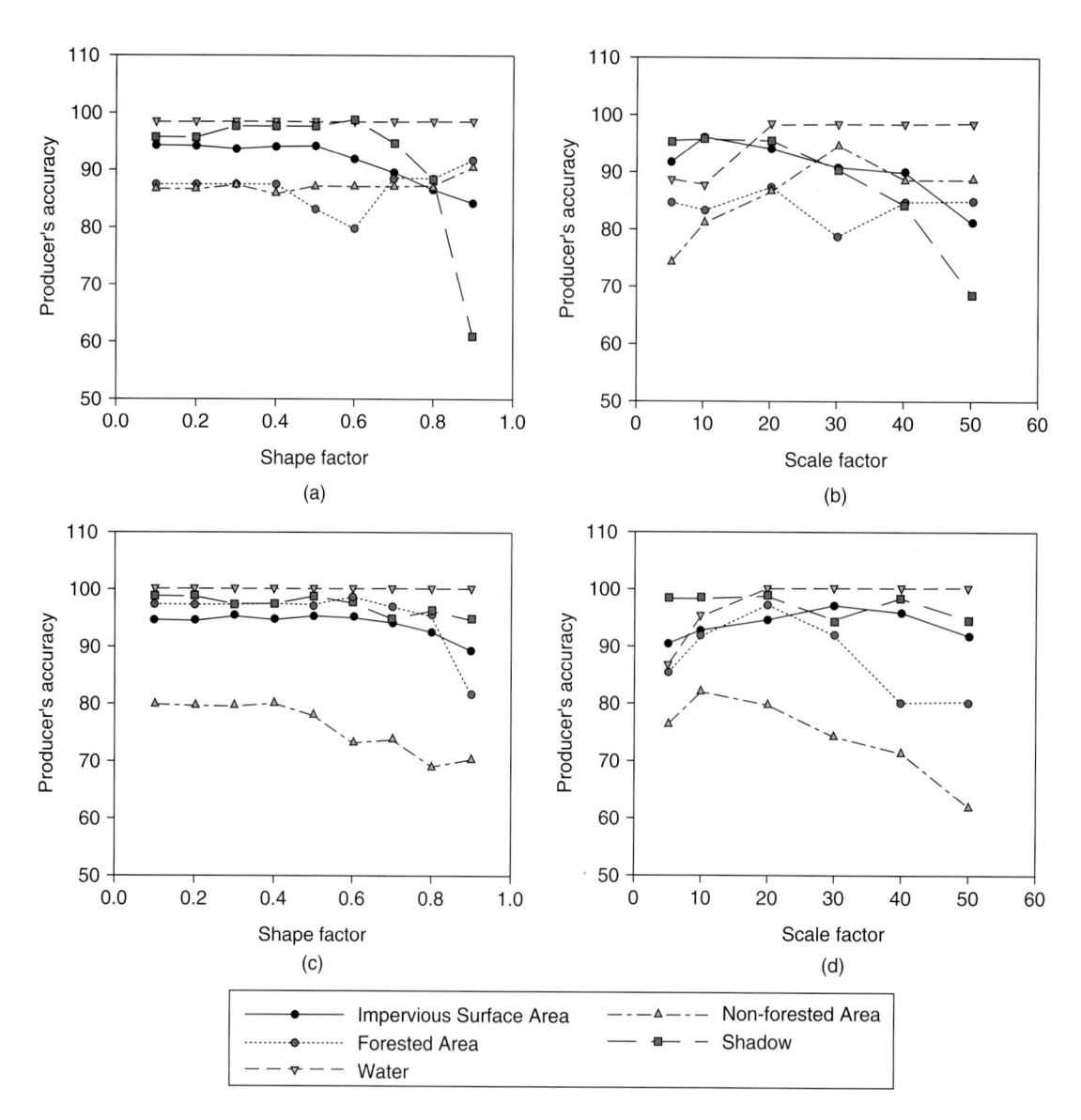

FIGURE 17.7 Effects of shape factor (a, b) and scale parameters (c, d) on the accuracy of object oriented classification.

TABLE 17.4 Summary of object oriented classifications accuracies (%) with different feature variables.

Land Cover Class	Mean brightness of all bands		Mean brightness of all bands; ratio of band 1		Mean brightness of all bands; ratio of band 4		Mean brightness of all bands; ratio of band 1; ratio of band 4	
	Producer	User	Producer	User	Producer	User	Producer	User
Impervious	94.4	93.2	92.1	94.4	94.2	93.7	94.2	94.5
Forest area	97	81.3	87.4	97.2	97	65.3	87.4	97.2
Water	98.5	100	98.5	94.7	98.5	100	98.5	100
Non-forested	71.9	83	86.6	76.7	55.3	76.7	86.6	79.5
Shadow	95.7	96.7	95.7	98.7	97.7	98.7	95.7	98.7
Overall	90.7		91.3		87.6		92.4	
Kappa	85.9		86.9		81.3		88.5	

17.4.3 Results

The results of the object-oriented classification change over the user-defined values of shape and scale factors (Fig. 17.7). Different land cover classes show different change trends. For example, the accuracy of water is consistent while the impervious class shows steadily decreasing accuracy with increases in the shape factor value. The overall classification accuracy reaches the highest point when the shape factor approximately equals to 0.3 and then it trends down to the lowest when the value is 0.9. On the other hand, the highest accuracy can be found for majority of the classes when scale parameter reaches about 20. Therefore, the optimal shape factor and scale parameter are set to 0.3 and 20, respectively for our classification.

Our analysis also demonstrates that adding the two extra feature variables of "Ratio of Band 4" and "Ratio of Band 1" did affect the classification accuracy. While the Kappa value increases 1% when adding the "Ratio of Band 1" only, the overall accuracy decreases 3% by adding the "Ratio of Band 4". This is probably because the "Ratio of Band 4" shows similar values for some forests and non-forested vegetation, which lead to decreased accuracy for forest and non-forested areas. Nonetheless, when both ratios of band 1 and band 4 were added, the classification results increased about 2% (Table 17.4).

The land cover map was classified based on the optimal shape and scale factor values and with the two extra feature variables included (Fig. 17.8a). Its accuracy has been assessed by creating error matrix using the same test area as reference data (Table 17.5). An overall accuracy of 92.5% was achieved for the

FIGURE 17.8 Object-oriented classification maps based on the QuickBird multispectral data. (a) General land covers classified by the object-oriented classifier, (b) Final impervious surface map after unmixing shadow class into impervious surface or other classes.

TABLE 17.5 Error matrix of object-based land cover classification.

Reference class	Object-oriented classification					
	Impervious	Forest	Water	Non-forest	Shadow	Producer's accuracy (%)
Impervious	7904	3	0	490	36	93.7
Forested area	0	1569	0	226	0	87.4
Water	0	0	935	0	14	98.5
Non-forested area	399	0	0	2752	0	87.3
Shadow	0	43	0	0	1799	97.7
User's accuracy (%)	95.2	97.2	100	79.4	97.3	
Overall accuracy (%) = 92.5 Kappa (%) = 88.7						

five classes. The final derived impervious surface classification map shows homogeneous patterns (Fig. 17.8b). Nevertheless, for small impervious patterns such as the single-family residential buildings, the object-based method tends to amalgamate impervious buildings and some surrounding lawn areas to the same object. A separate accuracy assessment using 200 randomly sampled impervious points indicated a 94% overall accuracy for the impervious/non-impervious map.

Conclusions

This chapter reviews the two groups of high resolution impervious surface estimation models: pixel-based and object-based methods. For pixel-based models, this chapter explored whether two popular impervious surface estimation methods developed for coarse-resolution remote sensing data can be transferred to IKONOS imagery. Further, to partially address the problems associated with the SMA and RT approaches, this paper developed an integrated model to improve estimation accuracy. Results suggest both SMA and RT models can be successfully applied to high-resolution remote sensing imagery with similar estimation accuracy, while the performance of the SMA model is slightly better. Moreover, the integrated model produces the best estimation accuracy, with the lowest *RMSE* and *MAE*, and the highest correlation coefficient between the modeled and 'true' impervious surface fractions. In addition, when hard classification is considered, the accuracy of overall classification is around 90%. These results suggest that these automatic methods have the potential to replace the labor-intensive and time-consuming digitizing process in creating high-resolution impervious surface information.

Although the results from pixel-based models are promising (with the *RMSE* about 10% and correlation coefficient about 0.9), it is still necessary to explore whether object-based methods can further improve the estimation accuracy. Through this case study, we found the object-based classification can produce accurate impervious surface map. The QuickBird imagery enabled mapping a complex urban area with high spatial variation. In addition, it is very convenient to refine the object-oriented classification result in eCognition™ since that the classification process is iterative. Nevertheless, the object-based method tends to amalgamate small buildings and surrounding lawn areas together. In object-oriented image analysis, multiresolution segmentation separates adjacent regions in an image as long as they have significant contrast. Successful segmentation should create image objects that have optimal information for further extraction of land cover information. Our study confirms the outcome of image segmentation is directly related to the user defined parameters of scale and shape. However, defining the optimal scale and shape factor is an intricate task since different land cover classes demonstrate different characteristics in relation to the modifications of these parameters. Moreover, this study indicates shadow problem can be addressed with class-related features in object-based classification. To obtain accurate impervious surface map, we have to differentiate tree shadows, mostly occurring in pervious areas, from building shadows that may occur in both pervious and impervious areas.

References

Adams, J.B., Sabol, D.E., Kapos, V., et al. (1995) Classification of multispectral images based on fractions of endmembers: Application to land cover change in the Brazilian Amazon. *Remote Sensing of Environment*, **52**, 137–154.

Baatz, M., Benz, U., Dehghani, S. et al. (2004) eCognition User Guide 4. Definiens Imaging.

Bauer, M.E., Loeffelholz B.C. and Wilson, B. (2007) Estimating and mapping impervious surface area by regression analysis of Landsat imagery, *Remote Sensing of Impervious Surfaces* (Q. Weng, editor), CRC Press, Boca Raton, pp. 3–20.

Benz, U.C., Hofmann, P., Willhauck, G., Lingenfelder, I. and Heynen, M. (2004) Multi-resolution object-oriented fuzzy analysis of remote sensing data for GIS-ready information. *ISPRS Journal of Photogrammetry and Remote Sensing*, **58**, 239–258.

Breiman, L., Friedman, J.H., Olshen, R.A. and Stone, C.J. (1984) *Classification and Regression Trees*, Wadsworth, Inc., Monterey.

Dougherty, M., Dymond, R.L., Goetz, S.J., Jantz, C.A. and Goulet, N. (2004) Evaluation of impervious surface estimates in a rapidly urbanizing watershed. *Photogrammetric Engineering and Remote Sensing*, **70** (11), 1275–1284.

Flanagan, M. and Civco, D.L. (2001) Subpixel impervious surface mapping. In *Proceedings of American Society for Photogrammetry and Remote Sensing Annual Convention*, St. Louis, MO, April 23–27.

Gillies, R.R., Box, J.B., Symanzik, J. and Rodemaker, E.J. (2003) Effects of urbanization on the aquatic fauna of the Line Creek watershed, Atlanta-a satellite perspective. *Remote Sensing of Environment*, **86**, 411–422.

Gitas, I.Z., Mitri, G.H. and Ventura, G. (2004) Object-based image classification for burned area mapping of Creus Cape, Spain, using NOAA-AVHRR imagery. *Remote Sensing of Environment*, **92**, 409–413.

Goetz, S.J., Wright, R.K., Smith, A.J., Zinecker, E. and Schaub, E. (2003) IKONOS imagery for resource management: tree cover, impervious surfaces, and riparian buffer analyses in the mid-Atlantic region. *Remote Sensing of Environment*, **88**, 195–208.

Green, A.A., Berman, M., Switzer, P. and Craig, M.D. (1988) A transformation for ordering multispectral data in terms of image quality with implications for noise removal, *IEEE Transactions on Geoscience and Remote Sensing*, **26**, 65–74.

Heinert, N.J. (2002) Impervious surface mapping using Landsat TM and IKONOS satellite remote sensing. MGIS (Master of Geographic Information Science) capstone paper. University of Minnesota.

Horne, J.H. (2003) A Tasseled Cap transformation for IKONOS images. In *ASPRS 2003 Annual Conference Proceedings*, May 2003, Anchorage, Alaska, USA.

Huang, C. and Townshend, J.R.G. (2003) A stepwise regression tree for nonlinear approximation: applications to estimating subpixel land cover, *International Journal of Remote Sensing*, **24** (1), 75–90.

Ji, C.Y. (2000) Land-use classification of remotely sensed data using kohonen self-organizing feature map neural networks. *Photogrammetric Engineering & Remote Sensing*, **66** (12): 1451–1460.

Kavzoglu, T. and P.M. Mather (2003) The use of backpropagating artificial neural networks in land cover classification. *International Journal of Remote Sensing*, **24** (23): 4907–4938.

Lu, D. and Weng, Q. (2004) Spectral mixture analysis of the urban landscape in Indianapolis city with Landsat ETM+ imagery. *Photogrammetric Engineering & Remote Sensing*, **70** (9), 1053–1062.

Lu, D. and Weng, Q. 2007. Mapping urban impervious surfaces from medium and high spatial resolution multispectral imagery. In *Remote Sensing of Impervious Surfaces* (Q. Weng, ed.), CRC Press, Boca Raton, pp. 59–77.

Lu, D. and Weng, Q. (2009) Extraction of urban impervious surfaces from IKONOS imagery. *International Journal of Remote Sensing*, **30** (5), 1297–1311.

Mohapatra, R.P. and Wu, C. (2007) Estimation with IKONOS imagery: an artificial neural network approach. In *Remote Sensing of Impervious Surfaces* (Q. Weng, ed.), CRC Press, Boca Raton, pp. 21–39.

Morton, T. and Yuan, F. (2009) Analysis of population dynamics using satellite remote sensing and U.S. census data. *Geocarto International*, **24** (2), 143–163.

Peterson, B. (2001) IKONOS relative spectral response and radiometric calibration coefficients. Document number SE-REF-016, Rev. A. Space Imaging, Inc., Thornton, CO.

Phinn, S., Stanford, M., Scarth, P., Murray, A.T. and Shyy, T. (2002) Monitoring the composition and form of urban environments based on the vegetation-impervious surface-soil (VIS) model by sub-pixel analysis techniques. *International Journal of Remote Sensing*, **23** (20), 4131–4153.

Pu, R., Gong, P., Michishita R. and Sasagawa, T. (2008) Spectral mixture analysis for mapping abundance of urban surface components from the Terra/ASTER data. *Remote Sensing of Environment*, **112**, 939–954.

Quinlan, J.R. (1993) Combining instance-based and model-based learning. In *Proceedings of the 10th International Conference of Machine Learning* (Utgoff, P., editor), Morgan Kaufmann Publishers, Amherst, MA, pp. 236–243.

Roberts, D.A., Gardner, M., Church, R., Ustin, S., Scheer, G. and Green, R.O. (1998) Mapping chaparral in the Santa Monica mountains using multiple endmember spectral mixture models. *Remote Sensing of Environment*, **65**, 267–279.

Sawaya, K., Olmanson, L., Heinert, N., Brezonik, P. and Bauer M. (2003) Extending satellite remote sensing to local scales: Land and water resource monitoring using high-resolution imagery. *Remote Sensing of Environment*, **88**, 144–156.

Schueler, T.R. (1994) The importance of imperviousness. *Watershed Protection Techniques*, **1** (3), 100–110.

Small, C. (2003) High spatial resolution spectral mixture analysis of urban reflectance. *Remote Sensing of Environment*, **88**, 170–186.

Song, C. (2005) Spectral mixture analysis for subpixel vegetation fractions in the urban Environment: how to incorporate endmember variations? *Remote Sensing of Environment*, **95** (2), 248–263.

Sun, X. (2003) Comparison of pixel-based and object-oriented approaches to land cover classification using high-resolution IKONOS satellite data. Master's Thesis, University of Minnesota.

Wang, L., Sousa, W. and Gong, P. (2004) Integration of object-based and pixel-based classification for mangrove mapping with IKONOS imagery. *International Journal of Remote Sensing*, **25** (24), 5655–5668.

Weng, Q. and Hu, X. (2008) Medium spatial resolution satellite imagery for estimating and mapping urban impervious surfaces using LSMA and ANN. *IEEE Transactions on Geoscience and Remote Sensing*, **46** (8), 2397–2406.

Willhauck, G. (2000) Comparison of object-oriented classification techniques and standard image analysis for the use of change detection between SPOT multispectral satellite images and aerial photos. *International Archives of Photogrammetry and Remote Sensing*, **XXXIII**, Part B3, 214–221.

Wu, C. (2004) Normalized spectral mixture analysis for monitoring urban composition using ETM+ imagery. *Remote Sensing of Environment*, **93**, 480–492.

Wu, C. (2009) Quantifying high-resolution impervious surfaces using spectral mixture analysis, *International Journal of Remote Sensing*, **30** (11), 2915–2932.

Wu, C. and Murray, A. (2003) Estimating impervious surface distribution by spectral mixture analysis. *Remote Sensing of Environment*, **84**, 493–505.

Wu, C and Murray, A.T. (2007) Population estimation using Landsat Enhanced Thematic Mapper imagery. *Geographical Analysis*, **39**, 26–43.

Xian, G. and Crane, M. (2005) Assessments of urban growth in the Tampa Bay Watershed using remote sensing data. *Remote Sensing of Environment*, **97**, 203–215.

Yang, L., Xian, G., Klaver, J.M. and Deal, B. (2003) Urban land-cover change detection through sub-pixel imperviousness mapping using remotely sensed data. *Photogrammetric Engineering and Remote Sensing*, **69** (9), 1003–1010.

Yang, X. (2006) Estimating landscape imperviousness index from satellite imagery. *IEEE Geoscience and Remote Sensing Letters*, **3** (1), 6–9.

Yang, X. and Liu, Z. (2005) Use of satellite-derived landscape imperviousness index to characterize urban spatial growth. *Computers, Environment and Urban Systems*, **29** (5), 524–540.

Yuan, F, Bauer, M.E., Heinert, N.J. and Holden, G. (2005) Multi-level land cover mapping of the Twin Cities (Minnesota) metropolitan area with multi-seasonal Landsat TM/ETM+ data. *Geocarto International*, **20** (2), 5–14.

Yuan, F. and Bauer, M.E. (2007) Comparison of impervious surface area and Normalized Difference Vegetation Index as indicators of surface Urban Heat Island effects in Landsat Imagery. *Remote Sensing of Environment*, **106** (3), 375–386.

Yuan, F., Wu, C. and Bauer, M.E. (2008) Comparison of various spectral analytical techniques for impervious surface estimation using Landsat imagery. *Photogrammetric Engineering and Remote Sensing*, **74** (8), 1045–1055.

18

Use of impervious surface data obtained from remote sensing in distributed hydrological modeling of urban areas

Frank Canters, Okke Batelaan, Tim Van de Voorde, Jarosław Chormański and Boud Verbeiren

While the increase of impervious surface cover in urbanized areas has a clear impact on urban hydrological processes, the relationship between flood conditions and urban development has been poorly studied. This chapter focuses on a case study demonstrating the impact of different remote sensing methods for characterizing the distribution of impervious surfaces on runoff estimation, and how this affects the assessment of peak discharges in an urbanized watershed in the Brussels Capital Region, Belgium. In the study use is made of WetSpa, a grid-based spatially distributed hydrological model adapted to incorporate information on the proportion of different types of land-cover at grid cell level. The study shows that use of detailed information on the spatial distribution of impervious surfaces, as obtained from remotely sensed data, strongly affects local runoff estimation and has a clear impact on the modeling of peak discharges. Little difference, however, is observed between results obtained with impervious surface maps derived from high-resolution remote sensing data (IKONOS, 4 m resolution) and sub-pixel estimates of impervious surface cover derived from satellite data matching the model's resolution (Landsat, 30 m resolution).

18.1 Introduction

More than half of the world's population lives in urbanized areas, and in the future cities will house an even increasing number of people in both absolute and relative terms (Martine, 2007). Many cities worldwide also grow faster spatially than demographically. A study of the European Environment Agency states that European cities have expanded on average by 78% since the mid-1950s, while during the same period the population increased by only 33% (EEA, 2006). One of the most obvious impacts of urban expansion is a substantial increase of impervious surfaces, i.e., man-made surface types that prevent direct infiltration of surface water into the soil, forcing it to travel downhill to sewers or to places where it can penetrate. Parking lots, pavements, asphalted streets and rooftops are all examples of urban surface types that make it impossible for precipitation to directly seep into the ground. The increase of impervious surfaces leads to more surface runoff, which in turn increases the risk for water pollution (Peters, 2009) and floods in urbanized watersheds, hampers the recharge of aquifers, and boosts erosion (Schueler, 1994; Brun and Band, 2000). Incorporating information on the spatial distribution of impervious surfaces is thus important in hydrological modeling of urbanized areas. Impervious surface cover is also increasingly used as a key indicator for surface water quality (Carlson, 2008), aquatic fauna (Gillies et al., 2003) and for the overall ecological condition of watersheds (Arnold and Gibbons, 1996; Sleavin et al., 2000).

The importance of impervious surface cover for storm drainage is well known to hydraulic engineers, who use since the second half of the 19th century the "rational method" for the prediction of peak discharge from small drainage catchments mostly for the design of urban drainage systems (Mulvany, 1850; Pilgrim and Cordery, 1992; TxDOT, 2009). In the rational method a runoff coefficient is used, which is defined as the fraction of rainfall that becomes runoff and which is assumed to be independent of rainfall intensity or volume. The runoff coefficient is typically selected from tables based on the type of land-use. For a catchment mostly a spatially constant or an area-weighted value based on the land-use distribution is determined. Since the seminal paper of Horton (1933) on the role of infiltration in the hydrological cycle much more awareness exists on the importance of the infiltration capacity of soils for runoff generation. Horton describes how the infiltration capacity of a drainage basin can be determined from runoff and rain-intensity data and how conversely, if the infiltration capacity is known, the surface runoff can be determined. He concludes that this is an improvement over the rational method (Horton, 1933). Both the rational method as well as the Hortonian runoff approach have influenced hydrologists for decades in applying spatially constant runoff concepts and models, although Beven (2006) remarks that Horton did show how different parts of a catchment with different infiltration capacities could be taken into account in the modeling.

In the 1960s a clear impact was identified of urban development on the hydrological regime in general and on peak discharges in specific (Carter, 1961; Anderson, 1968; Leopold, 1968). Since the 1980's various studies also pointed at the correlation between imperviousness and the health of drainage basins, leading to the use of imperviousness as a simple, easily measurable index of environmental disturbance (Schueler, 1994; Arnold and Gibbons, 1996; Moglen, 2009). One of the best-known applications linking watershed imperviousness to environmental quality is the Connecticut Nonpoint Education for Municipal Officials (NEMO) project, which was initiated in 1991 to assist communities in dealing with the complexities of polluted runoff management (Arnold et al., 1994). In this project remote sensing data were used to estimate current impervious surface cover as well as to determine zoning specific impervious surface cover coefficients, which were then applied to estimate future levels of imperviousness within watersheds. This gave town officials a prediction of the future status of their town in terms of impervious surface cover and health of their local water resources (Arnold and Gibbons, 1996). Schueler, Fraley-McNeal and Cappiella (2009) performed a review of recent research on the impervious cover model (ICM), which is based on the hypothesis that the behaviour of urban stream indicators can be predicted from the percentage of impervious cover in their contributing sub-watershed. Results of a meta-analysis of 65 new (since 2003) and 250 older studies showed that the majority of research confirmed or reinforced the basic premise of the ICM. The authors present a reformulated ICM model that can be readily understood by watershed planners, storm-water engineers, water quality regulators, economists, and policy makers. However, they conclude also that more information is needed to extend the ICM as a method to classify and manage small urban watersheds and decide on the optimum combination of best management practices to protect or restore streams within each sub-watershed classification (Schueler, Fraley-McNeal and Cappiella, 2009).

With the increasing awareness of the impact of impervious surface cover on the urban hydrological cycle there is also an increase in environmentally sensitive design practices. This requires that watershed professionals and storm water practitioners not only have a need to better understand the role of impervious cover types in urban settings, but also of pervious covers (Law, Cappiella and Novotney, 2009). Mejía and Moglen (2009) state that the relationship between flood conditions and the spatial distribution of urban development has been poorly studied, often because of limitations in streamflow data availability or because of the common use of lumped watershed models in urban hydrological modeling. They conclude that measures for minimizing the impact of urban development on water resources will have to take better account of spatial patterns and morphology of urban areas.

18.2 Spatially distributed hydrological modeling

Hydrological modeling has evolved enormously over the last four decades and has taken advantage of the development of computational power. Traditionally, storm runoff response is determined by the Rational Method and Soil Conservation Service (SCS) Curve Number (CN) method (USDA, 1986). These, and other simple empirical methods, have been widely used to predict the effect of urbanization on precipitation runoff processes (Carlson, 2004, 2008; Melesse and Wang, 2008). Although these methods can make use of area-weighted influence of different land-uses on the runoff coefficient or curve number they are not able to take spatial patterns into account.

An early example of the use of a simulation model for the estimation of the effect of urban development on peak discharges was an application of the Stanford Watershed Model (James, 1965). The Stanford Watershed Model (Crawford and Linsley, 1966) and its follow up HSPF (Bicknell *et al.*, 1997) utilize time series meteorology data to simulate hydrological processes in both pervious and impervious land segments. Although the integration of land-use related information in lumped watershed models like HSPF allows one to take account of differences in physical characteristics within the catchment, linked to different types of land-use, the lumped approach does not define physical processes in a spatially explicit way (NOAA, 2006). Spatial interaction between neigbboring locations is not considered in the modeling. This can be seen as a strong limitation, especially in urbanized areas, which are often characterized by highly complex land-use patterns with a strong impact on runoff processes.

The need to better deal with the spatial characteristics of hydrological processes has led to the development of distributed hydrological models, which explicitly take into account the physical characteristics of every location, within the limits of a certain discretization defined by the model's grid cell size. Hence, there is a strong tendency in research towards improving hydrological models to integrate as much as possible different hydrological processes taking place in the atmosphere, vegetation, land surface, soil and subsurface at the natural scale of a catchment. Most models opt for a physical description of the hydrological processes instead of e.g. statistical relationships. Physically based models have the obvious advantage that it can be assumed that predicted consequences of changed conditions hold within the calibrated range of the model and more importantly that predicted consequences can be explained in function of changed physical conditions. Early examples and trendsetters were TOP-MODEL (Beven and Kirkby, 1979) and the SHE model (Abbott *et al.*, 1986), while presently SWAT (Gassman *et al.*, 2007) is widely used. For a recent review we refer to Todini (2007). Developments in GIS technology combined with an increasing use of remote sensing as a data source for producing spatially distributed data are the backbone of this trend in hydrological modeling.

Remote sensing provides hydrologists spatial as well as temporal data on physical properties of the earth's surface that may be used for estimating hydrological model parameters (Schmugge *et al.*, 2002). Land-use datasets, which are usually obtained through visual interpretation of aerial photography or satellite data, are probably the most common remote sensing derived product used in hydrological modeling. Just like in lumped modeling approaches, also in distributed hydrological modeling relevant model parameters, such as vegetation type and vegetation fraction, leaf area index, interception capacity, root depth and Manning coefficient, are often directly inferred from land-use, based on general look-up tables found in the literature (Liu *et al.*, 2004). This, of course, is not an optimal approach. Because land-use types are in most cases quite broadly defined, and have a functional rather than a physical meaning (e.g., built-up area, agriculture, forest), surface related parameters that are relevant in hydrological modeling may show substantial variation within a particular land-use patch and among patches of the same land-use type. This is particularly so in urbanized areas, which are characterized by a strong landscape heterogeneity. When using land-use related look-up tables for hydrological parameter estimation in such areas, spatially-explicit detail on land surface properties that may be important in modeling and understanding of dynamic hydrological processes is discarded (Dams *et al.*, 2009).

In recent years, there has been a growing interest in the development of methods to derive various land-related parameters that are relevant for hydrological modeling directly from remote sensing data (Boegh *et al.*, 2004). A lot of work has been done on the mapping of impervious surface distribution, both at pixel level, using high-resolution multispectral data, as well as at sub-pixel level, using data from medium-resolution multispectral and, more recently, hyperspectral sensors. An overview of different approaches for mapping impervious surface cover from remotely sensed data is given below. Besides mapping of impervious surfaces to improve runoff prediction, current hydrological remote sensing research also aims at improving spatial and temporal parameterization of other hydrological processes. An overview of this research is beyond the scope of this chapter, but excellent reviews have been published on topics such as estimation of soil moisture (Anderson and Croft, 2009; Petropoulos *et al.*, 2009), mapping of evapotranspiration (Schmugge *et al.*, 2002; Kalma, McVicar and McCabe, 2008; Li *et al.*, 2009), and flood extent mapping (Schumann *et al.*, 2009).

The remainder of this chapter focuses on the use of information on the distribution of impervious surface cover derived from remote sensing in spatially distributed hydrological modeling. More in particular, a case study will be presented, demonstrating the impact of different remote sensing based methods for characterizing the distribution of impervious surfaces on runoff estimation in an urbanized watershed, as well as the effect the different methods have on the assessment of peak discharges at the outlet of the catchment. Study area is the Woluwe, a strongly urbanized watershed in the urban fringe of the Brussels Capital Region.

In Section 18.3 a short review of different approaches for mapping impervious surface cover from remote sensing data is provided. Section 18.4 gives an overview of the WetSpa model, a grid-based spatially distributed hydrological model that has been used in the case study on the Woluwe. Section 18.5 describes the set-up and the results of the case study, and demonstrates the need for detailed information on surface imperviousness for spatially explicit modeling of hydrological processes in urbanized areas.

18.3 Impervious surface mapping

Field inventorying and visual interpretation of large-scale, orthorectified aerial photographs are the most reliable methods to map impervious surfaces. However, because these methods are very time-consuming, they can in practice only be applied to relatively small areas. Satellite imagery, obtained from high-resolution multispectral sensors like IKONOS or Quickbird, offers an interesting alternative for producing maps of surface imperviousness. Although high-resolution imagery does not provide the same level of detail as large-scale aerial photographs, the use of automated or semiautomated image interpretation methods, exploiting the multispectral information content of the imagery, substantially reduces the effort that is required to produce reliable information on the distribution of impervious surfaces from this data.

Several studies have focused on the mapping of impervious surface cover from high-resolution satellite data. Because of the limited spectral resolution – most high-resolution sensors that are commonly used have only four spectral bands: blue, green, red, near infrared – a major challenge in these studies lies in separating impervious surface materials from other, spectrally similar land-cover types. Indeed, dark materials like asphalt or roofing may be difficult to distinguish from other low-albedo surface types like water and shadowed surfaces (Hodgson et al., 2003; Dare, 2005; Lu and Weng, 2007). Other impervious materials may be spectrally similar to bare soil (Thomas, Hendrix and Congalton, 2003; Van de Voorde, De Genst and Canters, 2007). Various approaches have been suggested to improve the separation of impervious from non-impervious surface types. Lu and Weng (2009) propose a hybrid approach that combines the use of a decision tree classifier with an unsupervised classifier and requires intervention from an image analyst to separate impervious surfaces from water and shadow in vegetated areas. De Roeck, Van de Voorde and Canters (2009) apply an object-based, hierarchical classification approach, using texture measures calculated at the level of image objects to complement the spectral data, and improve the distinction between impervious surfaces, shadowed areas and bare soil. Thomas, Hendrix and Congalton (2003) and Van de Voorde, De Genst and Canters (2007) suggest object-based post-classification approaches to improve the accuracy of an initial classification, using expert-based rules that rely on context information. Others have proposed the use of external data, such as elevation, to reduce the confusion between impervious and non-impervious surface types (Hodgson et al., 2003).

Despite the advantages of deriving impervious surface maps directly from high-resolution satellite imagery, the limited footprint and the relatively high cost of such images may pose difficulties if impervious surface maps are required for larger areas. A less expensive and more efficient alternative to map the spatial distribution of imperviousness in such cases is to develop models that allow estimation of the degree of imperviousness inside pixels of medium-resolution imagery (Landsat ETM+, ASTER). Over the past years many sub-pixel regression and sub-pixel classification methods have been proposed to derive the fraction of impervious surfaces within a medium-resolution pixel directly from the spectral data, including methods based on regression analysis (Gillies et al., 2003; Yang and Liu, 2005; Bauer, Loffelholz and Wilson, 2008), linear spectral unmixing (Ji and Jensen, 1999; Phinn et al., 2002; Rashed et al., 2003; Wu and Murray, 2003), artificial neural networks (Flanagan and Civco, 2001; Wang and Zhang, 2004; Pu et al., 2008) and regression trees (Huang and Townshend, 2003; Yang et al., 2003; Xian, 2006). For a comparison of different methods the reader is referred to Weng and Hu (2008), Yuan, Wu and Bauer (2008), and Van de Voorde, De Roeck and Canters (2009). Although the sub-pixel approach does not allow one to achieve the same degree of accuracy as when high-resolution data would be used, most studies report an average per pixel proportional error in the estimation of impervious surfaces that varies between 10% and 20%, depending on the characteristics of the study area and the way the validation is done. Recently, sub-pixel approaches have also been applied to high-resolution satellite imagery to improve the accuracy of impervious surface maps in areas with high proportions of mixed pixels (Mohapatra and Wu, 2007; Lu and Weng, 2009; Wu, 2009).

18.4 The WetSpa model

WetSpa was originally developed by Wang, Batelaan and De Smedt (1996) and rewritten with a focus towards flood prediction and water balance simulation by Liu et al. (2003). Batelaan and De Smedt (2007) derived from the original a quasi-steady state version, which is mainly used for recharge prediction.

Figure 18.1 (left) shows the hydrological processes simulated in the model. These processes are computed at grid cell level, and water balance calculations are performed in four control volumes: interception, depression, root zone and saturated subsurface. The processes are simulated in a cascading manner starting from a precipitation event, followed by interception, depression storage, surface runoff, infiltration, evapotranspiration, percolation, interflow and groundwater flow. In addition, the water balance of each grid cell is divided into impervious, vegetated, bare soil and open water parts to account for the non-uniformity in land-use/land-cover in individual cells. This aspect is especially for urban areas very important since on the typical resolution of remotely sensed data, land-cover shows strong heterogeneity (Fig. 18.1, right).

The main input of WetSpa, to describe various hydrological processes, is hydrometeorological data (precipitation, evapotranspiration and temperature) and raster maps (topography, land-use and soil). Figure 18.2 presents the relationship between the three main input maps, digital elevation model, land-use and soil map, and the WetSpa model parameters, which are derived by using physical and empirical equations. The main outputs of the model include river flow hydrographs, which can be defined for any location in the channel network, and spatially distributed hydrologic characteristics of the catchment such as evapotranspiration, recharge and runoff.

18.4.1 Surface runoff

One of the most important processes with respect to urban areas is surface runoff, which is therefore described in more detail. For the formulation of other processes we refer to Liu and De Smedt (2004). Surface runoff is calculated in the model by a modified rational method as:

$$R_s = C_p P_n (\theta/\theta_s)^\alpha \qquad (18.1)$$

where R_s [LT^{-1}] is the rate of surface runoff, C_p [−] is a potential runoff coefficient, P_n [LT^{-1}] is the rainfall intensity after canopy interception, θ and θ_s [L^{3}L^{-3}] are actual and saturated soil moisture content, and α [−] is an empirical exponent related to the rainfall intensity. The potential runoff coefficient C_p is a measure of rainfall partitioning capacity, depending upon slope, soil type and land-use combination. Default potential runoff coefficients for different slope, soil type and land-cover under the condition of near saturated soil moisture are interpolated from Kirkby (1978), Chow, Maidment and Mays (1988), Browne (1990), Fetter (1980) and Smedema and Rycroft (1988). A lookup table (Table 18.1) has been built to relate the potential runoff coefficient to different combinations of slope, soil type and land-use (Liu and De Smedt, 2004). To simplify this table, land-use classes are reclassified into five classes as forest, grass, crop, bare soil and impervious area. In addition, surface slope is discretized into four classes. But, in order to estimate the potential runoff coefficient on the basis of a continuous slope, a simple linear relationship between potential runoff coefficient for the discretized slope classes and surface slope is used. The potential runoff coefficients for impervious (including open water) surface are set to 1.

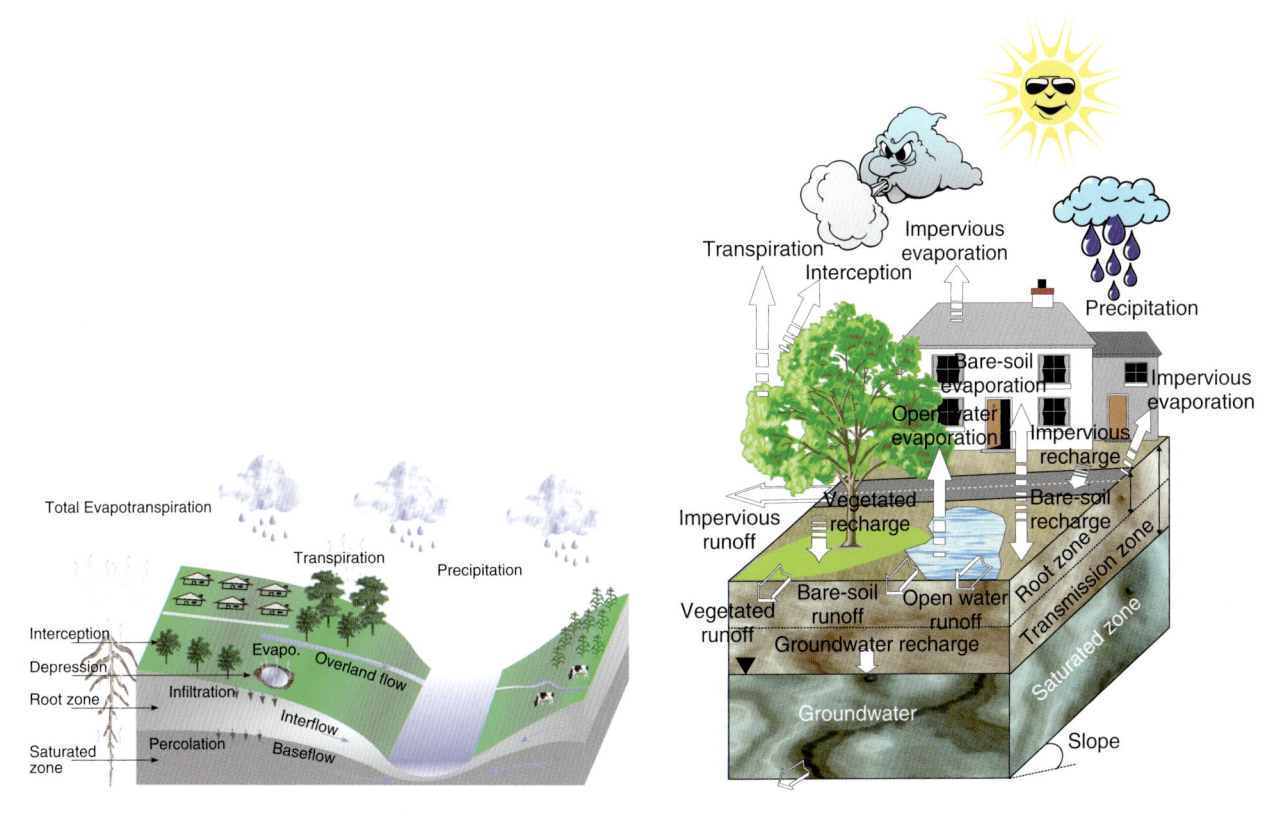

CHAPTER 18 USE OF IMPERVIOUS SURFACE DATA OBTAINED FROM REMOTE SENSING 259

FIGURE 18.1 Hydrological processes simulated in WetSpa model (Mohammed, 2009) (left), an example of a heterogeneous urban cell (right).

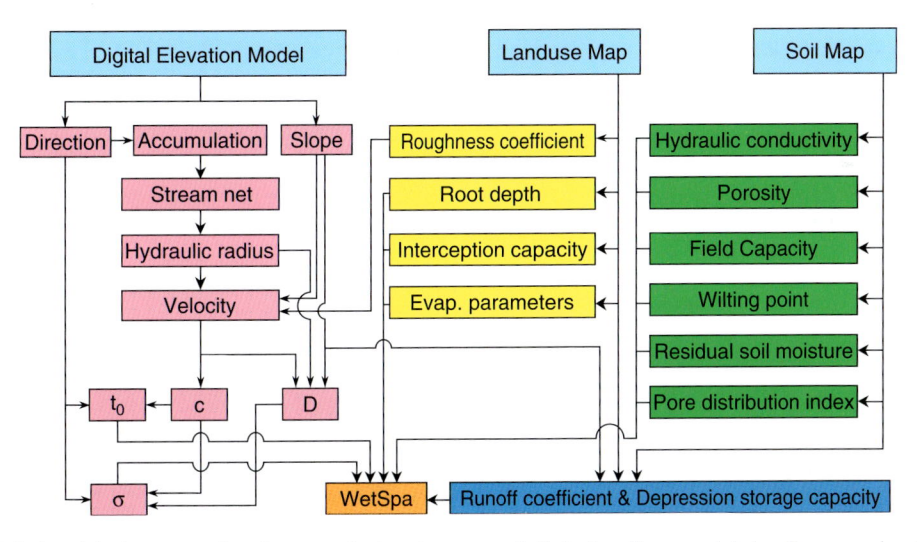

FIGURE 18.2 Relationship between the three main input maps, digital elevation model, land-use and soil map, and the WetSpa model parameters. t_0 is the mean flow time, c is the celerity of the diffusion wave, D is the dispersion coefficient coefficient, σ is the standard deviation of the flow time.

The influence of urban areas on storm runoff is self-evident. Due to the raster resolution, cells may not be 100% impervious. In WetSpa the remaining area is assumed to be pervious and covered by grass, and therefore, the potential runoff coefficient for urban areas is calculated as

$$C_u = IMP + (1 - IMP)C_{grass} \quad (18.2)$$

where C_u and C_{grass} are potential runoff coefficient for urban and grass raster cells, and IMP is the proportion of impervious area. Table 18.2 presents standard impervious cover percentages for different land-use categories. Imperviousness percentage for residential, commercial and industrial area is estimated based on the information in Chow, Maidment and Mays (1988). Other estimates are considered reasonable guesses. Zero impervious percentage is assumed for land-use categories not listed (i.e., agriculture, grassland, and forest land).

The effect of rainfall duration is also included in the runoff process, as more runoff is produced during a storm event due to

TABLE 18.1 Potential runoff coefficients of the WetSpa model for different land-use, soil type and slope.

Land-use	Slope (%)	Sand	Loamy sand	Sandy loam	Loam	Silt loam	Silt	Sandy clay loam	Clay loam	Silty clay loam	Sandy clay	Silty clay	Clay
Forest	< 0.5	0.03	0.07	0.10	0.13	0.17	0.20	0.23	0.27	0.30	0.33	0.37	0.40
	0.5–5	0.07	0.11	0.14	0.17	0.21	0.24	0.27	0.31	0.34	0.37	0.41	0.44
	5–10	0.13	0.17	0.20	0.23	0.27	0.30	0.33	0.37	0.40	0.43	0.47	0.50
	> 10	0.25	0.29	0.32	0.35	0.39	0.42	0.45	0.49	0.52	0.55	0.59	0.62
Grass	< 0.5	0.13	0.17	0.20	0.23	0.27	0.30	0.33	0.37	0.40	0.43	0.47	0.50
	0.5–5	0.17	0.21	0.24	0.27	0.31	0.34	0.37	0.41	0.44	0.47	0.51	0.54
	5–10	0.23	0.27	0.30	0.33	0.37	0.40	0.43	0.47	0.50	0.53	0.57	0.60
	> 10	0.35	0.39	0.42	0.45	0.49	0.52	0.55	0.59	0.62	0.65	0.69	0.72
Crop	< 0.5	0.23	0.27	0.30	0.33	0.37	0.40	0.43	0.47	0.50	0.53	0.57	0.60
	0.5–5	0.27	0.31	0.34	0.37	0.41	0.44	0.47	0.51	0.54	0.57	0.61	0.64
	5–10	0.33	0.37	0.40	0.43	0.47	0.50	0.53	0.57	0.60	0.63	0.67	0.70
	> 10	0.45	0.49	0.52	0.55	0.59	0.62	0.65	0.69	0.72	0.75	0.79	0.82
Bare soil	< 0.5	0.33	0.37	0.40	0.43	0.47	0.50	0.53	0.57	0.60	0.63	0.67	0.70
	0.5–5	0.37	0.41	0.44	0.47	0.51	0.54	0.57	0.61	0.64	0.67	0.71	0.74
	5–10	0.43	0.47	0.50	0.53	0.57	0.60	0.63	0.67	0.70	0.73	0.77	0.80
	> 10	0.55	0.59	0.62	0.65	0.69	0.72	0.75	0.79	0.82	0.85	0.89	0.92
IMP		1.00	1.00	1.00	1.00	1.00	1.00	1.00	1.00	1.00	1.00	1.00	1.00

TABLE 18.2 Average degree of imperviousness for different urban land-use classes: default values and estimates derived from IKONOS data and Landsat ETM+ data for the Woluwe case study.

Land-use classes	Default degree of imperviousness	IKONOS-derived value	Landsat-derived value
Low density built-up	0.30	0.12	0.12
High density built-up	0.50	0.57	0.61
City center	0.70	0.45	0.38
Infrastructure	0.50	0.58	0.60
Roads	0.50	0.36	0.31
Industry	0.70	0.84	0.86

increasing soil moisture content. In general, the model accounts for the effect of slope, soil type, land-use, soil moisture, rainfall intensity and its duration on the production of surface runoff in a realistic way.

18.4.2 Flow routing

In WetSpa, flow routing is carried out at a grid cell, flow path and catchment level as shown in Fig. 18.3(a). Assuming the grid cell as a reach with one-dimensional unsteady flow and neglecting the acceleration terms of the St. Venant equation and combining it with the continuity equation yields the diffusive wave equation, which is the basis for the flow routing in the grid cell (Miller and Cunge, 1975; Jobson, 1989):

$$\frac{\partial Q}{\partial t} + c_i \frac{\partial Q}{\partial x} - d_i \frac{\partial^2 Q}{\partial x^2} = 0 \quad (18.3)$$

where Q [$L^3 T^{-1}$] is the flow rate, x [L] is the distance along the flow direction, t [T] is the time, c_i [LT^{-1}] is an advective velocity and d_i [$L^2 T^{-1}$] is the diffusion coefficient. The advective velocity is a function of the flow velocity, which is calculated using Manning's equation. The diffusion coefficient is a function of the flow velocity, the hydraulic radius and the slope of the cell.

For a unit impulse input, the solution to Equation 18.3 at the grid cell outlet, assuming open boundary upstream and closed boundary downstream is a first passage time distribution as given by (Eagleson, 1970):

$$u_i(t) = \frac{l_i}{2\sqrt{\pi d_i t^3}} \exp\left[-\frac{(c_i t - l_i)^2}{4 d_i t}\right] \quad (18.4)$$

where $u_i(t)$ [T^{-1}] is the cell impulse response function at time t (Fig. 18.3b) and l_i [L] is the length of the cell. After determining the cell impulse response function using Equation 18.4, the flow path response is determined using a linear routing by successively applying the convolution integral. Assuming that the path response is also a first passage time distribution, an approximate numerical solution for the convolution integral is given by De Smedt, Liu and Gebremeskel (2000). Equation 18.5 shows the convolution integral and its numerical solution.

$$U_i(t) = \prod_{j=1}^{N} u_i(t) = \frac{1}{\sqrt{2\pi \sigma_i^2 t^3 t_i^{-3}}} \exp\left[-\frac{(t - t_i)^2}{2 \sigma_i^2 t t_i^{-1}}\right] \quad (18.5)$$

where $U_i(t)$ [T^{-1}] is the path response function (Fig. 18.3c), the subscript i denotes the cell where the input occurs, j refers to the cell connecting cell i with the outlet, N is the total number of cells along the flow path, t_i [T] is the mean flow time from the input cell to the end of the flow path, and σ_i [T] is the standard deviation of the flow time.

The flow path input response function is used to determine the flow response at the end of a flow path by convoluting the

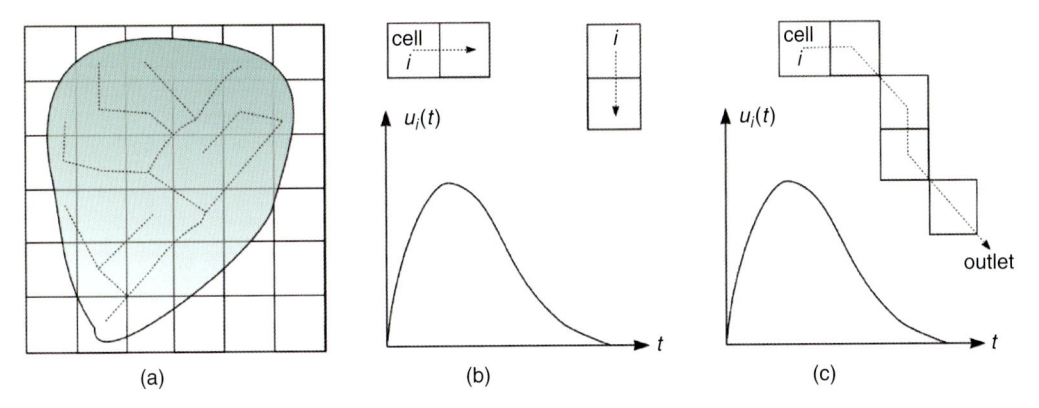

FIGURE 18.3 (a) Catchment and stream network, (b) unit impulse response function for a single cell, (c) path impulse response function as a convolution of unit impulse function of cells along a flow path (Mohammed, 2009).

input runoff volume. This is equivalent to decomposing the input into infinite impulses and summing all the responses to get a single response. Consequently, the outflow hydrograph to an arbitrary input can be determined as

$$Q_i(t) = \sum_{\tau=0}^{t-\tau} V_i(\tau) U_i(t-\tau) \qquad (18.6)$$

where $Q_i(t)$ [L^3T^{-1}] is the outflow at the end of a flow path produced by an arbitrary input in cell i, $U_i(t-\tau)$ is the flow path response function, τ [T] is the time delay, and $V_i(\tau)$ [L^3] is the input runoff volume at cell i and time τ. Flow response of the catchment is the sum of the response of the contributing cells as

$$Q(t) = \sum_{i=0}^{N_w} Q_i(t) \qquad (18.7)$$

where $Q(t)$ is the total flow at the catchment outlet and N_w is the number of cells in the catchment.

18.4.3 Water balance

In WetSpa, water balance computation is carried out in four control volumes: interception storage, depression storage, root zone and saturated subsurface. If we consider the root zone as an example, its main input is infiltration (F) and its main outputs are evapotranspiration (ET), lateral interflow (RI) and percolation to the groundwater storage (RG). Then, the water balance in the root zone with thickness D for a single cell i at time t can be written as

$$F_i(t) - ET_i(t) - Rl_i(t) - RG_i(t) = D_i[\theta_i(t) - \theta_i(t-1)] \qquad (18.8)$$

where $\theta_i(t)$ and $\theta_i(t-1)$ [L^3L^{-3}] are the soil moisture content of cell i at time t and $t-1$, D_i [L] is the root depth. The same principle is extended to the other control volumes based on the inputs and outputs in the control volumes. To keep track of water change in the entire catchment, an additional water balance is computed for the entire catchment using rainfall as the driving input and runoff and evapotranspiration as an output to yield a change in soil moisture and groundwater storage.

The WetSpa model, documentation, and examples are available from http://www.vub.ac.be/WetSpa/.

18.5 Impact of different approaches for estimating impervious surface cover on runoff calculation and prediction of peak discharges

The remainder of this chapter reports on a case study in which high-resolution and medium-resolution land-cover maps derived from remotely sensed data were used to estimate the distribution of impervious surfaces and other land-cover types within the urban classes of an existing land-use map for the Brussels Capital Region. The "remote sensing enhanced" versions of the land-use map were used as input for a rainfall-runoff simulation on the upper catchment of the Woluwe River, located in the southeastern part of the Brussels region. The impact of different methods for characterizing urban land-cover on estimated runoff patterns and on peak discharges at the outlet of the catchment has been examined. One of the objectives of the study was to examine how sub-pixel estimation of impervious surface cover based on medium-resolution data (Landsat ETM+, 30 m resolution) compares to detailed mapping of impervious surfaces using high-resolution satellite data (IKONOS, 4 m resolution) in terms of runoff and peak discharge prediction, using the WetSpa distributed hydrological model.

18.5.1 Study area and data

The Woluwe River is a tributary of the Zenne River, which runs through the center of Brussels, and is part of the Scheldt basin. The study area consists of the upper part of the Woluwe catchment and has an area of about 31 km². The upstream part of the study area is located in the protected zone of the Sonian forest, southeast of the city center of Brussels. In the downstream part, the Woluwe River flows partially through several vaulted stretches, parks and ponds. The elevation of the catchment varies between 49 and 129 m, with an average of 94 m above sea level. The elevation decreases gradually from south to north, with slopes varying between 0 and 22%, with an average of 6%. The long-term yearly precipitation and open-water evaporation is respectively 780 and 657 mm/year and is measured at the Royal Meteorological Institute at Ukkel, located close to the catchment border.

The land-use map of Flanders, produced by the Agency for Geographic Information Flanders (AGIV), was used as a reference data set to characterize land-use within the study area. The original map has a resolution of 20 m and was resampled to a 30 m cell size for modeling purposes (Fig. 18.4). A Digital Elevation Model (DEM) at 30 m resolution was interpolated from elevation contours for every 2.5 m and the river network. Both were digitized from topographic maps at a scale of 1:10 000. The DEM was generated in ArcInfo using TOPOGRID. A digital soil map of the study area was obtained from AGIV, rasterized and reclassified into 12 USDA soil texture classes based on textural properties.

Two image data sets were used: an ortho-rectified high-resolution image (IKONOS), acquired on 8 June 2000, to derive a detailed high-resolution land-cover map, and a medium-resolution image (Landsat ETM+), acquired on 18 October 1999, for estimating land-cover class proportions at sub-pixel level. The IKONOS image has four multispectral bands that capture the reflectance in the blue, green, red and near infrared part of the spectrum at 4 m spatial resolution, and an additional panchromatic band at 1 m resolution. The Landsat ETM+ multispectral image includes six spectral bands in the visual-to-mid-infrared range that were used in this study, and has a spatial resolution of 30 m. The Landsat image was geometrically co-registered to the IKONOS image by a first-order polynomial transformation. The RMS error on an independent set of control points was 5.78 m, which implies that on the average the geometric shift in x and y between IKONOS and ETM+ data corresponds to less than 4% of the area of an ETM+ pixel.

18.5.2 Impervious surface mapping

To produce a high-resolution impervious surface map of the study area, the IKONOS image was classified using a multiple layer perceptron classification approach (MLP). Eleven land-cover classes were distinguished: light and dark red built-up area (mainly red tiled roofs), light, medium and dark gray built-up area (consisting of building materials such as concrete, asphalt, slate), bare soil, water, crops, shrub and trees, grass and shadow. Input variables used for defining the classification model were the four multispectral bands, the PAN band, the NDVI and two texture measures (local variance, binary comparison matrix) calculated on the PAN band. Training data for the classification were obtained by digitizing about 200 training samples for each class. Transformations of the input bands and a selection according to their relative contribution to the overall information content were accomplished with NeuralWorks Predict®. The transformed input variables that were retained to perform the classification were chosen from a set of five mathematical transformations per original input band using a genetic-based variable selection algorithm embedded in the software.

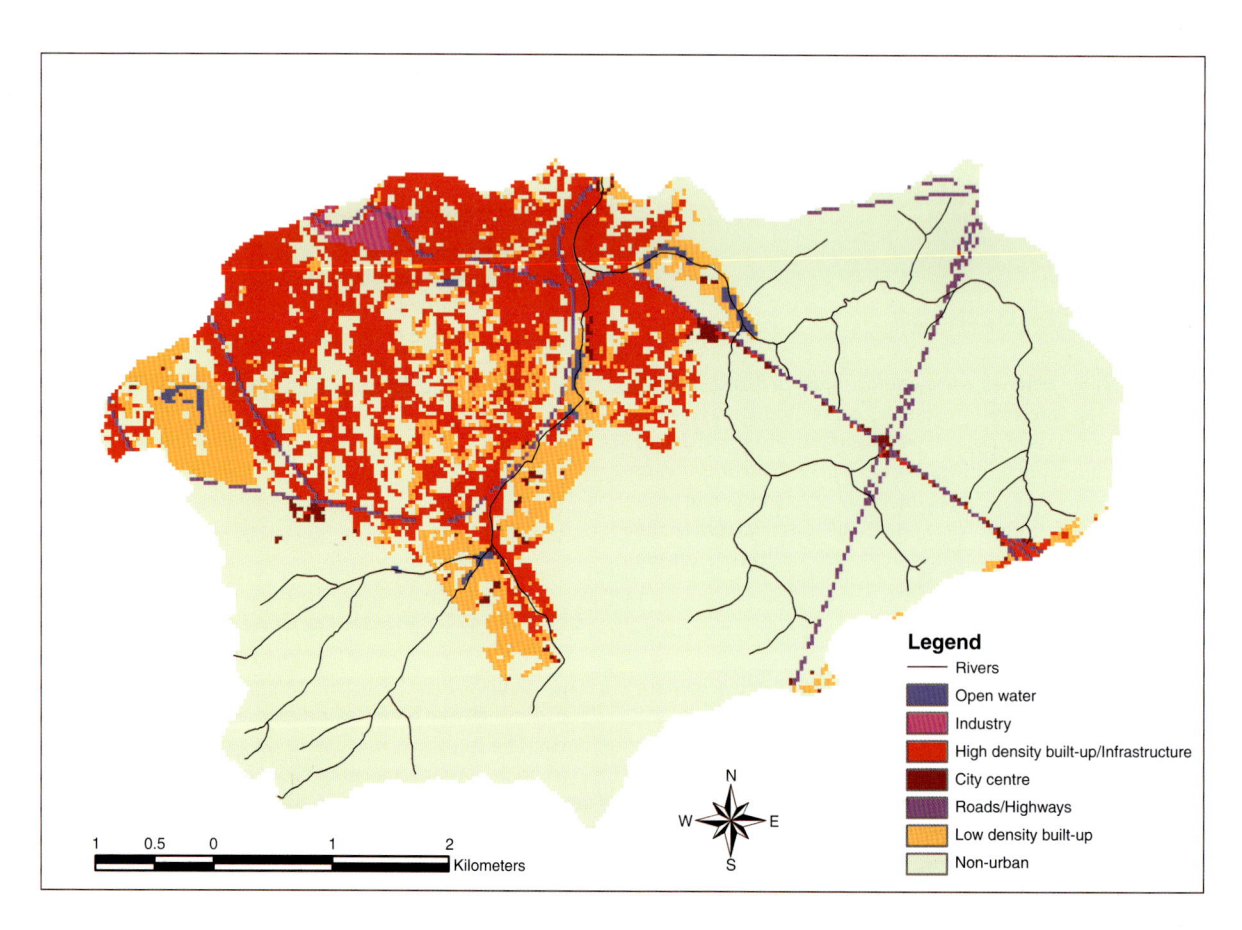

FIGURE 18.4 Land-use map for the Upper Woluwe River catchment.

After classification, the two red surface classes (light, dark) and the three gray surface classes (light, medium, dark) were aggregated into one red surface class and one gray surface class respectively, reducing the total number of classes to seven plus shadow. The overall performance of the classifier on an independent stratified random sample of validation pixels for these classes was characterized by a kappa index-of-agreement of 0.91. The accuracy and the spatial coherence of the classification was further improved by applying dedicated post-classification techniques to remove shadow, correct remaining classification errors and reduce structural noise typical for pixel-based classifications (Van de Voorde, De Genst and Canters, 2007). Post-classification improvement of the land-cover map resulted in an overall accuracy of 95.7% and increased the kappa index from 0.91 to 0.95 (Table 18.3). The seven remaining land-cover classes were grouped into a single vegetation class (including shrub and trees, grass, and crops), a single impervious class (including red, and gray surfaces), water and bare soil.

To characterize land-cover at the medium resolution, an MLP sub-pixel classification model was developed, estimating the proportion of the four major classes (impervious surfaces, vegetation, bare soil and water/shade) in each pixel of the Landsat ETM+ image. To train and validate the sub-pixel classifier, the land-cover classification derived from the IKONOS image was spatially aggregated to produce reference proportions for the four major land-cover classes for each ETM+ pixel in the overlapping zone between the two images. Because the images are not of the same date, precautions had to be taken not to include pixels in the training and validation phase with different land-cover in both images due to seasonal shifts (leaf condition, crop cycles) or due to a change in land-use (e.g., transition from non-built to built area). Since urban areas are dominated by impervious surfaces and vegetation, most changes in the 8-month period between the acquisition dates of both images are related to changes in the vegetation component of the pixels. Therefore, in order to detect anomalous pixels the ETM+ NDVI values of all pixels in the overlap with the IKONOS image were plotted against the mean NDVI value of the constituent IKONOS pixels, after converting the raw DNs for both images to at-satellite reflectance values. A strong linear relationship was observed between ETM+ NDVI values and IKONOS mean NDVI values. Pixels deviating significantly from the trend were considered as indicative of major changes in land-cover and were not used for training and validation (Van de Voorde, De Roeck and Canters, 2009).

Feature selection and neural network building were accomplished with NeuralWorks Predict® software, starting from a set of transformations of all multispectral Landsat ETM+ bands (1–5, 7) and all possible ratios between these bands. Application of the model to the part of the Landsat ETM+ image that overlaps the IKONOS image resulted in four proportion maps (vegetation, impervious surfaces, water and bare soil). Figure 18.5 (left) shows the proportion map obtained for impervious surfaces. A visual comparison with the proportion map of impervious surfaces, derived from the IKONOS image (Fig. 18.5, right), shows a good correspondence in the overall pattern of imperviousness, although the sub-pixel classification result is more generalized and includes less structural detail.

The performance of the sub-pixel classifier was assessed on an independent validation set by calculating per-class mean error (ME_C), as well as per-class mean absolute error (MAE_C):

$$ME_{Cj} = \frac{\sum_{i=1}^{N}(P'_{ij} - P_{ij})}{N} \quad (18.9)$$

$$MAE_{Cj} = \frac{\sum_{i=1}^{N}(P'_{ij} - P_{ij})}{N} \quad (18.10)$$

where N is the total number of pixels in the validation sample, C is the total number of target classes (4: vegetation, impervious surfaces, water and bare soil), C_j is target class j, P_{ij} is the proportion of class j inside validation pixel i, derived from the high-resolution land-cover map (ground truth), P'_{ij} is the proportion of class j inside validation pixel i, estimated by the sub-pixel classifier.

To assess the impact of cell aggregation on proportional accuracy, all error measures were calculated at the original 30 m resolution, as well as after aggregation of proportions to 60 m and 90 m. As shown in Table 18.4, impervious surfaces and bare soil are slightly underestimated, while vegetation and water are slightly overestimated. The slight underestimation of impervious surfaces (−1.8%) may be explained by the presence of shadow in urbanized areas. When examining the estimated proportion of water within the area, it turns out that small portions of

TABLE 18.3 Confusion matrix for the high-resolution land-cover map, obtained by MLP-classification and improved with post-classification techniques.

	Red surfaces	Bare soil	Water	Grass	Crops	Shrub and trees	Grey surfaces	Sum	User's accuracy
Red surfaces	87							87	1.00
Bare soil	1	146		1			1	149	0.98
Water		1	59				5	65	0.91
Grass			1	278	51	7		337	0.82
Crops					172			172	1.00
Shrub and trees				6	2	346		354	0.98
Gray surfaces		2				1	797	800	1.00
Sum	88	149	60	285	225	354	803	1964	
Producer's accuracy	0.99	0.98	0.98	0.98	0.76	0.98	0.99		PCC = 95.7%

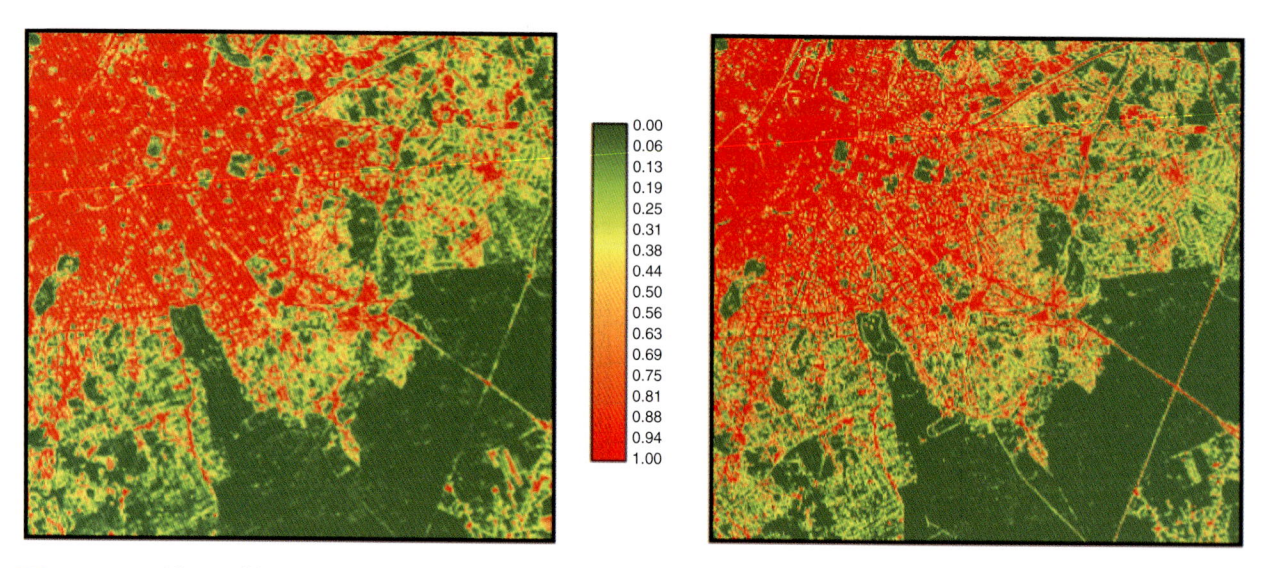

FIGURE 18.5 Maps of impervious surface fractions at the cell size of the WetSpa model (30 m), obtained by sub-pixel classification of Landsat ETM+ data (left) and by spatial aggregation of the IKONOS-derived land-cover map to the 30 m resolution (right).

TABLE 18.4 Mean error and mean absolute error of sub-pixel class proportions for the four major land-cover classes, for cell sizes of 30 m, 60 m and 90 m.

	Mean error				Mean absolute error			
Cell size	Impervious	Vegetation	Bare soil	Water	Impervious	Vegetation	Bare soil	Water
30 m	−0.018	0.015	−0.049	0.011	0.1030	0.1017	0.0593	0.0166
60 m	−0.017	0.014	−0.048	0.011	0.0752	0.0720	0.0514	0.0152
90 m	−0.019	0.015	−0.045	0.011	0.0611	0.0590	0.0467	0.0153

many urban pixels are assigned to water, although no water is present within these pixels. This phenomenon is most clearly observed in dense urban areas and is most likely caused by the presence of shadow, which is spectrally similar to water. As such, the proportion of water identified by the sub-pixel estimator should better be interpreted as water/shade, as it may point to the presence of both components. This may also explain the slight overestimation of water (+1.1%). The mean absolute error for impervious surfaces and vegetation, which are the two dominant classes in the image, is around 10%. Aggregation to cell sizes of 60 m and 90 m reduces this error to 7.5% and 6.1% for impervious surfaces, and to 7.2% and 5.9% for vegetation.

Figure 18.6 summarizes the different stages in the processing of the high-resolution (IKONOS) and medium-resolution (Landsat ETM+) imagery.

18.5.3 Impact of land-cover distribution on estimation of peak discharges

Starting with the land-use map of Flanders, three scenarios were defined to "enhance" the map, using the land-cover information derived from the IKONOS and Landsat ETM+ images. The three scenarios correspond to a gradual increase of information on the spatial distribution of land-cover within the urban area. These scenarios were used as a starting point for calculating runoff coefficients and for estimating peak discharges at the outlet of the catchment, using the WetSpa model. The spatial distribution of the WetSpa model parameters for the Woluwe study area were calculated based on the DEM, the soil map, the land-use map, and field measurements of hydraulic properties of the river channel. The period of simulation covers 4 days from 3 May 2005, 1.00 am till 6 May 2005, 9.00 am. This period was selected because of the typical occurrence of a number of spring storms. Since no continuous discharge measurements where available, no formal calibration of the model could be performed. However, some global parameters were adjusted during an initial trial and error calibration procedure, while spatial model parameters were kept constant. In the remainder of this section the effect of the three different land-cover input scenarios on the simulated runoff for the different storms in the period is examined. The three discussed scenarios correspond to a non-distributed, semidistributed and fully distributed approach for incorporating information on imperviousness in the runoff modeling.

18.5.3.1 Scenario 1: Non-distributed approach

The first scenario that was examined corresponds to a situation that is standard practice in hydrological modeling: the case where no information is available about different types of urban land-use, nor about the distribution of impervious surfaces within the urban area. To simulate runoff for this scenario all urban classes in the Flemish land-use map were aggregated into a single urban

CHAPTER 18 USE OF IMPERVIOUS SURFACE DATA OBTAINED FROM REMOTE SENSING

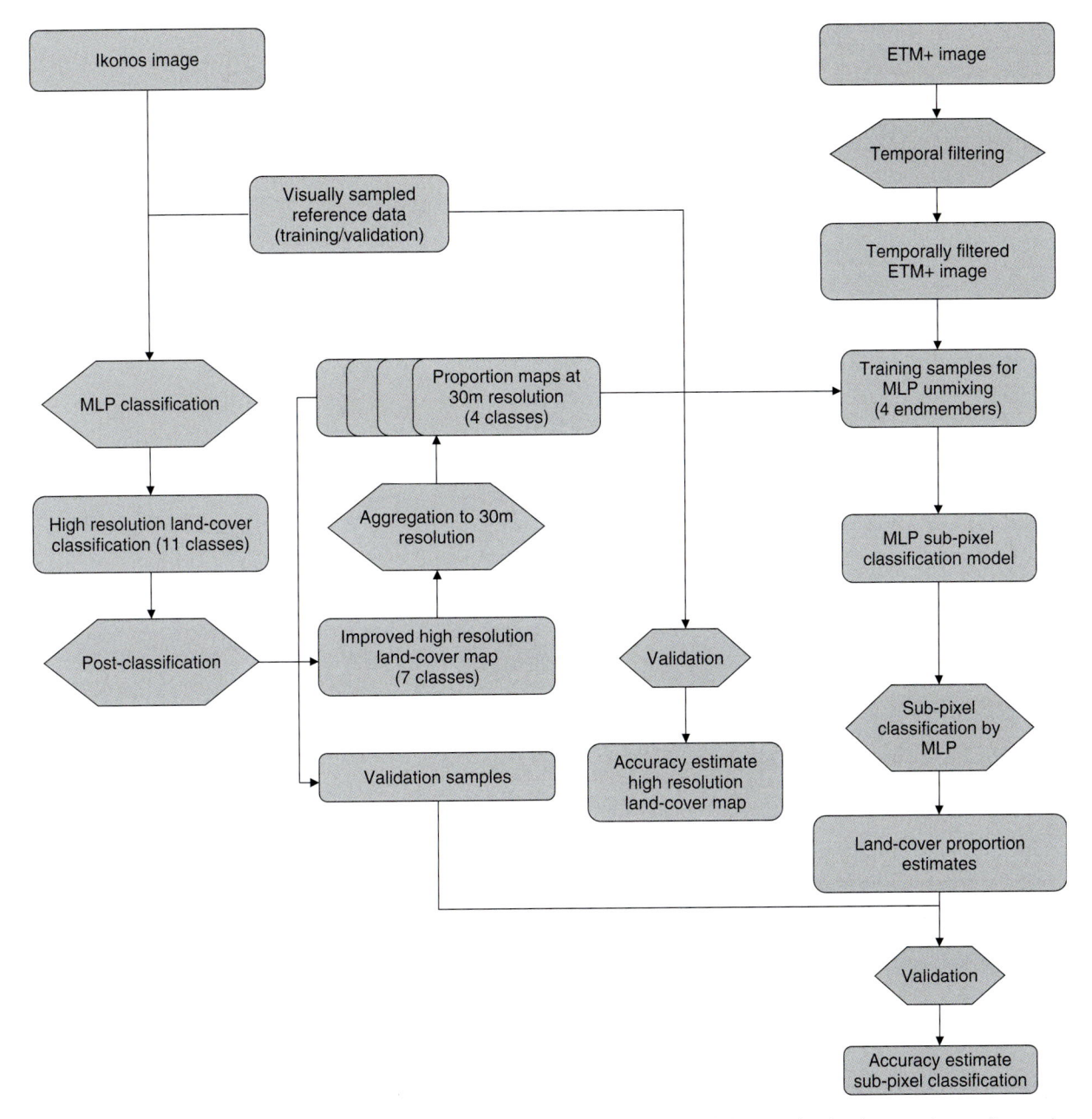

FIGURE 18.6 Overview of the procedure applied for determining fractions of four major land-cover types (impervious surfaces, vegetation, bare soil and water) at the cell size of the WetSpa model (30 m), using both high-resolution (IKONOS) and medium-resolution (Landsat ETM+) satellite data.

class. Next, according to the WetSpa default values (Table 18.2), 30% of each cell in the urban area was assumed to be impervious. The remaining part of each urban cell was assumed to be covered by grass. The surface runoff coefficient and depression storage were calculated as a weighted average of the parameters for the two classes. This basic scenario was improved by using the proportion maps, derived from the IKONOS land-cover classification and from the sub-pixel classification of Landsat ETM+ data, to estimate the average percentage of imperviousness for the urban area. The average degrees of imperviousness for the urban area as a whole obtained with IKONOS and Landsat proved to be very similar (44% and 46% respectively), but were substantially different from the default value (30%). Simulating runoff for different degrees of imperviousness indicated that peak discharges are very sensitive to the presence of impervious surfaces in urbanized watersheds. Estimated peak discharges for an average degree of imperviousness of 44% (IKONOS estimate) proved to be 10–20% higher than for the default scenario (Fig. 18.7).

18.5.3.2 Scenario 2: Semidistributed approach

In this scenario, each of the six built-up classes in the land-use map of Flanders were assigned a different level of imperviousness based on the IKONOS- and Landsat-derived proportion maps. Like in the first scenario, the remainder of each urban cell

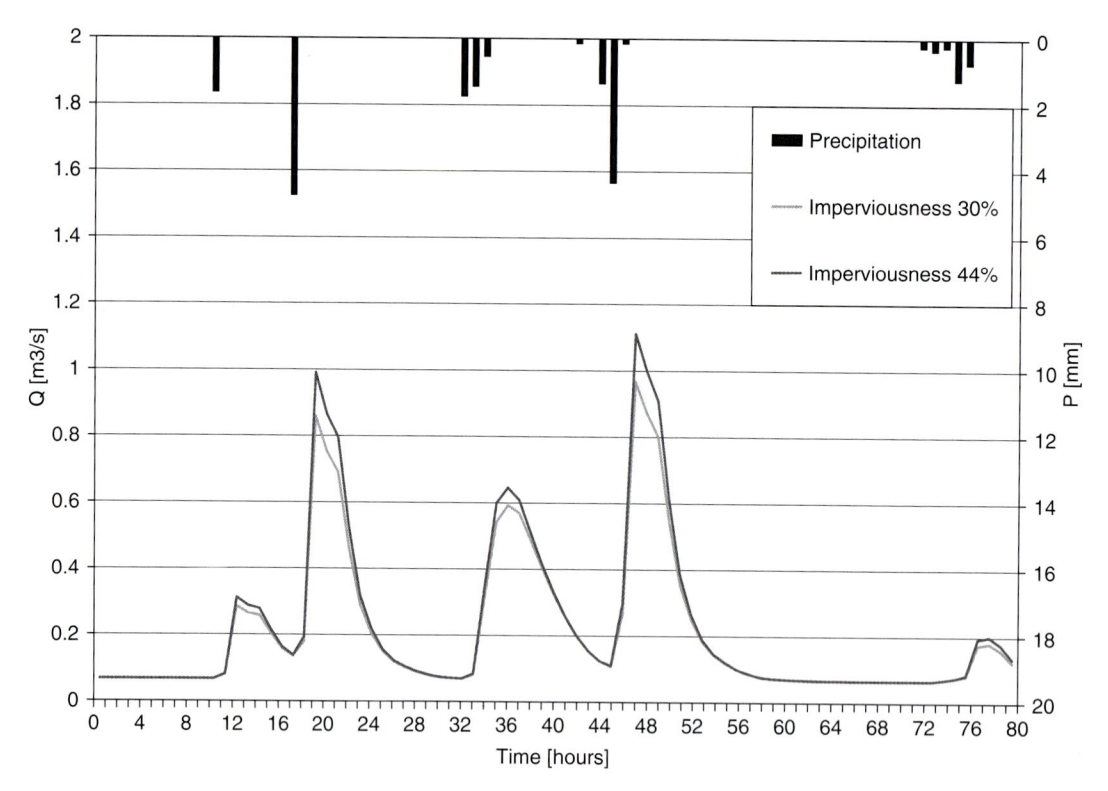

FIGURE 18.7 Hydrographs for scenario 1 with 30% imperviousness (default) and 44% imperviousness (IKONOS) for the period from 3 May 2005 1.00 am till 6 May 2005 9.00 am.

was considered as covered by grass. The average degree of imperviousness for different urban land-use classes obtained from IKONOS and Landsat data are quite similar (Table 18.2). For some classes, however, the values derived from the proportion maps strongly differ from default values found in the literature. This is especially the case for the low-density built-up and city center classes, where levels of imperviousness are much lower than assumed values. Also for roads and highways remote sensing estimates are lower than the default value. For the high density built-up class, for infrastructure and for industrial area remote sensing estimates are higher than the assumed level of imperviousness. The use of class-specific levels of imperviousness produces maximum peak discharges that are higher than the values obtained by defining one average level of imperviousness for the whole built-up area (between 5% and 10% higher for the two most important peaks in rainfall intensity, depending on the method used for estimating class-specific levels of imperviousness) (Fig. 18.8). The map of estimated runoff shows a clear increase in the value of the runoff coefficient in the urbanized area close to the river outlet in the northern part of the catchment, and a decrease in runoff values in areas located further away from the outlet, compared to scenario 1 (see Fig. 18.10). The more pronounced spatial variation in runoff coefficient values, however, seem to have only a minor effect on estimated peak discharges for less intense rainfall events (Fig. 18.8).

18.5.3.3 Scenario 3: Fully distributed approach

Finally, a fully distributed scenario was applied, where each cell in the urban area is assigned its proportion of vegetation, impervious surfaces, bare soil and water, as obtained from the IKONOS- and Landsat-derived proportion maps. The runoff coefficient and depression storage for each cell are then estimated as an area weighted average of the parameters for the four classes. Analysing and comparing the results for the three scenarios, the runoff calculated in scenario 3, and based on the IKONOS data, produces the highest peak discharges. The peak discharges are up to 15% higher than in the non-distributed simulation based on IKONOS data (scenario 1) (Fig. 18.8). This demonstrates that use of detailed information about the spatial distribution of land-cover within urban areas may have a clear impact on the modeling of peak discharges at the outlet of the catchment and, therefore, on flood prediction. The spatial variation in runoff is obvious from the map of runoff coefficient values obtained for scenario 3 (see Fig. 18.10). High values for the runoff coefficient, in the range of 0.8–1.0, prove to be linked to a closely connected pattern of impervious areas. In other words, the spatial variation in runoff and the connectivity between cells with high runoff coefficients in particular, proves to be a major factor influencing estimated discharge volume.

The hydrograph simulated in scenario 3, but based on Landsat ETM+ data, has lower peak values than the one based on IKONOS data (Fig. 18.9). This is most likely due to a smoothing effect in the distribution of impervious surface proportions caused by the sub-pixel classification process (see also Fig. 18.5), with a direct impact on the spatial variation of runoff coefficient values (Van de Voorde, De Roeck and Canters, 2009). As can be seen in Fig. 18.10, the spatial patterns of runoff for scenario 3, using either IKONOS or Landsat ETM+ data are very similar, yet for the Landsat scenario the pattern appears to be more generalized. This suggests that sub-pixel estimation of land-cover class proportions based on Landsat ETM+ data is less

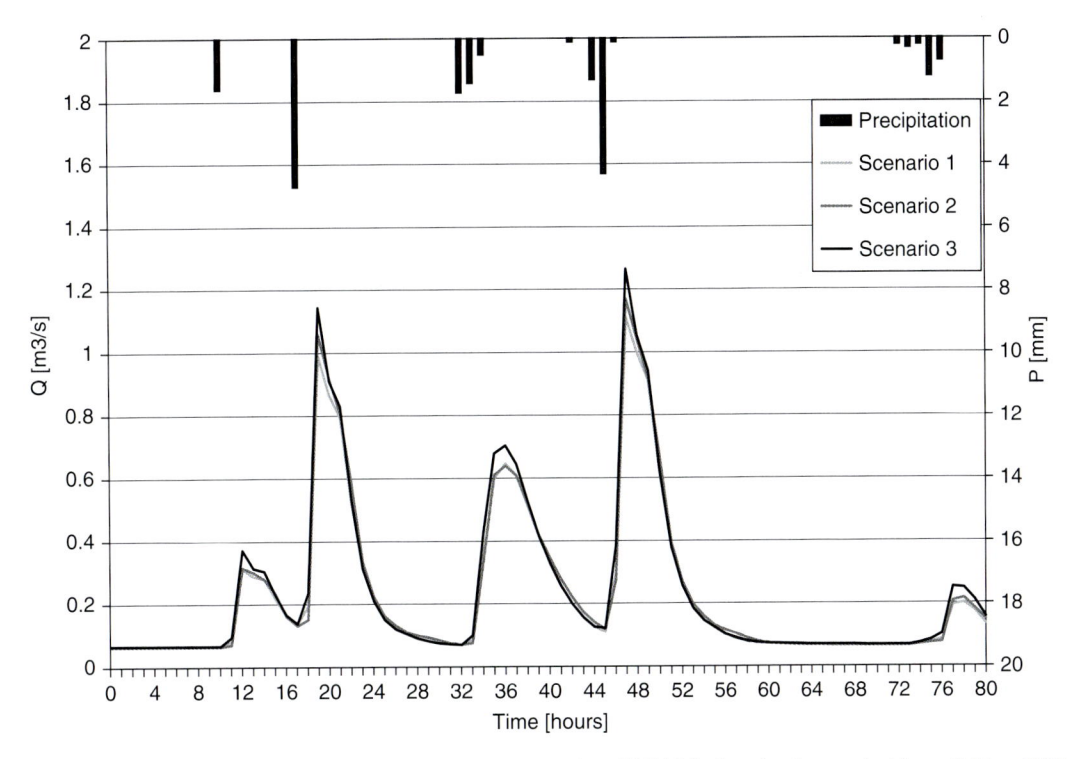

FIGURE 18.8 Comparison of hydrographs for the 3 scenarios using IKONOS data for the period from 3 May 2005 1.00 am till 6 May 2005 9.00 am.

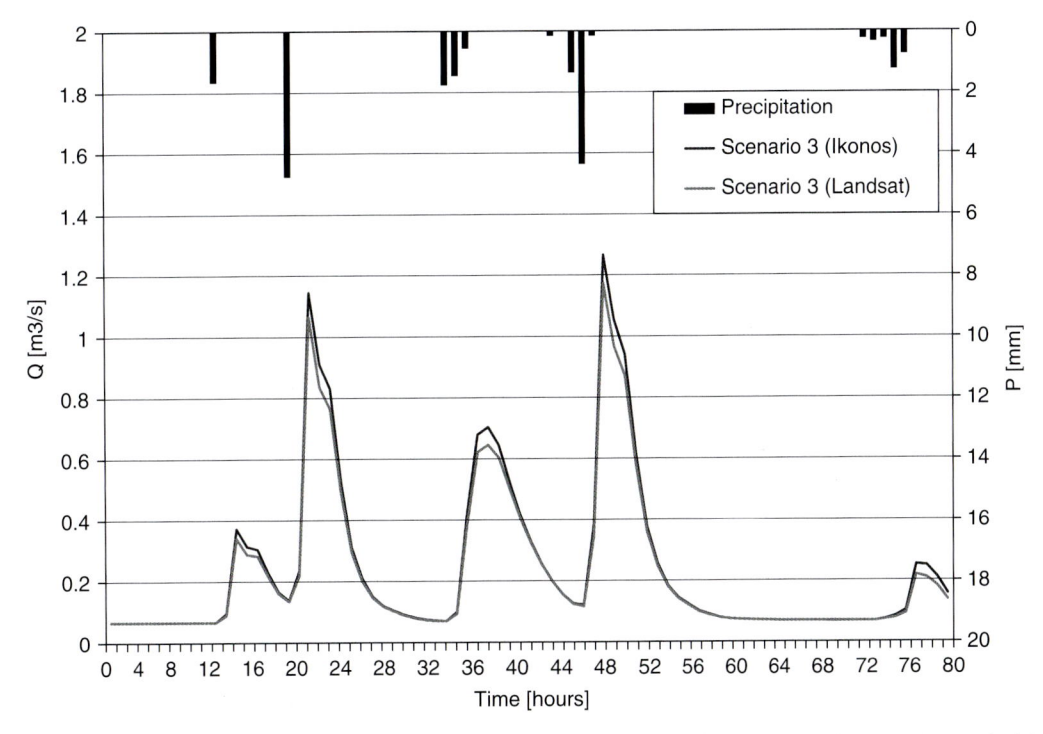

FIGURE 18.9 Comparison of hydrographs for scenario 3 using IKONOS data and Landsat data for the period from 3 May 2005 1.00 am till 6 May 2005 9.00 am.

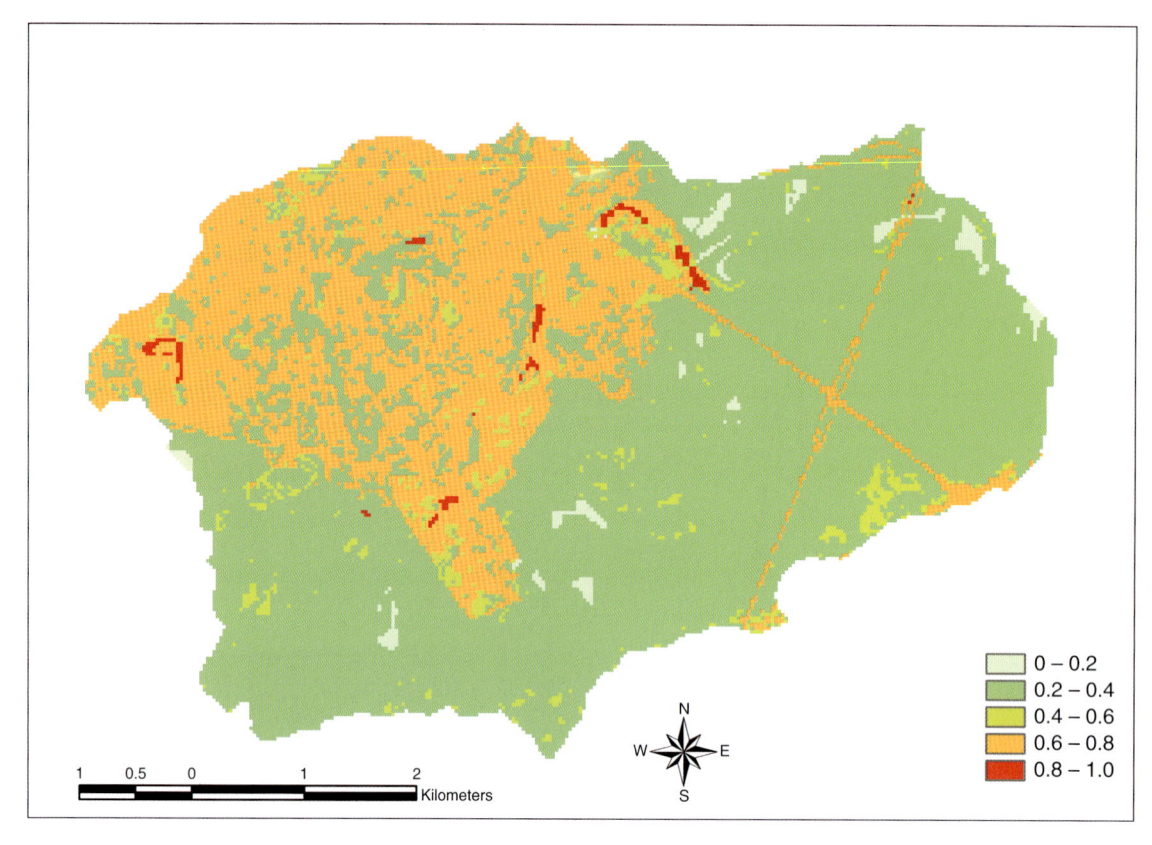

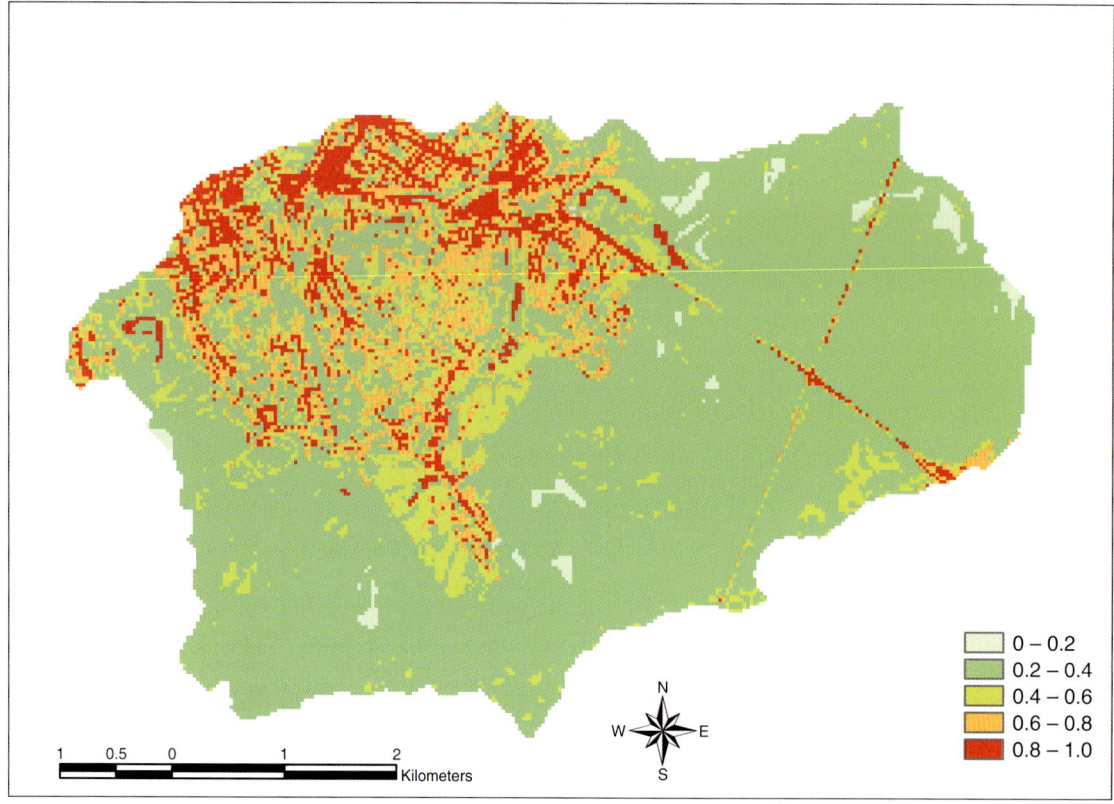

FIGURE 18.10 Runoff coefficient maps for scenario 1 (upper left), scenario 2 (upper right) and scenario 3 (lower left), based on IKONOS-derived land-cover fractions, and for scenario 3 (lower right), based on Landsat ETM+-derived land-cover fractions.

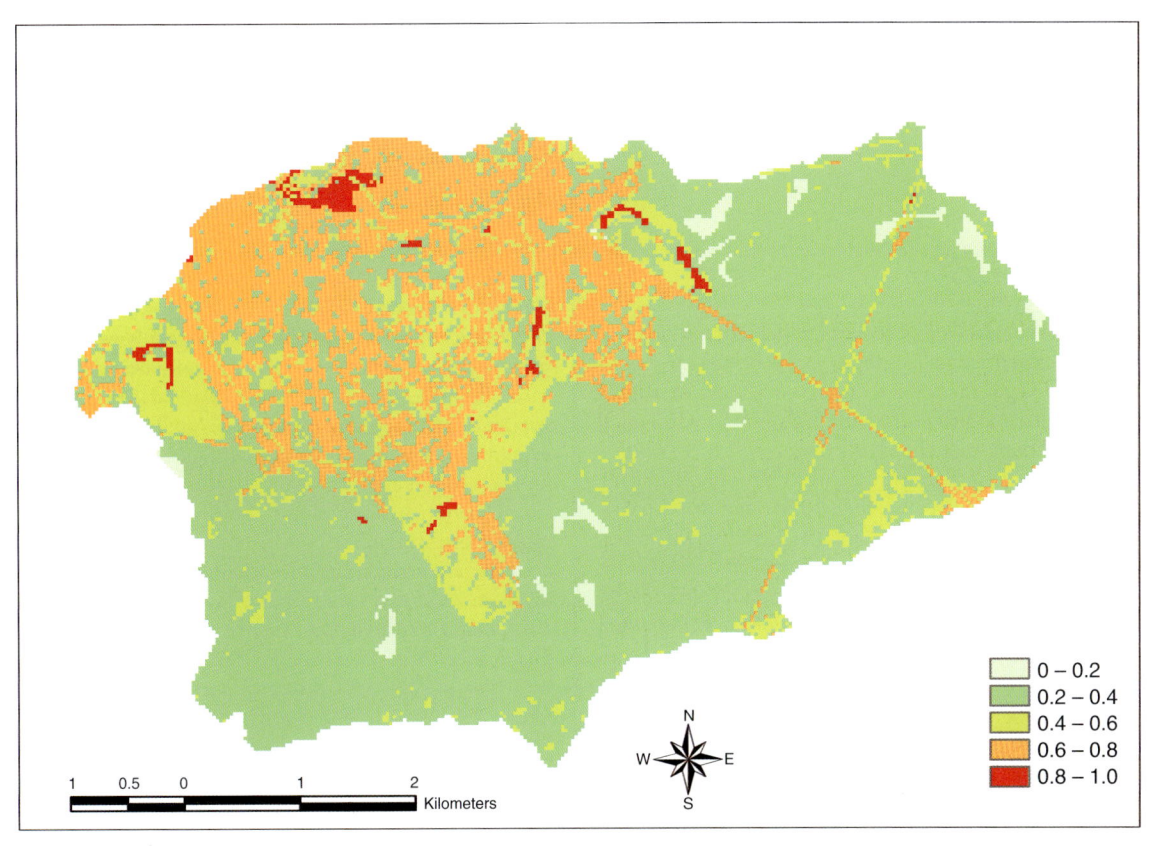

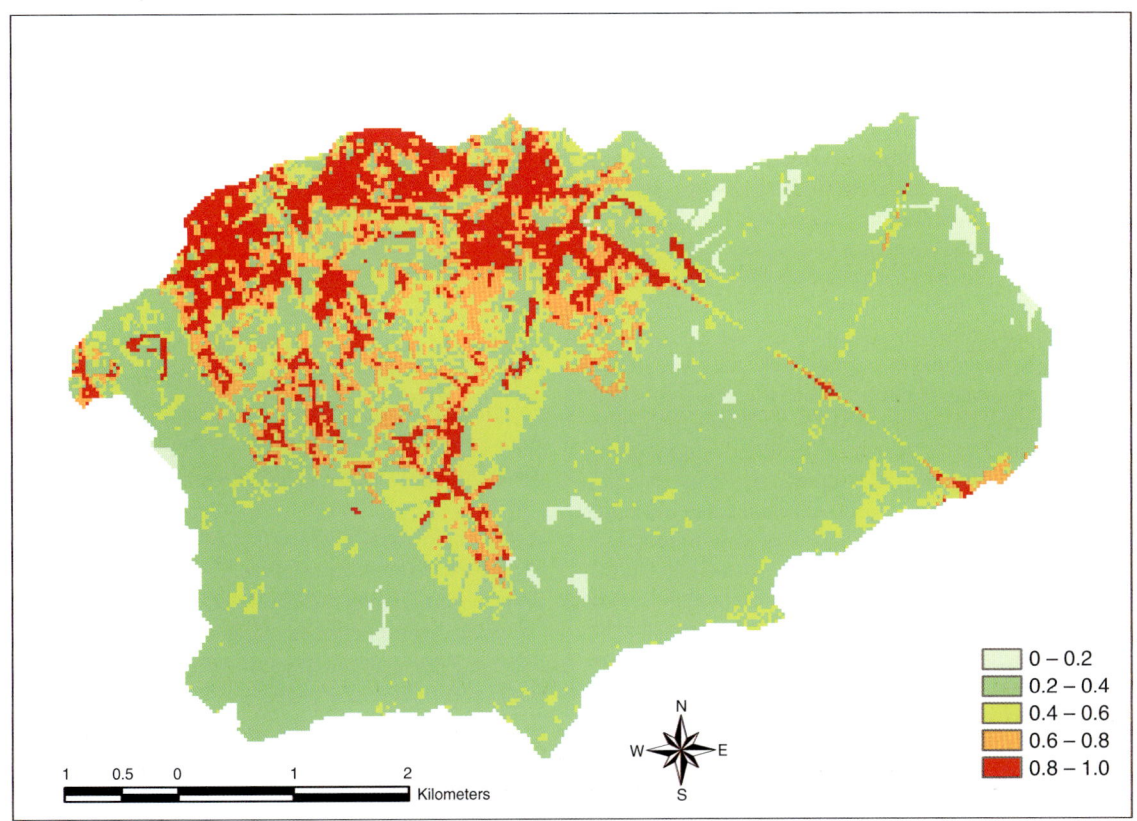

FIGURE 18.10 (*continued*).

accurate than high-resolution remote sensing for modeling the spatial distribution of runoff at a 30m cell size. However, if high-resolution data are not available or if resources are limited, the technique may be very useful to account for spatial variation in runoff.

Conclusions

Until recently the relationship between flood conditions and the spatial distribution of urban development has been poorly studied, partly because of the traditional use of lumped models in urban hydrological modeling. Presently, distributed hydrological models are becoming more popular because of their advantage of taking the spatial dynamics of hydrological processes explicitly into account. The fact that the spatial distribution of physical characteristics in hydrological models is mostly described by raster data makes the link to remotely sensed data obvious and highly interesting. One of the most important factors influencing runoff in urbanized areas is the extent and the spatial distribution of impervious surfaces within the catchment. In recent years, much research in urban remote sensing has been devoted to the development of methods for accurate mapping of impervious surfaces. Several studies addressed detailed impervious surface mapping from high-resolution remote sensing data. Other research focused on fractional mapping of impervious surface cover from medium-resolution data at the sub-pixel scale, offering the possibility to map impervious surface cover for large areas at reasonable cost.

In this study the impact of different methods for characterizing the distribution of impervious surfaces on the estimation of peak discharges was assessed for a strongly urbanized watershed in the Brussels Capital Region. The study shows that estimates of average imperviousness derived from satellite data may strongly differ from expert knowledge on the imperviousness of urban land-use classes documented in the literature and may lead to substantially different estimates of discharge at catchment level. The study also demonstrates that use of spatially detailed information on imperviousness obtained from high-resolution remote sensing leads to substantially higher estimates of peak discharge, compared to the use of average levels of imperviousness for different types of land-use, which can be considered the standard approach in hydrological modeling. The spatial variation in runoff and the connectivity between cells with high runoff coefficients in particular, proves to be a major factor influencing estimated discharge volume. Sub-pixel estimation of land-cover class proportions based on Landsat ETM+ data having the same resolution as the model's grid cell size (30 m) produces less structural detail, yet spatial patterns of imperviousness are well characterized. The higher level of generalization results in a smoothing of the runoff coefficient pattern and leads to slightly lower peak discharge estimates compared to the use of high-resolution impervious surface maps as input for the modeling. However, if high-resolution data are not available or if resources are limited, the technique may be an interesting alternative for high-resolution impervious surface mapping in rainfall-runoff modeling within urbanized catchments.

A lot of open research issues are recognized, which should allow future improvements in both impervious surface mapping as well as runoff modeling. New hyperspectral sensors allow a better discrimination between spectrally similar impervious and non-impervious cover types, both at high resolution (Franke et al., 2009) and medium resolution (Weng, Hu and Lu, 2008; Demarchi et al., 2010), leading to further improvement in the accuracy of impervious surface mapping. The increasing spatial resolution of hyperspectral sensors will be important as well for mapping of spatially heterogeneous urban land-cover. For the modeling of urban rainfall-runoff many challenges lie ahead. Methodologically, a strong focus should be given to the coupling of hydrological models for the estimation of runoff with hydraulic models for river and especially sewer flow. Also interaction of hydrological processes at the land surface, sewer and drinking distribution networks with the groundwater needs conceptual model improvements. Effects of climate change on hydrological processes in urban environments as well as demographic and economic changes will impact land and water use. New study and predictions approaches for adapting land and water management will have to be developed. Finally, the challenge remains, as already identified in the NEMO project in the 1990s, to translate scientific research results into actual urban water and land management or policy. Collaboration over the borders of scientific disciplines and with land and water policy/management will thus be required.

Acknowledgments

The research presented in this chapter is partly funded by the Belgian Science Policy Office in the frame of the STEREO programme – project SR/00/02 (SPIDER). J. Chormanski acknowledges the support of the Research in Brussels Programme for a postdoctoral fellowship.

References

Abbott, M., Bathurst, J., Cunge, J. *et al.* (1986) An introduction to the European Hydrological System – Systeme Hydrologique Europeen, "SHE", 2: Structure of a physically based, distributed modeling system. *Journal of Hydrology*, **87**, 61–77.

Anderson, D.G. (1968) Effects of urban development on floods in northern Virginia. *US Geological Survey Water-Supply Paper* 2001-C.

Anderson, K. and Croft, H. (2009) Remote sensing of soil surface properties. *Progress in Physical Geography*, 457–473.

Arnold, C.L., Jr. and Gibbons, C.J. (1996) Impervious surface coverage: the emergence of a key urban environmental indicator. *Journal of the American Planning Association*, **62** (2), 243–258.

Arnold, C.L., Jr., Crawford, H.M., Gibbons, C.J. and Jeffrey, R.F. (1994) The use of Geographic Information System images as a tool to educate local officials about the land use/water quality connection, in *Proceedings of the Watershed '93 Conference*, Alexandria, Virginia, March 1993, pp. 373–377.

Batelaan, O. and De Smedt, F. (2007) GIS-based recharge estimation by coupling surface-subsurface water balances. *Journal of Hydrology*, **337** (3–4), 337–355.

Bauer, M.E., Loffelholz, B.E. and Wilson, B. (2008) Estimating and mapping impervious surface area by regression analysis of Landsat imagery, in *Remote Sensing of Impervious Surfaces* (ed Q. Weng), CRC Press, Taylor & Francis Group, Boca Raton, pp. 3–19.

Beven, K. (2006) Streamflow generation processes. *Benchmark Papers in Hydrology 1*, IAHS Press, Wallingford, UK.

Beven, K. and Kirkby, M. (1979) A physical based, variable contributing area model of basin hydrology. *Hydrological Sciences Journal*, **24**, 43–69.

Bicknell, B.R., Imhoff, J.C., Kittle, J.L., Jr. et al. (1997) *Hydrological Simulation Program–Fortran: User's manual for version 11*, US Environmental Protection Agency, National Exposure Research Laboratory, Athens, GA, EPA/600/R-97/080.

Boegh, E., Thorsen, M., Butts, M. et al. (2004) Incorporating remote sensing data in physically based distributed agro-hydrological modeling. *Journal of Hydrology*, **287**(1–4), 279–299.

Browne, F.X. (1990) Stormwater management, in *Standard Handbook of Environmental Engineering* (ed. R.A. Corbitt), Chapter 7.1–7.135, McGraw-Hill, New York.

Brun, S.E. and Band, L.E. (2000) Simulating runoff behavior in an urbanizing watershed. *Computers, Environment and Urban Systems*, **24**, 5–22.

Carlson, T.N. (2004) Analysis and prediction of surface runoff in an urbanizing watershed using satellite imagery. *Journal of the American Water Resources Association*, **40** (4), 1087–1098.

Carlson, T.N. (2008) Impervious surface area and its effect on water abundance and water quality, in *Remote Sensing of Impervious Surfaces* (ed. Q. Weng), CRC Press, Boca Raton, pp. 353–367.

Carter, R.W. (1961) Magnitude and frequency of floods in suburban areas, in *Short Papers in the Geologic and Hydrologic Sciences, US Geol. Survey Professional Paper 424-B*, pp. B9–Bll.

Chow, V.T., Maidment, D.R. and Mays, L.W. (1988) *Applied Hydrology*, McGraw-Hill Inc., New York.

Crawford, N.H. and Linsley, R.K. (1966) *Digital Simulation in Hydrology: Stanford Watershed Model IV*, Technical Report No. 39, Department of Civil Engineering, Stanford University.

Dams, J., Batelaan, O., Nossent, J. and Chormanski, J. (2009) Improving hydrological model parameterisation in urbanised catchments: remote sensing derived impervious surface cover maps, in *Water and Urban Development Paradigms. Towards an Integration of Engineering, Design and Management Approach* (eds. J. Feyen, K. Shannon and M. Neville), Taylor & Francis Group, London.

Dare, P.M. (2005) Shadow analysis in high-resolution satellite imagery of urban areas. *Photogrammetric Engineering and Remote Sensing*, **71**, 169–177.

Demarchi, L., Canters, F., Chan, J.C.-W. and Van de Voorde, T. (2010) Mapping sealed surfaces from CHRIS/Proba data: a multiple endmember unmixing approach, in *Proceedings of the 2nd Workshop on Hyperspectral Image and Signal Processing: Evolution in Remote Sensing (Whispers 2010)*, Reykjavik, Iceland.

De Roeck, T., Van de Voorde, T. and Canters, F. (2009) Full hierarchic versus non-hierarchic classification approaches for mapping sealed surfaces at the rural–urban fringe using high-resolution satellite data. *Sensors*, **9**, 22–45.

De Smedt, F., Liu, Y.B. and Gebremeskel, S. (2000) Hydrological modeling on a catchment scale using GIS and remote sensed land use information, in *Risk Analyses II* (ed. C.A. Brebbia), WIT Press, Southampton, Boston.

Eagleson, P.S. (1970) *Dynamic Hydrology*, McGraw-Hill Inc., New York.

European Environment Agency (EEA) (2006) *Urban Sprawl in Europe – the Ignored Challenge*, EAA report 10, OOPEC, Brussels.

Fetter, C.W. (1980) *Applied Hydrogeology*, Charles E. Merrill Pub., Columbus, OH.

Flanagan, M. and Civco, D.L. (2001) Subpixel impervious surface mapping, in *Proceedings of the ASPRS 2001 Annual Conference*, American Society for Photogrammetry and Remote Sensing, Bethesda, MD, unpaginated CD-ROM.

Franke, J., Roberts, D.A., Halligan, K and Menz, G. (2009) Hierarchical multiple endmember spectral mixture analysis (MESMA) of hyperspectral imagery for urban environments. *Remote Sensing of Environment*, **113**, 1712–1723.

Gassman, P.W., Reyes, M.R., Green, C.H. and Arnold, J.G. (2007) The soil and water assessment tool: historical development, applications, and future research directions. *Transactions of the ASABE, American Society of Agricultural and Biological Engineers*, **50**, 1211–1250.

Gillies, R.R., Box, J.B., Symanzik, J. and Rodemaker, E.J. (2003) Effects of urbanization on the aquatic fauna of the Line Creek watershed, Atlanta – a satellite perspective. *Remote Sensing of Environment*, **86**, 411–422.

Hodgson, M.E., Jensen, J.R., Tullis, J.A. et al. (2003) Synergistic use of lidar and color aerial photography for mapping urban parcel imperviousness. *Photogrammetric Engineering and Remote Sensing*, **69**, 973–980.

Horton, R.E. (1933) The role of infiltration in the hydrological cycle. *Transactions of the American Geophysical Union*, **14**, 446–460.

Huang, C. and Townshend, J.R.G. (2003) A stepwise regression tree for nonlinear approximation: applications to estimating subpixel land cover. *International Journal of Remote Sensing*, **24** (1), 75–90.

James, L.D. (1965) Using a digital computer to estimate the effects of urban development on flood peaks, *Water Resources Research*, **1**(2), 223–234.

Ji, M. and Jensen, J.R. (1999) Effectiveness of subpixel analysis in detecting and quantifying urban imperviousness from Landsat Thematic Mapper imagery. *Geocarto International*, **14**, 33–41.

Jobson, H.E. (1989) Users manual for an open channel stream flow model based on the diffusion analogy, in *Techniques of Water Resources Investigations*, US Geological Survey, pp. 89–4133.

Kalma, J., McVicar, T. and McCabe, M. (2008) Estimating land surface evaporation: A review of methods using remotely sensed surface temperature data. *Surveys in Geophysics*, **29** (4), 421–469.

Kirkby, M.J. (1978) *Hill Slope Hydrology*. John Wiley & Sons, Ltd, Chichester.

Law, N.L., Cappiella, K. and Novotney, M.E. (2009) The need for improved pervious land cover characterization in urban watersheds. *Journal of Hydrologic Engineering*, **14**, 305–308.

Leopold, L.B. (1968) Hydrology for urban land planning – a guidebook on the hydrologic effects of urban land use. *US Geological Survey Circular 554*.

Li, Z.-L., Tang, R., Wan, Z. et al. (2009) A review of current methodologies for regional evapotranspiration estimation from remotely sensed data. *Sensors*, **9** (5), 3801–3853.

Liu, Y.B., Gebremeskel, S., De Smedt, F. et al. (2003) A diffusive transport approach for flow routing in GIS-based flood modeling. *Journal of Hydrology*, **283**, 91–106.

Liu, Y.B. and De Smedt, F. (2004) *WetSpa Extension, A GIS-based Hydrologic Model for Flood Prediction and Watershed Management, Documentation and User Manual*. Available http://www.vub.ac.be/WetSpa/ (accessed 18 November 2010).

Liu, Y., De Smedt, F., Hoffmann, L. and Pfister, L. (2004) Assessing land use impact on flood processes in complex terrain by using GIS and modeling approach. *Environmental Modeling and Assessment*. **9** (4), 227–235.

Lu, D. and Weng, Q. (2007) Mapping urban impervious surfaces from medium and high spatial resolution multispectral imagery, in *Remote Sensing of Impervious Surfaces* (ed. Q. Weng), CRC Press, Taylor & Francis Group, Boca Raton, pp. 59–77.

Lu, D. and Weng, Q. (2009) Extraction of urban impervious surfaces from IKONOS imagery. *International Journal of Remote Sensing*, **30** (5), 1297–1311.

Martine, G. (2007) *The State of the World Population 2007*. United Nations Population Fund, New York.

Mejía, A.I. and Moglen, G.E. (2009) Spatial patterns of urban development from optimization of flood peaks and imperviousness-based measures. *Journal of Hydrologic Engineering*, **14**, 416–424.

Melesse, A.M. and Wang, X. (2008) Impervious surface area dynamics and storm runoff response, in *Remote Sensing of Impervious Surfaces* (ed. Q. Weng), CRC Press, Boca Raton, pp. 369–386.

Miller, W.A., and Cunge, J.A. (1975) Simplified equations of unsteady flow, in *Unsteady Flow in Open Channels* (eds. K. Mahmood and V. Yevjevich), Water Resources Publications, Fort Collins, CO.

Moglen, G.E. (2009) Hydrology and impervious areas. *Journal of Hydrologic Engineering*, **14**, 303–304.

Mohammed, G.A. (2009) Modeling groundwater-surface water interaction and development of an inverse groundwater modeling methodology, Ph.D. thesis, Vrije Universiteit Brussel, Brussels. Available http://twws6.vub.ac.be/hydr/download/GetachewAdemMohammed.pdf (accessed 18 November 2010).

Mohapatra, R.P. and Wu, C. (2007) Estimation with IKONOS imagery: An artificial neural network approach, in *Remote Sensing of Impervious Surfaces* (ed. Q. Weng), CRC Press, Taylor & Francis Group, Boca Raton, pp. 21–39.

Mulvany, T.J. (1850) On the use of self registering rain and flood gauges. *Insttute of Civil Engineers Proceedings (Dublin)*, **4**, 1–8.

NOAA, The Great Lakes Environmental Research Laboratory (2006) *Distributed Large Basin Runoff Model*. Available http://www.glerl.noaa.gov/res/Programs/pep/dlbrm/home.html (accessed 18 November 2010).

Peters, N.E. (2009) Effects of urbanization on stream water quality in the city of Atlanta, Georgia, USA. *Hydrological Processes*, **23**, 2860–2878.

Petropoulos, G., Carlson, T., Wooster, M. and Islam, S. (2009) A review of Ts/VI remote sensing based methods for the retrieval of land surface energy fluxes and soil surface moisture. *Progress in Physical Geography*, 224–250.

Phinn, S., Stanford, M., Scarth, P. et al. (2002) Monitoring the composition and form of urban environments based on the vegetation–impervious surface–soil (VIS) model by sub-pixel analysis techniques. *International Journal of Remote Sensing*, **23**, 4131–4153.

Pilgrim, D. and Cordery, I. (1992) Flood runoff, in *Handbook of Hydrology* (ed. D.R. Maidment), McGraw-Hill, Inc., New York, pp. 9.1–9.41.

Pu, R., Gong, P., Michishita, R. and Sasagawa, T. (2008) Spectral mixture analysis for mapping abundance of urban surface components from the Terra/ASTER Data. *Remote Sensing of Environment*, **112**, 939–954.

Rashed, T., Weeks, J.R., Roberts, D. et al. (2003) Measuring the physical composition of urban morphology using multiple endmember spectral mixture models. *Photogrammetric Engineering and Remote Sensing*, **69**, 1011–1020.

Schmugge, T.J., Kustas, W.P., Ritchie, J.C. et al. (2002) Remote sensing in hydrology. *Advances in Water Resources*, **25**, 1367–1385.

Schueler, T.R. (1994) The importance of imperviousness. *Watershed Protection Techniques*, **1** (3), 100–110.

Schueler, T.R., Fraley-McNeal, L. and Cappiella, K. (2009) Is impervious cover still important? Review of recent research. *Journal of Hydrologic Engineering*, **14**, 309–315.

Schumann, G., Bates, P.D., Horritt, M.S. et al. (2009) Progress in integration of remote sensing-derived flood extent and stage data and hydraulic models. *Reviews of Geophysics*, **47**, RG4001–.

Sleavin, W.J., Civco, D.L., Prisloe, S. and Gianotti, L. (2000) Measuring impervious surfaces for nonpoint source pollution modelling, in *Proceedings of the ASPRS 2000 Annual Conference*, Washington DC, unpaginated CD-ROM.

Smedema, L.K. and Rycroft, D.W. (1988) *Land Drainage*, BT Batsford Ltd., London.

TxDOT, Texas Department of Transportation (2009) *Hydraulic Design Manual, revised version*. Available http://onlinemanuals.txdot.gov/txdotmanuals/hyd/hyd.pdf (accessed 18 November 2010).

Thomas, N., Hendrix, C. and Congalton, R.G. (2003) A comparison of urban mapping methods using high resolution digital imagery. *Photogrammetric Engineering and Remote Sensing*, **69**, 963–972.

Todini, E. (2007) Hydrological catchment modelling: past, present and future. *Hydrical Earth Systems Science*, **11**, 468–482.

USDA, United States Department of Agriculture (1986) *Urban Hydrology for Small Watersheds*. Technical Release 55 (TR-55) (2nd Ed.), Natural Resources Conservation Service, Conservation Engineering Division. Avaialble http://www.cpesc.org/reference/tr55.pdf (accessed 18 November 2010).

Van de Voorde, T., De Genst, W. and Canters, F. (2007) Improving pixel-based VHR land-cover classifications of urban areas with post-classification techniques. *Photogrammetric Engineering and Remote Sensing*, **73**, 1017–1027.

Van de Voorde, T., De Roeck, T. and Canters, F. (2009) A comparison of two spectral mixture modeling approaches for impervious surface mapping in urban areas. *International Journal of Remote Sensing*, **30** (18), 4785–4806.

Wang, Y. and Zhang, X. (2004) A SPLIT model for extraction of subpixel impervious surface information. *Photogrammetric Engineering and Remote Sensing*, **70**, 821–828.

Wang, Z.M., Batelaan, O. and De Smedt, F. (1996) A distributed model for water and energy transfer between soil, plants and atmosphere (WetSpa). *Physics and Chemistry of the Earth*, **21** (3), 189–193.

Weng, Q. and Hu, X. (2008) Medium spatial resolution satellite imagery for estimating and mapping urban impervious surfaces using LSMA and ANN. *IEEE Transactions on Geoscience and Remote Sensing*, **46** (8), 2397–2406.

Weng, Q., Hu, X. and Lu, D. (2008) Extracting impervious surfaces from medium spatial resolution multispectral and hyperspectral imagery: a comparison. *International Journal of Remote Sensing*, **29** (11), 3209–3232.

Wu, C. (2009) Quantifying high-resolution impervious surfaces using spectral mixture analysis. *International Journal of Remote Sensing*, **30**(11), 2915–2932.

Wu, C. and Murray, A.T. (2003) Estimating impervious surface distribution by spectral mixture analysis. *Remote sensing of Environment*, **84**, 493–505.

Xian, G. (2006) Assessing urban growth with sub-pixel impervious surface coverage, in *Urban Remote Sensing* (eds Q. Weng and D.A. Quattrochi), Taylor & Francis Group, Boca Raton.

Yang, L., Huang, C., Homer, C.G. et al. (2003) An approach for mapping large-area impervious surfaces: synergistic use of Landsat-7 ETM+ and high spatial resolution imagery. *Canadian Journal of Remote Sensing*, **29**, 230–240.

Yang, X. and Liu, Z. (2005) Use of satellite-derived landscape imperviousness index to characterize urban spatial growth. *Computers, Environment and Urban Systems*, **29**, 524–540.

Yuan, F., Wu, C. and Bauer, M.E. (2008) Comparison of various spectral analytical techniques for impervious surface estimation using Landsat imagery. *Photogrammetric Engineering and Remote Sensing*, **74** (8), 1045–1055.

19

Impacts of urban growth on vegetation carbon sequestration

Tingting Zhao

The last two centuries saw increasing human impacts on the global carbon cycle through land-cover/land-use conversions and the burning of fossil fuels, both highly associated with urban development. The accounting of urban carbon balance that considers both vegetation carbon sinks and human energy uses, however, remains largely incomplete at regional to global scales. One of the challenges facing the scientific community involves availability of data on carbon exchange between vegetation and the atmosphere at large geographic scales. Light-use efficiency (LUE) models provide a partial solution to this problem through integration of moderate- to coarse-resolution remote sensing imagery with the measured or modeled vegetation biophysical parameters. In this chapter, the LUE approach is briefly reviewed and applied to the estimation of gross primary production (GPP) in eight states and the District of Columbia in the eastern United States in 1992 and 2001. The estimated GPP was associated with four settlement densities identified based on Census 1990 and 2000 housing unit data. The LUE-based vegetation productivity estimates may be integrated with carbon emissions data prepared based on energy and transportation surveys, so as to provide a comprehensive view of net carbon exchange between land and the atmosphere owing to human urban development.

Urban Remote Sensing: Monitoring, Synthesis and Modeling in the Urban Environment, First Edition. Edited by Xiaojun Yang.
© 2011 John Wiley & Sons, Ltd. Published 2011 by John Wiley & Sons, Ltd.

19.1 Introduction

There is growing consensus that humans play an important role in modifying the global carbon cycle through land-cover/land-use conversions (e.g., deforestation) and the burning of fossil fuels, both highly associated with urban development (Boyle and Lavkulich, 1997; Nowak and Crane, 2002; Pouyat et al., 2002; Pataki et al., 2006; Churkina, 2008; Grimm et al., 2008; Churkina, Brown and Keoleian, 2010). Urban lands, accounting for approximately 3% of the total US land area in 1995, were estimated on average to be 1.6% less effective in absorbing carbon dioxide by green vegetation than their immediately adjacent non-urban surroundings (Imhoff et al., 2004). By contrast, an urban system may release a thousand more units of carbon dioxide to the atmosphere than a forest ecosystem of the same area, due to the burning of fossil fuels to produce adequate amount of energy for its sustainment (Odum, 1997).

The accounting of carbon balance in urban areas that considers vegetation productivities as well as human energy uses, however, remains largely incomplete at regional to global scales. Although the conceptual framework of the integrated urban ecological and socioeconomic analysis became mature by the end of the last century (Grimm et al., 2000; Pickett et al., 2001), quantitative assessment of carbon fluxes between urban areas and the atmosphere has been limited to a few metropolitan cities where either field measurements of vegetation productivities or detailed information on energy consumption was available (Kaye, McCulley and Burke, 2005; Pataki et al., 2009). The large-extent urban carbon accounting relies increasingly on the availability of data concerning carbon fluxes or stocks observed and modeled beyond the local scale.

Remote sensing technique offers a unique opportunity for the investigation of carbon dynamics of terrestrial lands at regional to global scales. Satellite imagery and aerial photos provide spatially and temporally continuous data on vegetation properties, including vegetation types, phenology, and biophysical characteristics such as leaf area index (LAI; Turner, Ollinger and Kimball, 2004). These data captured at the spatial resolution of a few meters to many kilometers have been applied widely to productivity and biomass estimation in natural vegetation and agricultural land uses (Prince and Goward, 1995; Haboudane et al., 2002; Zhao et al., 2009). Remote sensing also provides information on human settlement patterns (Herold, Scepan and Clarke, 2002), which may be used to infer carbon emissions from urban land uses, energy consumption, and transportation.

In this chapter, I will review measures and estimation of vegetation productivities, with emphasis given to light-use efficiency (LUE) models that utilize reflectance data gathered with optical remote sensing to estimate gross and net primary production (Section 19.2). Then, I will present a case study using the LUE approach to calculate changes in gross primary production (GPP) in the US Census South Atlantic division between 1992 and 2001; these changes will be related to urban growth identified based on changes in the US Census housing-unit density during the roughly same period (Sections 19.3 and 19.4). Finally, I will discuss the impacts of urban growth on vegetation productivities and outlook for future work (Section 19.5). By integrating the LUE-based vegetation productivity estimates with emission data prepared using energy and transportation surveys, a comprehensive view of net carbon exchange between land and the atmosphere owing to human urban development will be generated (Section 19.6).

19.2 Vegetation productivities and estimation

19.2.1 Vegetation productivities

The Earth consists of several massive carbon reservoirs including the atmosphere, plants and other living organisms, soils, marine sediments, and sedimentary rocks etc., where the element carbon takes different forms (e.g., CO_2 and carbonate) as it cycles through those reservoirs. The exchange of carbon between the atmosphere and terrestrial ecosystems (including plants, animals, soils, and soil microbial species) relies significantly on green vegetation photosynthesis, the process by which CO_2 is transformed to carbohydrate in the presence of solar radiation; and respiration, the process by which CO_2 is released back to the atmosphere.

The total amount of carbon that enters an ecosystem through photosynthesis during a certain time period, usually a year, is referred to as gross primary production (GPP; Chapin, Matson and Mooney, 2002). GPP indicates the maximum potential carbon uptake from the atmosphere by plants in an ecosystem. By subtracting the total plant respiration from GPP over the same time span, net primary production (NPP) is obtained. This is the net carbon obtained by plants through photosynthesis. GPP and NPP are indicators for carbon flows between vegetation and the atmosphere regardless of the presence of other living organisms or soil carbon processes; therefore, they are referred to as vegetation productivities in this chapter.

19.2.2 Estimation of vegetation productivities

The direct measurement of GPP and NPP proves difficult for large areas; therefore, alternative solutions rely on modeling these vegetation productivities at regional to global scales (Cramer et al., 1999; Zheng, Prince and Wright, 2003). For example, Running and Hunt (1993) developed the Biome-BGC model that simulates photosynthesis, respiration, and other ecosystem processes for areas up to hundreds of square kilometers. This model, combined with Landsat-based land-cover classification and leaf area index as well as meteorological variables measured at eddy covariance towers, was used to estimate GPP in two hardwood and boreal forest sites in the United States (Turner et al., 2003).

The light-use efficiency (LUE) approach, also known as production efficiency model, is a type of productivity models that relies extensively on remotely-sensed biophysical characteristics of land. The concept of the LUE approach resides in the correlation between optical properties of green leaves and the leaf's physiological functions (Tucker and Sellers, 1986). Green leaves are highly efficient in absorbing shortwave radiation between 400 and 700 nm, due mainly to the presence of chlorophylls in chloroplasts. Chloroplasts are where the primary processes of photosynthesis occur (Lambers, Chapin and Pons, 1998). The reduced reflectance of solar radiation in the 400–700 nm wavelength range over a vegetation canopy is associated with the light-absorbing photosynthesis processes. Such reduction can be captured with commonly used terrestrial remote sensing instruments such as IKONOS, Landsat ETM+, Terra MODIS, and NOAA AVHRR.

LUE models treat vegetation productivity (i.e., GPP or NPP) as a function of available light used by plants (Monteith, 1972; Field, Randerson and Malmström, 1995; Prince and Goward, 1995; Running et al., 2004). The rate of carbon conversion from inorganic form (i.e., CO_2) to organic form (i.e., carbohydrate measured by GPP or NPP) is associated with the total amount of absorbed photosynthetically active radiation (APAR; i.e., shortwave radiation between 400 and 700 nm). APAR is determined by the amount of incident solar radiation and types of plant species being exposed to light. Carbon conversion efficiency (light-use efficiency; ε) measures carbon production per unit energy captured (g C MJ^{-1}), and may be derived from field observation or ecological process models (Gower, Kucharik and Norman, 1999). This parameter varies by types of vegetation (e.g., tropical evergreen forest vs. boreal evergreen forest) and by species. Therefore, GPP or NPP is essentially the function of incoming solar radiation and types of vegetation, the latter determining both vegetation reflectance characteristics and carbon conversion efficiency ε (Fig. 19.1).

LUE models have been applied to the regional and global characterization of primary productions across various vegetation types (Field, Randerson and Malmström, 1995; Prince and Gower, 1995; Running et al., 2000; Zhao et al., 2005; Yang et al., 2007). Prince and Goward (1995) developed the GLO-PEM model and estimated GPP for 12 broad vegetation types at the global scale, using the normalized difference vegetation index (NDVI) derived from the red and near infrared reflectance collected by the NOAA AVHRR sensor. Similar approaches were applied to NDVI calculated from the Terra MODIS reflectance to estimate the global GPP (Running et al., 2000; Zhao et al., 2005). Validation of satellite-derived vegetation productivities has also been accomplished using actual measurements at eddy covariance flux towers (Turner et al., 2003; Heinsch, et al., 2006; Yang et al., 2007). These research efforts, however, presented divergent findings on productivity estimates due possibly to large variation in the geographic extent of these studies, time that the data were collected, parameter settings for ε, and the data sources used to define vegetation types (Table 19.1).

19.3 Data and analysis

Impacts of urban growth on vegetation productivities are illustrated using a case study of urban growth and changes in GPP in the US Census South Atlantic division between 1992 and 2001. The study area consists of eight states in the Eastern United States, including Delaware, Florida, Georgia, Maryland, North Carolina, South Carolina, Virginia, and West Virginia, and in addition the District of Columbia (Fig. 19.2). This region accounted for less than 8% of the total US territory, but had 18.4% of the nation's total population in 2000 (US Census Bureau, 2001a). Sixteen of the 100 largest US cities were located in the South Atlantic division, with an average rate of population growth between 1990 and 2000 of 9.4% in the cities and 22.9% in the suburbs (Berube, 2003).

19.3.1 Identifying urban growth

Urban growth was measured as the increase in housing unit density (HUD) at the scale of US Census county subdivisions.

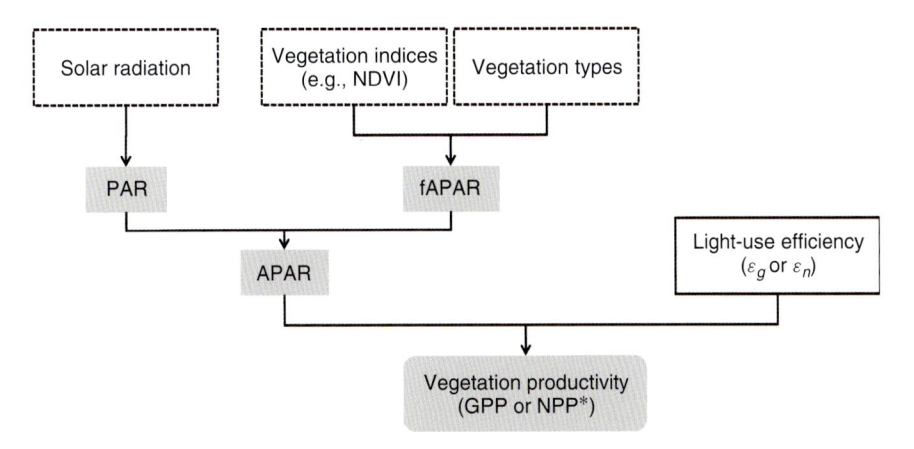

* Some LUE models estimate NPP directly as the product of APAR and ε_n (e.g., Field et al., 1995), while others derive NPP based on GPP (estimated from APAR and ε_g) minus modeled plant respiration (e.g., Prince and Gower, 1995; Running et al., 2000).

FIGURE 19.1 The light-use efficiency models estimate gross or net primary production based on the absorbed photosynthetically active radiation (APAR) and the light-use efficiency parameters of plants. APAR may be estimated using satellite-derived solar radiation, vegetation indices, and vegetation types (enclosed by dashed lines). Light-use efficiency (ε_g or ε_n) may be generated from field observation or ecological process models. The gray-shaded PAR, fAPAR, and APAR are intermediate model products. GPP or NPP is the end product of LUE models. NDVI is the normalized difference vegetation index. PAR is photosynthetically active radiation. fAPAR is the fraction of the absorbed photosynthetically active radiation. GPP is gross primary production. NPP is net primary production. ε_g is light-use efficiency (ε) for GPP. ε_n is ε for NPP.

TABLE 19.1 Light-use efficiency (ε_g, g C MJ^{-1}) and corresponding estimated gross primary production (GPP; g C m^{-2} year^{-1}) in previously published research at regional to global scales. Numbers in parentheses are standard deviation.

Vegetation Types	Prince and Gower (1995)[a] ε_g	GPP	Zhao et al. (2005)[b] ε_g	GPP	Yang et al. (2007)[c] ε_g	Turner et al. (2003)[d] ε_g
Evergreen broadleaf forest	0.71 (0.35)	1615 (475)	1.159	2699	-	-
Evergreen deciduous forest	0.66 (0.75)	1116 (552)	1.008	818	1.02 (0.40)	1.0 (0.5)
Broadleaf deciduous forest	0.41 (0.52)	736 (434)	1.044	1366	1.56 (0.36)	-
Mixed forest	0.64 (0.61)	1233 (532)	1.116	1125	1.31 (0.36)	1.8 (0.8)
Wooded grassland	0.43 (0.46)	865 (432)	0.768	1250	0.79 (0.59)	-
Grassland	0.40 (0.71)	377 (296)	0.068	396	0.86 (0.73)	1.7 (0.4)
Cropland	0.41 (0.53)	628 (314)	0.068	721	1.47 (0.53)	2.2 (0.7)

[a] Global estimation based on AVHRR NDVI in 1987.
[b] Global estimation based on MODIS NDVI during 2001–2003; ε_g came from Heinsch et al. (2003).
[c] Estimation for the conterminous United States based on MODIS NDVI in 2004.
[d] Estimation at four flux-tower-sized sites in the United States based on measured APAR for June-September only (two sites in 1997, one in 1999, and one in 2000).

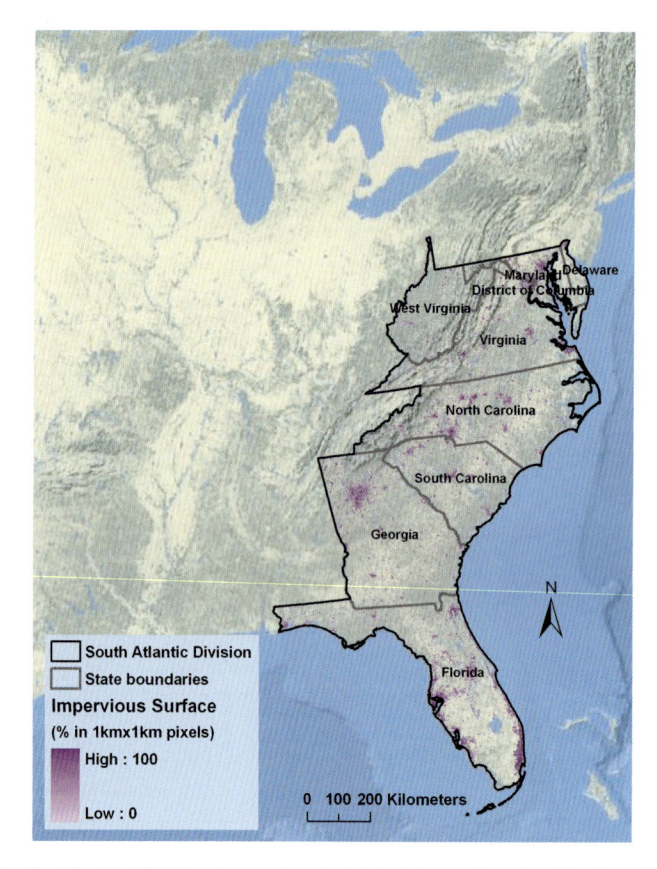

FIGURE 19.2 The Census South Atlantic Division is made of eight states, plus the District of Columbia, in the eastern US. Political boundaries came from the US Census Bureau. Proportions of the imperious surface came from Multi-Resolution Land Characteristics Consortium (MRLC). Base map is the World Physical Map obtained from ArcGIS Resource Centers.

HUD was calculated as the total number of housing units within a county subdivision divided by the area of that county subdivision. A housing unit refers to "a house, an apartment, a mobile home or trailer, a group of rooms, or a single room occupied as separate living quarters" (US Census Bureau, 2001b). The US decennial Censuses reported the total number of housing units at each summary scale ranging downward from the state level to the county, county subdivision, tract, and block group level. The scale of county subdivision was selected for this study following the rationale that patterns of changes in HUD should be captured at the scale of analysis, and that this scale should be manageable in handling data and analysis for a large geographic area. Previous research in the US Midwest showed that, compared with Census block groups, Census county subdivisions appeared to have the same efficiency in characterizing urban growth and its relationships to carbon changes (Zhao, 2007). The scale of county subdivision is practical for handling analysis of Census demographic changes across the South Atlantic division.

The housing-unit counts from the 1990 and 2000 U.S. Censuses (US Census Bureau, 1991, 2001a) were used to derive HUD for each county subdivision within the South Atlantic division in each of the two years separately. All county subdivisions were separated into four categories of settlement densities including urban (≤ 0.1 hectares per housing unit), suburban (0.1–0.69 hectares per housing unit), exurban (0.69–16.2 hectares per housing unit), and rural (> 16.2 hectares per housing unit) densities following the definition of Theobald (2005). This yielded maps of the four settlement densities for 1990 and 2000 with Census county subdivisions as mapping units in size ranging from 0.62 to 3069 km^2 (on average 214 km^2 or 21 437 hectares per county subdivision). Both maps were exported to raster format with the spatial resolution of 100 m, and used in change analysis that detects pixel-wise persistent or transitional status concerning settlement densities in 1990 and 2000. To minimize the sliver problem of GIS (Goodchild, 1978), the change data were clumped by types of settlement-density conversion using the four-neighbor rule and clumps smaller than 0.25 km^2 were merged into nearby larger clumps.

There were 16 possibilities of settlement-density conversion between 1990 and 2000, including four persistent types (i.e., Persistent Urban, Persistent Suburban, Persistent Exurban, and Persistent Rural) and 12 transitional types. Some of the transition was introduced by changes in boundaries of Census units between the two Census dates (Goodchild, Anselin and Deichmann, 1993). I attempted to minimize this type of transition, which obscures the actual changes in HUD, by excluding county subdivisions with area variation equal to or higher than 10% between 1990 and 2000. This resulted in a removal of 646 county subdivisions in the 2000 Census throughout the South Atlantic division, equivalent to 19.4% of the total count of county subdivisions and 16.4% of the total area of the South Atlantic division. Among the removed units, 414 county subdivisions were located within Virginia and West Virginia. These removed county subdivisions were masked out for further analysis.

19.3.2 Preparing vegetation maps and light-use efficiency parameters

The estimation of GPP based on LUE models requires data on light-use efficiency (ε_g), solar radiation, vegetation indices, and types of vegetation (Fig. 19.1). In this study, types of vegetation were inferred from the National Land Cover Dataset 1992 (Vogelmann et al., 2001) and National Land Cover Database 2001 (Homer et al., 2007), each providing the nationwide land-cover/land-use classification close to Census 1990 and Census 2000. The National Land Cover Dataset/Database (NLCD) land-cover/land-use classification was grouped into ten vegetation and land-use categories (Table 19.2), each differing from the other categories in terms of their ecological functions (Gower, Kucharik and Norman, 1999). This resulted in a map of 10 vegetation and land-use categories at the spatial resolution of 30 m across the South Atlantic study area in 1992 and 2001, respectively. The 30-m vegetation and land-use data were aggregated to 1 km, where the coarse pixel held proportions of individual vegetation and land-use categories by area within that pixel.

To determine values of light-use efficiency for individual vegetation and land-use categories, parameters documented in previous research (Yang et al., 2007) were applied to the forest, shrubland, grassland, and agriculture categories (Table 19.2). These documented values were used in estimating ε_g of Built-up and Wetland with a few assumptions applied to each category. For Wetland, ε_g was calculated as the average of ε_g for Deciduous Forest, Coniferous Forest, Mixed Forest, Shrubland, and Grassland, assuming equal distribution of wooded and herbaceous species given the absence of detailed species composition information in wetland areas. For Built-up, ε_g was estimated as the area weighted average of tree, lawn, impervious surface, and water. ε_g of impervious surface and water were assumed to be zero. ε_g of tree was calculated as the average of ε_g for Deciduous Forest and Coniferous Forest, which is equal to 1.29 g C MJ^{-1}. ε_g of lawn was assigned as the value for Grassland. The area proportion of tree, lawn, impervious surface and water for the Built-up category was estimated by examining percentage of canopy coverage and impervious surface within areas identified as Built-up in 2001. Canopy coverage and impervious surface were extracted from the NLCD 2001 Imperviousness and Tree Canopy data, which provide the pixel-wise Landsat-based estimation of canopy and impervious proportions at the 1-km resolution (Huang, Homer and Yang, 2003; Yang et al., 2003). The average tree (or canopy) cover was found to be 20.3% within the Built-up area across the South Atlantic division; the number was 21.5% for impervious surface. The area percentage of lawn or water was calculated as half of the rest area (i.e., 29.1% each). Therefore, ε_g of the Built-up type was estimated as the weighted average of tree (1.29 g C MJ^{-1}; 20.3%), lawn (0.86 g C MJ^{-1}; 29.1%), and impervious surface or water (0; 50.6%).

19.3.3 Estimating APAR, GPP and changes in GPP

To estimate GPP in 1992 and 2001, APAR was calculated for each year separately as the product of PAR and fAPAR (Fig. 19.1). PAR was the downward shortwave radiation, which came from the monthly average climate forcing data based on the NOAH land-surface model (Rodell et al., 2004), multiplied by a scalar 0.45. The 1-degree PAR data were re-sampled to 1-km resolution to be used jointly with vegetation maps and fAPAR data prepared at this finer resolution for the estimation of GPP. fAPAR was estimated based on the biweekly AVHRR NDVI and maps of vegetation and land uses (compiled in the previous step), using the look-up table approach (Knyazikhin et al., 1999). The NDVI data were processed to minimize cloud cover, sensor degradation, and atmospheric effects (USGS EROS Data Center, 2006) before being applied to this study. Broadleaf forest, needleleaf forest, shrubland, and grassland/cereal crops in Knyazikhin et al. (1999) approximated the Deciduous Forest, Coniferous Forest, Shrubland and Grassland categories in this study, whereas the NDVI-fAPAR look-up values for Mixed Forest was derived as the average between broadleaf forest and needleleaf forest. Agriculture was approximated as the average of broadleaf crop and grassland/cereal crops; and Wetland as the average of broadleaf forest, needleleaf forest, shrubland, and grassland/cereal crops. The NDVI-fAPAR look-up for the Built-up category was estimated as the area weighted average of Mixed Forest (20.3%), Grassland (29.1%), and impervious surface or water (50.6%), using the similar approach for the estimation of ε_g for the Built-up type.

TABLE 19.2 Vegetation and land-use categories generated by grouping the National Land Cover Dataset/Database (NLCD) land-cover/land-use classification (Vogelmann et al., 2001; Homer et al., 2007).

This study	Vegetation and land-use category NLCD 1992	NLCD 2001	ε_g (g C MJ^{-1})
1: Deciduous Forest	Deciduous forest (41)	Deciduous forest (41)	1.56
2: Coniferous Forest	Evergreen forest (42)	Evergreen forest (42)	1.02
3: Mixed Forest	Mixed forest (43)	Mixed forest (43)	1.31
4: Shrubland	Shrubland (51) Orchards, vineyards and other (61)	Shrub and scrub (52)	0.79
5: Grassland	Herbaceous grassland (71)	Herbaceous grassland (71)	0.86
6: Agriculture	Pasture and hay (81) Row crops (82) Small grains (83) Fallow (84) Urban and recreational grasses (85)	Pasture and hay (81) Cultivated crops (82)	1.47
7: Built-up	Low intensity residential (21) High intensity residential (22) Commercial, industrial and transportation (23)	Developed, open space (21) Developed, low density (22) Developed, medium density (23) Developed, high density (24)	0.51*
8: Wetland	Woody wetlands (91) Emergent herbaceous wetlands (92)	Woody wetlands (90) Emergent herbaceous wetland (95)	1.11*
9: Bare	Bare rock, sand and clay (31) Quarries, strip mines and gravel pits (32) Traditional barren (33)	Barren land (31)	n/a
10: Water	Open water (11) Perennial ice and snow (12)	Open water (11) Perennial ice and snow (12)	n/a

The corresponding light-use efficiency (ε_g) came from Yang et al. (2007), except for the Built-up and Wetland types. ε_g of Built-up was estimated as the weighted average of ε_g for deciduous and coniferous forests (1.29 g C MJ^{-1}; 20.3%), grassland (0.86 g C MJ^{-1}; 29.1%), and impervious surface or water (0; 50.6%); refer to the text for detailed explanation. ε_g of Wetland was estimated as the average of ε_g for Deciduous Forest, Coniferous FOREST, Mixed Forest, Shurbland, and Grassland.
*Estimated.

In the process of estimating GPP, temperature scalars were applied to ε_g, given that environmental constraints such as temperature, moisture, and nutrient availability influence the actual production of carbohydrate (Prince and Goward, 1995; Running et al., 2000). The rationale is that, when temperature drops below certain thresholds, plants would be less efficient in converting CO_2 to carbohydrate or completely shut down photosynthesis. The temperature scalar, ranging between 0 and 1, was determined by examining the minimum daily temperature (T_{min}). Its value is 0 for T_{min} below $-8°C$, 1 for T_{min} above the threshold temperature depending on types of vegetation and land uses, and linear function between 0 and 1 for T_{min} in between (Heinsch et al., 2003). The temperature data came from the monthly average climate forcing data based on the NOAH land-surface model (Rodell et al., 2004).

Changes in GPP were derived by subtracting the pixel-wise estimate in 1992 from the GPP estimate in 2001. Positive value indicates an increase in GPP at the later date, and negative value corresponds to the decline of GPP in 2001. These changes in GPP were associated with different types of urban growth (identified based on changes in Census HUD) throughout the South Atlantic study area.

19.4 Results

According to Census HUD, Rural densities (44.1%) predominated in the South Atlantic division by area in 1990, followed by the Exurban (39%), Suburban (14.9%), and Urban (2%) densities (Table 19.3; Fig. 19.3). By 2000, Exurban densities exceeded Rural densities by area and accounted for 42.7% of the total study area. The area of Urban densities increased by approximately 21% during the 1990–2000 period. Suburban densities grew at a higher rate (approximately 27%) than Urban densities. The area of Exurban densities also expanded (by approximately 9%) during the same time period. The growth of Urban, Suburban, and Exurban densities resulted in the loss of area at Rural densities, which dropped by approximately 19% between 1990 and 2000.

GPP was estimated in 1992 to be 4.11 g C m^{-2}day^{-1} and in 2001 to be 5.81 g C m^{-2} day^{-1} for the South Atlantic division as a whole. In both years, Exurban densities were found to be the

TABLE 19.3 Area and GPP of the four settlement densities.

Settlement densities	Area (km^2)			GPP (g C m^{-2} day^{-1})	
	1990	2000	Difference (%)	1992	2001
Urban	14172.58 {2.0}	17214.44 {2.4}	21.46	1.79 (1.11)	2.00 (1.57)
Suburban	106351.70 {14.9}	134873.06 {18.9}	26.83	3.44 (1.38)	4.80 (2.11)
Exurban	278966.30 {39.0}	304654.02 {42.7}	9.21	4.29 (1.02)	6.10 (1.50)
Rural	315563.94 {44.1}	257125.44 {36.0}	−18.52	4.20 (1.01)	6.03 (1.46)

Numbers in brackets represent percentage of the total study area excluding the removed county subdivisions due to boundary changes. Numbers in parenthesis are standard deviation.

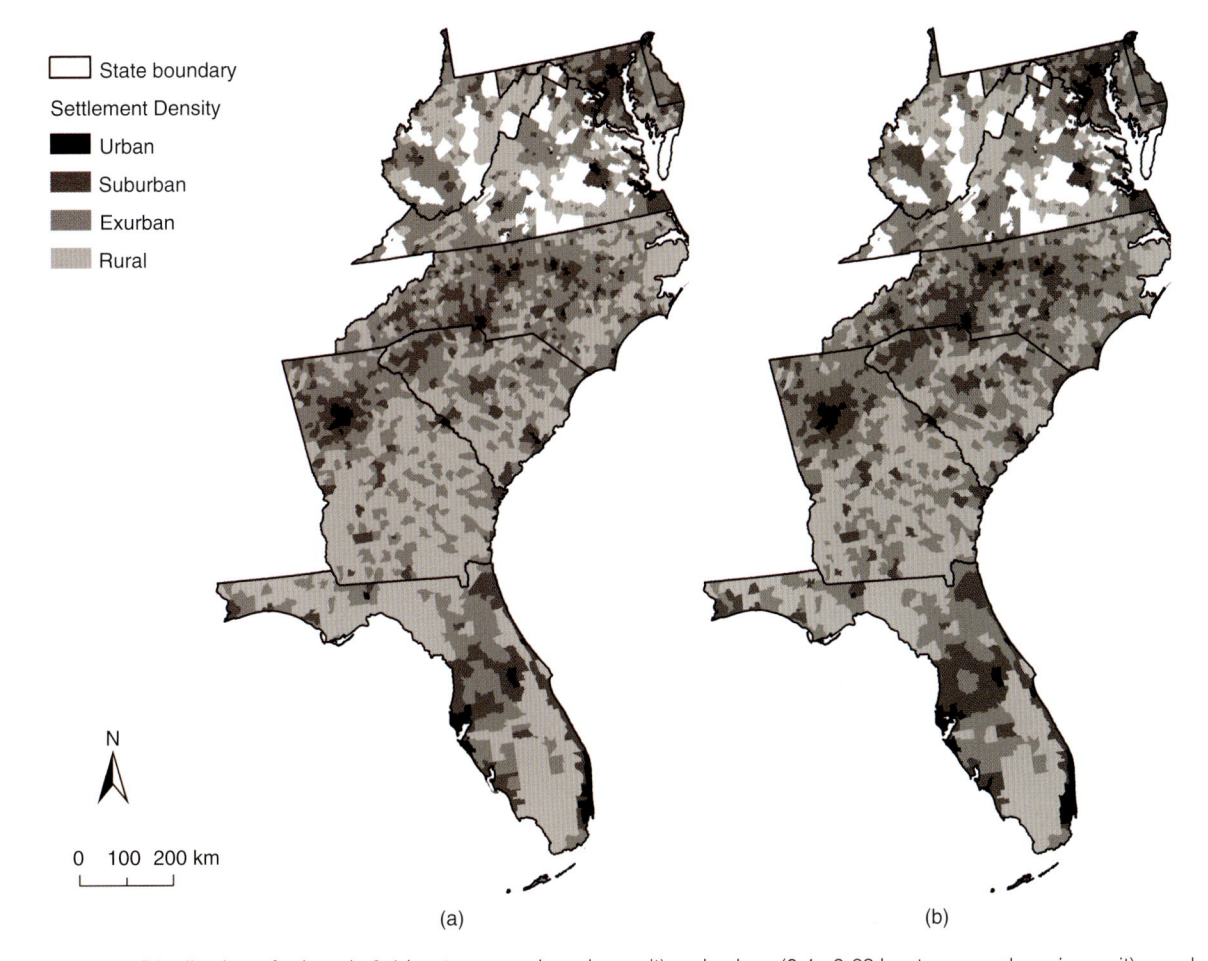

FIGURE 19.3 Distribution of urban (≤0.1 hectares per housing unit), suburban (0.1–0.69 hectares per housing unit), exurban (0.69–16.2 hectares per housing unit), and rural (> 16.2 hectares per housing unit) settlement densities in 1990 (a) and 2000 (b). The blank area indicates county subdivisions removed due to changes in Census boundaries.

most productive among the four settlement density categories; the values of GPP for Exurban densities were slightly (1–2%) higher than those for Rural densities (Table 19.3). Urban densities were associated with the lowest GPP estimates, accounting for 41.7% and 32.8% of the estimated GPP for Exurban densities in 1992 and 2001, respectively. Suburban densities were intermediate in carbon uptake through photosynthesis, accounting for approximately 80% of the GPP produced at Exurban densities in both years.

Approximately 14% of the South Atlantic study area, excluding county subdivisions whose boundaries were modified between 1990 and 2000, experienced transitions between settlement-density categories (Table 19.4; Fig. 19.4a). Among all types of categorical transition, the top transition (by area) was Rural densities converted to Exurban (R → E), which accounted for 8.53% of the total area. It was followed by Exurban densities converted to Suburban (E → S; 4.78%) and Suburban densities converted to Urban (S → U; 0.42%). The conversion from Exurban densities to Urban (E → U; 0.01%) or conversion from Rural densities to Suburban (R → S; 0.03%) was minor. Conversions from higher to lower densities, including six conversion categories, accounted for 0.21% of the study area in total.

The South Atlantic division saw an increase in GPP by 1.70 g C m^{-2} day^{-1} for the region as a whole between 1992 and 2001, which was equivalent to an increase by approximately 42% against the average GPP value estimated for this region in 1992.

282 PART V URBAN ENVIRONMENTAL ANALYSES

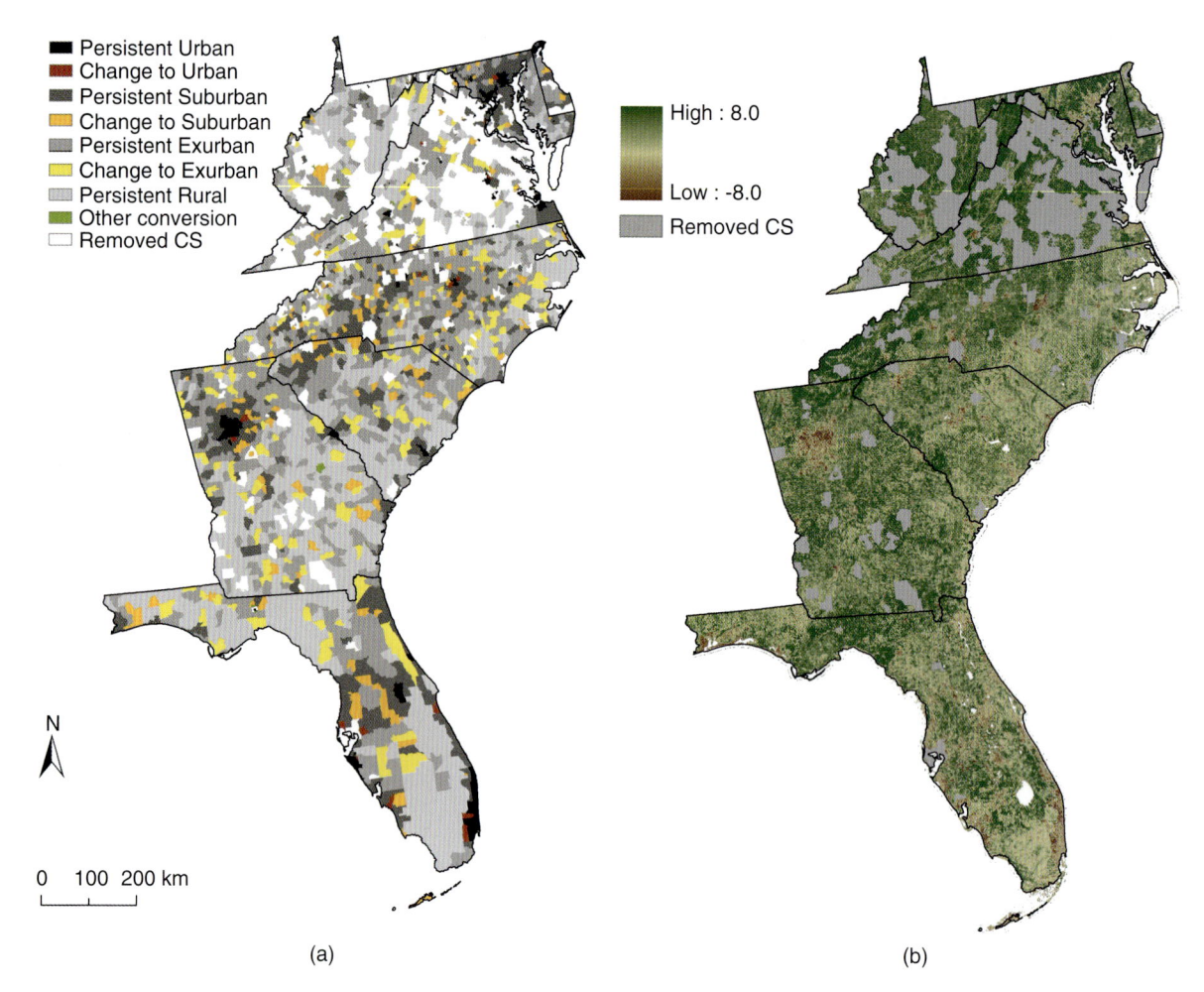

FIGURE 19.4 Changes in settlement-density categories (a) and in GPP (g C m^{-2} day^{-1}; b) between 1990 and 2000. The pixel-wise changes in GPP range between −5.31 and 7.94 g C m^{-2} day^{-1}. CS stands for county subdivision.

At the pixel level, the average value of changes in GPP varied between −5.31 and 7.94 g C m^{-2} day^{-1}, with the mountain areas to the north and plain areas around Florida Panhandle appearing to be related to the enhanced photosynthesis (i.e., positive values; Fig. 19.4b). Changes in GPP were found to be positive, on average, for all individual settlement-density transitions; however, the amount of change varied by transitional types (Table 19.4). The persistent Rural densities (R → R; 1.846 g C m^{-2} day^{-1}), Rural densities converted to Exurban (R → E; 1.813 g C m^{-2} day^{-1}), and persistent Exurban densities (E → E; 1.778 g C m^{-2} day^{-1}) were associated with the highest amount of increase in GPP, higher than the regional average difference (1.7 g C m^{-2} day^{-1}). They were followed by Exurban densities converted to Suburban (E → S; 1.442 g C m^{-2} day^{-1}), persistent Suburban densities (S → S; 1.125 g C m^{-2} day^{-1}), and Rural densities converted to Suburban (R → S; 1.071 g g C m^{-2} day^{-1}). Persistent Urban densities (U → U; 0.138 g C m^{-2} day^{-1}), Suburban densities converted to Urban (S → U; 0.096 g C m^{-2} day^{-1}), and Exurban densities converted to Urban (E → U; 0.04 g C m^{-2} day^{-1}) held the lowest values of increase in GPP.

Eight groups of settlement conversion were generated to characterize transitions among different settlement-density categories between 1990 and 2000 (Table 19.4). This grouping system separated the persistent types from those experiencing categorical changes. The persistent types referred to areas identified as the same settlement-density category in both Census years; for example, Persistent Rural represented areas identified as Rural densities in 1990 and remaining Rural densities in 2000. The actual categorical changes were divided into four groups: (1) Change to Urban represented conversions from any of the three lower-density categories to Urban densities; (2) Change to Suburban included conversions from Exurban or Rural densities to Suburban densities; (3) Change to Exurban was equivalent to the conversion from Rural to Exurban densities; and 4) Other Conversion referred to any conversion from higher to lower densities. From the perspective of photosynthesis, Change to Urban resulted in the lowest value of GPP increment (0.068 g C m^{-2} day^{-1}), followed by Persistent Urban (0.138 g C m^{-2} day^{-1}), Persistent Suburban (1.125 g C m^{-2} day^{-1}), and Change to Suburban (1.257 g C m^{-2} day^{-1}). All four types associated with persistent Urban/Suburban densities or being densified into Urban/Suburban densities yielded a smaller amount of GPP increment than the regional average, whereas the persistent Exurban densities (1.778 g C m^{-2} day^{-1}) and Rural densified into Exurban densities (1.813 g C m^{-2} day^{-1}) resulted in an enhanced amount of GPP increment than the regional average (Table 19.4).

TABLE 19.4 Conversions between settlement-density categories, their area proportions, and associated changes in GPP between 1990 and 2000.

Settlement-density conversion	Aggregated types of conversion	Percentage of the total area (%)	Changes in GPP (g C m^{-2} day^{-1}) Mean	Changes in GPP (g C m^{-2} day^{-1}) Std. Dev.
U → U	Persistent Urban	1.71	0.138	0.864
S → U	Change to Urban	0.42	0.096	1.085
E → U	Change to Urban	0.01	0.040	1.110
subtotal		2.13		
S → S	Persistent Suburban	14.91	1.125	1.158
E → S	Change to Suburban	4.78	1.442	1.088
R → S	Change to Suburban	0.03	1.071	1.040
subtotal		19.72		
E → E	Persistent Exurban	33.36	1.778	0.886
R → E	Change to Exurban	8.53	1.813	0.946
subtotal		41.89		
R → R	Persistent Rural	36.05	1.846	0.949
subtotal		36.05		
U → S	Other Conversion	0.03	0.994	1.196
U → E	Other Conversion	0.00	1.102	0.788
S → E	Other Conversion	0.05	1.685	0.977
U → R	Other Conversion	0.00	0.633	1.411
S → R	Other Conversion	0.01	1.214	1.079
E → R	Other Conversion	0.12	2.071	0.900
subtotal		0.21		

Positive values of changes in GPP indicate increase in GPP by 2001.

19.5 Discussion

19.5.1 Urban growth in the South Atlantic division

The US urban development has increased rapidly in the 20th century and was complicated by the decline of average household size since 1960. By 2000, the number of persons residing in Census urban areas reached 7.4 times the count of urban population in 1900 (US Census Bureau, 1995, 2001a). While the urban population kept growing from 69.9% to 79.0% of the total US population between 1960 and 2000, the average household size (i.e., numbers of people living in a housing unit) continuously shrank from 3.33 to 2.62 persons per housing unit during the same time period (US Census Bureau, 1995, 2009). The escalation of urban population and decline in the average household size accelerated urban sprawl at various development densities. In previous research, urban and suburban densities (≤0.69 hectares per housing unit) were found to expand by 31.5% in private lands by area across the conterminous United States between 1980 and 2000 (Theobald, 2005). The area of low-density development (0.4–16.2 hectares per housing unit) was found to be approximately 14 times greater than the area of higher-density settlement in the United States by 2000 (Brown et al., 2005).

This study found that the South Atlantic division, which consists of eight states and the District of Columbia in the eastern United States, experienced a similar settlement-density transition as patterns identified nationwide. Between 1990 and 2000, high-density urban and suburban development (≤0.69 hectares per housing unit) expanded their area by 20–30%. Although the exurban expansion bore a slower rate of growth (9%) during the same time period, the area extent of exurban densities was the highest among all settlement densities by 2001. In 2000, exurban densities (settled at 0.69–16.2 hectares per housing unit) were twice as much in area as suburban densities (settled at 0.1–0.69 hectares per housing unit) and were approximately 18 times the area of urban densities (settled at 0.1 hectares per housing unit and less). In terms of settlement densification between 1990 and 2000, exurbanization (the conversion from rural to exurban densities) occupied 1.8 times of the area of suburbanization (the conversion from exurban to suburban densities) and approximately 20 times of the area of urbanization (the conversion from suburban to urban densities).

19.5.2 Impacts of urban growth on vegetation productivities

Photosynthesis measured by GPP based on the LUE approach was found on average to be the highest at rural and low-density exurban development in South Atlantic division in 1992 and 2001. High-density urban and suburban development, compared to rural or low-density exurban development, was showed to reduce GPP by over 50% and approximately 20%, respectively, in both years. These numbers were higher than previous finding in four of the eight states located in my study area, where NPP (usually half the amount of GPP) of the urban land-cover class was estimated to be approximately 18% less than other natural vegetation combined (Milesi et al., 2003). The discrepancy may be introduced by different definition of "urban." In Milesi et al. (2003), the authors defined urban as a type of land cover based on remotely sensed nighttime light illumination, whereas in this study urban density was determined based on Census housing-unit density. Previous research (Imhoff et al., 2000) indicated that the urban land cover identified from night-time illumination imagery may include both urban and suburban densities as defined in this study.

The South Atlantic division for the region as a whole was found to be more productive in 2001 as measured by GPP, which increased by 1.70 g C m^{-2} day^{-1} comparing to the estimate in 1992 (4.11 g C m^{-2} day^{-1}). The monthly average of daily minimum temperature was not statistically different between 1992 and 2001 according to the paired-samples t test, which indicates minimum impacts of temperature on GPP. The region-wide increase in vegetation productivity in the latter year may partially resulted from the decline of incident solar radiation in the former year following the volcanic eruption of Mt. Pinatubo in June, 1991 (Ramachandran et al., 2000; Tucker et al., 2001; Zhao, Brown and Bergen, 2007). Although radiometric effects of this globally influential volcanic eruption was not explicitly corrected before estimating GPP in this study, variation of changes in GPP between 1992 and 2001 was still captured across different types of urban growth.

Exurbanization, the conversion from rural to exurban densities, was found to enhance the GPP increment (i.e., higher than the regional average increase of GPP) by 6.6%. Suburbanization, the conversion from exurban to suburban densities, was found to diminish the region-wide GPP increment (i.e., lower than the regional average increase of GPP) by 15.2%. Urbanization, the conversion from suburban to urban densities, was found to minimize the region-wide GPP increment by 94.4%, where the productivity gained through increased solar radiation in 2001 appeared to be nearly offset completely due to this high-density settlement development. Urbanization reduced GPP by a greater amount than the persistent urban densities, the latter reducing the region-wide GPP increment by 91.9%. Both exurbanization and suburbanization reduced GPP by a smaller amount, compared to the corresponding persistent types, due possibly to the remaining large proportions of vegetation cover in these transitional categories than the persistent exurban and suburban densities, respectively.

Spatial heterogeneity of changes in vegetation productivities associated with urban and urban growth has been documented in earlier research (Milesi et al., 2003; Imhoff et al., 2004; Zhao, Brown and Bergen, 2007). Relationships between vegetation productivities and human settlement patterns, especially those patterns connected to and measured by demographic and socioeconomic transitions, however, are not well understood yet. The case study in the Census South Atlantic division was among the first attempts to explore those relationships by examining urban growth at different intensities at local to regional scales. This research approach links Census demographic changes and landscape biophysical characteristics, and may be extended to large geographic areas (such as the entire United States) where demographic and vegetation data are publicly available. The study presented in this chapter focused mainly on analysis of GPP, whereas future work may be extended to the estimation of NPP by incorporating respiration from ecosystem models such as Biome-BGC (Running and Hunt, 1993). The modified approach may also incorporate other environmental variables such as soil moisture, soil nutrient availability, and vapor pressure deficit (Hargrove and Hoffman, 2004; Hashimoto et al., 2008) to simulate the real environmental condition for the analysis of vegetation carbon exchange with the atmosphere.

Conclusions

In this study, urban growth and its impacts on vegetation productivity measured by gross primary production (GPP) was examined throughout the US Census South Atlantic division between 1992 and 2001. In both years, GPP estimated based on the light-use efficiency (LUE) approach was found to be the highest in rural and low-density exurban areas. Between 1992 and 2001, GPP was estimated to increase by 1.70 g C m^{-2} day^{-1} or approximately 42% for the study area as a whole. Exurbanization, accounting for 8.53% of the study area, was showed to be slightly less productive than the persistent rural densities, slightly more productive than the persistent exurban densities, and 6.6% more productive than the regional average. Suburbanization and urbanization, accounting for 4.78% and 0.42% of the study area, were associated with decline in GPP increment compared to the regional average. The dampening effects of urban growth on vegetation photosynthesis, by the order of decreasing strength, was found to be urban densities developed from exurbs (E → U; 97.6% lower than the regional average GPP increment), urbanization (94.4% lower), persistent urban densities (U → U; 91.9% lower), suburbs developed from rural densities (R → S; 37% lower), persistent suburbs (S → S; 33.8% lower), and suburbanization (15.2% lower).

This study used publicly available Census demographic data, Landsat-based land-cover/land-use data, vegetation greenness index (i.e., AVHRR NDVI), climate variables on temperature and solar radiation, and light-use efficiency parameters based on empirical studies to identify urban growth and to estimate GPP. Research methods may be applied to other geographic areas at the comparable scale. The estimates of vegetation carbon fluxes may be incorporated with the measured or modeled anthropogenic CO$_2$ emissions from the burning of fossil fuels. Such integration would contribute to better understanding of human carbon impacts associated with cities and urban growth.

Acknowledgments

This work was supported by the 2008 FSU Council on Research and Creativity First Year Assistant Professor summer research program. The GLDAS climate data used in this study were acquired as part of the mission of NASA's Earth Science Division, and archived and distributed by the Goddard Earth Sciences (GES) Data and Information Services Center (DISC). The author would like to thank the two anonymous reviewers, Tao Zhang and Morton D. Winsberg.

References

Berube, A. (2003) Gaining but losing ground: Population change in large cities and their suburbs, in *Redefining Urban and Surburban America: Evidence from Census 2000* (v.1) (eds B. Katz and R.E. Lang), Brookings Institution Press, pp. 33–50.

Boyle, C.A. and Lavkulich, L. (1997) Carbon pool dynamics in the Lower Fraser Basin from 1827 to 1990. *Environmental Management*, **21**, 443–455.

Brown, D.G., Johnson, K.M., Loveland, T.R. and Theobald, D.M. (2005) Rural land-use trends in the conterminous United States, 1950–2000. *Ecological Applications*, **15** (6), 1851–1863.

Chapin, F.S., Matson, P. and Mooney, H.A. (2002) *Principles of Terrestrial Ecosystem Ecology*, Springer, Berlin.

Churkina, G. (2008) Modeling the carbon cycle of urban system. *Ecological Modelling*, **216**, 107–113.

Churkina, G., Brown, D.G. and Keoleian, G.A. (2010) Carbon stored in human settlements: The conterminous US. *Global Change Biology*, **16**, 135–143.

Cramer, W., Kicklighter, D.W., Bondeau, A. *et al.* (1999) Comparing global models of terrestrial net primary productivity (NPP): Overview and key results. *Global Change Biology*, **5** (Suppl. 1), 1–15.

Field, C.B., Randerson, J.T. and Malmström, C.M. (1995) Global net primary production: Combining ecology and remote sensing. *Remote Sensing of Environment*, **51**, 74–88.

Goodchild, M.F. (1978) Statistical aspects of the polygon overlay problem, in *Harvard Papers on Geographic Information Systems* (ed G. Dutton), Addison-Wesley, Reading.

Goodchild, M.F., Anselin, L. and Deichmann, U. (1993) A framework for the areal interpolation of socioeconomic data. *Environment and Planning A*, **25**, 383–397.

Gower, S.T., Kucharik, C.J. and Norman, J.M. (1999) Direct and indirect estimation of leaf area index, fAPAR and net primary production of terrestrial ecosystems. *Remote Sensing of Environment*, **70**, 29–51.

Grimm, N.B., Grove, J.M., Pickett, S.T.A. and Redman, C.L. (2000) Integrated approaches to long-term studies of urban ecological systems. *BioScience*, **50**, 571–584.

Grimm, N.B., Foster, D., Groffman, P. *et al.* (2008) The changing landscape: ecosystem response to urbanization and pollution across climatic and societal gradients. *Frontiers in Ecology and the Environment*, **6** (5), 264–272.

Haboudane, D., Miller, J.R., Tremblay, N., Zarco-Tejada, P.J. and Dextraze, L. (2002) Integrated narrow-band vegetation indices for prediction of crop chlorophyll content for application to precision agriculture. *Remote Sensing of Environment*, **81** (2–3), 416–426.

Hargrove, W.W. and Hoffman, F.M. (2004) The potential of multivariate quantitative methods for delineation and visualization of ecoregions. *Environmental Management*, **34**, S39–S660.

Hashimoto, H., Dungan, J.L., White, M.A. *et al.* (2008) Satellite-based estimation of surface vapor pressure deficits using MODIS land surface temperature data. *Remote Sensing of Environment*, **112**, 142–155.

Heinsch, F.A., Reeves, M., Votava P. *et al.* (2003) User's Guide, GPP and NPP (MOD 17A2/A3) Products, NASA MODIS Land Algorithm. Version 2.0, December 2, 2003.

Heinsch, F.A., Zhao, M., Running, S.W. *et al.* (2006) Evaluation of remote sensing based terrestrial productivity from MODIS using tower eddy flux network observations. *IEEE Transactions on Geoscience and Remote Sensing*, **44** (7), 1908–1925.

Herold, M., Scepan, J. and Clarke, K.C. (2002) The use of remote sensing and landscape metrics to describe structures and changes in urban land uses. *Environment and Planning A*, **34** (8), 1443–1458.

Homer, C., Dewitz, J., Fry, J. *et al.* (2007) Completion of the 2001 National Land Cover Database for the Conterminous United States. *Photogrammetric Engineering and Remote Sensing*, **73** (4), 337–341.

Huang, C., Homer, C. and Yang, L. (2003) Regional forest land cover characterization using Landsat type data, in *Methods and Applications for Remote Sensing of Forests: Concepts and Case Studies* (eds M. Wulder and S. Franklin), Kluwer Academic Publishers, Dordrecht, pp. 389–410.

Imhoff, M.L., Tucker, C.J., Lawrence, W.T. and Stutzer, D.C. (2000) The use of multisource satellite and geospatial data to study the effect of urbanization on primary productivity in the United States. *IEEE Transactions on Geoscience and Remote Sensing*, **38** (6), 2549–2556.

Imhoff, M.L., Bounoua, L., DeFries, R. *et al.* (2004) The consequences of urban land transformation on net primary productivity in the United States. *Remote Sensing of Environment*, **89**, 434–443.

Kaye, J.P., McCulley, R.L. and Burke, I.C. (2005) Carbon fluxes, nitrogen cycling, and soil microbial communities in adjacent urban, native and agricultural ecosystems. *Global Change Biology*, **11**, 575–587.

Knyazikhin, Y., Glassy, J., Privette, J.L. and others. (1999) MODIS leaf area index (LAI) and fraction of photosynthetically active radiation absorbed by vegetation (FPAR) product (MOD15), algorithm theoretical basis document. [Online] Available http://modis.gsfc.nasa.gov/data/atbd/atbd_mod15.pdf (accessed 17 November 2010).

Lambers, H., Chapin, F.S. and Pons, T.L. (1998) *Plant Physiological Ecology*, Springer, Berlin.

Milesi, C., Elvidge, C.D., Nemani, R.R. and Running, S.W. (2003) Assessing the impact of urban land development on net primary productivity in the southeastern United States. *Remote Sensing of Environment*, **86**, 401–410.

Monteith, J.L. (1972) Solar radiation and productivity in tropical ecosystems. *Journal of Applied Ecology*, **9**, 747–766.

Nowak, D.J. and Crane, D.E. (2002) Carbon storage and sequestration by urban trees in the USA. *Environmental Pollution*, **116**, 381–389.

Odum, E.P. (1997) *Ecology: A Bridge between Science and Society*, 3rd edn, Sinauer Associates, Sunderland, MA.

Pataki, D.E., Alig, R.J., Fung, A.S. *et al.* (2006) Urban ecosystems and the North American carbon cycle. *Global Change Biology*, **12**, 2092–2102.

Pataki, D.E., Emmi, P.C., Forster, C.B. *et al.* (2009) An integrated approach to improving fossil fuel emissions scenarios with urban ecosystem studies. *Ecological Complexity*, **6**, 1–14.

Pickett, S.T.A., Cadenasso, M.L., Grove, J.M. *et al.* (2001) Urban ecological systems: Linking terrestrial ecological, physical, and socioeconomic components of metropolitan areas. *Annual Review of Ecology and Systematics*, 32, 127–157.

Pouyat, R.V., Groffman, P., Yesilonis, I. and Hernandez, L. (2002) Soil carbon pools and fluxes in urban ecosystems. *Environmental Pollution*, **116**, S107–S118.

Prince, S.D. and Goward, S.N. (1995) Global primary production: A remote sensing approach. *Journal of Biogeography*, **22**, 815–835.

Ramachandran, S., Ramaswamy, V., Stenchikov, G.L. and Robock, A. (2000) Radiative impact of the Mount Pinatubo volcanic eruption: Lower stratospheric response. *Journal of Geophysical Research*, 105 (D19), 24409–24429.

Rodell, M., Houser, P.R., Jambor, U. *et al.* (2004) The Global Land Data Assimilation System, *Bulletin of American Meteorological Society*, **85** (3), 381–394.

Running, S.W. and Hunt, E.R. 1993. Generalization of a forest ecosystem process model for other biomes, Biome-BGC, and an application for global-scale models, in *Scaling Physiological Processes: Leaf to Globe* (eds J.R. Ehleringer and C.B. Field), Academic Press, San Diego, pp. 141–158.

Running, S.W., Thornton, P., Nemani, E.R. and Glassy, J.M. (2000) Global terrestrial gross and net primary productivity from the Earth Observing System, in *Methods in Ecosystem Science* (eds O.E. Sala, R.B. Jackson, H.A. Mooney and R.W. Howarth), Springer, New York, pp. 44–57.

Running, S.W., Nemani, R.R., Heinsch, F.A., Zhao, M., Reeves, M., and Hashimoto, H. (2004) A continuous satellite-derived measure of global terrestrial primary production. *BioScience*, **54**, 547–560.

Theobald, D.M. (2005) Landscape patterns of exurban growth in the USA from 1980 to 2020. *Ecology and Society*, **10** (1), 32.

Tucker, C.J. and Sellers, P.J. (1986) Satellite remote sensing of primary production. *International Journal of Remote Sensing*, 7 (11), 1395–1416.

Tucker, C.J., Slayback, D.A., Pinzon, J.E., Los, S.O., Myneni, R.B., and Taylor, M.G. (2001) Higher northern latitude normalized difference vegetation index and growing season trends from 1982 to 1999. *International Journal of Biometeorology*, **45**, 184–190.

Turner, D.P., Ritts, W.D., Cohen, W.B. *et al.* (2003) Scaling gross primary production (GPP) over boreal and deciduous forest landscapes in support of MODIS GPP product validation. *Remote Sensing of Environment*, **88**, 256–270.

Turner, D.P., Ollinger, S.V. and Kimball, J.S. (2004) Integrating remote sensing and ecosystem process models for landscape- to regional-scale analysis of the carbon cycle. *BioScience*, **54** (6), 573–584.

US Census Bureau (1991) *Census of Population and Housing, 1990: Summary Tape File 1*. The Bureau of the Census, Washington DC.

US Census Bureau (1995) Urban and Rural Population: 1900 to 1990. [Online]Available http://www.census.gov/population/censusdata/urpop0090.txt (accessed 17 November 2010).

US Census Bureau (2001a) Census 2000 Summary File 1; generated by T. Zhang and T. Zhao using American FactFinder. [Online] Avaialble http://factfinder.census.gov (accessed 17 November 2010).

US Census Bureau (2001b) Census 2000 Summary File 1 Technical Documentation. The US Census Bureau, Washington DC.

US Census Bureau (2009) Current Population Survey, March and Annual Social and Economic Supplements, 2008 and earlier. [Online] Available http://www.census.gov/population/www/socdemo/hh-fam.html#ht (accessed: 17 November 2010).

USGS EROS Data Center (2006) The Conterminous U.S. and Alaska Weekly and Biweekly AVHRR Composites, EROS Data Center, Sioux Falls, South Dakota.

Vogelmann, J.E., Howard, S.M., Yang, L., Larson, C.R., Wylie, B.K., and Van Driel, N. (2001) Completion of the 1990s National Land Cover Data Set for the Conterminous United States from Landsat Thematic Mapper Data and Ancillary Data Sources. *Photogrammetric Engineering and Remote Sensing*, **67**, 650–652.

Yang, F., Ichii, K., White, M.A. *et al.* (2007) Developing a continental-scale measure of gross primary production by combing MODIS and AmeriFlux data through Support Vector Machine approach. *Remote Sensing of Environment*, **110**, 109–122.

Yang, L., Huang, C., Homer, C.G., Wylie, B.K. and Coan, M.J. (2003) An approach for mapping large-area impervious surfaces: Synergistic use of Landsat-7 ETM+ and high spatial resolution imagery, *Canadian Journal of Remote Sensing*, **29** (2), 230–240.

Zhao, M., Heinsch, F.A., Nemani, R.R. and Running, S.W. (2005) Improvements of the MODIS terrestrial gross and net primary production global data set. *Remote Sensing of Environment*, **95**, 164–176.

Zhao, T. 2007. Changing Primary Production and Biomass in Heterogeneous Landscapes: Estimation and Uncertainty Based on Multi-Scale Remote Sensing and GIS Data. Dissertation. University of Michigan, Ann Arbor.

Zhao, T., Brown, D.G. and Bergen, K.M. (2007) Increasing Gross Primary Production (GPP) in the Urbanizing Landscapes of Southeastern Michigan. *Photogrammetric Engineering and Remote Sensing*, **73** (10), 1159–1167.

Zhao, T., Bergen, K. M, Brown, D.G. and Shugart, H.H. (2009) Scale dependence in quantification of land-cover and biomass change over Siberian boreal forest landscapes. *Landscape Ecology*, **24** (10): 1299–1313.

Zheng, D., Prince, S. and Wright, R. (2003) Terrestrial net primary production estimates for 0.5° grid cells from field observations – a contribution to global biogeochemical modeling. *Global Change Biology*, **9**, 46–64.

20

Characterizing biodiversity in urban areas using remote sensing

Marcus Hedblom and Ulla Mörtberg

Fauna and flora, and their diversity in cities have long been a neglected research area; instead, more natural environments or environments used for human production, such as forests or rural areas, have been prioritized. However, there has been a recent major increase in studies of urban green areas and their importance for species richness. The urbanization process has led to fragmentation of habitats, which has become one of the greatest threats to biodiversity worldwide. Remote sensing is a cost-efficient data source covering large areas, capturing information in a systematic manner and can provide data for spatiotemporal studies in urban environments. However, few studies have examined biodiversity in urban ecosystems using satellite images. Here, we review remote sensing techniques for the study of biodiversity in urban areas, different approaches for characterizing biodiversity with remote sensing and the effects of urbanization on biodiversity; we also discuss applications of remote sensing in planning and management, and past and future avenues for research. We conclude that urban biodiversity studies are still far from exploiting the full potential of advances in data capture, data interpretation and classification methods in combination with field studies for deriving ecologically meaningful information.

Urban Remote Sensing: Monitoring, Synthesis and Modeling in the Urban Environment, First Edition. Edited by Xiaojun Yang.
© 2011 John Wiley & Sons, Ltd. Published 2011 by John Wiley & Sons, Ltd.

20.1 Introduction

Urbanization, together with agriculture, is the most important threat to biodiversity worldwide (Ricketts and Imhoff, 2003). Urbanization will probably soon exceed agriculture as the dominant agent of loss, fragmentation and degradation of habitats as a result of the increasingly urbanized human population (Marzluff and Ewing, 2001). Urbanization will have a major effect on biodiversity, defined as the variability within species, between species and of ecosystems, since it is predicted that urban areas will increase by more than 600 000 km^2 in developing countries and 500 000 km^2 in industrialized cities between 2010 and 2030 (Angel, Sheppard and Civco, 2005). People traditionally settle in areas with highly productive ecosystems and abundant natural resources, therefore many cities are located in areas important for biodiversity conservation, so called ecological hotspots (Cincotta, Wisnewski and Engelman, 2000; Ricketts and Imhoff, 2003; Luck, 2007). Moreover, since the urban footprint extends far beyond municipal boundaries, urbanization may also reduce the diversity of native species at regional and global scales (Grimm et al., 2008). Accordingly, cities can, if managed properly, play an increasingly important role in sustaining the world's species (Rosenzweig, 2003; Dearborn and Kark, 2010). Interestingly, Grimm et al. (2008) considered cities to be microcosms of the kinds of changes that are happening globally, making them informative test cases for understanding socioecological system dynamics and responses to change.

Urban green areas, which act as habitats, can thus be of direct importance to the flora and fauna within cities but can also be beneficial to humankind. Natural remnant habitats or green areas in cities provide humans with a multitude of resources and processes that are supplied by ecosystems. Examples of such ecosystem services are rainwater drainage and noise reduction (Bolund and Hunhammar, 1999). Urban green areas have also been highlighted as being important for human well being (Grahn and Stigsdotter, 2010). Moreover, Fuller et al. (2007) showed an increase in psychological benefits with biodiversity in urban green areas.

In the scientific world, urban fauna and flora, and their diversity, have long been a neglected research area, while other more natural environments have been prioritized. However, during the last decade research into biodiversity in urban areas has increased rapidly, as seen in an increasing number of publications (McKinney, 2006; Chace and Walsh, 2006; Marzluff, 2008; McDonnell, Brueste and Hahs, 2009; Davies et al., 2009). Despite there being more studies on biodiversity in urban areas, remarkably few of them use remote sensing techniques such as satellite images (Gottschalk, Huettmann and Ehlers, 2005; Bino et al., 2008). This may be due to the fact that high spatial resolution satellite images have only been widely available since 1999 with the launch of IKONOS and NASA's Geocover dataset (Tucker, Grant and Dykstra, 2004). Prior to this there were very few freely available Landsat images globally (Bino et al., 2008). This situation is now changing rapidly. New data capture techniques using passive and active sensors, increased data availability and novel image interpretation methods, together with ongoing development of spatial ecological modeling approaches, mean that there is a greatly increased potential for identification, mapping and assessment of biodiversity components (e.g., Turner et al., 2003; Guisan and Thuiller, 2005; Gillespie et al., 2008).

Previously, many studies describing biodiversity in urban areas used the urban–rural gradient approach to study the ecology of cities and towns around the world (McDonnell et al., 1997). Although this approach has increased our knowledge of ecosystem processes and the distribution of species in relation to cities and urbanization, the use of remote sensor data in combination with previous gradient approaches has great potential for broadening our understanding of urban patterns, moving from a simplistic to a more complex, quantified and realistic view of urban–rural gradients and allowing us to develop a comprehensive terminology with respect to urban ecology (e.g., McDonnell and Hahs, 2008; Pickett et al., 2008). Moreover, combining field studies and remote sensing within a broad hierarchal perspective, including landscape analyses, such methods could be used to monitor long-term changes in urban landscapes. In the long run this approach could provide decision makers and landscape planners with important information about which urban green areas to prioritize. Finally, in this emerging research field of urban ecosystems and the relationships between urbanization and biodiversity, remote sensor data will make a crucial contribution and has yet to meet its full potential.

The following sections discuss capturing biodiversity information in urban areas and areas subject to urbanization using remote sensor data sources; we consider both direct and indirect approaches, the latter using spatial ecological modeling combined with field surveys. We also discuss their applications for monitoring, planning and management.

20.2 Remote sensing methods in urban biodiversity studies

The complexity of the biodiversity concept, as encapsulated in its definition, implies that deriving information relating to the wide array of biodiversity components, from genes to ecosystems, is complicated. Even if field surveys are necessary and give the most accurate information on vegetation, species, individuals and genes, the combined picture of many existing field surveys reveals a range of problems for comparative meta-analyses of, for example, the effects of urbanization on biodiversity. The different methods used may not be compatible with each other, there may also be extremely uneven coverage with a strong bias towards developed countries and existing knowledge centers, and, even in relatively well-studied regions, data is often spatially biased (e.g., Pautasso and McKinney, 2007). Furthermore, even if urban areas are largely accessible for conducting field surveys of biodiversity components, the increasing size of urban areas contributes to making the annual costs of detailed field studies very high, particularly since field surveys require skilled individuals and are very time consuming.

In the light of this, remote sensor data is especially attractive as a cost-effective source of information on biodiversity, since it offers a relatively inexpensive mean of providing complete spatial coverage of environmental information over large areas. Furthermore, remote sensing captures information in a systematic manner, can be used to examine inaccessible areas, and can be updated regularly (e.g., Luoto et al., 2004; Duro et al., 2007; Levin et al., 2007; Saatchi et al., 2008). However, even when high-resolution remote sensor data is available, it is difficult to determine species richness without conducting field surveys. In order to study biodiversity in urban areas (and, indeed, elsewhere) cost efficient combinations of the two methods are required; this has been the case in most previous studies of biodiversity using remote sensing (see Gottschalk, Huettmann and Ehlers, 2005).

Another main advantage of using remote sensing for biodiversity studies is the possibility of providing data for spatial-temporal analyses of landscapes. Aerial photographs are stored in many national archives and date from at least the early 1940s, while imaging from space has been widely collected since the 1970s. Furthermore, within the temporal domain provided by many satellite sensors, with repeat periods of between 15 minutes and a few weeks, it is also possible to undertake biodiversity studies of the monthly, seasonal and annual dynamics of landscapes (Groom et al., 2006).

Capturing biodiversity information through remote sensor can be divided into direct and indirect approaches (Turner et al., 2003, Duro et al., 2007). According to these authors, direct approaches are first-order analyses of the occurrence of species or species assemblages using remote sensing, while indirect approaches use remote sensor data to measure environmental variables that are related to biodiversity. The research into the two approaches is reviewed in the following subsections.

20.2.1 Direct approaches

Directly capturing information about biodiversity components from remote sensing is possible at a range of scales, from relatively coarse classifications of species assemblages or vegetation types to identifying single species at the scale of individual tree crowns (e.g., Foody et al., 2005). Land cover classification, although often used for deriving biodiversity-related information, can be considered to be more closely connected to the indirect methods, and therefore is considered under that heading. Direct methods tend to require high resolution imagery, from airplanes or satellites.

Aerial photographs have been collected since at least the early 1940s in many countries. Recently, the availability and quality of digital image data produced from aerial photographs has increased, often with national coverage and at resolutions of less than 1 m (Groom et al., 2006). Using color infrared (CIR) aerial photographs, methods have been developed for manual interpretation in order to obtain detailed vegetation maps with ecologically meaningful classifications, resulting in data with a very high information value for biodiversity studies (Ihse, 1995). These methods were further developed by Löfvenhaft, Björn and Ihse (2002), with the specific aim of supporting biodiversity issues relating to spatial planning in urban and suburban areas in Stockholm, Sweden. Apart from retrieving data pertaining to land cover and detailed vegetation types, they also captured data on habitat structures and vegetation cover within built-up areas. The accuracy compared to the field control was very high (93–95%) for the main land cover types and for vegetation type classes; for comparison, classes relating to habitat structure in hardwood deciduous forest were identified at an accuracy of 72–75%. This database has proved to be very valuable for the use in urban planning (Mörtberg and Ihse, 2006).

Since the late 1990s, there has been a major increase in the availability of digital satellite imagery, with very high spatial resolution, i.e., less than 5 m. Satellites that provide such high resolution data include Quickbird (0.6–2.5 m), IKONOS (1–4 m), and SPOT (2.5 m). There is great potential for manually or digitally identifying tree species and canopy attributes from these sources (e.g., Clark et al., 2004; Levin et al., 2009). In a comparison between manual and automated interpretation of CIR aerial images and IKONOS satellite images, the manual interpretation of CIR aerial images delivered data with the highest accuracy in classes with a very high information value for biodiversity studies (Groom et al., 2006). However, with the lower cost and greater availability of high-resolution satellite images and the rapid development of new classification techniques, in combination with other data sources, their value for biodiversity studies can be expected to increase significantly. Promising research on new classification techniques embrace, for example, computerized approaches and the use of multilayer perception and neural networks that significantly improve accuracy (e.g., Boyd, Sanchez-Hernandez and Foody, 2006; Sesnie et al., 2008) and object-based image segmentation (e.g., Burnett and Blaschke, 2003), thus moving away from the pixel-based approach.

New developments in retrieving biodiversity-related information from remote sensor data are facilitated by major improvements in data capture. Data from active sensors, such as radar, can provide information on elevation, e.g., from the Shuttle Radar Topography Mission which has almost global coverage; these data are valuable for biodiversity studies. Radar can also provide high-resolution data, for example, Saatchi et al. (2008) used radar backscatter from QSCAT to improve models by providing information on vegetation structure. Taft, Haig and Kiilsgaard (2003) and Lang et al. (2008) used radar remote sensor data for wetland mapping, with resolutions of 5.6 to 68 cm in the bands used. Another type of active sensor is airborne lidar, which has been used to improve species distribution models by quantifying vegetation structure within a landscape (Hill and Thompson, 2005; Goetz et al., 2007) and for tree species classification with a resolution of 0.2 to 2 m (Hyyppä et al., 2008). Other advances are associated with techniques such as multi-angle viewing (e.g., Baltsavias et al., 2008) and hyperspectral remote sensing (e.g., Ben-Dor, Levin and Saaroni, 2001; Foody et al., 2004) that have considerable potential relevant to biodiversity studies. Finally, combinations of data from different sensors can be combined to enhance the information derived from them, e.g. laser scanning and multispectral remote sensing (e.g., Holmgren et al., 2008).

20.2.2 Indirect approaches

Despite the advances in extracting biodiversity-relevant information directly from remote sensor data, given the extremely wide definition of biodiversity, we can only expect a fraction of all biodiversity components to be detected directly by remote sensing. However, remote sensor data provide an invaluable and yet underused information source relating to biodiversity and ecological processes, which could be accessed through indirect approaches including spatial ecological modeling (e.g., Turner et al., 2003). In this way, it is possible to explore relationships between biodiversity components, ecological processes and predictive environmental variables derived from remote sensing and other data sources. Furthermore, once relationships between response variables and predictive variables have been established, the derived models can be used to predict outcomes of alternative future scenarios.

20.2.2.1 Predictive environmental variables in indirect approaches

The predictive environmental variables that can be derived from remote sensor (and other) data for use in indirect approaches can include proxies such as coarse land cover classes, or more or less direct measures of the ecological niche, such as climate, topography, hydrology, detailed vegetation types, vegetation structure, measures of productivity, and disturbance metrics (e.g., Elith *et al.*, 2006; Duro *et al.*, 2007; Saatchi *et al.*, 2008). Precipitation data has also been used, for example at 0.1° from NOAA satellites (Pearson *et al.*, 2007) and 0.25° from the Tropical Rainfall Mapping Mission (Saatchi *et al.*, 2008). Building on such variables, landscape patterns can be quantified via landscape metrics, using, for example, measures of fragmentation and heterogeneity, which can be related to biodiversity response variables (e.g., Luoto *et al.*, 2004).

A common method for obtaining environmental variables related to biodiversity is land cover classification. Information on land cover does not correspond directly to biodiversity components such as ecosystems, but it can provide very useful basic information such as the distribution of broad types of forest and grasslands. However, additional information is often needed in order to gain ecologically meaningful information for biodiversity studies, such as vegetation types or habitat quality for individual species, populations or guilds (Groom *et al.*, 2006; Mücher *et al.*, 2009). A wide range of land cover classification studies based on remote sensor data have been undertaken, both for research purposes, during the collection of environmental data relating to biodiversity studies, and management (e.g. Gillespie *et al.*, 2008; Mücher *et al.*, 2009).

Low resolution sensors, such as MODIS and AVHRR, with spatial resolutions of 250 m and 1.1 km respectively, can contribute to biodiversity studies mainly by helping us to understand ecological processes at regional to global scales, for example climate-associated vegetation growth patterns (e.g., Lotsch *et al.*, 2003). These sensors also provide data on temperature, precipitation and fire, which can add to the information derived from land cover classifications. For urban studies, they have potential to provide data on global urbanization patterns in relation to this type of information (Fig. 20.1).

Sensors with intermediate resolutions in the range 10–100 m include Landsat TM/ETM, SPOT HRV and IRS LISS. Of these, the NASA Landsat series is the most widely used for biodiversity related studies due to the ease of obtaining the data, the long time series and low cost (Gillespie *et al.*, 2008). Remote sensing studies using medium resolution sensors with relevance to biodiversity include classification of landscape types, classification and monitoring of vegetation types, monitoring vegetation degradation and optimization of land cover information for ecological purposes, from local to regional scales (Groom *et al.*, 2006; Gillespie *et al.*, 2008; Tomppo *et al.*, 2008; Mücher *et al.*, 2009). Data from these sensors have often been used in large-scale land cover mapping with a general aim to attain 85% accuracy across all mapping classes (Franklin and Wulder, 2002).

Furthermore, urbanization has complicated wide-ranging ecological implications (Grimm *et al.*, 2008) and in order to understand and model the ecological processes involved and their relation to biodiversity, it is necessary to quantify ecologically significant components of the urban system. Studies that have used remote sensing to quantify urban land cover in more detail for this purpose are listed in Section 20.3.2.

20.2.2.2 Response variables in indirect approaches

Biodiversity components that can be modeled through indirect approaches, using remote sensor data, range from detailed process models, for example, relating to individual life histories, to the modeling of broad patterns of biodiversity hotspots (Gontier, Balfors and Mörtberg, 2006). Individual-based models (e.g., Topping *et al.*, 2003) and population viability models (e.g., Akcakaya, 2001) will gain in precision as a result of improvements in remote sensor data capture and interpretation, but such detailed models also require a range of parameters to be calibrated before they can be applied. The modeling of individual movements and population viability in urban areas and areas becoming urbanized is of great importance for improving our knowledge of biodiversity in this context. However, the application of such complicated models may be hampered by a lack of information on parameter values and thresholds, especially in relation to urban environments (e.g., Gontier, Balfors and Mörtberg, 2006).

The modeling of the occurrence of single species is often called ecological niche or habitat suitability modeling and has developed rapidly in the field of ecological research (e.g., Scott *et al.*, 2002; Guisan and Thuiller, 2005). These models build on empirical data pertaining to species in the form of presence, presence–absence or abundance data, or expert knowledge on, for example, species' habitat preferences, in relation to environmental predictors (Gontier, Balfors, and Mörtberg, 2006). Such models have been used for predicting habitat suitability for single species at the regional level (e.g., Peterson *et al.*, 2006; Soberón and Peterson, 2009) and in areas subject to urbanization (e.g. Mörtberg, Balfors and Knol, 2007; Hepinstall, Alberti and Marzluff, 2008).

In addition, connectivity, indicating the movement potential for different species on a landscape scale, can be spatially modeled using cost-distance modeling (e.g., Adriansen *et al.*, 2003) and graph theory (Townsend *et al.*, 2009; Zetterberg, Mörtberg and Balfors, 2010). Advances have also been made in the modeling of species richness and diversity on different scales, in relation to remote sensor data such as land cover classifications, measures of productivity, measures of heterogeneity and vegetation indices (e.g., Luoto *et al.*, 2004; Leyequien *et al.*, 2007; Levin *et al.*, 2007).

The development of spatial ecological models involves increasingly sophisticated statistical and spatial analyses to study the distribution of biodiversity components, from univariate or multiple regression models to general linear models, general additive models, machine learning methods and geographically weighted regression analyses, each associated with specific improvements in accuracy and interpretability (e.g., Scott *et al.*, 2002; Foody, 2005; Mörtberg and Karlström, 2005; Guisan and Thuiller, 2005). All spatial ecological models, once established, can provide probability maps that can be regarded as spatial predictions of species distributions and patterns of diversity within landscapes and regions. These can be applied to the

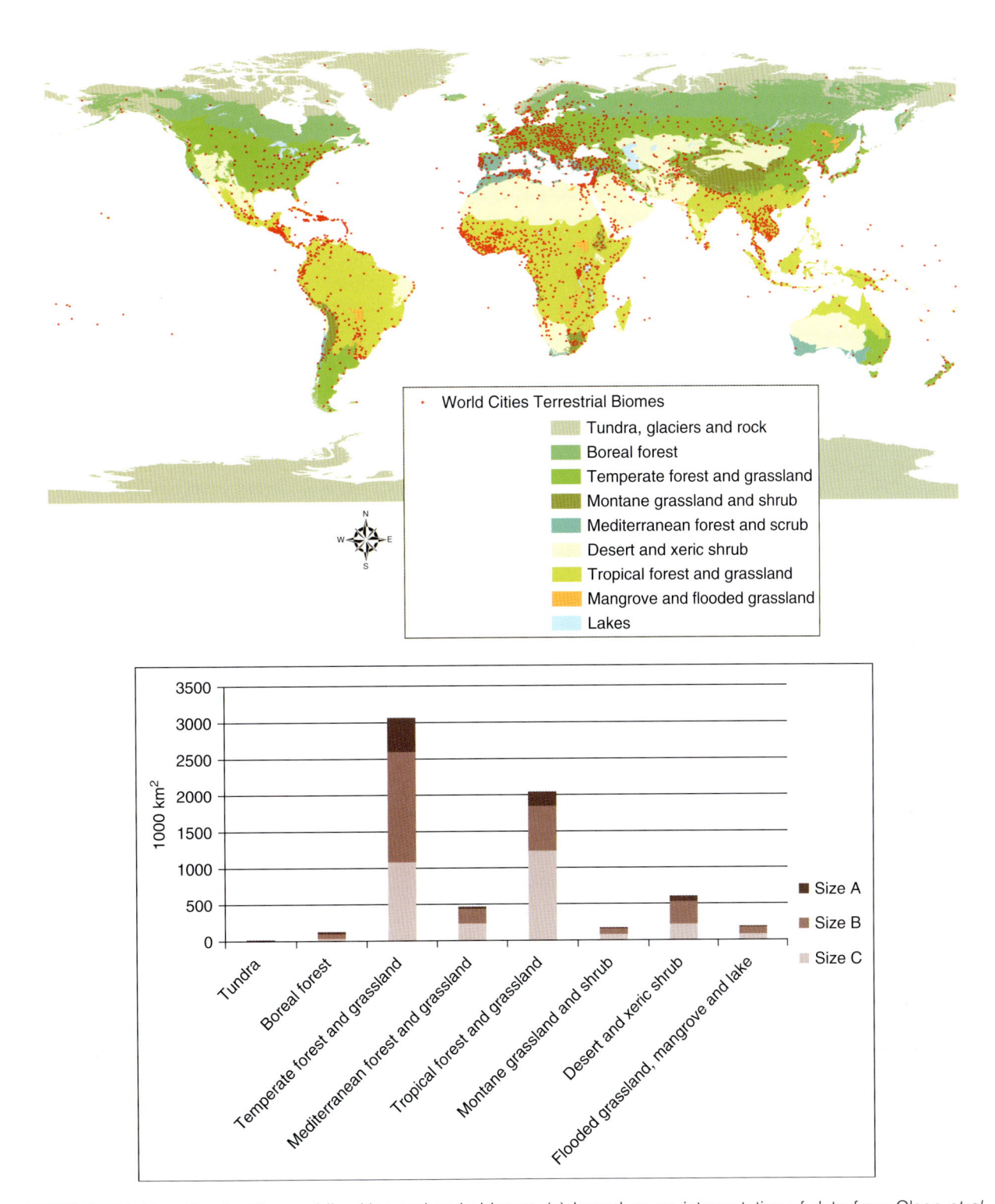

FIGURE 20.1 Map showing the world's cities and main biomes (a) based on an interpretation of data from Olson *et al.* (2001). The diagram (b) shows the size of periurban area within each biome for cities of different size, where A has >1 million inhabitants, B has 0.1–1 million inhabitants and C has <0.1 million inhabitants assuming that for larger cities the periurban area reaches up to 100 km, for smaller down to 10 km. This assumption is a rough estimate of the influence of the urban metabolism, based on the ideas of, for example Coelho and Roth (2006) on integrating the urban metabolism in urban ecology. As can be seen, temperate forest and grasslands are most affected by periurban areas, followed by tropical forest and grasslands. Basic geographical data from ESRI (2008).

present situation and to future scenarios, such as urbanization scenarios with different intensities and patterns. In addition, the use of agent-based models has been explored both for modeling biodiversity components such as individuals (e.g., Topping *et al.*, 2003) and for modeling urbanization and land use change (e.g., Fontaine and Rounsevell, 2009).

20.3 Hierarchical levels and definitions of urban ecosystems

Urbanization affects ecosystems and biodiversity in ways that have not been fully studied. According to Breuste, Niemelä and Snep (2008) urban ecosystems are highly complex, even more complex than many natural systems; this highlights the need to define clearly the components of urban ecosystems when comparing different cities and their biodiversity across the globe. There is, however, a lack of consistent definitions and terminology relating to the urban landscape and urban green habitats, making it difficult to compare related studies. For instance, Florgård (2007) found more than 10 different definitions of urban forest fragments. Nevertheless, from a broad perspective, certain terminology describing urban ecosystems is used frequently in the literature although not always with exactly the same meaning.

Moreover, different mechanisms affect species at different scales. It can be difficult to state categorically whether the quality of the local habitat or features of the adjacent landscape have the greatest effect on species richness of a population breeding in a fragmented urban habitat (Hedblom and Söderström, 2010). By subdividing urban areas into hierarchical levels (multiple geographical scales) with different ecological functions, a more accurate explanation for recorded species distributions could be found. Thus, as suggested by Clergeau, Jokimäki and Snep (2006), local habitats between houses, larger conglomerations of parks or houses, the whole town and differences between the northern and southern parts of continents are examples of hierarchical levels.

Furthermore, natural areas that become incorporated into urban or suburban settlements will change, even if they are protected, as a result of recreation pressure, urban predators, air pollution, traffic noise or other disturbances (Mörtberg, 2009). The joint impacts of such urban disturbances have only recently been studied in detail, but are receiving increasing research attention (McDonnell, Breuste and Hahs, 2009). Two major general patterns of urbanization affect ecological processes in cities. The first is expansion of the city into surrounding areas; when the area of the cities increases faster than the urban population growth this is known as urban sprawl. The second is fragmentation of existing green areas within the city, which is known as compaction, infill or urban condensation. The impacts of these urbanisation patterns are yet to be explored.

20.3.1 Flora and fauna along urban gradients

A large number of studies of urban areas have compared how biodiversity changes from the city center to the surrounding countryside (McDonnell and Hahs, 2008). These types of studies are known as urban gradient or urban–rural gradient studies (McDonnell *et al.*, 1997). The basic idea is to compare species composition in different parts of the city along transects that could be of any size stretching from city centers to the more natural or rural areas surrounding them. In this way differences in, for example, bird species richness in a city could be compared to that in more pristine areas. However, the ecological definitions of urban-to-rural areas differ between studies and vary depending on perspective. Some studies refer to urban, suburban and rural areas, but in regions of urbanization the land outside the urban and suburban areas tends to be heavily influenced by the proximity of the city, which also affects biodiversity; these areas have, therefore, sometimes been referred to as periurban areas (e.g., Mörtberg, 2009).

Previous studies has revealed a general pattern of increased urbanization leading to the presence of fewer species (Jokimäki and Kaisanlahti-Jokimäki, 2003; Luck, 2007; Garaffa, Filloy and Belloq, 2009) and higher densities of birds (Chace and Walsh, 2006). McKinney (2006) argued that cities are great homogenizing forces, where some species become so called urban adaptors (Blair, 1996) and are more common in cities worldwide; it was also suggested that subsets of native species become locally and regionally abundant at the expense of indigenous species. However, the picture is more complex than a simple negative relationship between human density and native bird diversity (Turner, 2003; McDonnell and Hahs, 2008). Thus, the general patterns differ depending on definitions of scale, species and habitats that are compared. McKinney (2006) concluded that species richness, biotic interactions and ecosystem complexity decline with increasing urbanization whereas biomass, total organism abundance and ecosystem reliance on external subsidies all increase with increasing urbanization.

Blair (1996) suggested that the highest species richness of birds was at the border of the city where the rural landscape meets the urban landscape, i.e. suburbia. Blair referred to this as the intermediate-disturbance theory (Hansen *et al.*, 2005). It means that there are more heterogeneous landscapes in the suburban area (compared to periurban and rural areas, open grassland, golf courses, parks and town center) that allow a diversity of possible habitats for species such as birds. Plant species have been shown to exhibit the opposite trend from vertebrates and invertebrates in that species richness and evenness increases in cities compared to periurban areas. This is probably because of the highly heterogeneous patchwork of habitats coupled with human introductions of native and exotic species and the human preference for few individuals of many species in close proximity (Grimm *et al.*, 2008; McDonnell and Hahs, 2008).

One further example of the complexity associated with the effects of urbanization on biodiversity is the different responses of different vegetation types, as illustrated by an example from Stockholm, the capital of Sweden (Mörtberg, 2009). The total volume of forest per hectare has been found to decrease along the urbanization axis from periurban to urban, as can be seen in Fig. 20.2. Along the same axis, the volume of deciduous forest only decreased slightly, and the volume of oak (*Quercus robur*, a native species) per hectare even exhibited a small increase. Similar patterns have been reported, for example, from Finland (Jokimäki and Suhonen, 1998). Possible explanations are: that conifers are particularly sensitive to air pollution; that deciduous trees and shrubs are favoured in urban planning and park management;

settlements (Falkenmark and Chapman, 1989; Luck, 2007). In Stockholm, the remnant native deciduous trees and woodlands support red-list bird species (Mörtberg and Wallentinus, 2000) and the relatively large oak stands support the very high diversity of invertebrates associated with this habitat (Mörtberg and Ihse, 2006).

According to a review by McDonnell and Hahs (2008), the responses of organisms to urbanization gradients are not predictable and could be anything from negative to positive, or even indistinguishable. Furthermore, urban ecosystems and species are affected in different ways according to the size of the city. Garaffa, Filloy and Bellocq (2009) found thresholds with respect to the size of cities, where towns with over 7000 inhabitants were negatively related to species richness (see also Jokimäki and Kaisanlahti-Jokimäki, 2003). However, Hedblom and Söderström (2010) found no threshold with respect to size of cities and abundance of species. Few studies have compared species richness and green area cover in more than one city at the time. Although Clergeau *et al.* (2006) found differences along the urban–rural gradient when comparing bird species richness in 19 European cities, it was difficult to explain the species distribution since they did not use multiple geographical scales with comparable well-described habitats.

20.3.2 Using remote sensing to quantify urbanization patterns

As mentioned earlier, urbanization gradient studies have been criticized for presenting gradients that are too simplistic (Alberti, Botsford and Cohen, 2001; McKinney, 2006). Not all cities are circular and therefore transects are not always the best way of describing differences. By using remote sensing it may be possible to describe true gradients within urban areas and thus investigate patterns within cities (McDonnell and Hahs, 2008). Patterns within cities could be divided according to the hierarchical levels suggested by Clergeau, Jokimäki and Snep (2006), including, for example, patterns within a single residential area or between residential areas. Such approaches could change some of the previous results from gradient analyses.

However, in order to find appropriate large scale ecological and ecosystem functions, it might be of interest to start comparing broad-scale studies, for example, looking at patterns between cities. In order to facilitate comparative studies between cities we suggest quantitative measures of urbanization that try to encapsulate the global process of urbanization and its impacts on ecosystems and biodiversity (McDonnell and Hahs, 2008). Quantitative measures of landscapes undergoing urbanization, urban land use and urban disturbances will be necessary in order to support studies that start to disentangle how ecological processes are affected by urbanization. Examples of quantitative measures that can provide proxy variables for such studies are landscape metrics, urban form parameters, demographic measures, the proportion of impervious surfaces, building density, road density, traffic noise, night-time light emissions, air pollution and other physical and chemical measures but also socio-economic factors and the presence of informal settlements (e.g., Small, 2001; Luck and Wu, 2002; Wu and Murray, 2003; Small, Pozzi and Elvidge, 2005; Pauleit, Ennos and Golding, 2005; Hahs and McDonnell, 2008; Hepinstall, Alberti and Marzluff, 2008; Mörtberg, 2009).

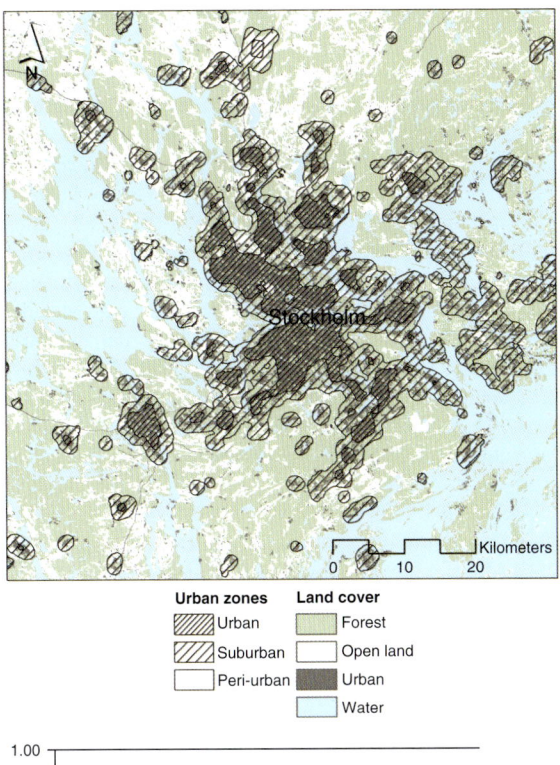

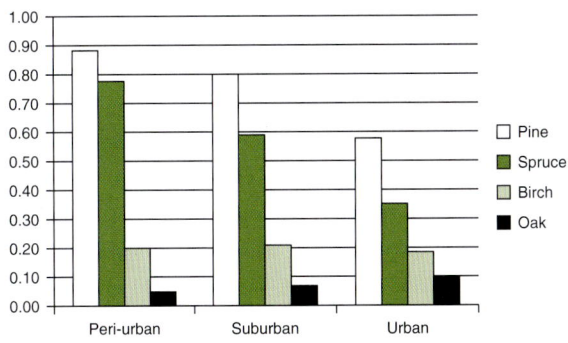

FIGURE 20.2 Differences in the mean volume of different tree species ($m^3 \times 10^{-1}/m^2$), mirroring the forest proportion and composition along an urbanization gradient, here expressed as urban, suburban and periurban zones (see map for location of zones). The urban zone is dominated by dense urban areas with small green areas, the suburban zone is dominated by urban areas with larger green areas and industrial areas, while the periurban zone is dominated by single houses, forest and open land, according to Swedish Landcover Data (National Landsurvey of Sweden, 2006). The proportion of coniferous and deciduous tree species in the periurban zone reflect the situation in much of the hemiboreal forest, while in the urban zone the proportion of deciduous trees is much higher, especially of oak. Data on forest variables was obtained from Holmgren *et al.*, 2000). Developed from Mörtberg (2009).

and that deciduous trees are present as a result of vegetation succession on temporarily abandoned open land. Furthermore, the concentration of oak and other hardwood deciduous trees in urban areas of the hemiboreal zone may coincide with the preference for good soils and microclimate, which also characterize long-established cities built on or in proximity to early human

However, measures of urbanization tend to be highly redundant. Therefore, Hahs and McDonnell (2006) created an index that explained variability along the urbanization gradient. It was based on two measures: density of people working in non-agricultural employment (obtained from census information), and the proportion of the landscape covered by impervious surfaces (determined from satellite imagery). They suggested that this index could constitute a broad measure of urbanization, suitable for comparative studies. Remote sensor data plays a key role here for capturing urbanization measures and their changes over time, especially when urban growth is rapid and/or unplanned, causing topographic maps to be misleading with respect, for example, to the extent of a city (Angel, Sheppard and Civco, 2005). Thus, in order to allow comparisons, future studies trying to explain species richness in a local urban habitat need to combine both field studies and remote sensing techniques and examine different hierarchical levels.

20.4 Using remote sensing to interpret effects of urbanization on species distribution

In order to explain species abundances in urban areas it is necessary to select field data at different hierarchical levels (multiple geographical scales) from the local habitat of urban green spaces to differences between cities, nations and regions (Clergeau, Jokimäki and Snep, 2006; McDonnell and Hahs, 2008). Thus, it is costly and time consuming to have field personnel working in many green spaces and cities at the same time (sometimes large sampling efforts are required over short time periods, for example when conducting breeding bird surveys). Moreover, it is difficult to use only remote sensing without complementary field surveys to explain species distribution. For example, Clergeau *et al.* 2006 compared cities by collecting data from inventories compiled in 19 European cities. The results supported previous findings that indicated a decline in species richness from rural areas to urban centers. However, the species data had been collected using different methods without detailed knowledge of the habitats. Therefore it was difficult to draw conclusions about whether it was the distance to the center, the surrounding landscape, or the quality of the habitat that had the main effect on species distribution. In a detailed study conducted in one city by Blair (1996) there were clear differences in species richness depending on whether the habitat studied was a golf course, park, lawn or natural area. The author described the dominant landscape character surrounding the study areas (e.g., residential area) but did not quantify them making it difficult to interpret possible fragmentation effects.

In northern Europe remnant forests are under severe threat from development infilling the remaining natural spaces and as a result of weak legal protection (Carlborg, 1991). These urban woodlands also have high potential for biodiversity and are popular for recreation (Hedblom and Söderström, 2008). Studies by Hedblom and Söderström (2008, 2010) revealed that valuable information on how landscape fragmentation affects birds in local urban woodlands could be provided if different geographical scales were examined.

As pointed out, landscape fragmentation due to urbanization is considered to be a major cause of the decline in species richness. Andrén (1994) suggested a theoretical landscape level threshold of 20–30% of remaining habitat below which the loss of species, or populations of species, is higher than the combined effects of habitat loss and fragmentation. Few empirical studies of landscape thresholds exist (but see Betts *et al.*, 2007). In Sweden, 1–34% (average 20%) of the area of cities consists of urban woodlands in the form of remnant forests (Hedblom and Söderström, 2008). In comparison, the proportion of urban woodland averages 7% in the 22 largest cities in Holland. Thus, with this range in extent of urban green species, it is possible to undertake studies of thresholds in Swedish urban landscapes.

Hedblom and Söderström (2008, 2010) used remote sensing to select 34 cities out of 100 in which the proportion woodland varied from 1–34%. The cities were also chosen on the basis of other criteria such as: being in different regions of Sweden (regional differences with respect to climate, etc.), cities surrounded predominantly by agricultural land or forests (the two most common landscape characteristics in Sweden) and whether the city was mainly suburban, an industrial city, a major city, etc. (see Hedblom and Söderström, 2008, for details). The woodlands were chosen using remote sensing and thereafter habitat structure (whether deciduous or coniferous woodland dominated and the age structure of the trees) and bird species were inventoried. Along a gradient from center to the surrounding, woodlands were selected in triplets of equal size (i.e., in three size intervals: >1 to ≤3 ha, >3 to ≤8 ha or >8 ha, see Hedblom and Söderström, 2010). After using remote sensing to find woodlands fulfilling these criteria, field assistants were despatched primarily to confirm that the woodlands had the characteristics indicated by the remote sensor data (namely, the tree species configuration, age structure and tree size).

Bird surveys were conducted by skilled ornithologists in 474 urban woodlands in 34 cities (see Hedblom and Söderström, 2010, for details). Since the digitalized remote sensor data adequately represented the habitat characteristics (as confirmed by the ground truth collected by the field assistants), digital maps were used to quantify the proportion of woodland in cities and in the surrounding area. Thereafter, species abundance in local urban woodlands was compared with the proportions of woodland at the city level and in surrounding areas. The results revealed that some bird species breeding in local urban woodlands were affected by the total proportion of woodland at the city level, indicating possible thresholds. Moreover, some species breeding in local urban woodlands were affected by the proportion of woodland in the surrounding landscape; this was especially the case in cities surrounded by agricultural land (see Hedblom and Söderström, 2010). Thus, urban planners and decision makers could take these results into consideration when examining the effects of urban sprawl; in addition, they should carefully consider applications to exploit local urban woodlands in cities where they constitute 20–30% of the original landscape cover. However, to understand fully species distribution and landscape effects it is also important to include temporal studies of species fluctuations.

20.5 Long-term monitoring of biodiversity in urban green areas – methodology development

A long term goal of urban ecology should be to uncover the factors regulating the success or failure of species in inhabited areas and use these factors to develop principles for the design of urban landscapes compatible with nature (Turner, 2003). In order to reveal why species richness and abundance fluctuate in urban ecosystems long-term data relating to species abundance, habitat quality and habitat quantity are needed. Long-term data could provide valuable information not only about the condition of single species, large ecosystems and the general environment but also indirect effects on humans such as through ecosystem services. A number of existing monitoring programs are biodiversity-oriented, although most of them started fairly recently. One of the programs that has been operational for a long time is the British Countryside Survey (Haines-Young *et al.*, 2003; Barr *et al.*, 2003). Monitoring programs have demonstrated a decrease in many species of farmland birds in Europe due to the intensification of agriculture (Wretenberg *et al.*, 2006). This in turn has led to a demand that the EU needs to introduce policies to protect or enhance bird diversity in agricultural areas (Vickery, 2001; Pan European Common Bird Monitoring Scheme, 2007). In Europe alone, historical photographs reveal that urbanization has been the predominant agent of land cover change since 1950 (Mücher, 2009). This process probably affects both species richness and habitat quality. However, to date, no national biodiversity monitoring program for urban green areas exists.

National monitoring programs covering certain habitats have existed for a long time, for example the Swedish forest monitoring program that began in 1923 (Fransson, 2009). In Europe a coordinated program (EBONE) for monitoring biodiversity at the national, region and European levels is under development (Jongman *et al.*, 2006). In Sweden a nationwide environmental protection program called NILS (National Inventory of Landscape in Sweden) has monitored the conditions and changes in the landscape and recorded how these changes influence the conditions for biodiversity since 2003 (Ståhl *et al.*, 2010). The program is founded by the Swedish Environmental Protection Agency and is supposed to conduct data from year 2003 into an indefinite future. The NILS program uses both remote sensing and field studies and covers approximately 600 landscape squares (5 × 5 km) that are randomly scattered (stratified in certain regions) all over Sweden. Every fifth year an aerial picture is taken and field studies are conducted (see Ståhl *et al.*, 2010 for details). The aerial pictures are manually interpreted (see Fig. 20.3). However, despite this recent increase in monitoring programs covering whole landscapes and regions, only a tiny proportion of the urban landscapes is included (Hedblom and Gyllin, 2009). This small proportion of urban cover is not enough to provide decision makers or urban planners with information about the effects of fragmentation on biodiversity, not even in a national context (Hedblom and Gyllin, 2009). There is an increased awareness of the effects of fragmentation on biodiversity in urban areas and also the value of urban green spaces for human recreation (Grahn and Stigsdotter, 2010) and well-being (Fuller *et al.*, 2007). Thus, urban green areas and their biodiversity are included in the Swedish Environmental Quality Objectives (Governmental bill to Parliament, Prop. 2009/10:155).

The NILS methodology, based on hierarchical structure (see Fig. 20.3), has been used as a model for the development for a specific urban monitoring program called NILS Urban (National Inventory of Landscape in Swedish urban areas, Fig. 20.3, Hedblom and Gyllin, 2009). A questionnaire was sent out by the NILS Urban program to municipalities and governmental organizations asking what data could be important with respect to the urban landscape; the responses revealed a great interest in biodiversity-related topics such as long-term data on urban green habitats, species richness and fragmentation changes. Interestingly, information about human requirements for urban green areas was particularly emphasized (Hedblom and Gyllin, 2009). Apparently, monitoring urban data relating to green areas have many potential users, more than in, for example, forested and agricultural areas, where one parcel of land could have one or a few owners and users; urban areas might, therefore, be more complex to monitor. The method developed for NILS Urban illustrated some constraints and possibilities associated with an urban monitoring program.

NILS Urban suggests a hierarchical level including remote sensing and field inventories (Fig. 20.3). From a national perspective, the previous method used in NILS with random 5 × 5 km landscape squares within cities would be enough to highlight spatial–temporal differences. However, from a local (municipal) perspective, it is of greater interest to cover the whole urban landscape in order to get detailed data relating to urban condensation (infill development) and urban sprawl, to provide local planners with information. However, to interpret aerial pictures manually over large urban areas, for example, the city of Stockholm covering 400 km^2, would be costly and time consuming. It is, therefore, more cost-effective to use satellite images with automated interpretation but lower resolution (see Fig. 20.3). Statistics Sweden (2008) used the satellite SPOT 5 with a 10 m resolution and automated interpretation of urban green areas, while NILS uses manually interpreted infrared aerial photographs with a ground resolution between 0.5 and 1.5 m (Allard *et al.*, 2003). Thus, using manual interpretation is advantagous since it allows special focus on features such as vulnerable biodiversity hotspots, for example small ponds.

Although not yet used in monitoring, a different kind of remote sensor data source could provide measurements of species movements from space. These possibilities are provided by the ARGOS satellite tracking system, using relatively small transmitters for studying larger animals, including mammals and reptiles. For smaller animals (less than ~300 g), Wikelski *et al.* (2007) proposed a small-animal satellite tracking system that would enable the global monitoring of animals down to the size of the smallest birds, mammals, marine life and eventually large insects. The technique would utilize satellite-mounted radio-receivers for tracking radio-tags weighing less than 1 g. This type of technique will provide important new data that will allow quantitative assessment of dispersal and migration and has the potential to provide new insights into population dynamics and persistence. This technique has great potential for analyzing how species use the urban landscape.

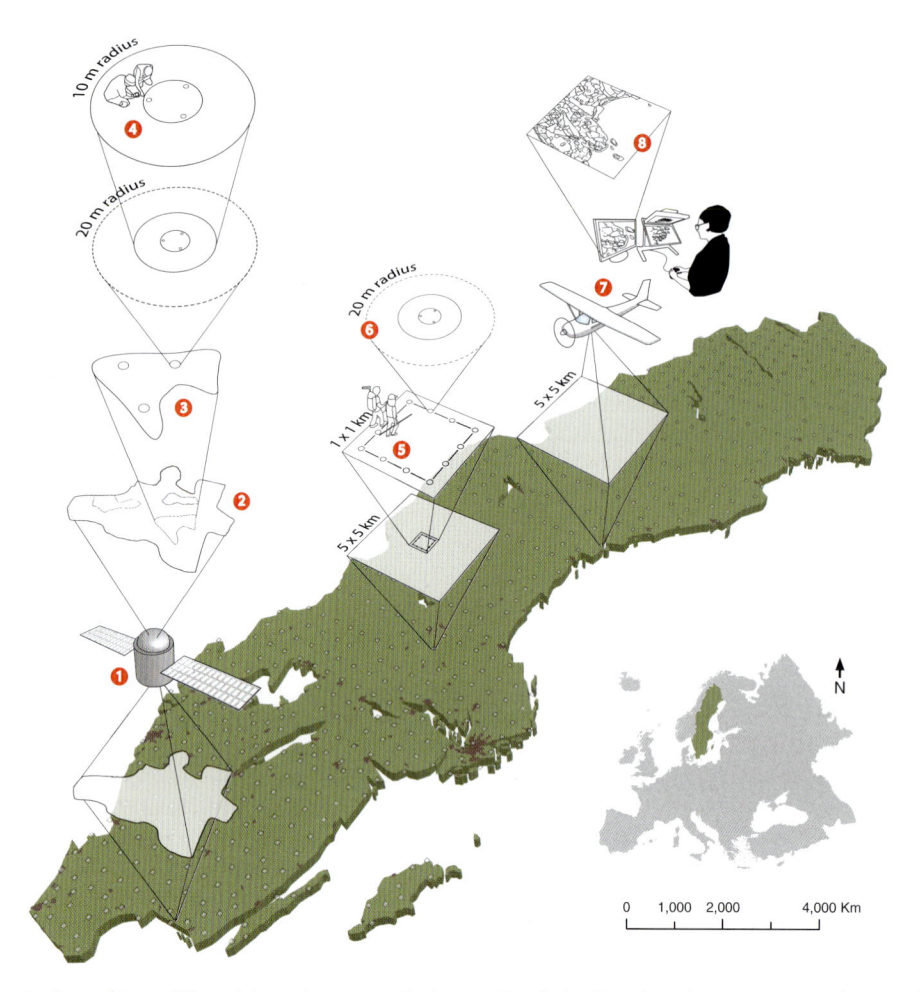

FIGURE 20.3 Illustration of two different long-term monitoring methods in Sweden. 1–4 represent the monitoring of urban areas within the NILS Urban project; 4–7 represent the NILS (National Inventory of Swedish Landscapes) monitoring program. (1) A satellite image that covers the whole city and the periurban (surrounding) area. (2) Green areas, e.g., parks, churchyards, recreation areas, golf courses, urban woodlands, etc. that are accessible to the public are chosen on the basis of the satellite image. (3) One field plot or a number of field plots (depending on the size of the green area) with a 20 m radius are chosen within each green area; and (4) field surveys of biodiversity and perception values are conducted within each of the plots in circles with different radii (Esseen et al., 2007). (5) In the NILS program, 12 sample plots (radius 20 m) in each central landscape square are chosen for survey. (6) Field surveyng is conducted the same way as in the Urban NILS except recreational values are not recorded. (7) Aerial photographs are taken for each 5 × 5 km landscape square. (8) Digital aerial photos are interpreted with the aid of a polarized screen and special glasses. It is possible to see a three dimensional picture using two aerial photos in a so called stereo model (Allard et al., 2003). (Figure created by Erik Cronvall, 2009).

Human perception of biodiversity has mainly been ascertained through interviews (e.g., Bjork et al., 2008), a method that is difficult to use in monitoring because it is costly and time consuming. Indeed, it is always difficult to measure human perceptions due to individual and cultural preferences. Within the NILS Urban project a framework developed by the European project Visualand (Tveit, Ode and Fry, 2006; Ode, Tveit and Fry, 2008) was used and consisted of nine descriptions of nine landscape characteristics (Ode, Tveit and Fry, 2010). Preliminary studies indicate that it is possible to use field assistants to estimate the nine landscape characteristics in a systematic way, thus collecting information during the same field visit that the biodiversity sampling is conducted. This suggests that it might be possible to collect valuable data in the field more efficiently than is possible through interviews.

Tools that could rationalize urban monitoring are developing rapidly; these include new satellites, computer based interpretation of images and 3D landscape images (e.g., Olsson, Scheider and Koukal, 2008). Thus, current monitoring programs need to be under constant development.

20.6 Applications in urban planning and management

Remote sensing data are important for applications that aim to integrate biodiversity conservation and development into urban planning and management. Such applications involve deriving ecologically meaningful information and mapping it over large areas using both direct and indirect approaches, and reusing the data for prioritizing areas for biodiversity conservation in urban planning and management contexts.

One example of a modeling approach using remote sensor data for mapping biodiversity components is the modeling of the spatial distribution of Natura 2000 habitats across Europe (Mücher et al., 2009). The Natura 2000 sites are protected under the Habitats Directive (92/43/CEE) and represent habitats with high biodiversity values; the spatial distribution of these habitats outside the protected areas was, however, unknown. Datasets derived from remote sensor data, such as CORINE land cover (European Environment Agency, 2000) and GTOPO30 (NASA, 2009) were used in combination with other data, such as the distributions of indicator plant species. The data were combined through spatial distribution modeling using expert knowledge and machine-learning methods. The result was a series of maps of 27 rather specific habitats across Europe; these maps are very valuable for biodiversity studies, including urbanization studies.

Recently, spatial prioritizing models have been used in urban areas to identify urban forests that are of high priority for conservation and species richness (De Wan et al., 2009). In Australia, prioritization methods have incorporated species-specific connectivity into multi-species conservation planning within the urban context (Gordon et al., 2009). These prioritization tools could be used for strategic decision-making by land-use planners.

In the Stockholm region, long-term research has been conducted concerning methods for integrating biodiversity issues into urban planning and management, ranging from mapping detailed vegetation types and habitat structures (Löfvenhaft, Björn and Ihse, 2002) to spatial ecological modeling and the development of planning process tools (Löfvenhaft, Runborg and Sjögren-Gulve, 2004; Gontier, Balfors and Mörtberg 2006; Mörtberg, Balfors and Knol 2007; Gontier, Mörtberg and Balfors, 2010; Zetterberg, Mörtberg and Balfors, 2010).

These methods have been applied in case studies in several real-world planning situations within the region (Mörtberg and Ihse, 2006; Mörtberg, Zetterberg and Balfors, 2009; Mörtberg et al., 2010a). All case studies used remote sensor data (in combination with other data) for spatially complete coverage of vegetation and land cover parameters. Vegetation data derived by manual interpretation of CIR aerial photographs (Löfvenhaft, Björn and Ihse, 2002), covering the Stockholm municipality and including the city center, were used in two of the case studies. The other three case studies covered larger areas and therefore had to rely on less detailed data derived from the classification of Landsat images (Holmgren et al., 2000).

One of the case studies concerned Hanveden, a large forest-dominated suburban and periurban area in south Stockholm. Three municipalities, Haninge, Huddinge and Botkyrka, are responsible for the area. Furthermore, large areas are owned by a fourth municipality, the city of Stockholm. Part of the rationale of the study was concern about policy changes imposed by this big land owner. The overall aim of the project was to create a plan and a common strategy for nature conservation, recreation and multi-objective forestry within Hanveden. With respect to planning, there were three main scenarios: urban exploitation of land parcels; intensive commercial forestry; and a management plan for the whole Hanveden area with biodiversity and recreation objectives.

The biodiversity targets were derived via a participatory process, involving stakeholders, and aimed at finding priority habitat types and ecological profiles for the study area. The targeted habitat types were coniferous forest and deciduous forest with several ecological profiles for each. Habitat networks for the priority ecological profiles were derived through GIS-based habitat modeling, using both empirical models (Gontier, Balfors and Mörtberg, 2006, Mörtberg, Balfors and Knol 2007; Gontier, Mörtberg and Balfors, 2010) and expert models (Mörtberg et al., 2010a; Zetterberg, Mörtberg and Balfors, 2010). The study resulted in habitat networks that were composed of reproduction areas and connectivity zones for invertebrates, as well as areas likely to supply suitable conditions for breeding birds (Fig. 20.4).

The results were used to evaluate the scenarios. The urban exploitation scenario would have strong negative impacts on biodiversity targets. The forest management scenarios examined forest growth 40 years ahead with different management plans for different forest stands (see Fig. 20.4). The commercial forestry scenario would have relatively strong negative impacts on coniferous forest targets, given that hardwood deciduous forest was conserved in this scenario. The scenario with forest management adapted to biodiversity needs would have positive effects on biodiversity targets, as expected. The results provided decision support for joint strategies for the municipalities with respect to multi-purpose forestry with integrated biodiversity, recreation and economic objectives. The method has great potential for incorporating biodiversity issues into the creation and evaluation of development and management scenarios. Furthermore, the study indicated that there may be particular opportunities for management to enhance biodiversity close to urban areas that may not be available elsewhere.

Conclusions

Urbanization has major effects on biodiversity and this, in combination with the tendency for cities to be leading centers of policy and innovation, means that implementation of, for example, biodiversity-informed planning tools, infrastructure guidelines, education and public awareness will have influence at wider scales than within the cities themselves. For such planning tools and guidelines, remote sensing has considerable potential as a source of information on biodiversity at landscape and regional spatial scales.

Remote sensor data is increasingly being used in biodiversity studies, which take advantage of recent rapid advances in data capture, data interpretation and classification methods for deriving ecologically meaningful information, and spatial ecological modeling. Important biodiversity research directions and monitoring schemes involve systematic combinations of remote sensor data and field studies. Furthermore, there are great possibilities for quantifying the urbanization gradient using remote sensor data; this will be useful for further research into urban ecosystem processes and the impacts of urbanization on biodiversity. However, urban biodiversity studies are far from using the full capacity of these rapidly expanding databases and research directions.

In order to reach the full potential of urban biodiversity studies, high-quality data need to be readily available, derived data and new methods effectively communicated and shared, and the necessary inter- and transdisciplinary research must also incorporate demographic, social and economic issues. Finally, for future knowledge of how to manage urban habitats and assure

298 PART V URBAN ENVIRONMENTAL ANALYSES

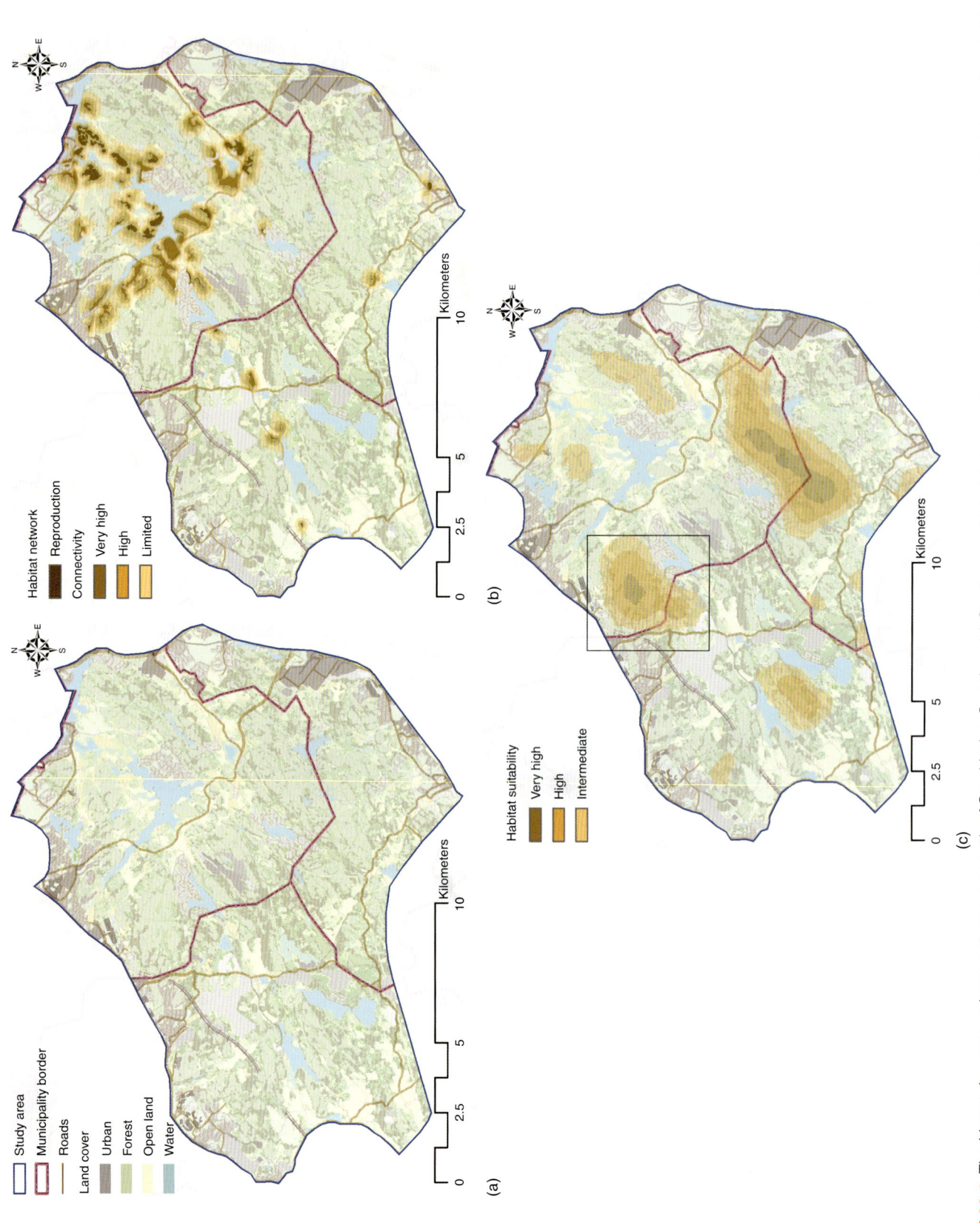

FIGURE 20.4 The Hanveden area, a large nature area south of Stockholm, Sweden: (a) Study area, (b) habitat network for species tied to hardwood deciduous forest, (c) habitat suitability for area demanding species of coniferous forest, (d) detail of scenario with normal forest management and (e) detail of scenario with forest management supporting biodiversity target of coniferous forest (Mörtberg et al., 2010b). The rectangle in (c) embraces the area of scenarios in (d) and (e).

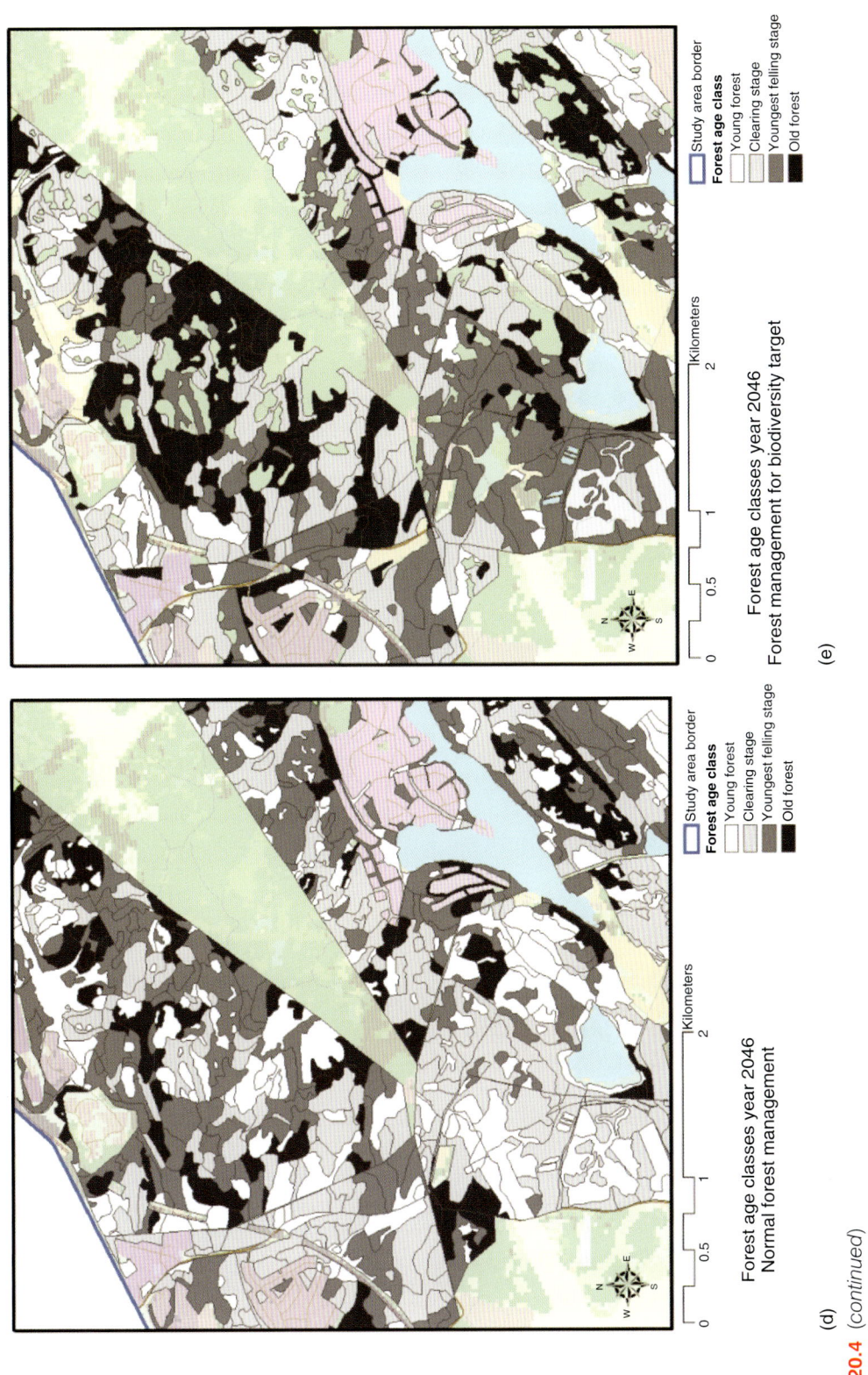

FIGURE 20.4 *(continued)*

species survival, urbanization processes and their interaction with ecological processes and biodiversity need to be explored within urban landscapes, urban regions, and at a global level.

Acknowledgments

The authors would like to thank the volume editor and three anonymous reviewers for valuable suggestions that help improve the scholarly quality of the text.

References

Adriaensen, F., Chardon, J.P., De Blust, G. et al. (2003) The application of 'least-cost' modelling as a functional landscape model. *Landscape and Urban Planning*, **64**, 233–247.

Akcakaya, H.R. (2001) Linking population-level risk assessment with landscape and habitat models. *Science of the Total Environment*, **274**, 283–291.

Alberti, M., Botsford, E. and Cohen, A. (2001) Quantifying the urban gradient: linking urban planning and ecology. In *Avian Ecology and Conservation in an Urbanizing World* (eds J.M. Marzluff R. Bowman and R. Donnelly), Kluwer Academic Publishers, Norwell, MA, pp. 89–115.

Allard, A., Nilsson, B., Pramborg, K., Ståhl, G. and Sundquist, S. (2003) Manual for aerial photo interpretation in the national inventory of landscapes in Sweden. Swedish University for Agricultural Sciences, SLU, Umeå, Sweden.

Andrén, H. (1994) Effects of habitat fragmentation on birds and mammals in landscapes with different proportions of suitable habitat-a review. *Oikos*, 71, 355–366.

Angel, S., Sheppard, S.C., and Civco, D.L. (2005) The dynamics of global urban expansion. Transport and Urban Development Department, World Bank, Washington DC, 205 pp.

Baltsavias, E., Gruen, A., Eissenbeiss, H., Zhang, L. and Waser, L. (2008) High-quality image matching and automated generation of 3D tree models. *International Journal of Remote Sensing*, **29**, 1243–1259.

Barr C.J., Bunce R.G.H. and Clarke R.T. et al. (2003) Methodology of Countryside Survey 2000 Module 1: Survey of broad habitats and landscape features, Final Report. Contract report to Department for Environment, Food and Rural Affairs from the Centre for Ecology and Hydrology, Cumbria, England.

Ben-Dor E., Levin N. and Saaroni H. (2001) A spectral based recognition of the urban environment using the visible and near-infrared spectral region (0.4–1.1 μm). A case study over Tel-Aviv, Israel. *International Journal of Remote Sensing*, **22**, 2193–2218.

Betts, M.G., Forbes, G.J., Diamond, A.W. and Bêty, J. (2007) Thresholds in songbird occurrence in relation to landscape structure. *Conservation Biology*, **21**, 1046–1058.

Bino, G., Levin, N., Darawshi, S., van der Hal, N., Reich-Solomon, A. and Kark, S. (2008) Landsat derived NDVI and spectral unmixing accurately predict bird species richness patterns in an urban landscape. *International Journal of Remote Sensing*, **29**, 3675–3700.

Bjork, J., Albin, M., Grahn, P. et al. (2008) Recreational values of the natural environment in relation to neighbourhood satisfaction, physical activity, obesity and wellbeing. *Journal of Epidemiology and Community Health*, **62**, 4.

Blair, R.B. (1996) Land use and avian species diversity along an urban gradient. *Ecological Applications*, **6**, 506–519.

Bolund, P. and Hunhammar, S. (1999) Ecosystem services in urban areas. *Ecological Economics*, **29**, 293–301.

Boyd, D.S., Sanchez-Hernandez, C. and Foody, G.M. (2006) Mapping a specific class for priority habitats monitoring from satellite sensor data. *International Journal of Remote Sensing*, **27**, 2631–2644.

Breuste, J., Niemelä, J., and Snep, R.H. (2008) Applying landscape ecological principles in urban environments. *Landscape Ecology*, **23**, 1139–1142.

Burnett, C. and Blaschke, T. (2003) A multi-scale segmentation/object relationship modeling methodology for landscape analysis. *Ecological Modelling*, **168**, 233–249.

Carlborg, N. (1991) Tätortsnära skogsbruk. The national board of forestry, Report 1, Jönköping (In Swedish).

Chace, J.F. and Walsh, J.J. (2006) Urban effects on native avifauna: a review. *Landscape and Urban Planning*, **74**, 46–69.

Cincotta R.P., Wisnewski. and Engelman R. (2000) Human population in the biodiversity hotspots. *Nature*, **404**, 990.

Clark, D.B., Read, J.M., Clark, M.L., Cruz, A.M., Dotti, M.F., and Clark, D.A. (2004) Application of 1-M and 4-M resolution satellite data to ecological studies of tropical rain forests. *Ecological Applications*, **14**, 61–74.

Clergeau, P., Jokimäki. J. and Snep. R. (2006) Using hierarchical levels for urban ecology. *Trends in Ecology and Evolution*, **21**, 660–661.

Clergeau, P., Croci, S., Jokimäki, J., Kaisanlahti-Jokimäki, M.J. and Dinetti, M. (2006) Avifauna homogenisation by urbanisation: Analysis at different European latitudes. *Biological Conservation*, **127**, 336–344.

Davies, Z.G., Fuller, R.A., Loram, A., Irvine, K.N., Sims, V. and Gaston, K.J. (2009) A national scale inventory of resource provision within domestic gardens. *Biological Conservation*, **142**, 761–771.

De Wan, A.A., Sullivan, P.J., Lembo, A.J. et al. 2009. Using occupancy models of forest breeding birds to prioritize conservation planning. *Biological Conservation*, **142**, 982–991.

Dearborn, D. and Kark, S. (2010) The motivation for conserving urban biodiversity. *Conservation Biology*, **24** (2), 432–440.

Duro, D., Coops, N. C., Wulder, M. A., and Han, T. (2007) Development of a large area biodiversity monitoring system driven by remote sensing. *Progress in Physical Geography*, **31**, 235–260.

Elith, J., Graham, C. H., Anderson, R. P. et al. (2006) Novel methods improve prediction of species' distributions from occurrence data. *Ecography*, **29**, 129–151.

ESRI 2008. ESRI Data & Maps. ArcGIS 9.3, Environmental Systems Research Institute, Inc., Redlands, CA.

Esseen P.A., Glimskär, A,, Ståhl, G. and Sundquist, S. (2007) Field instruction for the National inventory of the landscape in Sweden, NILS. Swedish University of Agricultural Sciences, Department of Forest Resource Management, Umeå, Sweden.

European Environment Agency (2000) CORINE land cover database 2000. The European Topic Centre on Terrestrial Environment. Online. Available http://www.eea.europa.eu/data-and-maps/data/corine-land-cover-2000-clc2000-seamless-vector-database (accessed 18 November 2010).

Falkenmark, M. and Chapman, T. (1989) Comparative hydrology. An ecological approach to land and water resources, UNESCO, Paris.

Florgård, C. (2007) Preserved and remnant natural vegetation in cities: A geographically divided field of research. *Landscape Research*, **32**, 79–94.

Fontaine, C.M. and Rounsevell, M.D.A. (2009) An agent-based approach to model future residential pressure on a regional landscape. *Landscape Ecology*, **24**, 1237–1254.

Foody, G.M. (2005) Mapping the richness and composition of British breeding birds from coarse spatial resolution satellite sensor imagery. *International Journal of Remote Sensing*, **26**, 3943–3956.

Foody, G.M., Atkinson, P.M., Gething, P.W., Ravenhill, N.A., and Kelly, C.K. (2005) Identification of specific tree species in ancient semi-natural woodland from digital aerial sensor imagery. *Ecological Applications*, **15**, 1233.

Foody, G.M., Sargent, I.M.J., Atkinson, P.M. and Williams, J.W. (2004) Thematic labeling from hyperspectral remotely sensed imagery: trade-offs in image properties. *International Journal of Remote Sensing*, **25**, 2337–2363.

Franklin, S.E. and Wulder, M.A. (2002) Remote sensing methods in medium spatial resolution satellite data land cover classification of large areas. *Progress in Physical Geography*, **26**, 173–205.

Fransson, J. (2009) Forestry statistics 2009 Official Statistics of Sweden. Swedish University of Agricultural Sciences Umeå 2009. Online. Avaialble www.srh.slu.se (accessed 18 November 2010).

Fuller, R.A., Irvine, K.N., Devine-Wright, P., Warren, P.H. and Gaston, K.J. (2007) Psychological benefits of greenspace increase with biodiversity. *Biology Letters*, **3**, 390–394.

Garaffa, P.I., Filloy, J. and Bellocq. M.I. (2009) Bird community responses along urban–rural gradients: Does the size of the urbanized area matter? *Landscape and Urban Planning*, **90**, 33–41.

Gillespie, T.W., Foody, G.M., Rocchini, D., Giorgi, A.P., and Saatchi, S. (2008) Measuring and modelling biodiversity from space. *Progress in Physcial Geography*, **32**, 203–221.

Goetz, S., Steinberg, D., Dubayah, R. and Blair, B. (2007) Laser remote sensing of canopy habitat heterogeneity as a predictor of bird species richness in an eastern temperate forest, USA. *Remote Sensing of Environment*, **108**, 254–263.

Gontier, M., Balfors, B., and Mörtberg, U. (2006) Biodiversity in environmental assessment – current practice and tools for prediction. *Environmental Impact Assessment Review*, **26**, 268–286.

Gontier, M., Mörtberg, U. and Balfors, B. (2010) Comparing GIS-based habitat models for applications in EIA and SEA. *Environmental Impact Assessment Review*, **30**, 8–18.

Gordon, A., Simondson, D., White, M., Moilanen, A., and Bekessy, S.A. (2009) Integrating conservation planning and land use planning in urban landscapes. *Landscape and Urban Planning*, **91**, 183–194.

Gottschalk, T.K., Huettmann, F. and Ehlers, M. 2005. Thirty years of analyzing and modeling avian habitat relationships using satellite imagery data: a review. *International Journal of Remote Sensing*, **26**, 2631–2656.

Grahn, P. and Stigsdotter, U.K. (2010) The relation between perceived sensory dimensions of urban green space and stress restoration. *Landscape and Urban Planning*, **94**, 264–275.

Grimm, N.B., Faeth, S.H., Redman, C.L. et al. (2008) Global change and the ecology of cities. *Science*, **319**, 756–760.

Groom, G., Mücher, C.A., Ihse, M. and Wrbka, T. (2006) Remote sensing in landscape ecology: experiences and perspectives in a European context. *Landscape Ecology*, **21**, 391–408.

Guisan, A. and Thuiller, W. (2005) Predicting species distribution: offering more than simple habitat models. *Ecology Letters*, **8**, 993–1009.

Hahs, A.K. and McDonnell, M.J. (2006) Selecting independent measures to quantify Melbourne's urban-rural gradient. *Landscape and Urban Planning*, **78**, 435–448.

Haines-Young, R., Barr, C.J, Firbank, L.G. et al. (2003) Changing landscapes, habitats and vegetation diversity across Great Britain. *Journal of Environmental Management*, **67**, 267–281.

Hansen A.J., Knight R.L., Marzluff J.M. et al. (2005) Effects of exurban development on biodiversity: patterns mechanisms, and research needs. *Ecological Applications*, **15**, 1893–1905.

Hedblom, M. and Gyllin, M. (2009) Övervakning av biologisk mångfald och friluftsliv i tätorter – en metodstudie. (In Swedish with short summary in English). Swedish National Protection Agency, Report 5974.

Hedblom, M. and Söderström, B. (2008) Woodlands across Swedish urban gradients: status, structure and management implications. *Landscape and Urban Planning*, **84**, 62–73.

Hedblom, M. and Söderström. B. (2010) Landscape effects on birds in urban woodlands: an analysis of 34 Swedish cities. *Journal of Biogeography*, **37** (7), 1302–1316.

Hepinstall, J.A., Alberti, M. and Marzluff, J.M. (2008) Predicting land cover change and avian community responses in rapidly urbanizing environments. *Landscape Ecology*, **23**, 1257–1276.

Hill, R.A. and Thomson, A.G. (2005) Mapping woodland species composition and structure using airborne spectral and LiDAR data. *International Journal of Remote Sensing*, **26**, 3763–3779.

Holmgren, J., Joyce, S., Nilsson, M. and Olsson, H. (2000) Estimating stem volume and basal area in forest compartments by combining satellite image data with field data. *Scandinavian Journal of Forest Research*, **15**, 103–111.

Holmgren, J., Persson, Å and Söderman, U. (2008) Species identification of individual trees by combining high resolution LiDAR data with multi-spectral images. *International Journal of Remote Sensing*, **29**, 1537–1552.

Hyyppä, J., Hyyppä, H., Leckie, D., Gougeon, F., Yu, X. and Maltamo, M. (2008) Review of methods of smallfootprint airborne laser scanning for extracting forest inventory data in boreal forests. *International Journal of Remote Sensing*, **29**, 1339–1366.

Ihse, M. (1995) Swedish agricultural landscapes – patterns and changes during the last 50 years, studied by aerial photos. *Landscape and Urban Planning*, **31**, 21–37.

Jokimäki, J. & Kaisanlahti-Jokimäki, M.L. (2003) Spatial similarity of urban bird communities: a multiscale approach. *Journal of Biogeography*, **8**, 1183–1193.

Jokimäki, J. and Suhonen, J. (1998) Distribution and habitat selection of wintering birds in urban environments. *Landscape and Urban Planning*, **39**, 253–263.

Jongman, R.H.G., Bunce, R.G.H., Metzger, M.J. et al. (2006) Objectives and applications of a statistical environmental stratification of Europe. *Landscape Ecology*, **21**, 409–419.

Lang, M.W., Kasischke, E.S., Prince, S.D. and Pittman, K.W. (2008) Assessment of C-band synthetic aperture radar data for mapping coastal plain forested wetlands in the mid-Atlantic region USA. Remote Sensing of Environment 112: 4120–4130.

Levin N., Shmida A., Levanoni O., Tamari H. and Kark, S. (2007) Predicting mountain plant richness and rarity from space using satellite-derived vegetation indices. *Diversity and Distributions*, **13**, 692–703.

Levin, N., McAlpine, C., Phinn, S et al. (2009) Mapping forest patches and scattered trees from SPOT images and testing their ecological importance for woodland birds in a fragmented agricultural landscape. *International Journal of Remote Sensing*, **30**, 3147–3169.

Leyequien, E., Verrelst, J., Slot, M., Shaepman-Strub, G., Heitkönig, I.M.A. and Skidmore, A. (2007) Capturing the fugitive: applying remote sensing to terrestrial animal distribution and diversity. *International Journal of Applied Earth Observation and Geoinformation*, **9**, 1–20.

Löfvenhaft, K., Björn, C. and Ihse, M. (2002) Biotope patterns in urban areas: a conceptual model integrating biodiversity issues in spatial planning. *Landscape and Urban Planning*, **58**, 223–240.

Löfvenhaft, K., Runborg, S. and Sjögren-Gulve, P. (2004) Biotope patterns and amphibian distribution as assessment tools in urban landscape planning. *Landscape and Urban Planning*, **68**, 403–427.

Lotsch, A., Tian, Y., Friedl, M.A. and Myneni, R.B. (2003) Land cover mapping in support of LAI and FPAR retrievals from EOS-MODIS and MISR: classification methods and sensitivities to errors. *International Journal of Remote Sensing*, **24**, 1997–2016.

Luck, G.W. (2007) A review of the relationships between human population density and biodiversity. *Biological Reviews*, **82**, 607–645.

Luck, M. and Wu, J. (2002) A gradient analysis of urban landscape pattern: a case study from thePhoenix metropolitan region, Arizona, USA. *Landscape Ecology*, **17**, 327–339.

Luoto, M., Virkkala, R., Heikkinen, R.K., and Rainio, K. (2004) Predicting bird species richness using remote sensing in boreal agricultural-forest mosaics. *Ecological Applications*, **14**, 1946–1962.

Marzluff, J.M. (ed.) (2008) *Urban Ecology: An International Perspective on the Interaction Between Humans and Nature*, Springer, New York.

Marzluff, J.M. and Ewing, K. (2001) Restoration of fragmented landscapes for the conservation of birds: A general framework and specific recommendations for urbanizing landscapes. *Restoration Ecology*, **9**, 280–292.

McDonnell, M.J., Pickett, S.T.A., Groffman, P. et al. (1997) Ecosystem processes along an urban-to-rural gradient. *Urban Ecosystems*, **1**, 21–36.

McDonnell, M.J. and Hahs, A.K. (2008) The use of gradient analysis studies in advancing our understanding of the ecology of urbanizing landscapes: current status and future directions. *Landscape Ecology*, **23**, 1143–1155.

McDonnell, M.J., Breuste, J. and Hahs, A.K. (eds) (2009) *Ecology of Cities and Towns: A Comparative Approach*, Cambridge University Press, Cambridge.

McKinney, M.L. (2006) Urbanization as a major cause of biotic homogenization. *Biological Conservation*, **127**, 247–260.

Mörtberg, U. (2009) Landscape ecological analysis and assessment in an urbanizing environment, in *Ecology of Cities and Towns: A Comparative Approach* (eds M.J. McDonnell, J. Breuste and A.K. Hahs), Cambridge University Press, Cambridge, pp 439–455.

Mörtberg, U. and Ihse, M. (2006) Landskapsekologisk analys av Nationalstadsparken. Underlag till Länsstyrelsens program för Nationalstadsparken. Länsstyrelsen i Stockholms län, Rapport 2006: 13. [In Swedish].

Mörtberg, U., and Karlström, A. (2005) Predicting forest grouse distribution taking account of spatial autocorrrelation. *Journal for Nature Conservation*, **13**, 147–159.

Mörtberg, U.M. and Wallentinus, H.-G. (2000) Red-listed forest bird species in an urban environment – assessment of green space corridors. *Landscape and Urban Planning*, **50**, 215–226.

Mörtberg, U.M., Balfors, B. and Knol, W.C. (2007) Landscape ecological assessment: a tool for integrating biodiversity issues in strategic environmental assessment and planning. *Journal of Environmental Management*, **82**, 457–470.

Mörtberg, U., Zetterberg, A. and Balfors, B. (2009) Regional landscape strategies – integrating biodiversity, recreation and cultural history in an urbanising area, in (eds J. Breuste, M. Kozová and M. Finka), *European Landscapes in Transformation. Challenges for Landscape Ecology and Management. Proceedings of the European IALE Conference 2009, 70 Years of Landscape Ecology in Europe*, 12–16 July 2009, Salzburg, Austria, pp. 66–70.

Mörtberg, U., Balfors, B., Zetterberg, A. and Gontier, M. (2010a) Implementation of landscape ecological methods in urban planning, in *Implementation of Landscape Ecological Knowledge in Practice* (eds A. Maciasand A. Mizgajski), Proceedings of the 1st IALE-Europe Thematic Symposium, Poznan 16–19 June 2010, pp 39–51.

Mörtberg, U., Braide, A., Gontier, M. and Balfors, B. (2010b) Prognosverktyg för biologisk mångfald i planering – fallstudie Hanveden. Forskargruppen för miljöbedömning och – förvaltning, Kungl. Tekniska högskolan. TRITA-LWR.REPORT 3027. [In Swedish].

Mücher, C.A., Hennekens, S.M., Bunce, R.G., Schaminée, J.H., and Schaepman, M.E. (2009) Modelling the spatial distribution of Natura 2000 habitats across Europe. *Landscape and Urban Planning*, **92**, 148–159.

Mücher, C.A. (2009) Geospatial modelling and monitoring of European landscapes and habitats using remote sensing and field studies. Doctoral thesis, Wageningen University, ISBN: 978-90-8585-453-1.

NASA (2009) Shuttle Radar Topography Mission. Online. Available http://www2.jpl.nasa.gov/srtm/ (accessed 18 November 2010).

National Landsurvey of Sweden 2006. GSD-Landcover Map [GSD Data]. National Landsurvey of Sweden, ([grant I 2008/1818] https://butiken.metria.se/digibib/(accessed 30 November 2010).

Ode, Å., Tveit, M. and Fry, G. (2008) Capturing landscape visual character using indicators – touching base with landscape aesthetic theory. *Landscape Research*, **33**, 89–118.

Ode, Å., Tveit, M.S. and Fry, G. (2010) Advantages of using different data sources in assessment of landscape change and its effect on visual scale. *Ecological Indicators*, **10**, 24–31.

Olsson, H., Schneider, W. and Koukal, T. (2008) 3D remote sensing in forestry. *International Journal of Remote Sensing*, **29**, 1239–1242.

Pan European Common Bird Monitoring Scheme (2007) http://monitoring.sor.ro/Download/StateEuropeCommonBirds 2007.pdf (accessed 1 December 2010).

Pauleit, S., Ennos, R. and Golding, Y. (2005) Modeling the environmental impacts of urban land use and land cover change – a study in Merseyside, UK. *Landscape and Urban Planning*, **71**, 295–310.

Pautasso, M., and McKinney, M.L. (2007) The botanist effect revisited: plan species richness, county area, and human population size in the United States. *Conservation Biology*. **21**, 1333–1340.

Pearson, R.G., Raxworthy, C.J., Nakamura, M. and Townsend Peterson, A. (2007) Predicting species distributions from small numbers of occurrence records: a test case using cryptic geckos in Madagascar. *Journal of Biogeography*, **34**, 102–107.

Peterson, A.T., Sanchez-Cordero, V., Martinez-Meyer, E. and Navarro-Sigüenza, A.G. (2006) Tracking population extirpations via melding ecological niche modeling with land-cover information. *Ecological Modelling*, **195**, 229–236.

Pickett, S.T.A., Cadenasso, M.L. Grove, J.M. *et al.* (2008) Beyond urban legends: An emerging framework of urban ecology as illustrated by the Baltimore Ecosystem Study. *BioScience*, **58**, 139–150.

Ricketts, T. and Imhoff, M. (2003) Biodiversity, urban areas, and agriculture: locating priority ecoregions for conservation. *Conservation Ecology*, **8**, 1.

Rosenzweig, M.L. (2003) *Win-win Ecology: How earth species can survive in the midst of human enterprise*, Oxford University Press, New York, USA.

Saatchi, S., Buermann, W., Mori, S., ter Steege, H., and Smith, T. (2008) Modeling distribution of Amazonian tree species and diversity using remote sensing measurements. *Remote Sensing of the Environment*, **112**, 2000–2017.

Scott, J.M., Heglund, P.J., Morrison, M.L. *et al.* (eds.) (2002) *Predicting Species Occurrences: Issues of accuracy and scale*, Island Press, Washington.

Small, C. (2001) Estimation of urban vegetation abundance by spectral mixture analysis *International Journal of Remote Sensing*, **22**, 1305–1334.

Small C., Pozzi, F. and Elvidge, C.D. (2005) Spatial analysis of global urban extent from DMSP-OLS night lights. *Remote Sensing of Environment*, **96**, 277–291.

Sesnie, S.E., Gessler, P.E. Finegan, B. and Thessler, S. (2008) Integrating Landsat TM and SRTM-DEM derived variables with decision trees for habitat classification and change detection in complex neotropical environments. *Remote Sensing of Environment*, **112**, 2145–2159.

Soberón, J. and Peterson, A.T. (2009) Monitoring biodiversity loss with primary species-occurrence data: Toward national-level indicators for the 2010 target of the Convention on Biological Diversity. *Ambio*, **38**, 29–34.

Ståhl, G., Allard, A., and Esseen, P.-A. *et al.* (2010) National Inventory of Landscapes in Sweden (NILS) – Scope, design, and experiences from establishing a multi-scale biodiversity monitoring system. *Environmental Monitoring and Assessment*, doi: 10.1007/s10661-010-1406-7 Online First™.

Statistics Sweden (2008) Grönytor i tätort. Satellitdata som stöd vid kartering av grönytor i och omkring tätorter. DNR (Rymdstyrelsen): 180/06.

Taft, O.W., Haig, S.M., and Kiilsgaard, C. (2003) Use of radar remote sensing (RADARSAT) to map winter wetland habitat for shorebirds in an agricultural landscape. *Environmental Management*, **32**, 268–281.

Tomppo, E., Olsson, H., Ståhl, G., Nilsson, M., and Katila, M. (2008) Creation of forest data bases by combining National Forest Inventory Field Plots and Remote Sensing. *Remote Sensing of Environment*, **112**, 1192–1999.

Topping, C.J., Hansen, T.S., Jensen, T.S., Jepsen, J.U., Nikolajsen, F. and Odderskaer, P. (2003) ALMaSS, an agent-based model for animals in temperate European landscapes. *Ecological Modelling*, **167**, 65–82.

Townsend, P.A., Lookingbill, T.R., Kingdon, C.C. and Gardner, R.H. (2009) Spatial pattern analysis for monitoring protected areas. *Remote Sensing of Environment*, **113**, 1410–1420.

Tucker, C.J., Grant, D.M. and Dykstra, J.D. (2004) NASA's global orthorectified Landsat data set. *Photogrammetric Engineering and Remote Sensing*, **70**, 313–322.

Turner, W., Spector, S., Gardiner, N., Fladeland, M., Sterling, E. and Steininger, M. (2003) Remote sensing for biodiversity science and conservation. *Trends in Ecology and Evolution*, **18**, 306–314.

Turner, W.R. (2003) Citywide biological monitoring as a tool for ecology and conservation in urban landscapes: the case of the Tucson Bird Count. *Landscape and Urban Planning*, **65**, 149–166.

Tveit, M., Ode, Å. and Fry, G. (2006) Key concepts in a framework for analyzing visual landscape character. *Landscape Research*, **31**, 229–255.

Vickery, J. (2001) Consequences of EU expansion for farmland birds in eastern Europe. *Trends in Ecology and Evolution*, **16**(4), 176.

Wikelski, M., Kays, R.W., Kasdin, N.J., Thorup, K., Smith, J.A., and Swenson, G.W. (2007) Going wild: what a global small-animal tracking system could do for experimental biologists. *Journal of Experimental Biology*, **210**, 181–186.

Wretenberg, J., Lindstrom, A., Svensson, S., Thierfelder, T. and Pärt, T. (2006) Population trends of farmland birds in Sweden and England: similar trends but different patterns of agricultural intensification. *Journal of Applied Ecology*, **43**, 1110–1120.

Wu, C. and Murray, A.T. (2003). Estimating impervious surface distribution by spectral mixture analysis. *Remote Sensing of Environment*, **84**, 493–505.

Zetterberg, A., Mörtberg, U. and Balfors, B. (2010) Making graphy theory operational for landscape ecological assessments, planning, and design. *Landscape and Urban Planning*, **95**, 181–191.

21

Urban weather, climate and air quality modeling: increasing resolution and accuracy using improved urban morphology

Susanne Grossman-Clarke, William L. Stefanov and Joseph A. Zehnder

Expansion of cities to accommodate increasing population has global, regional and local effects on weather, climate and air quality, subsequently influencing human comfort and health. Atmospheric models are increasingly being employed to improve understanding of the impact of urban development on atmospheric processes. In this chapter we will give an overview of the recent development of physical approaches for the representation of urban areas in regional atmospheric models, input data, and urban characterization. These model developments were enabled by improvements in computer capacity and the accompanying increase in spatial resolution of atmospheric models. The demands on model input parameters characterizing urban land use and building morphology, along with the methods to derive them from remotely sensed data from various platforms will be discussed. We will review the findings of studies for several cities around the globe that have investigated the influence of urban land use and land cover changes on urban meteorology, climate and air quality. In a specific case study for the rapidly urbanizing Phoenix (Arizona, USA) area, we will demonstrate how remotely sensed data are used to study the effect of historic land use changes on near-surface air temperature during recent extreme heat events.

Urban Remote Sensing: Monitoring, Synthesis and Modeling in the Urban Environment, First Edition. Edited by Xiaojun Yang.
© 2011 John Wiley & Sons, Ltd. Published 2011 by John Wiley & Sons, Ltd.

21.1 Introduction

Expansion of cities to accommodate increasing population has global, regional and local effects on weather and climate due to land use and land cover (LULC) changes and accompanying effects on physical processes governing energy, momentum, and mass exchange between land surfaces and the atmosphere (Cotton and Pielke, 2007). Urbanization significantly impacts regional near-surface air temperatures, wind fields, the evolution of the planetary boundary layer, and precipitation, subsequently influencing air quality, human comfort, and health. Increased scientific interest in capturing the details of meteorological fields at the urban scale results from both, the need to understand and forecast the environmental conditions in cities where most humans live, and by the improved computing ability to resolve heterogeneity and physical characteristics of urban areas in regional meteorological and air quality modeling (Brown, 2000; Martilli, 2007).

The urban ecosystem is imbedded in, and responds to, climatic conditions that vary on a wide range of temporal scales. The earth's climate system is driven by solar forcing and global scale forcing such as anomalies in sea-surface temperatures associated with El Niño and the Pacific Decadal Oscillation. The local response to larger-scale meteorological forcing is determined in large part by land surface characteristics such as albedo, emissivity, thermal capacity, available moisture and surface roughness – all influencing the energy, mass, and momentum exchange between the earth surface and the atmosphere. In urban areas the energy balance characteristics differ as the building materials and morphology lead to an increased surface volumetric heat-storage capacity and thermal conductivity; short- and long-wave radiation trapping because of the vertical structure of buildings; and anthropogenic heat release.

The extent of many urban areas around the globe is large enough to affect mesoscale phenomena on the 2–200 km scale (Orlanski, 1975) such as thunderstorms, convection, complex terrain flows, sea and land breezes. One of the well-known effects of cities on the atmosphere is the urban heat island (UHI) effect, i.e. a warming of the near-surface atmosphere in cities in comparison to rural areas. Comprehensive overviews of the specifics of the urban energy balance and the effects of cities on weather, climate and air quality are given in Oke (1987), Arnfield (2003), Collier (2006), Masson (2006), Cotton and Pielke (2007) and Fisher et al. (2005).

Regional atmospheric models aim to capture mesoscale atmospheric phenomena. Very good overviews of the basics of mesoscale meteorological and air quality modeling are given in Pielke (2002), Seinfeld (2006) and Byun and Schere (2006), respectively. The term model, as used in the context of meteorology and air quality, refers to a complex computer code that numerically solves a set of differential equations that govern the evolution of the state of the atmosphere in space and time in terms of air temperature, pressure, specific humidity, chemical constituents and wind speed. The evolution is determined in part through the interaction between the model variables, but also through external forcing (e.g., solar radiation) and interactions with the earth's surface through fluxes of heat, moisture, momentum and emissions. Physical properties of the earth's surface that influence the exchange with the atmosphere depend on LULC characteristics. Therefore the accurate characterization of LULC and corresponding physical properties is important for atmospheric modeling.

Regional meteorological modeling is a prerequisite for air quality modeling since it predicts the transport, mixing and conditions for emissions, deposition and chemical reactions of trace gases and aerosols. Particularly biogenic emission and deposition modeling are directly influenced by LULC characterization and enhancements of the latter for urban areas improve quantification of those processes in air quality models. In current state of the art atmospheric models the chemistry, transport and mixing is either simulated simultaneously with the meteorology (as for example in the Weather Research and Forecasting (WRF) – chem. Model); or the meteorological fields are generated by a meteorological model independently first (for example WRF), then converted by a Meteorology-Chemistry Interface Processor as input for the air quality model (for example the Community Multi-scale Air Quality Modeling system CMAQ). Therefore improved representation of meteorological fields for urban areas potentially leads to a direct improvement of simulated trace gas and aerosol concentrations, emissions and deposition processes.

In Section 21.2 we will give an overview of the recent development of physical approaches for the representation of urban areas in regional atmospheric models along with necessary model input parameters. Section 21.3 addresses remote sensing platforms and methods that support the acquisition and derivation of urban model input parameters such as urban land use and associated physical characteristics. In Section 21.4 we will review the findings of studies for several cities around the globe that have investigated the influence of urbanization on urban meteorology and air quality. In a specific case study for the rapidly urbanizing Phoenix (Arizona, USA) area, we will demonstrate how remotely sensed data are used to study the effect of historic land use and land cover changes on near-surface air temperature during recent extreme heat events.

21.2 Physical approaches for the representation of urban areas in regional atmospheric models

In order to improve forecast model performance in urban areas for weather, climate and air quality applications several physical approaches of varying complexity were developed that describe the energy and matter exchange between urban surfaces and the atmosphere. During the past decade computer capacity has increased significantly, allowing spatial resolutions of regional atmospheric models as high as a few hundred meters. However, the airflow around individual buildings and roads cannot be spatially resolved. Therefore in analogy to vegetation canopies, the terms "urban parameterization" or "urban canopy model" (UCM) are widely used in the scientific community to mathematically describe the average effects of a configuration of buildings and streets on the atmosphere. "Urban canopy parameters" (UCPs) provide values for UCM input parameters (Ching et al. 2009). An excellent review of principal physical approaches and the efforts to include them in regional meteorological models is given by Brown (2000), while an update on more recent developments and challenges of urban regional modeling is discussed in Masson (2006) and Martilli (2007).

In order to resolve the air flow around individual buildings computational fluid dynamics (CFD) models such as Reynolds-averaged Navier–Stokes and Large-Eddy Simulation models with spatial resolutions of ∼1 m are applied to areas of interest within cities. Those models can resolve urban aerodynamic features such as street level flow. Increasingly output from regional atmospheric models is used to provide initial and boundary conditions to the CFD models while on the other hand CFD model output on can be aggregated and transferred back to the regional model (Chen et al., 2010).

The main physical approaches to represent urban areas in regional atmospheric models are: roughness approach, single-layer urban canopy approaches, and multi-layer urban canopy approaches.

21.2.1 Roughness approach

A relatively simple urban parameterization to account for the influence of urban areas on the atmosphere is the "roughness approach", in which the urban surface is treated physically like a soil surface, but with an adjusted increased roughness, heat capacity and conductivity and with modified albedo, emissivity and water availability for evapotranspiration. The roughness approach was successfully enhanced by Taha (1999) by means of including an empirical Objective Hysteresis Model by Grimmond, Cleugh and Oke (1991) thereby accounting for the increased heat storage in cities.

The advantage of the roughness approach is the relatively low demand on input parameters and simplicity of coupling the approach with regional atmospheric models. Usually, remote sensing-derived LULC classes with certain associated physical characteristics are used as model input. The roughness approach, however, does not resolve vertically the effects of buildings on the urban canopy air, which is important for some model applications, usually related to air quality and human comfort (Masson 2006).

21.2.2 Single-layer urban canopy approaches

The single-layer UCMs incorporate urban geometry into the surface-energy balance and wind-shear calculations by using the geometry of a generic street canyon that is characterized by road width, building width and building heights. The surface energy balance and a multilayer heat conductivity equation are solved for roof, wall, and road surfaces. Within-urban canopy profiles of air temperature, humidity and wind speed can be diagnosed from the predicted variables above roof level and empirical profile functions. The demand on input parameters is significantly larger than for the roughness approach and must include parameters defining the average urban geometry, together with physical characteristics of roof, road, and wall materials such as heat conductivity, heat capacity, albedo and emissivity. The application of single-layer UCMs with regional atmospheric models increases their computational needs and demands by coupling the urban surface parameters to the atmospheric models. Examples of single layer UCMs are the Town Energy Balance Scheme (TEB; Masson, 2000), the Noah Urban Canopy Model (Noah UCM; Kusaka and Kimura, 2004) and models by Mills (1997) and Oleson et al. (2008).

21.2.3 Multilayer urban canopy approaches

The multilayer UCM is currently the most complex among the urban approaches used in regional atmospheric modeling. Examples of multilayer UCMs include the Building Parameterization Scheme (BEP) by Martilli et al. (2002) and the scheme by Dupont, Otte and Ching (2004). In a multilayer UCM, exchanges with the atmosphere occur at multiple vertical levels within the urban canopy by directly modifying the prognostic differential equations of the regional atmospheric model to include additional terms such as urban drag force, heating, turbulent kinetic energy production and dissipation terms. At each level within the urban canopy the urban surface energy balance is solved for roofs and walls and, at the bottom of the model, for roads. Hence, within urban canopy profiles of air temperature, humidity and wind speed can be predicted by the regional model and therefore the environmental conditions were humans live. However, the models are difficult to couple with regional atmospheric models and significantly increase their computational time. The coupling of UCMs with regional atmospheric models requires the preprocessing of LULC and urban parameters; to enable land surface models to average calculated energy and momentum fluxes and surface temperatures from natural and urban fractions of cover of a model grid cell; and in case of multilayer UCMs the modification of the prognostic equations of the atmospheric model.

Single-layer UCMs make use of the geometry of a generic two-dimensional street canyon as an abstraction of the true heterogeneous urban geometry. The geometry is enhanced in the multilayer UCMs by considering subgrid scale features such as street canyons of several street directions and vertical changes of building density. However, for both approaches there are limitations particularly when investigating micrometeorological or neighborhood-scale characteristics of meteorological variables where site-specific details may become important. This applies for example when studying mitigation strategies for the urban heat island.

The demand on model input parameters in terms of urban geometry is also increased. Table 21.1 gives an overview of input parameters for the different urban model types (from Ching et al., 2009). Burian and Ching (2009) present a detailed derivation of UCPs for the example of Houston. In the next section we will discuss how those parameters can be derived from remotely sensed data.

21.3 Remotely sensed data as input for regional atmospheric models

Remotely sensed data is an important tool for obtaining spatially extensive, detailed and temporally repeatable data that records the physical character of the urban land surface that is a result

TABLE 21.1 Input parameters for urban physical representations of different degrees of complexity in regional atmospheric models (from Ching et al., 2009). For an explanation of parameters see Burian and Ching (2009).

Parameters	Roughness approach	Single-layer model	Multi-layer model
Albedo	x		
Emissivity	x		
Roughness length for momentum and heat above the urban canopy layer	x	x	
Zero-plane displacement height above the urban canopy layer	x	x	
Heat capacity and heat conductivity of urban surface	x		
Anthropogenic heat	x	x	x
Surface fractions of vegetation, water, soil, road, roof		x	x
Building height		x	
Albedo of roof, road and wall surfaces		x	x
Emissivity of roof, road and wall surfaces		x	x
Heat capacity of roof, wall and road surfaces		x	x
Heat conductivity of roofs, roads and walls		x	x
Roughness length of heat and momentum of roof, road and wall		x	x
Mean and standard deviation of building height		x	x
Plan area weighted mean building and vegetation height		x	
Building height histograms			x
Plan area fraction and frontal area index			x
Plan area density			x
Rooftop area density			x
Frontal area density			x
Complete aspect ratio		x	x
Building area ratio		x	x
Building height-to-width ratio		x	x
Distribution of street orientation and width			x
Drag coefficient of buildings		x	

of complex interactions between the land cover and land use (Jansen, 2009). Land cover is the observed surface material, whereas land use is a description of the human activity taking place at the surface; the former is detectable by remote sensing systems, but the latter typically requires additional information from other sources to make a correct determination. Throughout the text, we use "LULC" in cases where combined land use and land cover classifications are discussed; cases where land use or land cover alone are used are explicitly noted. Appropriate urban LULC classifications (i.e., using land cover, land use, or a combination of both) and building characteristics derived from remotely sensed data for atmospheric modeling depend on the model formulation and parameterization; spatial grid resolution; any simplifying assumptions, and the urban physical approach provided with the regional model; and application in terms of processes of interest. We focus our discussion on optical satellite-based sensor systems, as these provide the most extensive historical and spatial coverage of urban areas.

21.3.1 Urban land use and land cover data

For many applications LULC and biophysical measurements derived from remotely sensed data – such as vegetation cover, albedo, emissivity – are appropriate inputs for atmospheric models (see Stefanov and Brazel, 2007 for a recent overview of relevant orbital and airborne sensors and their data specifications; also see chapters in Part II of this volume). However, because of the relatively recent emphasis on detailed urban atmospheric modeling, there are efforts going on to update the spatial extent and consider heterogeneity of urbanizing regions in the LULC data sets provided with the standard release versions of atmospheric models.

For example, in order to adjust to the increasing resolution of atmospheric models Masson et al. (2003) derived a new global database for land surface parameters including LULC at the 1 km scale and associated parameters of 215 ecosystems derived by combining existing LULC, climate maps and Advanced Very High Resolution Radiometer (AVHRR) satellite data. For the European continent the cover types of the Coordination of Information of the Environment (CORINE) land use classification are used (Heymann et al., 1994), including 11 urban types (continuous urban fabric, discontinuous urban fabric, industrial or commercial units, road and rail networks and associated land, port areas, airports, mineral extraction sites, dump sites, construction sites, green urban areas, sport and leisure facilities). The CORINE data were derived from medium-resolution satellite images such as Landsat Thematic Mapper (TM) and Système Probatoire d'Observation de la Terre multispectral (Spot XS) scanner images and complementary information from ancillary data sets. The data base is currently available with the regional atmospheric models used by the French and German weather services, MESO-NH and COSMO respectively.

Lemonsu *et al.* (2006) present a general methodology of urban classification for typical North-American cities that is currently applied to update LULC information in the regional forecast model of the Meteorological Service of Canada. The methodology is based on the joint processing of medium-resolution Advanced Spaceborne Thermal Emission and Reflection Radiometer (ASTER) and Landsat-7 Enhanced Thematic Mapper Plus (ETM+) imagery, digital elevation models (DEMs) derived from the Shuttle Radar Topography Mission (SRTM-DEM) and national elevation DEMs for both the USA and Canada. A decision tree model is applied to identify urban LULC classes. First a 15 m resolution unsupervised classification of preprocessed remotely sensed data (ASTER or ETM+) is performed to obtain 11 generalized urban/non-urban classes. Building heights at 15 m spatial resolution are determined based on the difference of SRTM-DEM and the national DEMs, and used as additional criteria to aggregate the 15 m classifications to a 60 m resolution fraction of natural surface and urban cover. The decision-tree model is then applied to derive twelve urban LULC classes: high buildings, mid-high buildings, low buildings, very low buildings, sparse buildings, industrial areas, roads and parking, road mix, dense residential, medium-density residential, low-density residential, mix of natural and built. This methodology was refined by Leroux *et al.* (2009) to create a fully automated geospatial database processing approach for generation of urban LULC. Designed for use with Canadian cities the LULC output can be used directly in mesoscale atmospheric models. The automated system incorporates vector-based land use and land cover data, SRTM-DEMs, Canadian Digital Elevation Data DEMs, and census data to derive urban LULC classifications. Such an automated system may be applicable in other countries where similar data are available.

For the American community Weather Research and Forecasting (WRF) model (Skamarock *et al.*, 2005) Moderate Resolution Imaging Spectroradiometer, or MODIS, 1-km global data were made available recently by Boston University (Friedl *et al.*, 2002). Those data are classified according to a 20-class LULC classification developed by the International Global Biosphere Program (Lambin and Geist, 2006). Since the classification is based on 2001 images the representation of urban areas in terms of extent has been improved. However, only one urban land use class is considered ("urban built-up"). For WRF applications to urban areas users are currently still encouraged to provide their own updated urban LULC dataset.

The current generation of very high (1–4 m/pixel) to ultra high (<1 m/pixel) spatial resolution sensors on board Earth-observing satellites (e.g., IKONOS, Quickbird, Orbview) has enabled the development of sophisticated approaches to classification of urban LULC and measurement of biophysical variables (also see chapters in Part III of this volume). Very high to ultra high resolution multispectral data is now available for numerous urban centers around the world. These data provide spatial resolution similar to or better than traditional aerial photography, but with increased spectral information and geometric accuracy. Urban/suburban land cover classes such as residential lawns, driveways and parking lots, and even individual tree canopies can be fully resolved in these data. This allows avoidance of sub-pixel spectral mixing of built and natural surface materials that is a problem with coarser resolution data such as that obtained from the Landsat sensors or MODIS (Woodcock and Strahler, 1987; Small, 2007). Current atmospheric and weather prediction models operate at grid scales two to three orders of magnitude larger than the scale of land cover classifications produced from very high to ultra high resolution data, but as computational power continues to increase it may soon become possible to utilize such fine-grained land cover and biophysical information (such as vegetation indices) directly. At present, however, fine spatial resolution land cover information can be aggregated to larger grid scales as model inputs.

The majority of urban and suburban LULC classification techniques using remotely sensed data have been developed for use with medium to high resolution (50–4 m/pixel; Ehlers, 2004) multispectral and hyperspectral data collected by satellite-based and airborne sensors. Classification approaches include unsupervised/supervised clustering algorithms and fuzzy membership classifiers (Jensen, 1996; Zhang and Foody, 1998); techniques that integrate both remotely sensed and ancillary georeferenced data such as expert systems (Gong and Howarth, 1990; Stefanov, Ramsey and Christensen, 2001; Wentz *et al.*, 2008) and neural networks (Civco, 1993; Foody, McCulloch and Yates, 1995; Weng and Hu, 2008); decision trees (Mahesh and Mather, 2003); image spectral analysis (Lu and Weng, 2006; Rashed *et al.*, 2003 Ridd, 1995, Small, 2005; Wu and Murray, 2003), and image spatial analysis (Bian, 2003; Myint, Lam and Tyler, 2004). Many of the above-listed techniques, particularly spectral analysis, rely on multiple bands of information spanning the visible through midinfrared (or thermal infrared) portions of the electromagnetic spectrum; this condition is satisfied by most of the current generation of high to low resolution orbital and airborne sensors.

Direct application of these spectrally-based techniques to very high and ultra high resolution orbital and airborne data has been problematic due to the relatively low spectral information content of these systems, which typically consists of visible red, green, blue (and/or a panchromatic band in these wavelengths), and a near infrared band. Some of these spectral limitations can be overcome by adding other information such as light detection and ranging (lidar) data (e.g., Tooke *et al.*, 2009), or extracting spatial information from the fine resolution data such as texture (e.g., Su *et al.*, 2008) to provide additional information for LULC classification. Newer satellite-based systems such as DigitalGlobe's WorldView-2 (launched in 2009) and the RapidEye constellation of satellites (launched in 2008) exceed the multispectral capabilities of previous very high to ultra high resolution sensors by adding addition bands in the visible and near-infrared wavelengths (Ehlers, 2009). The site revisit times of these new sensor systems are also reduced to 1–4 days by off-nadir pointing capability (WorldView 2) and multiple satellite platforms (RapidEye). An extensive review of existing and planned Earth observing satellite systems is presented in Petrie and Stoney (2009).

A relatively recent classification approach known as object-based image analysis (OBIA) takes advantage of the high spatial information content of very high to ultra high resolution data, and is particularly well-suited for land use and land cover classification of urban and suburban areas (Banzhaf and Netzband, 2004; Blaschke and Hay, 2001; Blaschke, Burnett and Pekkarinen, 2006; Hofmann *et al.*, 2008; also see Chapter 7, this volume). OBIA uses image segmentation to identify homogeneous image objects at several different scales, rather than the classical pixel-based (and single-scale) classification approaches. A particular pixel included in an image segment might have simultaneous membership in several hierarchical levels depending on the spatial scale of segmentation (e.g., a Vegetation superclass, a Canopied Vegetation class, and a Deciduous Tree subclass), with membership in each determined by fuzzy classification algorithms using spectral character, shape, and neighborhood relationships across the hierarchical levels as inputs. Built materials can be identified in similar fashion (e.g., image objects with high length to width ratios, low reflectance in the visible to near infrared wavelengths), and location in or near an "urban" superclass image object could be identified as asphalt roadways. The OBIA approach has also been used for LULC classification of medium resolution (15 m/pixel) ASTER data of Phoenix, AZ and Las Vegas, NV (Schöpfer and Moeller, 2006), suggesting that it may be useful for semiautomated classification using historical Landsat datasets.

The High Ecological Resolution Classification for Urban Landscapes and Environmental Systems (HERCULES) classification scheme of Cadenasso, Pickett and Schwarz, (2007) uses OBIA of very high resolution aerial photography and lidar data to classify six urban land cover types: coarse-textured vegetation, fine-textured vegetation; bare soil, pavement, buildings, and building typology (spatial arrangement, general structural type, and general height classes). The classification scheme is informed by the vegetation-impervious – soil, or VIS classification model of Ridd (1995). While the HERCULES classification scheme is specifically designed for urban ecological studies, the output can easily be converted into building height and vegetation percentage format for input into climate models.

21.3.2 Building data

As mentioned in Section 21.2 for the application of the single-layer and multilayer UCMs, building-related model input parameters are necessary. Three-dimensional building data sets and derived building statistics have become readily available for many cities around the world (Ratti, Sabatino and Britter, 2002, Brown et al., 2002, Burian et al., 2006, 2008, Martilli, 2009). Print stereographic aerial photography and manual photogrammetric techniques have been used extensively in urban areas to derive such parameters as building heights and dimensions for planning purposes; an extensive review of the subject is outside the scope of this chapter, but Jensen (2000) provides a good introduction to the subject. More recently, photogrammetric analysis of aerial orthophotographs derived from digital camera data has become the state of the art. The availability of photogrammetry software, global positioning system (GPS) and ground survey data, and increased desktop computing capabilities has facilitated the use of digital aerial photography for creation of both digital terrain models (DTMs) and digital elevation models (DEMs) that allow for the extraction of building locations and dimensions at very high spatial resolution.

Data acquired from both synthetic aperture radar (SAR) and interferometric synthetic aperture radar (InSAR) satellite and airborne systems has been used to map urban building form, distribution, density, and in some cases to extract building heights by measuring reflected energy from the land surface and roughness elements such as buildings, trees, etc. (Brenner and Roessing, 2008; Dell'Acqua and Gamba, 2006; Gamba, Houshmand and Saccani, 2000; Gens and Van Genderen, 1996; also see Chapter 5, this volume). Both SAR and InSAR systems are active in the sense that data take geometry and power/wavelength output can be varied. This is in contrast to passive optical remote sensing systems that typically have a stable viewing and illumination geometry that relies on the Sun for energy input to the land surface. The spatial resolution of orbital SAR sensors such as ENVISAT, the European Remote Sensing (ERS) satellite, and TerraSAR-X – 1 to more than 10 m per pixel – has limited their use for urban building form and height extraction as unambiguous interpretation of the data is difficult. Airborne systems such as TopoSAR and the Phased Array Multifunctional Imaging Radar (PAMIR) can provide subdecimeter resolutions that greatly facilitate mapping of urban building form and height, as well as provide information on the composition of the urban materials themselves (Brenner and Roessing, 2008; Dell'Acqua, 2009).

Three-dimensional building data sets are increasingly derived from airborne lidar systems. Lidar provides DEMs and DTMs, which allow the derivation of building size, shape, orientation, relative location to other buildings and other urban morphological features (trees, highway overpasses, etc.; also see Chapter 6, this volume). The finest resolution is typically on the order of 1–5 m. Lidar provides a high-resolution representation of urban morphological features for entire metropolitan areas, with a minimal set of airplane flyovers. However, lidar is costly and presents a data management challenge given the massive size of datasets (Ching et al., 2009). Using building data in conjunction with a geographic information system (GIS) Brown et al. (2002) developed scripts for automating the calculations of UCM building-related input parameters (see Table 21.1). Those data can then be correlated with LULC data sets, or building attributes can be accepted by the atmospheric model in a spatially explicit manner, i.e. they do not need to be associated with a particular land use or land cover class.

Based on the need for advanced treatments of high resolution urban morphological features in meteorological and air quality modeling systems, the "Community"-based National Urban Database and Access Portal Tool (NUDAPT) was developed (Ching et al., 2009). NUDAPT is currently sponsored by the US Environmental Protection Agency (USEPA) but involves collaborations and contributions from federal and state agencies, and from national and international private and academic institutions. NUDAPT contains archived copies of lidar DEM and DTM data currently being acquired by the National Geospatial Agency (NGA; formerly the National Imagery and Mapping Agency). When completed, NGA will have obtained data from as many as 133 urban areas. NUDAPT provides a web-based portal technology which facilitates the customization of data handling and retrievals to generate gridded fields of urban canopy parameters for UCMs. Data are currently available for Houston (Texas, USA) and New York (New York, USA).

Techniques to build georeferenced three-dimensional building models using aerial and ground-based imagery, together with traditional ground maps, have been developed (Gatzidis, Liarokapis and Brujic-Okretic, 2007). A typical workflow is to use ground mapping data to accurately position building footprints, followed by extrusion to correct building height using ancillary vector data. Lastly, ground or aerial imagery can be used to construct a photorealistic representation of a given building. Similar tools are now publicly available through Web-based geospatial browsers, e.g. Google Building Maker (http://sketchup.google.com/3dwh/buildingmaker.html) and provide an alternative means of generating three-dimensional urban models. The three-dimensional building representations must be created on a building-by-building basis; for modeling of an entire metropolitan area it may still be more effective to use remotely sensed data as discussed above.

21.4 Case studies investigating the effects of urbanization on weather, climate and air quality

21.4.1 Studies investigating effects of urban land use and land cover on meteorology and air quality

The application of detailed remote sensing derived urban LULC data in conjunction with UCMs and associated input parameters in meteorological and air quality modeling studies for cities around the world has strongly increased in recent years. Several studies investigated the effects of urbanization and accompanying LULC changes on various meteorological and air quality phenomena such as precipitation, local circulations, the UHI, and near-surface ozone formation. Other studies underline the potential use of regional atmospheric modeling for urban air quality regulatory and planning purposes. A very recent application focuses on high-resolution downscaling of global climate model output in order to investigate interactions of global climate change and urbanization on meteorology and air quality. Here we give an overview of some of the studies and their findings.

A series of studies showed an improvement of the quality of the simulations for urban areas in terms of capturing meteorological fields and phenomena when applying single- or multi-layer UCMs in conjunction with detailed high-resolution urban morphological and LULC data instead of using the simpler roughness approach. For example, Lemonsu and Masson (2002) simulated the UHI circulation over Paris using the MESO-NH model with the single-layer UCM TEB. Zhang *et al.* (2008) coupled the urban canopy model of Kusaka and Kimura (2004) into Regional Atmospheric Modeling System (RAMS) and improved the representation of the UHI in Chongqing (China). Chen, Tewari and Ching, (2007), Kusaka *et al.* (2005) and Miao *et al.* (2008) used the single-layer Noah-UCM (Kusaka and Kimura, 2004) with WRF and found that the simulation of UHI and PBL characteristics was improved in Houston (USA), Tokyo (Japan) and Beijing (China), respectively. The multilayer UCM by Dupont, Otte and Ching (2004) was incorporated in the Mesoscale Meteorological Model MM5 (Otte *et al.* 2004) and its performance evaluated for Philadelphia (USA). The study showed that the simulated results including the wind velocity, friction velocity, turbulent kinetic energy, and potential temperature vertical profiles were more consistent with the observations than that using the roughness approach.

By applying state-of-the-art urban mesoscale meteorological and air quality modeling as well as UCPs Taha (2008b, 2008c) demonstrated the potential use of regional atmospheric modeling for urban planning (regarding UHI mitigation) and air quality regulatory purposes. In this study MM5 was used in conjunction with an enhanced version of the multi-layer UCM by Dupont, Otte and Ching (2004) for Sacramento (California, USA) and the potential for mitigating the UHI in response to increased albedo of roof, walls and roads (finest model resolution 1 km) and urban vegetation cover was investigated. It was shown that air temperature could potentially decrease by up to 3°C with a subsequent reduction in ozone concentration in most areas of the city. The study also showed that the application of the multilayer UCM and consideration of effects of vertical changes in building density on the urban planetary boundary layer enabled the atmospheric model to capture vertical variations in turbulent kinetic energy budget components.

Regional meteorological models are an important research tool to investigate the influence of urbanization on convection (Shepherd, 2005). For example, using the Noah-UCM with WRF Lin *et al* (2008) analyzed the impact of urbanization on precipitation over Taipei (Taiwan). Rozoff, Cotton and Adegoke (2003) used TEB coupled with RAMS to simulate the urban atmosphere and its role in deep, moist convection over St. Louis (Missouri, USA) for summer precipitation. Sensitivity experiments show that the UHI was the most important factor in initiating deep, moist convection downwind of the city and that there is a large sensitivity of simulated urban-enhanced convection to the details of the urban surface model. Shem and Shepherd (2009) applied WRF with an urbanized roughness approach by Liu *et al.* (2006) in order to investigate for a case study the impact of urbanization on summertime thunderstorms in Atlanta. The results show that WRF with the urban enhancements captured timing and amount of convective rainfall reasonably well and that urban characteristics in the model affected rainfall.

Several studies investigated the effects of urbanization and accompanying LULC changes on air quality near-surface ozone formation. For example, the Houston metropolitan region was the focus of two air quality experiments (2000 and 2006 Texas Air Quality Studies I and II) and modeling efforts due to the fact of frequent exeedances of the 8-hour US National Ambient Air Quality Standard for ozone and the need to better understand the contribution of regional transport and local sources to the exceedances (Kemball-Cook *et al.*, 2009). Cheng and Byun (2008) and Cheng, Kim and Byun (2008) investigated how the application of high resolution LULC data for coupled WRF and CMAQ modeling in the Houston-Galveston area affected simulated meteorology and air quality. In those model simulations the Texas Forest Service LULC dataset established with Landsat satellite imagery correctly represented the Houston-Galveston-Brazoria area as mixtures of urban, residential, grass, and forest LULC types thereby improving emissions, air quality and meteorological modeling. Houston 2001 LULC and lidar derived building data (650 000 buildings) were incorporated in the NUDAPT data base (Ching *et al.*, 2009) together with sets of gridded daughter products (UCM model parameters), anthropogenic heat fluxes, and day-night population data. Regional atmospheric modeling was carried out for the TEXAS 2000 intensive field study using the detailed LULC and building morphological characteristics. This improved the accuracy of the simulation of the Galveston Bay bay-land breeze flow reversal in the Houston area and therefore of the meteorological variables (Chen, Tewari and Ching, 2007, Taha 2008a) and ozone concentrations.

Wang *et al.* (2009) investigated the impacts of urban expansion from pre-urbanization to current LULC on surface ozone for the Pearl and Yangtze River Delta regions (China). Results show that urbanization increases day- and night-time 2 m air temperatures, leads to a reduction in the wind speed and higher PBL. These modified meteorological conditions lead to detectable ozone-concentration increases of 2.9–4.2% and 4.7–8.5% for day- and night-time surface-ozone concentrations.

Civerolo et al. (2007) quantified the effects of increased urbanization on surface meteorology and ozone concentrations in the New York City metropolitan region. A land use change model was used to extrapolate urban LULC over this region from "present-day" (ca. 1990) conditions to a future year (ca. 2050), and these projections were subsequently integrated into MM5 and CMAQ simulations. The results suggest that extensive urban growth in this metropolitan area has the potential to increase afternoon near-surface temperatures by more than 0.6°C, ozone by about 1–5 ppb, and episode-maximum 8-hour ozone levels by more than 6 ppb across much of the New York City area.

The focus of this study is on regional atmospheric modeling. There are also efforts going on to better represent urban areas in climate models. For example Oleson et al. (2008) included a single layer UCM in the land model for the Community Climate System Model (CCSM). However to fully make use of the recent development of improvements in regional atmospheric modeling high-resolution dynamical downscaling to urban areas is necessary that can resolve heterogeneity of urban LULC. A good example of such a study is Jiang et al. (2008) who applied WRF-chem with the single-layer UCM to investigate the impacts of climate and LULC changes on surface ozone in the Houston area for August of current (2001–2003) and future (2051–2053) years. The model was forced by downscaled 6-hourly Community Climate System Model (CCSM) version 3 outputs. High-resolution current year land use data from National Land Cover Database (NLCDF) and future year land use distribution based on projected population density for the Houston area were used. The simulations show in the urban area, the effect of climate change alone accounts for an increase of 2.6 ppb in daily maximum 8-h O3 concentrations, and a 62% increase of urban land use area exerts more influence than does climate change. The combined effect of the two factors on O3 concentrations can be up to 6.2 ppb.

21.4.2 Case study for Phoenix

The impact of 1973–2005 LULC changes on near-surface air temperatures during four recent summer extreme heat events (EHEs) are investigated for the arid Phoenix metropolitan area using WRF.

During three decades of rapid urbanization the population of the Phoenix metropolitan region increased by ~45% per decade from about 971 000 to nearly 4 million currently. It is expected that the population will grow to ~10 million by 2050 (Maricopa Association of Governments, 2007). Understanding the effects of urbanization on near-surface air temperature is particularly important in this arid region in order to enhance the adaptive capacity of a city that regularly experiences high average daily summertime temperatures and extended periods without rainfall. Multiday EHEs strongly influence human comfort and health. Between the years 1993–2002 Arizona led the United States in deaths from heat exposure (CDC, 2005). EHEs were identified based on criteria developed by Huth, Kysely and Pokorna (2000) and Meehl and Tebaldi (2004) using the distribution of maximum recorded summer time air temperatures between 1961 and

TABLE 21.2 EHE and the highest recorded maximum and minimum daily temperatures during each period for Phoenix Sky Harbor station (1961–2008) based on Huth et al. (2000) criteria.

EHE	Highest observed daily temperature (°C)	
	Maximum	Minimum
25–28 June 1979	47.2	26.7
07–09 June 1985	46.1	27.8
21–23 June 1988	46.7	31.7
03–05 July 1989	47.8	30.6
25–28 June 1990	50.0	33.9
26–29 July 1995	49.5	31.7
12–16 July 2003	46.7	35.6
12–17 July 2005	46.7	33.9
21–24 July 2006	47.8	35.0
03–06 July 2007	46.7	33.9

1991. Table 21.2 lists EHEs identified for the Phoenix metropolitan region between 1961 and 2008 plus the highest recorded maximum and minimum temperatures. WRF simulations were carried out for each EHE using LULC classification data for the years 1973, 1985, 1998 and 2005. The scientific question asked was whether urban development caused an intensification and expansion of the area characterized by extreme temperatures. Landsat Multispectral Scanner (MSS), Landsat Thematic Mapper (TM) and Landsat ETM+ derived LULC data for 1973 (Moeller, 2005), 1985, 1998 (Stefanov, 2000) and 2005 (Buyantuyev, 2005) are used to provide the basis for model parameter values. Landsat MSS, TM, and ETM+ – based 12-category LULC data (rescaled to 30 m pixels) for 1973, 1985, 1998 and 2005 were analyzed using the procedure of Stefanov, Ramsey and Christensen (2001). The data were incorporated by recoding classes to conform to the 33-category 30 s global USGS Land Use/Land Cover System currently used in WRF. The classification is part of the USGS "North America Land Cover Characteristics Data Base Version 1.2" (USGS, 2008) and contains 24 of the 37 categories of the Anderson et al. (1976) Level II classification (U.S. Geological Survey, 2008). Additional categories in WRF's global 30-s classification are: "Playa" (25), "Lava" (26), "White Sand" (27), "Unassigned" (28–30), "Low intensity Residential" (31), "High intensity Residential" (32) and "Industrial or Commercial" (33).

WRF in conjunction with a single-layer UCM (Kusaka and Kimura, 2004) as applied in this study includes three urban land use classes characterized as commercial, high-intensity and low-intensity residential. Here, these have been adjusted: commercial/industrial, urban mesic residential and urban xeric residential, which are distinguished by the type of vegetation and irrigation (no vegetation, well-watered flood or overhead sprinkler irrigated, and drought-adapted vegetation with drip irrigation, respectively).

For each LULC class we adjusted the urban fraction according to detailed vegetative cover measurements that were obtained from an extensive field survey carried out in 30 × 30 m field plots at 200 randomly selected sites across the entire urban area (Grossman-Clarke et al., 2005). Development across the Phoenix area is largely suburban in nature, with an urban core of very limited spatial extent, unlike many older cities in more temperate regions. The fraction of the surface cover types in the residential urban categories is remarkably uniform across the entire metropolitan area. The fractional cover of built, vegetation and soil surfaces comprises 0.73, 0.10 and 0.17 in the xeric residential and 0.60, 0.23 and 0.17 in the mesic residential categories, respectively. The building morphology is very similar in the residential classes with average building heights, roof and road widths of 6 m, 10 m and 15 m.

The commercial/industrial land use category is comprised almost entirely of anthropogenic surfaces (fractional cover of built surfaces 0.95). The average building height of commercial and industrial buildings is 10 m. The large variation in road width (10 m to 100 m) and building area complicates the assignment of building geometry. Here for the commercial/industrial class the building width (10 m) and street width (10 m) of the commercial land use class provided with WRF's urban parameter table (version 3.0.1.1) were applied. This highlights the observation that three urban land use classes as available with the standard WRF might be too few to represent the heterogeneity of urban form in high resolution WRF applications to urban areas. Similarly, the WRF standard values for heat capacity, conductivity, albedo, emissivity roughness length for heat and momentum of roof, road and wall surfaces were used.

Figure 21.1 shows the 1973, 1985, 1998 and 2005 LULC classification data as processed for the WRF model simulations. The predominant land cover category in rural areas is shrubland which represents desert vegetation in the USGS classification.

WRF simulations were conducted for each EHE (12–16 July 2003, 12–17 July 2005, 21–24 July 2006 and 3–7 July 2007) with four sets of LULC: 1973, 1985, 1998 and 2005 (16 runs in total) with a spatial resolution of 2 km. Comparison of measured 2 m air temperatures T_{2m} and 10 m wind speeds for 18 surface stations located in rural and urban parts of the region show a good agreement between observed and simulated data for all simulation periods (Grossman-Clarke et al., 2010).

Here the focus is on the regional scale effects of LULC changes. Hence local scale effects on a sub-neighborhood scale might not be captured by WRF (highest resolution in this study: 2 km). Also, a dominant land use class is assigned to each model grid cell with its fixed vegetation fraction and morphological characteristics. It would potentially enhance the quality of the simulations in some areas of the urban region if spatially explicit urban vegetation cover and building characteristics are used for each grid cell rather than urban land use classes. As discussed previously there are limitations to the single-layer UCM because of the assumed urban geometry of a street canyon. The single-layer UCM was chosen for the simulations over the available multilayer UCM option since development across the Phoenix area is largely suburban in nature and one-story single-family homes are the predominant building type. Therefore the vertical extent of the city is generally low and the benefits of a multilayer UCM such as the direct interaction with the planetary boundary layer are relatively small. Differences in the simulated maximum and minimum daily 2 m air temperatures, T_{2m}, between the land use scenarios were used as a measure of LULC classification effects on regional near-surface air temperature. To illustrate the effects of land use changes on minimum and maximum daily temperatures over time, the effects were averaged over the 18 simulated days that constituted the four EHEs. Difference maps allow comparison between the regional T_{2m} simulated with the 2005 LULC classification to the T_{2m} obtained with the 1973, 1985 and 1998 LULC classifications, respectively at 0500 Local Standard Time (LST) and 1700 LST 2005 (Fig. 21.2). Based on the simulations, new urban development caused an intensification and expansion of the area experiencing extreme temperatures. As expected, 1998 with the shortest time for LULC changes, show the smallest but still detectable differences.

For the minimum temperatures (0500 LST) Fig. 21.2 (upper) shows that with urban development in formerly agricultural as well as desert areas the temperatures increased on average by 5–8°C, leading to an expansion of the area experiencing elevated night-time temperatures. The largest changes in minimum night-time temperatures occur in the center of the formerly agricultural areas to the southeast of Phoenix and on some nights approach 10°C. Smaller changes in minimum T_{2m} of 4–6°C are detected when desert land underwent the transformation to urban land use. Night-time temperatures in the existing urban core show only relatively small changes of up to 1–2°C with the ongoing LULC changes. Comparing changes in minimum temperatures between land use scenarios showed that strong changes in T_{2m} were relatively localized where new urban development took place. The expansion of the built-up area also increased minimum temperatures in the previously developed fringe region by 1–3°C, as in the center of the metropolitan region. For the urban land use categories the night-time net radiation amounts to up to ~−200 W m^{-2} leading to higher surface temperatures and a warming of the atmosphere in the city relative to the rural surroundings. With maximum night-time values of ~150 W m^{-2} the ground heat flux for the urban land use types is significantly higher than for the rural land use types. The ground heat flux for the urban land use categories also accounts for the heat storage in roofs and walls. Positive night-time sensible heat fluxes (QH) are calculated for the urban categories while they are negative for the rural ones. Heat fluxes are positive when directed away from the surface into the atmosphere and negative when directed towards or into the surface as for the ground heat flux.

Daytime temperatures (1700 LST) are little affected when urban development replaces desert (Fig. 21.2 lower). In the urban center the increase over time in T_{2m} amounted to about 0.5°C. As the main conversion of agricultural to urban land use concluded in the mid 1990s the changes in average maximum T_{2m} between 1998 and 2005 were relatively small. However, when irrigated agricultural land was converted to suburban development, maximum T_{2m} increased on average by 2–4°C. The increase in sensible heat fluxes with urban development led to the daytime warming when agricultural land was replaced with urban development. Maximum daytime latent heat fluxes (QE) for irrigated agriculture land are of the order of 500 W m^{-2} compared to about 20 W m^{-2} and 75 W m^{-2} for commercial/industrial and xeric residential areas, respectively. Due to the low soil moisture content in the natural desert QE is simulated to be ~20 W m^{-2}. The magnitude of maximum daily QH was ~100 W m^{-2} for irrigated agricultural, ~450 W m^{-2} for desert and ~400 W m^{-2} for commercial/industrial areas. The increase in sensible heat fluxes with urban development led to the daytime warming when agricultural land was replaced with urban development. Because of the large daytime heat storage flux for the commercial/industrial land use category during the hours before noon of up to ~−450 W m^{-2} the simulated QH for the desert areas is larger than for the urban area. However net radiation is significantly larger for urban vs. desert LULC classes due to the lower albedo and the effects of the sky view factor of the urban classes.

The influence of LULC changes on 10 m wind speed during the EHE episodes is of interest as increased urban roughness leads potentially to an increase in vertical momentum fluxes thereby

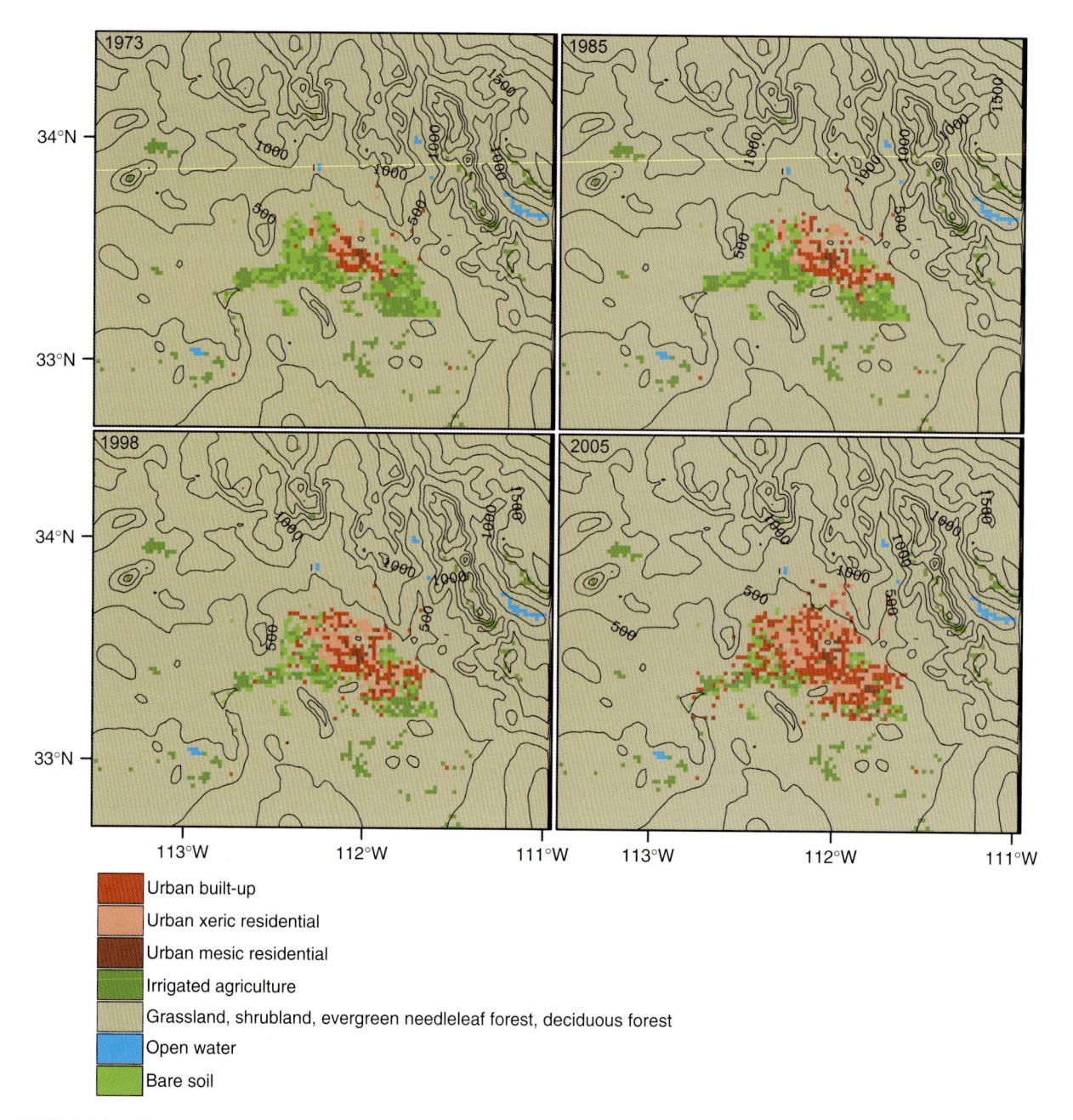

FIGURE 21.1 Topography contours (from 0 to 3000 m interval 250 m) and land use/land cover for 1973, 1985, 1998, 2005 produced by WRF's preprocessing system for the inner modeling domain (2 × 2 km grid resolution) based on Landsat MSS, TM, and ETM+ image-derived land use/cover data and three urban land use/cover classes. In order to emphasize the urban land use changes the colors are grouped together for the rural land use classes (grassland, shrubland, deciduous broadleaf forest, evergreen needle leaf forest). The extent of the area is ~ 240 × 198 km.

reducing mean flow (Brown, 2000). Urban development has led to a reduction of the already relatively weak night-time winds by 1 to 2 m s^{-1} and therefore a reduction of advection of cooler air into the city.

Based on the simulations, we conclude that land use and land cover characteristics that evolved over the past 35 years in the Phoenix metropolitan region have had a significant impact on near-surface air temperatures occurring during EHEs in the area. Future intensification and temporal expansion of EHE episodes is also possible due to global warming. Although the interactions of urbanization and global climate change and their effect on the severity of summer EHEs in the region were not investigated here, it can be assumed that the region will be highly sensitive to any additional warming.

CHAPTER 21 URBAN WEATHER, CLIMATE AND AIR QUALITY MODELING 315

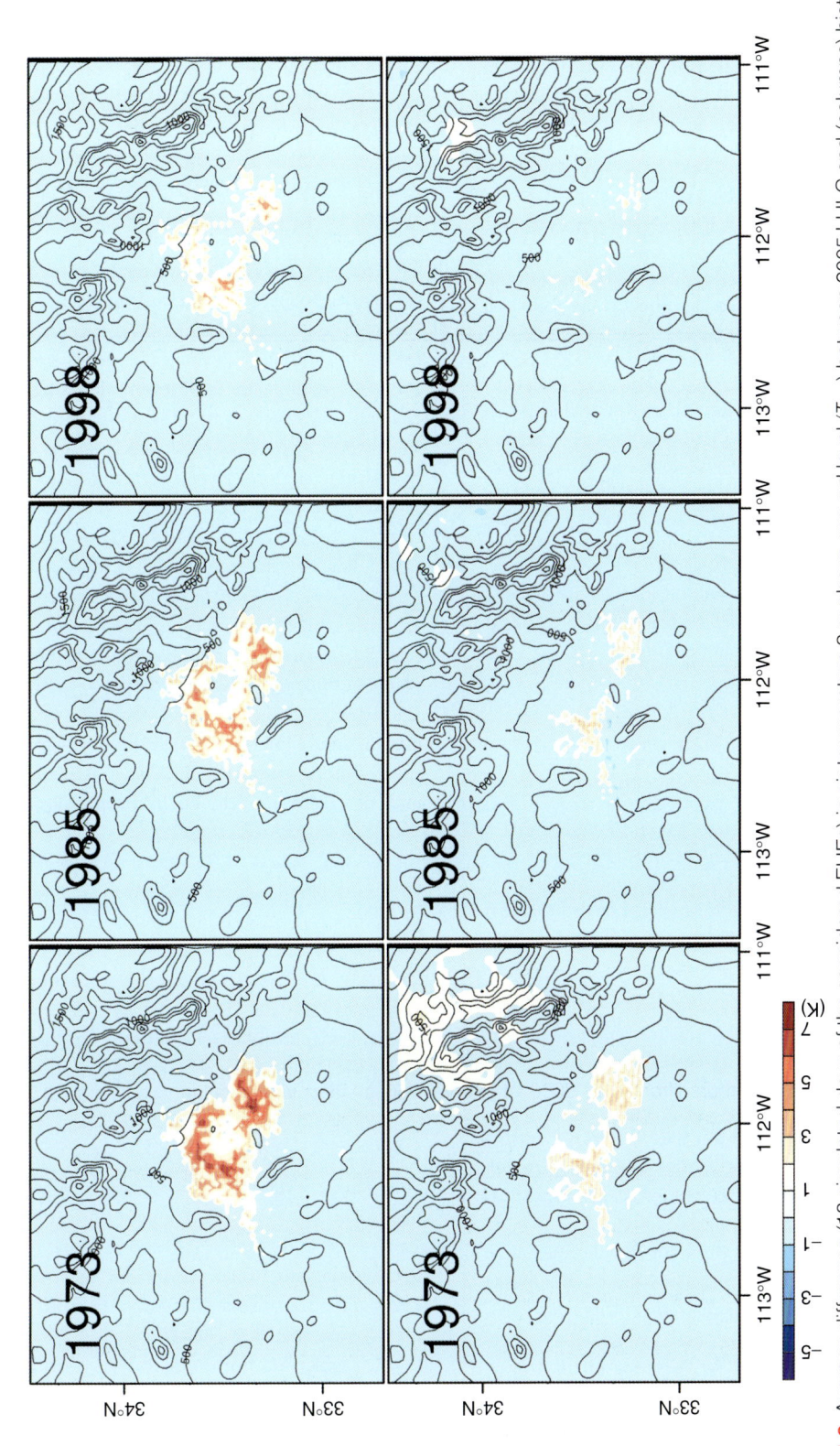

FIGURE 21.2 Average difference (18 simulated days of the considered EHEs) in air temperature 2 m above ground level (T_{2m}) between 2005 LULC and (columns) historic LULC data (1973, 1985, 1998) for (top row) 0500 LST (~ daily minimum temperatures) (bottom) 1700 LST (~ daily maximum temperatures). Also included are topography contours (from 0 to 3000 m interval 250 m).

Conclusions

Here we gave an overview of the recent development of physical approaches for the representation of urban areas in regional atmospheric models, input data, and urban characterization along with the methods to derive them from remotely sensed data from various platforms. The review of meteorological, climate and air quality modeling studies for cities, in which those recent model developments were applied, showed an improvement of the quality of the simulations in terms of capturing meteorological fields and phenomena and subsequently air quality.

An example was given in how LULC are used in regional atmospheric modeling. The potential contribution of LULC changes to daily minimum and maximum near-surface air temperatures, during four recent summer EHEs in the Phoenix metropolitan area, were studied using WRF with Noah Urban Canopy Model and remotely sensed LULC classification data for 1973, 1985, 1998 and 2005. Simulations were carried out for each EHE with the four LULC classification data sets. Results show that urban land use characteristics that have evolved over the past ~35 years in the Phoenix metropolitan region have had a significant impact on extreme near-surface air temperatures occurring during EHEs in the area. Simulated maximum daytime and minimum night-time temperatures were notably higher due to the conversion of agricultural to urban land use (by ~2–4 K and 8–10 K, respectively). The conversion of desert to urban land use led to a significant increase in night-time air temperatures (6–7 K) and no significant changes in daytime temperatures.

Acknowledgments

This material is based upon work supported by the National Science Foundation under Grants No. ATM-0710631, No. GEO-0816168, No. DEB-0423704, Central Arizona – Phoenix Long-Term Ecological Research (CAP LTER). Any opinions, findings and conclusions or recommendation expressed in this material are those of the authors and do not necessarily reflect the views of the National Science Foundation.

References

Anderson, J.R., Hardy, E.E., Roach, J.T. and Witmer, R.E. (1976) *A Land Use and Land Cover Classification System for Use with Remote Sensor Data.* Geological Survey Professional Paper 964, US Government Printing Office.

Arnfield, J.A. (2003) Two decades of urban climate research: a review of turbulence, exchanges of energy and water, and the urban heat island. *International Journal of Climatology*, **23**, 1–26.

Banzhaf, E. and Netzband, M. (2004) Detecting urban brownfields by means of high resolution satellite imagery, *The International Archives of the Photogrammetry, Remote Sensing, and Spatial Information Sciences*, **25**, 6 p. (on CDROM).

Bian, L. (2003) Retrieving urban objects using a wavelet transform approach. *Photogrammetric Engineering and Remote Sensing*, **69**, 133–141.

Blaschke, T. and Hay, G. (2001) Object-oriented image analysis and scale-space: Theory and methods for modeling and evaluating multi-scale landscape structure. *International Archives of Photogrammetry and Remote Sensing*, **34** (4/ W5), 22–29.

Blaschke, T., Burnett, C. and Pekkarinen, A. (2006) Image segmentation methods for object-based analysis and classification, in (eds de Jong, S.M. and F.D. van der Meer), *Remote Sensing Image Analysis: Including the Spatial Domain*, Springer, Berlin.

Brenner, A.R. and Roessing, L. (2008) Radar imaging of urban areas by means of very high-resolution SAR and Interferometric SAR. *IEEE Transactions in Geoscience and Remote Sensing*, **46(10)**, 2971–2982.

Brown, M. (2000) Urban Parameterizations for Mesoscale Meteorological Models, in *Mesoscale Atmospheric Dispersion* (ed. Z. Boybeyi), WIT Press, Southampton, UK, pp. 193–255.

Brown, M.J., Burian, S.J., Linger, S.L., Velugubantla, S.P. and Ratti, C. (2002) An overview of building morphological characteristics derived from 3D building databases. *Preprint Proceedings of the American Meteorological Society's Fourth Symposium on the Urban Environment*, 20–24 May 2002, Norfolk, VA.

Burian, S.J., Brown, M.J., McPherson, T.N. *et al.* (2006) Emerging urban databases for meteorological and dispersion modeling. *Preprints, AMS Annual Meeting*, 6th Symposium on the Urban Environment, 29 January – 2 February 2006, Atlanta, Georgia.

Burian, S.J., Augustus, N., Jeyachandran, I. and Brown M. (2008). *National Building Statistics Database, Version 2.* Final Report. LA-UR-08-1921, Los Alamos National Laboratory.

Burian, S. J. and Ching, J. (2009) Development of gridded fields of urban canopy parameters for advanced urban meteorological and air quality models, Environmental Protection Agency, Washington, EPA/600/R-10/007.

Byun, D. and Schere, K.L. (2006) Review of the governing equations, computational algorithms, and other components of the models-3 Community Multiscale Air Quality (CMAQ) modeling system. *Applied Mechanics Reviews*, **59**, 51–77.

Buyantuyev, A. (2005) Land cover classification using Landsat Enhanced Thematic Mapper (ETM) data – year 2005. Online. Available http://caplter.asu.edu/home/products/showDataset.jsp?keyword=Land-Use%20and%20Land-Cover%20Change&id=377_1 (accessed on 18 November 2010).

CDC (Centers for Desease Control) (2005) Heat-related mortality – Arizona, 1993–2002, and United States, 1979–2002. *Morbidity & Mortality Weekly Report*, **54**, 628–630.

Cadenasso, M.L., Pickett, S.T.A. and Schwarz, K. (2007) Spatial heterogeneity in urban ecosystems: reconceptualizing land cover and a framework for classification. *Fronteirs in Ecology and Environment*, **5** (2), 80–88.

Chen F., Kusaka, H., Bornstein, R. *et al.* (2010) The integrated WRF/urban modeling system: development, evaluation, and applications to urban environmental problems. *International Journal of Climatology*, doi: 10.1002/joc.2158.

Chen, F., Tewari, M. and Ching, J. (2007) Effects of high resolution building and urban data sets on the WRF/urban coupled model simulations for the Houston-Galveston areas. Seventh

Symposium on the Urban Environment, San Diego, CA, Sep 10–13, American Meteorological Society, Boston, Paper 6.5.

Cheng, Y.Y. and Byun, D.W. (2008) Application of high resolution land use and land cover data for atmospheric modeling in the Houston-Galveston metropolitan area, Part I: Meteorological simulation results. *Atmospheric Environment*, **42**, 7795–7811.

Cheng, F.Y., Kim, S. and Byun, D.W. (2008) Application of high resolution land use and land cover data for atmospheric modeling in the Houston-Galveston Metropolitan area: Part II Air quality simulation results. *Atmospheric Environment*, **42**, 4853–4869.

Ching, J., Brown, M., Burian, S. *et al.* (2009) National Urban Database and Access Portal Tool NUDAPT. *Bulletin of the American Meteorological Society*, 10.1175/2009BAMS2675.1.

Civco, D.L. (1993) Artificial neural networks for land-cover classification and mapping. *International Journal of Geographical Information Systems*, **7**(2), 173–186.

Civerolo K., Hogrefe C., Lynn B. *et al.* (2007) Estimating the effects of increased urbanization on surface meteorology and ozone concentrations in the New York City metropolitan region. *Atmospheric Environment*, **41**, 1803–1818.

Collier, C.G. (2006) The impact of urban areas on weather. *Quarterly Journal of the Royal Meteorological Society*, **132**, 1–25.

Cotton, W.R. and Pielke, R.A. (2007) *Human Impacts on Weather and Climate*, Cambridge University Press, Cambridge.

Dell'Acqua, F. (2009) The role of SAR sensors, in *Global Mapping of Human Settlement: Experiences, Datasets, and Prospects* (eds. P. Gamba and M. Herold), CRC Press, Boca Raton, Florida, 364 pp.

Dell'Acqua, F. and Gamba P. (2006) Discriminating urban environments using multiscale texture and multiple SAR images. *Interantional Journal of Remote Sensing.*, **27**(18), 3797–3812.

Dupont, S., Otte, T.L. and Ching, J.K.S. (2004) Simulation of meteorological fields within and above urban and rural canopies with a mesoscale model (MM5). *Boundary-Layer Meteorology*, **113**, 111–158.

Ehlers, M. (2004) Remote sensing for GIS applications: New sensors and analysis methods, in *Remote Sensing for Environmental Monitoring, GIS Applications, and Geology III*, Proceedings of SPIE 5239, 1–13.

Ehlers, M. (2009) Future EO sensors of relevance – integrated perspective for global urban monitoring, in *Global Mapping of Human Settlement: Experiences, Datasets, and Prospects* (eds. P. Gamba and M. Herold), CRC Press, Boca Raton, FL, 364 pp.

Fisher, B., Joffre, S., Kukkonen, J. *et al.* (2005) *Meteorology Applied to Urban Air Pollution Problems – Final Report COST Action 715*, European Commission in Science and Technology, Demetra Ltd Publishers.

Foody, G.M., McCulloch, M.B. and Yates, W.B. (1995) Classification of remotely sensed data by an artificial neural network: Issues related to training data characteristics. *Photogrammetric Engineering and Remote Sensing*, **61**(4), 391–401.

Friedl, M.A., McIver, D.K., Hodges, J.C.F. *et al.* (2002) Global land cover mapping from MODIS: algorithms and early results. *Remote Sensing of Environment*, **83**, 287–302.

Gatzidis, C., Liarokapis, F., and Brujic-Okretic, V. (2007) Automatic modelling, generation and visualization of realistic 3D virtual cities for mobile navigation, in *9th International Conference on Virtual Reality*, April 18–22, 2007, Laval, France.

Gamba, P., Houshmand, B. and Saccani, M. (2000) Detection and extraction of buildings from Interferometric SAR data. *IEEE Transactions on Geoscience and Remote Sensing*, **38**(1), 611–617.

Gens, R. and Van Genderen, J.L. (1996) SAR interferometry: Issues, techniques, applications. *Interantional Journal of Remote Sensing*, **17**(10), 1803–1836.

Gong, P. and Howarth, P.J. (1990) The use of structural information for improving land-cover classification accuracies at the rural-urban fringe. *Photogrammetric Engineering and Remote Sensing*, **56**(1), 67–73.

Grimmond, C.S.B., Cleugh, H.A. and Oke, T.R. (1991) An objective urban heat storage model and its comparison with other schemes. *Atmospheric Environment*, **25B**, 311–32.

Grossman-Clarke, S., Zehnder, J.A., Stefanov, W.L., Liu, Y. and Zoldak, M.A. (2005) Urban modifications in a mesoscale meteorological model and the effects on surface energetics in an arid metropolitan region. *Journal of Applied Meteorology*, **44**, 1281–1297.

Grossman-Clarke, S., Zehnder, J.A., Loridan, T. and Grimmond, C.S.B. (2010) Contribution of land use changes to near surface air temperatures during recent summer extreme heat events in the Phoenix Metropolitan area. *Journal of Applied Meteorology and Climatology*, **49**, 1649–1664.

Heymann, Y. Steenmans, Ch., Croisville, G. and Brossard, M. (1994) CORINE land cover: Technical guide. Environment, Nuclear Safety and Civil Protection Series. Commission of the European Communities, Office for Official Publication of the European Communities, Luxembourg, 144 pp.

Hofmann, P., Strobl, J., Blashke, T. and Kux, H. (2008) Detecting informal settlements from QuickBird data in Rio de Janeiro using an object-based approach, in *Object-Based Image Analysis* (eds. T. Blaschke, S. Lang and G.J. Hay), Springer, Berlin, Germany, 817 pp.

Huth, R., Kysely, J. and Pokorna, L. (2000) A GCM Simulation of heat waves, dry spells, and their relationship to circulation. *Climatic Change*, **46**, 29–60.

Jansen, L.J.M. (2009) Semantic characterization of human settlement areas: Critical issues to be considered, in *Global Mapping of Human Settlement: Experiences, Datasets, and Prospects* (eds. P. Gamba and M. Herold), CRC Press, Boca Raton, FL, 364 pp.

Jensen, J.R. (1996) *Introductory Image Processing: A Remote Sensing Perspective* (2nd edn), Prentice Hall, Upper Saddle River, New Jersey.

Jensen, J.R. (2000) *Remote Sensing of the Environment: An Earth Resource Perspective*, Prentice Hall, Upper Saddle River, New Jersey.

Jiang, X.Y., Wiedinmyer, C., Chen, F., Yang, Z.L. and Lo, J.C.F. (2008) Predicted impacts of climate and land use change on surface ozone in the Houston, Texas, area. *Journal of Geophysical Research-Atmospheres*, **113**, D20312.

Kemball-Cook, S., Parrish, D., Ryerson, T. *et al.* (2009) Contributions of regional transport and local sources to ozone exceedances in Houston and Dallas: Comparison of results from a photochemical grid model to aircraft and surface measurements. *Journal of Geophysical Research – Atmospheres*, **114**, D00F02.

Kusaka, H., Chen, F., Tewari, M. and Hirakuchi, H. (2005) Impact of the urban canopy model in the Next-Generation Numerical Weather Prediction Model, WRF. *Environmental Systems Research*, **33**, 159–164 (in Japanese).

Kusaka H. and Kimura F. (2004) Thermal effects of urban canyon structure on the nocturnal heat island: numerical experiment using a mesoscale model coupled with an urban canopy model. *Journal of Applied Meteorology*, **43**, 1899–1910.

Lambin, E. F. and H. J. Geist (Eds) (2006) *Land-Use and Land-Cover Change. Local Processes and Global Impacts*, The IGBP Series, Springer-Verlag, Berlin, 222 pp.

Lemonsu, A. and V. Masson (2002) Simulation of a summer urban Breeze over Paris. *Boundary-Layer Meteorology*, **104**, 463–490.

Lemonsu, A., Leroux, A., Bélair, S., Trudel, S., and Mailhot, J. (2006) A general methodology of urban cover classification for atmospheric modelling, in *The 86th American Meteorological Society Annual Meeting (Sixth Symposium on the Urban Environment)*, January 28–February 3, 2006, Atlanta, Georgia.

Leroux, A., Gauthier, J.-P., Lemonsu, A., Bélair S. and J. Mailhot (2009) Automated urban land use and land cover classification for mesoscale atmospheric modeling over Canadian cities. *Geomatica*, **63**, 13–24.

Lin, C.-Y., Chen, F., Huang, J.C., Chen, W.-C., Liou, Y.-A., Chen, W.-N. and Liu, Shaw-C. (2008) Urban heat island effect and its impact on boundary layer development and land–sea circulation over northern Taiwan. *Atmospheric Environment*, **42**, 5635–5649.

Liu, Y., F. Chen, T. Warner, and J. Basara (2006) Verification of a mesoscale data-assimilation and forecasting system for the Oklahoma City area during the Joint Urban 2003 Field Project. *Journal of Applied Meteorology*, **45**, 912–929.

Lu, D. and Weng, Q. (2006) Spectral mixture analysis of ASTER imagery for examing the relationship between thermal features and biophysical descriptors in Indianapolis, Indiana. *Remote Sensing of Environment*, **104**(2), 157–167.

Mahesh, P. and Mather, P.M. (2003) An assessment of the effectiveness of decision tree methods for land cover classification. *Remote Sensing of Environment*, **86**, 554–565.

Maricopa Association of Governments, (2007): Arizona's Future – Population Projections From 2000–2050 [Online] Available http://www.mag.maricopa.gov/pdf/cms.resource/IS_2007_ArizonasFuture_2000-2050_LRG73684.jpg (accessed 19 November 2010).

Martilli, A. (2007) Current research and future challenges in urban mesoscale modelling. *International Journal of Climatology*, **27**, 1909.

Martilli, A. (2009) On the derivation of input parameters for urban canopy models from urban morphological datasets. *Boundary-Layer Meteorology*, **130**, 301–306.

Martilli, A, Clappier, A. and Rotach, M.W. (2002) An urban surface exchange parameterization for mesoscale models. *Boundary-Layer Meteorology*, **104**, 261–304

Masson, V. (2000) A physically-based scheme for the urban energy budget in atmospheric models, *Boundary Layer Meteorology*, **94**, 357–397.

Masson V. (2006) Urban surface modeling and the meso-scale impact of cities. *Theoretical and Applied Climatology*, **84**, 35.

Masson, V., Champeaux, J.-L., Chauvin, F., Meriguet, C. and Lacaze R. (2003) A global database of land surface parameters t 1-km resolution in meteorological and climate models. *Journal of Climate*, **16**, 1261–1282.

Meehl, G.A. and Tebaldi C. (2004) More intense, more frequent, and longer lasting heat waves in the 21st century. *Science*, **305**(5686), 994–997.

Miao, S., Chen, F., Lemone, M.A. *et al.* (2008) An Observational and modeling study of characteristics of urban heat island and boundary layer structures in Beijing. *Journal of Applied Meteorology and Climatology*, **48**, 484–501.

Mills, G. (1997) An urban canopy-layer climate model. *Theoretical and Applied Climatology*, **57**, 229.

Moeller, M. (2005) Land cover classification using Landsat (MSS) data – year 1973. Online. Available http://caplter.asu.edu/home/products/showDataset.jsp?keyword=Land-Use%20and%20Land-Cover%20Change&id=286_1 (accessed 18 November 2010).

Myint, S.W., Lam, N. and Tyler, J.M. (2004) Wavelets for urban spatial feature discrimination: comparisons with fractal, spatial autocorrelation, and spatial co-occurrence approaches. *Photogrammetric Engineering and Remote Sensing*, **70**, 803–812.

Oke, T. R. (1987) *Boundary Layer Climates*, Routledge, and John Wiley & Sons, Chichester.

Oleson, K.W., Bonan, G.B., Feddema, J., Vertenstein, M. and Grimmond, C.S.B. (2008) An urban parameterization for a global climate model. 1. Formulation and evaluation for two cities. *Journal of Applied Meteorology and Climatology*, **47**, 1038–1060.

Orlanski, I. (1975) A rational subdivision of scales for atmospheric processes. *Bulletin of the American Meteorological Society*, **56**(5), 527–530.

Otte, T. L., Lacser, A., Dupont, S. and Ching, J. K. S. (2004) Implementation of an urban canopy parameterization in a mesoscale meteorological model. *Journal of Applied Meteorology*, **43**, 1648–1665.

Petrie, G. and Stoney, W.E. (2009) The current status and future direction of spaceborne remote sensing platforms and imaging systems, in *Earth Observing Platforms & Sensors, Manual of Remote Sensing*, 3rd edn., Volume 1.1 (ed. M.W. Jackson), American Society for Photogrammetry and Remote Sensing, Bethesda, MD, 520 pp.

Pielke Sr, R.A. (2002) *Mesoscale Meterorological Modeling*, Academic Press, San Diego.

Rashed, T., Weeks, J.R., Roberts, D.A., Rogan, J. and Powell, R. (2003) Measuring the physical composition of urban morphology using multiple endmember spectral mixture models. *Photogrammetric Engineering and Remote Sensing*, **69**(9), 1011–1020.

Ratti, C., Di Sabatino S. and Britter R. (2002) Analysis of 3-d urban databases with respect to air pollution dispersion for a number of European and American cities. *Water, Air and Soil Pollution*, **2**, 459–469.

Ridd, M.K. (1995) Exploring a V-I-S (vegetation-impervious surface-soil) model for urban ecosystem analysis through remote sensing: comparative anatomy for cities. *International Journal of Remote Sensing*, **16**, 2165–2185.

Rozoff, C.M., Cotton, W.R. and Adegoke, J.O. (2003) Simulation of St. Louis, Missouri, land use impacts on thunderstorms. *Journal of Applied Meteorology*, **42**, 716–738.

Schöpfer, E. and Moeller, M.S. (2006) Comparing metropolitan areas – a transferable object-based image analysis approach, *Photogrammetrie, Fernerkundung, Geoinformation*, **4**, 277–286.

Seinfeld, H. J. (2006) *Atmospheric Chemistry and Physics: From Air Pollution to Climate Change*, John Wiley & Sons, Inc, New York.

Shem, W. and Shepherd, J.M. (2009) On the impact of urbanization on summertime thunderstorms in Atlanta: Two numerical model case studies. *Atmospheric Research*, **92**, 172–189.

Shepherd, J.M. (2005) A review of current investigations of urban-induced rainfall and recommendations for the future. *Earth Interactions*, **9**, 1–27.

Skamarock, W.C., Klemp, J.B., Dudhia, J. *et al.* (2005) A Description of the Advanced Research WRF Version 2. NCAR Technical Note-468+STR.

Small, C. (2005) A global analysis of urban reflectance. *International Journal of Remote Sensing*, **26**(4), 661–681.

Small, C. (2007) Spatial analysis of urban vegetation scale and abundance, in *Applied Remote Sensing for Urban Planning, Governance and Sustainability* (eds. M. Netzband, W.L. Stefanov and C. Redman), Springer, Berlin, Germany, 278 pp.

Stefanov, W.L. (2000) 1985, 1990, 1993, 1998 Land Cover Maps of the Phoenix, Arizona Metropolitan Area, Geological Remote Sensing Laboratory, Department of Geological Sciences, Arizona State University, Tempe, Arizona, 4 Plates, scale 1:115,200.

Stefanov, W.L. and Brazel, A.J. (2007) Challenges in characterizing and mitigating urban heat islands – a role for integrated approaches including remote sensing, in *Applied Remote Sensing for Urban Planning, Governance and Sustainability* (eds. M. Netzband, W.L. Stefanov and C. Redman) Springer, Berlin, Germany, 278 pp.

Stefanov, W.L., Ramsey, M.S. and Christensen, P.R. (2001) Monitoring urban land cover change: An expert system approach to land cover classification of semiarid to arid urban centers. *Remote Sensing of Environment*, **77**, 173–185.

Su, W., Li, J., Chen, Y. *et al.* (2008) Textural and local spatial statistics for the object-oriented classification of urban areas using high resolution imagery. *International Journal of Remote Sensing*, **29**(11), 3105–3117.

Taha, H. (1999) Modifying a mesoscale meteorological model to better incorporate urban heat storage: A bulk-parameterization approach. *Jurnal of Applied Meteorology*, **38**, 466–473.

Taha, H. (2008a) Urban surface modification as a potential ozone air-quality improvement strategy in California: A mesoscale modeling study. *Boundary Layer Meteorology*, **127**, 219–239.

Taha, H. (2008b) Episodic performance and sensitivity of the urbanized MM5 (uMM5) to perturbations in surface properties in Houston TX. *Boundary Layer Meteorology*, **127**, 193–218.

Taha, H. (2008c) Meso-urban meteorological and photochemical modeling of heat island mitigation. *Atmospheric Environment*, **42**, 8795–8809.

Tooke, T.R., Coops, N.C., Goodwin, N.R. and Voogt, J.A. (2009) Extracting urban vegetation characteristics using spectral mixture analysis and decision tree classifications. *Remote Sensing of Environment*, **113**(2), 398–407.

US Geological Survey, (2008): North America Land Cover Characteristics Data Base Version 1.2. [Online] (Updated 23 June 2008) Available http://edc2.usgs.gov/glcc/nadoc1_2.php (accessed 26 March 2010).

Wang, X.M., Chen, F., Wu, Z.Y. *et al.* (2009) Impacts of weather conditions modified by urban expansion on surface ozone: comparison between the Pearl River Delta and Yangtze River Delta regions. *Advances in Atmospheric Sciences*, **26**, 962–972.

Weng, Q. and Hu, X. (2008) Medium spatial resolution satellite imagery for estimating and mapping urban impervious surfaces using LSMA and ANN. *IEEE Transactions in Geoscience and Remote Sensing*, **46**(8), 2397–2406.

Wentz, E.A., Nelson, D., Rahman, A., Stefanov, W.L. and Roy, S.S. (2008) Expert system classification of urban land use/cover for Delhi, India. *Interantional Journal of Remote Sensing*, **29**(15), 4405–4427.

Woodcock, C.E. and Strahler, A.H. (1987) The factor of scale in remote sensing, *Remote Sensing of Environment*, **21**, 311–332.

Wu, C. and Murray, A.T. (2003) Estimating impervious surface distribution by spectral mixture analysis. *Remote Sensing of Environment*, **84**(4), 493–505.

Zhang, J. and Foody, G.M. (1998) A fuzzy classification of sub-urban land cover from remotely sensed imagery. *Interantional Journal of Remote Sensing*, **19**, 2721–2738.

Zhang, H., Sato, N., Izumi, T., Hanaki, K. and Aramaki, T. (2008) Modified RAMS-Urban canopy model for heat island simulation in Chongqing, China. *Journal of Applied Meteorology and Climatology*, **47**, 509–524.

vi

URBAN GROWTH AND LANDSCAPE CHANGE MODELING

Developing dynamic modeling techniques to simulate and predict future urban growth and landscape changes has been an active research area in urban remote sensing. The role of remote sensing is indispensible in the entire model development process from model conceptualization to implementation. Part VI (Ch. 22–26) examines some most exciting developments in this aspect. Chapter 22 discusses the evolution from pixel-matrix structures towards cell and agent-based models, the challenge of integrating spatial and a-spatial data structures and models, and a new data structure and modeling approach. Chapter 23 reviews the developments in calibration and validation of urban cellular automata models emphasizing on models tasked with simulating human−environment interactions. Chapter 24 introduces the agent-based urban modeling technique, followed by a case study to demonstrate the utilities of this type of modeling technique for urban growth and landscape change simulation. Chapter 25 discusses the utilities of ecological modeling for predicting changes in biodiversity in response to future urban development. The last chapter (Ch. 26) included in Part VI shifts the discussion from technical aspects to the underlying root metaphors embedded in various modeling efforts.

22

Cellular automata and agent base models for urban studies: from pixels to cells to hexa-dpi's

Elisabete A. Silva

The age of digital data collection and analysis is pivotal to land related fields. The contribution of remote sensing towards the development of one of the most important data structures (raster data structure) and its associated matrix/pixel structure, impacted diverse scientific fields and allowed for multiple practical applications and use. Also, the contribution of research in the remote sensing arena towards the development of powerful algorithm-classifiers that allowed interpretation of the available digital data pointed the direction, among other things, towards data mining and pattern identification. This chapter explores how such important developments in remote sensing, together with developments in physics, maths, chemistry, and computer science take center stage in one of the most important revolutions in the geographical sciences and towards the study of complexity. This chapter will explore the evolution from pixel–matrix structures towards cell and agent base models, and why we are facing an important challenge (the integration of spatial and a-spatial data structures and models) that will ultimately led us towards the integration of multiple approaches. A new data structure and modeling approach is proposed (the hexa-dpi) and an emphasis is placed on remote sensing by its potential for data collection, processing and as an integrator of modeling approaches such as CA, ABM and GA.

22.1 Introduction

This chapter focus is on the development of digital data structures and main models during the last decades; in particular it will emphasize the importance of the development of raster data structures though remote sensing that were pivotal in the development of key artificial intelligence models, such as cellular automaton (CA), agent base models (ABM) and genetic algorithms (GA).

It will emphasize the interaction between remote sensing and CA-ABM; in particular it will focus on the trilogy pixels–cells–hexa-dpi, it will explain why this is important for CA and ABM and how it can be enhanced by remote sensing. It will present today's pixel–cell interaction and potentialities and it will also point out future directions by exploring the opportunity of integrating remote sensing, CA, ABM, and other AI approaches.

In order to fully understand where we are today in terms of the multiple attempts to integrate different approaches, a set of phased steps will tell the "history" from the pixel to the cell, from the cell and agent's individual development to its recent integrated approaches; this will be followed by a discussion of one of the main challenges at the present moment (integration of spatial and aspatial data structures and modeling environments), the chapter finishes with a discussion towards a proposal of a hexa-dpi data structure that will close the circle of dpi–pixel–cell–hexa-dpi and will allow to integrate spatial and aspatial dimensions in time and space. A hexa-dpi structure is a hexagonal spatial matrix that will work as the "virtual" magnetic field where spatial cell-based-dots can interact in different time/scale dimensions. Allowing spatial and aspatial attributes to converge and synchronize in dynamic spatial representations.

22.2 Computation: the raster–pixel aproach

Aerial photos, "dots in a paper," had a profound impact on cartography and its affiliated subjects. Gray tones or false colors that represented different land attributes and the development of stereoscopic pairs that would allow three-dimensional visualization transformed the science of cartography, geography, urban, and environmental planning.

Remote sensing as:

> the practice of deriving information about earth's land and water surfaces using images acquired from an overhead perspective, using electromagnetic radiation in one or more regions of the electromagnetic spectrum, reflected of emitted from the earth's surface
>
> *(Campbell, 1996, p. 5)*

was at the basis of one of the most important revolutions in data collection and analysis – plains and satellites sensed the earth for multiple purposes. Less known among the research community is the importance of remote sensing to the development of matrix-base data structures and associated algorithms (Fig. 22.1).

Gathering information at a distance using satellites with spectrometers that surveyed the earth by means of reflected or emitted electromagnetic energy and that would send information to earth codified as raster images, allowed increasing the pace of surveying and the geographic wideness it could cover (Fig. 22.1). Each image was treated as an array of values corresponding to the spectral channels, and the pixel was the basic/minimum unit of the image, each pixel's location was identified by row–column coordinates (Fig. 22.2).

If image acquisition in remote sensing allowed to define new data structures (raster/matrix/pixel), image classification of remotely sensed data through the use of image classifiers (computer programs that would implement processes to classify images – convert numeric images to classified images) started to develop a new field that would allow to make sense of the vast amounts of data arriving in a matrix of numbers.

The classified images would then feed other computer applications that would describe attributes, perform statistical analysis, spatial descriptors, search spatial relationships, and perform spatial analysis and simulation. These new opportunities tended to link cartographic production of base maps resulting from these remotely sensed data to the production of thematic maps for multiple uses. A name that started to be associated with the process of producing/analyzing digital maps was commonly known as geographic information systems (GIS). A toolbox:

> **a GIS can be seen as a set of tools for analyzing spatial data. These are, of course, computer tools, and a GIS can then be thought as a software package containing the elements necessary for working with spatial data**
>
> *(Clarke, 1999).*

From this initial definition GIS, migrated towards more integrated approaches, linking basic statistic analysis with more sophisticated data mining approaches, and advanced modeling approaches (Maguire, Batty and Goodchild, 2005; Haining, 2003), contributing towards what is nowadays considering a geographic information science (Goodchild, 2009).

Therefore, new technologies, new methodologies and a new language started to emerge from the previous developments. For the purpose of this chapter, the development of the pixel–matrix data structure and the development of computer algorithms that would explore pixels in a matrix in order to assign meanings to numbers and classify patterns allowed, in a subsequent phase, the implementation of new emerging theoretical developments and the creation of the "cell."

22.3 Cells: migrating from basic pixels

If the pixel was the basic unit of a matrix and this allowed to scan/convert and produce digital maps, with each pixel having a specific numeric attribute (i.e. resulting from a spectral signature, or from a gradation of tones in a color/grayscale table) and if computation was already allowing moving windows to operate simple/complex operations in a matrix or between matrixes, why not allow a pixel interaction at a deeper level? And why would this be needed?

Starting to answer this question: why would this (interaction between pixels) be needed? There are at least three obvious reasons for this: (1) computer supported technology was able to do more than simple scans of the Earth and re-produce base-cartography; (2) the development of analytical thematic cartography required more that a simple overlap of maps; (3)

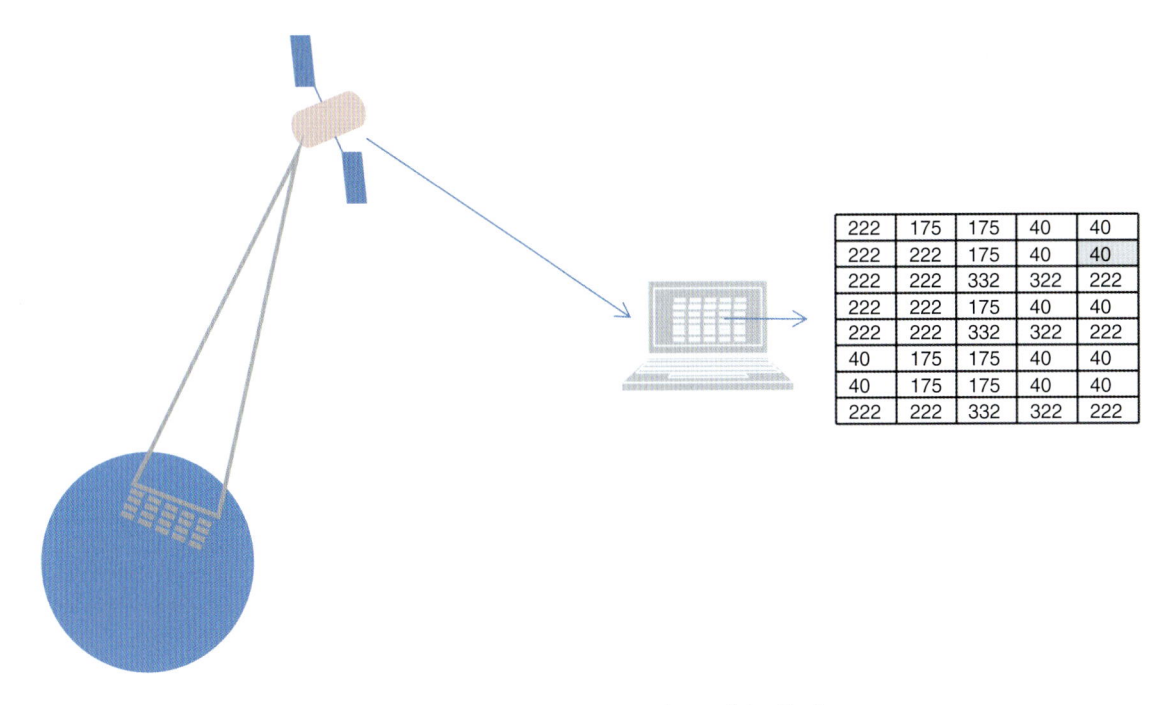

CHAPTER 22 CELLULAR AUTOMATA AND AGENT BASE MODELS FOR URBAN STUDIES: FROM PIXELS TO CELLS TO HEXA-DPI'S

FIGURE 22.1 Scanning the surface of the Earth.

FIGURE 22.2 Snapshot of remote sensed data for a given moment (T1) and a uniform scan spatial unit (SU).

analysis of dynamics of different phenomena required dynamic "pixels" self-aware of their 'value' in time-1, but also being sensitive to change in time-2; being sensitive to its immediate neighbors, or to specific action at a distance.

In order to perform analysis that included some of the previous needs, this would require more sophisticated pixel analysis, linking it with the concepts of self-awareness, neighborhood interaction (sensitivity to local conditions), but also sensitivity to regional/global inference/interference. Consequently this would mean allowing a "pixel" (and the matrix) to self-organize, to change. These needs departed substantially from the traditional pixel–matrix static concepts, and therefore, if these goals were achievable, it would make sense to name this "new pixel" as a "cell" and the new kind of science dealing with all, the science of complexity.

The founding fathers that are internationally recognized as the ones contributing to the birth of the cell, self-organization, and complexity are Von Neumann and Morgenstern (1944), Von Neumann (1966) Ulam (1960, 1974), Prigogine (1999), Prigogine and Stengers (1984) and Prigogine and Nicolis (1977), Tobler (1979), Kauffman (1984, 1993), Wolfram (1994), Holland (1995, 1999), and Crutchfield and Mitchell (1995).

During the 1940s Von Neumann and Morgenstern's theory of games and economic behaviour (1944) demonstrated that it is indeterminism, probability, and discontinuous changes of state that control the behaviour of most systems, in particular the behaviour of the elements of systems and their contribution to the overall patterns and processes in a system. Therefore, the emphasis is in the representation and understanding of the 'behaviour of the elements'.

The proposal of cells as these new "elements" happened in the 1950s with joint work by Von Neumann's and Standislav Ulam that linked game-theory and microbehavior, through the development of a simple CA. CA provided local rules which could generate mathematical patterns in two-dimensional and three dimensional space. Therefore global order could be produced from local action. A similar conclusion was found through the observation of particles behaviour, Prigogine's work on molecular interactions and the resulting large-scale spatial structure demonstrated the importance of this micro-macro interaction in pattern identification (Prigogine and Nicolis 1977; Prigogine and Stengers, 1984; Prigogine, 1999).

By the 1950s the concept of cells as dynamic entities that could be simulated in a computer gave rise to the CA. A computer generated being, CA can be seen in its more primary state, as a discrete model of spatiotemporal dynamics obeying local laws.

If cells are not about deterministic behavior, if the behavior of cells varied and their random behavior had to be studied and classified, not as an absolute value but as a probability, the next obvious development had to be the understanding of how random behavior could generate regular patterns. The works of Kauffman (1984), Crutchfield (1995), and Holland (1999) have been important in pointing out the emergent properties of random complex automata.

Tobler and Burks (1979) work is particularly important in the field of CA as they were the first to clearly define what was its structure and basic principles: (i) A grid or raster space – organized by cells that are the smallest units in that grid/space; (ii) cell states – cells must manifest adjacency or proximity – the state of a cell can change accordingly to transition rules, which are defined in terms of neighborhood functions; (iii) the neighborhood and dependency of the state of any cell on the state and configuration of other cells in the neighbourhood of that cell; (iv) transition rules that are decision rules or transition functions of the CA model and can be deterministic or stochastic; (v) sequences of time steps. When activated, the CA proceeds through a series of iterations (Silva and Clarke, 2002, 2005; Langton 1986, 1995). For each iteration (time step), the cells in the grid are examined. Based on the composition of cells in the neighborhood of that central cell, transition rules are applied to determine the central state of the cell in the next iteration.

As seen in Fig, 22.3, multiple grids can be used as complementary information/attributes that will help the development of behavior in the seed layer(s) allowing for a vertical interaction among different cells (most basic CA only need one matrix (seed-layer), increasing the number of matrixes will increase the levels of complexity). This interaction (being vertical or horizontal) was, and still is in many cases, an extremely important feature of these models, as there are some difficulties in finding the appropriate data structure that would embed all attributes in one cell. As a consequence these models require matrixes that perfectly overlay (geographically referenced) and where the same cell might have different attributes in each of the layers.

Once vertical interaction (at each individual cell, but accordingly the attributes its has in each layer/matrix) and horizontal interaction (neighborhood effects among nearby cells – usually four or eight neighboring cells) happens among cells, a final synchronous update warps the matrix to a new moment in time with a new configuration of the matrix (synchronous update), where cells can assume different values and different spatial configurations. This local self-organization of cells allows the identification of different regional patterns, sometimes allowing the development of new emergent behavior where original conditions would not anticipate the formation of new/different patterns, but also the existence of phase-transitions where conditions can promote for instance boom and bust phases (i.e. sudden increase of speed of urban cells in a certain area of the matrix).

With the study of random complex CA came an understanding of its basic patterns: as they appear to fall into four qualitative classes, in one-dimension (1-D) CA evolution leads to: (i) a homogeneous state; (ii) a set of separated simple stable or periodic structures; (iii) a chaotic pattern; (iv) complex localized structures, sometimes long-lived (Wolfram, 1984:5).

In 1994 Wolfram outlined a number of characteristics that CA possesses: (i) The correspondence between physical and computational processes is clear; (ii) CA models are much simpler than complex mathematical equations, but produce results that are

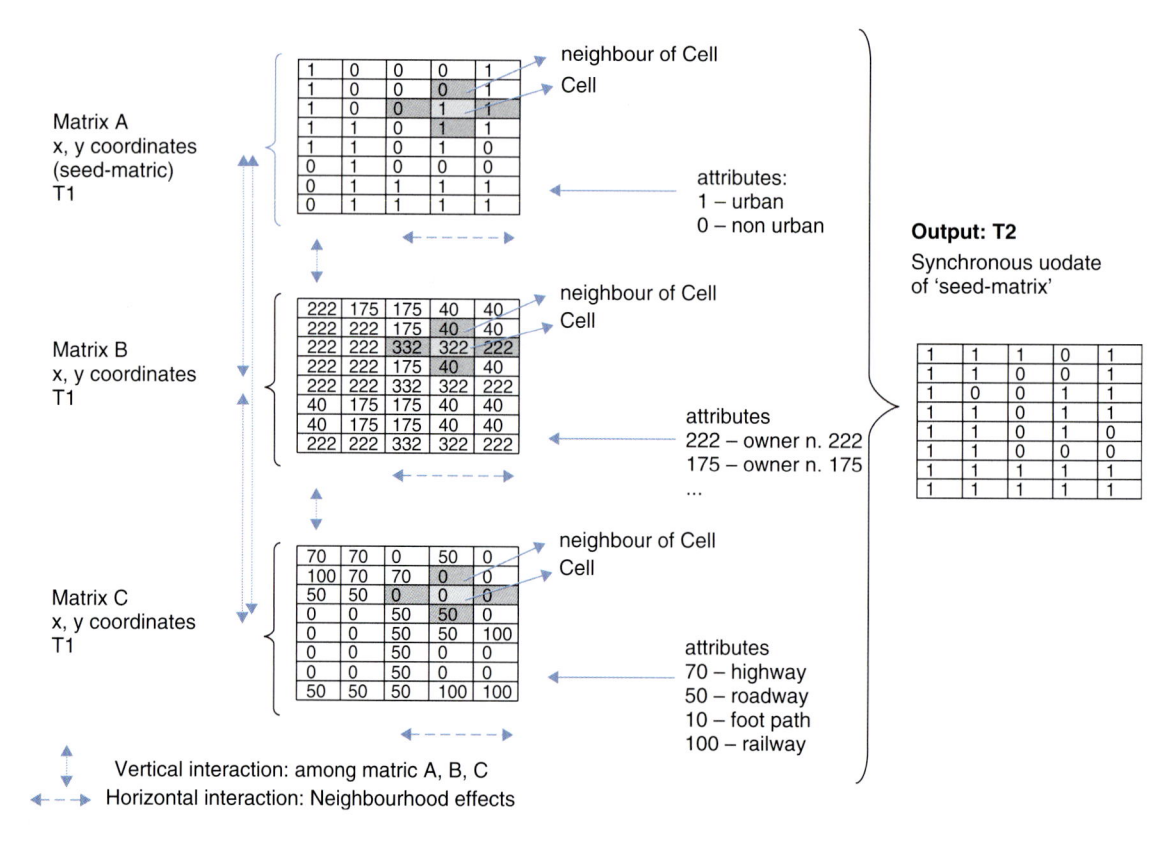

FIGURE 22.3 The cellular environment.

more complex; (iii) CA models can be modeled using generating precise results (degree of closeness with real world systems); (iv) CA can mimic the actions of any possible physical system; (v) CA models are irreducible.

Conways's game of life, popularized by Gardner in 1970, was pivotal in the promotion of these ideas through other fields of science (Wolfram, 2002). Simulations of death and life in a game showed the striking similarities between real life and simulated life in a computer. Cells would survive or die in competition for space, crowdedness, minimum number of neighbors required in order to live in society, and reproduction. Emergent behavior, not possible to justify directly by the input data, was one of the topics of most curiosity among researchers.

In the field of urban and environmental planning, Waldo Tobler, in contact with Arthur Burks, was exposed to Von Neumann's works, and published "Cellular geography" (1979). At NCGIA-Santa Barbara, Helen Couclelis and Keith Clarke, published respectively "Cellular worlds" (Couclelis, 1985) and develop the first fully operational and implementable CA (Clarke and Gaydos, 1998). Michael Batty, initially at NCGIA-Buffalo and afterwards at CASA-UCL, developed the theory and practice that culminated in the publication of the seminal books *Fractal Cities* (Batty and Longley, 1994) and *Cities and Complexity* (Batty, 2005). Recently, Wolfram's book *A New Kind of Science* (2002) explores the importance of having a new science that does not dismiss complexity, instead that tries to understand it and apply it to a pan disciplinary development, that the geographic information sciences obviously incorporate.

From there, links to fractals and bifurcations were an obvious development as it became apparent that in some cases transition was not constant and continuous, but the development of patterns was less than randomized in time and space and some patterns seemed to be repeating and multiple time/space dimensions (Silva, 2010a, Silva and Clarke 2002; Batty, Xie and Sun 1999).

As previously seen, starting with a smart-cell in a matrix called a CA, it was now possible to identify/quantify processes and patterns through computation, it was now possible to identify/quantify phase transitions, and detect emergent behaviour. And it was possible to do all of this not only across space in a matrix of cells for a specific moment in time, but also along systematic moments in time.

22.4 Agents: joining with cells

If physical structures (spatial structures) and their (self) organization in the physical world at multiple scales and through time could be measured computationally using CA. What about immaterial structures (i.e., socioeconomic conductions), how are they structured in order to produce an action? How do immaterial structures (aspatial structures) self-organize into meaningful actions accordingly to a change in conditions? Why would immaterial structures and their organization be meaningful to physical structures and to the material world?

Starting to answer the question: Why would immaterial structures be important to the material world? Because a thought is capable of directing a physical action (i.e., command and control of the brain that will activate movement in a hand). That is to say that, while the physical world (without humankind's in the equation) can be represented/explained by pixels and cells, if one wants to represent a physical world where humankind intervenes, it would be unreasonable to assume only a pure physical (spatial) representation, and this means the need of including the immaterial world (aspatial) in these new models and theories. The challenge was to create a data structure that would allow that.

Therefore while CA developments were decisive to the understanding of complexity in spatial settings, at the same time (during the 1940s and 1950s) a second branch of research was also being developed that would be determinant to the development of aspatial organization.

The founding fathers that are internationally recognized as the ones contributing to the birth of the developments in mimicking the human decision-making process were: Albert W. Tucker (1957, 1960) Merrill Flood and Melvin Dresher (RAND, 1952), and RAND's (www.rand.org) and Santa Fe Institute's (www.santafe.edu) developments during the 1980s and 1990s.

One of the initial developments was linked with GA. In the case of GAs, the goal was to focus on behavioral and social systems. This second stream of research focuses on the different behavioral options that human beings are faced with, using for instance decision trees and neuronal nets, and extrapolating such concepts to a new modeling environment called GAs.

While GAs usually do not have a direct representation in space as it is the case with CA, they are very important as representations of phenomena in terms of explaining what is at the basis of decisions and options (i.e., why do we choose to perform a specific option). Therefore GAs play a major role in explaining decision-making processes that lead to specific spatial actions. The core theory of GA is in the work developed by John Nash exploring research results by Merrill Flood and Melvin Dresher at RAND Corporation in the 1950s; in doing so he opened up a new field of computational exploration of human behaviour (Nash 1950, 1953).

The research developed by Flood and Dresher at RAND Corporation are at the basis of the understanding of human behaviour and how to extrapolate that behavior to understandable rules. Their most known finding is "the prisoner's dilemma:" if two prisoners (players) have a choice each on whether to betray the other, and thus to decrease one's own jail time, while increasing the jail time for the other, common sense tell us that in this competitive environment a cooperative strategy should be the best option; nevertheless, because of the uncertainty both will tend to choose to "betray." This leads to what is commonly known as a "non-zero-sum game." The current set of strategy choices and the corresponding payoffs constitute a sate of equilibrium (the "Nash equilibrium").

Consequently, human behavior could be classified according to patterns in a more quantifiable way and Nash's equilibrium represents the first clear demonstration of the outcome of self-organization of agents accordingly to a set of new conditions. Therefore, the post Second World War explanations of behavior in a mathematical way, by attempting to incorporate the complexity associated to individual behaviour, using game theory and strategizing techniques, are at the basis of another important element of theory by focusing on public–individual choice and option.

One of the most common approaches towards the simulation of the 'genetics' of human thinking is ABM. In contrast to CA's abilities to model the spatial dynamics of land change, ABMs proven to be most effective with aspatial dynamics. Advantages of ABMs include their ability to model individual decision-making entities and their interactions, to incorporate social processes on decision-making, and their dynamic socioeconomic environmental linkages (Matthews *et al.*, 2007).

In the development of ABM models, the focus is on the moment of taking a decision, the moment of moving from one place to another; in other words the moment of the "command" that will lead to a decision plays an important role (in these micro-simulation "command and control" models where each agent has its own unit of command and control) – that is to say that, for instance, in the decision tree the "nodes" that allow the ramification of the decision tree will be pivotal in order to enable a new change as a result of that decision/command. As a consequence most of these ABM should rely on the simulation capabilities of GAs and representations of phenomena in terms of explanations of what is at the basis of decisions and options (i.e., why do we decide to choose a specific option). Therefore ABM-GAs can play an explanatory/experimental role in formalizing the decision-making processes that leads to specific spatial actions.

For the purpose of this chapter, we will keep addressing ABM-GA, as we feel that this will be one of the most important modelling approaches to optimize the decision trees of each individual agent (or groups of agents).

ABM-GAs are constituted of: (i) agents that do not have the constraints of neighborhood effects; (ii) behavioral roles among agents and the environment itself; (iii) independence from central command/control, but able to act if action at a distance is required; (iv) states of agents tend to represent behavioral forms.

The most basic model environment of an ABM-GA will have a set of attributes per agent (or group of agents), (one)a set of decision trees and trigger points that will allow to set the context for a new movement (upgrade of the spatial/temporal environment) in time/space (Fig. 22.4).

These models are characterized by behaving as populations of simple agents that interact locally with each other and their environment and produce swarm intelligence, that is to say, machine resulting intelligence, or emergent behaviour in machines not directly resulting from code development (Fig. 22.4). This emergent behaviour in GA, ABM and also CA are one of the most interesting developments of these new theories, models and data structures, pointing out to what is called 'the intelligence in the machine', 'artificial intelligence', or 'artificial life' as described in the next points of this chapter.

While today research starts to integrate both approaches of CA and GA, ABM during several decades these were studied apart. While CA has a spatial explicit representation that makes it very apt to model urban and environmental systems (i.e., land parcels, transportation infrastructures), ABM-GAs are important for their behavioural roles that are very apt to model individual agents and their behaviour (i.e., households, vehicles and pedestrians).

From the 1950s to the 1980s the basic elements of complexity analysis both in the natural and social sciences were drafted. Some of them were at that time still conjectures that are now being explored with more detail. During the 1990s it started to become clear that these two approaches of modeling with CA or with ABMs were the two most used approaches to work with complexity in a quantitative formulation.

Nevertheless, researchers would tend to separate both approaches. CA and ABM researchers tended to be in opposing fields, defending one or another approach. These were almost hermetic areas of research where little information would cross-fertilize the development of new methodologies. For some of us it started to be clear that without this cross-fertilization it would be difficult to select the best methodological option (CA, ABM, GA) or to develop hybrid methodologies that included CA, ABM and GA and fully integrate aspatial and spatial structures.

22.5 Cells and agents in a computer's ``artificial life''

Artificial life is the study of man-made systems that exhibit behavioral characteristics of natural living systems. CA and intelligent agents (ABMs that rely in technological developments such as the previously described GAs) are typical artificial life techniques, and in the past decades they have been increasingly implemented in urban studies (Wu and Silva, 2010).

It is now commonly accepted that CA are very capable of simulating complex spatial phenomena. CA models include the following capabilities: (1) representation of complexity and dynamics in systems (Silva, 2010a, b); (2) spatial integration (Batty, Xie and Sun, 1999; Piyathamrongchai and Batty 2007; Silva and Clarke 2002, 2005) and self-organizing mapping (Castilla and Blas *et al.* 2008); 3. Extensibility and adaptability (Dragicevic 2007; Kocabas and Dragicevic 2007; Jenerette and Wu 2001); (4) both simplicity and complexity: (Torrens and O'Sullivan 2000 (5) visibility: through the lattice structure (cells in CA model correspond to the grids in a raster image), and the link to geographic data makes CA models highly visual: CA are popular as land change models (Clarke and Gaydos, 1998; Batty, Xie and Sun, Straatman and Engelen, 2004; Silva, Wileden and Ahern, 2008).

Ligtenberg *et al.* (2004) described an ABM to explore different ways in which decisions regarding land use could be made. Ligmann and Jankowski (2007) focused on using ABM for spatially explicit modeling of real-world policy scenarios. Benenson, Martens and Birfir, (2007) simulated urban parking policy scenarios and analyzed their impacts from the user and public policy perspective. The simulation of social-economic interactions in urban systems is another important application of ABM. Milner-Gulland *et al.* (2006) described the use of an ABM to investigate the trade-offs in allocation of wealth by households. Ettema *et al.* (2007) presented a multi-agent based urban model (PUMA) concerning land conversion problems.

The phenomena of land change needs to include spatial and aspatial dynamics. During the past, the stationary transition probabilities limited the model's ability to reflect feedbacks to global changes (action at a distance) to influence transitions at the cellular level. Some exceptions include, SLEUTH, by enabling and disabling boom and bust phases (Clarke *et al.*, 1998, Silva and Clarke, 2002), and CVCA, by allowing localized changes accordingly to proximity, area issues and globally by allowing changes to the landscape shape index (Silva, Wileden and Ahern 2008). Nevertheless, if CA are incorporated with ABM, it expands its potentiality by integrating two important components: cellular models are used to describe spatial dynamics and ABM represent a-spatial, social interactions (Wu and Silva, 2010).

Recently, integrated approaches using hybrid modelling systems are providing promising solutions for urban land studies. Agent-based cellular automata (AB-CA) is one such hybrid systems, that integrates agent based modelling and cellular automata. This is particularly important when one needs to include both

Modelling Environment T1

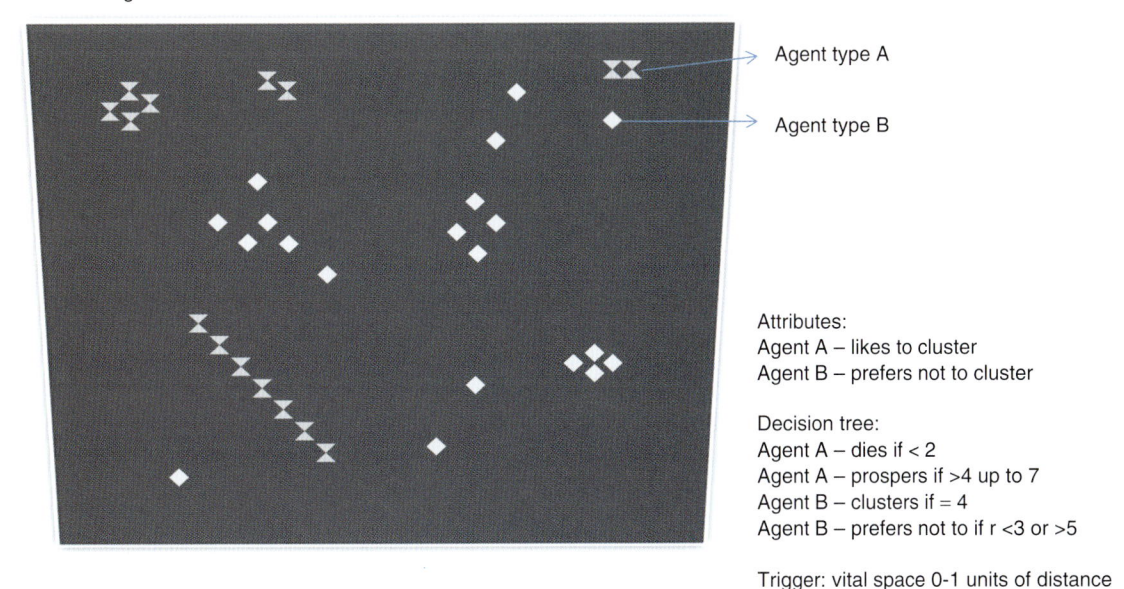

Agent type A
Agent type B

Attributes:
Agent A – likes to cluster
Agent B – prefers not to cluster

Decision tree:
Agent A – dies if < 2
Agent A – prospers if >4 up to 7
Agent B – clusters if = 4
Agent B – prefers not to if r <3 or >5

Trigger: vital space 0-1 units of distance

Modelling Environment T2

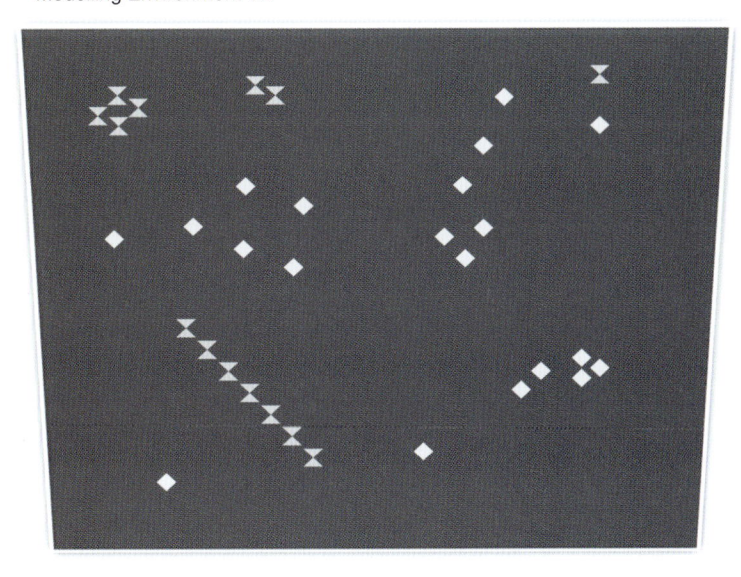

FIGURE 22.4 The agent's environment.

spatial and a-spatial approaches in the same model, but they can also be used as helpful tools to calibrate models. For example, Sudhira et al. (2005) proposed an AB-CA framework for urban sprawl simulation, where the ABM is used to define the transitional rules of the CA. Wu and Silva compared 11 recent AB-CA hybrid systems in urban studies (Wu and Silva, unpublished work). In these 11 models, agents usually represented the behaviour of pedestrians, vehicles, drivers, and passengers in urban traffic systems, or householders, stakeholders, and residents in urban systems. Geographical changes were represented by CA. Nevertheless, the spatial context was not provided in an explicit way, and no attempt is made to merge or integrate the data structures.

While there are not many loose coupled models exploring individually CA, ABM and GA. Wu and Silva (2010) already implemented it in a fully coupled CA-ABM/GA (model:

DG-ABC). the authors present an integrated model that incorporates ABM, CA and a GA in order to include both spatial and a spatial dynamics supplying a new solution for urban studies. In the model (DG-ABC stands for "developing genetic-agent based cells"), the social economic behaviors of heterogeneous agents (resident, property developer, and government) will be regulated by GA and the theory of planning behaviour (TpB).

A simplification of integration of CA-ABM-GA models can be seen in Fig. 22.5. Agents and cells are still confined to the seed-matrix and the modeling environment, spatial/temporal synchronization of cells and agents allows adjusting the variability in scales/time (i.e., agent's actions happening in minutes and action in cells of 100 ha changing every year). In these synchronization tables horizontal interaction pays a pivotal role, as it will be the most favourable environment for cells and agents

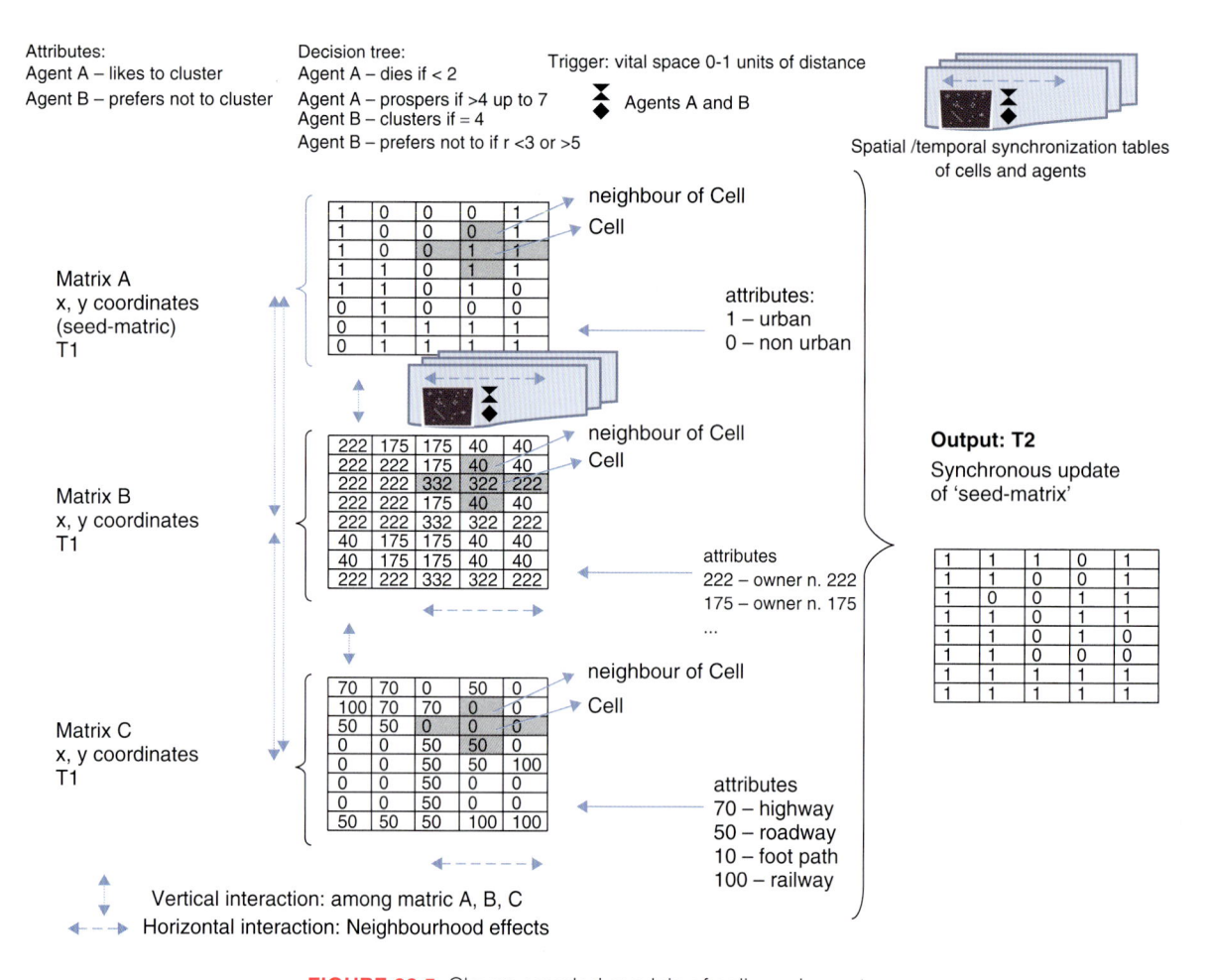

FIGURE 22.5 Clouse coupled models of cells and agents.

to interact. The result is still a synchronous update of the entire matrix in a subsequent time.

At this point we face a dilemma: the spatial models require specific data structures (i.e., matrix-cell based) and the aspatial models required substantially different data structures (i.e. decision trees/neuronal nets) how to integrate both data structures? How to integrate data, data structures, theories, models, algorithms? How to integrate the past 50 years of outstanding developments into a unifying theory?

22.6 The hexa-dpi: closing the cycle in the digital age

The integration of computational simulation methods for CA and ABM seems to be the most promising approach in order to include both spatial and aspatial dynamics. Yet both CA and ABM rely on data structures that make this integration difficult to achieve. Modeling spatial dynamics often requires long timespans for successful calibration and to capture the slower pace of decadal change in land processes such as urban growth. Within these models, space is usually represented by cells (or polygons if one considers vector data structures, not detailed in this chapter), and land use change tends to be heavily dependent on local conditions, that shows strong spatial autocorrelation. Models of aspatial dynamics include characteristics that might not have an immediate reference to space (i.e., socioeconomic condition) and so do not rely as much on proximity characteristics as they do on action at a distance or other contextual information. Integrating these properties is a serious challenge for the future of land-related computer simulation modeling.

It is important to build on previous work that developed robust, but standalone or close coupled, CA and ABM models and link these two different but common-heritage models by using innovative approaches to data structures.

The answer to the need of integrating spatial and aspatial models can be clearly understood by the following question: Why does the investigation of individual people's movement and changes in land ownership parcels require two different modeling and planning approaches? By answering this question, it would be possible to solve the problem of linking spatial and a-spatial modeling approaches (and data structures) by exploiting a hybrid of ABM and CA.

The concept that best describes such a modeling approach is the goal of specifying two data structures that automatically interact, but that are constrained by different time frames, and different behavioral and ontological schema. It is important to examine both the revealed effect, e.g., representation of change in land parcels (i.e., from undeveloped to built-up land) and at the same time the agents (people or groups) and their choices that enabled that change. While major changes in urban growth and land use change tend to take months to years, agents action

The hexa-dpi structure:

- As traditional cell structures the hexa-dpi-structure base is a matrix, in this case an irregular matrix (the hexagon is the basic unit of every hexa-matrix) – therefore no uniform spatial unit (SU)

- Each hexa-cell accommodates a dot – the grouping of dots in patterns will define an object
- Dots will ensure 3D structure
- The interplay of dots-per-inch and hexa-matrix will define a scale and units of analysis
- Zooming-in/out will immediately redefine the scale of analysis and the units
- Objects resulting from dot-pattern identification will have different mobility rates (i.e microseconds-seconds-minutes-days-years-centuries-millennia), this will be indexed to scale and units of analysis

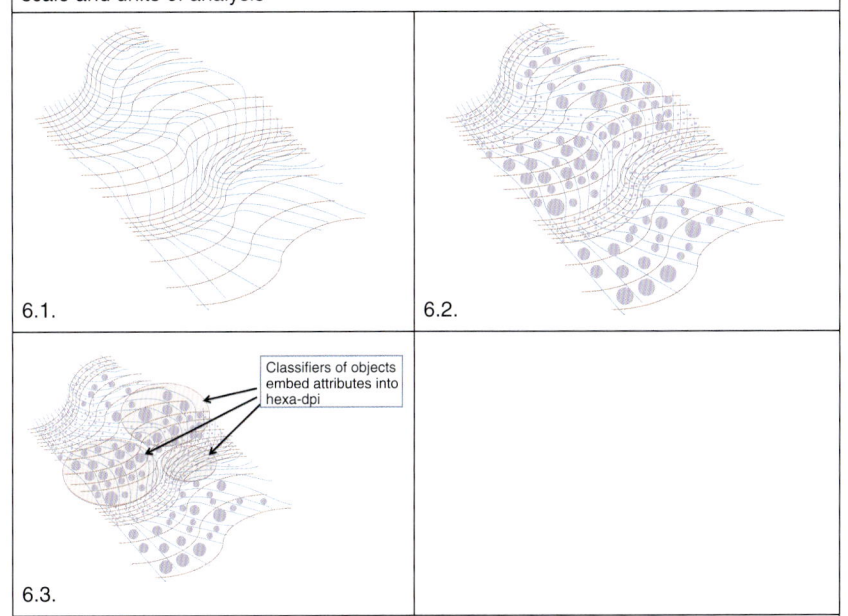

6.1.

6.2.

6.3. Classifiers of objects embed attributes into hexa-dpi

The Cellular Environment
-Dot size and hexa-matrix will define the unit of analysis and scale
-Scalability will be a proportion of the object (i.e. 2 times current scale or phenomena's fractal dimension)
-Scale will be variable in for each T-time accordingly to object zoomed in
-Image recognition of dots/patterns will form objects (that can feed attribute tables)
-The matrix will grant horizontal integrity of the same perspective
-A zoom-in will immediately re-scale dots-patterns to new perspective and new objects or more detail will be unveiled and allow scalability
-The hexa-matrix will also allow vertical integrity as it will allow to 'filter' to lower levels fine-grained dots allowing to build the next zoom-in level (and the next scale and unit of analysis)
-Obviously scalability to upper levels isn't a problem as there are many algorithms that already to that.

Having the Cellular Environment defined in such a way makes it easy to incorporate **the agent's behavioural/environment**

-Agents will be the 'materialization' of some of the 'mobile'-physical structures into mobile objects (i.e. aggregation of so me of the dots) across the matrix for different time spans, or for the same time/space when zooming in. This materialization for a sequence of time-frames for objects will depend on the observable hexa-dpi structure.

In theory, this new data-structure and new-cartographic representation will allow to chart from planetary/galaxy to the particle and sub-particle levels (and, in between, the surface of the earth), for both mobile/immobile objects in a specific Time. Because everything is mobile across time (depending of the scale of analysis), but immobile in each time-frame this seems to be the best compromise to answer to the needs to these two different data structures.

FIGURE 22.6 The hexa-dpi structure, and the cellular agents environments.

tend to be more important at a timescale of weeks to months. At the same time people move freely, or even independently, of space (grid) while land uses tend to be heavily dependent on local characteristics.

By integrating CA-cell based models (spatial models) and ABM it will be possible to represent more realistic processes where spatial/land and a-spatial/socioeconomic dynamics interact and produce the feedbacks that re-produce emergent and complex land changes. In order to achieve this integration goal one of the primary tasks will be to solve the different data-structures and data requirements of the two model types.

Urban land dynamics are self-organizing, stochastic, catastrophic and chaotic (i.e., Barredo et al. 2003; Batty 1997; Silva, 2004; Silva and Clarke 2005; de Roo and Silva, 2010). Spatial and temporal dynamics are two important driving forces of the complex adaptive process. The key to integrating new computational approaches into urban dynamic models is to understand the interaction and synchronization of spatial and temporal processes, and presenting innovative solutions of scalability effects (i.e., MAUP effects – Martinez, Viegas and Silva, 2009; and Wu and Silva's spatial/temporal synchronization tables, 2010).

The obvious question is: is this possible? A straight forward answer would be: Yes – it already exists: (1) we have the computational capabilities; (2) we have the modeling technology studied and understood (for instance we are now coupling modeling approaches to common platforms of GA and CA); (3) we are now exploring massive amounts of data and unveiling new relations and new models; (4) we have personnel from different fields working together (increasing multidisciplinarity).

Remote sensing can/will play an important role for two main reasons: (1) as the main input data source into the spatial dimension (though satellites that scan the surface of the earth), and into the a-spatial dimension (using other satellites that capture the movement of mobile agents in shorter time spans and link it with GA base models); (2) as a base aggregator where spatial and aspatial dimensions find a common environment/platform.

What would be the new data structure that would link CA, ABM, GA – that would integrate matrix and vector data structures, spatial and aspatial dimensions in a dynamic evolving self-organizing environment.

The hexa-dpi structure has the potential to overcome some of the existent limitations, by proposing a new hexagonal spatial matrix that will work as the "virtual" magnetic field where spatial cell-based-dots can move freely and aggregate, vary its dependence on local and global intervention, and exist as a material entity or as an immaterial attribute. No hexagonal-cell will be "empty" at any point of space, the scale of analysis and units will play an important role in defining the objects capable to observe/classify.

Figure 22.6 synthesizes some of the main characteristics of the hexa-dpi structure and its associated cellular and agent's environment. In the figure, box. 6.1 represents the basic structure (that was simplified for the purpose of this chapter), it should be approximately hexagonal, doesn't need to be a physical representation as such, but will be a fluid physical reference (a kind of magnetic field where other structures will congregate accordingly to the dimension of the net); still as part of the cellular/spatial environment the dots represent an important addition (explored in box. 6.2) they have different sizes and will have higher/lower mobility depending of the scale and unit of analysis (they are mutable/transformable, never disappearing only changing to lower/higher scales/nets). The interplay of the hexagonal structure and dots in the 'cellular environment' will grant among other things, scalability. Finally, Figure 22.6.3 represents the 'materialization' of some of the spatial features into objects that can be identified by us as recognizable objects, allowing classifiers of objects to embed attributes into the hexa-dpi structure and build objects (i.e. cars, houses, urban land uses, farmland).

In a way we are coming full circle, from initial dots in a picture, to rigid pixel–matrix structures and its associated deterministic models, to more flexible, changing and emergent behaviours of cells and agents and their initial stochastic models. To today's hybrid models that have the goal of aggregating the benefits of both worlds. The future requires full integration of spatial–aspatial data structures.

Conclusions

Computers and the opportunities derived from computation allowed the application or simultaneous development of new theories. What is now known as the study of complexity has its roots during the 1950s. Pre and post Second World War developments allowed remote surveying of the land, first as photos of dots per inch, latterly as pixels in a matrix, scanned from satellites. These vast matrixes with an individual number per pixel were important not only to analyze existent data but to derive new data. With time deterministic analysis gave place to probabilistic analysis and more complex data mining techniques that needed to rely on more elaborate algorithms. The acknowledgment that stochastic analysis could be more accurate in the representation of the phenomena, allowed the development of CA models and GA/ABM models. In turn, the observation that real life does not separate aspatial and spatial patterns and processes is now directing new developments in modeling that avoid the spatial vs. aspatial divide and this requires new data models and new data structures (termed here as hexa-dpis).

References

Batty, M. (2005) *Cities and Complexity*, The MIT Press, Cambridge MA.

Batty, M. and Longley, P. (1994) *Fractal Cities: Geometry of form and function*, Academic Press, New York.

Batty, M. (1997) Cellular Automa and Urban Form: A primer. *Journal of the American Planning Association*, **63** (2), 266–274

Batty, M., Xie, Y., and Sun, Z. (1999) Modelling urban dynamics through GIS-based CA. *Computers, Environment and Urban Systems*, **23**, 205–233

Barredo, J., I, Marjo, K., McCormick, N. and Lavalle C. (2003) Modelling dynamic spatial processes: simulation of urban future scenarios through cellular automata, *Landscape and Urban Planning*, **64**: 45–160

Benenson, I., K. Martens and S. Birfir (2007) Agent-based model of driver parking behavior as a tool for urban parking policy evaluation. Paper presented at the 10th AGILE International

Conference on Geographic Information Science, 8–11 May 2007, Aalborg University, Aarhus, Denmark.

Campbell, J. (1996) *Introduction to Remote Sensing*, The Guilford Press, New York.

Castilla, A., and Blas, N. (2008) Self-organizing Map and Cellular Automata Combined Technique for Advanced Mesh Generation in Urban and Architectural Design, International Journal Information Technologies and Knowledge, **2**, 354–359

Clarke, K.C. (1999) *Getting Started with Geographic Information Systems*, Prentice Hall, New Jersey.

Clarke, K.C. and Gaydos, L. (1998) Loose coupling a cellular automaton model and GIS: long-term growth prediction for San Francisco and Washington/Baltimore. *International Journal of Geographical Information Science*, **12**, (7), 699–714.

Couclelis, H. (1985) Cellular Worlds: a framework for modelling micro-macro dynamics, *Environment and Planning A*, (**17**), 585–596

Crutchfield, J. and Mitchell, M. (1995) The evolution of emergent computing. *Proceedings National Academy of Sciences – USA*, **92**, 10742–10746

de Roo, G., and Silva, E.A. (2010) *A Planners' Encounter with Complexity*, Ashgate Publishers Ltd, Aldershot.

Dragicevic, S. (2007) Embedding spatial agents into irregular cellular automata models of urban land use change to improve scenario exploration and decision making. Extended abstract in Proceedings of GeoComputation'07 Conference, Maynooth, 3–5 September, Ireland.

Ettema, D.F., De Jong, K., Timmermans, H.J.P. and Bakema, A. (2007) PUMA: multi-agent moddeling of urban systems, in Moddeling Land-Use Change (eds H.J. Scholten, E. Kommen and J. Stilwell), Springer Verlag, Berlin. pp. 237–257

Goodchild, M. (2009) Geographic information systems and science: today and tomorrow. *Annals of GIS*, **15**(1), 3–9.

Haining, R. (2003) *Spatial Data Analysis: Theory and Practice*, Cambridge University Press, Cambridge, UK.

Herold, M., Goldstein, N., Menz, G. and K.C. Clarke (2002): Remote sensing based analysis of urban Dynamics in the Santa Barbara region using the SLEUTH urban growth model and spatial metrics. Proceedings of the 3rd Symposium on Remote Sensing of Urban Areas. Istanbul, Turkey, June 2002.

Holland, J. (1995) *Hidden Order: How Adaptation Builds Complexity*, Helix Books, Reading MA.

Holland, J. (1999) *Emergence: From Chaos to Order*, Helix Books, Reading MA.

Jenerette, G.D. and Wu, J.G. (2001) Analysis and simulation of land-use change in the central Arizona-Phoenix region USA. *Landscape Ecology*, **16** (7), 611–626.

Kauffman, S. (1993) *Origins of Order: Self-Organization and Selection in Evolution*, Oxford University Press, Oxford.

Kocabas, V., Dragicevic, S. (2007) Enhancing a GIS cellular automata model of land use change: Bayesian networks, influence diagrams and causality. *Transactions in GIS*, **11** (22), 681–702.

Langton, C. (1986) Studying artificial life with cellular automata, *Physica D*, **22**, 120–149.

Langton, C. (1995) *Artificial Life. An Overview*, Bradford & The MIT Press, Cambridge MA.

Ligmann Z.A. and Jankowski, P. (2007) Agent-based models as laboratories for spatially explicit planning policies. *Environment and Planning B: Planning and Design*, **34** (2), 316–335.

Ligtenberg, A., Wachowicz, M., Bregt, A.K., Beulens, A. and Kettenis, D.L. (2004) A design and application of a multi-agent system for simulation of multi-actor spatial planning. *Environmental Management*, **72**, 43–55.

Martinez, L.M., Viegas, J.M. and Silva, E.A. (2009), Modifiable areal unit problem effects on traffic analysis zones delineation. *Environment and Planning B – Planning and Design*, **36** (4), 625–643.

Maguire, D., Batty, M. and Goodchild, M. (2005) *GIS, Spatial Analysis and Modelling*, ESRI Press, Redlands CA.

Matthews, B., Nigel, G., Gilbert, A., Roach, J., Polhill, C., and Gotts, N. (2007) Agent-based land-use models: a review of applications. *Landscape Ecology*, **22**, 1447–1459.

Milner-Gulland, E.J, Kerven, C., Behnke, R., Wright, I.A., and Smailov. A. (2006) A multi-agent system model of pastoralist behaviour in Kazakhstan. *Ecological Complexity*, **3** (1), 23–36.

Nash, J. (1953) Two-person Cooperative Games, *Econometrica* **21**, 128–140.

Nash, J. (1950) Equilibrium Points in N-person Games, *Proceedings of the National Academy of Sciences*, **36**, 48–49.

Piyathamrongchai, K. and M. Batty (2007) Integrating cellular automata and regional dynamics using GIS Modelling land-use Change **90**, 259–277.

Prigogine, I. (1999) Laws of nature, probability and time symmetry breaking, *Physica D*, **263**, 528–539.

Prigogine, I. and Nicolis, G. (1977). *Self-Organization in Non-Equilibrium Systems*, John Wiley & Sons, Ltd, Chichester.

Prigogine, I. and Stengers, I. (1984) *Order out of Chaos. Man's New Dialogue With Nature*, Bantam Books, Toronto.

RAND (1952) On Game-Learning Theory and some decision-making experiments. Rand Research Paper.

Silva, E.A. and Clarke, K. (2005) Complexity, Emergence and Cellular Urban Models: Lessons Learned from Appling SLEUTH to two Portuguese Cities. *European Planning Studies*, **13** (1), 93–115.

Silva, E.A. (2004) The DNA of our regions: artificial intelligence in regional planning. *Futures*, **36** (10), 1077–1094.

Silva, E.A. (2010a) Waves of complexity. Theory, models, and practice, in *A Planners' Encounter with Complexity* (eds G. de Roo, and E.A. Silva), Ashgate Publishers Ltd, Aldershot.

Silva, E.A. (2010b) Complexity and CA, and application to metropolitan areas, in *A Planners' Encounter with Complexity* (eds G. de Roo, and E.A. Silva), Ashgate Publishers Ltd, Aldershot.

Silva, E.A. and Clarke, K. (2002) Calibration of the SLEUTH Urban Growth Model for Lisbon and Porto, Portugal. *Computers, Environment and Urban Systems*, **26** (6), 525–552.

Silva, E.A. and Clarke, K. (2005) Complexity, emergence and cellular urban models: lessons learned from applying SLEUTH to two Portuguese cities. *European Planning Studies*, **13** (1), 93–115.

Silva, E.A., Wileden, J. and Ahern, J. (2008) Strategies for landscape ecology in metropolitan planning: applications using cellular automata models. *Progress in Planning*, **70** (4), 133–177

Sudhira, H.S., Ramachandra, T.V., Wytzisk, A. and Jeganathan, C. (2005) Framework for integration of agent-based and cellular automata models for dynamic geospatial simulations. Technical Report: 100,Centre for Ecological Sciences, Indian Institute of Science, Bangalore.

Tobler, W. (1979) Cellular geography, in Philosophy and Geography (eds S. Gale and G. Olosson), D. Reidel, Dordrecht, pp. 279–386.

Torrens, P. and O'Sullivan, D. (2000) Cities, cells, and complexity: Developing a research agenda for urban geocomputation, in *Proceedings of the 5th International Conference on Geocomputation*, University of Greenwich, London.

Tucker, A.W. (1957) Linear and Nonlinear Programming, *Operations Research*, **5**, 244–257.

Tucker, A.W. (1960) Solving a Matrix Game by Linear Programming, *IBM Journal of Research*, **4**, 507–517.

Ulam, S. (1960) *A Collection of Mathematical Problems*, Interscience Publishers, New York.

Ulam, S. (1974) *Sets, Numbers and Universes*, MIT Press, Cambridge, Massachusetts.

Wolfram (1984) Universality and complexity in cellular automata, *Physica D* (**10**), 1–35

Wolfram, S. (2002) *A New Kind of Science*, Wolfram Media Inc, Champaign.

Von Neumann, J. (1966) *Theory of Self-Reproducing Automata*, University of Illinois Press, Urbana.

von Neumann, J. and Morgenstern, O. (1944). Theory of games and economic behavior. Princeton, NJ: Princeton University Press.

White, R., Straatman, B. and Engelen G. (2004) Planning scenario visualization and assessment – a cellular automata based integrated spatial decision support system, in (eds.: Spatially Integrated Social Science, edited by M.F. Goodchild, and D. Janelle), Oxford University Press, New York, USA, pp. 420–442

Wu, N. and Silva, E.A. (2009) Artificial Intelligence and 'waves of complexity' for urban dynamics, *inWSEAS. Recent Advances in Artificial Intelligence, Knowledge Engineering and Databases*, pp. 459–469.

Wu, N. and Silva, E.A. (2010). Artificial intelligence solutions for urban land dynamics: a review. *Journal of Planning Literature*, **24**, 246–265

Wu, N. and Silva, E.A. (2010) Integration of genetic agents and cellular automata for dynamic urban growth modelling: pilot studies. WCTR-World Congress of Transportation Research, Lisbon, Portugal. Conference program: page 61; Book of Abstracts: (digital format paper 1858 (page 119)

Wu, N. and Silva, E.A. *(unpublished work) Dynamic Models in urban land studies – exploring sophisticated modelling for urban simulation, JPL*

23

Calibrating and validating cellular automata models of urbanization

Paul M. Torrens

Automata-based modeling of urban dynamics is an active research enterprise that has gathered momentum at an increasing pace. Despite its popularity, research and development in this domain remains in a state of relative stasis, largely constrained by persistent problems in registering simulation scenarios to the real world that model developers seek to simulate. To a degree, these problems are indicative of modeling as a scientific endeavor more broadly, but application of automata models to human–environment contexts raises some particularly challenging issues for calibration and validation that the field could benefit from addressing. This chapter reviews the state of the art in applying cellular automata models to human–environment interactions in complex, messy, dynamic urban systems, with all of its associated problems and promise.

23.1 Introduction

All models are abstractions of a complicated reality and in simulation there is an imperative to assess the match between modeled phenomena and the real world. *Empirically* determining this correspondence is always somewhat of a losing proposition because of limits to what might be observed, the partiality inherent in most datasets, discrepancy between modeled spatial and temporal scales and those of natural and social phenomena, and the non-uniqueness of any hypothesis or assumption (Oreskes, Shrader-Frechette and Belitz, 1994, Batty and Torrens, 2005). These problems are exacerbated when models are used to speculate about futures, about which we can know almost nothing with certainty and they become especially problematic when we fashion models of complex systems (Allen and Torrens, 2005). Regardless of how problematic issues of mismatch might be, models are nevertheless used as research, experimental, and diagnostic tools and so treatment of their "fit" to the real world or to a purpose is significant, particularly when models are tasked to test theories or to support plans, decisions, or policies. Calibration and validation exercises are particularly significant in urban applications of models to human-environment interactions, where there has been a tradition of using computer models as planning support systems (Brail and Klosterman, 2001), often for use in determining the potential impact of infrastructure developments, human activities, and land-uses upon the environment, or even for assessing compliance with legislative rulings relating to transportation efficiency and air emissions standards (Southworth, 1995).

The terminology surrounding model-matching exercises is infamously vague and confused (Oreskes, Shrader-Frechette and Belitz, 1994), but is perhaps best summarized as follows. "Verification" exercises serve to register a model (generally) to a particular application, system, place, or time, or to fit a particular purpose (normative modeling, conceptual modeling, decision support). "Calibration" involves (specifically) adjusting model parameters so that simulations (where simulation is the act of "running" a model on data or applying it to a given scenario) can be performed with a level of fitness or sufficiency for their intended purpose. "Validation" involves assessing the success of a model or simulation run in achieving its (specific) intended goals. In all cases, these exercises usually involve comparing the performance of the model to some properties of the real system being simulated. In a relative minority of cases, similar schemes are used to compare models to other models, or individual runs of a single model in varying places and times (Pontius Jr. *et al.*, 2008, Wegener, 1994). Comparisons are commonly made against known or observed conditions.

Urban automata models, usually formed as cellular automata, individual-based models, agent-based models, or multi-agent systems (Benenson and Torrens, 2004) have become increasingly popular in human-environment research, in large part owing to their flexibility in representing an almost limitless variety of phenomena and systems. Among their advertised advantages, automata model-builders often tout the ability of automata to simulate complex dynamic systems that defy easy analysis by more traditional forms of modeling (Grimm, 1999; Manson, 2001; Parker *et al.*, 2003; Batty, 2005). Verifying the utility of such models for the systems and uses to which they are applied in urban and environmental studies has, however, proven to be a thorny issue (Torrens and O'Sullivan, 2001; Batty and Torrens, 2005; Manson, 2007; Pontius Jr. *et al.*, 2008).

Nevertheless, work in this area is active and it is advancing. Much of the current research emphasizes statistical procedures, which are employed in gauging uncertainty or sensitivity in empirical analysis, providing formal methodology for analyzing relationships and the strength of association between simulation output and data about the system being simulated (Kocabas and Dragićević, 2006; Dragićević and Kocabas, 2007; Pontius Jr. *et al.*, 2008). Remotely-sensed imagery has also emerged as an important source of "dataware" for model validation and calibration, feeding models with data and providing the catalyst that allows model data output to be contextualized into information (Herold and Clarke, 2002; Torrens, 2006a; Bone *et al.*, 2007). Increasingly, model developers are relying on spatial analysis and image processing techniques that are more usually applied to the analysis of real cities to analyze patterns and regularities in simulated cities: to extract features, examine geographic trends, analyze spatial structures, etc. (Herold, Clouclesi and Clarke, 2005). Some of the advances in the computer sciences that enabled the introduction of automata techniques to the geographical sciences have also been employed in the calibration and validation of automata models using techniques from artificial intelligence (AI) (Almeida *et al.*, 2008).

In his chapter, I review the state of the art in calibration and validation of urban cellular automata models, with particular emphasis on models tasked with simulating human–environment interactions through land-use and land cover change, largely because it is in these application domains that development of automata models is advancing most rapidly. The chapter continues in the next sections with discussion of calibration mechanisms – conditional transition, weighted transition, and state-based constraints. Discussion of the derivation of values for calibration parameters follows, with particular attention to visual calibration, statistical tests, the use of historical data, regression of parameter values, the use of exogenous models, and automatic calibration procedures. The focus of this chapter then moves to consideration of model validation routines through use of visual inspection, pixel-matching, feature and pattern recognition, and the analysis of complexity signatures. Various procedures for sweeping the parameter space of models are then discussed before the chapter draws to a close with some concluding discussion.

23.2 Calibration

Cellular automata models are generally calibrated by tailoring the parameters that control transition rules. In addition, calibration may be performed on a state-basis or cell-basis, allowing special conditions by which the normal state transition procedure for those states or cells might be specially treated. For system properties that might sit beyond the range of an automata simulation, external models may be used to calibrate automata parameters.

23.2.1 Conditional transition rules

Conditional transition invokes calibration by determining circumstances under which transition may or may not take place. One classic example is a so-called stopping rule, which freezes state transition to prevent a simulation from evolving beyond a particular time-step, over a geographical boundary, or outside specified state parameters. Stopping rules do have some theoretical basis. For example, in urban applications they are synonymous with holding capacity for built infrastructure or the environment; they may be associated with the notion that cities can accommodate people, vehicles, only up until a certain point or up to a particular threshold of sustainability. The actual stopping procedure can be introduced, *a priori*, or it can be context-specific, subject to evolving conditions in a simulation. In their model of urbanization in Dongguan, China, for example, Yeh and Li (2002) used population totals as a stopping rule for simulation runs. A simulation is halted when the total target ("known") population for a simulation run is reached. Alternatively, Ward, Murray and Phinn (2000a) used geographical extent as the basis for a stopping rule in their models of urbanization; once a simulation reaches a target area, urbanization ceased to proceed further.

Thresholds are another example of conditional calibration, used to establish the lower- and upper-limits for certain properties or events in a simulation run. Often, attainment of a threshold value stimulates a particular action in a simulation (a particular state transition, introduction of a new rule, initiation of a feedback loop). Transition thresholds have strong analogies with properties of real urban systems (trip-making is often governed by capacity thresholds, for example). White, Engelen and Uljee (1997) used thresholds to specify a quota of total cell transitions in their simulations of urbanization. Their models were run as normal, transitioning between states on a cell-by-cell basis, until the threshold quota is reached. At that point, transition was constrained by setting the transition potential of all cells in the model to zero.

The use of transition hierarchies is another way to calibrate a model. Often, the sequence of urbanization adheres to a particular formal sequence. Vacant lots, for example, have the potential to be used for any activity; a site that was previously occupied by a heavy chemical industry, on the other hand, has limited potential for re-development because of the considerable effort required to "reclaim" it. Hierarchies are generally implemented as rankings on state transition potentials in urban models. This process operates as follows. Particular uses are rank-ordered: states that are observed in great abundance in the system (residential land-use, for example) may be ranked on one (higher) extremity of a scale. States that feature less frequently (some form of locally unwanted land-use such as waste disposal, for example) will be ranked on another (lower) extremity of that scale. State transition can then be specified in such a way that states may only transition to other states with *higher rankings*. White and Engelen (1993) used this approach to condition land-use transition in their models. In their examples, a vacant cell may transition to *any* other use (housing, industry, commerce). A cell with a housing land-use can transition either to industrial or commercial uses. An industrial cell may transition *only* to a commercial state. The use of hierarchies thereby establishes a mechanism for plausible urban momentum.

23.2.2 Weighted transition rules

Weights are used to calibrate automata models by "elasticizing" transition rules; this allows models to amplify or dampen processes by determining the *likelihood* of an event occurring in a simulation run, or the *intensity* of its influence. Weights are commonly specified as transition potentials, controlling the likelihood of transition to a particular state. Often, the influence of several such potentials can be combined, as a *weighted sum*. Transition weights have analogies in real urban systems. Agglomeration effects are the most obvious example: colocation of particular land uses (car dealerships, for example) may establish mutual further potential for additional collocation of the same activity, say, through economies of scale, for example. The phenomenon also works inversely; certain activities act with a repellant influence: power plants, hazardous industries, disposal facilities are rarely collocated with residential land-uses (they have a negative elasticity).

Weights are particularly useful for calibrating rates of change and transition potentials for automata (Xie, 1996; Batty and Xie, 1997). Static weights remain constant over the course of a simulation run. Dynamic weights, on the other hand, may change in influence as a simulation run evolves. The most common exemplar in automata modeling is the use of growth rates. By weighting growth (or decline), the speed of evolution in a model run can be hastened, placed in relative stasis, or slowed. In a sense, growth rates set the overall metabolism for a simulation. In their model of urbanization in Cincinnati, OH, White and colleagues (1997) introduced a static, universal growth rate, which held constant over a simulation. In the SLEUTH models developed by Clarke and colleagues (Clarke, 1997; Clarke, Hoppen and Gaydos, 1997; Clarke and Gaydos, 1998; Clarke *et al.*, 2007), the rate of urbanization in their simulations is linked, dynamically, to the (evolving) size (area extent) of the simulated city. As the city grows, its rate of development accelerates. As growth slows toward a standstill, the growth rate is dampened. Weighting by distance is a related approach. White and colleagues (1997) made use of a distance-dependent weight on state transition in their cellular automata models. Transition is weighted more heavily – specifically, it has a higher probability of successful transition – for cells that are closer to a specific cell. Essentially, this establishes a mechanism to temper transition by distance-decay. Elsewhere, White and colleagues (1997) introduced weights designed to mimic the attraction-repulsion effects of various land-uses on one another, designing weights to mimic agglomeration and nuisance effects and the distance-decay of those influences. Weights can also be used to introduce stochastic perturbation to model dynamics, as a proxy for unexplained (or uncertain) factors in a model. For example, White and Engelen (1993) used a weighting parameter to control the overall level of perturbation in their models. They found that their simulation runs were highly sensitive to the perturbation parameter, concluding that it was an important factor in calibrating a model to *real* applications. Yeh and Li (2002) used a random variable for stochastic perturbation, scaling (weighting, really) it to restrict it within a "normal range of fluctuation" (plus or minus 10%).

23.2.3 Seeding and initial conditions

Urbanization often exhibits trends of path dependence around a particular condition or event. Once this pattern is "locked-in" (Arthur, 1990), it becomes relatively difficult to change the trajectory of the system to a new phase. The canonical example in urban studies is the initial settlement site for a given city, which loses significance as the city develops but retains a central position in its metropolitan hierarchy because of inertia (Hall, 1988). Similar mechanisms are used to calibrate models to known (historical) conditions. Clarke and colleagues (1997), for example, used historical maps and other data sources to identify and place the initial position of settlements in the San Francisco Bay Area. This information was then used to set seed cells for a cellular automata simulation of the area's urbanization; in a sense, it ensured that the right stuff grew in the right places. They also introduced a layer of road states, employed such that once a simulation run transitioned through specific temporal markers, the appropriate road data features were read into the model. Torrens (2006b) similarly used the initial settlement pattern of the system of cities around Lake Michigan and known population totals to establish initial conditions in his model of sprawl formation in the Midwestern megalopolis. Seed constraints can also be applied in the opposite direction: the SLEUTH model, for example, has been run from contemporary seed conditions, and reversed back in time over a historical period for the Santa Barbara Region, beginning with conditions in 1998 and running back to 1929 (Goldstein, Candau and Clarke, 2004).

Exclusion mechanisms can also be used to constrain simulations by withholding certain cells from transition consideration over a model run, or over a particular time period in a simulation. White, Engelen, and colleagues, for example, used this procedure to distinguish between "fixed" and "functional" cells in their cellular automata models. "Fixed" cells correspond to areas of a city that remain exempt from urbanization: permanent land-uses such as rivers, parks, and railways. "Functional" cells are consistent with sites that are open to urbanization; they are subject to the full range of transition rules specified in the model (White and Engelen, 2000). Clarke and colleagues similarly enjoyed a layer of "excluded areas" in their models (Clarke *et al*., 2007). This layer was used to represent cells that are immune to growth processes in a simulation: ocean, lake, and "protected areas". They also introduced topography constraints that prohibited urbanization above a given slope threshold. A similar scheme was employed in the Dynamic Urban Evolution Model (DUEM) developed by Xie and Batty (Batty, Xie and Sun, 1999). Li and Yeh (2002) varied this standard approach slightly in their applications: they associated development with the potential environmental costs of urban growth, excluding areas of cropland, forest, and wetland from their model. In a similar exercise, they developed mechanisms to freeze urban growth in environmentally sensitive and important agricultural areas of their simulated area, relating the constraint to ideas of sustainability (Li and Yeh, 2000).

23.2.4 Specifying the value of calibration parameters

There are a variety of methods by which the empirical value of calibration parameters such as weights and constraints might be derived. These methodologies include visual calibration, statistical tests, the use of historical data, and statistically regressing parameter values. In addition, exogenous models may be used to supply parameter values. In recent years, several automated calibration procedures have also been developed for urban automata models.

Visual calibration involves the manual tweaking of rules by comparing simulation runs visually (Aerts *et al*., 2003). Essentially, a model designer or user acts as an expert system, adjusting weights and parameter values based on her observation of model dynamics: "The eye of the human model developer is an amazingly powerful map comparison tool, which detects easily the similarities and dissimilarities that matter, irrespective of the scale at which they show up. And the model developer, with all his knowledge of spatial form and spatial interaction processes, is a very able (if often unwilling) "calibration machine"" (Straatman, White and Engelen, 2004). On other occasions, the values of calibration parameters have been derived from historical data sources. Most often, land-use maps are used. White and colleagues calibrated weights in their model of Cincinnati against land-use maps dating back to 1970 (White, Engelen and Uljee, 1997), for example.

Statistics may also be used to calculate parameter values. Clarke and Gaydos ran descriptive statistical tests to determine the parameter vales in their models of San Francisco, Washington DC, and Baltimore (Clarke and Gaydos, 1998). Wu and Webster (1998) used a series of logistic regressions to calibrate the probabilities for transition rules in their model of land development in Guangzhou, China: the beta-values from a regression equation were used, directly, as the value for model weights. Similarly, in his model of polycentric urbanization, Wu (1998) introduced a transition rule (which he termed as an "action function") that calibrated the model's decision- and preference-based transition rules. This function was specified using a regression equation that was itself calculated based on real choice observation data. Arai and Akiyama (2004) also used regression to derive transition weights in their model of urbanization in the Tokyo Metropolitan area.

23.2.5 Coupling automata and exogenous models

Strictly speaking, automata models should be closed systems: they should not be open to external influence. However, in some instances (and often born of a necessity to better ally with the reality that they try to emulate), urban automata models have been coupled with exogenous simulations: model output, routines, or equations external to an automata model are used to influence transition in an automata simulation, either by initiating state transition directly, or by determining the value of constraints and weights. There are a few rationales for using exogenous models in this way. Simple convenience is one reason: exogenous models can be employed to introduce mechanisms that may not be treated (or that a model-builder may not wish to treat because of disjointed time scales, for example). There are also theoretical justifications. Exogenous models can be used to represent phenomena that are external to the system of interest: non-urban subsystems such as climate, hydrology, geology, or external (but locally relevant) phenomena such as the operation of cities in a regional system, national boom-and-bust cycles, and geopolitical dynamics. The MURBANDY model developed by White, Engelen, and colleagues (Engelen, White and Uljee, *et al*., 2002) used three exogenous models to establish the level of demand for cell state transition in its automata-based microsimulation of urbanization. Sea-level rises, as simulated in a natural environment model, may trigger a conversion in state variables from active land-use to inactive sea uses at low elevations, for example. Many top-down systems, such as the evolution of public policy, are very difficult to "get right" in simulation and are perhaps best handled as a given, but external, influence on system dynamics. Indeed, a model might actually have been built to evaluate the nitty-gritty system dynamics that follow on from the use of various policy "levers" on a system's trajectory.

Growth rates may also be specified exogenously, for example, when growth is assumed to originate outside a given urban system, for example, through in-migration or as a function of economic and demographic links with other systems (Torrens, 2006b). White and Engelen (1993), for example, treated growth as external to their cellular automata models because they regarded growth in a real city as being dependent on the position of a city in its regional or national economy, rather than its internal spatial structure. However, problems may arise if exogenously-specified constraints conceal the actual interactions between local states in a CA (Straatman, White and Engelen, 2004).

23.2.6 Automatic calibration

Urban automata models may be specified with a bewildering array of inter-dependent parameters and widgets. Calibrating those mechanisms often requires the use of high-performance computing (Hecker *et al.*, 1999; Bandini, Mauri and Serra, 2001; Nagel and Rickert, 2001; Guan, Clarke and Zhang, 2006). Several authors have developed schemes for automatic calibration, typically adding calibration modularly as a step in a simulation run. A range of techniques have been employed to achieve this, including brute-force approaches, optimized search routines, and self-modification schemes.

Brute-force calibration (Silva and Clarke, 2002) is a procedure for essentially throwing computer power at a model to run it over many permutations and combinations of parameter values. Several developments to the SLEUTH model have been undertaken around these goals (Guan, Clarke and Zhang, 2006). In such instances, results of varying parameters are sorted according to some metric, and the highest-scoring results are fed into the next iteration of the procedure.

Optimized search procedures are similar to the brute-force approach, but introduce an element of machine intelligence. Unlike the brute-force approach, which rank-selects parameter values rather simply, optimized search schemes use a variety of guidelines (decision rules) to target parameter adjustments. This approach begets related issues: how to gauge error; how to formulate a decision rule (adjust up, adjust down, aim for the optimal solution, go for a less-than-optimal solution); how to choose between adjustments that appear to yield the same improvement; and what to do if the adjustment decision yields an unsatisfactory result that manifests only after several subsequent transitions have taken place. In the example developed by Straatman and colleagues (2004), a procedure was introduced that targeted searches based on benchmarks for maximum error (the neighborhood with the greatest difference between desired potential and undesired potential) and total error (the number of neighborhoods, or cell count, that resulted in a wrong state). Adjustment was carried out by means of "length searches" (adjustments that yield a result that is more likely to convene on a desired state, or less likely to result in an undesired state) and "width searches" (adjustment guided by the relative ability of the adjustment options to reduce total error). To avoid having to make random choices between similarly-optimal adjustments, a variety of techniques were proposed, including making the length and width searches inter-dependent, varying those approaches, and making use of "backtracking" (setting the adjustment procedure back by a number of transition steps, and relaxing the adjustment to allow for less-than-optimal solutions).

An alternative method for automatic calibration involves the use of self-modifying mechanisms: transition rules are allowed to change (in relative importance or weighting) as a simulation evolves. Under self-modifying calibration schemes, changes in parameter values are often linked to evolving conditions within the model itself. In the SLEUTH model, for example, the rate of growth serves as a trigger for adaptation in the application of rules (Clarke, 1997; Clarke, Hoppen and Gaydos, 1997; Clarke and Gaydos, 1998; Candau, Rasmussen and Clarke, 2000; Silva and Clarke, 2002; Goldstein *et al.*, 2004;). Under conditions of rapid growth, growth control parameters in the model are exaggerated. Essentially, this acts as a brake on the growth metabolism or momentum of the model. If a simulation exhibits "little or no growth", the growth parameters are adjusted to dampen growth. Without self-modification of this nature, simulation runs would produce only linear or exponential growth (Silva and Clarke, 2002).

23.3 Validating automata models

For the most part, model calibration takes place before a simulation run. Validation relates to the assessment of a model's performance, and this generally takes place after a simulation has been run. In abstract terms, we may differentiate between qualitative validation and quantitative validation. The former refers to the evaluation of general agreement between a simulation and observed conditions; the latter denotes the assessment of empirical goodness-of-fit between simulation outputs and observed conditions. We can also distinguish between validation mechanisms designed to assess model performance through analysis of the patterns and outputs that a simulation generates and mechanisms that analyze a simulation run itself, as a simulation of the system being considered.

23.3.1 Pixel matching

Validation by visual inspection is one of the most straightforward (but subjective) techniques for assessing the performance of a model. In a sense, model designers or users act as their own test case in studying the system as it unfolds in simulation. For example, one might evaluate whether a model generated plausible urban forms (Wu, 1998), or whether the simulated processes operated at sensible rates and with appropriate consequences (Torrens, 2006b). In their models of urban growth in the San Francisco Bay Area, Clarke and colleagues used visual validation to determine estimates for parameter settings in the model. Simulations were run, their performance was evaluated visually, and parameters were adjusted based on those evaluations if necessary (Clarke, Hoppen and Gaydos, 1997). They looked, in particular, at whether their model generated realistic patterns of historical growth, and realistic urban extents. They also looked at the plausibility of area, edge, and cluster attributes in the simulation output.

One might opt, alternatively, to have a *computer* inspect the visual match between a model's output and a real-world pattern. This is usually done by pixel matching or by pattern recognition. Pattern recognition implies a targeted search – analysis to determine the presence of specific features, artifacts, or configurations that are known or spelled-out *a priori*. Pixel-matching, by contrast, is a more mechanical approach, involving the analysis of the composition of a particular scene and usually this involves matching the pixel images generated by a model to those in a digitized map or remotely sensed image, but – and this is important – ignoring the configurations of those elements in relation to each other (Torrens, 2006a). Pixels (picture elements) in one scene are simply registered to another scene and the level of coincidence between the two is gauged.

The most straightforward approach to pixel matching is to fashion an inventory of pixel attributes within a simulated scene and to compare the results to corresponding attributes in a remotely sensed image. In analyzing output from their models, Clarke and colleagues (1997) collated information regarding the total number of urban pixels (urban extent), the number of edge pixels on the boundary of a simulated landscape, and the amount of pixel clusters at various stages in the course of a simulation run. They then performed a validation exercise by determining the correlation between those results and observed "knowns" from historical maps of the area. Pixel matching can also be performed on a pixel-by-pixel basis. These techniques originate in image processing, where they are used to evaluate classification accuracy. Coincidence matrices are commonly used to register cell-by-cell comparisons between simulated and remotely sensed scenes: pixels are evaluated to determine whether they are identical on both scenes, in terms of states. In some instances, error data is also calculated. Mismatches may be recorded, and moving filters can be used to discern displacement in the mismatch. There is a problem, however, with spatial autocorrelation (Moran, 1950), that weakens the reliability of correlation statistics of this form when applied over geographically-coincident pixels, and with serial autocorrelation when matching is performed between temporal snapshots (Berry, 1993). Similarly, results at one scale of observation or matching may not hold at other scales (Qi and Wu, 1996).

The kappa-statistic is commonly used in conjunction with pixel matching coincidence matrices. Low values of kappa indicate conditions in which there is little correspondence between observed and expected scenes; high values indicate a "good" match between the two. The kappa-statistic has been used in these contexts in a number of urban cellular automata models (White, Engelen and Uljee, et al., 997; Wu, 1998), but there are some important complications associated with the technique. The statistic is almost unsuitably sensitive. If a simulated scene is similar to an observed scene, but the correspondence is mismatched by just one pixel in a few places, the overall accuracy of the match may suffer considerably (Wu, 1998). Dislocation of this variety may be particularly problematic in sparsely-populated areas of a scene (Wu and Webster, 1998). Correspondence between observed and expected results may be highly susceptible to variation when different resolutions are used. Coarse resolutions essentially "average out" disagreement between scenes. In the context of urban models, there may be difficulties associated with particular state variables. States designed to represent transport infrastructure are an excellent example. White and colleagues (1997) noted a displacement – of one cell in value – owing to the rasterization of road and rail in their model. This may seem minor, but it ended up becoming (statistically) significant in the context of coincidence matrices. Moreover, the problem, in this example, was transport-specific, and therefore introduced error to a significant feature of the simulation by misrepresenting transport land-use specifically.

23.3.2 Feature and pattern recognition

Recognition exercises focus on identifying particular (and theoretically well-understood and/or significant) features and patterns within model output. An advantage of pattern generation in urban simulations is that the patterns to be sought are often robust to changes in rules and parameter values (Andersson *et al.*, 2002). Whereas pixel-matching techniques deal with relatively simple compositional correspondence between scenes, pattern recognition techniques are used to measure specific (space–time) structures in model output.

For example, edge detection may be used to measure the morphology of urban boundaries: the perimeter of an entire city, or the boundary of zones of activity within a city, or well-understood morphologies such as downtown, suburbs, exurbs, and urban–rural fringe (Torrens, 2008). Fractal analysis (Batty and Longley, 1994) can be employed in determining space-filling of urban (or landscape) processes and this has the advantage of being quantifiable across model variables and can help in avoiding thorny scale issues. Cluster-frequency spectra may be used to compare hierarchies in urban clustering, to look at central-place structures, for example. White and colleagues (1997) have performed extensive analysis of their models using these techniques, as have Batty and Xie (Xie, 1996). A similar approach is employed by Torrens (2006b) in modeling suburbanization within the American midwestern megalopolis, where clustering-frequency is used to distinguish between the varying growth trajectories of cities within that urban system.

Another approach to pattern-based validation is to study urban "patchiness." In landscape ecology (Forman and Gordon, 1986), a patch is representative of a spatial object (with homogenous characteristics or states) situated within a broader landscape, e.g., the coverage of a particular vegetative type within a larger forest composed of several vegetative species. One can easily see how this concept might be related to urban contexts: we could distinguish clusters of retail activity amidst a sea of urban development, "ecologically," for example. There is a variety of spatial analysis techniques associated with landscape ecology, each designed to support the quantification of composition and configuration properties of patches in landscapes (Gustafson, 1998). Several such landscape metrics have been used to validate urban automata models. Goldstein and colleagues (2004), for example, have calculated a variety of spatial metrics to explore the robustness of their SLEUTH simulation of the Santa Barbara region. These metrics are also employed by Torrens in exploring the results of his automata simulations (Torrens, 2006b), as well as in measuring patterns of sprawl in real-world contexts (Torrens, 2008).

Fuzzy pattern recognition directly addresses the weaknesses of pixel-matching techniques, by introducing additional functionality. First, maps, or images that are being registered for agreement are processed to yield "soft," fuzzy or transitional, boundaries between pixels, rather than the "hard," crisp, discrete boundaries used in basic matching methodologies (Heikkila, Shen and Kaizhong, 2003). Second, "fuzzy logic" (Kosko, 1993) is used to determine the agreement between scenes, using AI-inspired pattern recognition rather than brute force matching. In addition to determining the *overall* agreement between scenes, steps are added to the procedure to determine *localized* agreement (Liu and Phinn, 2003), using, for example, linguistic membership functions. Power and colleagues (2000) used fuzzy validation procedures, in this context, to assess the performance of automata models developed by Engelen and colleagues (1995), enabling disagreement between observed and expected scenes to be broken down into state-specific (land-use types) details. This has obvious advantages, beyond remedying the weaknesses of brute-force pixel-matching; it could, for example, highlight whether particular state variables were subject to systematic agreement errors (which could point to data problems or could indicate areas where amendments to a model's rules might be appropriate).

23.3.3 Running models exhaustively

Sweeping the parameter space of a model involves exploring the complete range of outcomes possible with a particular model specification, or looking at its "space of possibilities" (Couclelis, 1997). One approach is to map those possibilities using graphs. Finite state transition graphs can be used to examine the *global evolution* – step-by-step or transition-by-transition – of an automata simulation, relying on graphs (networks) to plot the "state space" of an automaton (Wolfram, 1994). Essentially, the complete trajectory of a model can be visualized. Early inroads are being made toward such a scheme, beginning with data-mining procedures for automata models (Hu and Xie, 2006).

Using stochastic (probabilistic) constraint parameters creates an interesting problem: different results can be produced from identical parameterizations; there are often a near infinite number of micro-states that might determine macro-conditions, even for a small set of model parameterizations (Wilson, 1970; Oreskes, Shrader-Frechette and Belitz, 1994). One way to "smooth out" this sort of variation and to narrow the candidate configurations to a more manageable set size is to employ Monte Carlo averaging. Simulations can be run from identical conditions or using the same parameter values (variation then comes from different random number draws used in simulation); they may also be run repeatedly, using different combinations of parameter values (Li and Yeh, 2000), or using variable start conditions. Monte Carlo averaging is also useful for generating probability maps for use in prediction. Simulations with the SLEUTH model, for example, have been run using Monte Carlo Averaging in an application to the Santa Barbara Region, enabling the selection of locations based on a cut-off rate of 90% success (Goldstein, Candau and Clarke, 2004).

Conclusions

This chapter has presented a review and discussion of the state-of-the-art in calibration and validation of automata models to urbanization applications in the context of complex and dynamic city-systems. A variety of techniques have been introduced and assessed. It is important to realize, however, that many of the models described in this chapter are in very early stages of development. Also, in the context of calibration and validation, it is noteworthy that the intentions of many of the modeling projects and exercises discussed here differ from those that characterize work in land-use and transport modeling traditions common in support of municipal planning. Automata modeling represents somewhat of a paradigm shift in urban simulation (Albrecht, 2005), away from thinking of models as diagnostic or prescriptive tools, toward a conceptualization of urban models as artificial laboratories for experimenting with ideas about urban dynamics. Consequently, many models may not be validated at all; they may be developed as pedagogic instruments, or as "tools to think with". Accordingly, we might consider a broad spectrum of models (Torrens and O'Sullivan, 2001), ranging from very simply-parameterized models in the tradition of Wolfram's CA designed to test universality (Wolfram, 1984) to "fuller" planning support systems designed to assist planning management, and policy exercises (Torrens, 2002; Engelen, White and Uljee, 2002). Depending on their position on this spectrum, models may have different calibration and validation requirements.

Nevertheless, there are some important issues that relate to calibration and validation across that spectrum, including simplicity in model specification, data issues, the generality of models, and the relationship between models and urban theory. Simplicity is one of the most commonly advertised advantages of automata models. This stems from their association with the idea of generative emergence – the concept that simple rules can generate surprising and intricate complexity and that, unlike chaos, the path from simplicity to complexity can be traced through a causal relationship (Batty and Torrens, 2005). Those ideas are often taken at face value when automata models are developed: simple parameters are often introduced to models – and choice may be a function of what data are to hand – and simulations are run as "blue skies" experiments to see, essentially, what will come out. A problem with using automata in this regard is that as simulations runs evolve beyond initial conditions, automata (and particularly urban cellular automata) have a strong tendency to "go exponential" and must often be tightly constrained in order to produce patterns that resemble real cities. The resulting "soup" is sometimes confused with emergent phenomena (Faith, 1998; Epstein, 1999).

Another issue relates to the role of automata modeling in the experimental process. Urban automata models are often borrowed from the physical sciences; while automata methodologies are usually similar regardless of application (an automaton is an automaton after all), urban automata *experiments* often bear little resemblance to those in fields such as computer science, physics, and chemistry (Oreskes, Shrader-Frechette and Belitz, 1994), where experiments deal with systems that are relatively well-understood in comparison to urban systems. The idea of simple models generating surprising complexity is a powerful one, but is predicated on the notion that an (often small) set of the "right" rules are out there, that the simple ingredients to a complex system can be uncovered. As Wolfram wrote, a model developer should:

> 'attempt to distill the mathematical essence of the process by which complex behavior is generated. ... To discover and analyze the mathematical basis for the generation of complexity, one must identify simple mathematical systems that capture the essence of the process'
>
> (Wolfram, 1994) (p.411).

These sorts of sentiment are wonderfully romantic in their tractability and parsimony. Alas, the reality is quite different: indeed, Wolfram spent the next two decades searching for sets of simple rules (Wolfram, 2002) and many believe the issue to be a red herring (Horgan, 1995). Of course, in the context of urban systems, we often have no idea what the "right" rules are (if we did, planning cities would be easy). The rules are always changing and there are myriad confounding influences on any given urban process. Nevertheless, urban automata models can be used as exploratory tools, experiments for evaluating what the "right" rules might actually be, what dynamics might result given a particular set of candidates for the "right" rules, or where we might efficiently devote our intellectual efforts in searching for them. Recent emphasis in the literature on calibration and validation is certainly taking the field in that direction and provides opportunities for it to evolve from early, experimental phases.

Validation and calibration verification exercises are inextricably intertwined with data. This relationship can be synergetic in some cases, and terribly constraining in others. One criticism that could be made of several of the parameter-based calibration mechanisms that I have discussed in this chapter is that they are data-specific; models may become "captives of their data sets" (Ward, phin and Murray, 2000). Also, many automata models require individual-scale data, of individual households, parcels of land, and so on. Some detailed models have been developed where model developers have access to entity-scale databases (Benenson, Omer and Hatna, 2002), but such data may not always be available. A confounding issue is that calibration and validation with data from particular locations may result in a model that is "fit" for use in that location, but is not applicable to other cities or times. The danger is that, after careful calibration and validation, you may end up, not with a model of sprawl, for example, but with a model of "sprawl in Bloomington, Indiana."

A way to dodge this problem is to design urban automata models for general use, with general rules of urbanization that transcend the specifics of a particular location, problem, or period. The models developed by Roger White, Guy Engelen, and colleagues at the Research Institute for Knowledge Systems (RIKS) are a good example of this approach; the models have been applied to a diverse range of scenarios in various locations, including the Netherlands, Saint Lucia, Dublin, and Cincinnati (White and Engelen, 1993, 1994, 1997, 2000; Engelen *et al.*, 1995; White, Engelen and Uljee, 1997; White, 1998; Engelen, White and Uljee, 2002; Power, Simms and White, 2000). Keith Clarke's SLEUTH model has also seen a rich range of applications. It has been used to model urbanization in Santa Barbara, the San Francisco Bay Area, Washington DC and Baltimore, Lisbon, and Porto (Clarke, 1997; Clarke, Hoppen and Gaydos, 1997; Clarke and Gaydos, 1998; Candau, Rasmussen and Clarke, 2000; Herold, 2002; Silva and Clarke, 2002; Goldstein, Candau and Clarke, 2004). In addition, Benenson and colleagues at Tel Aviv University's Environmental Simulation Laboratory have developed a general all-purpose modeling environment (Benenson, Birfur and Kharbash, 2006) for building simulations of *any* description, based on the idea of reconfigurable and malleable Geographic Automata Systems (Torrens and Benenson, 2005). However, general models lose some fidelity to detail, almost by definition, and how to reconcile more specific models in this sort of ecosystem is a question that remains to be answered.

Ultimately, several of these issues discussed above relate to a central concern: the relationship between models and theory. Models can be calibrated with vast quantities of detailed data, and using sophisticated procedures. They can be validated for historical time periods with high degrees of success. However, a model is only as good as the rules that drive its behavior. Good rules require good theory. The relationship is symbiotic: good theory often relies on good models to test the theory. Interestingly, the emergence of automata models has facilitated the exploration of new question about urban systems, and the evaluation of hypotheses that were hitherto inaccessible with conventional simulation methodology (Batty, 2005). But, in many instances, theory has been found wanting, particularly at microscales and in relation to phenomena that operate across scales. Conventional simulation methodology is inadequate, in many respects, for exploring ideas about urban dynamics in terms of complex adaptive systems. Nevertheless, these issues were present in previous generations of urban models and remain unresolved. While still in its relative infancy as a field, automata models, with their almost limitless malleability, are the best option we have for reconciling theory, plans, and reality through simulation.

Acknowledgments

This material is based upon work supported by the National Science Foundation under Grants Nos. 1002519 and 0643322.

References

Aerts, J.C.J.H., Clarke, K.C. and Keuper, A.D. (2003) Testing popular visualization techniques for representing model uncertainty. *Cartography and Geographic Information Science*, **30**, 249–261.

Albrecht, J. (2005) A new age for geosimulation. *Transactions in Geographic Information Science*, **9**, 451–454.

Allen, P.M. and Torrens, P.M. (2005) Knowledge and complexity. *Futures*, **37**, 581–584.

Almeida, C.M., Gleriani, J.M., Castejon, E.F. and Soares-Filho, B.S. (2008) Using neural networks and cellular automata for modelling intra-urban land-use dynamics. *International Journal of Geographic Information Science*, **22**, 943–963.

Andersson, C., Steen, R. and White, R. (2002) Urban settlement transitions. *Environment and Planning B: Planning and Design*, **29**, 841–865.

Arai, T. and Akiyama, T. (2004) Empirical analysis of the land use transition potential in a CA based land use model: application to the Tokyo Metropolitan Region. *Computers, Environment and Urban Systems*, **28**, 65–84.

Arthur, W.B. (1990) Positive feedbacks in the economy. *Scientific American*, February, 80–85.

Bandini, S., Mauri, G. and Serra, R. (2001) Cellular automata: From a theoretical parallel computational model to its application to complex systems. *Parallel Computing*, **27**, 539–553.

Batty, M. (2005) *Cities and Complexity: Understanding Cities with Cellular Automata, Agent-Based Models, and Fractals*, Cambridge, MA, The MIT Press.

Batty, M. and Longley, P. (1994) *Fractal Cities*, London, Academic Press.

Batty, M. and Torrens, P.M. (2005) Modeling and prediction in a complex world. *Futures*, **37**, 745–766.

Batty, M. and Xie, Y. (1997) Possible urban automata. *Environment and Planning B*, **24**, 175–192.

Batty, M., Xie, Y. and Sun, Z. (1999) Modeling urban dynamics through GIS-based cellular automata. *Computers, Environment and Urban Systems*, **23**, 205–233.

Benenson, I., Birfur, S. and Kharbash, V. (2006) Geographic automata systems and the OBEUS software for their implementation. In: PORTUGALI, J. (ed.) *Complex Artificial Environments: Simulation, Cognition and VR in the Study and Planning of Cities*, Springer-Verlag, Berlin..

Benenson, I., Omer, I. and Hatna, E. (2002) Entity-based modeling of urban residential dynamics: the case of Yaffo, Tel Aviv. *Environment and Planning B: Planning and Design*, **29**, 491–512.

Benenson, I. and Torrens, P.M. (2004) *Geosimulation: Automata-Based Modeling of Urban Phenomena*, John Wiley & Sons, Chichester..

Berry, W.D. (1993) *Understanding Regression Assumptions*, Sage, Newbury Park, CA..

Bone, C., Dragićević, S. and Roberts, A. (2007) Evaluating forest management practices using a GIS-based cellular automata modeling approach with multispectral imagery. *Environmental Modeling and Assessment*, **12**, 105–118.

Brail, R.K. and Klosterman, R.E. 2001. *Planning Support Systems in Practice: Integrating Geographic Information Systems, Models, and Visualization Tools*, ESRI Press and Center for Urban Policy Research Press, Redlands, CA and New Brunswick, NJ..

Candau, J.T., Rasmussen, S. and Clarke, K.C. (2000) A coupled cellular automata model for land use/land cover change dynamics, in *4th International Conference on Integrating GIS and Environmental Modeling (GIS/EM4): Problems, Prospects and Research Needs*, Banff, Alberta, Canada.

Clarke, K. 1997. Land transition modeling with deltatrons. Department of Geography, University of California, Santa Barbara. Online.Available: http://www.geog.ucsb.edu/~kclarke/Papers/deltatron.html (accessed 19 November 2010).

Clarke, K. C. and Gaydos, L. (1998) Loose coupling a cellular automaton model and GIS: long-term growth prediction for San Francisco and Washington/Baltimore. *International Journal of Geographical Information Science*, **12**, 699–714.

Clarke, K.C., Gazulis, N., Dietzel, C. and Goldstein, N.C. (2007) A decade of SLEUTHing: Lessons learned from applications of a cellular automaton land use change model, in *Classics in IJGIS: Twenty years of the International Journal of Geographical Information Science and Systems* (ed. P. Fisher, P.), CRC Press, Boca Raton, FL.

Clarke, K.C., Hoppen, S. and Gaydos, L. (1997) A self-modifying cellular automaton model of historical urbanization in the San Francisco Bay area. *Environment and Planning B*, **24**, 247–261.

Couclelis, H. (1997) From cellular automata to urban models: New principles for model development and implementation. *Environment and Planning B*, **24**, 165–174.

Dragićević, S. and Kocabas, V. (2007) Enhancing GIS cellular automata model of land use change: Bayesian networks, influence diagrams and causality. *Transactions in GIS*, **11**, 679–700.

Engelen, G., White, R. and Uljee, I. (2002) *The MURBANDY and MOLAND models for Dublin*. Research Institute for Knowledge Systems (RIKS) BV, Maastricht..

Engelen, G., White, R., Uljee, I. and Drazan, P. (1995) Using cellular automata for integrated modelling of socio-environmental systems. *Environmental Monitoring and Assessment*, **30**, 203–214.

Epstein, J.M. (1999) Agent-based computational models and generative social science. *Complexity*, **4**, 41–60.

Faith, J. (1998) Why gliders don't exist: anti-reductionism and emergence, in *Artificial Life VI: Proceedings of the Sixth International Conference on Artificial Life* (ed. C. Adami) MIT Press, Cambridge, MA.

Forman, R. and Gordon, M. (1986) *Landscape Ecology*, John Wiley & Sons, Inc., New York..

Goldstein, N.C., Candau, J.T. and Clarke, K.C. 2004. Approaches to simulating the &March of Bricks And Mortar". *Computers, Environment and Urban Systems*, **28**, 125–147.

Grimm, V. (1999) Ten years of individual-based modelling in ecology: what have we learned and what could we learn in the future. *Ecological Modelling*, **115**, 129–148.

Guan, Q., Clarke, K. C. and Zhang, T. (2006) Calibrating a parallel geographic cellular automata model. AutoCarto, Vancouver, WA.

Gustafson, E.J. (1998) Quantifying landscape spatial pattern: what is the state of the art? *Ecosystems*, **1**, 143–156.

Hall, P. (1988) *Cities of Tomorrow: An Intellectual History of Urban Planning and Design in the Twentieth Century*, Blackwell, Oxford.

Hecker, C., Roytenberg, D., Sack, J.-R. and Wang, Z. (1999) System development for parallel cellular automata and its applications. *Future Generation Computer System*, **16**, 235–247.

Heikkila, E.J., Shen, T. and Kaizhong, Y. (2003) Fuzzy urban sets: theory and application to desakota regions in China. *Environment and Planning B*, **30**, 239–254.

Herold, M. (2002) Remote sensing based analysis of urban dynamics in the Santa Barbara Region using the SLEUTH urban growth model and spatial metrics, in *Third Symposium on Remote Sensing of Urban Areas*, Istanbul, Turkey.

Herold, M. and Clarke, K.C. (2002) The use of remote sensing and landscape metrics to describe structures and changes in urban land uses. *Environment and Planning A*, **34**, 1443–1458.

Herold, M., Couclelis, H. and Clarke, K.C. (2005) The role of spatial metrics in the analysis and modeling of urban land use change. *Computers, Environment and Urban Systems*, **29**, 369–399.

Horgan, J. (1995) From complexity to perplexity: Can science achieve a unified theory of complexity systems? Even at the Santa Fe Institute, some researchers have their doubts. *Scientific American*, **284**, 104–109.

Hu, G. and Xie, Y. (2006) An extended cellular automata model for data mining of land development data, in *First IEEE/AICS International Workshop on Component-Based Software Engineering, Software Architecture and Reuse*, Institute of Electrical and Electronics Engineers Computer Society, Honolulu, HI. pp. 201–207.

Kocabas, V. and Dragićević, S. (2006) Assessing cellular automata model behaviour using a sensitivity analysis approach. *Computers, Environment and Urban Systems*, **30**, 921–953.

Kosko, B. (1993) *Thinking Fuzzy: The New Science of Fuzzy Logic*, Hyperion Press, New York.

Li, X. and Yeh, A.G.-O. (2000) Modelling sustainable urban development by the integration of constrained cellular automata and GIS. *International Journal of Geographical Information Science*, **14**, 131–152.

Liu, Y. and Phinn, S.R. (2003) Modeling urban development with cellular automata incorporating fuzzy-set approaches. *Computers, Environment and Urban Systems*, **27**, 637–658.

Manson, S. M. (2001) Simplifying complexity: A review of complexity theory. *Geoforum*, **32**, 405–414.

Manson, S.M. (2007) Challenges in evaluating models of geographic complexity. *Environment and Planning B*, **34**, 245–260.

Moran, P.A.P. (1950) Notes on continuous stochastic phenomena. *Biometrika*, **37**, 17–23.

Nagel, K. and Rickert, M. (2001) Parallel implementation of the TRANSIMS micro-simulation. *Parallel Computing*, **27**, 1611–1639.

Oreskes, N., Shrader-Frechette, K. and Belitz, K. (1994) Verification, validation, and confirmation of numerical models in the earth sciences. *Science*, **263**, 641–646.

Parker, D.C., Manson, S.M., Janssen, M.A., Hoffmann, M.J. and Deadman, P. (2003) Multi-Agent System models for the simulation of land-use and land-cover change: a review. *Annals of the Association of American Geographers*, **93**, 314–337.

Pontius Jr., R.G., Boersma, W., Castella, J.-C., *et al.* (2008) Comparing the input, output, and validation maps for several models of land change. *Annals of Regional Science*, **42**, 11–47.

Power, C., Simms, A. and White, R. (2000) Hierarchical fuzzy pattern matching for the regional comparison of land use maps. *International Journal of Geographical Information Systems*, **15**, 77–100.

Qi, Y. and Wu, J. (1996) Effects of changing spatial resolution on the results of landscape pattern analysis using spatial autocorrelation indices. *Landscape Ecology*, **11**, 39–49.

Silva, E.A. and Clarke, K.C. (2002) Calibration of the SLEUTH urban growth model for Lisbon and Porto, Portugal. *Computers, Environment and Urban Systems*, **26**, 525–552.

Southworth, F. (1995) *A technical review of urban land use: Transportation models as tools for evaluating vehicle reduction strategies*. Oak Ridge National Laboratories, Oak Ridge, TN.

Straatman, B., White, R. and Engelen, G. (2004) Towards an automatic calibration procedure for constrained cellular automata. *Computers, Environment and Urban Systems*, **28** (1–2), 149–170.

Torrens, P.M. (2002) Cellular automata and multi-agent systems as planning support tools, *Planning Support Systems in Practice* (eds S. Geertman and J. Stillwell), Springer-Verlag, London.

Torrens, P.M. (2006a) Remote sensing as dataware for human settlement simulation, in *Remote Sensing of Human Settlements* (eds M. Ridd and J.D. Hipple), American Society of Photogrammetry and Remote Sensing, Bethesda, MA.

Torrens, P.M. (2006b) Simulating sprawl. *Annals of the Association of American Geographers*, **96**, 248–275.

Torrens, P.M. (2008) A toolkit for measuring sprawl. *Applied Spatial Analysis and Policy*, **1**, 5–36.

Torrens, P.M. and Benenson, I. (2005) Geographic Automata Systems. *International Journal of Geographical Information Science*, **19**, 385–412.

Torrens, P.M. and O'Sullivan, D. (2001) Cellular automata and urban simulation: where do we go from here? *Environment and Planning B*, **28**, 163–168.

Ward, D.P., Murray, A.T. and Phinn, S.R. (2000) A stochastically constrained cellular model of urban growth. *Computers, Environment and Urban Systems*, **24**, 539–558.

Ward, D.P., Phinn, S.R. and Murray, A.T. (2000) Monitoring growth in rapidly urbanizing areas using remotely sensed data. *The Professional Geographer*, **52** (3), 371–386.

Wegener, M. (1994) Urban/regional models and planning cultures: lessons from cross-national modelling project. *Environment and Planning B*, **21**, 629–641.

White, R. (1998) Cities and cellular automata. *Discrete Dynamics in Nature and Society*, **2**, 111–125.

White, R. and Engelen, G. (1993) Cellular automata and fractal urban form. *Environment and Planning A*, **25**, 1175–1199.

White, R. and Engelen, G. (1994) Urban systems dynamics and cellular automata: fractal structures between order and chaos. *Chaos, Solitions, and Fractals*, **4** (4), 563–583.

White, R. and Engelen, G. (1997) Cellular automata as the basis of integrated dynamic regional modelling. *Environment and Planning B*, **24**, 235–246.

White, R. and Engelen, G. (2000) High-resolution integrated modelling of the spatial dynamics of urban and regional systems. *Computers, Environment and Urban Systems*, **24**, 383–400.

White, R., Engelen, G. and Uljee, I. (1997) The use of constrained cellular automata for high-resolution modelling of urban land use dynamics. *Environment and Planning B*, **24**, 323–343.

Wilson, A.G. (1970) *Entropy in Urban and Regional Modelling*, Pion, London.

Wolfram, S. (1984) Universality and complexity in cellular automata. *Physica D*, **10**, 1–35.

Wolfram, S. (1994) *Cellular Automata and Complexity*, Addison-Wesley, Reading, MA.

Wolfram, S. (2002) *A New Kind of Science*, Wolfram Media, Inc., Champaign, IL.

Wu, F. (1998) An experiment on the generic polycentricity of urban growth in a cellular automatic city. *Environment and Planning B*, **25**, 731–752.

Wu, F. and Webster, C.J. (1998) Simulation of land development through the integration of cellular automata and multicriteria evaluation. *Environment and Planning B*, **25**, 103–126.

Xie, Y. (1996) A generalized model for cellular urban dynamics. *Geographical Analysis*, **28**, 350–373.

Yeh, A. G.-o. and Li, X. (2002) A cellular automata model to simulate development density for urban planning. *Environment and Planning B: Planning and Design*, **29**, 431–450.

24

Agent-based urban modeling: simulating urban growth and subsequent landscape change in suzhou, china

Yichun Xie and Xining Yang

Agent-based modeling (ABM), as a computational model for simulating the actions and interactions of autonomous individuals in a network, has attracted growing attention in urban modeling arena in recent years. The ABM can integrate natural environments (the agents' physical space) with policy making rules (the agents' intelligence), combine bottom-up actions with global interactions, and simulate processes of urban growth that are locally determined but moderated by higher-level macro economy. In this chapter, we introduce agent-based urban modeling and survey current literature. We then describe four key elements of ABM: model design, model construction, model calibration, and model validation. Furthermore, we provide a case study to illustrate the aforementioned modeling procedures by applying them to the desakota development in the Suzhou region (China) between 1990 and 2000. Two types of agents are designed: (1) the township policy agents who control global development policies and competitiveness of towns and villages in the model; and (2) the developer agents who determine where open land is to be converted into urban use to seek economic profits. The simulation outcomes produce spatial patterns of urban growth close to those observed over the calibration period 1990 to 2000.

Urban Remote Sensing: Monitoring, Synthesis and Modeling in the Urban Environment, First Edition. Edited by Xiaojun Yang.
© 2011 John Wiley & Sons, Ltd. Published 2011 by John Wiley & Sons, Ltd.

24.1 Introduction

Urban landscape changes are usually outcomes of urban growth. It has been a challenge for both researchers and practitioners to capture the dynamics of urban growth and the causal relationships between urban landscape changes and underlying socioeconomic driving forces. Rigorous efforts have been made in developing computational and mathematical models to explore, explain, and predict urban growth during the past half a century. However, urban modeling has been a controversial field since its reception. The critics claim that conventional urban modeling is either based on aggregated numerical analysis, or generalized behavior approach. These models are either averaging out the characteristics of study units of interest, or assuming that urban drivers behave in a rational way (Ligmann-Zielinska and Jankowski, 2007). True urban complexity may be lost in the process of generalization forced by mathematical tractability (Axtell, 2003). Moreover, the differential equation-based approach may unnecessarily make the model mathematically complicated and therefore the potential user perceives it as a black box (Lee, 1973). Urban modeling was regarded as a misunderstanding of the role of modeling in real-world planning (Lee, 1973).

Only very recently have the conceptual and mathematical foundations for substantive inquiry into urban dynamics been made possible due to the growing understanding of open systems structures and human decision processes. The applications of systems theory to urban dynamics have been made possible by fundamental advances in the theories of nonlinear systems, including dissipative structures, synergetics, chaos and bifurcation in the physical sciences. In fact many of the originators of these new ways of articulating how complex systems work, have seen cities as being a natural and relevant focus for their work. Prigogine's work on dissipative structures, for example, has been applied to urban and regional systems by Allen (1997) while Haken's work on self-organization has been implemented for city systems by Portugali (2000) and Weidlich (2000). Ideas about how life can be created artificially have guided many of these developments and in this context, highly disaggregate dynamic models based on cellular automata (CA) have become popular as the metaphor for a complex system (Batty and Xie, 1994; Xie, 1996). CA models articulate a concern that systems are driven from the bottom up, in which local rules generate global patterns, and are good indicators that there is no hidden hand in the form of top-down control. Again cities are excellent exemplars of these kinds of systems (Holland, 1975). More recently CA has been elevated to the status of a 'new science', articulated as the basis for taking a new look at a wide range of applications to scientific inquiry (Wolfram, 2002).

In fact, CA focuses on physical processes of urban systems and simulates land use changes through rules usually acting upon immediate neighboring cells or at best some wider set of cells in which the notion of a restricted neighborhood for spatial influence is central (Batty, 1998; Batty, Xie and Sun, 1999; Clarke and Gaydos, 1998; Li and Yeh, 1998; White and Engelen, 1993; Wu and Webster, 1998; Wu, 2002; Xie, 1996). Though many innovative ideas such as genetic algorithms, neural network methods, and stochastic calibration for determining weights and parameters have been proposed and successfully integrated, such CA models are essentially heuristic and simplistic. Hence, agent-based models (ABMs) and simulations hold a promise to remedy these modeling weaknesses. ABMs (also termed multi-agent systems or individual-based models) are increasingly used to represent decision-making in urban land-change context (Parker, Manson and Janssen, 2003; Xie, Batty and Zhao, 2007). ABMs have been generally employed as a computational model for simulating the actions and interactions of autonomous individuals in a network. Though what agents actually are remains a topic of debate, they are generally regarded to be goal-driven, autonomous, and adaptive. In the context of urban and geographic studies, these agents are 'geo', because they interact with scale-dependent geographic environment in given contexts and have the ability to reason spatially (Yu and Peuquet, 2009). Therefore, these features of ABM are salient to supplement some weaknesses in the commonly used urban models. For instance, the ABM can integrate natural environments (the agents' physical space) with policy making rules (the agents' intelligence), combine bottom-up actions with global interactions, and simulate processes of urban growth that are locally determined but moderated by higher-level macro economy.

24.2 Design, construction, calibration, and validation of ABM

From the implementation point of view, ABM involves four important steps: design, construction, calibration, and validation, which are four key elements of an ABM. Moreover, ABMs are primarily applied as a tool-box for computer simulation at present. Computer simulation (or computational modeling) is one of three "symbol systems" available to social scientists for developing models in addition to the familiar verbal argumentation and mathematics (Ostrom, 1988). The underlying motivation of using computer simulation in developing models is analogous to the use of more familiar statistical methods in modeling. In either case, there is a driving issue that the researchers want to understand and investigate, which is called the "research question". A model exploring this research question is often built through a theoretically motivated process of abstraction (Gilbert and Terna, 2000). However, the approaches that are deployed to assess behaviors of various models are significantly different between statistical models and computer simulations. If the model is a statistical equation, it is run through a statistical analysis program such as SPSS. If the model is a computer simulation, its behavior is assessed by "running" a computational program iteratively to evaluate the effect of different input parameters (Gilbert and Terna, 2000). It is the task of model design to parameterize the behavior of a computer simulation program.

In the urban modeling context, the design task of an ABM is to enable "agents" to represent complex behavior of interacting entities such as households, businesses, planners, developers, or decision-makers in a given (metropolitan) region. Model researchers have to provide theoretical and methodological foundations as well as practical solutions to the identified decision-making task that drives urban land change or urban development policy. Thus the heart of urban ABM design lies in how to abstracting and computerizing a decision-making problem when it is identified. Agents are designed to be capable of making decisions concerning an identified research question, such as choosing locations for any given production activity like urban growth in a simulated landscape according to land suitability (Manson, 2005). The decision making is commonly formulated as a form of multicriteria evaluation (Collins, Steiner and Rushman, 2001; Xiao, Bennett and Armstrong, 2007). However, a key challenge in designing software agents that represent real decision-makers is to determine how each agent behaves over the simulated landscape. In other words, it is a task of how to computerize the behavior of each agent in the context of solving a multicriteria evaluation problem.

There are several approaches employed in current literature to enable agents to make intelligent decisions. From the perspective of dynamic systems concept, evolutionary programming is employed to provide agents with intelligence to solve this multi-criteria evaluation problem as an optimization problem (Bennett and Tang, 2006; Manson, 2006; Xiao, Bennett and Armstrong, 2007). Agents are also empowered with primitive intelligence like swarms in natural environments (Parunak et al, 2006; Alexandridis and Pijanowski, 2007) or random walkers in built environments (Batty, 2001). In a spatially-explicit context, agent-based models can be considered as generalized forms of cellular automata where agents are not restricted to the cells of a raster environment (Goodchild, 2005). Cellular automata can be integrated as a physical infrastructure, over which agents move according to specified rules to generate spatial patterns of urban land uses (Li and Liu, 2008; Liu, Li and Liu, 2008; Torrens and Benenson, 2005; Xie and Batty, 2005). Local neighborhood-based analyses are frequently treated as important inputs for agents to explore or examine rules of decision-making (Deadman and Robinson, 2004; Torrens and Benenson, 2005; Alexandridis and Pijanowski, 2007). Under the consideration of urban planning, probabilities derived from logistic regression models are reformulated as quantitative measurements from which agents make decisions (Waddell, 2002). Other statistical and quantitative methods, such as regression model (Xie, Batty and Zhao, 2007) and discrete-choice model (Parker and Meretsky, 2004; Jepsen, Leisz and Rasmussen, 2006), are deployed as the foundations on which agents make decisions among choices of various land developments, or compare model outputs with observed land-use patterns.

Model construction is an implementation process that codes the model design by using existing tools or programming from scratch for agent models. There are many tools available for constructing ABMs and they are generally grouped into four categories (The Center for the Study of Complex Systems, 2009). The first group is open-source, among which Repast (Recursive Porous Agent Simulation Toolkit [Online] Available at: http://repast.sourceforge.net/ [accessed 19 November 2010]) and Swarm ([Online] Available at: http://www.swarm.org/index.php/Main_Page [accessed 19 November 2010]) are the most known. The second category is freeware. NetLogo ([Online] Available at: http://ccl.northwestern.edu/netlogo/ [accessed 19 November 2010]) and StarLogo ([Online] Available at: http://education.mit.edu/starlogo/ [accessed 19 November 2010]) are the representatives of this group. Proprietary tools, such as AgentSheets and iGEN, belong to the third group. The fourth group includes a good number of tool boxes that have been developed by researchers and many of those are prototypes and hard to share or reuse for a generic purpose (Karssenberg, De Jong and Van Der Kwast 2007).

Model calibration and validation deal with the initiation and the correctness of a model. In other words, calibration means finding out the right values for the parameters contained in a model, while validation aims at proving that a model is built correctly. Hence, calibration is a process of initializing a model such that the model parameters are consistent with the data used to create the model and they are the best fit with the real-world data (Verburg et al., 2006). A rigorous and comprehensive illustration is provided in a paper dealing with the calibration of a SLEUTH model (Dietzel and Clarke, 2007). However, validation is a challenging task in the domain of urban studies. In reality, urban land use changes are spatially specific and temporally dependent. Spatial variations and temporal changes are also impacted by macro-scale policies and local socioeconomic conditions, which are hard to capture in agent models. Moreover, not all variables used in agent models are numerical. Texts and if–then rules are often deployed to describe agent behaviors. Therefore, validation is a critical but challenging task of agent-based modeling.

There are several published papers that provide a comprehensive review of validation techniques designed for spatial models (Turner, Costanza and Sklar, 1989; Kocabas and Dragicevic, 2009). Based on the summary provided by Kocabas and Dragicevic (2009), there exist three positions on validation with regards to complex systems models applied to geographical contexts. First, spatial models cannot be validated in a rigorous way (Oreskes, Shrader-Frechette and Belitz, 1994), or cannot be used for prediction (Batty, 2005). Second, outcomes of ABMs are sensitive to the initial conditions of the model (Parker, Manson and Janssen, 2003). Different initial conditions will lead to varied evolution pathways and result in different structural patterns. Therefore, path dependence analysis (Brown, Page and Riolo, 2005) and structural validation have to be performed to verify an ABM. Third, it is possible to carry out ABM validation, but specific procedures have to be followed.

The heart of these specific procedures relies on comparison of simulation results with real-world observations. It involves running an ABM with a variety of input parameters and observing the program's outputs (Bratley, Fox and Schrage, 1987). The values of the output variable derived from the model runs are compared with the corresponding values of the independent variable measured in observation (Gilbert and Troitzsch, 1999). If the output from the model and the data collected from the real-world observations are sufficiently similar, this will be a good evidence in support of the validity of the model (Gilbert and Terna, 2000).

The most common validation procedure is the map-comparison, which compares the modeled output against the map representing the reality. The relative operating characteristic (ROC) method is used for assessing model validity in the context of land-use-cover changes (Pontius and Schneider, 2001). Chi-square and kappa statistics are often used to quantify the raster-by-raster map comparisons, though there are obvious limitations in the use of the kappa index and coincidence matrix (Barredo, Kasanko and McCormick, 2003; Straatman, White and Engelen, 2004; Hargrove, Hoffman and Hessburg, 2006). There are many other methods used in model validation. For instance, model validation on the basis of multiscale approaches was proposed by Kok and colleagues (2001), and by comparison at pixel-to-pixel-level was tested by Boots and Csillag (2006). Map comparisons are based on landscape metrics (Lei et al., 2005) and vector polygons using the goodness-of-fit at various spatial configurations (Xie and Ye, 2007). A hierarchical fuzzy-pattern-matching method to measure agreement between the model output and the real-world data was proposed by Power, Simms and White (2001) and White (2006). Kocabas and Dragicevic (2009) developed a validation method for agent models based on Bayesian network and vector spatial data.

24.3 Case study – desakota development in Suzhou, China

Suzhou, the cradle of Wu Culture, is a city well-known for its history (tracing back to the later Shang Dynasty more than 2500 years ago) and its classical gardens, called "Paradise on the Earth" or the "East Venice of the World." Suzhou is one of 13 prefecture cities of Jiangsu Province and the study area covers the entire prefecture city, which includes Suzhou and six county-level cities under its jurisdiction: Changshu, Zhangjiagang, Kunshan, Taicang, Wujiang and Wuxian, with the urban district (Shiqu) of Suzhou located within the administrative territory of Wuxian (Fig. 24.1). The municipality of Suzhou has a population of 6.244 million (Suzhou Municipality Statistical Bureau, 2008). Suzhou has served as a model for the development of rural industries based on diversified collective enterprises run by the local municipality model (Tan, 1986), and its network of pre-existing towns and cities (Chen, 1988; Marton, 2000). Today over 50% of GDP is provided by secondary industries with another 30% by tertiary (service) activities, though agriculture still accounts for most employment of the rural labor force (Xie, Yu and Bai, 2006).

There is a remarkable link between rapid urban growth and rural industrialization and subsequent investments in non-farm enterprises (Kirkby, 2000). New towns and urban development have been observed across the entire study area, which has been called "desakota." The desakota emergence issue surrounding rural urbanization stems from complex factors (Lin, 2001; Xie and Batty, 2005). This process to a large degree is local, spontaneous and unplanned (McGee, 1989, 1991, 1998; Marton, 2000; Kirkby, 2000). The phenomenon of desakota conforms with the notion that a structure emerges from a bottom-up process through which local actions and interactions produce global patterns. The nature of desakota fits well with the capabilities of complexity system based modeling approach, like agent-based modeling. Thus, Suzhou Municipality is a good site for applying an agent-based model to study the dynamics of urban growth.

The data for this model was derived from diverse sources. Population, household and related socio-economic data were collected from the statistical yearbooks (Suzhou City Statistical Bureau, 1991, 1996, 2001). Data on urban and rural construction (which we assume proportional to household change) was generated from remote sensed imagery using Landsat Thematic Mapper (TM) images for 1990, 1995 and 2000. The pixel resolution of Landsat TM images is 30 m. This resolution suffices to differentiate urban construction patches since they are large and contiguous. Patches of rural construction have a non-contiguous impervious surface and are small in size. Rural construction is

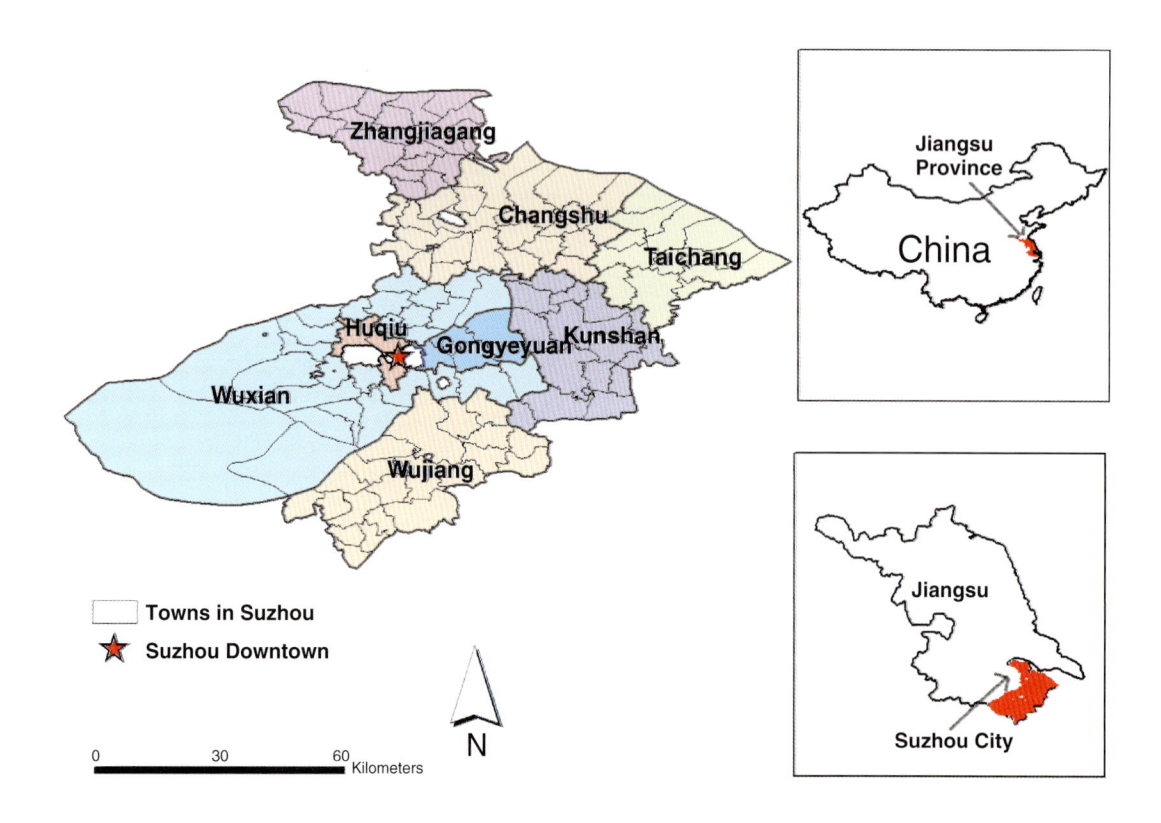

FIGURE 24.1 The study area: Suzhou Prefecture City.

a rather fuzzy type of land use, based on the official Chinese statistical yearbooks (Suzhou City Statistical Bureau, 2001).

These images were classified into a dozen land use categories used to derive the transition matrices indicating the amount of each land use which was converted to any other during the two periods in question: 1990 to 1995 and 1995 to 2000: $T \rightarrow T + 1$ and $T + 1 \rightarrow T + 2$. From these images, land parcel data was extracted and then converted to vector data sets, complemented by data associated with topography, geomorphology, vegetation, precipitation and temperature used as the ancillary data in the interpretation process. Further details on Landsat classification are given in Liu, Liu and Deng (2002). The method adopted here to extract dynamic changes in the vector land use datasets was based on post-classification image comparison complemented by field sampling to ensure quality control in the resulting classifications. Control was executed by checking the identities and the boundaries of sample land use patches with manual adjustment to decrease the incidence of major errors. The TM images in 1995 were used to interpret the dynamic change vector data by comparing with the vector data derived from interpreting TM images in 1990, while the same method was applied for the period 1995 to 2000. In fact over both time periods comparing 1990 with 2000 data, the overall classification accuracy of measured areas of all land types is about 97% (Liu, Liu and Deng, 2002), which gives us a high level of confidence in the extracted change data and its allocation into land use categories.

The main changes of land use in the study area took place with the construction land and the paddy fields. In Table 24.1, we have extracted the changes from an aggregated set of classes into urban and rural construction (which we take to be urban/household unit development in this context). The total construction volume was significantly reduced during 1995 to 2000 (40.7%) and in particular urban construction (new development) was dramatically trimmed down to around 55.4%. In contrast, rural construction was reduced 19.5% in 1995 to 2000 compared to the period 1990 to 1995. This conversion confirms that the policy of protecting agriculture land and curbing the overheated economy enacted in 1995 has had a notable impact on the desakota process through subsequent land use changes and patterns. Moreover, there has been a sharp increase in land use being converted from paddy fields and a consequent drop in conversion from drylands between the first and second time periods. This indicates that the preferred land supply – drylands – for urban development, has been severely diminished, which in turn has forced people to take more productive farmland – paddy fields. Although strict measures for controlling investment and urban expansion are noticeable, it has been hard for government to discourage rural residents from building more spacious houses due to increasing affluence. As conversion from paddy fields is the largest category in both periods, we can also examine the extent to which paddy fields are converted to other land uses which is shown in Table 24.2. Urban and rural construction still dominates taking some 77% and 79% of paddy field land in the two respective time periods with factory and transportation uses taking 3.5% and 4.7%. Therefore the main focus of the ABM in this chapter is to simulate the transition from rural (paddy fields) to urban growth (including urban construction, rural construction and special construction – factory and transportation).

24.4 The Suzhou Urban Growth Agent Model

24.4.1 The model design

The formal structure of the Suzhou Urban Growth Agent Model (SUGAM) is primarily derived from the Wuxian urban growth agent model developed by Xie, Batty and Zhao (2007). There are two types of agents considered in SUGAM: the township

TABLE 24.1 Percentage conversion of land use to urban and rural construction over the two macro-time periods.

Land use cover type	1990–1995	1995–2000
Converted hectares of urban and rural construction	27860(=16504+11356) 16514(=7374+9140)	
Shrub and loose forest	0.30	0.65
Other forest including orchards	0.06	1.12
Highly-covered grassland	0.00	0.36
Lake, reservoir, and pond	0.41	1.56
Shoal	0.00	0.16
Hill and plain paddy field	67.71	88.69
Hill and plain drylands	31.65	5.58

TABLE 24.2 Percentage conversion of paddy fields to other land uses over the two macro-time periods.

Land use cover type	1990–1995	1995–2000
Lost hectares of paddy fields	24362	18546
Dense forest	0.12	0.00
Shrub forest	0.03	0.00
Sparse forest	0.18	0.00
Orchard	0.14	0.00
Dense grassland	0.05	0.00
River	0.12	0.00
Lake	2.66	0.00
Reservoir and pond	14.02	16.33
Shoal	0.01	0.00
Urban construction	39.33	35.31
Rural construction	38.10	43.66
Large factory and transportation	3.51	4.70
Plain dry land	1.72	0.00

policy agents and the local developer agents. The policy agents, including town-village governments, municipality governments, and external developers such as investors from abroad and other China municipalities as well as national policy-making organizations, are acting at the macro level. The developer agents are the collection of individual entrepreneurs, small corporations, town and village owned enterprises, and privately operated businesses at the micro level. Therefore, SUGAM is specified at two levels. The township policy agents who are making decisions and policies at townships and over higher spatial units are functioning at the global level to determine the rates of growth in each of the townships measured by changes in households which can be converted into developable units. The rate of change, $R_k(\Delta T)$ is defined as

$$R_k(\Delta T) = f\{X_k^1(T), X_k^2(T), \ldots\} \quad (24.1)$$

where $X_k^l(T), \ell = 1, 2, \ldots, L$ are socioeconomic drivers associated with economic development and regional policy appropriate to the township level. Equation 24.1 is in fact the basis for the estimation of the importance of exogenous variables to the rates of change fitted using linear regression methods in a later section with these rates determining the amount of growth over the macro-time period ΔT. To generate a total growth over a period, they are applied directly to the total households (as developable units) at the initial time

$$P_k(T+1) = [1 + R_k(\Delta T)]P_k(T). \quad (24.2)$$

We can then factor this rate into a rate per unit time period $\Delta t = [t+1] - [t]$ by discounting the cumulative rate as

$$\widetilde{r}_k(t) = \left\{\frac{P_k(T+1)}{P_k(T)}\right\}^{\frac{1}{\tau}} = 1 + r_k(t). \quad (24.3)$$

When applied cumulatively to the population $P_k(t) = \widetilde{r}_k(t)P_k(T)$, Equation 24.3 updates the totals at each time period t to meet the constraint that $P_k(t + \tau) = P_k(T+1)$. In each macro-time period ΔT, the total change $\Delta P_k = P_k(T+1) - P_k(T)$ is broken into its finer temporal units using Equation 24.3 and each subtotal $\Delta P_k(t), \Delta P_k(t+1), \ldots, \Delta P_k(t+\tau)$ forms the control for the detailed urban development process at the cellular level.

At the micro level, the local developer agents take into consideration accessibilities (to economic center and to transportation), land cost reflected through suitability, and growth management policies to allocate future growth. Land suitability in the fine cell i in township k is defined as $C_{ik}(t)$, accessibility to economic centers as $E_{ik}(t)$, and accessibility to transportation facilities as $T_{ik}(t)$. The interaction or feedback between the local developer agents and the township policy agents is realized through a policy index $S_k(t)$, which is in effect a "Township Competition Index" derived from the rate of change in $k, R_k(T), \forall_{i,t}$. This links the effects of accessibility and suitability with respect to the growth management and economic policies set at the township level with the index being set in proportion to the rate of growth of each township (Xie, Batty and Zhao, 2007; Xie and Batty, 2005). These factors are used by the developer agents to determine a probability for development $\rho_{ik}(t)$ which is a form of utility given as

$$\rho_{ik}(t) = \mu T_{ik}(t) + \lambda E_{ik}(t) + \psi S_{ik}(t). \quad (24.4)$$

In general, land is converted to urban uses by the developer agent j who for each cell i in township k evaluates the probability of development, subject to the suitability of the land in question as reflected in the measure $C_{ik}(t)$. In principle, what each agent is doing is converting the land in question to an urban use, to $P_{ik}(t)$ by maximizing $\rho_{ik}(t)$ subject to the constraint posed by the land suitability $C_{ik}(t)$.

An agent's behavior of maximizing development probability is realized through intelligent search in two levels of spaces. The first level of space is a township, which is ranked by the agent according to the policy (township competition) index, $S_k(t)$. A higher rank of the township competition index allows the agent to choose a "maximized" probability of development, which means that the agent can break the limitation of growth rate (the rate of change), converting more open or agricultural land to urban uses in that township. This also indicates that townships with lower policy indices will not be able to keep growth rates that were determined by their socioeconomic factors. Thus, a competition among townships is generated. The second search space is an extended neighborhood, which is taken in this case study as a neighborhood of 20 × 20 cells (2 × 2 km) surrounding each economic center. A developer agent will look first in this neighborhood for conjunctions of high accessibilities to both economic centers, $E_{ik}(t)$ and transportation facilities, $T_{ik}(t)$. The developer agent will assign maximized probabilities of development to these conjunctions as new sources of development. The agent will then look at high accessibilities to either economic centers or transportation facilities, but assign these locations much lower development probabilities in comparison with the former.

The implementation of maximizing development probability in this case study is very close to the notion of solving agents' multi-criteria evaluation problem as an optimization problem (Bennett and Tang, 2006; Manson, 2006; Xiao, Bennett and Armstrong, 2007). However, the focus of this case study is on township competitiveness, location advantage, and easy-to-implement in simulation. Moreover, it is worth pointing out that the maximization (or optimization) decision processes taken by the developer agents in this model distinguish themselves from the traditional cellular automata approaches. These processes also provide heuristic methods to assign values to the three parameters μ, λ, and ψ in equation 4.

24.4.2 The model construction

The SUGAM model is implemented in the open source modeling language *RePast 3* (Collier, Howe, and North 2003 [Online] Available at: http://repast.sourceforge.net [accessed 19 November 2010]). One thing that needs to be pointed out is that *Repast* is using a different time concept in its simulation, which is referenced as ticks. These ticks do not match the real times t and T for they are essentially used to track the movement of agents across the space as they search for suitable cells to transform and as such, reflect the various iterations that are used to achieve the control totals from the global level.

24.4.3 The model calibration

The model calibration is done in two steps: first calibrating urban growth rate in township k; and second calibrating the probability of development in terms of the local parameters μ, λ,

TABLE 24.3 Demographic and socioeconomic variables considered in the stepwise regressions.

Description	Unit
Agricultural population	IND*
Non-Agricultural population	IND
Total population	IND
Land area in cultivation	MU*
Total output value of agriculture, forestry, animal husbandry and fishery	MY*
Gross domestic product value	MY
Gross product value of primary and secondary industries	MY
Gross product value of tertiary industries	MY
Total value of fixed assets investment	MY
Total income of rural economy	MY
Income in agriculture, forestry, animal husbandry and fishery	MY
Income in non-agriculture, non-forestry, non-animal husbandry and non-fishery	MY
Total expense in rural economy	MY
Total income of the farmers	MY
Total value of industrial assets	MY
Net value of the fixed assets	MY
The number of factories	CNT*
The number of employed people at the year end	IND
Total tax value	MY
Sold ratio of the product value	Percent

IND*: individual count of all people; MU*: 1 mu = 1/15 hectare = 1/6 acre. MY*: million Chinese yuan CNT*: count of all factories."

and ψ associated with Equation 24.4. For calibrating township growth rate, twenty socioeconomic variables were collected in 1990, 1995, and 2000 (Table 24.3). Stepwise regression was employed to find out the best variables that could explain urban growth at township level. The results are given below on the basis of best fit:

$$R_k(1990 \rightarrow 1995)$$
$$= \left. \begin{array}{l} 15.68 + 0.627\,RP_k(1990) + 7.69\,P_k(1990) \\ \quad(0.89)\quad\ (2.48)\qquad\qquad\quad(4.54) \\ -5.02\,E_k(1990) \\ (-3.49) \end{array} \right\} \quad (24.5)$$

$RP_k(1990)$ is the rural (non-urban population), $P_k(1990)$ the urban population, and $E_k(1990)$ the employment (labor force) total at 1990. The *t* statistics (**bold**) under each weight and variable make clear that the parameters β_k^l are all significantly different from zero at the 5 percent level. The amount of variance explained by this equation is 72 percent which is acceptable for driving the simulation from its start point.

$$R_k(1995 \rightarrow 2000)$$
$$= \left. \begin{array}{l} 0.99 + 0.24\,TAX_k(1995) - 0.28\,INA_k(1995) \\ (-0.17)\ \ (8.58)\qquad\qquad\ (-4.65) \\ +0.17\,GDP_k(1995) - 0.14\,FA_k(1995) + 0.25\,RE_k \\ (4.35)\qquad\qquad\ (-4.04)\qquad\qquad\ (2.37) \end{array} \right\} \quad (24.6)$$

$TAX_k(1995)$ is the total tax levied, $INA_k(1995)$ income in the non-agricultural sectors, $GDP_k(1995)$ gross domestic product, $FA_k(1995)$ the net value of fixed total assets, and $RE_k(1995)$ the expenditure in the rural economy, all at 1995, and defined in each township *k*. The *t* statistics (**bold**) under each variable imply that the β_k^l parameters are all significantly different from zero at the 5% level. The variance explained by this equation is 88 percent which is particularly high, given the aggregation and uncertainties posed by the quality of the data. Moreover, different explanatory variables were chosen in Equation 24.6 from those in Equation 24.5. The main cause was the distinct policy shift between 1990–1995 and 1995–2000 (Xie and Batty, 2005).

Equations 24.5 and 24.6 are those used in the global model which in terms of the simulation provides the parameters determining the overall rates of change from 1990 to 1995 and 1995 to 2000 in the 122 townships. These are used to compute the rates input to Equation 24.2 that in turn is used to factor the total urban change into its constituent components, which is then allocated to the cells by the lower-level agent model.

The values of the three parameters μ, λ, and ψ in Equation 24.4 are determined by using our judgment and controlled by the land suitability $C_{ik}(t)$ as the scores used in obtaining a probability of development (conversion) from rural to urban land use. Thus it is their relative values that are important. In fact, during this process because land suitability is taken into account, developers will not develop a cell if the land suitability is less than a certain threshold $\Xi_k(T)$, that is, if $C_{ik}(t) < \Xi_k(T)$. The reasons for this initial allocation step which is different from the subsequent steps within the macro-time period, rests on the fact that there was a strong shift in policy between 1995 and 2000 in this region and this needs to be reflected in the initial placement. We call this first process *random allocation* but in subsequent time periods, the master developer agents are used to "spawn" additional agents which add up to the total required in subsequent micro-time periods. These agents begin by considering development in the cellular neighborhood of each master agent activating a process we call *neighborhood allocation*. It is at this point that the probabilities defined in Equation 24.4 are considered in neighborhood order: that is, the developer agent begins by considering cells in the immediate band of eight cells around the master agent – in the Moore neighborhood – and if no suitable cell is found, then the agent considers the next band of cells, and so on until a suitable cell is located. The reason for this somewhat convoluted process is to ensure that development remains "close" to existing development which reflects the need for connectivity in the urbanizing system.

24.4.4 The model validation

The validation procedure used in this paper is a variation of the map-comparison method. The steps of validation are following those developed in the Wuxian city agent based model (Xie, Batty and Zhao, 2007). The validation has been conducted over the entire period from 1990 to 2000 but aggregated to the 122 townships. The goodness of fit is based on the comparison between the urban cells that were converted from rural predicted by the model and those observed from the remote imagery maps, which is formally formulated as,

$$\Phi(\mu,\lambda,\psi) = \sum_{k=1}^{122} \left[P_k(T \rightarrow T+2) - \sum_{\tau=1}^{10} \sum_{i \in Z_k} P_{ik}(\tau) \right]^2 \Big/ 122 \quad (24.7)$$

where the temporal summations are over the period from 1990 to 2000 and the spatial summations over the number of cells in each township. These parameters values have produced simulations of spatial patterns of urban growth close to those that we observe over the calibration period 1990 to 2000 (Fig. 24.2). These parameters are also used to predict urban growth in Suzhou Prefecture in 2005 and 2010 (Fig. 24.3). It seems that some overestimation took place in the northern sections of Changshu and Taichang. Although it cannot be verified whether there is an overestimation or not because of the lack of ancillary data at this moment, the official statistics observation confirm that lot of urban growth has happened in these areas in recent years (Suzhou Municipality Statistical Bureau, 2008).

Summary and conclusion

In this chapter, we introduced general technical steps of developing agent-based models in the context of urban studies, and illustrated an agent-based urban growth simulation (desakota development) in Suzhou, China. In the illustration of SUGAM, the developer agents are rational in making decisions about where to convert land use from rural to urban. The developer agents take a comprehensive review of government policies, socioeconomic conditions and external factors as a baseline to determine whether new development is feasible. First, the developer agents communicate with the township policy agents to determine the rates of development, which are quantified through stepwise regression analysis. Second, the developer agents compare the development rates and rank a township competitiveness measurement, which is quantified as the policy index. Third, the developer agents examine the accessibilities to either economic center or transportation or both within a localized area to determine another measure of development probability. The developer agents evaluate those three aforementioned quantities to determine location with maximized or optimized development probabilities. Finally, the developer agents have to check the land suitability to see if new developments are supported by available land. Therefore the developer agents are strategic (in terms of considering competition among 122 townships), task-oriented, and sensitive to

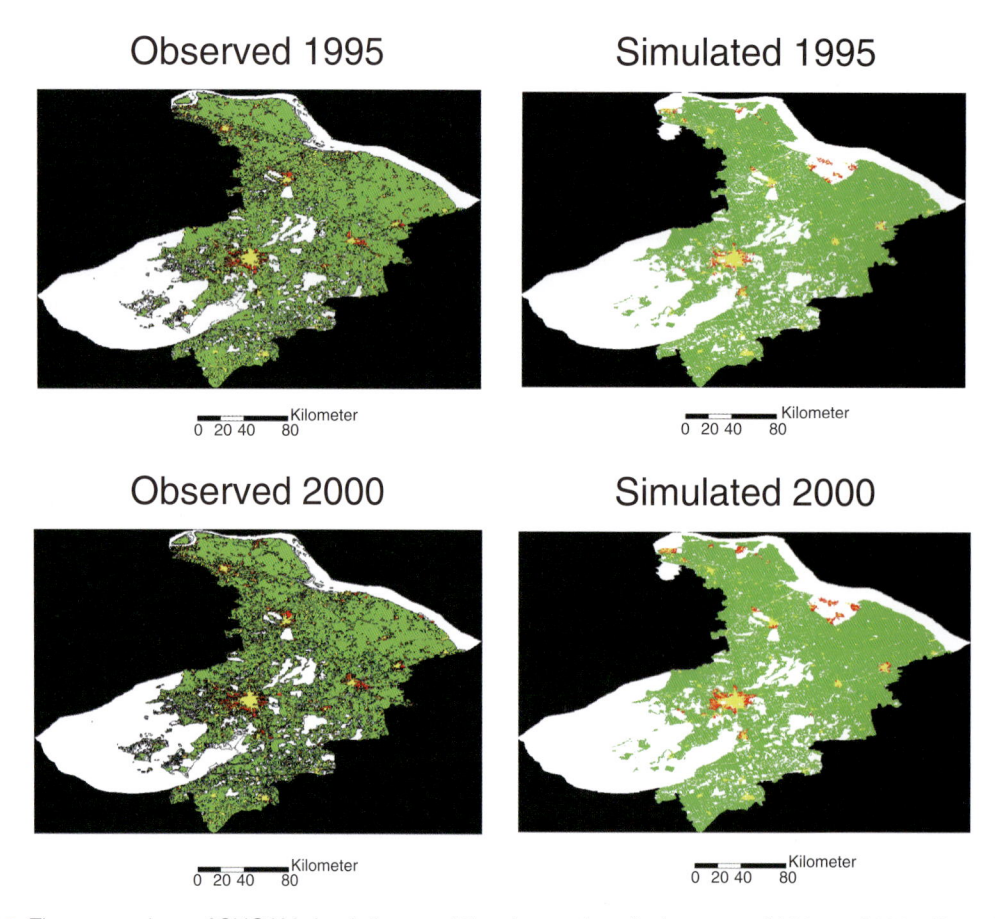

FIGURE 24.2 The comparison of SUGAM simulations and the observations in the years of 1995 and 2000 (the yellow patches are land uses in 1990).

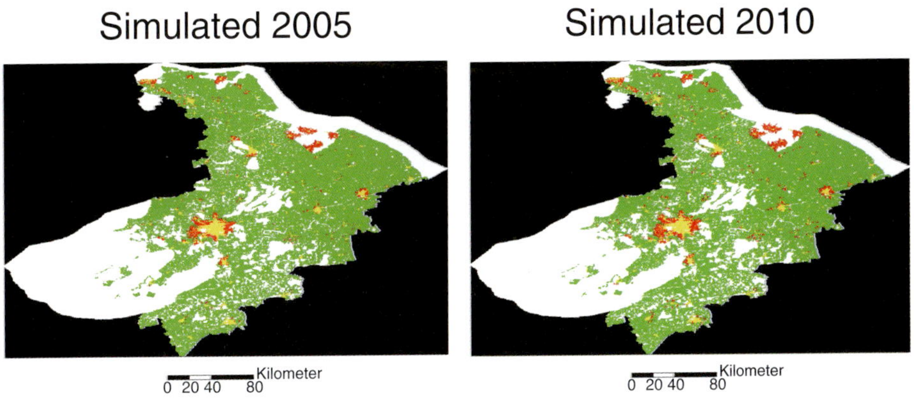

FIGURE 24.3 The predicted urban growth from SUGAM in the years of 2005 and 2010 (the yellow patches are land uses in 1990).

resource constraints. They combine the top-down approach (the government policy) with the bottom-up momentum (local development opportunity) to decide on locations and rate of urban growth.

As shown, the agent-based models are not only well suited to disaggregate systems but can also be used to integrate different levels and scales essential to simulating what at first sight appears to be bottom-up phenomenon such as desakota. Of course, enabling agents to search and to extract information beyond their locality is a challenging task. Our handling of the agents at the local neighborhoods and at the level of the townships in this chapter is a considerate attempt to understand the interactions of geographic agents at different scales. However, as we pointed out earlier, the consideration of agents acting in a competitive environment both from regional and national perspectives in a world of economic globalization leads to more realistic simulations.

On the other hand, there are several areas where further explorations and improvements are needed. The most challenging one is the verification of the simulation results. Currently, the verification of the simulation results is carried out between the simulation results and the regression analysis. Further verifications are needed to check errors of the total count and sub-counts of land development by land categories. Attentions should be paid to the verification and validation of spatial distribution and patterns and the statistics derived from landscape metrics (Xie, Yu and Bai, 2006) and computation geometry (Xie and Ye, 2007).

Finally, the simulation results raise a number of questions on rural urbanization in China. An urbanization policy that emphasizes rural areas continues to be successful, because rural urbanization shows a very positive response to economic prosperity. Less restrictive migration policies and less parochial town and county governance are leading to more efficient allocations of resources and greater economic growth in the long run. However, the rural industrialization is facing many challenges. Compared with large cities, lack of high-tech and trained workers often lead to the quality concerns and thus slowing demand for rural township and village enterprise (TVE) produced goods. Moreover, due to relatively loosened environmental protection or control in rural areas, environmental consequences of uncontrolled rural industrialization are serious. Furthermore, new competitive demands are putting pressure on China's industries that may eventually reward the economies of scale and locational advantages of large metropolitan areas. Perhaps continued rural urbanization is neither economically sound, nor environmentally wise. The pattern changes of land use identified in these agent-based simulations support some of these concerns. For instance, hilly and plain dry-lands, environmentally sensitive ecosystems, were almost wholly converted to urban land use. In other word, these concerns of sustainable development and environmental quality should be integrated in agent-based urban modeling.

References

Alexandridis, K. and Pijanowski, B.C. (2007) Assessing multi-agent parcelization performance in the MABEL simulation model using Monte Carlo replication experiments. *Environment and Planning B: Planning and Design*, **34**, 223–244.

Allen, P.M. (1997) *Cities and Regions as Self-Organizing Systems: Models of Complexity*, Taylor & Francis, London.

Axtell, R.L. (2003) The New Coevolution of Information Science and Social Science: From Software Agents to Artificial Societies and Back or How More Computing Became Different Computing, Working paper, Brookings Institution. Avaialble http://www.econ.iastate.edu/tesfatsi/compsoc.axtell.pdf (accessed 19 November 2010).

Barredo, J.I., Kasanko, M. and McCormick, N. *et al.* (2003) Modelling dynamic spatial processes: simulation of urban future scenarios through cellular automata. *Landscape and Urban Planning*, **64**, 145–160.

Batty, M. (1998) Urban evolution on the desktop: simulation using extended cellular automata. *Environment and Planning A*, **30**, 1943–1967.

Batty, M. (2001) Agent-based pedestrian modelling. *Environment and Planning B: Planning and Design*, **28**, 321–326.

Batty, M. (2005) *Cities and Complexity: Understanding Cities with Cellular Automata, Agent-based Models, and Fractals*, MIT Press, Cambridge, MA.

Batty, M. and Xie, Y. (1994) From Cells to Cities. *Environment and Planning B*, **21**, 31–48.

Batty, M., Xie, Y. and Sun, Z. (1999) Modeling urban dynamics through gis-based cellular automata. *Computers, Environments and Urban Systems*, **233**, 205–233.

Bennett, D. A. and Tang, W. (2006) Modelling adaptive, spatially aware, and mobile agents: Elk migration in Yellowstone. *International Journal of Geographical Information Science*, **20**, 1039–1066.

Boots, B. and Csillag, F. (2006) Categorical maps, comparisons, and confidence. *Journal of Geographical Systems*, **8**, 109–118.

Brately, P., Fox, B. and Schrage, L.E. (1987) *A guide to simulation*, Springer, New York.

Brown, D. G., Page, S. and Riolo, R. et al. (2005) Path dependence and the validation of agent-based spatial models of land use. *International Journal of Geographical Information Science*, **19**, 153–174.

Chen, H. (1988) Jiangsu expands its economic horizons. *Beijing Review*, **23**, 18–25.

Clarke, K., and Gaydos, L. (1998) Loose-coupling a cellular automaton model and GIS: long-term urban growth prediction for San Francisco and Washington/Baltimore. *International Journal of Geographical Information Science*, **12**, 699–714.

Collier, N., Howe, T. and North, M. (2003) Onward and upward: The transition to Repast 2.0. *Proceedings of the First Annual North American Association for Computational Social and Organizational Science Conference*. Electronic Proceedings, Pittsburgh, PA: 241–268; http://repast.sourceforge.net/ (accessed 19 November 2010).

Collins, M.G., Steiner, F.R. and Rushman, M.J. (2001) Land-use suitability analysis in the United States: historical development and promising technological achievements. *Environmental Management*, **28**, 611–621.

Deadman, P. and Robinson, D. (2004) Colonist household decision making and land-use change in the Amazon Rainforest: an agent-based simulation. *Environment and Planning B: Planning and Design 2004*, **31**, 693–709.

Dietzel, C. and Clarke, K.C. (2007) Toward Optimal Calibration of the SLEUTH Land Use Change Model. *Transactions in GIS*, **11**, 29–45.

Gilbert, N. and Terna, P. (2000) How to build and use agent-based models in social science. *Mind & Society*, **1**, 57–72.

Gilbert, N. and Troitzsch, K.G. (1999) *Simulation for the Social Scientist*, Open University Press, Milton Keynes, UK.

Goodchild, M.F. (2005) GIS, spatial analysis, and modelling overview, in *GIS, Spatial Analysis, and Modeling* (eds D. J. Maguire, M. Batty and M. F. Goodchild), ESRI Press, Redlands, CA.

Hargrove, W.W., Hoffman, F.M. and Hessburg, P.F. (2006) Mapcurves: a quantitative method for comparing categorical maps. *Journal of Geographical Systems*, **8**, 187–208.

Holland, J. H. (1975) *Adaptation in Natural and Artificial Systems*, University of Michigan Press, Ann Arbor, MI.

Jepsen, M. R., Leisz, S. and Rasmussen, K. et al. (2006) Agent-based modelling of shifting cultivation field patterns, Vietnam. *International Journal of Geographical Information Science*, **20**, 1067–1085.

Karssenberg, D., De Jong, K. and Van Der Kwast, J. (2007) Modelling landscape dynamics with Python. *International Journal of Geographical Information Science*, **21**, 483–495.

Kirkby, R. (2000) *Urban land reform in China*, Oxford University Press, Oxford, UK.

Kocabas, V. and Dragicevic, S. (2009) Agent-based model validation using Bayesian networks and vector spatial data. *Environment and Planning B: Planning and Design*, **36**, 787–801.

Kok, K., Farrow, A. and Veldkamp, A. et al. (2001) A method and application of multi-scale validation in spatial land use models. *Agriculture Ecosystems and Environment*, **85**, 223–238.

Lee, D.B. (1973) Requiem for large-scale models. *Journal of the American Institute of Planners*, **39**, 163–177.

Lei, Z., Pijanowski, B.C. and Alexandridis, K.T. et al. (2005) Distributed modeling architecture of a multi-agent-based behavioral economic landscape (MABEL) model. *Simulation-transactions of the Society for Modeling and Simulation International*, **81**, 503–515.

Li, X. and Liu X. (2008) Embedding sustainable development strategies in agent-based models for use as a planning tool. *International Journal of Geographical Information Science*, **22**, 21–45.

Li. X. and Yeh. A. (1998) Principal Components Analysis of Stacked Multi-Temporal Images for Monitoring Rapid Urban Expansion in the Pearl River Delta. *International Journal of Remote Sensing*, **19**, 1501–1518.

Ligmann-Zielinska, A. and Jankowski, P. (2007) Agent-based models as laboratories for spatially explicit planning policies. *Environment and Planning B: Planning and Design*, **34**, 316–335.

Lin, C. (2001) Metropolitan development in a transitional socialist economy: spatial restructuring in the Pearl River Delta, China. *Urban Studies*, **38**, 383–406.

Liu, J., Liu, M., and Deng, X. (2002) The land-use and land-cover change database and its relative studies in China. *Journal of Geographical Sciences*, **12**, 275–82.

Liu, X., Li, X. and Liu, L. et al. (2008) A bottom-up approach to discover transition rules of cellular automata using ant intelligence. *International Journal of Geographical Information Science*, **22**, 1247–1269.

Manson, S.M. (2005) Agent-based modeling and genetic programming for modeling land change in the Southern Yucatan Peninsula Region of Mexico. *Agriculture, Ecosystems and Environment*, **111**, 47–62.

Manson, S.M. (2006) Bounded rationality in agent-based models: experiments with evolutionary programs. *International Journal of Geographical Information Science*, **20**, 991–1012

Marton, A. (2000) *China's Spatial Economic Development: Restless landscapes in the Lower Yangtze Delta*, Routledge, New York.

McGee, T.G. (1989) Urbanisasi or Kotadesasi? Evolving patterns of urbanization in Asia, in *Urbanization in Asia* (eds F.J. Costa, A.K. Dutt, L.J.C. Ma and A.G. Noble), University of Hawaii Press, Honolulu, pp. 93–108.

McGee, T.G. (1991) The Emergence of desakota regions in Asia: expanding a hypothesis, in *The Extended Metropolis: Settlement Transition in Asia* (eds N. Ginsburg, B. Koppel and T.G. McGee), University of Hawaii Press, Honolulu, pp. 3–25.

McGee, T.G. (1998) Five decades of urbanization in Southeast Asia: A personal encounter, in *Urban Development in Asia: Retrospect and Prospect*, Chinese University of Hong Kong, Hong Kong, pp. 55–91.

Oreskes, N., Shrader-Frechette, K. and Belitz, K. (1994) Verification, validation, and confirmation of numerical-models in the earth-sciences. *Science*, **263**, 641–646.

Ostrom, T. (1988) Computer simulation: the third symbol system, *Journal of Experimental Social Psychology*, **24**, 381–392.

Parker, D.C. and Meretsky, V. (2004) Measuring pattern outcomes in an agent-based model of edge-effect externalities using spatial metrics. *Agriculture, Ecosystems & Environment*, **101**, 233–250.

Parker, D.C., Manson, S.M. and Janssen, M.A. *et al.* (2003) Multi-agent systems for the simulation of land-use and land-cover change: a review. *Annals of the Association of American Geographers*, **93**, 314–337.

Parunak, H.V., Brueckner, S.A., Matthews, R. and Sauter, J. (2006). Swarming methods for geospatial reasoning. *International Journal of Geographic Information Science*, **20**, 945–964.

Pontius, R.G. and Schneider, L.C. (2001) Land-cover change model validation by an ROC method for the Ipswich watershed, Massachusetts, USA. *Agriculture Ecosystems and Environment*, **85**, 239–248.

Portugali, J. (2000) *Self-Organization and the City*, Springer-Verlag, Berlin.

Power, C., Simms, A. and White, R. (2001) Hierarchical fuzzy pattern matching for the regional comparison of land use maps. *International Journal of Geographic Information Science*, **15**, 77–100.

Straatman, B., White, R. and Engelen, G. (2004) Towards an automatic calibration procedure for constrained cellular automata. *Computers, Environment and Urban Systems*, **28**, 149–170.

Suzhou Municipality Statistical Bureau (1991) *Statistical yearbook of Suzhou Municipality – 1990*. Suzhou Municipality, Jiangsu Province.

Suzhou Municipality Statistical Bureau. (1996) *Statistical yearbook of Suzhou Municipality – 1995*. Suzhou Municipality, Jiangsu Province.

Suzhou Municipality Statistical Bureau (2001) *Statistical yearbook of Suzhou Municipality – 2000*. Suzhou Municipality, Jiangsu Province.

Suzhou Municipality Statistical Bureau (2008) *Statistical yearbook of Suzhou Municipality – 2006*. Suzhou Municipality, Jiangsu Province.

Tan, K. (1986) Revitalized small towns in China. *Geographical Review*, **76**, 138–148.

The Center for the Study of Complex System (2009) Agent- and Individual-based Modeling Resources [Online] Available at: http://www.swarm.org/index.php/Tools_for_Agent-Based_Modelling (accessed 19 November 2010).

Torrens, P.M. and Benenson, I. (2005) Geographic automata systems. *International Journal of Geographical Information Science*, **19**, 385–412.

Turner, M.G., Costanza, R. and Sklar, F.H. (1989) Methods to evaluate the performance of spatial simulation models. *Ecological Modelling*, **48**, 1–18.

Verburg P.H, Kok, K., Pontius, G.R. and Veldkamp, A. (2006) Modeling land-use and land-cover change, in *Land-use and Land-cover Change: Local Processes and Global Impacts* (eds E.F. Lambin and H.J Geist), Springer-Verlag, Berlin, pp. 117–137.

Waddell, P. (2002) UrbanSim: modeling urban development for land use, transportation, and environmental planning. *The Journal of the American Planning Association*, **68**, 297–314.

Weidlich, W. (2000) *Sociodynamics: A Systematic Approach to Mathematical Modelling in the Social Sciences*, Harwood Academic Publishers, Amsterdam, The Netherlands.

White R. (2006) Pattern based map comparisons. *Journal of Geographical Systems*, **8**, 145–164.

White, R., and Engelen, G. (1993) Cellular automata and fractal urban form: a cellular modeling approach to the evolution of urban land-use patterns. *Environment and Planning A*, **25**, 1175–1189.

Wolfram, S. (2002) *A New Kind of Science*, Wolfram Media, Inc., Urbana, IL.

Wu, F. (2002) Calibration of stochastic cellular automata: the application to rural–urban land conversions. *International Journal of Geographical Information Science*, **16**, 795–818.

Wu, F. and Webster, C.J. (1998) Simulation of land development through the integration of cellular automata and multicriteria evaluation. *Environment and Planning B*, **25**, 103–126.

Xiao, N., Bennett, D.A., and Armstrong, M.P. (2007) Interactive evolutionary approaches to multiobjective spatial decision making: A synthetic review, *Computers, Environment and Urban Systems*, **31**, 232–252.

Xie, Y. (1996) A generalized model for cellular urban dynamics. *Geographical Analysis*, **28**, 350–373.

Xie, Y., and Batty, M. (2005) Integrated urban evolutionary modelling, in *Geodynamics* (eds. P.M. Atkinson, G.M. Foody, S.E. Darby, and F. Wu), CRC Press, Boca Raton, FL, pp. 273–293.

Xie, Y., and Ye, X. (2007) Comparative tempo-spatial pattern analysis: CTSPA. *International Journal of Geographic Information Science*, **21**, 49–69.

Xie, Y., Batty, M., and Zhao, K. (2007) Simulating emergent urban form using agent-based modeling: desakota in the Suzhou-Wuxian Region in China. *Annals of Association of American Geographers*, **97**, 477–495.

Xie, Y., Yu, M. and Bai, Y. (2006) Ecological analysis of an emerging urban landscape pattern-desakota: a case study in Suzhou, China. *Landscape Ecology*, **21**, 1297–1309.

Yu, C. and Peuquet, D.J. (2009) A GeoAgent-based framework for knowledge-oriented representation: Embracing social rules in GIS. *International Journal of Geographic Information Science*, **23**, 923–960.

25

Ecological modeling in urban environments: predicting changes in biodiversity in response to future urban development

Jeffrey Hepinstall-Cymerman

Human population growth is increasingly transforming the landscape into urbanized land uses as more than 50% of humans worldwide now live in urban areas. The intensification of land use caused by urbanization threatens the long-term provisioning of ecosystem services. Urban remote sensing provides the opportunity to develop data to support ecological modeling specifically through the high resolution mapping of land cover over space and time. Integrated modeling of land cover, land use, and ecological systems is required to predict possible changes in future ecological conditions. I present an example of urban ecological modeling where I combine the output from a land use change model with output from a land cover change model to predict avian species richness 25 years into the future.

25.1 Introduction

Humans have been transforming the Earth's land area increasingly in the past decades with nearly half showing evidence of human manipulation (Houghton, 1994; Lambin et al., 2001). As human populations increase, there is increasing pressure to convert land to urban land uses (Meyer and Turner, 1992). More than 50% of humans now live in cities; up from only 10% in 1900 (Sadik, 1999). Urbanization, defined here as the conversion of undeveloped or agricultural lands to build environments, dramatically alters landscapes and substitutes natural processes with human ones (Grimm et al., 2000; Alberti et al., 2003). The intensification of land use caused by urbanization threatens the long-term provisioning of ecosystem services (Foley et al., 2005). Specifically, altering land use of a parcel of land inevitably alters the land cover (e.g., vegetation, impervious surface, etc.) present on a site, thereby changing the species that can occupy that land and changing many other processes (i.e., surface water flow, groundwater penetration) as well.

In this chapter, I am specifically interested in understanding how future urbanization may change the biodiversity of a region. To address this question, I must be able to: (1) develop models that predict future land cover; (2) develop models that estimate biodiversity; (3) integrate the output from land cover change models into models estimating biodiversity. Although none of these steps are trivial on their own and the integration of multiple models makes this process even more challenging, such integrated modeling is the goal of many modeling efforts (Alberti and Waddell, 2000). I will first introduce the major steps required to address this issue and then provide an example drawn from my work in western Washington State, USA.

25.1.1 Using urban remote sensing to develop land cover maps for ecological modeling

As urban remote sensing is the focus of this book, I focus my discussion here on specific techniques useful for creating data necessary for ecological modeling. The use of satellite imagery and remote sensing techniques, while employed by both urban planners and ecologists for over 30 years, has increased greatly within the both fields in the last twenty years (Wilson et al., 2003; Sudhira, Ramachandra and Jagadish, 2004; Croissant and York, 2005; Fassnacht, Cohen and Spies, 2006; Jat, Garg and Khare, 2008; Munroe, Tole, 2008). The utility of remotely sensed data to detect patterns and change in patterns over time is well suited for both fields and is discussed in detail in other portions of this book. Landsat sensors, in particular, have played an important role in land cover change (Yuan et al., 2005; Taubenböck et al., 2009) and ecological studies (Cohen and Goward, 2004). Urban and regional planners utilize remotely sensed data in many ways including tracking changes in land use and land cover (LULC) over time. Ecologists in general, and landscape ecologists in particular, primarily have been interested in the patterns and changes in pattern of vegetation or land cover, often as a surrogate for other features of interest such as wildlife habitat or primary productivity (Xu et al., 2007).

Urban environments, because of their inherent fine-grained heterogeneity of land use (i.e., the specific use(s) of the parcel) and land cover (i.e., the vegetation, soil, rock, or built material present) present a special problem when attempting to map and model vegetation and land cover (Yang and Lo, 2002, Yuan et al., 2005), often requiring a specialized classification system to be useful to ecologists and urban planners alike (Cadenasso, Pickett and Schwarz, 2007). Previous attempts at mapping urban areas with fine-grained detail have generally used either high resolution imagery (e.g., IKONOS, Quickbird, SPOT), which necessarily limits the spatial extent of a study area, or spectral mixture analysis of medium resolution imagery (e.g., Landsat Thematic Mapper [TM], MODIS). Landsat TM data in particular, with its synoptic view of the earth, high (30 m) spatial resolution, multidecade data stream, multispectral sensors designed specifically for detecting vegetation, and low cost (now free), makes it especially attractive to both ecological and urban and regional planning studies. However, to make Landsat TM data useful for classifying urban areas, specialized classification techniques are required. Other chapters in this book go into more detail on satellite sensors and classification methods, but in my example below I will briefly discuss several image classification techniques (i.e., image segmentation, supervised classification, and spectral mixture analysis) that are useful for deriving data to be used in urban ecological modeling.

Several recent studies have used multi-date satellite imagery to track changes in urban extent and form over time. For example, Taubenböck et al. (2009) used a Landsat Multispetral Scanner (MSS) and TM time-series to identify urban footprints and spatial patterns of development for India's 12 largest urban agglomerations from the 1970s until 2000. They used metrics of landscape composition and configuration to classify different urban growth types (e.g., sprawling or increased density, mono- or polycentric, laminar or punctual). Tole (2008) used 1986 and 2001 Landsat TM satellite imagery and spectral mixture analysis to document changes in the extent and pattern of urban land in the greater Toronto area, Canada. Tole concluded that during both time periods the area had a highly dispersed, sprawling pattern of built-up land (i.e., urban development). Similarly, Rashed (2008) used spectral mixture analysis and fuzzy logic to classify two time periods of Landsat TM imagery for Los Angeles, California, and landscape metrics to analyze changes in the extent and shape of urban areas. Jat, Garg and Khare (2008) used similar methods and determined that between 1977–2002, developed land increased three times as fast as the growth in population for Amjer, India. Each of these studies (and many others) used satellite imagery to examine patterns and changes in patterns of urban land cover over time which is often the first step in understanding the status of ecological systems of a region.

25.1.2 One example of ecological modeling: modeling species habitat

Urban ecology, as a branch of ecology, has recently emerged as an interdisciplinary field interested in understanding how urban systems function (Grimm et al., 2000; Pickett et al., 2001; Alberti et al., 2003; Alberti, 2008). The resilience of urban systems is linked to the dynamic interactions between social, economic, and biophysical processes operating over multiple spatial and temporal scales (Alberti and Marzluff, 2004). As land cover changes with urbanization, one proposed indicator of urban resilience, the biodiversity of a site or geographic area, is often negatively affected as habitats are altered or removed entirely. Human activities change the selective forces acting upon the biota existing on a site, causing some species to decline and eventually go extinct while others expand, colonize, and thrive in these same areas (Marzluff, 2001; McKinney, 2002).

Many researchers are interested in how the biodiversity of an area will change with urbanization (Schumaker et al., 2004; Hepinstall, Marzluff and Alberti, 2009). To determine how an area's biota may respond to land cover changes, several steps are required. First, the habitat requirements of a species must be known, generally the result of field studies conducted to document existing species occupancy in relation to vegetation types and structure (or more generally, land cover). From these habitat relationships we can develop a habitat model. Modeling a species' habitat, however, is not a trivial task as each species has specific habitat requirements that may differ: (1) at different times of the year (i.e., breeding versus non-breeding season); (2) in different portions of the species range; and (3) depending on if the species is declining and not occupying all "suitable" habitat or is increasing and occupying marginal habitat. Second, once the habitat requirements of a species are determined, we must be able to map these habitats across large areas. Third, we must be able to predict and map how these habitats will likely change in the future so that we may predict how species will likely change in response to future habitat availability.

While these steps may seem simple, the reality is much more complicated. For example, species occupancy of a site (MacKenzie et al., 2006) while determined by many factors, is primarily related to the amount of habitat present on a site. Terrestrial vertebrate habitats can be mapped through the use of remotely sensed data and while the focus of such work has generally been on non-urban areas (Villard, Trzcinski and Merriam, 1999), increasingly studies have focused on urban ecological relationships (Blair, 1996, 2004; Donnelly and Marzluff, 2004, 2006; Marzluff, 2005). Many studies have clearly documented that as land cover composition and configuration change, habitat for species changes as well. However, some species respond to the direct loss of habitat while others are also sensitive to change in the spatial arrangement of remaining habitat (e.g., fragmentation, isolation, distance between patches of habitat, amount of edge versus interior habitat). While there is a rich literature of studies investigating the effects of habitat loss and fragmentation on species occupancy (McGarigal and McComb, 1995; Fahrig, 1999, 2003; Villard, Trzcinski and Merriam, 1999; Betts et al., 2006), the specific form (i.e., statistical distribution) of these relationships is often unknown.

Once empirical data relating species with habitat are available, models of species occupancy can be developed. Here I discuss two basic types of spatially explicit models relating species to habitat. With one type, species-habitat association models, species are associated with different vegetation types through a set of rules such as those used in habitat suitability index (HSI) models and Gap analyses (Scott et al., 1993) where simple deductive relationships between habitat elements (e.g., canopy closure, tree height, tree species) and a species use of these elements is scaled to an index ranging from 0–100. These rule-based models are quite general and suffer from many problems (Roloff and Kernohan, 1999). The second general category of models is those that develop statistical relationships between field observations and explanatory variables (i.e., inductive models). The statistical form of the model can be quite varied depending on the response variables. Many different statistical techniques are in common use today including logistic regression (Betts, Forbes and Diamond, 2007, Hepinstall, Marzluff and Alberti, 2009), classification and regression trees and random forests (Prasad, Iverson and Liaw, 2006; Cutler et al., 2007), and climate envelope models (Pearson and Dawson, 2003), to name but a few of the more common techniques. Several review articles have been published that categorize modeling approaches (Guisan and Zimmermann, 2000; Guisan, Edwards and Hastie, 2002,). These two broad types of modeling approaches each have their place. Rule-based models may often be preferable to statistical models because of the ease of interpretation and the ability to illustrate basic principles. Statistical models, conversely, may allow the exploration of specific hypothesis such as the relative influence of habitat amount versus habitat configuration in determining species presence on a site. Statistical models, however, also suffer from errors and error propagation (Conroy et al., 1995) and difficulty in interpreting model results. For example, spatially explicit population models (SEPMs) attempt to relate species habitat requirements to demographic rates (survival, reproduction, dispersal) to model populations (Dunning Jr. et al., 1995). The added complexity of SEPMs necessary to model demographic processes requires many more parameter estimates and therefore includes many more potential sources of error. While these types of models represent an improvement in understanding the complexity of ecological systems, often the data requirements preclude their application for more than a few select species.

Models that are derived from empirical data, but do not attempt to model specific demographic processes, are more likely to be available for coupling with land cover change models. For example, Schumaker et al. (2004), developed species-habitat models for 279 species present in their western Oregon study area that were subsequently used to rank different future scenarios of land cover change (Hulse, Branscob and Payne, 2004). They used rule-based models of species-habitat associations because it was impossible to develop more specific habitat models for each species from existing data. Instead, expert knowledge was used to develop suitability rankings from 1 to 10 for each of 34 habitats for each species. These rankings were then modified by up to 50 different adjacency rules to incorporate the importance of landscape context.

If the goal is to model a smaller subset of species, it may be possible to develop more refined species models to use as input into predictions of land cover change. For example, Hepinstall, Marzluff and Alberti (2009) used 6437 point count surveys across 992 locations in 139 study sites to develop linear regression models predicting community measures (species richness, guild-specific community richness) and population measures (relative abundance of individual species) for a subset of the avian community found on the sites. The regression models used predictor variables that were output from land use and land cover models that had been developed concurrently (Hepinstall, Alberti and Marzluff, 2008).

25.1.3 Predicting future land use and land cover

The next step in the process is to develop a model that will produce a spatially explicit map of the future. Many different models have been developed to predict future land use (Schneider and Pontius, 2001; Parker *et al.*, 2003; Verburg *et al.*, 2004; Tang *et al.*, 2005; Verburg, Eickhout and Vanmeijl 2008), land cover (Hepinstall, Alberti and Marzluff, 2008), or combined land use and land cover. Most ecological processes are influenced by land cover, so models predicting future land cover are needed. However, most urban development and transportation models predict future land development (e.g., land use, development type and intensity) and not land cover (Parker *et al.*, 2003; Verburg *et al.*, 2004). For example, UrbanSim, an urban development model based on simulating the interactions between land use, transportation, and the economy, predicts specific development events and intensity (i.e., new square feet of commercial land use, new residential units). While these future events are highly relevant to the change in urbanization of a geographic region, they do not directly relate to how the land cover will change in the amount of impervious area and vegetation (e.g., species, type, or areal extent) present on a site. Another commonly used land use model, CLUE-S (Verburg *et al.*, 2002; Verburg and Veldkamp, 2004), uses spatial allocation rules to simulate competition between different land uses for a specific location on the ground. A third model, GEOMOD2 (Pontius, Huffaker and Denman, 2004) models future land use based on biophysical and climate factors (lifezone or ecoregion, elevation, soil moisture, precipitation) and potential land use. With each of these models, the output is land use or a combination of land use and land cover. These outputs must be translated to changes in land cover, vegetation, or species habitat to be used as input into species habitat models.

The Land Cover Change Model (LCCM) represents one model that has been developed to integrate predictions of future land use (derived from another model: UrbanSim) into predictions of land cover change (Hepinstall, Alberti and Marzluff, 2008). Specifically, LCCM uses the output of the UrbanSim economic development model discussed above as one of many inputs into a model that predicts changes in land cover. The outputs from LCCM and UrbanSim are then available as inputs into biodiversity models (see details below).

25.1.4 Integrating models to predict future biodiversity

Studies that have documented land cover change over time often link these changes to changes in available habitat for wildlife species (Pearson, Turner and Drake, 1999; Tworek, 2002; Turner *et al.*, 2003; Gude, Hansen and Jones, 2007). Linking land cover and land use change models to biodiversity can be accomplished using spatially explicit habitat data and either the coarse habitat-association models or more sophisticated statistical models discussed above. Often constraints (on gathering empirical data, computing power, storage) will dictate either few statistically based high spatial and thematic resolution models for few species (Dunning Jr. *et al.*, 1995; Schumaker *et al.*, 2004; Hepinstall, Marzluff and Alberti, 2009) or simpler, lower spatial resolution models for many species (White *et al.*, 1997; Schumaker *et al.*, 2004).

25.2 Predicting changes in land cover and avian biodiversity for an area north of Seattle, Washington

This section details an example an integrated modeling approach that uses classified remotely sensed data to predict likely changes in ecological systems in an urban and urbanizing environment. This process entailed the multiple steps discussed above and depicted in Fig. 25.1. Observed land cover change was modeled into the future with LCCM. UrbanSim (www.urbansim.org) was used to simulate land use change (e.g., change in land use type, number of residential units, and area of industrial or commercial buildings). Models of avian biodiversity were developed from field data using predictor variables that were output by UrbanSim and LCCM. Future biodiversity was predicted by applying the biodiversity model to the predicted future land cover and land use. Below I go into more detail regarding each step in the process.

25.2.1 Land cover maps

Land cover maps depicting 14 different land cover classes were available for 1995, 1999, and 2002 for six counties in western Washington, USA (Hepinstall-Cymerman, Coe and Alberti, 2009). These data were developed using Landsat TM imagery with 30 m cell resolution. Images were segmented into vegetated, unvegetated, and shadow using spectral mixture analysis (Rashed *et al.*, 2003; Powell *et al.*, 2007). Supervised classification was used on the different image segments. Multi-season (i.e., leaf-on and leaf-off) data were used within each year to separate deciduous and mixed forest from coniferous forest. Spectral mixture analysis was used to extract three land cover classes of urban (heavy [> 80% impervious surface within a 30 m pixel], medium [50–80% impervious surface], and light [20–50% impervious surface]). Temporal trajectories were used to separate bare soil into separate classes of agriculture, clearcut forest, and cleared for development. Overall classification accuracy ranged from 80% to 89% (Hepinstall-Cymerman, Coe and Alberti, 2009).

25.2.2 Land use change model

Both the LCCM (below) and the avian richness models require variables measuring urban development to run. I derived variables associated with land use change and intensity of new development from UrbanSim (Waddell, 2002) using simulations for 2003–2027 (i.e., 25 years into the future) developed for the Puget Sound Regional Council (P. Waddell, personal communication) as input to both the LCCM and the avian species richness

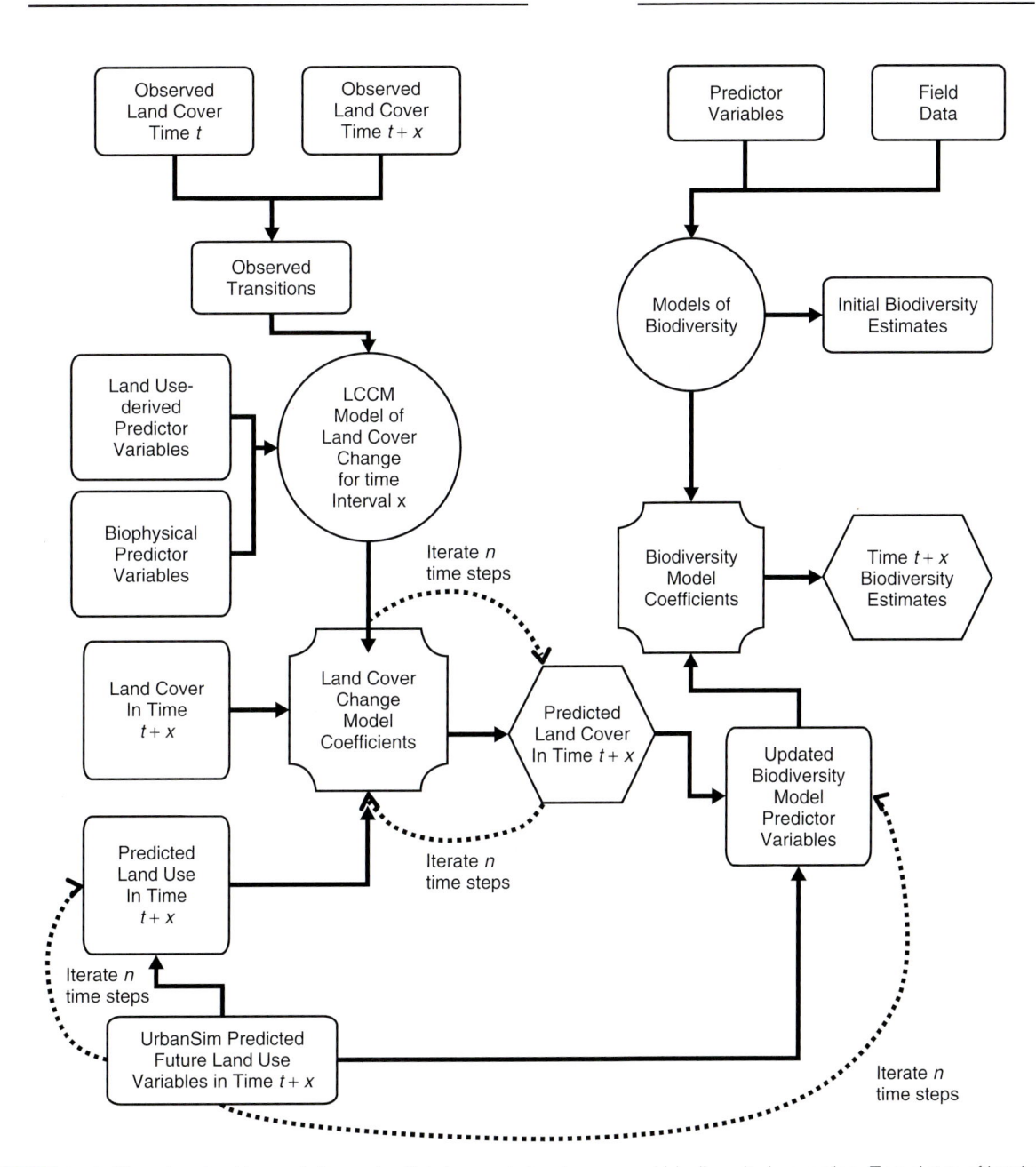

FIGURE 25.1 Steps involved in modeling potential change in land cover and biodiversity in a region. Two dates of land cover (t and $t + x$) are used to build a database of observed transitions separated by x years. LCCM (a multinomial logit-based land cover change modeling framework) is paired with land use and biophysical predictor variables to develop equations of land cover change. These coefficients are then used to predict future land cover using land use predictions derived from UrbanSim. The biodiversity model consists of linear regression models derived from field data that relate avian species richness to land use and land cover characteristics in 1 km^2 units. The biodiversity model coefficients are then applied to predicted future land use and land cover to derive estimates of future biodiversity. The land use and land cover models iterate forward 'x' years and predict the next time interval which is then fed into the biodiversity model to predict biodiversity at that next date.

model. UrbanSim is designed to model and predict many different land use and development variables, several of which are important in predicting land cover change in the LCCM including: number of residential units built, area of commercial or industrial land use added, and occurrence of a development event each year (Hepinstall, Alberti and Marzluff, 2008). Other variables such as change in land use type, density of residential land use, density of commercial land use and age of development can be derived from these variables and are important drivers of land cover change.

25.2.3 Land cover change model

I used LCCM (Hepinstall, Alberti and Marzluff, 2008) applied to the Central Puget Sound, Washington, to predict land cover change for a 15.5 × 33.8 km swath of Snohomish county, Washington (Fig. 25.2; north of the city of Seattle). This area includes a portion of the city of Everett as well as the towns of Mill Creek, Snohomish, and Monroe, and fertile agricultural lands surrounding the Snohomish River.

The basic LCCM specification allows the user to develop (i.e., specify and parameterize) multinomial logit equations of land cover change from empirically derived land cover and potential explanatory variables that are believed to drive land cover using the time-step interval of input data (Hepinstall, Alberti and Marzluff, 2008). For my example here, I selected two dates of land cover (1995 and 1999; 4-year time step) to derive estimates of plausible land cover transitions (including no-change). I developed multinomial equations for each transition using 65 candidate explanatory variables measuring landscape composition, configuration, contagion/distance, development intensity, and indicator variables defining land use, parcel value, parcel size. A full listing of the variables available for the Puget Sound implementation of the LCCM can be found in Table 2 of Hepinstall, Alberti and Marzluff (2008). I used 1999 land cover as the base land cover to predict to 2027 (the last date of available UrbanSim predictions for the study area). The predictive accuracy of LCCM

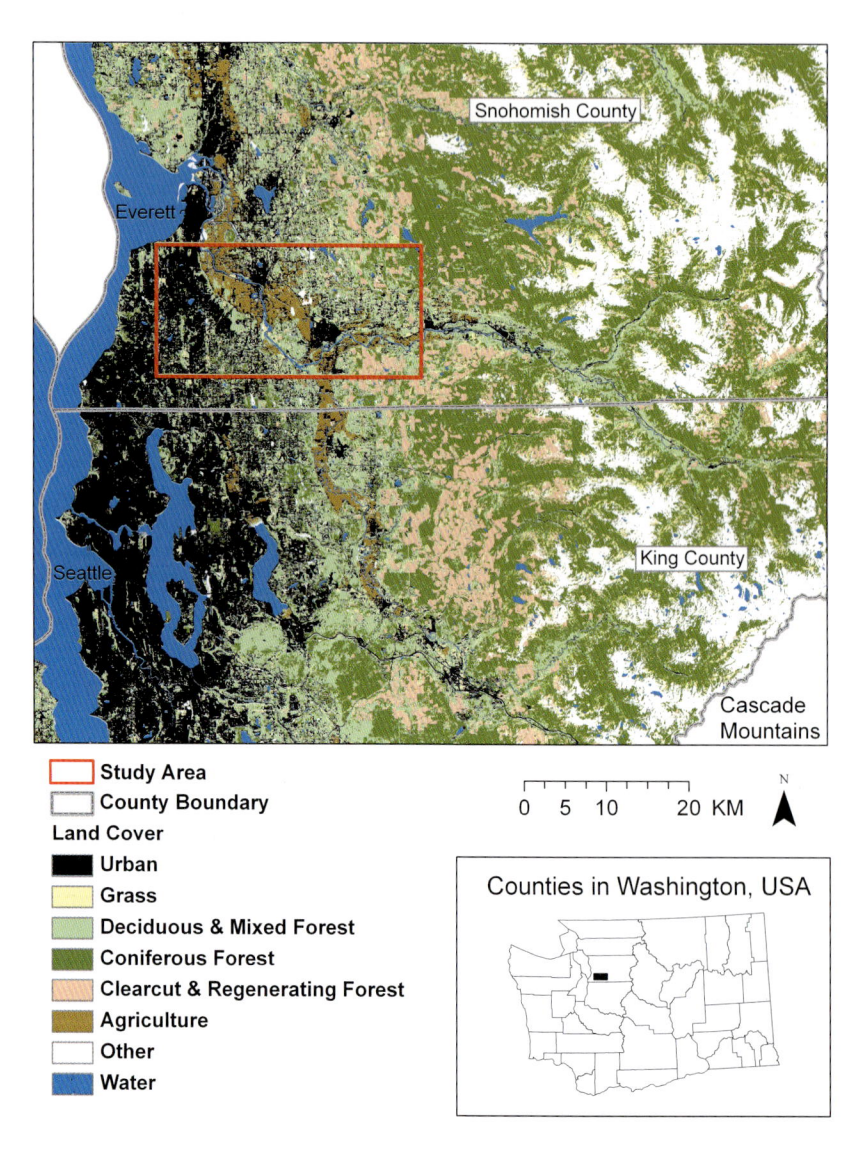

FIGURE 25.2 Land cover in 2002 for example 508 km^2 study area and for portions of Snohomish and King Counties, Washington, USA.

TABLE 25.1 Land cover (km² and percent of study area) observed (2002) and predicted (2003, 2027), change in percent of the study area from 2002 to 2003 and 2027, and percent increase by 2027 from 2002 land cover.

	Observed 2002 km²	Observed 2002 %	Predicted 2003 km²	Predicted 2003 %	Predicted 2003 Change in %	Predicted 2027 km²	Predicted 2027 %	Predicted 2027 Change in %	Predicted 2027 % Increase
Heavy Urban	29.2	5.7	30.6	6.0	0.3	103.8	20.4	14.7	255.9
Medium Urban	66.0	13.0	51.0	10.0	−3.0	53.9	10.6	−2.4	−18.4
Light Urban	99.1	19.5	113.5	22.3	2.8	144.8	28.5	9.0	46.1
Grass	33.3	6.6	64.6	12.7	6.2	57.8	11.4	4.8	73.7
Agriculture	49.5	9.7	34.2	6.7	−3.0	0.1	0.0	−9.7	−99.9
Deciduous and Mixed Forest	118.3	23.3	123.1	24.2	0.9	67.2	13.2	−10.1	−43.2
Coniferous Forest	55.7	11.0	43.3	8.5	−2.4	23.1	4.6	−6.4	−58.4
Regenerating Forest	37.3	7.3	28.7	5.7	−1.7	38.3	7.5	0.2	2.5
Other	19.7	3.9	19.1	3.8		19.1	3.8		

applied to various spatial partitions of the Central Puget Sound was reported in Hepinstall, Alberti and Marzluff (2008) and was calculated by comparing predictions for 2003 with observed land cover for 2002 (the closest available date of land cover data); kappa values varied between 0.63 and 0.77. Land cover can be predicted for as many time steps as desired, but research by Pontius, Huffaker and Denman (2004) has shown that the accuracy of these models likely converge on the level of accuracy derived from random models after multiple time steps. Prior runs of LCCM in the Central Puget Sound indicated the accuracy of model output likely approached the accuracy of random models after 8–10 time steps (Hepinstall-Cymerman, unpublished data).

25.2.4 Avian biodiversity model

In this example, avian biodiversity, as indexed by expected species richness (total and for native forest species) on a site, was modeled using published models for the Central Puget Sound (Hepinstall, Marzluff and Alberti, 2009) and are described in detail below. These models were developed from extensive field data designed to quantify species richness on sites representing different locations along an urban-to-wildland gradient (Donnelly and Marzluff, 2004, 2006; Marzluff, 2005; Marzluff et al., 2007). I used linear regression equations that related the percentage forest, aggregation of residential land use, and mean development age per km² to: (a) total species richness (for a subset of 57 common species); and (b) species richness of native forest species (19 species typically found in intact, mature forests) (Hepinstall, Marzluff and Alberti, 2009). These variables were selected in the avian community models because they were correlated with vegetation characteristics to which birds were responding and were variables output by the UrbanSim and LCCM models, and therefore available as spatially explicit predictions of future conditions.

25.2.5 Predicted land cover change for study area

The study area in 2027 is predicted to have more heavy urban, light urban, and grass and less medium urban (converting to heavy urban), agriculture (converting to urban or grass), and forest (deciduous and mixed forest and coniferous forest both converting to urban classes) (Table 25.1). The greatest increase is predicted to be in heavy urban and this pattern is visible in Fig. 25.3.

25.2.6 Predicted changes in avian biodiversity for study area

The predicted changes in land cover obviously have great potential for changing the avian species able to inhabit this landscape. Indeed, the total species richness is expected to decline across 94% of the study area (Table 25.2). The losses in total species richness (Fig. 25.4) clearly follow the spatial pattern of conversion of forests and agriculture to heavy urban (Fig. 25.3). For native forest birds, already low in this area because of the dominance of urban and agriculture, 59% of the area is predicted to harbor fewer of these species. As expected by the area totals, the overall pattern for native forest birds is approximately equal between areas of loss where forest is predicted to convert to urban and stable where agriculture is predicted to convert to urban since native forest birds currently do not occupy these agricultural lands (Fig. 25.5). Interestingly, some of the greatest predicted losses of species are in large patches; patterns that would not be easily discerned by looking only at the predicted land cover maps. This results points to the importance of modeling specific biodiversity responses rather than relying on land use or land cover changes alone to tell the story. Other published studies have found similar results with differential effects being predicted depending on what aspects of biodiversity were being considered (Schumaker et al., 2004; Gude, Hansen and Jones, 2007)

Conclusions

I have provided a brief overview of one example of integrated land use, land cover, and biodiversity modeling in an urbanizing environment. Such integrated modeling efforts provide the opportunity to simulate the interrelated components of urban development (Alberti and Waddell, 2000). I have shown how

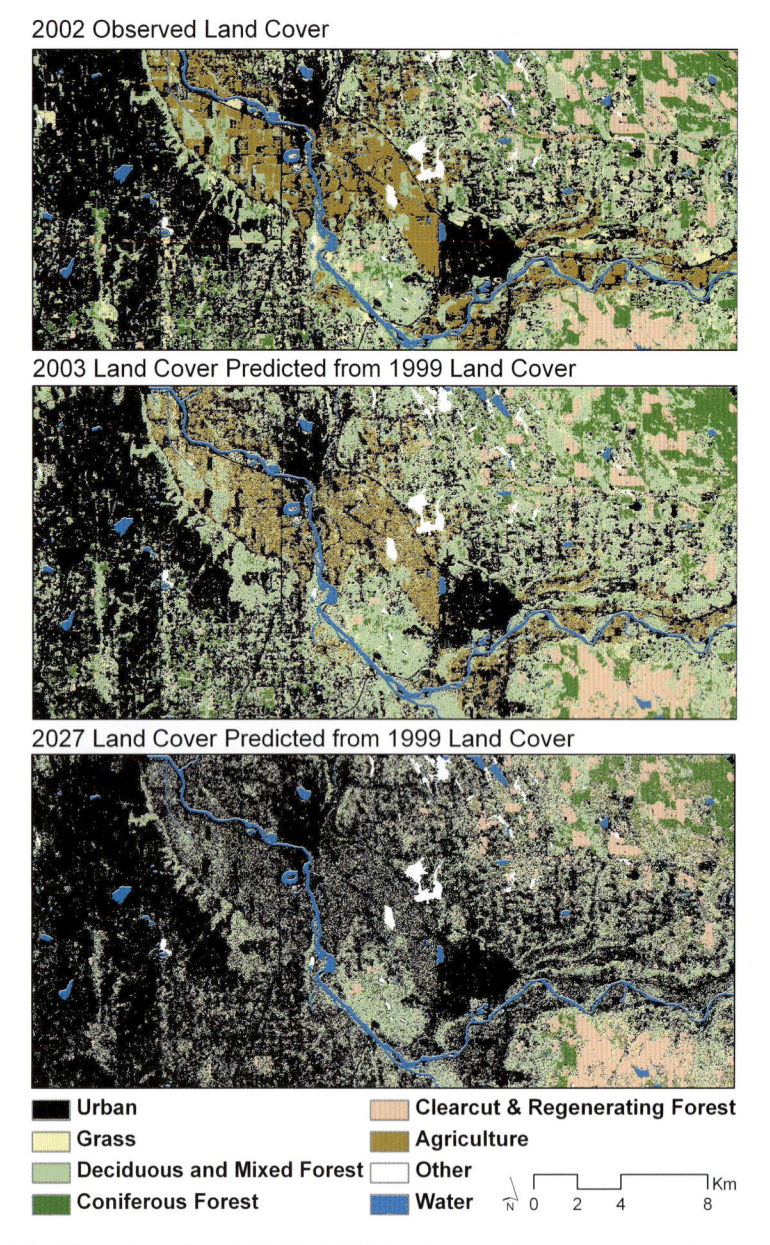

FIGURE 25.3 Observed (2002) and predicted (2003, 2027) land cover classes for example study area north of Seattle, Washington.

economic and land cover models predicting future changes in land use and land cover can be integrated with models predicting avian biodiversity to predict how this biodiversity may change in the future. All of these steps are required in some form to be able to relate ecological systems (in my example avian biodiversity) to land cover and land cover in the future. To map land cover requires remote sensing and advanced image classification techniques. To predict future land cover in urban environments requires modeling land use, since urban development is the main driving factor behind changes in land use and changes in land use translate into changes in land cover. The output from these models can be incorporated into many different natural resource planning and conservation applications. For example, site selection studies (e.g., selecting which set of sites to protect to conserve the most biodiversity) can use the outputs from predictive land use and land cover models as "threats" or "hazards" to weight the site selection process (Kiesecker et al., 2009).

Many factors are now favorable for the use of remotely sensed data of urban environments to support environmental modeling.

TABLE 25.2 Area (km^2) and percent of study area expected to gain, remain stable, or lose bird species richness as a result of changes in the landscape predicted to occur by 2027.

	All birds		Native forest birds	
	km^2	%	km^2	%
Gain	0.5	0.1	0.0	0.0
Stable	28.7	5.6	209.4	41.2
Loss	479.0	94.3	298.8	58.8

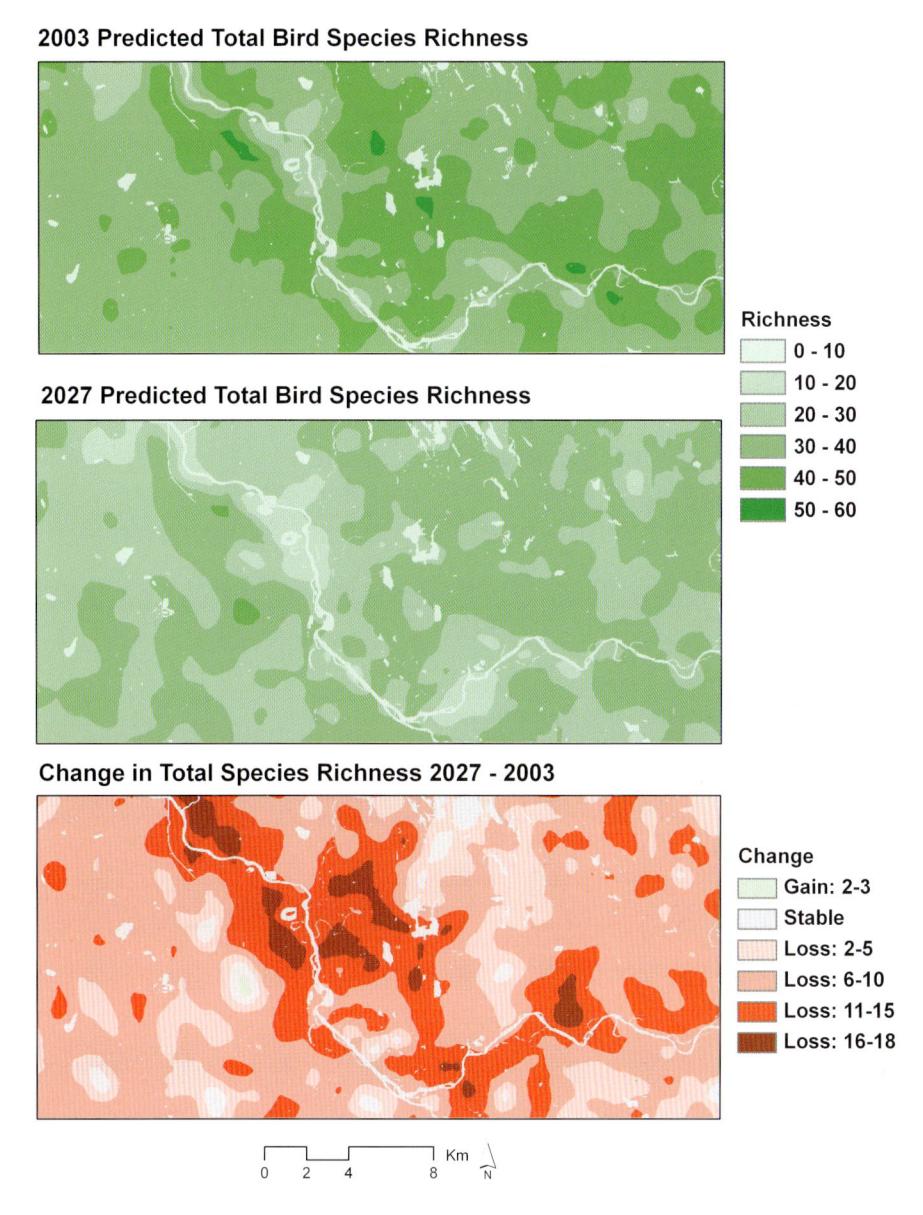

FIGURE 25.4 Predicted total bird diversity for 2003 and 2027 and the difference in species richness with time (2027–2003).

However, many limitations still exist. In the example provided, there was no attempt to model how the error in the initial land cover classifications may have influenced the results. Such errors would have propagated through: (1) the LCCM model specification, leading to erroneous associations between land cover change and predictor variables and ultimately erroneous predictions of future land cover; (2) erroneous avian biodiversity models since they were derived from the land cover data; and (3) erroneous predictions of future biodiversity change.

Additionally, each model had its own set of assumptions. For example, the LCCM assumes that observed change will continue into the future; specifically that current drivers of land cover change will be the same and of the same magnitude into the future. Because of these changes, future land cover change (rates and types of change) are likely to be different than what was observed in the past. Other studies have demonstrated how changes in data used to specify the LCCM and application to different areas have lead to different future predictions (Hepinstall, Alberti and Marzluff, 2008). Avian community models represent only one aspect of the biodiversity of the area and only one way of modeling avian diversity. Looking at individual species will likely yield different results with more complicated patterns (Hepinstall, Marzluff and Alberti in press).

We are now in an era where increased remotely sensed data, other geospatial data, computer power, decreased storage costs, and improved training at our undergraduate institutions will combine to provide many opportunities to model the past, present, and future effects of urbanization on ecological systems.

Acknowledgments

I thank Marina Alberti and the Urban Ecology Research Laboratory staff, and Paul Waddell and the Center for Urban Simulation and Policy Analysis staff, and John Marzluff. I thank Krista Merry

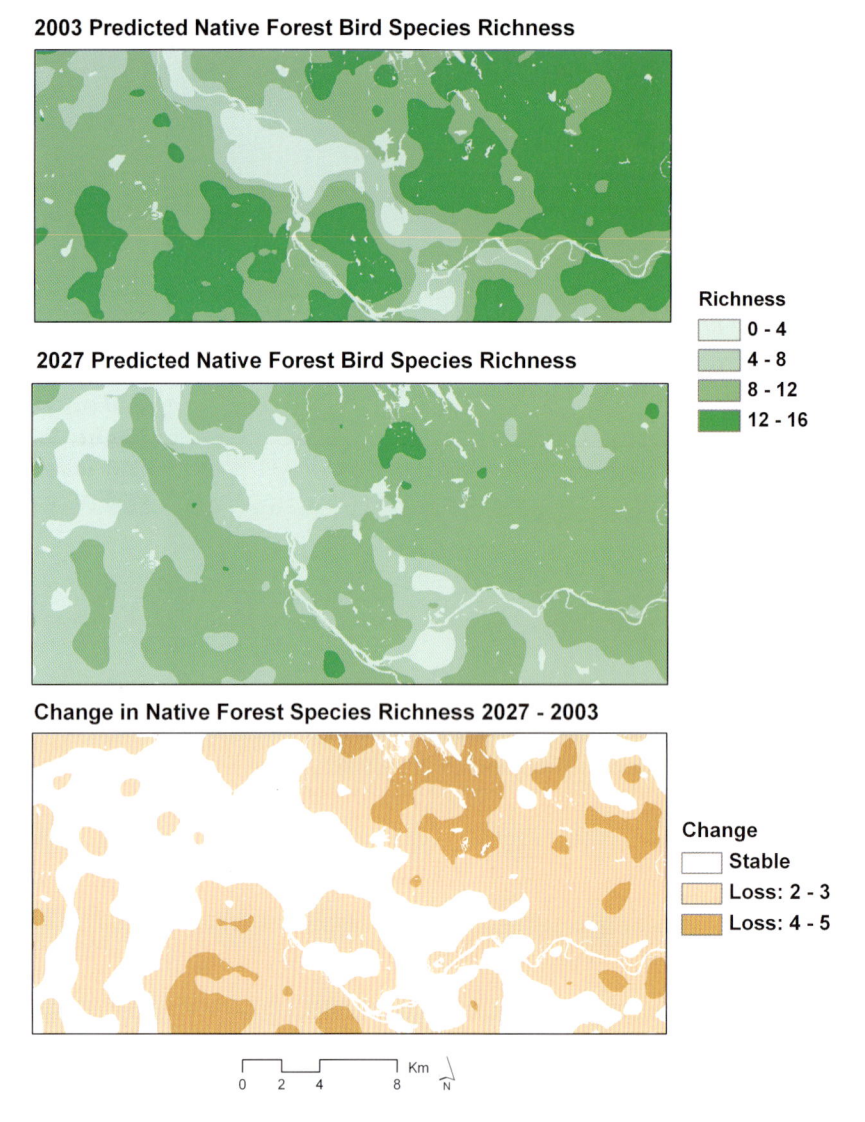

FIGURE 25.5 Predicted native forest bird diversity for 2003 and 2027 and the difference in species richness with time (2027–2003).

for comments on earlier drafts. Two anonymous reviewers and Xiaojun Yang also provided insightful comments that substantially improved the text. This research was supported by the University of Georgia's Warnell School of Forestry and Natural Resources, the National Science Foundation (DEB-9875041, BCS 0120024, IGERT-0114351), the University of Washington (Tools for Transformation Fund), and the University of Washington's College of Forest Resources.

References

Alberti, M. (2008) *Advances in Urban Ecology: Integrating Humans and Ecological Processes in Urban Ecosystems*, Springer, New York.

Alberti, M. and Marzluff, J.M. (2004) Ecological resilience in urban ecosystems: Linking urban patterns to human and ecological functions. *Urban Ecosystems*, **7**, 241–265.

Alberti, M. and Waddell, P. (2000) An integrated urban development and ecological simulation model. *Integrated Assessment*, **1**, 215–227.

Alberti, M., Marzluff, J.M., Shulenberger, E., et al. (2003) Integrating humans into ecology: Opportunities and challenges for studying urban ecosystems. *Bioscience*, **53**, 1169–1179.

Betts, M.G., Forbes, G.J., Diamond, A.W. and Taylor, P.D. (2006) Independent effects of fragmentation on forest songbirds: An organism-based approach. *Ecological Applications*, **16**, 1076–1089.

Betts, M.G., Forbes, G.J. and Diamond, A.W. (2007) Thresholds in songbird occurrence in relation to landscape structure. *Conservation Biology*, **21**, 1046–1058.

Blair, R.B. (1996) Land use and avian species diversity along an urban gradient. *Ecological Applications*, **6**, 506–519.

Blair, R.B. (2004) The effects of urban sprawl on birds at multiple levels of biological organization. *Ecology and Society*,

9, (5), 2–. [Online]. Available http://www.ecologyandsociety.org/vol9/iss5/art2/(accessed 22November 2010).

Cadenasso, M.L., Pickett, S.T.A. and Schwarz, K. (2007) Spatial heterogeneity in urban ecosystems: Reconceptualizing land cover and a framework for classification. *Photogrammetric Engineering and Remote Sensing*, **71**, 169–177.

Cohen, W.B. and Goward, S.N. (2004) Landsat's role in ecological applications of remote sensing. *Bioscience*, **54**, 535–545.

Conroy, M.J., Cohen, Y., James, F.C., et al. (1995) Parameter estimation, reliability, and model improvement for spatially explicit models of animal populations. *Ecological Applications*, **5**, 17–19.

Cutler, D.R., Edwards, T.C., Jr., Beard, K.H., et al. (2007) Random forests for classification in ecology. *Ecology*, **88**, 2783–2792.

Donnelly, R. and Marzluff, J.M. (2004) Importance of reserve size and landscape context to urban bird conservation. *Conservation Biology*, **18**, 733–745.

Donnelly, R. and Marzluff, J. (2006) Relative importance of habitat quantity, structure, and spatial pattern to birds in urbanizing environments. *Urban Ecosystems*, **9**, 99–117.

Dunning Jr., J.B., Stewart, D.J., Danielson, B.J., et al. (1995) Spatially explicit population models: Current forms and future uses. *Ecological Applications*, **5**, 3–11.

Fahrig, L. (1999) Forest loss and fragmentation: Which has the greater effect on persistence of forest-dwelling animals? in *Forest Fragmentation: Wildlife and Management Implications* (eds J.A. Rochelle, L.A. Lehmann, J. Wisniewski and L. Fahrig), Brill Academic Publishing, Leiden, The Neatherlands..

Fahrig, L. (2003) Effect of habitat fragmentation on biodiversity. *Annual Review of Ecology, Evolution, and Systematics*, **34**, 487–515.

Fassnacht, K.S., Cohen, W.B. and Spies, T.A. (2006) Key issues in making and using satellite-based maps in ecology. *Forest Ecology and Management*, **222**, 167–181.

Foley, J.A., Defries, R., Asner, G.P., et al. (2005) Global consequences of land use. *Science*, **309**, 570–574.

Grimm, N.B., Grove, J.M., Pickett, S.T.A. and Redman, C.L. (2000) Integrated approaches to long-term studies of urban ecological systems. *BioScience*, **50**, 571–584.

Gude, P.H., Hansen, A.J. and Jones, D.A. (2007) Biodiversity consequences of alternative future land use scenarios in greater yellowstone. *Ecological Applications*, **17**, 1004–1018.

Guisan, A. and Zimmermann, N.E. (2000) Predictive habitat distribution models in ecology. **135**, 147–186.

Guisan, A., Edwards, T.C. and Hastie, T. (2002) Generalized linear and generalized additive models in studies of species distributions: Setting the scene. *Ecological Modelling*, **157**, 89–100.

Hepinstall-Cymerman, J., Coe, S. and Alberti, M. (2009) Using urban landscape trajectories to develop a multi-temporal land cover database to support ecological modeling. *Remote Sensing*, **1**, 1353–1379.

Hepinstall, J.A., Alberti, M. and Marzluff, J.M. (2008) Predicting land cover change and avian community responses in rapidly urbanizing environments. *Landscape Ecology*, **28**, 1257–1276.

Hepinstall, J.A., Marzluff, J.M. and Alberti, M. (2009) Modeling bird responses to predicted changes in land cover in an urbanizing region, in *Models for Planning Wildlife Conservation in Large Landscapes* (eds J.A. Hepinstall, J.M. Marzluff and M. Alberti), Academic Press, San Diego.

Hepinstall, J.A., Marzluff, J.M. and Alberti, M. (in press) Predicting avian community responses to increasing urbanization. *Studies in Avian Biology*.

Houghton, R.A. (1994) The worldwide extent of land-use change. *BioScience*, **44**, 305–313.

Hulse, D.W., Branscomb, A. and Payne, S.G. (2004) Envisioning alternatives: Using citizen guidance to map future land and water use. *Ecological Applications*, **14**, 325–341.

Jat, M.K., Garg, P.K. and Khare, D. (2008) Monitoring and modelling of urban sprawl using remote sensing and gis techniques. *International Journal of Applied Earth Observation and Geoinformation*, **10**, 26–43.

Kiesecker, J.M., Copeland, H., Pocewicz, A., et al. (2009) A framework for implementing biodiversity offsets: Selecting sites and determining scale. *BioScience*, **59**, 77–84.

Lambin, E.F., Turner, B.L., Geist, H.J., et al. (2001) The causes of land-use and land-cover change: Moving beyond the myths. *Global Environmental Change*, **11**, 261–269.

Mackenzie, D.I., Nichols, J.D., Royle, J.A., et al. (2006) *Occupancy Estimation and Modeling: Inferring Patterns and Dynamics of Species Occurrence*, Elsevier, Amsterdam.

Marzluff, J.M. (2001) Worldwide urbanization and its effects on birds, in *Avian ecology and conservation in an urbanizing world* (eds J.M. Marzluff, R. Bowman and R. Donnelly), Kluwer Academic Publishers, Norwell, MA.

Marzluff, J.M. (2005) Island biogeography for an urbanizing world: How extinction and colonization may determine biological diversity in human-dominated landscapes. *Urban Ecosystems*, **8**, 157–177.

Marzluff, J.M., Withy, J.C., Whittaker, K.A., et al. (2007) Consequences of habitat utilization by nest predators and breeding songbirds across multiple scales in an urbanizing landscape. *Condor*, **109**, 516–534.

McGarigal, K. and McComb, W.C. (1995) Relationships between landscape structure and breeding birds in the oregon coast range. *Ecological Monographs*, **65**, 235–260.

Mckinney, M.L. (2002) Urbanization, biodiversity, and conservation. *BioScience*, **52**, 883–890.

Meyer, W.B. and Turner, B.L., (1992) Human population growth and global land-use/cover change. *Annual Review of Ecology and Systematics*, **23**, 39–61.

Munroe, D.K., Croissant, C. and York, A.M. (2005) Land use policy and landscape fragmentation in an urbanizing region: Assessing the impact of zoning. *Applied Geography*, **25**, 121–141.

Parker, D.C., Manson, S.M., Janssen, M.A., et al. (2003) Multi-agent systems for the simulation of land-use and land-cover change: A review. *Annals of the Association of American Geographers*, **93**, 314–337.

Pearson, R.G. and Dawson, T.P. (2003) Predicting the impacts of climate change on the distribution of species: are climate envelope models useful? *Global Ecology and Biogeography*, **12**, 361–371.

Pearson, S.M., Turner, M.G. and Drake, J.B. (1999) Landscape change and habitat availability in the southern Appalachian Highlands and Olympic Peninsula. *Ecological Applications*, **9**, 1288–1304.

Pickett, S.T.A., Cadenasso, M.L., Grove, J.M., *et al.* (2001) Urban ecological systems: Linking terrestrial ecological, physical, and socioeconomic components of metropolitan areas. *Annual Review of Ecology and Systematics*, **32**, 127–157.

Pontius Jr, R.G., Huffaker, D. and Denman, K. (2004) Useful techniques of validation for spatially explicit land-change models. *Ecological Modelling*, **179**, 445–461.

Pontius, J.R.G., Cornell, J. and Hall, C. (2001) Modeling the spatial pattern of land-use change with geomod2: Application and validation for costa rica. *Agriculture, Ecosystems and Environment*, **85**, 191–203.

Powell, R.L., Roberts, D.A., Dennison, P.E. and Hess, L.L. (2007) Sub-pixel mapping of urban land cover using multiple endmember spectral mixture analysis: Manaus, Brazil. *Remote Sensing of Environment*, **106**, 253–267.

Prasad, A.M., Iverson, L.R. and Liaw, A. (2006) Newer classification and regression tree techniques: Bagging and random forests for ecological prediction. *Ecosystems*, **9**, 181–199.

Rashed, T., Weeks, J.R., Roberts, D., *et al.* (2003) Measuring the physical composition of urban morphology using multiple endmember spectral mixture models. *Photogrammetric Engineering and Remote Sensing*, **69**, 1011–1020.

Rashed, T. (2008) Remote sensing of within-class change in urban neighborhood structures. *Computers, Environment and Urban Systems*, **32**, 343–354.

Roloff, G.J. and Kernohan, B.J. (1999) Evaluating reliability of habitat suitability index models. **27**, 973–985.

Sadik, N. (1999) The state of world population 1999 – 6 billion: A time for choices, United Nations Population Fund, New York, USA.

Schneider, L.C. and Pontius, R.G. (2001) Modeling land-use change in the Ipswich Watershed, Massachusetts, USA. *Agriculture, Ecosystems and Environment*, **85**, 83–94.

Schumaker, N.H., T. Ernst, D. White, *et al.* (2004) Projecting wildlife responses to alternative future landscapes in Oregon's Willamette Basin. *Ecological Applications*, **14**, 381–400.

Scott, J.M., Davis, F., Csuti, B., *et al.* (1993) Gap analysis – a geographic approach to protection of biological diversity. *Wildlife Monographs*, 1–41.

Sudhira, H.S., Ramachandra, T.V. and Jagadish, K.S. (2004) Urban sprawl: Metrics, dynamics and modelling using GIS. *International Journal of Applied Earth Observation and Geoinformation*, **5**, 29–39.

Tang, Z., Engel, B.A., Pijanowski, B.C. and Lim, K.J. (2005) Forecasting land use change and its environmental impact at a watershed scale. *Journal of Environmental Management*, **76**, 35–45.

Taubenböck, H., Wegmann, M., Roth, A., *et al.* (2009) Urbanization in India – spatiotemporal analysis using remote sensing data. *Computers, Environment and Urban Systems*, **33**, 179–188.

Tole, L. (2008) Changes in the built vs. Non-built environment in a rapidly urbanizing region: A case study of the greater toronto area. *Computers, Environment and Urban Systems*, **32**, 355–364.

Turner, M.G., Pearson, S.M., Bolstad, P. and Wear, D.N. (2003) Effects of land-cover change on spatial pattern of forest communities in the southern Appalachian Mountains (USA). *Landscape Ecology*, **18**, 449–464.

Tworek, S.A. (2002) Different bird strategies and their responses to habitat changes in an agricultural landscape. *Ecological Research*, **17**, 339–359.

Verburg, P.H. and Veldkamp, A. (2004) Projecting land use transitions at forest fringes in the philippines at two spatial scales. *Landscape Ecology*, **19**, 77–98.

Verburg, P.H., Soepboer, W., Limpiada, R., *et al.* (2002) Land use change modelling at the regional scale: The CLUE-S model. *Environmental Management*, **30**, 391–405.

Verburg, P.H., Schot, P.P., Dijst, M.J. and Veldkamp, A. (2004) Land use change modelling: Current practice and research priorities. *GeoJournal*, **61**, 309–324.

Verburg, P.H., Eickhout, B. and Vanmeijl, H. (2008) A multi-scale, multi-model approach for analyzing the future dynamics of european land use. *Annals of Regional Science*, **42**, 57–77.

Villard, M.-A., Trzcinski, M.K. and Merriam, H.G. (1999) Fragmentation effects on forest birds: Relative influence of woodland cover and configuration on landscape occupancy. *Conservation Biology*, **13**, 774–783.

Waddell, P. (2002) Urbansim: Modeling urban development for land use, transportation and environmental planning. *Journal of the American Planning Association*, **68**, 297–314.

White, D., Minotti, P.G., Barczak, M.J., *et al.* (1997) Assessing risks to biodiversity from future landscape change. *Conservation Biology*, **11**, 349–360.

Wilson, J.S., Clay, M., Martin, E., *et al.* (2003) Evaluating environmental influences of zoning in urban ecosystems with remote sensing. *Remote Sensing of Environment*, **86**, 303–321.

Xu, C., Liu, M., An, S., *et al.* (2007) Assessing the impact of urbanization on regional net primary productivity in Jiangyin County, China. *Journal of Environmental Management*, **85**, 597–606.

Yang, X. and Lo, C.P. (2002) Using a time series of satellite imagery to detect land use and land cover changes in the Atlanta, Georgia metropolitan area. *International Journal of Remote Sensing*, **23**, 1775–1798.

Yuan, F., Sawaya, K.E., Loeffelholz, B.C. and Bauer, M.E. (2005) Land cover classification and change analysis of the Twin Cities (Minnesota) metropolitan area by multitemporal landsat remote sensing. *Remote Sensing of Environment*, **98**, 317–328.

26

Rethinking progress in urban analysis and modeling: models, metaphors, and meaning

Daniel Z. Sui

> It has been well said that analogies may help one *into the saddle*, but are encumbrances on a long journey.
>
> A. Marshall

Previous assessment on urban analysis and modeling efforts has concentrated primarily on technical and methodological details without probing the underlying root metaphors embedded in the diverse modeling efforts. This chapter presents a comprehensive review of four major traditions in urban modeling during the past 50 years – spatial morphology, social physics, social biology, and spatial events. Using Pepper's world hypotheses as a guiding framework, this chapter argues that the root metaphors embedded in the four urban modeling traditions correspond to those in Pepper's world hypotheses – the world as forms, machines, organism, and arenas. It is argued that what we traditionally regard as progress is, in fact, a shift of metaphors used for conceptualizing cities. In this context, what we must recognize is the process whereby meaning is produced from metaphor to metaphor, rather than, as it was often assumed by urban modelers, between model and the world. We need not only to check the validity of our models from the technical perspective in terms of data accuracy and consistency with empirical results, but also to scrutinize the driving conceptual metaphors deeply embedded in the models. Only then can we weave the insights gained from the urban modeling efforts with other urban narratives to have a more sensible urban future.

26.1 Introduction

According to a new estimate by the United Nations, starting in 2009 more than half of the world's 6.7 billion people are living in urban areas starting in 2009, marking the beginning of the first urban century in human history (http://esa.un.org/unup). The pace of the world's urbanization has been continuing to accelerate at an astonishing rate (Cohen, 2004). By 2030, nearly 5 billion people will be urban residents (www.sciencemag.org/cities). Many of the challenges facing humanity today, from the economy, environment, and education to health, security, and development, are all manifesting (though not exclusively) in the world's urban areas. Thus, a better understanding of diverse urban forms and processes will be of paramount importance. Not surprisingly, we have witnessed renewed interests by both the academic world and the general public on issues related to urban development in recent years (Ash et al., 2008).

As we enter the very first urban century in human history, it is fitting to consider what progress has been made in understanding and modeling urban phenomena and how well equipped we are to deal with issues that are likely to arise. Over the past 50 years, the geospatial community in general and urban modelers in particular have made significant contributions to improving our understanding of various complex issues related to urban development. As reflected from the diverse and informative chapters included in this and the related volume (Rashed and Juergens, 2006), many exciting interdisciplinary advances in urban analysis and modeling have taken place during the first decade of the 21st century, from data acquisition, spatial metrics, and dynamic visualization to model construction, calibration, and applications. At the time this chapter was written (November, 2009), a simple query of "urban analysis and modeling" in Google had more than 6.6 million returns, with the same query in Google Scholar having 1.3 million returns. To better understand such a voluminous literature, we need a more robust, inclusive framework. The primary goal of this chapter is to initiate such an endeavor to take a synoptic overview of the recent development of urban analysis and modeling. Moving away from technical and methodological discussions, as most the chapters in this volume have so eloquently presented, this chapter aims to present a panoramic overview of urban analysis and modeling from a conceptual/theoretical perspective. I hope this chapter, not intended to be exhaustive due to space limitations, does provide enough food for thoughts in future discussions.

The particular angle this chapter takes for looking at the urban analysis and modeling efforts is to probe the dominant metaphors used in various urban analyses and modeling efforts. Human languages are metaphorical by nature, and rhetorical devices are instrumental in creating meaning. Metaphors assert a similarity between two or more different things while converting the unfamiliar to the familiar. As historians and philosophers of science have demonstrated (Black, 1962; Hesse, 1963, 1980), analogical thinking and metaphoric description play a pivotal role in the formation of scientific ideas. Throughout history, scientific knowledge has advanced through the creative insights emanating from analogical connections between ostensibly distinct realms of phenomena (Rothbart, 1997). Intractable obstacles to understanding complex processes can be overcome by analogical comparisons to ostensibly distinct domains. Such comparisons often span different disciplinary boundaries.

The field of urban analysis and modeling is no exception. Urban modelers have relied on analogies and metaphors borrowed from many other fields to construct models for understanding the complexity of cities. A few modelers have made explicit acknowledgement of the use of metaphors/analogies in their work (Wilson, 1969a, b; Isard, 1999) while the majority of them have been silent on this issue, often taking some of the fundamental assumptions in their urban models for granted. Reflecting on the general trend of interests in the role of metaphors in scientific process (Leatherdale, 1974; Lakoff and Johnson, 1980), social scientists in general (Mirowski, 1988, 1994; Verma, 1993) and geographers in particular have probed their role in developing theoretical discourses.

Geographers have examined the use of metaphors in economic geography (Barnes, 1992, 1996), cultural geography (Tuan, 1978; Duncan, 1980), health geography (Kearns, 1997), environmental geography (Mills, 1982; Norwick, 2006), geographic methodology (Livingston and Harrison, 1981), and geographical thought in general (Sui, 2000). Although both Buttimer (1993) and Barnes (1996) have touched on the role of metaphors related to urban analysis and modeling indirectly, so far our collective knowledge about urban analysis and modeling efforts are still dominated by technical discussions on the methodological details without probing deeply the root metaphors in urban models. There exist few systematic, comprehensive works examining the role of metaphors in urban modeling except Couclelis's (1984) early work, which drew little immediate attention among urban modelers.

This chapter aims to fill in this void in the literature by presenting some preliminary discussions on analogies and metaphors in various urban models. By doing so, it aims to raise the sensitivity level of both modelers and users about the power (and also the constraints) each metaphor brings, and to better appreciate the intended and unintended consequences in the real world when policies are made according to the results of these modeling efforts. Unlike the post-modernists or post-structuralists, my goal here is not to denigrate or discard urban models, but to make both modelers and users more keenly aware of the theoretical commitment they made prior to their number-crunching modeling efforts, and the impacts of these metaphorical commitments on their understanding of the urban reality. I also hope that this chapter will open new dialogs about possible alternative metaphors we can rely on for launching our next generation of operational models.

The chapter is composed of five sections. After a brief introduction, Stephen Pepper's world hypotheses and their related root metaphors are introduced in Section 26.2. Using Pepper's framework as a guide, the third section identifies four traditions in urban modeling and the driving metaphors urban modelers created in each tradition. Section 26.4 presents further discussions on the meaning of progress in urban modeling from a metaphorical perspective, followed by a summary and concluding remarks in the last section[1].

26.2 Pepper's world hypotheses: the role of root metaphors in understanding reality

Inspired by Buttimer's (1993) early work on the topic, Pepper's world hypotheses will be used as the guiding conceptual framework for this chapter. In his classic volume on metaphysics, Berkeley philosopher Stephen Pepper (1942) postulated four world hypotheses, each having its own root metaphor, theory of truth, world picture, and narrative style (Table 26.1). As Buttimer (1993) has demonstrated, Pepper's world hypotheses can serve as an inclusive framework for our understanding of reality in general, and the models in social sciences and humanities in particular.

Formism grounds itself on a common sense experience based in similarity; its cognitive claims rest on a correspondence theory of truth. A proposition is considered as true if it corresponds to a fact or some portion of reality. Its world picture is a dispersed one; each form may be analyzed and explained in terms of its own nature and appearance. Its practice motto is to "get to the top of things."

Mechanism takes a common sense experience with the machine as its root metaphor. Its claim to cognitive validity rests on a causal adjustment theory of truth, which is based on the idea that a proposition is considered as true only if there is an appropriate causal connection between the states of affairs. Mechanism offers an integrated world view while also affording guidelines for detailed analysis. Its practice motto is to "get to the bottom of things."

Organism also offers an integrated picture of the world, but it aims at synthetic understanding of the whole rather than the analysis of its parts. It implicitly assumes that every event in the world follows a concealed process, all eventually reaching maturation (and transcendence) in an organic whole. It sees all events in life cycles, more or less following specific organic forms of growth. Cognitive claims of the organic world view rest on a coherent theory of truth. The truth of any new proposition consists in its coherence with some specified set of propositions. Its practices motto is to "get to whole of things."

Contextualism draws inspiration from the common-sense experience of unique events. It sees the world as an arena of unique occurrences for some local contexts. It seeks to unravel the texture and strands of processes operating in, or associated with, particular events. Its world-view is a dispersed one with a synthetic goal. Contextualism espouses an operational theory of truth. What is considered as true is something that works or can be operationalized (at a time) in the real world. Its practice motto is to "get to each individual thing itself."

Although published over 65 years ago, Pepper's world hypotheses are still highly relevant today to help understanding the diverse strands of urban analysis and modeling efforts.

26.3 Progress in urban analysis and modeling: metaphors urban modelers live by

There have been continuing efforts to review progress in urban modeling during the past 50 years (Harris, 1965, 1985; Batty, 1976, 1994, 2007; de la Barra, 1989; Klosterman, 1994, Wilson, 1998; Wegener, 2004). In general, a consensus seems to exist among urban modelers that there are at least two generations of urban models (Sui, 1997). The first generation corresponding largely to those published prior to Lee's (1973) paper, are typically top-down models focusing on land-use and transportation interaction according to principles drawn from neoclassic economics, regional science, and spatial interaction models. The second generation favors a more bottom-up approach, focusing on individual behaviors using either cellular automata or agent-based modeling approaches. Although classifying models through this two-generation-approach serves a pedagogic purpose for students of urban modeling to understand fundamental changes in urban modeling, it doesn't do justice to the diverse literature on urban analysis modeling.

Looking through the glass of Pepper's four world hypotheses, the efforts of urban analysis and modeling during the past 50

TABLE 26.1 Pepper's world hypotheses, root metaphors, and theories of truth.

World hypothesis	Root metaphors	Theories of truth	World picture/ narrative style	Practice motto
Formism	Mosaic/similarity of forms	Correspondence	Dispersed/analytical	"Get to the top of things"
Mechanism	Machine	Causal adjustment	Integrated/analytical	"Get to the bottom of things"
Organism	Organic whole	Coherence	Integrated/synthetic	"Get to the whole of things"
Contextualism	Arena/spontaneous (historic) events	Operational/pragmatic/ anarchistic	Dispersed/synthetic	"Get to each individual thing itself"

TABLE 26.2 Four urban modeling traditions: root metaphors and operational models.

Urban analysis and modeling traditions	Related pepper world hypothesis	Dominant metaphors	Operational models/ measurements
Spatial morphology	Formism	Cities as Fractals	Fractals Lacunarity Spatial metrics
Social physics	Mechanism	Cities as Machines	The Lowry Model CUFM, ITLUP, MEPLAN, RURBAN, URBANSIM
Social biology	Organism	Cities as Organisms	Cellular Automata, Agent-based Models (ABM), Life Cycle Analysis (LCA)
Spatial events	Contextualism	Cities as Arenas	Field-based Time Geography; Sequence Alignment; Urban Social Tapestry Analysis

years can better be grouped into four major traditions – spatial morphology, social physics, social biology, and spatial events. Embedded in each of these traditions is a different driving metaphor urban modelers live by (Table 26.2), corresponding to one of Pepper's four root metaphors in his world hypotheses.

Viewed from a metaphorical angle, the first generation of urban models was primarily motivated by conceptualizing cities as machines following the social physics tradition, whereas the second generation of urban models conceives cities as organisms following the social biology tradition. In addition to these two dominant types of models, urban modelers have never ceased their efforts to search for innovative ways to better describe complex urban forms. In addition, the recent emergence of geographical information systems (GIS) and science (GIScience), Web 2.0, and user-generated content have also promoted a nascent paradigm of studying cities as events. As shown in Table 26.2, Pepper's four world hypotheses can provide us with a more holistic perspective on diverse urban analysis and modeling efforts (Wilson, 1967; Batty, 1976; Harris, 1985; Batty and Longley, 1994; Benenson and Torrens, 2004; Hudson-Smith et al., 2009).

26.3.1 Cities as forms – the spatial morphology tradition

The study of urban forms has a long interdisciplinary history linked to geography, planning, architecture, economics, and sociology (Morris, 1979; Vance, 1990; Kostof, 1991). This tradition focuses on the description, analysis and modeling of existing or ideal urban forms. It corresponds to Pepper's formism world hypothesis, with a root metaphor focusing on the various mosaics of geometric forms.

The spatial morphology tradition is perhaps the oldest among the four. Good city form is often linked to the stars in ancient China and India (Lynch, 1981). In classic locational theories, von Thünen's concentric rings for agriculture, Weber's triangle for manufacturing, Christaller's hexagons for services, and Alonso's theory for residential location all entail certain geometric forms. In urban geography, Burgess's (1925) concentric rings, Hoyt's (1939) sectoral radiation, and Harris and Ullman's (1945) multinuclei, and Garreau's (1992) edge city are widely accepted as standard descriptions of urban spatial structure based on geometric forms, which in turn have been transplanted to social area analysis from a factorial ecology perspective. Recent debates on new urbanism and whether urban development should follow more compact or dispersed/sprawling patterns are also, perhaps indirectly, a reflection of our obsession for the pursuit of formism (the "ideal" form) in understanding cities.

Although non-Euclidean geometry has been occasionally mentioned in the literature (Muller, 1982) to depict functional areas in cities, until recently descriptions of urban forms have been predominantly couched in Euclidean geometry. Exceptions like Wilson (1981) attempted to apply Thom's (1975) catastrophe/bifurcation theory to study cities generated few followers.

One of the major breakthroughs in the spatial morphology tradition is perhaps the conceptualization of cities as fractals according to Benoit Mandelbrot's fractal geometry. A fractal is a geometric form that has the property of self-similarity – some part of the form has the same properties of the form as a whole. In urban studies, cities possess fractal properties in a statistical sense, meaning that a part of the form has the same statistical properties as the whole. Batty and Longley (1994) argued that much of our pre-existing urban theory is a special case of the fractal city, and they see fractal as the best hope for a holistic understanding for cities by linking the micro with macro, local with the global, intra- to inter-urban levels of analysis. To a larger extent, recent advances in our understanding of the size, scale, and shape of cities are all closely related to conceiving cities as fractals (Batty, 2008; Rozenfeld et al., 2008).

By conceptualizing cities as fractals, we can now better understand the complexity of urban forms in ways that are impossible using the Euclidean geometry, such as multi-scale measurements and dimensions between 1 and 2. The fractal dimension is quickly becoming a major component of spatial metrics for describing landscape patterns and urban forms (Sudhira, Ramachandra and Jagadish, 2004; Herold, Couclelis and Clarke, 2005). We can calculate fractal dimensions easily using public domain data and software like Fragstat.

Bolstered by new insights into the complexity of a geometric kaleidoscope from the fractal perspective, both quantitative urban modelers and qualitative urban social theorists have gained new insights on urban spatial structure. For example, Sui and Wu (2006) used lacunarity analysis – a new multiscale residential segregation measure inspired by fractal geometry – to critically re-examine the relative significance of class vs. race in shaping the residential segregation patterns in a prismatic metropolis. Social theorist Ed Soja (2000) chose the fractal dimension as a metaphor to describe the restructured social mosaic in Los Angeles. Fractal cities were used as one of the conceptual pillars by Soja (2000) to characterize the "metropolarities" in what he called the "Post-metropolis."

By introducing fractals into urban studies, we have gradually shifted from viewing cities as mosaics of geometric forms as defined in Euclidean geometry to viewing cities as a recursive world as described by the fractal geometry (Batty and Hudson-Smith, 2007), echoing Berry's (1964) early synthesis of "cities as systems within systems of cities." The significance of this fundamental change in the spatial morphology tradition is still unfolding. Undoubtedly, fractals have enabled urban researchers to better understand the complex relationship between the parts and the whole. Conceptually, fractals also imply unsettled complexity and instability among the multiple social and spatial relationships within cities, placing significant constraints on confident generalization and accurate cartographic depiction as classic urbanists have done.

26.3.2 Cities as machines – the social physics tradition

The social physics tradition of urban modeling concentrates on studying mechanisms found in urban development. This tradition can be equated to Pepper's mechanism world hypothesis. The driving metaphor of the social physics tradition has been conceiving cities as machines, aiming to model interaction among its different components, like the interactions between land use and transportation. The large scale-models developed around the mid-20th century can be classified as following this tradition (Batty, 1976).

The application of basic laws from Newtonian physics to social domains has a long history. They were earlier advocated by Carey (1858), then applied in migration studies by Ravenstein (1885) and in retail studies by Reilly (1931) and Lakshmanan and Hansen (1965). After the principles of social physics were proposed shortly after World War II (Stewart, 1947, 1950), urban modelers have attempted to use all major concepts and techniques developed in physics, such as entropy maximization in statistical mechanics (Wilson, 1967), information theory (Webber, 1979), synergetics (Haken, 1983), dissipative structure (Straussfogel, 1991), self-organizing criticality (Bak, 1996), self-organizing systems (Allen, 1997), small world/complex networks (Urry, 2004), Brownian agents of active particles (Schweitzer and Farmer, 2007), and even quantum physics (Arida, 2002). But as far practical applications are concerned, the types of models developed by Lowry (1964) were perhaps those having the biggest impacts. Essentially, the Lowry model is an aggregated large-scale urban model based upon economic base theory with a built-in spatial interaction submodel to predict journey-to-work and journey-to-shop flows. These types of models rely on the implicit assumption that cities are like simple systems, usually involving a finite number of individual elements. Consequently, the entities in the models have to be aggregated to predefined spatial units.

Although mounting criticisms in the 1970s (Lee, 1973; Sayer, 1976) of large-scale urban modeling approaches shook both their foundation and practice, urban modeling efforts following the social physics tradition in general and Lowry models in particular did not die [perhaps Forrester's (1969) model is an exception], but continued to evolve across several different continents (Wegener, 1994; Wegener and Furst, 1999). For instance, modified versions of the Lowry models have been revitalized by behavioral modifications through discrete choice models based upon random utility theory (Anas, 1982; Wrigley and Longley, 1984; Ben-Akiva and Lerman, 1985; Roy and Thill, 2004), or through their integration with GIS (Landis, 1995; Sui, 1998). Instead of the crude predictions of spatial interaction modeling, new models can predict choices between alternatives as function of attributes of the alternatives, subject to stochastic dispersion constraints taking account of unobserved attributes of the alternatives, such as differences in taste between decision makers and uncertainty/lack of information.

In general, urban models following the social physics tradition tend to be aggregated, cross-sectional, and non-dynamic. Gregory (1980) argued that the failure of this modeling tradition "... is strategic: it allows for the uncontested mobilization of its discursive element to secure the reproduction of specific structures of domination (p.341)." Perhaps inspired by physicists' dream of a final theory for the ultimate law of nature (Weinberg, 1994), Wilson (1995) also dreamed of a final theory in locational analysis following the social physics tradition, but so far such a final theory remains a dream. The urban modeling efforts following the social physics tradition have been stagnant/dormant in recent years and seem to have been overtaken by the next two traditions.

26.3.3 Cities as organisms – the social biology tradition

The social biology tradition of urban modeling conceptualizes cities as organisms. This tradition is related to Pepper's organism world hypothesis, as it aims to understand how parts of the city (or the city as a whole) grow/evolve. In contrast to the aggregated, static, non-behavioral large-scale urban models following the social physics tradition, models following the social biology tradition tend to be disaggregated, dynamic, and more behaviorally oriented, simulating patterns at the individual levels[2].

Similar to the social physics tradition, conceiving cities as an organism or some other biological entity has a long interdisciplinary history (Jacobs, 1961; Alexander, 1964; 2004; Steadman, 1979; Larkham, 1995; Khalil, 1998). Cities have been studied by invoking ecological metaphors for understanding their resilience (Pickett, Cadenasso, and Grove, 2004), metabolism metaphor for exploring their internal material and energy flows (Olson, 1982), cancer/epidemic metaphors for explaining urban systems growth (Hern, 2008), or brain neural net metaphors for modeling traffic networks within a city (Changizi and Destefano, 2009).

Modeling cities as an organism has also been greatly facilitated by the biologically inspired computing paradigm, ranging from neural computing and genetic algorithms to evolutionary programming. Due to sharing common data structure, cellular automata (CA) and agent-based (AB) models have increasingly stepped out of the artificial world (Besussi and Cecchini, 1996), and been integrated with GIS (Takeyama and Couclelis, 1997; Liu, 2008) as an integral part of geosimulation tools (Benenson and Torrens, 2004). Viewed from a larger context, CA and AB models are increasingly becoming part of the emerging generative social science (Epstein, 2006) aiming to simulate diverse social processes across space.

In many ways similar to the dominance of Lowry models in the social physics tradition, the social biology tradition has been dominantly by models developed from cellular automata (CA) and agent-based models (ABM) – what Portugali (1999) called FACS (free agent in a cellular space) model. As demonstrated in some chapters in this book, there have been major methodological accomplishments during the past 15 years. The driving metaphor behind this new modeling paradigm using cellular automata and agent-based modeling is more biologically than mechanically motivated, by conceiving cities as growing cells[3].

Different from the Lowry and Forester types of urban models, CA and AB models conceive cities as complex systems involving a large but finite number of intelligent and adaptive agents. The behaviors of these agents are contingent on the availability of information and subject to modification of their rules of action based upon new flows of information. This continual and dynamic change of agents behavior makes prediction and measurement using old rules of science impossible. CA and AB-based urban models have been characterized by new concepts and theories based on non-linear dynamics. After about a decade of theoretical reflection (Couclelis, 1985, 1988, 1997), CA and AB models have been applied to simulate complex real world issues (Clarke and Gaydos, 1998; Sui and Zeng, 2001; Waddell, 2002; Yang and Lo, 2003; Brown *et al.*, 2004; Torrens, 2006; Liu and Seto, 2008).

Just like the physical/mechanical metaphors, conceiving cities as organism also has unintended consequences. For example, the idea of life cycles is embedded in all organism metaphors. According to Roberts (1991), the uncritical use of a life cyclical approach allowed the idea of city to slip from being an image to being a cause in accounts of urban decline, and has served a predominantly conservative urban policy. The use of the biological metaphor in general and the life cycle metaphor in particular, is profoundly ideological as it espouses inevitability, fatalism, determinism, and inexorability (Furbey, 1999). According to the policies motivated by biological metaphors, any government intervention has been criticized as tinkering with some natural order of things, which has led to policies of planning shrinkage, and managing decline and dispersion. A deterministic, biologically motivated urban triage often directly attacks empowering views of polity and place.

Nonetheless, CA and ABM-based simulations are emerging as part of the generative social science, which is changing the new frontier of science (Casti, 1997; Wolfram, 2002). For instance, Batty (2009) argued that we can develop a digital breeder to simulate urban growth according to different scenarios. Spiller (2009) even suggested that:

> we can use the natural imperatives of plants and maybe animal cells as a means to 'rewire' them to create huge rafts of new architectural flora and fauna. We might be able to make truly sustainable and green materials whose biodegradability is simply a natural side effect of these technologies (p. 131). Remote sensing technologies are going to play increasingly important roles in the urban modeling following the social biology tradition. Instead of simply being tools to study cities from the sky (Campanella, 2001), remote sensing now is regarded as "the X-ray crystallography" for urban "DNA"
>
> *(Wilson, 2009)*

.

26.3.4 Cities as arenas – the spatial event tradition

The spatial event tradition is relatively new, perhaps the least developed and least noticed among the four traditions, but perhaps has the greatest potential and momentum for new growth. This new urban modeling tradition conceptualizes cities as spatial events (Batty, 2002). Conceptually, this tradition is related to Pepper's contextualism world hypothesis. Consistent with Hägerstrand's time geography, the spatial event tradition is closely allied with the development of people-based GIS (Miller, 2007) and the concept of a real-time city (Foth, 2008).

Attempting to move away from any pre-conceived notion of how cities work, this tradition aims to understand how individual events occur spontaneously within a city. According to Batty (2002), this tradition grew out of a need for a more temporal emphasis in our theories and models. For quite some time, we have placed too much emphasis on finding equilibrium instead of on understanding the dynamics of urban change despite Wilson's (1981) early attempt to do so from the perspective of catastrophe and bifurcation. By conceiving cities as arenas for spontaneous events, we can better understand rhythms of urban life that take place in both time and space. Nowadays we have the technologies to analyze events by their location, duration, intensity, and sometimes volatility. More than in any other tradition, humans are becoming the censors (Goodchild, 2007) as a part of the participatory sensing (http://research.cens.ucla.edu), and urban modeling has become closer to story-telling and narrative development (Guhathakurta, 2002).

The metaphor of "cities as arena" has meanings not only in a physical sense, but also increasingly in a virtual sense. The growth of web 2.0 and related user-generated content in the form of spatially and temporally tagged information may greatly facilitate the study of cities as spatial events (Hudson-Smith *et al.*, 2008, 2009). For example, Girardin, Vaccari, and Gerber (2009) used geotagged photos tourists voluntarily posted on flickr.com as the primary data source to reconstruct tourist flows. Similarly, Dykes *et al.* (2008) used volunteered geographic information to describe the diversity of places.

The emerging participatory sensing has enabled citizens to record and transit data about the sourrounding environment using means like their cell phones. High school students can develop a mashup in Google Earth using realistic three-dimensional imageries for showing potential impacts of global climate change in their hometown. The US Centers for Disease Control and Prevention (CDC) has been tracking the diffusion of H1N1 flu across the United States by closely monitoring tweets on Twitter. Urban planners have also started using Second Life, simulating potential planning scenarios and policy impacts

(Foth *et al.*, 2009). The emerging web-based technologies have also enabled a new level of synthesis of a heterogeneous geospatial information and lead to a better understanding of urban forms and processes at multiple scales (http://www.londonprofiler.org).

26.4 Models, metaphors, and the meaning of progress: further discussions

As an organizing framework, Pepper's world hypotheses undoubtedly can help us understand the diverse urban modeling efforts in a more systematic manner. But I must emphasize that although the four traditions and their dominant/driving metaphors can be identified, urban modelers seldom rely on one metaphor exclusively; instead, most have used a mixture of multiple metaphors or modeling traditions. For example, Michael Batty, who has been active in the field for over four decades, has contributed to all four modeling traditions in his work. Most of Batty's work has been motivated by a combination of mathematical, physical, biological, and computation metaphors. Alan Wilson, the leading urban modeler following the social physics tradition in the 1960s and 1970s, has invoked DNA, genetic, and ecological analogies in his recent work (Wilson, 2008), although he still relies primarily on techniques from math and physics to implement and operationalize his models.

Each modeling tradition is also conceptually and methodologically permeable. In morphological studies linked to the concept of fractal cities, urban development has been designed to follow the diffusion-limited aggregation (DLA) model of physics, which generates tree-like clusters of population around some focal points. Modeling and simulating individual events are often relying on CA/AB models, whereas the new field-based time geography (Miller and Bridwell, 2009) is based upon concepts from physics. Lefebvre (2004)'s approach for rhythm analysis could also be used as a new operational approach within the spatial event tradition; his 'alignments' of urban rhythms – arrhythmia, polyrhythmia, eurhythmia, and isorhythmia – were all borrowed from medical literature describing rhythms in the human body. DNA sequence alignment techniques have also been adapted to analyze daily urban events in space and time (Shoval and Isaacson, 2007).

If all these four traditions of urban modeling are so permeable, it is natural to ask whether they will continue to diverge or to converge some unified theory of cities? It seems to me that complexity, or polyplexity as Couclelis (2009) called it, is a concept that could serve as a general umbrella that can tie all four traditions together. It is beyond the scope of this chapter to discuss this unified framework for linking the four traditions, but basic ground work has already been laid out by Batty (2004), Wilson (2000), and Allen, Strathern, and Baldwin (2007). Technological convergence to the new web-based digital environment, coupled with continued efforts to address global critical issues, could also facilitate further synthetic development of urban modeling. Otherwise, urban modeling could suffer the same fate as regional science (Barnes, 2004).

By looking at urban modeling efforts through a metaphorical perspective, we can better understand what Couclelis (1984) called the "prior structure" of the models. As the driving metaphors in urban models, regardless whether you are drawn from mathematics, physics, biology or computational science, some "prior structure" is already embedded in the conceptual foundation of every model. Precisely because of the baggage with various metaphors, modelers and critics alike have called for alternative grounding of our modeling efforts. O'Kelly (2004) called for a move from crude physical or biological metaphors into grounding our modeling efforts more in social/behavioral sciences. The difficulty is that some of the leading theories in social and behavior sciences are also embedded in physical or biological metaphors (Mirowski, 1991; Levine, 1995; McCloskey, 1995; Gaziano, 1999; Phillips, 2007). Social scientists have never ceased their efforts to ground their work in various natural sciences (Phillips, 1992). As Solow (2005) pointed out, economics got the way it is primarily due to influences from natural sciences, especially physics and biology. In this context, what I believe urban modelers need to realize is that whatever insights we may gain from understanding the modeling efforts do not necessarily represent (much less correspond to, as most urban modelers seem to believe) a better match to the external urban reality, but will reflect primarily the internal logics of the metaphor employed. In other words, urban modeling results are not necessarily a testimony to the empirical robustness of what is physically inside cities, but primarily to the power of the various metaphors in urban modelers' imagination. In other words, urban modeling is constitutive, not simply reflective.

Previous reviews on urban modeling efforts have concentrated primarily on the technical and methodological details without probing deeply the underlying metaphors embedded in the diverse modeling efforts. What we traditionally consider as progress is, in fact, nothing but a shift in the driving metaphors we used for conceptualization of cities. In this context, what we should recognize is the process whereby meaning is produced from metaphor to metaphor, rather than, as it was often assumed by urban modelers, between model and the world. We need not only to check the validity of our models from the technical perspective in terms of data accuracy and consistency with the previous data (Yeh and Li, 2006), but also, and perhaps more importantly, to scrutinize the driving conceptual metaphors deeply embedded in the models. Only then can we weave the insights gained from the urban modeling efforts with other urban narratives (Finnegan, 1998) to have a more sensible and sensitive urban life.

Summary and concluding remarks

> We have an incapacity for proving anything which no amount of dogmatism can overcome; we have an idea of truth which no amount of skepticism can overcome
>
> *B. Pascal*

Using Stephen Pepper's world hypotheses as a guiding framework, this chapter painted a broad-brush picture of urban analysis and modeling during the past 50 years. As revealed from this broad overview, urban analysis and modeling have made enormous strides drawn from the powerful methodological techniques as well as conceptual metaphors in mathematics, physics, biology, and computational science. For quite some time, metaphors have been regarded as frivolous, ornamental, obfuscatory, and even logically perverted. We now understand metaphors are not mere ornamentations or decorative constructs, but are central in formulating the problem and finding solutions. Although the correspondence between its theoretical category and understanding a geographical reality has been questioned, there seems to have no better alternatives – something we are struck with and stuck to (Barnes, 1996) in our intellectual enterprise.

It is interesting to observe that humanity's understanding of cities have evolved from conceiving them as celestial manifestations on earth to thinking of cities as machines to simulating cities as organisms to regarding cities simply as spatial events. Chronologically, we have drawn inspiration from astronomical, physical, biological, and computation metaphors to operationalize urban models. With the dramatic changes that are anticipated in the human conditions in our cities, it is more urgent than ever that we develop more robust urban models to better understand the urban forms and process. Much of the progresses reported in this book related to data acquisition, analysis, and modeling may serve as a springboard to lunch our new round of search for new models and metaphors. It is between Pascal's two extremes that we have to come to terms with our urban models and our new urban reality.

Acknowledgments

The author would like to thank Morton O'Kelly, Ed Malecki, Larry Brown, Kevin Cox, and Jose Gavinha for their critical comments on an earlier draft. Comments from four anonymous reviewers and the editor of this volume also substantially improved this chapter. Research assistance by Jay Knox, Wenqin Chen, and Li Li is gratefully acknowledged. The author is solely responsible for any remaining errors or omissions.

Notes

1 Throughout this chapter, metaphors and analogies are used interchangeably. No distinctions are made among metaphors, analogies, and similes in this chapter, but literary scholars do assign different meanings to these three terms. In the geographic literature, Downs (1981) made a distinction between metaphor (a means of expression) and analogy (a means of explanation) in the context of understanding mental maps.

2 One reviewer pointed that spatial system simulation, rooted in system ecology (Odum and Odum, 2000; Costanza and Voinov, 2004; Huang, Kao and Lee, 2007; Lee, Huang, and Chan, 2008), is an emerging approach for modeling cities, but this approach is dominated by deterministic, top-down models. Although it is suitable for the motto of organism world hypothesis, this modeling approach focuses more on the ecological rather than social processes, thus discussions on spatial system simulation have been excluded in this section.

3 I am aware that Batty (2007) regarded CA and AB models as the new social physics. As discussed in section four of this chapter, this tradition is closely related to physics, but its driving metaphor is biologically motivated, thus I put all CA and AB models in the social biology modeling tradition.

References

Alexander, C. (1964) *Notes on the Synthesis of Form*, Harvard University Press, Cambridge.

Alexander, C. (2004) *The Nature of Order*, Taylor & Francis, Berkeley.

Allen, P.M. (1997) *Cities and Regions as Self-organizing Systems: Models of Complexity*, Taylor & Francis, London.

Allen, P.M., Strathern, M. and Baldwin, J. (2007) Complexity: The integrating framework for models of urban and regional systems, in *The Dynamics of Complex Urban Systems: An Interdisciplinary Approach* (eds S. Albeverio, D. Andrey, P. Giordano and A. Vancheri), Springer, Berlin, pp. 21–41.

Anas, A. (1982) *Residential Location Models and Urban Transportation: Economic Theory, Econometrics, and Policy Analysis with Discrete Choice Models*, Academic Press, New York.

Arida, A. (2002) *Quantum City*, Architectural Press, Oxford.

Ash, C., Jasny, R., Roberts, L. et al. (2008) Reimagining cities. *Science*, **319**, 739.

Bak, P. (1996) *How Nature Works: The Science of Self-organized Criticality*, Springer-Verlag, New York.

Barnes, T.J. (1992) Reading the text of theoretical economic geography: The role of physical and biological metaphors, in *Writing Worlds: Discourse, Text, and Metaphor in the Representation of Landscape* (eds T. J. Barnes and J. S. Duncan), Routledge, London, pp. 118–135.

Barnes, T.J. (1996) *Logics of Dislocation: Models, Metaphors, and Meanings of Economic Space*, Guilford Press, New York.

Barnes, T.J. (2004) The rise (and decline) of American regional science: Lessons for the new economic geography. *Journal of Economic Geography*, **4**, 107–129.

Batty, M. (1976) *Urban Modelling: Algorithms, Calibrations, Predictions*, Cambridge University Press, Cambridge.

Batty, M. (1994) A chronicle of scientific planning: The Anglo-American modeling experience. *Journal of the American Planning Association*, **60**, 7–16.

Batty, M. (2002) Thinking about cities as spatial events. *Environment and Planning B: Planning and Design*, **29**, 1–2.

Batty, M. (2004) *Cities and Complexity: Understanding Cities with Cellular Automata, Agent-Based Models, and Fractals*, The MIT Press, Cambridge.

Batty, M. (2007) Fifty years of urban modelling: Macro statics to micro dynamics, in *The Dynamics of Complex Urban Systems: An Interdisciplinary Approach* (eds S. Albeverio, D. Andrey, P. Giordano and A. Vancheri), Physica-Verlag, Heidelberg, pp. 1–20.

Batty, M. (2008) The size, scale, and shape of cities. *Science*, **319**, 769–771.

Batty, M. (2009) A digital breeder for designing cities. *Architectural Design*, **79**, 46–49.

Batty, M. and Hudson-Smith, A. (2007) Imagining the recursive city: Explorations in urban simulacra, in *Societies and Cities in the Age of Instant Access* (ed. H. J. Miller), Springer, Dordrecht, pp. 39–55.

Batty, M. and Longley, P. (1994) *Fractal Cities: A Geometry of Form and Function*, Academic Press, London and San Diego.

Ben-Akiva, M. E. and Lerman, S. R. (1985) *Discrete Choice Analysis: Theory and Application to Travel Demand*, The MIT Press, Cambridge.

Benenson, I. and Torrens, P.M. (2004) *Geosimulation: Automata-based Modelling of Urban Phenomena*, John Wiley & Sons Ltd., Chichester.

Berry, B.J.L. (1964) Cities as systems within systems of cities. *Papers in Regional Science*, **13**, 146–163.

Besussi, E. and Cecchini, A. (eds) (1996) *Artificial Worlds and Urban Studies*, DAEST, Venice.

Black, M. (1962) *Models and Metaphors: Studies in Language and Philosophy*, Cornell University Press, Ithaca.

Brown, D.G., Page, S.E., Riolo, R. and Rand, W. (2004) Agent-based and analytical modeling to evaluate the effectiveness of greenbelts. *Environmental Modelling & Software*, **19**, 1097–1109.

Burgess, E.W. (1925) The growth of the city: An introduction to a research project, in *Urban Ecology: An International Perspective on the Interaction Between Humans and Nature* (eds J. M. Marzluff, E. Shulenberger, W. Endlicher et al.), Springer, pp. 71–78.

Buttimer, A. (1993) *Geography and the Human Spirit*, Johns Hopkins University Press, Baltimore.

Campanella, T. (2001) *Cities from the Sky: An Aerial Portrait of America*, Princeton Architectural Press, New Jersey.

Carey, H.C. (1858) *Principles of Social Science*, J.B. Lippincott, Philadelphia.

Casti, J. L. (1997) *Would-Be Worlds: How Simulation Is Changing the Frontiers of Science*, John Wiley & Sons, Inc., New York.

Changizi, M.A. and M. Destefano (2009) Common scaling laws for city highway systems and the mammalian neocortex. *Complexity*, **15**, 1–8.

Clarke, K.C. and Gaydos, L.J. (1998) Loose-coupling a cellular automaton model and GIS: Long-term urban growth prediction for San Francisco and Washington/Baltimore. *International Journal of Geographical Information Science*, **12**, 699–714.

Cohen, B. (2004) Urban growth in developing countries: A review of current trends and a caution regarding existing forecasts. *World Development*, **32**, 23–51.

Costanza, R. and Voinov, A. (2004) *Landscape Simulation Modeling*, Springer, New York.

Couclelis, H. (1984) The notion of prior structure in urban modelling. *Environment and Planning A*, **16**, 319–338.

Couclelis, H. (1985) Cellular worlds: A framework for modeling micro-macro dynamics. *Environment and Planning A*, **17**, 585–596.

Couclelis, H. (1988) Of mice and men: What rodent populations can teach us about complex spatial dynamics. *Environment and Planning A*, **20**, 99–109.

Couclelis, H. (1997) From cellular automata to urban models: New principles for model development and implementation. *Environment and Planning B: Planning and Design*, **24**, 165–174.

Couclelis, H. (2009) Polyplexity: A complexity science for the social and policy sciences, in *Complexity and Spatial Networks* (eds A. Reggiani and P. Nijkamp), Springer, Heidelberg, pp. 75–88.

de la Barra, T. (1989) *Integrated Land Use and Transport Modelling*, Cambridge University Press, Cambridge.

Downs, R.M. (1981) Maps and metaphor. *The Professional Geographer*, **33**(3), 287–293.

Duncan, J. (1980) The superorganic in American cultural geography. *Annals of the Association of American Geographers*, **70**, 181–198.

Dykes, J., Purves, R. S., Edwardes, A. and Wood, J. (2008) Exploring volunteered geographic information to describe place: visualization of the 'Geography of British Isles' collection. In *Proceedings of GIS Research UK*, UNIGIS UK, pp. 256–267.

Epstein, J. M. (2006) *Generative Social Science: Studies in agent-based computational modeling*, Princeton University Press, Princeton

Finnegan, R. H. (1998) *Tales of the City: A Study of Narrative and Urban Life*, Cambridge University Press, New York.

Forrester, J. W. (1969) *Urban Dynamics*, MIT Press, Cambridge.

Foth, M. (2008) *Handbook of Research on Urban Informatics: The Practice and Promise of the Real-Time City*, Information Science Reference, Hershey.

Foth, M., B. Bajracharya, R. Brown, and G. Hearn (2009) The Second Life of urban planning? Using NeoGeography tools for community engagement. *Journal of Location Based Services*, **3**(2), 97–117.

Furbey, R. (1999) Urban "regeneration": Reflections on a metaphor. *Critical Social Policy*, **19** (4), 419–445.

Garreau, J. (1992) *Edge City: Life on the New Frontier*, Anchor, New York.

Gaziano, E. (1999) Ecological metaphors as scientific boundary work: Innovation and authority in the interwar sociology and biology. *American Journal of Sociology*, **101**, 874–907.

Girardin, F., Vaccari, A. and Gerber, A. (2009) Quantifying urban attractiveness from the distribution and density of digital footprints. *International Journal of Spatial Data Infrastructures*, **4**, 175–200.

Goodchild, M.F. (2007) Citizens as sensors: The world of volunteered geography. *GeoJournal*, **69**, 211–221.

Gregory, D. (1980) The ideology of control: Systems theory and geography. *Tijdschrift voor Economische en Sociale Geografie (TESG)*, **71**, 327–342.

Guhathakurta, S. (2002) Urban modeling as storytelling: Using simulation models as narrative. *Environment and Planning B*, **29**, 895–911.

Haken, H. (1983) *Synergetics: An Introduction*, Springer-Verlag, Berlin.

Harris, B. (1965) Urban development models: A new tool for planners. *Journal of the American Institute of Planners*, **31**, 90–95.

Harris, B. (1985) Urban simulation models in regional science. *Journal of Regional Science*, **25**, 545–567.

Harris, C.D. and Ullman, E.L. (1945) The nature of cities. *The Annals of the American Academy of Political and Social Science*, **242**, 7–17.

Hern, W. M. (2008) Urban malignancy: Similarity in the fractal dimensions of urban morphology and malignant neoplasm. *International Journal of Anthropology*, 23 (1–2), 1–19.

Herold, M., Couclelis, H. and Clarke, K.C. (2005) The role of spatial metrics in the analysis and modeling of urban land use change. *Computers, Environment and Urban Systems*, **29**, 369–399.

Hesse, M. B. (1963) *Models and Analogies in Science*, Sheed & Ward, London.

Hesse, M. B. (1980) *Revolutions and Reconstructions in the Philosophy of Science*, Harvester Press, Brighton.

Hoyt, H. (1939) *The Structure and Growth of Residential Neighborhoods in American Cities*, Federal Housing Administration, Washington, DC.

Huang, S. L., Kao, W. C., and Lee, C. L. (2007). Energetic mechanisms and development of an urban landscape system. *Ecological Modelling*, **201** (3-4), 495–506.

Hudson-Smith, A., Milton, R., Dearden, J. and Batty, M. (2008) The neogeography of virtual cities: Digital mirrors into a recursive world, in *Handbook of Research on Urban Informatics: The Practice and Promise of the Real-Time City, Information Science Reference* (ed M. Foth), Idea Group Inc, Hershey, pp. 270–290.

Hudson-Smith, A., Crooks, A., Gibin, M. *et al.* (2009) NeoGeography and Web 2.0: Concepts, tools, and applications. *Journal of Location Based Services*, **3**, 118–145.

Isard, W. (1999) Regional science: Parallels from physics and chemistry. *Papers in Regional Science*, **78**, 5–20.

Jacobs, J. (1961) *The Death and Life of Great American Cities*, Vintage Books, New York.

Kearns, R.A. (1997) Narrative and metaphor in health geographies. *Progress in Human Geography*, **21**, 269–277.

Khalil, E.L. (1998) The five careers of the biological meaphor in economic theory. *Journals of Socio-Economics*, **27**, 29–52.

Klosterman, R.E. (1994) Large-scale urban models: Retrospect and prospect. *Journal of the American Planning Association*, **60** (1), 3–6.

Kostof, S. (1991) *The City Shaped: Urban Patterns and Meanings Through History*, Thames and Hudson, London.

Lakoff, G. and Johnson, M. (1980) *Metaphors We Live By*, University of Chicago Press, Chicago.

Lakshmanan, J.R. and Hansen, W.G. (1965) A retail market potential model. *Journal of the American Institute of Planners*, **31** (2), 134–143.

Landis, J.D. (1995) Imagining land use futures: Applying the California futures model. *Journal of the American Planning Association*, **61**, 438–457.

Larkham, P.J. (1995) Organic thought in urban geography: The "evolution" of towns. *Australian Geographical Studies*, **30**, 3–8.

Leatherdale, W. H. (1974) *The Role of Analogy, Model and Metaphor in Science*, North-Holland Publishing Co., Amsterdam.

Lee, C.L., Huang, S.L., and Chan, S.L. (2008) Biophysical and system-based approaches for simulating land-use change. *Landscape and Urban Planning*, **86** (2): 187–203.

Lee, D.B. (1973) Requiem for large scale models. *Journal of the American Institute of Planners*, **39**, 163–178.

Lefebvre, H. (2004) *Rhythm Analysis: Space, time, and everyday life*, Contium, London.

Levine, D.N. (1995) The organism metaphor in sociology. *Social Research*, **62**, 239–265.

Liu, W. and Seto, K.C. (2008) Using the ART-MMAP neural network to model and predict urban growth: A spatiotemporal data mining approach. *Environment and Planning B: Planning and Design*, **35**, 296–317.

Liu, Y. (2008) *Modelling Urban Development with Geographical Information Systems and Cellular Automata*, CRC Press, Boca Raton.

Livingston, D.N. and Harrison, R.T. (1981) Meaning through metaphor: Analogy as epistemology. *Annals of the Association of American Geographers*, **71**, 95–107.

Lowry, I.S. (1964) *A Model of Metropolis*, Rand Corporation, Santa Monica.

Lynch, K. (1981) *A Theory of Good City Form*, MIT Press, Cambridge.

McCloskey, D.N. (1995) Metaphors economists live by. *Social Research*, **62**, 215–237.

Miller, H.J. (2007) Place-based versus people-based geographic information science. *Geography Compass*, **1**, 503–535.

Miller, H.J. and S.A. Bridwell (2009) A field-based theory for time geography. *Annals of the Association of American Geographers*, **99**, 49–75.

Mills, W.J. (1982) Metaphorical visions: Changes in Western attitudes toward the environment. *Annals of the Association of American Geographers*, **72**, 237–253.

Mirowski, P. (1988) *Against Mechanism: Protecting Economics from Science*, Rowman & Littlefield Pub Inc, Totowa.

Mirowski, P. (1991) *More Heat than Light: Economics as Social Physics, Physics as Nature's Economics*, Cambridge University Press, Cambridge.

Mirowski, P. (ed.) (1994) *Natural Images in Economic Thought: "Markets read in tooth and Claw,"* Cambridge University Press, New York.

Morris, A. E. J. (1979) *History of Urban Form: Before the Industrial Revolution*, Longmans Scientific and Technical, London.

Muller, J.C. (1982) Non-Euclidean spaces: Mapping functional distances. *Geographical Analysis*, **14**, 189–203.

Norwick, S.A. (2006) *The History of Metaphors of Nature: Science and Literature From Homer to Al Gore*, Edwin Mellen Press, Lewiston.

O'Kelly, M.E. (2004) Isard's contributions to spatial interaction modeling. *Journal of Geographical Systems*, **6**, 43–54.

Odum, H.T. and Odum, E.C. (2000) *Modeling for All Scales*, Academic Press, San Diego.

Olson, S. (1982) Urban metabolism and morphogenesis. *Urban Geography*, **3**, 87–109.

Pepper, S.C. (1942) *World Hypotheses: A Study in Evidence*, University of California Press, Berkeley.

Phillips, DC. (1992) *The Social Scientist's Bestiary: A guide to fabled threats to, and defense of, naturalistic social science*, Pergamon Press, New York.

Phillips, P.J. (2007) Mathematics, metaphors, and economics of visualisability. *Quarterly Journal of Austrian Economics*, **10**, 281–299.

Pickett, S.T.A., Cadenasso, M.L. and Grove, J.M. (2004) Resilient Cities: Meaning, models, and metaphor for integrating the ecological, socioeconomic, and planning realms. *Landscape and Urban Planning*, **69**, 369–384.

Portugali, J. (1999) *Self-organization and the City*, Springer Verlag, Berlin Heidelberg.

Rashed, T. and Juergens, C. (eds) (2006) *Remote Sensing of Urban and Suburban Areas*, Springer, Berlin and New York.

Ravenstein, E.G. (1885) The laws of migration. *Journal of the al Statistical Society*, **48**, 167–235.

Reilly, W.J. (1931) *The Law of Retail Gravitation*, W.J. Reilly, New York.

Roberts, S. (1991) A critical evaluation of the city life cycle idea. *Urban Geography*, **12**, 431–449.

Rothbart, D. (1997) *Explaining the Growth of Scientific Knowledge: Metaphors, Models, and Meanings*, Edwin Mellen Press, Lewiston.

Roy, J.R. and Thill, J.C. (2004) Spatial interaction modelling. *Papers in Regional Science*, **83**: 339–361.

Rozenfeld, H.D., Rybski, D., Andrade, J.S. *et al.* (2008) Laws of population growth. *Proceedings of the National Academy of Sciences*, **105**, 18702–18707.

Sayer, R.A. (1976) A critique of urban modelling from regional science to urban and regional political economy. *Progress in Planning*, **6**, 187–254.

Schweitzer, F. and Farmer, J.D. (2007) *Brownian Agents and Active Particles: Collective Dynamics in the Natural and Social Sciences*, Springer Verlag, Berlin.

Shoval, N. and Isaacson, M. (2007) Sequence alignment as a method for human activity analysis in space and time. *Annals of the Association of American Geographers*, **97**, 282–297.

Soja, E. (2000) *Postmetropolis: Critical studies of cities and regions*, John Wiley & Sons, Inc., New York.

Solow, R.M. (2005) How did ecnomics get that way & what way did it get? *Dædalus: Journal of the American Academy of Arts & Sciences*, **134** (4), 87–100.

Spiller, N. (2009) Parallel biological futures. *Architectural Design*, **79**, 130–131.

Steadman, P. (1979) *The Evolution of Designs: Biological Analogy in Architecture and the Applied Arts*, Cambridge University Press, Cambridge.

Stewart, J.Q. (1947) Suggested principles of "social physics". *Science*, **106**, 179–180.

Stewart, J.Q. (1950) The development of social physics. *American Journal of Physics*, **18**, 239–253.

Straussfogel, D. (1991) Modeling suburbanization as an evolutionary system dynamic. *Geographical Analysis*, **23**(1), 1–24.

Sudhira, H. S., Ramachandra, T. V. and Jagadish, K. S. (2004) Urban sprawl: Metrics, dynamics and modelling using GIS. *International Journal of Applied Earth Observations and Geoinformation*, **5**, 29–39.

Sui, D. Z. (1997) The syntax and semantics of urban modeling: Versions vs. visions. Proceedings of USGS-NCGIA Workshop in Landuse Modeling, Sioux Falls, SD., June 4–7.[Online] Available http://www.ncgia.ucsb.edu/conf/landuse97/papers/sui_daniel/paper.html (accessed 22 November 2010).

Sui, D.Z. (1998) GIS-based urban modeling: Practice, problems, and prospects. *International Journal of Geographical Information Science*, **12**(7), 651–671.

Sui, D.Z. (2000) Visuality, aurality, and the shifting metaphors in geographic thought in the late 20th century. *Annals of the Association of American Geographers*, **90** (2), 322–343.

Sui, D.Z. and Zeng, H. (2001) Modeling the dynamics of landscape structure in Asia's emerging desakota regions: A case study in Shenzhen. *Landscape and Urban Planning*, **53**, 37–52.

Sui, D.Z. and X.B. Wu (2006) Changing residential segregation patterns in a prismatic metropolis: The case of Houston. *Environment and Planning B*, **33**, 559–579.

Takeyama, M. and Couclelis, H. (1997) Map dynamics: Integrating cellular automata and GIS through Geo-Algebra. *International Journal of Geographical Information Science*, **11**, 73–91.

Thom, R. (1975) *Structural Stability and Morphogenesis*, Benjamin, Reading.

Torrens, P.M. (2006) Simulating sprawl. *Annals of the Association of American Geographers*, **96**, 248–275.

Tuan, Y.F. (1978) Sign and metaphor. *Annals of the Association of American Geographers*, **68**, 363–372.

Urry, J. (2004) Small worlds and the new 'social physics'. *Global Networks*, **4**, 109–130.

Vance, J. E. (1990) *The Continuing City: Urban morphology in Western Civilization*, John Hopkins University Press, Baltimore.

Verma, N. (1993) Metaphor and analogy as elements of a theory of similarity for planning. *Journal of Planning Education and Research*, **13**, 13–25.

Waddell, P. (2002) UrbanSim: Modeling urban development for land use, transportation, and environmental planning. *Journal of the American Planning Association*, **68**, 297–314.

Webber, M.J. (1979) *Information Theory and Urban Spatial Structure*, Croom Helm, London.

Wegener, M. (1994) Operational urban models: State of the art. *Journal of the American Planning Association*, **60**, 17–29.

Wegener, M. (2004) Overview of land use transport models, in *Transport Geography and Spatial Systems. Handbook 5 of the Handbook in Transport* (eds D. A. Hensher and K. K. Button), Pergamon/Elsevier Science, Oxford, pp. 127–146.

Wegener, M. and Furst, F. (1999) *Land-Use Transport Interaction: State of the Art*, IRPUD, Dortmund.

Weinberg, S. (1994) *Dreams of a Final Theory*, Vintage Books, New York.

Wilson, A.G. (1967) A statistical theory of spatial distribution models. *Transportation Research*, **1**, 253–269.

Wilson, A.G. (1969a) Notes on some concepts in social physics. *Papers of the Regional Science Association*, **22**, 159–193.

Wilson, A.G. (1969b) The use of analogies in geography. *Geographical Analysis*, **1**, 225–233.

Wilson, A.G. (1981) Catastrophe theory and bifurcation, in *Quantitative geography: A British view* (eds N. Wrigley and R. J. Bennett), Routledge, London, pp. 192–202.

Wilson, A.G. (1995) Simplicity, complexity, and generality: Dreams of a final theory in locational analysis, in *Diffusing Geography: Essays for Peter Haggett* (eds. A.D. Cliff, P.R. Gould, A.G. Hoare and N.J. Thrift), Blackwell, Oxford, UK, pp. 342–352.

Wilson, A.G. (1998) Land use/transport interaction models: Past and future. *Journal of Transport Economics and Policy*, **32**, 3–23.

Wilson, A.G. (2000) *Complex Spatial Systems: The modelling foundations of urban and regional analysis*, Pearson Education, London and Harlow.

Wilson, A.G. (2008) Urban and regional dynamics–3:'DNA'and 'genes' as a basis for constructing a typology of areas. *Centre for Advanced Spatial Analysis Working Paper*, **130**, 1–11.

Wilson, A.G., 2009. Remote sensing as the 'X-Ray crystallography' for urban 'DNA'. [Online] Available http://www.isibang.ac.in/~sirarm/Paper-1-Alan-Wilson-Seminar-18-March-09.pdf (accessed 22 March 2010).

Wolfram, S. (2002) *A New Kind of Science*, Wolfram Media Inc., Champaign.

Wrigley, N. and Longley, P.A. (1984) The use of discrete choice modeling in urban analysis, in *Geography and the Urban Environment* (eds D.T. Herbert and R.J. Johnston), John Wiley & Sons, New York, pp. 45–95.

Yang, X. and Lo, C.P. (2003) Modelling urban growth and landscape changes in the Atlanta metropolitan area. *International Journal of Geographical Information Science*, **17**, 463–488.

Yeh, A. G.-O. and Li, X. (2006) Errors and uncertainties in urban cellular automata. *Computers, Environment and Urban Systems*, **30**, 10–28.

INDEX

A

AAG (Association of American Geographers) 4, 5
Agent-based modeling 9, 292, 321, 323, 324, 327, 328, 329, 330, 332, 336, 347, 348, 349, 350, 351, 373, 376, 377
Absolute radiometric correction 29
Abundance 54, 56, 58, 112, 117, 118, 124, 172, 213, 290, 292, 293, 294, 295, 337, 361
Accessibility 167, 170, 171, 173, 177, 197, 352
Accuracy 7, 9, 17, 19, 21, 22, 28, 36, 44, 46, 66, 68, 69, 80, 83, 86, 87, 89, 93, 100, 101, 107, 111, 112, 114, 115, 117, 118, 119, 123, 124, 125, 129, 130, 132, 135, 136, 137, 141, 156, 186, 187, 191, 196, 197, 198, 199, 201, 205, 234, 242, 243, 247, 249, 251, 258, 265, 289, 290, 305, 309, 311, 340, 362, 364, 371, 377
Activation function 96, 98, 99, 100, 101, 103, 104, 105
Aerial photography 16, 118, 156, 159, 186, 229, 249, 257, 309, 310
Airborne 6, 7, 13, 38, 50, 56, 66, 67, 75, 77, 79, 81, 83, 85, 87, 89, 143, 160, 183, 184, 186, 234, 289, 308, 309, 310
Algorithm parameters 7, 93, 96
ALOS (Advanced Land Observation Satellite) 65
Analytical models 8, 160
Ancillary data 8, 19, 21, 39, 68, 101, 107, 156, 170, 175, 176, 195, 196, 197, 198, 199, 200, 201, 202, 203, 205, 229, 234, 249, 351, 354
Ancillary information 68, 156
APAR (Active Phased Array Radar) 277, 278, 279
Areal interpolation 8, 163, 195, 196, 197, 198, 199, 203, 205, 206, 207, 228, 229
A-spatial data structures 9, 321, 323
ASTER (Advanced Spaceborne Thermal Emission and Reflection Radiometer) 228, 258, 309, 310
Atlanta metropolitan area 6, 18, 19, 22, 25, 28, 101, 196, 200
ATLAS (Advanced Thermal and Land Applications Sensor) 19, 226
Atmospheric properties 29
Austin, Texas 185
Automated identification 54
AVHRR (Advanced Very High Resolution Radiometer) 212, 277, 278, 279, 284, 290, 308
Avian biodiversity 362, 365, 366, 367

B

Back-propagation algorithm 96, 98, 99, 100, 106, 107
Batty, Michael 4, 168, 169, 172, 200, 324, 327, 328, 336, 337, 338, 340, 341, 348, 349, 350, 351, 352, 353, 354, 373, 374, 375, 376, 377
Bayesian classification 159
Belfast, Northern Ireland 157, 158, 159
Bifurcation 327, 348, 374, 376
Binary method 199, 202, 203, 204

Biodiversity 6, 8, 9, 239, 287, 288, 289, 290, 291, 292, 293, 294, 295, 296, 297, 298, 299, 321, 359, 360, 361, 362, 363, 365, 366, 367
Biomass 50, 51, 276, 292
Bird habitat models 361, 362
Bird species richness 292, 293, 366
Boundary errors 21
Broad-band sensors 27
Broyden, Fletcher, Goldfarb, and Shanno (BFGS) algorithm 100, 105
Brussels 255, 257, 261
Building boundary 77, 78, 81, 82, 83, 84
Building characteristics 63, 189, 308, 313
Building damage assessment 6, 7, 43
Building detection 75, 76, 77, 78, 89, 185, 186, 187, 188, 189, 191
Building footprint 45, 76, 186, 187, 189, 191, 200, 310
Building reconstruction 7, 67, 75, 76, 77, 78, 79, 81, 83, 85, 87, 89
Building volume 183, 184, 186, 187, 189, 190, 191, 200

C

Calibration 8, 9, 119, 123, 130, 189, 190, 199, 264, 321, 330, 335, 336, 337, 338, 339, 341, 347, 348, 349, 352, 354, 372
Calibration parameter values 9
Capability of convergence 105, 106, 107
Carbon cycle 275, 276
Carbon dioxide 276
Carbon flux 276, 284
Cellular automata model (CA) 9, 321, 323, 325, 327, 328, 329, 331, 335, 336, 337, 338, 339, 340, 341, 348, 349, 352, 373, 374, 376
Census block groups 201, 227, 228, 278
Census blocks 183, 185, 186, 187, 191, 201, 203, 227, 228, 278
Census county subdivision 277, 278, 279
Census data 176, 186, 196, 199, 226, 228, 229, 230, 234, 309
Census South Atlantic division 276, 277, 278, 284
Census tracts 172, 196, 197, 227, 229, 234
Central Puget Sound, Washington 362, 364, 365
Change detection 6, 15, 19, 21, 27, 28, 29, 36, 39, 40, 43, 46, 50, 68
Change of support problem 196
Chaos 341, 348
Choropleth map 196, 197, 229, 230, 231
Class adaptiveness 39
Classification accuracy 100, 102, 103, 104, 106, 107
Clustering parameters 21
Coefficient of determination 99, 118, 187
Coefficient of efficiency (CE) 98
Common validation procedure 349
Compactness 131, 133, 134, 135, 139, 157, 176

Urban Remote Sensing: Monitoring, Synthesis and Modeling in the Urban Environment, First Edition. Edited by Xiaojun Yang.
© 2011 John Wiley & Sons, Ltd. Published 2011 by John Wiley & Sons, Ltd.

COMPAS 157, 159
Complexity 5, 9, 20, 64, 67, 76, 77, 98, 100, 103, 105, 107, 115, 117, 118, 119, 120, 121, 122, 124, 125, 157, 160, 174, 196, 198, 243, 288, 292, 306, 308, 323, 325, 326, 327, 328, 336, 341, 348, 350, 361, 372, 374, 375, 377
Component substitution 141, 142, 144, 145, 146, 147
Conjugate gradient method 99, 100
Context-based decision 143, 144
Contextualism 373, 374, 376
Conventional pattern classifiers 7
Convergence value 21
CORINE 297, 308
Correlation 65, 68, 69, 78, 82, 117, 118, 121, 124, 184, 187, 189, 230, 231, 248, 249, 256, 276, 340
COSMO/SkyMed 64, 65, 68, 69
Curve Number (CN) 256

D

Damage assessment 6, 7, 43, 44, 67, 68
Dasymetric mapping 8, 163, 195, 196, 197, 198, 199, 200, 201, 202, 203, 204, 205, 207, 228, 229, 234
Data acquisition 16, 17, 19, 27, 29, 36, 52, 96, 107, 372
Data fusion 4, 6, 143
Data pre-processing 107
Decision rule 7, 81, 129, 132, 133, 134, 136, 137, 139, 144, 326, 339
Decision support system 132
Decision tree 243, 258, 309, 327, 328, 329, 330
Deductive models 156
Defense Meteorological Satellite Program (DMSP) 184, 211, 212, 213, 214, 215, 216, 217, 218, 219, 220, 221, 222
Definiens 129, 130, 131, 132, 133, 134, 140
Dempster-Shafer 188
Denver, Colorado 17, 112, 113, 122, 123, 124, 125, 228
Desakota 347, 350, 351, 354
Digital elevation model (DEM) 36, 39, 66, 79, 258, 259, 262, 264, 309, 310
Digital image processing 4, 28
Digital terrain model (DTM) 76, 186, 310
Dimensionality reduction 54, 56, 57
Discrete measurement 77, 79, 80, 86
Discrete -choice model 349
Distributed hydrological model 255, 257, 261
DMSP (Defense Meteorological Satellites Program) 184, 211, 212, 213, 214, 215, 216, 217, 218, 219, 220, 221, 222
DOE (Department of Energy) 211, 212, 214
Double bounce 64, 69
Douglas-Peuker 78, 81, 87
Drivers 6, 7, 119, 121, 159, 227, 329, 348, 352, 364, 367
Dwelling type 184, 189
Dwelling unit 185, 186, 191
Dynamic modeling 5, 6, 9, 321

E

Earthquakes 43, 46, 65, 234
eCognition 130, 131, 186, 188, 189, 249
Ecological indicator 60,
Ecological modeling 9, 288, 289, 297, 321, 359, 360, 363, 365, 36
Edge cities 169, 374
Electrification rate estimation 211
Endmembers selection 56, 114, 116, 118, 123, 249
Enhanced Thematic Mapper Plus (ETM+) 16, 101, 228, 309

Entropy-maximization 375
Environmental amenity 228, 234
Environmental equity 226, 227, 228, 230
Environmental hazard 225, 226, 227, 228, 229, 230, 234
Environmental justice 8, 163, 197, 225, 226, 227, 228, 229, 230, 231, 233, 234
Environmental racism 226
Environmental risk 226, 227, 229, 234
ENVISAT 310
Evaluation protocol 144
Evapotranspiration 122, 257, 258, 259, 261, 307
Evolutionary programming 349, 376
Exurbanization 283, 284

F

fAPAR(Fraction of Absorbed Photosynthetically Active Radiation) 277, 279
Feature function 53, 54
Feature space 53, 55, 56, 57, 58, 79, 81, 133, 245, 246
Feed-forward networks 96, 97
Field spectra 55
Filtering 78, 79, 144, 145, 146, 265
Fletcher-Reeves (CGF) algorithm 99, 100, 105, 106, 107
Flood 327
Flow routing 260
Foreshortening 66
Fractal city 374
Fractal geometry 157, 172, 374, 375
Fraction images 115, 117, 120, 121, 124, 246, 248
Fuzzy logic 341, 360

G

Gaussian maximum likelihood (GML) 101, 103, 105
Gauss-Newton method 100
Genetic algorithm (GA) 142, 324, 348, 376
Geographic information science (GISci) 324
Geographic information systems (GIS) 4, 6, 21, 27, 45, 46, 68, 69, 159, 163, 176, 198, 225, 226, 324, 374, 375, 376
Geometric rectification 20, 28
Geosimulation 376
Global gains 147
GLP-CBD algorithm 144
Google building maker 310
Google Earth 19, 68, 101, 213, 376
GPP (Gross Primary Production) 6, 8, 239, 275, 276, 277, 278, 279, 280, 281, 282, 283, 284
Gradient descent with momentum (GDM) 99, 100, 101, 105, 106, 107
Grafton, Wisconsin 241, 242, 243, 244, 246, 248
Gwinnett, Georgia 25, 196, 200, 201, 202, 203, 204, 205, 206, 207

H

Habitat structure 8, 289, 294, 297
Hexa-dpi data structure/model 324
Hidden layers 96, 97, 98, 100, 101, 103, 104, 105
Hierarchical classification 258
Hierarchical fuzzy-pattern-matching 349, 350
High Ecological Resolution Classification for Urban Landscapes and Environmental Systems (HERCULES) 310
High resolution remote sensing 183, 242, 243
High-density urban 20, 21, 22, 23, 24, 25, 26, 27, 29, 101, 105, 199, 283, 284
High-performance computing 339

Histogram matching 29, 142, 145
Homogeneous areas 7, 29, 116
Housing unit 157, 170, 177, 184, 187, 200, 244, 275, 277, 278, 279, 281, 283
Housing unit density 277
Human-environment interactions 336
Hydrological modeling 255, 256, 257, 264
Hyperspectral remote sensing 6, 49, 50, 52, 289

I

IEA (International Energy Agency) 8, 56, 212, 213, 214, 215, 216, 217, 218, 219, 220, 221, 222
IHS transform 142
IKONOS 6, 36, 37, 43, 50, 64, 142, 147, 150, 151, 156, 157, 158, 159, 160, 184, 234, 242, 243, 244, 245, 246, 248, 255, 257, 260, 261, 262, 263, 264, 265, 266, 267, 268, 277, 288, 289, 309, 360
Image classification 19, 103, 105, 107
Image fusion 7, 93, 141, 142
Image preprocessing 19, 20, 27, 28
Image resolution 15, 27, 28, 29
Image segmentation 78, 79, 86, 129, 130, 131, 134, 241, 243, 289, 309, 360
Image spectra 52, 54, 55
Image-to-image comparison 29
Imaging sensors 6, 7, 15, 16
Imaging spectroscopy 50, 52
Impervious Cover Model (ICM) 256
Impervious surface 8, 112, 114, 116, 117, 119, 121, 137, 138, 186, 200, 228, 229, 234, 239, 241, 242, 243, 244, 245, 246, 247, 248, 249, 250, 251, 255, 256, 257, 258, 259, 261, 262, 263, 264, 265, 266, 267, 269, 279, 280, 350, 360, 362
In situ data 9, 118
Inductive models 361
Infiltration capacity 256
Injection model 141, 143, 144, 146, 151
InSAR(Interferometric Synthetic Aperture Radar) 36, 66, 67, 234, 310
Integrated adaptive spatial approach 7, 35
Integrated modeling environment 9
Intensity image 41, 145
Inter-class variability 28
Intra-class variability 29
ISODATA(Iterative Self Organizing Data Analysis) 21, 43, 45
ISPRS (International Society for Photogrammetry and Remote Sensing) 4

J

JURSE (Joint Urban Remote Sensing Event) 4, 5, 36

K

Kappa statistic 103, 187, 349

L

Land cover change modeling 363
Land cover types 28, 50, 51, 52, 79, 155, 157, 243, 245, 246, 249, 289, 310
Landsat Data Continuity Mission (LDCM) 17
Landsat ETM+ (Enhanced Thematic Mapper Plus) 16, 17, 21, 29, 101, 103, 105, 106, 119, 123, 126, 230, 243, 244, 258, 261, 262, 263, 265, 266, 270, 277, 209, 312
Landsat MSS (Multispectral Scanner) 17, 19, 28, 312, 314
Landsat TM (Thematic Mapper) 20, 28, 113, 119, 121, 186, 229, 241, 244, 290, 350, 360, 362

Landscape changes 5, 6, 7, 8, 9, 13, 15, 16, 321, 348
Landscape ecology 27, 172, 173, 340
Landscape metrics 17, 173, 176, 290, 293, 340, 349, 360
Land use classification 186, 189
Land use/cover change 7, 8, 17, 21, 24, 29, 239
Laplacian pyramids 143
Leapfrog development 167, 169
Learning process 96, 98
Least square template matching 78, 81, 84, 87
Levenberg-Marquardt (LM) algorithm 100, 105, 106, 107, 232
LIDAR (light detection and ranging) 6, 7, 8, 13, 36, 66, 67, 75, 76, 77, 78, 79, 80, 81, 82, 83, 85, 86, 87, 89, 163, 183, 184, 185, 186, 187, 189, 191, 200, 234, 289, 309, 310, 311
Linear concentration
Linear features 80, 81, 83, 89
Linear regression 104, 145, 352, 361, 363, 365
Lo, C.P. 4, 6, 7, 8, 9, 16, 17, 18, 20, 27, 28, 29, 107, 157, 159, 175, 184, 197, 198, 200, 205, 228, 229, 360, 376
Local features 141
Local runoff estimation 8, 255
Logistic sigmoid function 105
Low-density urban 20, 21, 22, 23, 24, 25, 26, 27, 28, 29, 101, 199
LUE (Light-Use Efficiency) models 8, 275, 276, 277, 279, 280, 283, 284

M

Macro urban remote sensing 155, 156
Mankato, Minnesota 241, 242, 249
Man-made materials 50, 51, 52, 54
Map-to-map comparison 29
Material mapping 38, 50, 51, 52, 54, 56, 57, 58
Mathematical morphology 7, 39, 46
Matrix analysis 25,
MAUP (Modifiable Areal Unit Problem) 196, 227
Maximum difference 136, 137, 138, 139
Maximum noise fraction (MNF) transformation 55, 245, 246
Mean average error (MAE) 118, 121, 124, 187, 191, 248, 249
Mean squared error (MSE) 20, 98, 106, 145, 146
Mean squared relative error (MSRE) 98
Medium resolution 28, 65, 230, 241, 243, 263, 290, 310, 360
MESMA (Multiple Endmember Spectral Mixture Analysis) 7, 56, 58, 111, 112, 114, 115, 116, 117, 118, 119, 123, 124, 125, 246
Metaphors 9, 321, 371, 372, 373, 374, 375, 376, 377
Micro urban remote sensing 156
Mis-registration 142, 147
Mixed pixel 50, 54, 56, 58, 112, 242, 258
Model calibration 8, 339, 347, 349, 352
Model conceptualization 8, 321
Model design 96, 347, 348, 349, 351
Model implementation 112, 125
Model validation 8, 9, 187, 336, 347, 349, 353
MODIS (Moderate Resolution Imaging Spectroradiometer) 212, 277, 278, 290, 309, 360
Modular transfer function 117
Monitoring 4, 6, 13, 15, 16, 17, 27, 28, 29, 35, 38, 49, 50, 65, 75, 130, 234, 241, 255, 275, 287, 288, 290, 295, 296, 297, 323, 335, 347, 376
Monte Carlo averaging 341
Multi-criteria evaluation problem 349, 352
Multi-layer perceptron (MLP) 56, 95, 96, 97, 98, 99, 100, 101, 105, 262, 263, 265

Multi-layer urban canopy model 307
Multi-nomial logit regression 364
Multinucleated pattern 27
Multiple bounce 66, 67
Multi-resolution analysis 43, 186
Multispectral 17, 28, 38, 43, 49, 50, 118, 125, 129, 130, 132, 134, 141, 142, 143, 144, 145, 146, 147, 149, 151, 159, 242, 249, 251, 257, 262, 263, 289, 308, 309, 312, 360
Multitemporal 17, 29, 36, 39, 40, 68, 156, 159

N

NASA (National Aeronautics and Space Administration) 16, 17, 212, 284, 288, 290, 297
NCGIA (National Center for Geographic Information and Analysis) 327
NDVI (Normalized Difference Vegetation Index) 132, 186, 225, 228, 230, 231, 232, 234, 242, 262, 263, 277, 278, 279, 284
Nearest neighbor 7, 129, 130, 132, 133, 134, 136, 137, 138, 139, 159, 249
Neighborhood allocation 353
NLCD (National Land Cover Database) 201, 202, 203, 204, 205, 228, 243, 246, 279, 280, 312
NEMO (Naval EarthMap Observer) 256
Net carbon exchange 8, 275, 276
Neural network configuration 107
Neurons 96, 97, 98, 100, 101, 103, 105
Nighttime lights 221
NOAA (National Oceanic and Atmospheric Administration) 212, 257, 277, 290
Non-algorithm parameters 95, 96
Normal distribution 132
NSMA (Normalized Spectral Mixture Analysis) 243
NUDAPT (National Urban Database and Access Portal Tool) 310, 311

O

OBIA (Object-Based Image Analysis) 7, 38, 39, 93, 129, 130, 131, 309, 310
Object-oriented classification 87, 130, 249, 251
Operational Linescan System (OLS) 184, 212, 220, 221, 225, 227, 231, 232, 233
Orthogonal constraint 84, 87, 89
Orthophotographs 310

P

Paddy fields 351
PAMIR (Phased Array Multifunctional Imaging Radar) 310
Panchromatic sensor 38
Pan-sharpening 7, 141, 143, 146, 147, 148, 150
PAR 228, 277, 279
Parametric statistics 7, 29
Parcel data 57, 200, 201, 202, 203, 204, 351
Partial memberships 29
Patches 66, 130, 172, 173, 175, 257, 340, 350, 351, 354, 361, 365
Pattern classification 7, 29, 30
Pattern matching 350
Patterns 5, 8, 9, 16, 17, 25, 39, 66, 159, 160, 166, 167, 168, 169, 172, 175, 176, 225, 226, 227, 234, 242, 248, 256, 257, 261, 266, 276, 278, 283, 284, 288, 290, 292, 293, 324, 325, 326, 327, 331, 336, 339, 340, 341, 347, 348, 349, 350, 351, 354, 360, 365, 367, 374, 375

PCA (Principal Component Analysis) 54, 55, 134, 135, 136, 137, 138, 139, 142, 228
Peak discharge 8, 239, 255, 256, 261, 264, 265, 266
Pepper's world hypotheses 9, 371, 372, 373, 377
Per-pixel 7, 36, 38, 46, 58, 93, 111, 112, 114, 117, 118, 121, 124, 129, 130, 139, 186, 187, 189, 243
Phenology 123, 276
Philadelphia 199, 225, 226, 229, 230, 231, 232, 311
Phoenix metropolitan area 6, 8, 312
Photo analysis 142
Photogrammetry 4, 5, 159, 310
Photosynthesis 276, 277, 280, 281, 282, 283, 284
Pixel purity index 116
Pixel-based models 241, 242, 243
Pixel-by-pixel classification 28
Planning support systems 336, 341
Polak-Ribiere (CGP) algorithm 100
Population density 8, 163, 167, 170, 171, 172, 174, 177, 184, 187, 196, 198, 199, 200, 201, 202, 203, 205, 212, 228, 229, 230, 231, 232, 234, 312
Population estimation 8, 163, 183, 184, 185, 186, 187, 189, 190, 191, 196, 234
Post-classification comparison (see map-to-map comparison) 27, 39
Postmodern urban dynamics 19
Powell-Beale (CGB) algorithm 99
Predictive models 9
Processes 5, 6, 8, 9, 17, 76, 87, 95, 97, 121, 122, 124, 130, 156, 157, 159, 160, 166, 168, 170, 174, 197, 201, 227, 234, 255, 256, 257, 258, 259, 276, 277, 288, 289, 290, 292, 293, 297, 305, 306, 308, 324, 325, 326, 327, 328, 330, 337, 338, 339, 340, 347, 348, 352, 360, 361, 362, 372, 373, 376, 377
Proprietary tools 349
Public policy 328, 339

Q

Q index 144, 147
Quality assessment 143
Quantitative measures 8, 112, 165, 293
Quasi-Newton method 99, 100
Quickbird 7, 36, 37, 38, 40, 43, 44, 45, 118, 123, 124, 129, 130, 132, 133, 134, 142, 147, 148, 149, 156, 184, 234, 242, 243, 249, 251, 257, 289, 309, 360

R

RADARSAT 64, 65, 66
Radiometric normalization 19, 20, 29
Radiometric resolution 17, 21, 28, 134
Radiometry 29, 142
RAND 327
Random sample consensus 78, 82
Random utility theory 375
Range data 75, 85
RapidEye 309
Raster data structures 323, 324
Rational method 256, 258
RBV (Return Bean Vidicon) 16
Real-time city 376
Recurrent networks 96
Region growing 82, 130, 188, 189
Regional atmospheric modeling 307, 311, 312
Regression modeling 205, 242

Regression tree 241, 242, 243, 246, 247, 248
Relative operating characteristic 349
Relative radiometric normalization 19, 20, 29
RePast 352
Residential land use 172, 185, 186, 364, 365
Resilient propagation (RP) 99, 100
RMSE (Root Mean Square Errors) 20, 203, 248, 249
Rule set 129, 130, 132, 134, 135, 136, 137, 138, 139
Rule-based classifiers 30
Runoff coefficient 259, 268
Rural 21, 66, 168, 198, 213, 222, 228, 230, 231, 242, 244, 248, 249, 279, 280, 281, 282, 283, 284, 287, 288, 292, 293, 294, 306, 313, 314, 340, 350, 351, 353, 354

S
Sampling design 5
Satellite imagery 6, 7, 8, 16, 19, 27, 36, 95, 175, 176, 195, 196, 198, 199, 205, 234, 242, 257, 258, 276, 289, 294, 311, 360
Satellite remote sensing 6, 13, 15, 16, 27, 184, 225, 226
Scale of analysis 8, 115, 124, 155, 156, 157, 165, 169, 175, 227, 278, 331
Scaled conjugate gradient (SCG) algorithm 99, 100
Scan Line Corrector (SLC) 17
Seattle, Washington 362, 364, 366
Self-organization 325, 326, 348
Sensitivity analysis 336
Sensor variations 29
Sensor-target-illumination geometry 29
Shade normalization 118
Shanghai, China 67
Shape 9, 41, 43, 56, 69, 76, 77, 78, 81, 82, 130, 131, 132, 133, 134, 135, 139, 143, 169, 171, 172, 174, 175, 186, 187, 189, 196, 249, 250, 251, 310, 328, 360, 374
SHE 257
Simulation 9, 89, 159, 256, 258, 261, 264, 266, 307, 311, 313, 321, 324, 327, 328, 329, 330, 335, 336, 337, 338, 339, 340, 341, 347, 348, 349, 352, 353, 354
Single-layer networks 96
Single-layer urban canopy model 307
SLEUTH model 338, 339, 341, 349
SMA (Spectral Mixture Analysis) 7, 56, 93, 111, 112, 113, 114, 115, 116, 118, 121, 123, 125, 228, 234, 242, 243, 244, 245, 246, 247, 248, 249, 360, 362
Smoothness 131, 134
Social biology 9, 371, 374, 375, 376
Social media 226
Social physics 9, 371, 374, 375, 376, 377
Sociodemographic data 8, 163, 195, 196, 197, 198, 199, 200, 201, 203, 205, 207
Socioeconomic data 7, 227, 229, 230
'Soft' classifiers 29
Soil moisture 228, 257, 258, 259, 260, 261, 284, 313, 362
Space-borne radar 6, 13, 63
Spatial analysis 4, 6, 21, 41, 170, 172, 309, 324, 336, 340
Spatial autocorrelation 227, 230, 231, 232, 330, 340
Spatial configuration 58, 157
Spatial consequences 4, 8
Spatial data structures 9, 321, 323
Spatial dynamics 27, 197, 327, 328, 329, 330
Spatial econometric model 231, 232
Spatial error model 232, 233
Spatial events 9, 371, 374, 376
Spatial geometry 170, 172
Spatial lag model 232
Spatial metrics 157, 160, 170, 176, 340, 372, 374
Spatial modeling 330
Spatial morphology 9, 371, 374, 375
Spatial quality 148
Spatial reclassification 21
Spatial resolution 7, 9, 17, 27, 28, 29, 36, 38, 42, 43, 46, 50, 64, 65, 66, 101, 105, 112, 114, 116, 118, 121, 123, 124, 125, 129, 130, 134, 138, 141, 142, 143, 144, 146, 156, 157, 159, 183, 184, 186, 195, 199, 201, 227, 234, 244, 262, 276, 279, 288, 289, 305, 309, 310, 313, 360, 362
Spatial scale of analysis 8, 165
Species richness 87, 289, 290, 292, 293, 294, 295, 297, 359, 361, 362, 363, 365, 366, 367
Spectral resolution 38, 43, 50, 56, 117, 123, 225, 234, 257
Spectral Angle Mapper (SAM) 56, 116
Spectral characteristics 9, 50, 54
Spectral complexity 117, 118, 119, 121, 124
Spectral confusion 21, 27, 86, 121
Spectral library 52, 53, 57, 58, 112, 114, 115, 116, 119, 123
Spectral Mixture Analysis (SMA) 7, 56, 58, 93, 111, 112, 113, 114, 115, 117, 119, 121, 123, 125, 228, 241, 242, 243, 245, 246, 247, 248, 360, 362
Spectral unmixing 54, 56, 57, 58, 59, 258
Spectrally pure pixel 57, 58
SPOT (Landsat Système Probatoire d'Observation de la Terre) 16, 18, 50, 184, 242, 289, 290, 295, 308, 360
SRTM (Shuttle Radar Topography Mission) 67, 212, 234, 309
Stanford Watershed Model 257
Steepest gradient descent (SGD) 99, 100, 105
Street network 177, 197, 199, 200
Sub-pixel classification 265
Suburbanization 18, 27, 159, 169, 234, 283, 284, 340
Suzhou, China 347, 350, 351, 354
Supervised training 99
Surface materials 6, 7, 13, 49, 50, 51, 52, 54, 55, 56, 57, 58, 59, 112, 258, 309
Surface temperature 6, 306, 307, 312, 313
Surface water quality 256
SWAT 257
Sweden 5, 217, 222, 289, 292, 293, 294, 295, 296, 298
Synthetic Aperture Radar (SAR) 63, 64, 65, 67, 69, 310
Systems theory 348

T
Tangent sigmoid function 98
Tasseled Cap transformation 20
Technological integration 9
Temporal lag 7, 93, 155, 156, 157, 159, 160
Temporal resolution 6, 195
TerraSAR-X 64, 65, 67, 310
TopoSAR 310
Texture 42, 68, 69, 77, 86, 89, 130, 132, 133, 149, 205, 229, 249, 258, 262, 309, 373
Thematic accuracy assessment 19, 20, 21, 22
Thermal remote sensing 6
Three-class method 199
Three-dimensional building reconstruction 7
Time-geography 376, 377
Tobler, Waldo 197, 199, 325, 326, 327
TOPMODEL 257

Township and village enterprise 355
Training algorithm 7, 96, 101, 105, 106, 107
Training efficiency 106
Transportation 8, 18, 20, 25, 27, 101, 163, 167, 168, 169, 173, 184, 186, 226, 242, 244, 248, 275, 276, 280, 328, 336, 351, 352, 354, 362, 373, 375
Tree cover 122, 124, 228, 229, 234

U

Unsupervised classification 21
Urban air quality 311
Urban areas 4, 5, 6, 7, 8, 9, 13, 16, 17, 20, 22, 24, 26, 28, 36, 38, 39, 40, 42, 43, 44, 45, 46, 49, 50, 52, 54, 56, 58, 63, 64, 65, 66, 67, 68, 69, 75, 76, 78, 79, 80, 81, 82, 84, 86, 88, 111, 112, 121, 141, 142, 143, 145, 147, 149, 151, 155, 156, 157, 160, 169, 174, 175, 176, 184, 227, 228, 229, 230, 234, 239, 242, 245, 246, 248, 255, 256, 258, 259, 263, 264, 266, 276, 283, 287, 288, 289, 290, 291, 292, 293, 294, 295, 297, 299, 305, 306, 307, 308, 309, 310, 311, 312, 313, 359, 360, 361, 372
Urban atmospheric modeling 308
Urban canopy model 306, 307, 311
Urban climate 50, 122, 242
Urban complexity 9, 348
Urban dynamics 8, 16, 19, 335, 341, 348
Urban environments 6, 9, 112, 114, 116, 117, 118, 124, 130, 228, 234, 287, 290, 359, 360, 361, 363, 365, 366, 367
Urban expansion 8, 121, 125, 176, 256, 311, 351
Urban feature extraction 4, 6, 7, 13
Urban forms 18, 339, 372, 374
Urban geography 4, 27, 157, 160, 200, 374
Urban gradients 292
Urban growth 5, 6, 8, 9, 15, 16, 17, 18, 19, 21, 27, 28, 50, 123, 156, 160, 166, 167, 168, 275, 276, 277, 278, 279, 280, 281, 283, 284, 294, 312, 321, 324, 326, 328, 330, 336, 338, 339, 340, 347, 348, 349, 350, 351, 352, 353, 354, 360, 362, 364, 366, 372, 376
Urban heat island 6, 229, 234, 306, 307
Urban impervious covers 8, 230, 242, 243
Urban mapping 6, 7, 38, 46, 93, 129

Urban model 306, 307, 328, 375
Urban morphology 4, 112, 119, 159, 305
Urban physical structure 4
Urban sprawl 8, 63, 159, 163, 165, 166, 167, 168, 169, 171, 172, 173, 175, 177, 184, 197, 283, 292, 294, 295, 329
Urbanization 4, 5, 6, 8, 16, 239, 242, 256, 283, 284, 287, 288, 290, 292, 293, 294, 295, 297, 306, 311, 312, 314, 335, 337, 338, 339, 341, 350, 359, 360, 361, 362, 367, 372
UrbanSim 362, 363, 364, 365, 374
USDA (US Department of Agriculture) 249, 256, 262
User-generated content 374, 376
USGS EROS Data Center 16, 17, 279

V

Validation 8, 9, 43, 44, 96, 107, 118, 119, 121, 187, 190, 258, 263, 265, 277, 321, 335, 336, 339, 340, 341, 347, 348, 349, 350, 353, 354
Vector data structures 330
Vegetation index 8, 132, 225, 228, 242, 277
Vegetation productivities 276, 277, 283, 284
Verification 336, 342, 355
Very-high spatial resolution 35, 36
Very-high spectral resolution 38
V-I-S model 112
Voting districts 201, 203, 205, 206

W

Water balance 258, 261
Water quality 8, 212, 242, 256
Watershed health 8
Wavelets 142, 143, 144
Web 2.0 374, 376
Weighted links 96, 97, 98
WEO (World Energy Outlook) 212, 222
WetSpa 255, 257, 258, 259, 260, 261, 264, 265
Worldview-2 36, 38, 156, 309

Z

ZIP codes 196, 197, 203, 206, 227

Hinduism
for AS students

Roger J. Owen, Series Editor

Roger J. Owen was Head of RE in a variety of schools for thirty years, as well as being a Head of Faculty, advisory teacher for primary and secondary RE, Section 23 Inspector and 'O' Level and GCSE Chief Examiner. Author of seventeen educational titles, he is currently an education consultant and WJEC Religious Studies AS and A2 Chair of Examiners.

Acknowledgements

The author and publishers would like to thank the following for permission to reproduce copyright material in this book:

Akg-images: p53(t); By kind permission of Barry Geller: p21; © The Bhaktivedanta Book Trust-International, Inc. www.krishna.com: p3, 17, 19, 23; Dalit Solidarity: p29; Dinodia Photo Library: p15, 31, 32 (all images), 33(b), 34(t); with thanks to Hashita, North Wales: p56, 61; India 280 'R. Eime/ Travel-Images.com': p34(c); India 239 'W.Allgower/Travel-Images.com': p53(b); Topfoto: p39, 58; V&A Images/Victoria and Albert Museum: p45; World Religions/Christine Osborne: 43, 47; World Religions/Chantal Boulanger: p13; World Religions/Nick Dawson: p36, 41; World Religions/Paul Gapper: p16; World Religions/Louise B. Duran: p10 (all images), 35, 38, 48, 49, 55(t), 55(b); World Religions/Claire Stout: p 46.

Every effort has been made to contact copyright holders of material reproduced in this publication. Any omission will be rectified in subsequent printing if notice is given to the publisher.
While the information in this publication is believed to be true and accurate at the date of going to press, neither the author nor the publisher can accept any legal responsibility for any errors or omissions that may have been made.

Published by UWIC Press
UWIC, Cyncoed Road,
Cardiff CF23 6XD
cgrove@uwic.ac.uk
029 2041 6515

Design by *the info group*
Picture research by Sue Charles
Printed by *HSW Print*

ISBN 978-1-905617-19-7

Sponsored by the Welsh Assembly Government

© 2007 Huw Dylan Jones

These materials are subject to copyright and may not be reproduced or published without the permission of the copyright owner

Front cover: Ganesh with his father Shiva and his mother Parvati, World Religions/Claire Stout

Hinduism
for AS students

by Huw Dylan Jones
Series Editor: Roger J Owen

Contents

Introduction — i

Section 1 — Beliefs about deity and humanity

Chapter 1	The Concept of God	1
Chapter 2	The Trimurti	7
Chapter 3	Shaivism and Vaishnavism	15
Chapter 4	Karma and Reincarnation	21

Section 2 — Some key beliefs and practices

Chapter 5	The Four Varnas and the Status of Dalits	27
Chapter 6	Varnashramadharma	31
Chapter 7	Rites of Passage	35
Chapter 8	Gods and Goddesses	43
Chapter 9	Festivals	51
Chapter 10	Worship	57

Bibliography — 64

Note about dates

This book uses the abbreviations CE and BCE for Common Era and Before the Common Era. In some books you will find AD (Anno Domini) for CE, and BC (Before Christ) for BCE. The actual years are the same, only the tag is different.

Hinduism

Introduction

This book assumes no prior knowledge about Hinduism and presents the religion in such a way as to meet the requirements of the WJEC AS specification. However under no circumstances should this book be used as the sole textbook for the Hinduism course, since advanced study requires the skills of wide reading and the analysis of a range of scholarly views on different issues.

The book is designed to be used in tandem with the teachers' book, which provides more detailed background information on some of the topics covered and assistance with the tasks that appear in the text.

AS level candidates are expected to demonstrate not only knowledge and understanding but also certain skills, such as the ability to sustain a critical line of argument to justify a point of view, and to relate elements of their course of study to their broader context, as well as to specified aspects of human experience. Some of the tasks that appear below are designed to assist in developing those skills. Teachers and students will doubtless think of others. It is important to remember at all times, however, to go beyond the simple facts and stories within Hinduism, which are relatively easy to learn, and to respond to all aspects of the religion in an open, empathetic, and critically aware manner. Being able to appreciate different points of view is crucial in this regard.

This book, and the accompanying teachers' book, is constructed with Key Skills in mind. Students are asked to develop communication skills by taking part in discussions, gathering information and writing. They are asked to develop ICT skills through encouragement to make critically aware use of the Internet, and to present findings in the form of project dossiers and class presentations. They are asked to solve problems through making cases for particular viewpoints, and to work with others on joint research projects.

The students' and teachers' books both attempt to reflect the huge diversity and variety found within Hinduism. Not only is this a requirement of the WJEC AS Specification, but it is also crucial to a proper, rounded understanding of the religion. All religious traditions contain a range of viewpoints and practices and students should be able to demonstrate a critical, yet non-judgemental, awareness of this fact. Teachers and students should take every opportunity to demonstrate the diversity of Hinduism: to do otherwise would be to risk a partial view of the religion, which could lead to unhelpful stereotyping.

When studying Hinduism candidates must be aware of the vast denominational and regional differences within the religion. In their personal study they will encounter different versions of the stories, rites, practices and festivals from those found in this book. No one version should be regarded as the defining version, but all should be accepted as being part of the rich diversity within Hinduism.

Candidates will also come across a number of important terms, which, depending on the tradition they come from within Hinduism, are not always spelt in the same way. This is sometimes a little confusing: it is important, however, for students to become familiar with these terms because, for every term learnt, something important about Hinduism is also learnt.

Section 1

Beliefs about deity and humanity

Aim of the section

This section asks you to consider the diversity of beliefs about the divine found within Hinduism and the way those beliefs affect the relationship between the deity and humanity.

This means you will have to consider six key matters:

1. Different concepts of God found within Hinduism;
2. The relationship between Brahman and atman;
3. The main features of Vaishnavism and Shaivism and the understanding of God within those traditions;
4. The notion of avatar and its implication in the God/humanity relationship;
5. The importance and significance of the tradition of Bhakti;
6. The doctrine of karma and reincarnation.

Chapter 1

The Concept of God

> **Aim**
>
> After studying this chapter you should be able to show kowledge and understanding about the different concepts of God found within Hinduism – polytheism, monotheism, henotheism and monism. You should also have knowledge and understanding of the concepts of Brahman and atman (soul) and the relationship between them. You should be able to explain your reaction to these beliefs using critical arguments to justify your view.

In Hinduism there are a diversity of beliefs about the divine and it is impossible to fit Hinduism into any particular system of beliefs – monotheistic, pantheistic, henotheistic or monistic. Hinduism can be described as all of these or none of these. Hinduism accepts that, since people are different, then it is completely natural for them to have different views of God. Each individual has a different approach to God. It is illogical in Hinduism to say that everyone has to have the same view.

Monotheism: the belief in one single, universal God. God is personal, exhibits qualities, displays a form and performs activities. Within Hinduism Vaishnavism, which regards Vishnu as God, and Shaivism, which regards Shiva as God, could be argued to be monotheistic. The Rig Veda (an ancient collection of Sanskrit hymns) contains evidence of emerging monotheistic thought:

> 'To what is One, sages give many a title.' (Pada 1.164.46c)

There are different types of monotheism within Hinduism. Inclusive monotheism regards the different deities as just different names for the single monotheistic God. Smartism, a division of Hinduism, argues that God is one but has different aspects and can be called by different names (see hard and soft polytheism below). This view of Hinduism is very popular amongst non-Hindus in the West. Exclusive monotheism believes that the deities are false and distinct from the one God, either invented or simply incorrect, e.g. Vaishnavism regards the worship of anyone other than Vishnu in this way.

Monism: the belief that everything is made up of one essential essence – atman. The soul is one with God in all respects and the numerous deities are aspects of the formless, all pervading world-soul, Brahman. God is impersonal, without qualities, without form, although represented in human form for our own understanding. Hindu belief in Brahman as the Supreme absolute from which everything, living or not, comes can be seen as a monistic belief. Brahman is in all things and each thing is a part of Brahman.

Another type of monism found in Hinduism is called Monistic Theism. This is a belief in a personal God who is a universal, omnipotent, Supreme being who is both immanent (in all things) and transcendent (above all things).

Chapter 1

Henotheism: this is devotion to a single God while accepting the existence of other gods. The term henotheism was created by Max Muller, who said henotheism was 'monotheistic in principle and polytheistic in practice.' While a monotheist would worship only one god, a henotheist might worship any god within the pantheon (all the gods), but be devoted more to one supreme deity. Again some forms of Hinduism could be argued to be henotheistic such as sects who follow Bhakti (loving devotion).

Polytheism: the belief in, or worship of, many gods or divinities. The word comes from the Greek words 'poly' and 'theoi', literally meaning 'many gods'. Most ancient religions were polytheistic, and had pantheons of different deities. However the belief in many gods does not preclude the belief in an all-powerful, all-knowing Supreme Being. In polytheistic belief systems gods are seen as complex personalities, of greater or lesser importance, with individual skills, needs, desires and stories. Usually such gods are not omnipotent or omniscient but are portrayed as similar to humans in their personalities, but with additional individual powers, abilities, knowledge and perceptions.

In a polytheistic pantheon, the gods have multiple names, each with its own significance in specific roles, and they have authority over specified areas of life and the cosmos.

Hard and soft polytheism: hard polytheists believe that the gods are separate and distinct beings. Soft polytheists, also known as inclusive monotheists, regard their many gods as representing different aspects of a single Supreme God. This can be seen in Smarta Hinduism: although perceived as polytheistic, it is also a form of inclusive monotheism, where one God is perceived as having many forms. A hard polytheist thinks that two Gods are different, but a soft polytheist such as a Smarta Hindu thinks that Vishnu and Shiva are different aspects of the same God. A follower of Smarta Hinduism would have no problem worshipping every imaginable deity with equal devotion, because they view them as being manifestations of the same God.

> **For reflection**
>
> *'Is it really possible to have more than one concept of God within a religion?'*

Brahman

It is a basic understanding in Hinduism that all reality is saturated with the divine. The physical world is created and sustained through the breath of God, or Brahman. Brahman is the name given to Ultimate Reality that, in fact, cannot be sufficiently named. In addition to recognising that God cannot be adequately named or described, Hindus believe that the 'real one' is known by different names. In Hinduism, although God (or the Ultimate, or Truth) is One, that One may be experienced in many different forms.

There is a conversation recorded in which a famous Hindu guru (teacher) is asked how many gods (devas) there are and he answers 'three hundred and three and three thousand and three.' Unsatisfied, the questioner repeats the question. The teacher replies that, actually, there are thirty-three gods. He is asked again and the teacher concedes that there are six; that there are three; that there are two. Finally, he states that there is one God. The questioner then wants to know which god is that One god and the guru replies, 'The prana (breath, life). The Brahman. He is called tyat (that)'. Brahman is considered to be eternal, genderless, omnipotent, without form and indescribable.

A representation of Brahman

Brahman is the origin of all things and Brahman is in all things and each thing is a part of Brahman. Each god is an aspect of Brahman or Brahman itself. This can be compared to the sun and its rays. We cannot experience the sun itself but can experience its rays and the qualities those rays have. There are many rays but only one source, the sun.

There is a tension between the one and the many in the Hindu understanding of the ultimate. On the one hand, the gods of the Hindu pantheon – the elevated gods and goddesses – are many, and their stories and qualities are varied. On the other hand, many Hindu scriptures, especially from the Upanishads, emphasise the oneness of the Ultimate. Additionally, while some Hindus experience the Ultimate in the form of a single deity, or god, other Hindus insist that any name or form for god is inadequate and that the Ultimate will always remain nameless.

Brahman is described as being composed of three qualities: sat, cit and ananda. Sat is pure being; cit is pure consciousness or awareness; and ananda is pure bliss. As such, Brahman can encompass all of creation (as pure being), can enfold all consciousness, and is the only source of eternal bliss. Of course, the best way of realising God has always been a source of controversy. The influential 9th-century philosopher, Shankara, suggested that humans contemplate Brahman on one of two levels. On a lower level, one may think of Brahman as Saguna Brahman, or 'God with attributes'. However, on a higher spiritual level, one realises Nirguna Brahman, or 'God without attributes'. Not all Hindus agree with Shankara that one approach to God is necessarily higher than another approach.

Saguna Brahman (God with attributes)

Brahman is sometimes conceived of as having attributes, such as those of the Supreme Person, and is given titles like Ishvara or Bhagavan (Lord). In the bhakti tradition, the person who worships and adores God personally as Lord Vishnu or the Goddess Kali is an example of someone who understands God with attributes. The bhakti tradition

Chapter 1

maintains that adoring God demands that one associate God with the proper attributes (of mercy and omnipotence, etc.), because it is God's overwhelming character of infinite attributes which saves humans in their limitedness.

Nirguna Brahman (God without attributes)

Brahman is also described as having no attributes and no names, since all descriptions are incomplete and limiting. Some devotees maintain that, because God is ultimately beyond all the words and thoughts which humans posses,s then representations of God only lead the person astray and limit God's true scope.

> **Discussion topic**
>
> 'It's impossible to worship a god without form or description.'

Atman

Humans possess a divine spark or atman within their physical body and this is ultimately associated with Brahman. In the Upanishads the atman is said to be found in all living beings, not only in humans, and is described as being 'identical with Brahman'. Every being has an eternal soul and that eternal soul is God. A popular greeting in India is 'Namaste', which means 'I greet the divinity within you.'

Hinduism differentiates between matter and spirit. Spirit is divided into two main categories:

- the individual self or soul (jiva-atman)
- the Supreme Self or God (paramatman).

To Monists the soul and god are the same, while the monotheist emphasises the distinction between them.

> **For reflection**
>
> 'Do all living beings have a soul? How might this affect the way you regard other forms of life?'

The majority of Hindus believe that the atman is eternal and never changes, as distinct from the temporary body made of matter and the mind. The soul is caught in the cycle of samsara and suffers repeated birth and death. Each soul creates its own destiny according to the law of karma. The goal of most Hindus is moksha, liberation from this perpetual cycle and re-identification with the Supreme Brahman.

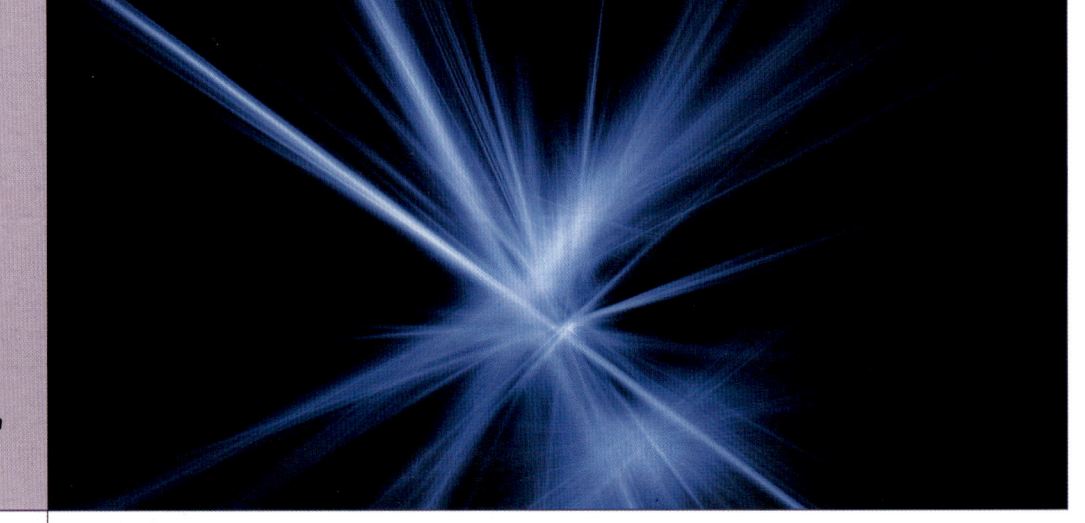

> **Seminar topic**
>
> How does this definition of the soul compare to ideas to be found in other religions?

A representation of atman, the soul in all living beings

A representation of the relationship between Brahman and atman

Brahman and Atman

To some Hindus the relationship between Brahman and atman is dual – the atman only being a part of Brahman and not wholly identifiable with it. Others hold a monistic view, in which Brahman and atman are one.

Many comparisons have been used to explain the relationship between Brahman and atman. In the Upanishads a father called Uddalaka teaches his son, Svetaketu, about the nature of atman and its relationship to Brahman. Uddalaka uses a number of images to make his teaching understood :

'As the bees, my dear Svetaketu, make honey by collecting the juices of distant plants and reduce the juices into one form, and as these juices have no discrimination, so that they might say, I am the juice of this tree or that, in the same manner, my son, all these creatures, when they have become merged in Being, do not know that they are merged in Being. Whatever these creatures are here, whether a lion or a wolf or a boar or a worm or a midge or a gnat or a mosquito, they are all identical with that Being which is subtle essence. The whole universe is identical with that subtle essence, which is none other than the self [Atman]. And you are That, Svetaketu. Place this salt in water, and then come to me in the morning.'

Svetaketu did as he was asked, and the next morning his father said to him:
'Bring me the salt which you placed in the water last night.' The son looked for it, but could not find it because it had dissolved completely.

'Sip the water from the surface,' said the father. *'How is it?' 'It is salt.' 'Sip it from the middle,'* said the father. *'How is it?' 'It is salt.' 'Sip it from the bottom,'* said the father. *'How is it?' 'It is salt.'*

Chapter 1

> **Seminar topic**
>
> *Read the conversation between Svetaketu and his father. How successful are the comparisons used in explaining the relationship between Brahman and atman?*

His father explained: 'In the same manner, my dear son, you do not see Being. But it is there. It is this subtle essence. The whole universe is identified with it, and it is none other than the self. And you are That, Svetaketu.'

Another comparison is space in a jar and space outside it. The space in the jar is temporarily confined as the atman is confined by the body. The space in the jar is still the same as the space outside it and when the jar breaks, the space is one just as atman and Brahman are really one.

Tasks

Writing tasks	a) Explain what Hindus believe about Brahman and atman.
	b) Assess the view that Hinduism is a polytheistic religion.
Writing and presentation task	Think of metaphors or comparisons that would appropriately describe the relationship between Brahman and atman and present them to the rest of the class.

Glossary

monotheism	belief in one single universal god
monism	belief that everything is made up of one essential essence
henotheism	devotion to a single god while accepting the existence of other gods
polytheism	belief in the worship of many gods or divinities
Brahman	The Supreme Spirit in Hinduism
Saguna Brahman	God with attributes
Nirguna Brahman	God without attributes
atman	the animating energy in any living creature, usually referred to as the soul
jiva-atman	the individual self or soul
paramatman	the supreme self or God

Chapter 2

The Trimurti

Aim

After studying this chapter you should be able to show knowledge and understanding about the three gods of the Trimurti and the beliefs, stories, symbolism and practices associated with them. You should also be able to convey your reaction to some of these stories and beliefs.

The Trimurti – Brahma, Vishnu and Shiva

In Hinduism Brahma, Vishnu and Shiva are called the Trimurti and are regarded as three different aspects of God – Brahma the creator, Vishnu the Preserver and Shiva the destroyer. Sometimes the three are represented as one God with three heads. Of the three, Vishnu and Shiva are by far the most important, since Brahma is very rarely worshipped independently.

As well as being regarded as the creative, preservative, and destructive aspects of God, many believe that they also represent earth, water and fire: Brahma, the earth, the originator of life; Vishnu, water, the sustainer of life; and Shiva, fire, which destroys and consumes life.

Many also believe that the Trimurti represent various stages in an individual's life. The student stage, or ashrama, is represented by Brahma, because during this stage an individual seeks knowledge and knowledge is represented by Saraswati, who is the consort of Brahma. The householder stage is represented by Vishnu, because during this stage an individual fulfils his family responsibilities such as ensuring that he has enough wealth to sustain his family; wealth is represented by Lakshmi, Vishnu's consort. The third stage is retirement, when an individual renounces all material possessions; this is represented by Shiva, who leads a homeless life with only the most basic possessions.

Chapter 2

Seminar topic

Is it possible to worship more than one god with equal devotion?

So, although the Trimurti is male in nature, the three gods are associated with female energy (shakti) – Saraswati, the goddess of knowledge and learning, is the shakti or consort of Brahma; Lakshmi, the goddess of beauty and wealth, is the consort of Vishnu; and Parvati, daughter of Himavat, the god of the Himalayan mountains, is the consort of Shiva.

Brahma, the creator god

Brahma

Brahma is the Hindu creator god. He is usually depicted with four heads, four faces and four arms. An interesting legend explains how he got his four heads: while he was creating the universe he created a female deity called Shatarupa and immediately became infatuated with her. In order to avoid his gaze Shatarupa moved in different directions, but whichever direction she chose Brahman developed a head. In all he developed five heads, one on each side, and one above the others. Therefore in order to control Brahma, Shiva cut off the top head. Since Shatarupa was created by Brahma, Shiva decided it was wrong for Brahma to become obsessed with her and ordered that there should be no proper worship of him in India.

There is another legend that explains why Brahma is not worshipped: once a great fire-sacrifice was being organised on earth by the high priest, Bhrigu. It was decided that the greatest amongst all the gods would preside over the sacrifice. Bhrigu set off to find the greatest among the Trimurti. When he went to Brahma he was so busy listening to Saraswati's music that he did not hear Bhrigu's calls. Bhrigu was so angry that he cursed Brahma, so that no person on earth would invoke or worship him again.

Therefore only Vishnu and Shiva are the focuses of popular worship, with Brahma being ignored. Ever since, Brahma has been reciting the four Vedas (sacred books) in repentance. Although Brahma is prayed to in almost all Hindu religious rites, only two temples are dedicated to him in India. The more important one is at Pushkar, close to

Jaipur. Once a year a religious festival is held in Brahma's honour, with thousands of pilgrims coming to bathe in the holy lake next to the temple.

Brahma holds no weapons in his four arms. In one of his hands he holds a sceptre in the shape of a spoon, which represents the pouring of holy ghee or oil into a sacrificial pyre, representing Brahma as lord of sacrifices. In another hand he holds a water pot, sometimes shown as a coconut shell, containing water which represents the beginning of creation. Brahma has a string of rosary beads which he uses to keep track of the Universe's time. He also holds the Vedas and sometimes a lotus flower.

Bahma's consort is Saraswati, the goddess of learning, and his vehicle is the swan, which has the ability to separate milk from water when they are mixed. This represents justice being given to all creatures in every situation and also the separation of good from evil.

Brahma is self-born, in the lotus which grew from the navel of Vishnu at the beginning of the universe. In another legend he is said to have created himself by first creating water. In this he planted a seed which became the golden egg and from this egg Brahma the creator was born. He is also said to be the son of the Supreme Being, Brahman, and his female energy, Maya.

> **Seminar topic**
>
> *How difficult is it to accept that the gods are all different aspects of the one god, Brahman?*

Vishnu

Vishnu is known as the preserver and is the second aspect of the Trimurti. For the followers of Vaishnavism he is the Supreme God and not just one aspect of him (see chapter 3). He is mainly associated with his avatars or incarnations especially Krishna and Rama. His consort is Lakshmi, goddess of wealth, and his vehicle is Garuda, the eagle.

The name Vishnu comes from the root 'vis', which means 'to spread in all directions, to pervade', and Vishnu is regarded as the core or nucleus through which everything exists.

Vishnu is usually depicted either reclining over the waves of the ocean on the coils of the serpent deity called Shesh Nag, or standing upright on a lotus flower, often with Lakshmi beside him. He is shown as having four arms which symbolise his all-powerful nature. The two front arms represent his physical existence, while the two arms at the back represent his presence in the spiritual world.

In each hand he holds one of his four main attributes :

a) In his lower right hand he holds a conch shell, which represents the five elements – water, fire, air, earth and sky - and is a symbol of creativity. When blown, it produces a sound that is associated with the sound 'Aum' from which creation developed.

b) The upper right hand holds a discus or wheel. The discus has six spokes and symbolises a lotus with six petals. It represents the power that controls all six seasons (of the Hindu calendar) and the fearful weapon that cuts off the heads of all demons.

Chapter 2

Vishnu and two of his avatars - Narasimha and Krishna

c) In the upper left hand is a lotus which symbolises the power from which the universe emerges and represents the truth from which Dharma and knowledge come.

d) A mace is held by the lower left hand and represents the force from which all physical and mental powers derive.

Vishnu is pictured as having blue skin, the colour representing his all-pervasive nature like the infinite sky and the infinite ocean. On his chest is the mark of sage Bhrigu's feet, as well as the srivasta mark, symbolizing his consort Lakshmi. Around his neck he wears the 'Kaustubha' jewel and a garland of flowers. A crown should adorn his head as a symbol of his supreme authority. He is also shown wearing two earrings, representing the opposites in creation – knowledge and ignorance, happiness and unhappiness, pleasure and pain.

Hindus believe that Vishnu incarnates from time to time in a variety of ways, to restore and promote good and destroy evil. Most Hindus believe that he incarnates as a living being in ten avatars:

- **Matsya** (fish) – According to Hindu mythology the universe is in a cycle of destruction and creation. Before the creation of the present universe, the four Vedas were lost in the waters. As they were needed to instruct Brahma about the work of creation, Vishnu was given the task of bringing them up from the deep. He took the form of a giant fish and brought the sacred books up from the water.

- **Kurma** (tortoise) – In this form Vishnu took the newly created earth upon his back, to give it stability in the 'churning of the ocean.'

- **Varaha** (boar) – Once, when the earth sank into the deep waters, Vishnu as Varaha went into the waters and pulled the earth back up with his tusks.

- **Narasimha** (half man, half lion) – In this form Vishnu killed the demon ruler Hiranyakashipu. He had pleased the god Brahma with his sacrifice and Brahma in return had given him the gift of not being able to be killed by man or animal, in the day or night, on earth or in the heavens, by fire, water or any weapon.

Therefore Vishnu, in order to kill the tyrant, took the form of Narasimha who was neither man nor animal, and killed Hiranyakashipu in the evening, which was neither day nor night.

- **Vamana** (dwarf) – Bali, the demon king, had defeated Indra, the king of heaven and had become invincible, ruling over the three worlds – heaven, earth and the underworld. Indra appealed to Vishnu for protection. Vishnu took the form of a dwarf to protect the world from the tyrant. Vamana went to see Bali in the company of other Brahmin priests to ask for gifts. Bali was amused by such a small holy man and promised that he would give him whatever he asked for. Vamana asked only for as much land as he could measure in three steps. Bali laughingly agreed. Vamana (Vishnu) covered heaven with one stride, then the earth with the second stride, and with his foot on Bali's head pushed him into the underworld where he was allowed to rule out of respect for his grandfather Prahalad's great virtues.
- **Parashurama** – Vishnu took the form of a brahmin who was armed with an axe-like weapon to protect the brahmin from the tyranny of the power mad Kshatriya caste at the request of the goddess Prithvi (Mother-Earth)
- **Rama** – The story of Rama and his battle with Ravana, the demon king of Lanka, is described in the epic Ramayana (see chapter 9).
- **Krishna** – Krishna and Rama are the best known avatars of Vishnu and are the most popular gods amongst Hindus everywhere. Krishna means black or dark in Sanskrit, refering to Krishna's black/bluish complexion.

The Mahabharata is the earliest text that describes the actions of Krishna. In it he is described as the incarnation of Vishnu and is one of the most important characters of the epic. In the sixth book there are eighteen chapters which make up the Bhagavad Gita, the best known of all Hindu texts. The chapters contain the advice given by Krishna to Arjuna on the battlefield. In the epic, Krishna is already an adult and therefore an appendix was added later to describe Krishna's childhood and youth. This appendix is called the Harivamsa.

Krishna was born to restore the balance of good in the world and therefore many of the stories involving Krishna describe him overcoming evil demons and people. He was born into the royal family of Mathura, the eighth son of Vasudeva and princess Devaki. His parents had been imprisoned by the evil demon king Kamsa, because a prophecy had predicted that the king would be killed by Devaki's eighth child. Krishna was born in prison but was smuggled out to safety to be raised by his foster parents, Yashoda and Nanda, in the village of Gokula (cow-village). Nanda was the head of a community of cow-herders. Kamsa heard about the child's escape and sent various demons to try to kill him.

The stories of Krishna's childhood and youth are very popular amongst Hindus of all ages. Although he is a divine being, and Krishna was fully aware at all times of his identity with Vishnu, his childhood and youth take place amongst ordinary people. He is described as a confident, fun loving character who plays tricks on others of his age in the village. He is also shown to be full of love. Although all the gopis (cowgirls) were in

Chapter 2

love with Krishna, he had a favourite, Radha. To many Hindus this love affair is a symbol of the love of the human soul for God and God's love for humanity. The love of the gopis for Krishna is a symbol of bhakti (devotion to God) where the worshipper loves God to such an extent that the self is lost. Krishna was a Kshatriya, while Radha was a low caste gopi, therefore bhakti is above class. However, in practice this is not always the case.

As a young man Krishna returned to Mathura and killed Kamsa, freed his parents and set up Ugrasena as the rightful king. Krishna himself became a leading figure at the court and during this time his friendship with Arjuna grew. He married Rukmini, daughter of King Bhishmaka, but also had seven other wives.

In the war between the Pandavas and Kauravas, Krishna was a cousin to both sides and therefore he gave them a choice of having his army or himself on their side. The Kauravas chose the army and therefore Krishna sided with the Pandavas. In the battle between the two sides he became the charioteer for Arjuna.

Krishna is worshipped in many forms – as a baby, youth, lover, lord or as the great God Vishnu.

- **Buddha** – The ninth avatar of Vishnu is the founder of Buddhism.

- **Kalki** – The last avatar of Vishnu is yet to appear, but is expected to come at the end of the present dark age. He is expected to appear seated on a white horse, blazing like a comet, and will finally destroy the wicked, restart the new creation and restore purity in people's lives.

For reflection

How difficult is it for Hindus to see Vishnu as Brahman or Rama as Vishnu?

Shiva

Shiva, which means 'kindly or auspicious', is the third member of the Trimurti. Shiva is the destroyer, but although he represents destruction he is viewed as a positive force – the destroyer of evil but also the creator following destruction.

Shiva is also called the Lord of Ascetics (Mahadeva); Rudra the god of storms; and Nataraja, lord of the dance. As Nataraja, he controls the movement of the universe. He is the god of reproduction and sexuality and is worshipped in the form of the linga, the male organ of reproduction. He is also the god of opposites – static and dynamic, old and young, gentle and fierce and, as Ardhanarishwara, both man and woman. Shiva is the source of fertility in all things and protects the good.

Shiva's consort is Devi, the mother goddess, who comes in many forms, two of the most popular being Parvati and Kali. Shiva and Parvati are the parents of Ganesh, the elephant-headed god of wisdom. Shiva is often shown riding a snow white bull called Nandi, which is often found in Shaivite (followers of Shiva – see chapter 3) temples as a doorkeeper. The bull is a symbol of sexual impulse, which Shiva controls.

Shiva is pictured in various ways, but his most distinctive features are:

- **Sitting on the skin of a tiger** - the tiger is the vehicle of Shakti, the goddess of power and force. Since Shiva is beyond and above any kind of force, he is the master of Shakti and the victor over every force. The tiger also represents lust and by sitting on it Shiva shows that he controls lust.

- **The cobra necklace** – Shiva is beyond the power of death, although he is surrounded and encircled by death. He is blue-necked, having consumed poison to save the world from destruction. The cobras also represent the spiritual energy within life.

Shiva

- **Long matted hair tied into a mop on his head** – this represents him as the lord of the wind or Vayu, who is the breath in all living things. Shiva is the lifeline of all living beings.

- **The crescent** – on his head Shiva has the crescent of the fifth day moon. It is placed near the third eye and shows the power of offering sacrifices. It shows that Shiva has the power of procreation as well as the power of destruction. The moon is also a measure of time and the crescent represents his control over time.

- **The Sacred Ganga** – The Ganga, the most sacred river in Hinduism, flows from the matted hair of Shiva. According to tradition, Shiva allowed the Ganga to fall on the earth from his own hair to save the destruction it would have caused had nothing broken its fall. It was also an outlet for the great river to bring purifying water to humanity. The flowing water is one of the five elements which compose the universe and from which the earth comes. The Ganga also represents fertility, which is one of the creative aspects of Shiva.

- **The third eye** – this is situated on Shiva's forehead and is the eye of wisdom which can see beyond the obvious. It represents his all-knowing nature. It is also associated with his energy which destroys evil-doers and sins. The Hindu epic, the Mahabharata, describes how Shiva got his third eye. One day his consort Parvati playfully went behind him and placed her hands over his eyes. Suddenly the whole world was covered in darkness and every living being was full of fear because the lord of the universe had closed his eyes. Suddenly a tongue of flame came from the forehead of Shiva and a third eye appeared to give light to the world.

- **The trident** – the three heads of Shiva's trident represent the three aspects of the Trimurti – creation, preservation and destruction. Since the trident is in the hands of Shiva it suggests that they are all in his control. Another interpretation of the three heads is that they represent the past, present and future, which are all under Shiva's control. As a weapon, the trident represents the punishment that evil-doers will suffer.

Chapter 2

- **The drum** – from the drum in the hand of Shiva comes the universal word 'Aum' – the source of all languages and expression.
- **The Vibhuti** – the three lines of ashes drawn on the forehead that represent the essence of being, after all the impurities have been burnt in the fire of knowledge.
- **The ashes** – Shiva smears his body with cremation ashes, which represents the philosophy of life and death.

Unlike Vishnu, Shiva does not traditionally have avatars and is worshipped directly by his followers. Shiva-Ratri, or Shiva's Night, is a festival held in honour of Lord Shiva. It is held on the fourteenth night of the dark half moon in the month of Magh (January-February). During the night Shiva's image is covered by showers of green leaves. This custom is based on a story about a hunter who lost his way in a thick forest and decided to spend the night under a tree. It was cold and he could not sleep properly and kept on changing position. His movements caused the leaves to fall on a Shiva-Linga which was placed at the foot of the tree. Shiva was so pleased with him that he blessed him with good luck.

> *Seminar topic*
>
> **'Hindus don't worship Brahman, they worship Shiva and Vishnu'.**

Tasks

Writing tasks	a) Explain the main characteristics of Vishnu. b) 'Vishnu is the most important god in Hinduism.' Assess this view.
Research and presentation task	Use the internet to find different stories and practices associated with the Trimurti and present them to the rest of the class in a powerpoint presentation.

Glossary

Trimurti	literally 'three-form', the Hindu trinity of three deities
Brahma	the creator aspect of Brahman in the Trimurti
Vishnu	the preserver aspect of Brahman in the Trimurti
Shiva	the destructive aspect of Brahman in the Trimurti
Shakti	female energy
Saraswati	Brahma's consort
Lakshmi	Vishnu's consort
Parvati	Shiva's consort
avatar	earthly incarnation of a deity

Chapter 3

Shaivism and Vaishnavism

Aim

After studying this chapter you should be able to show knowledge and understanding of the two major devotional traditions of Shaivism and Vaishnavism and the differences between them. You should also be able to show knowledge and understanding of the concept of avatar and the tradition of Bhakti and be able to react to them in a critical manner.

Shaivism is a branch of Hinduism that worships Shiva as the Supreme God. The followers of Shaivism are called Shaivas or Shaivites. Shaivism has appeal all over India but is particularly strong in south India and Sri Lanka. Shaivas acknowledge the existence of other gods, but as expressions of the Supreme God. This view is called Monistic Theism. Shaivas believe that God cannot be limited to any form or body and often worship Shiva in the form of a linga (a phallic symbol) symbolising the entire universe.

Shaivism is a deep, devotional and mystical denomination of Hinduism. It is a very broad religion which contains diverse ideas, rituals, legends and traditions. However the ultimate aim of Shaivas is to reach Moksha and break the cycle of birth, death and rebirth.

Nataraja Temple, dedicated to Shiva: Chidambaram, Tamil Nadu, India

The main features of Shaivism are :

1. Shaivas believe that Shiva is the Supreme God of the Trimurti.
2. Shiva is worshipped in two forms – as a linga and in human form.
3. Shaivas also worship his consort Parvati and his sons Ganapathi and Murugan.
4. There are many temples dedicated to Shiva but worship can also take place at home. Shaivas have idols of natural linga shaped Salagramam stones to which they offer flowers and food during worship.
5. Shaivas place great emphasis on certain parts of the Vedas, such as Rudram and Chamakam, which praise Shiva.
6. Sacred ash is an important part of Shaivite worship. Shiva is bathed in it and Shaivas wear it on their foreheads and other parts of the body as a mark of respect and reverance.

Chapter 3

7. The sacred syllable 'Aum' is an important part of shaivite worship, as is the five syllable word Na-ma-si-va-ya which Shaivas consider holy and regard it as their duty to repeat several times.
8. In Shaivite temples priests are called Shivacharyas. Brahmins are not allowed to perform the worship in the inner sanctum of Shaivite temples. This is a privilege held only by the Shivacharyas.
9. The holiest of Shiva shrines is the Nataraja temple in Tamilnadu.
10. Benares is considered the holiest city by all Hindus, including Shaivas.

In the past Shaivism has been in conflict with Vaishnavism, the other major denomination within Hinduism. However today there is more dialogue and discussion between them.

Seminar topic

'It is much easier to worship a personal god than an impersonal one.'

Followers of Vaishnavism in the streets of London

Vaishnavism

Vaishnavism is the other major denomination within Hinduism. Vaishnavites, as followers of Vishnu are called, worship Vishnu or one of his avatars as the Supreme God. Vaishnavism is principally monotheistic but also includes aspects which could be described as pantheistic. The worship of Vishnu dates back to vedic times.

Vaishnavites believe that Vishnu is the one Supreme God and that all other gods serve him. The heroes of both the great Indian Epics – the Ramayana and the Mahabarata – are believed to be avatars of Vishnu. Around half the world's one billion Hindus are Vaishnavites.

The main features of Vaishnavism are:

1. Vaishnavism is very devotional, and stresses the personal aspects of God and complete devotion to Vishnu or his avatars.
2. Vaishnavite worship is full of ecstatic dancing and chanting of holy names such as Rama and Krishna.
3. Temple worship and festivals are elaborately observed.
4. Vaishnavites believe that god and the soul are distinct from each other.
5. The main aim of Vaishnavites is Moksha, liberation, which is only attainable after death when the soul realises union with Vishnu's body, as part of him yet maintaining its individual personality. They believe that Vishnu is the soul of the universe.
6. The highest path to attaining Moksha is the path of Bhakti. To Vaishnavites performing Bhakti is very important, since it is a way of communicating with and receiving the grace of Vishnu through the temple or home deity.
7. The most important Vaishnavite scriptures are the Vedas and Puranas.

Seminar topics

'Vaishnavism and Shaivism are not denominations within Hinduism but religions of their own.'

'Brahman is not important to the followers of Vaishnavism or Shaivism.'

Avatar

The term avatar or avatara in Hinduism refers to the incarnation of an immortal being or God. It derives from the Sanskrit 'avatara', which means 'descent' and usually refers to God descending or coming down to the world in bodily form for a special purpose, usually the destruction of evil and the promotion of good. Hindu scriptures do not describe any appearance as an avatar by Brahma or Shiva. The term is usually used for an incarnation of Vishnu, the Preserver, who is worshipped as God by many Hindus. He is said to have ten great avatars (see chapter 2).

A representation of the ten great avatars of Vishnu

Chapter 3

Vaishnavites believe that Vishnu takes a special form whenever evil is on the rise and dharma (righteousness) is on the decline. Krishna, an avatar of Vishnu, explains his purpose as an avatar in the Bhagavad Gita 'for the protection of the good, for destruction of evil and for the establishment of righteousness, I come into being from age to age.' (Bhagavad Gita ch4 vs8).

Hindus believe that there is no difference between the worship of Vishnu and of his avatars, since it all leads back to him.

The importance of avatars in the Hindu religious life is that to many Hindus they are forms in which God can be truly appreciated e.g. Rama and Krishna have been loved and adored by Hindus for thousands of years. The idea of the avatar has also helped ordinary Hindus to understand the concept of the impersonal Brahman and the aid he gives to humanity in difficult times.

According to Vaishnava teaching there are two types of avatar – primary and secondary. In his primary avatars Vishnu directly descends e.g. Rama or Krishna, but in his secondary avatars he does not directly descend. Only the primary avatars are worshipped and in practice only Narasimha, Rama and Krishna.

Some Hindus regard the central characters of various non-Hindu religions as avatars e.g. Jesus and Muhammad, but others completely reject the idea of avatars outside traditional Hinduism.

> **For reflection**
>
> *Have the avatars of Vishnu, such as Rama and Krishna, become more important than Vishnu himself.*

Bhakti

Bhakti is the belief that a personal relationship with God is possible, based on love and devotion expressed through action (service). The term Bhakti comes from the root 'Bhaj' which means 'to be attached to' and refers to the intense loving devotion which the worshipper shows towards God. The worshipper (Bhakta) totally surrenders all aspects of himself/herself to his/her personal deity.

It is difficult to pinpoint exactly the origin of the Bhakti movement or movements in Hinduism. Some aspects can be traced to Vedic times in the worship of the Vedic god Varna, but it is in the Bhagavad Gita that the first full statement of liberation and fulfilment through devotion to a personal God is found. The Bhagavad Gita teaches that God's love is available to anyone who comes to him. Bhakti is reciprocal, in that the worshipper shows love towards God and also receives love from him. It is this personal relationship that is at the heart of Bhakti worship. Although Bhakti is usually associated with devotion to Brahman in the form of Vishnu or Shiva, the term has developed to refer to devotion in general within Hinduism.

Bhakti is a path that leads to liberation of the soul (moksha) in its own right and stresses the way of inner feelings rather than an institutionalised form of religion. The nature of the Bhakti movement is rebellious, disregarding such things as Scriptures and class distinctions. This is one reason why Bhakti is so popular. Worshippers have the freedom to choose their favourite deity, while not excluding all others. They can also worship God as child, parent, friend, master or lover.

Bhakti worship

There are various kinds of Bhakti –

- **Sakamya Bhakti** – This is devotion with desire for material gains. The worshipper wants something specific such as wealth, health or a certain career, and believes that if his Bhakti is intense and his prayers are sincere then God will grant his wish.
- **Nishkamya Bhakti** – In this form of Bhakti the worshipper seeks to become one with God and receive spiritual blessings such as wisdom and power.
- **Apara-Bakhti** – This is for beginners in yoga. The worshipper decorates an image with flowers and garlands, rings the bell, offers Naivedya (food offerings), waves lights and observes rituals and ceremonies. In this form of Bhakti the worshipper regards God as Supreme, who is present in the image and can only be worshipped in that form.
- **Para-Bhakti** – This is the highest form of Bhakti. The worshipper sees God and feels his power everywhere around him.

Although Bhakti is based on the doctrine 'God is love and love is God', it is more than simple emotion. It is a thorough disciplining and training of the will and mind. Everything must be done to create an environment which develops the feeling of Bhakti. The Puja room must be kept clean and decorated. Incense must be burnt, a lamp lit and a clean seat kept. Worshippers should bathe, wear clean clothes and apply sacred ash on the forehead. They should concentrate the mind on the deity they wish to worship.

Many other actions also help to develop Bhakti, such as right conduct, prayer, worship, service of saints, service of the poor and sick, and keeping to Varnashramadharma duties.

Chapter 3

When the worshipper grows in devotion, the self is completely forgotten. This is called Bhava. It is a true relationship between the worshipper and God.

There are five kinds of Bhava in Bhakti –

- **Shanta Bhava** – The worshipper is Shanta or peaceful. He is not highly emotional but his heart is filled with love and joy.
- **Dasya Bhava** – This is where the worshipper has the attitude of a servant and serves God whole-heartedly, pleasing him in all possible ways.
- **Sakhya Bhava** – This is where God is regarded as a friend, on equal terms, just as Arjuna and Krishna used to sit, eat, talk and walk together as friends.
- **Vatsalya Bhava** – The worshipper looks at God as his child. The worshipper serves, feeds and looks on God as a mother does her child.
- **Madhurya Bhava** – This is the highest form of Bhakti, where the worshipper regards God as his lover and they become one. However it must be stressed that they are not like earthly lovers.

For reflection

How can Bhakti be described as a religious experience?

Tasks

Writing tasks	a) Explain the main features of Shaivism.
	b) 'Shaivism is a monotheistic religion.' Assess this view.
	a) Explain what is meant by Bhakti.
	b) 'Bhakti is the highest form of worship in Hinduism.' Assess this view.
Research task	Use the internet and other resources to note the differences between Vaishnavism and Shaivism. Present your findings in a table

Glossary

Shaivism	the worship of Shiva as the Supreme God
Vaishnavism	the worship of Vishnu as the Supreme God
linga	phallic symbol used to represent Shiva
Shivacharyas	name for priests in Shaivite temples
Bhakti	the path of loving devotion

Chapter 4

Karma and Reincarnation

Aim

In this section you will learn and come to understand the doctrine of Karma and reincarnation and its influence on the everyday lives of Hindus. You should also be able to react in a critical manner to these doctrines.

Karma literally means 'deed' or 'act' and is the force that drives reincarnation. It is the principle of cause and effect and operates on a moral basis, that is a good action, mental or physical, leads to a good effect and vice-versa. In the Vedic religion our situation in this life is thought to be the fruit or result of our karma, action and reaction, in our past life or lives, as karma is accumulated throughout our reincarnated lives.

It is then possible to purify our karma and make it good, leading the atman to return to Moksha to be united with God. Therefore there is 'good karma' and 'bad karma', which are stored reactions that determine each soul's destiny. The concept of karma has been taken and adapted by other religions such as Buddhism and Jainism.

The law of karma is used in Hinduism to explain the problem of evil that persists in spite of an all powerful God. Karma is not fate, since each soul determines its own destiny through free will. However, this is exercised only in human form. In lower species the atman takes no moral decisions and is bound by instinct. It generates no new karma, only burning up bad karma and gradually rising towards a human birth. Karma is generated only in human life.

Seminar topics

'Karma is fatalistic.'

Discuss the cartoon. Is it a fair representation of the law of karma in Hinduism?

Chapter 4

The principle of karma is that those who sow goodness will reap goodness and those who sow evil will reap evil. Therefore it has been said that we are punished by our wicked actions and not for them. Some actions take a long time to come to fruition and may come about in later existences. Actions which lead to good karma are called punya (merit) and include activities such as following the principle of varnashramadharma, giving to charity and going on pilgrimage. Actions which lead to bad karma are called papa (sin) and include activities such as avoiding one's duty and the neglect and abuse of five sections of society – women, children, animals (especially cows), saintly people and the elderly.

There are many misunderstandings concerning the law of karma. One popular misunderstanding is that Hindus blame suffering on karma. This is not necessarily so, since blame and responsibility are different. Karma means understanding that everyone is responsible for their own lives. However this does not mean being indifferent to suffering, either our own or that of others. It stresses the principle of helping others to help themselves.

> **Seminar topic**
>
> *The teaching of the principle of karma is just a way of ensuring that people try to live a good life.*

Reincarnation

'As the embodied soul continually passes in this body from boyhood to youth to old age, the soul similarly passes into a new body at death.'
(Bhagavad Gita 2.13)

The process of the soul transmigrating into a new body is called reincarnation. All Hindus believe that the individual soul (atman) exists in a cycle of birth into a body, followed by death and then rebirth into a new body, although not necessarily a human one. This cycle is called Samsara. The ultimate aim of the soul is to be freed from this cycle altogether by attaining liberation (Moksha).

> **For reflection**
>
> *Do you think any part of a person is eternal?*

> **Seminar topic**
>
> *Is it reasonable to believe in reincarnation? If so, why? If not, why not?*

The quality of the life into which the soul is reborn depends on the previous life, on karma. Karma is not the same as the process of judgement in some other religions such as Christianity: it is automatic and impersonal. Hindus therefore aim to live their lives in a way that will earn them good karma and eventually free them from rebirth altogether.

They have four legitimate aims in life :

- **dharma** – living in an appropriate way. This means doing what is right for the individual, family, and social class. Hindus believe that following one's dharma is necessary to sustain cosmic order and therefore going against dharma results in bad karma. The Bhagavad Gita points out that it is better to perform your own duties poorly than another person's well.

A representation of the cycle of Samsara or reincarnation

- **artha** – the pursuit of wealth through lawful means. Instead of condemning the material aspects of life, Indian culture has given it a specific place as part of its religious goals. The attainment of wealth and the maintaining of peace and prosperity was, for the classical Indian king (the raja), part of his dharma towards his people. He was responsible for developing sound policies and had to be skilled as a ruler. On a wider level, the householder was expected to work for material gain to provide for his household. The pursuit of artha, however, is also subject to the moral structure of a person's dharma. Immoral actions to gain wealth will only produce bad karma.

- **kama** – the delight of the senses. This is also part of the second ashrama, and is encouraged for the householder. It is thought that the right experience of pleasure creates a well-developed character. It is also good preparation for the soul's devotion to God. Longing and loving adoration has always been a mark of religious aspirations.

Chapter 4

- **moksha** – release from rebirth. This is the ultimate aim of all Hindus. Each time a soul is born into a better life, it has the oppurtunity to improve itself further, and get closer to liberation. When a soul attains moksha it loses its individual identity and becomes part of Brahman. The soul is often compared to a drop of water, which at liberation falls into the ocean (the Supreme Soul, Brahman). It is a state of being where the soul no longer desires anything at all. However other Hindus believe that the soul and God are eternally distinct and that any merging is only apparent. The individual soul is compared to a green bird that enters a green tree (God). It appears to have merged but retains its separate identity. Liberation means entering into God's presence.

Each Hindu also has four daily duties – to show respect to the deities; to respect ancestors; to respect all beings; and to honour all humankind.

Tasks

Writing tasks	a) Explain Hindu beliefs about karma and reincarnation.
	b) 'The Hindu belief in reincarnation makes no sense in the modern world.' Assess this view.
Research task	Use the internet to find evidence to support the belief in reincarnation and use it in a class debate: 'There is no such thing as reincarnation.'

Glossary

karma	the theory of cause and effect, action and reaction
punya	actions which lead to good karma (merit)
papa	actions which lead to bad karma (sin)
reincarnation	the transmigration of the soul from one body to another
dharma	living in an appropriate way
artha	the pursuit of wealth through lawful means
kama	delight of the senses
moksha	release from rebirth

Section 2

Some key beliefs and practices

Aim of the section

This section focuses first on some of the key beliefs found within Hinduism and how they affect the daily lives of Hindus. During discussion of the key teachings it is important to consider how you might explore and illustrate them, and how you can express them in your own words.

Second, it explores ways in which Hindus practise their beliefs, focusing on the significance of festivals and some of the practices associated with worship. It will be important for you to consider how these practices reflect Hindu belief.

During your study of this section, remember that Hinduism is a diverse religion containing numerous different beliefs and practices

Chapter 5

The Four Varnas and the Status of Dalits

Aim

In this chapter you will learn about the varna system within Hinduism and its effect on Hindu society. You will also learn about the lives of those placed outside the system – the Dalits. You will have to show critical understanding of the system and be able to react in a balanced way to its strengths and weaknesses.

The word 'varna' literally means 'kind', although it is sometimes translated as 'colour' and is sometimes referred to in English as 'caste'. The term caste comes from the Portuguese meaning 'breed or race'.

The origin of the varna system can be traced to the Aryan invasion of India in the second millenium BCE. They devised a class system to organise the new society created by their arrival. Initially they created a system of three varnas, later expanded to include a fourth – the Shudra. The initial purpose of the varna system was to create and maintain rigid social boundaries between the invaders and the previous inhabitants. The top three varnas were the invaders and the fourth consisted of the Dravidian inhabitants.

The four varnas are called :

VARNA	OCCUPATION	STATUS
Brahmin	Priests/religious officials	Aryan/Twice-born
Kshatriya	Rulers/warriors	Aryan/Twice-born
Vaishya	Farmers/merchants/craftsmen	Aryan/Twice-born
Shudra	Servants of upper castes	Non-Aryan/not Twice-born

Diagram of varna system

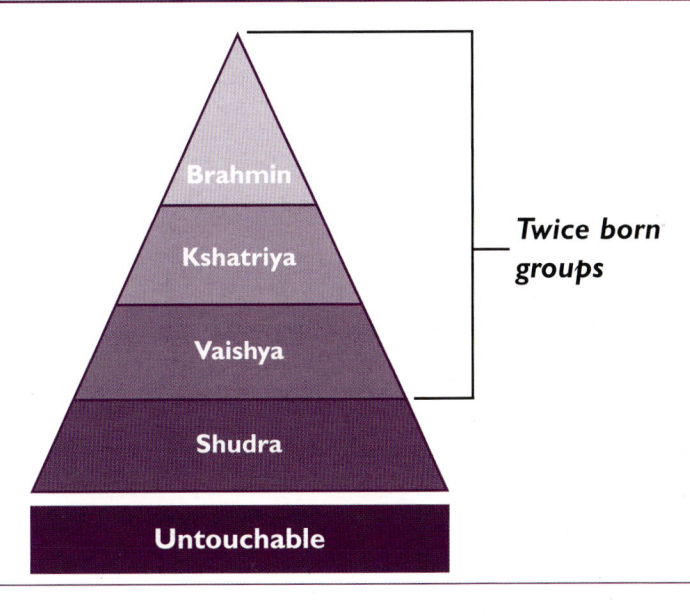

Brahmin
Kshatriya — Twice born groups
Vaishya
Shudra
Untouchable

For reflection

How would society function successfully if there were no orderly divisions within it?

Chapter 5

Hindus would also point to the religious origin of the varna system which is its divine justification. The Rig Veda refers to the four principal varnas. In a hymn the classes are compared to the body of a huge primeval man called Purusa:
'The Brahmin was his mouth, of both his arms was the Kshatriya made. His thighs became the Vaishya, from his feet the Shudra was produced.' (RV 10:90:12)
In the hymn the word 'varna' is not used and it is the only hymn in the Rig Veda in which the words Vaishya and Shudra are used.

Since the Brahmin came from the creator's mouth, they are the purest. Since the Kshatriya emerged from his arms, they are strong and meant to be soldiers. The Vaishya, because they came from his abdomen/thighs, are meant for craft and responsible for keeping society's stomach full. The Shudra, because they emerged from his feet, are considered as an impure or dirty part of a person's body and are therefore meant to be menial workers.

The Rig Veda also stressed that caste dharma needs to be strictly followed, as the soul climbs the class ladder with every birth. A soul is born into a varna as punishment/reward for its karmic influences. People are also born into the varna of their parents and cannot change varna during their lifetime.

Each varna is divided into a number of castes called jati, which provide a social hierarchy within a varna. The varna system serves a number of important functions :

1. It assigns occupations – The varna and jati to which a person belongs is usually linked to an occupation. Within the Vaishya there are jatis of bakers, sheep herders, metal workers etc.
2. The system separates the different varnas according to purity and impurity.
   The higher a person is in the system, the higher level of purity they must maintain. The lower a person is, the more likely they are to transmit impurity.
   The four areas where the purity restrictions appear most frequently are: marriage; drink; food; and touch. Marriage is only possible between members of the same varna. A touch by a Shudra on a Brahmin requires the Brahmin to undergo extensive rites of purification.
3. It was believed that the system was central in ensuring social harmony and order.

The top three varnas have a status which the fourth does not have: this is the status of being 'twice-born', which means that the Vedic religion applies only to them. 'Twice-born' refers to the rite of initiation which members of these varnas go through when they reach maturity. This rite brings them into the religion, where they are reborn as a Hindu and not just as a varna member. The Shudras, however, though are excluded from worship in the Vedic religion and are not even permitted to hear the Vedas read out loud. They therefore have their own priests and religious rites.

The varna system has been very stable in India for over two thousand years. It is only in recent times, after the state of India was formed, that the system has come under scrutiny. It is now forbidden by law, but many of the practices and attitudes are still strongly ingrained in Hindu society. There are many reasons for this. Rules for social behaviour expect Hindus to fulfil duties associated with the varnas. Also in the system of samsara and reincarnation, rebellion against class expectations will result in a lower rebirth in the next life.

Seminar topics

Is the varna/caste system fair? What are the advantages and disadvantages of the system?

'There is a class system in every society.'

Dalits

A Dalit, formerly called 'untouchable', is a person outside the four varnas and considered below them. When the Aryan conquest of India became more widespread and more people came into their society, the Aryans created a new class - the Untouchables. This new class was put outside the caste system altogether – they were outcastes. The purity requirements were such that not even the Shudras would relate to them and they were given occupations regarded as ritually impure, such as handling dead bodies, disposing of refuse or human waste. These activities were considered to be polluting to the individual who performed them and this pollution was considered to be contagious. Therefore Untouchables were banned from fully participating in Hindu religious life. They were considered unfit to have any social contact with the 'pure' sections of society.

The Untouchables suffered from extreme social restrictions. They used to live outside the limits of villages and townships, not allowed temple worship or water from the same sources. They were not allowed any contact with people from higher varnas. If a member of a higher varna came into physical or social contact with an untouchable, the member of the higher varna was polluted and had to wash repeatedly and engage in long and rigorous rituals to cleanse himself of the impurity. This was done even if the shadow of an Untouchable fell on a member of a higher varna.

Organised opposition to the caste system began with the Bhakti movement. Ghandi called the Untouchables 'Harijans' – the children of God, and wanted to place them

A Dalit family

Chapter 5

within the fourth varna. The term Dalit is now used, as the term Harijan is regarded as patronising. Another important reformer was Ranji Ambedkar, who was an Untouchable who succeeded in attaining a scholarship to study law. He disagreed with Ghandi over the future status of Untouchables and wanted a classless society. He was one of the main architects of the new Indian constitution of 1950, which outlawed untouchability and gave equal status to all citizens.

However, in practice caste differences continue, particularly in rural areas. In extremely traditional villages Dalits are still not allowed to let their shadow fall upon Brahmins for fear of polluting them and are still required to sweep the ground where they walk, to remove the contamination of their footsteps. In urban areas the concept of a strong caste system no longer exists, though most Indians still hold on to their varna origins. Positive discrimination is securing the Dalits more equal opportunities.

One religious question concerning Dalits is whether they can be regarded as Hindus. Traditionally Dalits have been barred from many activities central to Vedic religion and Hindu practice. Many Dalits continue to debate whether they are Hindu or non-Hindu. Some have succeeded in integrating into urban Indian society where class origins are less obvious and less important in public life. In rural India it is not as easy, because class origins are more apparent and Dalits remain excluded from local religious life. Because of this, and because many Dalits nevertheless feel the need for a formal religion, they have been drawn to other religions such as Buddhism. By moving away from a religious environment where they are left out and insulted, they have improved their social and economic standing.

Tasks

Writing tasks	a) Explain the concept of varna in Hinduism. b) 'Dalits are no longer discriminated against.' Assess this view.
Writing and presentation task	Write a diary entry for one day in the life of a Dalit and present it to the rest of the class

Glossary

varna	a social category
Brahmin (priests)	a member of the first group in the varna system
Kshatrya (warriors)	a member of the second group in the varna system
Vaishya (merchants)	a member of the third group in the varna system
Shudra (workers)	a member of the fourth group in the varna system
jati (caste)	sub-division within each varna, usually linked to occupation
Dalit	person outside the varna system who used to be called Untouchable

Chapter 6

Varnashramadharma

Aim

After studying this chapter you should be able to show knowledge and understanding of the role played by varnashramadharma in the daily lives of many Hindus. You should also be able to assess critically the importance of this concept within Hinduism.

Hindus believe that the universe is ordered and that each person has a role to play within it. If people fulfil their roles, then the universe operates harmoniously. However if they act against or outside their given role they threaten cosmic order. Their role is their duty or dharma. Dharma is roughly translated as 'religious duty' and in practical terms refers to leading a righteous life.

Varnashramadharma defines duties for the individual according to their class (varna) and by their stage of life (ashrama). Each varna and ashrama has its own specified dharma and what may be acceptable to one section of society might be totally unacceptable to another. Absolute non-violence is essential to a member of the Brahmin class, but unacceptable to a member of the Kshatriya (warrior) class; generating wealth and having children is essential for someone in the householder stage of life, but totally unacceptable for those in the retired stage.

Many people wrongly believe that caste and varnashramadharma are synonymous. Mahatma Ghandi, the most famous opponent of caste abuse, actually believed in the original principle of varnashramadharma. The system of four varnas was based on mutual support and service, allowing for upward and downward mobility. Krishna teaches in the Bhagavad Gita that people are allocated to a specific varna according to two criteria – gena/personal qualities and karma/aptitude for a type of work. He makes no mention of varna being determined by birth. This makes the original varnashramadharma different from the current caste system, which is based on descent. It is this hereditary nature that is the main difference between them.

The four varnas and dharma

Every Hindu must follow general moral codes and each has duties according to his or her varna.

Shudras (workers)

Only Shudras are able to be employed by others. The other varnas must be self-employed and financially self-sufficient. The Shudras' duties are:

- to render service to others
- to take pride in their work and be loyal
- to follow general moral principles
- to marry (the only compulsory rite of passage).

Shudra blacksmiths

Chapter 6

Vaishya shopkeeper

Vaishyas (farmers, merchants and business people)

This is the productive class. They are twice-born, indicating that they accept the sacred thread as spiritual initiation and must perform certain rituals and rites of passage. Their duties are:

- to protect animals (especially cows) and the land
- to create wealth and prosperity
- to produce food, clothes etc. for the workers
- to trade ethically
- to give taxes to the Kshatriya (ruling class).

Kshatriyas (warriors, police, administrators)

These are the nobility, the protectors of society, who are expected to display strength of body and character. They are also twice born. Their duties are:

- to protect citizens from harm, especially women, children, cows, brahmins and the elderly
- to ensure that others perform their dharma and move on spiritually
- to be first into battle and never surrender

Kshatriya policemen

- to keep their word
- to accept all challenges
- to develop noble qualities such as power, chivalry and generosity
- to levy taxes from the Vaishyas and never accept charity
- to take advice from the Brahmin
- to know the scriptures
- to deal strongly with crisis and lawlessness
- to ensure an heir.

Brahmin (priests, teachers and intellectuals)

These are the providers of education and spiritual leadership. Their basic needs are fulfilled, so that they can dedicate themselves to their spiritual tasks. Their duties are:

- to study and teach the Vedas
- to perform sacrifices and religious ceremonies and to teach others how to perform rituals
- to accept and give charity
- to offer guidance, especially to Kshatriyas
- to provide medical care and advice free of charge
- to never accept paid employment
- to develop ideal qualities – honesty, integrity, cleanliness, purity, knowledge and wisdom.

Brahmin priests

> **Seminar topics**
>
> *'Varnashramadharma leads to a well ordered society.'*
>
> *'Following varnashramadharma ensures that everybody within society is valued.'*

The four ashramas and dharma (ashramadharma)

Hinduism recognises four main stages of life called ashramas – the student stage; the householder stage; the retired person stage; and the ascetic stage, also known as a sannyasa. Each of the four ashramas has its own specific duty or dharma.

The student stage (brahmacharya)

During this stage a boy is traditionally expected to live away from home and study with a teacher (guru) for several years, to foster spiritual values. Today only a few Brahmin families follow this tradition to the full extent. A boy enters into student-hood at adolescence and spends most of his maturing years studying. This stage begins for members of the three upper castes after the ritual of the sacred thread, when they are reborn. This ceremony symbolises the entrance of the boy into Hinduism. Originally it was at this point that he would be first permitted to hear the words of the Vedas and to learn his first mantra.

Duties would include:

- studying the Vedas and other texts
- living a celibate and simple life
- serving the guru and collecingt alms for him
- learning how to set up and maintain household worship
- developing appropriate qualities such as humility, discipline, simplicity etc.
- understanding and performing various rituals.

The student stage (brahmacharya)

The householder stage (grihastha)

This stage is usually entered into when a Hindu decides to marry and accept family responsibilities. During this stage a man has children, forms a family, establishes himself in a career or job, and becomes an active member of his community. He is also expected to ensure that rituals of domestic life are carried out at their proper time and in a proper way..

Duties would include:

- making money and enjoying pleasure in an ethical manner
- performing sacrifice and observing religious rituals
- protecting and nourishing family members
- teaching children spiritual values
- giving to charity.

The householder stage (grihastha)

Chapter 6

The retired person stage (vanaprastha)

The retired person stage (vanaprastha)

When a man reaches old age and his son has a family and is ready to take over the leadership of the household, he and his wife will retire. Their household responsibilities diminish significantly and they are free to meditate on the meaning of death and rebirth. Some choose to withdraw into a secluded area, perhaps become a hermit, or they may involve themselves more with bhakti of a god or goddess. Some go on a pilgrimage, when they may be accompanied by their wife, but all sexual relations are forbidden.

Duties would include:

- devoting more time to spiritual matters
- going on pilgrimage.

The ascetic stage (sannyasa)

This is the fourth stage of life. Traditionally it is only available to men who exhibit the qualities of a Brahmin. The Sannyasin become wandering hermits, leaving their family and living a life dependant on God alone. They are conspicuous in their saffron dress, eating when they acquire food but never working to get it - food must be given or found. They seek spiritual enlightenment and power.

Duties would include:

- controlling the mind and senses, fixing the mind on the Supreme
- becoming detached and fearless, fully dependant on God as protector
- becoming aware of the self and of God.

The ascetic stage (sannyasa)

Tasks

Writing tasks	a) Explain the concept of varnashramadharma in Hinduism.
	b) 'Varnashramadharma is the most important concept in Hinduism.' Assess this view.
Writing and presentation task	Imagine that you belong to different varnas and are in different ashramas in life. Present your varnashramadharma to the rest of the class.

Glossary

Bhagavad Gita	Hindu scripture
varnashramadharma	duties of the individual according to his varna and stage in life (ashrama)
brahmacharya	the student stage of life
grihastha	the householder stage in life
vanaprastha	the retired person stage in life
sannyasa	the ascetic stage in life

Chapter 7

Rites of Passage

Aim

After studying this chapter you should be able to show knowledge and understanding of the key rites of passage within Hinduism and how they reflect Hindu beliefs. You will also have a critical perception of their value within Hinduism.

Hindu rites of passage are called 'samskaras'. They are not merely formalities or social observances, but serve to mark the various stages of life and to purify the soul at these critical junctions in life's journey. They are considered essential for members of the three highest varnas. Most traditions say that there are a total of 16 samskaras.

Four sets of rites are more popular than the others:

1. **Jatakarma** – birth ceremonies
2. **Upanayana** – initiation, Sacred Thread ceremony
3. **Vivaha** – marriage
4. **Antyeshti** – funeral and rites for the dead

Birth ceremonies

The first five samskaras of a Hindu's life take place before and after a child's birth. The first samskara, called purification of the womb, takes place before the child is conceived, when the Hindu couple pray for the kind of child they hope for. The second samskar takes place during the third month of pregnancy and the third during the fifth month. It is believed the second and third samskaras protect the child and gives strength to the mother.

Shortly after the birth, the fourth samskara takes place. The Jatakarma ceremony welcomes the baby into the world. The baby is washed and the father places a small amount of ghee and honey on the baby's tongue and writes the holy syllable 'Aum' on his tongue with a golden pen. This signifies the hope of long life. He also whispers the name of God in the child's ear, signifying intelligence and wisdom. He also touches the baby's shoulders while reciting a vedic verse, signifying strength.

The Jatakarma ceremony - the fourth samskara

Chapter 7

A boy's first haircut - the ninth samskara

On about the eleventh day after birth the child receives a name. This is the fifth samskara – Namakarana. It is a religious and social occasion. Rich Hindus, especially in cities, invite many relatives and friends to witness the ceremony and then enjoy a lavish meal afterwards. To a Hindu, the name chosen is very important, since it can bring good luck to the child throughout life. The child receives two names – the public name that is known to everyone and the private name that is used on special religious occasions.

In strictly religious families the naming ceremony is led by the priest. Usually the priest will draw a horoscope according to the time, day, month, place, and the position of the stars and planets on the day the baby was born. According to this horoscope he will suggest a number of suitable names. The baby is placed on his mother's lap, as she sits on the right of her husband. In front of them rice grains are spread on a metal plate. Then the father, with a gold ring or piece of gold wire, writes the name of the family deity, the date of birth of the child and the name chosen. The baby receives the name in a very simple way: the father whispers the name in the baby's right ear, then recites a number of mantras which ask for strength, intelligence and wisdom for the child. In other families the ritual is much more informal and very often attended only by women.

Other childhood samskaras

There are a number of important childhood samskaras which are carried out by many Hindus:

The sixth samskara
This is the child's first outing and takes place in the third or fourth month after birth. The baby is carried to see the sun for the first time after receiving its name. To Hindus the sun is very important since its power is necessary for life. The child is only shown to the sun for a few seconds, so that no harm can come to it. The purpose of the ritual is for a child to become aware of its surroundings.

The ninth samskara
This is a boy's first haircut. It usually takes place after his first birthday and is regarded as an important celebration. The priest usually leads the ceremony. The hair is shaved, apart from a small tuft which is left at the front. The shaved hair is then buried or put into a stream. This is to ensure that a child begins his life clean, deleting any evil influences from past lives.

Upanaya – the Sacred Thread ceremony
Of the first ten samskaras, the Sacred Thread ceremony is the most important. It is a ritual for boys of the three highest varnas only – Brahmin, Kshatriyas and Vaishyas. Usually, however, it is only boys who belong to the highest varna or have been born into a rich family who take part.

The ceremony can take place from nine years of age onwards. Once it has been completed the boy will be regarded as being born again and having come of age. During the ceremony the priest and the boy sit near the fire, which has been consecrated to Agni (god of fire). Hymns and prayers are chanted. When the thread has been placed over his left shoulder and under his right arm, the boy will repeat the prayers of the priest. Before the end of the ceremony the boy will receive a personal mantra, which is to be recited every time he prays.

The boy will wear the thread for the rest of his life. A special ceremony is held each year to change the thread for a new one, which is always put on before the old one is removed, so as to ensure that the boy is never without it. Often men meet on the banks of the Ganges to wash and change the thread.

The thread itself is made of wool or cotton and there are three strands: white, red and yellow. It is tied with a special knot known as 'Brahma Granthi', a spiritual knot. The purpose of the knot is to remind the boy of three duties during his life:
- to God for everything that sustains him throughout life
- to his parents for giving him life and teaching him about Hinduism
- to his guru (religious teacher) for giving him information and wisdom.

The three strands can also represent the Trimurti.

> **For reflection**
>
> *How important have any rites of passage been in your life?*

Chapter 7

Upanaya – the sacred thread ceremony

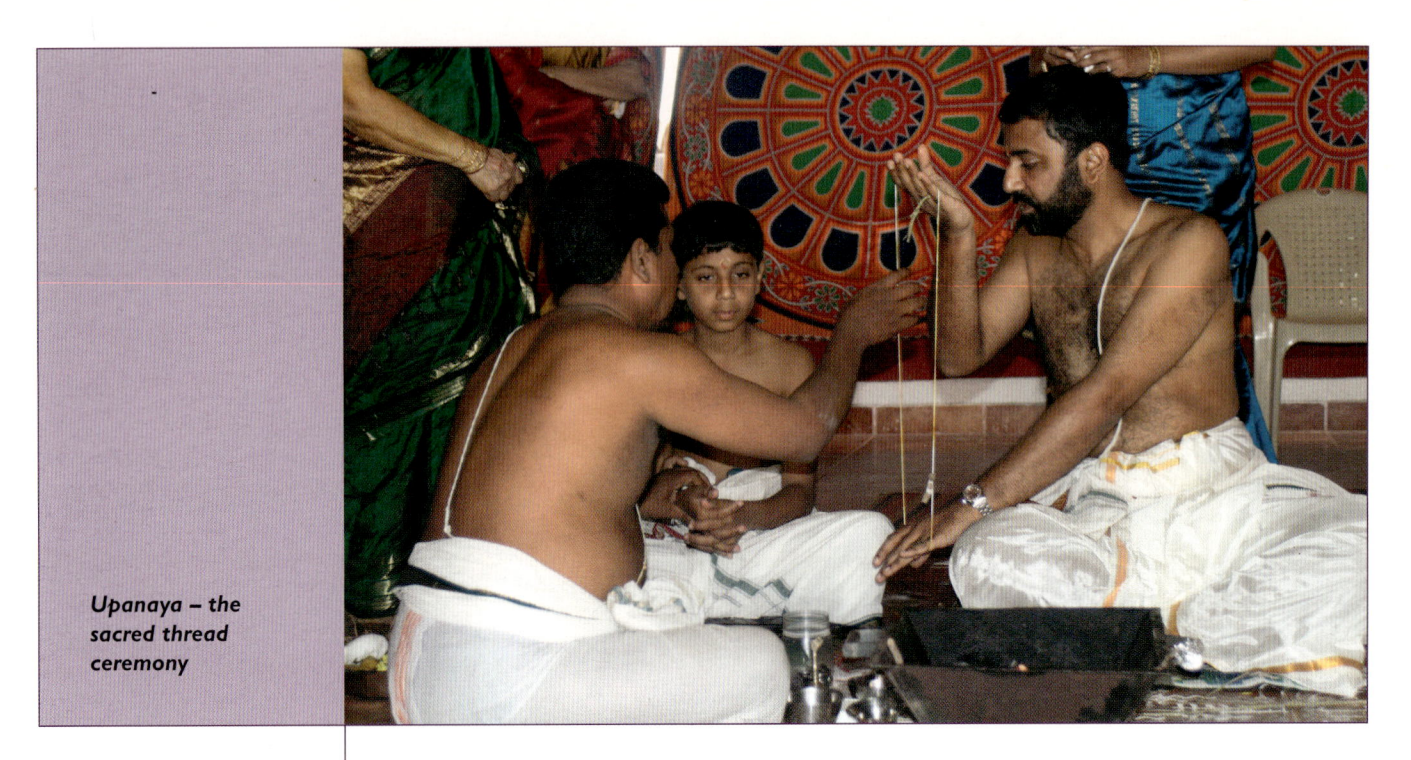

To a Hindu, receiving the Sacred Thread is a very important milestone in the life of a boy of the highest varna. After the ceremony the boy will be expected to pray three times a day, perform puja ceremonies and read, study and teach the holy scriptures.

In the past, after receiving the Sacred Thread a boy used to travel around the country, depending on others for food and shelter. Today the parents present him to a guru for spiritual training. It is the beginning of his formal education.

Vivaha - marriage

Marriage is perhaps the most important samskara. It is a duty of the higher varnas in Hinduism, since secure family life is regarded as being of the utmost importance. Traditionally marriage is regarded as being for life, or until the husband takes to the path of renunciation. Divorce used not to be allowed and those who left their partners were often shunned by society.

Marriage in Hinduism is basically an arrangement between two families and unites the families as well as the individuals. When a son/daughter reaches the age to marry, the usual practice is for the elders of the family to look for a suitable partner. Very often advertisements are placed in newspapers. Many factors are taken into consideration when assessing the suitability of a prospective partner – wealth, occupation, their respective horoscopes – but the most important is class, and marriage is usually between members of the same varna. Marriage outside of the religion is something which is not encouraged.

Marriages based on love are still the exception. Today, however, the couple are allowed to meet and if they are happy with the choice then preparations are made for the marriage. If not, then the whole process will begin again. It is said that a Hindu youth does not marry the girl he loves, but loves the girl he marries. Despite modern and western attitudes towards this practice, statistics suggest that these marriages work relatively well and that the divorce rate in such marriages is lower.

Bride and groom in traditional clothes and decoration for their marriage ceremony

In Hinduism marriage customs and traditions vary enormously. Usually the couple become engaged about a month before marriage and the bride's family pays for the wedding. In some areas the bride and groom will be anointed with oil and turmeric for several days before the ceremony. Puja will include not only the family deities but also Ganesh, the deity of good fortune. The bride often offers prayers to Parvati and Shiva and asks for blessings of prosperity, long life, health and children (especially sons).

Marriages are usually held in the home or in a hall, in a specially erected booth or canopy called a 'Mandap'. The bride wears a red sari, decorated with gold.
As already stated, there is much regional variation, but certain features are common:

1. **Welcoming the bridegroom**

    When the bridegroom arrives at the place of the wedding with his parents, he is welcomed by the bride's mother. The bridegroom's face is covered with decorations, which is associated with ancient practice when the couple used to see each other for the first time on the day of the marriage. The bridegroom is then welcomed in a worshipful manner by the bride's father, because on this day the groom is treated as a god and the bride as a goddess. The bride's father then gives him a mixture of water, yoghurt and honey to sweeten the welcome. It is also a symbol of hospitality, and hope that their future lives together will be full of pleasure.

2. **The daughter being given in marriage**

    This is the formal giving away of the bride by her father to the groom and his father. Hindus believe that one of the greatest gifts a man can give to another is the gift of a virgin daughter in marriage. If he does, he will obtain good karma or merit. It is also a way of showing publicly that the marriage is taking place with the parents' consent and blessing. The couple and their parents stand facing each other, holding hands, and take a vow of friendship.

Chapter 7

3 **Sacred Fire Ceremony**

The couple sit in front of the sacred fire and the groom makes offerings of darbha grass, sacred woods and ghee, while mantras and vedic verses are recited. As he makes these offerings the bride touches his right hand. The darbha grass is offered to the fire and is meant for the god Skanda (son of Shiva). Sacred woods are offered to Agni, the god of fire, as is most of the ghee. The bride then puts rice in the hand of the bridegroom, who throws the rice into the fire. This is done four times, while the priest recites words which stress that the woman is leaving her parents and joining the family of her husband, and has accepted this way of life.

4 **Holding Hands**

The bridegroom takes the bride's right hand and repeats a mantra after the priest. The mantra refers to the hope that they have good fortune and grow old together as man and wife. He also vows to carry out his duties as a married householder.

5 **Circumambulation of the sacred fire**

The bride's wrist is tied with a thread and she places her foot three times on the groom's family's grinding stone. This represents being faithful and is the action which seals the marriage. A white cord is used to tie the bride's sari to the groom's scarf and they walk around the sacred fire, to show that they accept each other in understanding and love. On the last round the bride moves from sitting with her own family to sit with her new in-laws.

6 **Marking the bride's hair with kumkum**

The bridegroom puts colour in the hair of the bride, as a sign that she is to be regarded as a married woman.

7 **Taking seven steps together**

To many Hindus, this is the most important part of the marriage. More ghee is poured on the fire to make it glow. According to the Scriptures the ritual should be performed to the north of the fire in a straight line. Usually the priest makes seven small heaps of rice, about 30cm apart.

The bridegroom puts his right hand on the bride's right shoulder and they take seven steps. After each step they stop to recite the seven vows:

a) The first step, follow me in my vows and let God lead you;

b) Take the second step for strength;

c) Take the third for wealth;

d) Take the fourth for joy;

e) Take the fifth for children;

f) Take the sixth for pleasure together;

g) Take the seventh for a lifelong union of friendship.

8 **Viewing the Pole Star**

If the ceremony is in the evening, the couple might go out to see the Pole Star. The bride will promise to be constant like the star within her new family and not to stand in the way of the good and just actions of her husband.

9 ***Blessings***

   The ceremony ends with prayers for good fortune and peace. The couple are blessed by the priest for a long and prosperous married life.

The couple have now reached a new position in their lives. As a married woman the girl will have to fulfil her religious duties in the home by:

- praying in front of the home shrine;
- making offerings of food to the family deity and sharing the food with the family as Prasad;
- teach the children about the family religion.

The bridegroom also accepts the responsibility of being a good husband – faithful and supportive. These are duties associated with the second ashrama.

Funeral and Rites for the Dead

Funeral rites are followed by most Hindus and follow similar patterns. Most Hindus cremate their dead: the exceptions are small children and saints, whose bodies are considered pure and are therefore buried, and also sannyasins who have no relatives to perform the cremation rites. Cremation is chosen because it enables the soul to abandon its attachment to its previous body and move quickly forward to the next. Funeral ceremonies are therefore performed as soon as possible. In India this means within hours of death, but elsewhere it usually takes longer.

The corpse is washed, anointed with sandalwood paste, wrapped in a cloth (white for men and red for women), and then carried to the cremation ground in a procession led by the eldest son, who also lights the funeral pyre. Funeral processions move as quickly as possible, chanting the name of God. Funeral pyres are often built by rivers, so that the ashes can be scattered in them later on. On the funeral pyre the corpse is placed so that the feet point to the south, towards the home of Yama, god of the dead, and the head towards the north, towards the home of Kubera, Lord of riches. Sometimes a pot is broken by the body's head, as a symbol that the soul is being released. After cremation the youngest member of the family leads the procession home.

Seminar topic

All societies have, and need, rites of passage of some kind.

A funeral pyre

After the funeral there is a period of mourning, lasting about thirteen days, although this varies according to varna and other considerations. During this time the family are considered impure. They will not attend religious functions or eat certain foods such as sweets. This period gives people a chance to come to terms with their loss and to grieve, so that they can then carry on with their lives. However, these rites are more for the benefit of the deceased than of the bereaved. They are thought essential to ensure the smooth passing of the soul to a higher level of existence. The most important are the Sraddha ceremonies, where balls of cooked rice are offered to God and the deceased, to sustain their body in the afterlife. This ceremony also takes place on the first anniversary of death.

When the period of death rites is over, the house has to be thoroughly cleaned and all the linen is given to be washed. The household deities are only returned when this has taken place.

Tasks

Writing tasks	a) Explain the significance in Hinduism of two key rites of passage. b) 'Modern life is so changeable and confusing that rites of passage have no meaning.' Assess this view.
Research and presentation task	Use the internet to find out more about different practices within any one rite of passage and use a Powerpoint presentation to present your findings to the rest of the class.
Writing task	Write a newspaper report on one Hindu rite of passage.

Glossary

samskaras	Hindu rites of passage
Jatakarma	birth ceremonies
Upanayana	Sacred Thread ceremony
Vivaha	marriage ceremony
Antyeshi	funeral and rites for the dead
Namakarana	fifth samskara when a child receives a name
Skanda	son of Shiva
Agni	god of fire
Yama	god of the dead

Chapter 8

Gods and Goddesses

Aim

In this chapter you will learn about seven gods and goddesses – Rama, Ganesh, Lakshmi, Kali, Durga, Parvati, Sita – together with some of the symbolism and stories associated with them, and will be able to critically assess their importance to Hindus.

As we have already seen in chapters 1 and 2, contrary to popular understanding, Hinduism recognises one God, Brahman, who is the cause and founder of all existence. All the other gods and goddesses of Hinduism are different manifestations and expressions of Brahman and different ways of approaching him. Many Hindu communities have their own divinities whom they worship. It is impossible to say how many Gods and Goddesses there are within Hinduism, as some Gods are worshipped under different names in different parts of India. There are also local deities which are not known on a national level. Yet in spite of this, the majority of Hindus would not refer to themselves as being polytheistic. One reason for this, as we have already noted, is that all the gods are manifestations of the Ultimate, Brahman, but many Hindus also distinguish between major and minor deities. Hindus usually worship one major deity and only worship the other deities on certain occasions. Major deities are those who are regarded as being Brahman and are referred to as 'Isvara' – Lord. The minor deities are regarded as just divine beings, and are called 'deva' if they are male and 'devi' if they are female.

Rama

Rama is a great favourite amongst Hindu deities and has many followers. He is regarded as being the perfect avatar of the Supreme Protector, Vishnu, and a popular symbol of chivalry and virtue – 'the embodiment of truth, of morality, the ideal son, ideal husband and ideal king.' He is the seventh incarnation of Lord Vishnu, born to destroy the evil forces of the age, and his exploits form the Ramayana, the great Hindu epic. To many Rama is not very different in looks from Lord Vishnu or Krishna. He is most often represented as a standing figure, with an arrow in his right hand, a bow in his left and a quiver on his back. He is also usually accompanied by his wife Sita, brother Laksmana, and Hanuman, the

Rama, with Sita and Laksmana

monkey king. Rama is depicted in princely adornments with a 'tilak' or mark on the forehead. His affinity with Vishnu and Krishna is shown by his dark, almost bluish, complexion.

Chapter 8

The Ramayana

The Ramayana is one of the two great Indian epics, telling about life in India around 1000 BCE and offering models in dharma. Rama, the hero of the epic, lived his whole life by the rules of dharma and that is why Hindus regard him as a hero and teach their young people to 'be as Rama' or 'be as Sita.'

Prince Rama was the oldest of the four sons of King Dasaratha of Ayodha and was chosen by his father as his heir. However his stepmother, Kaikeyi, wanted to see her son Bharata, Rama's younger brother, become king. She reminded the king that he had once promised to grant her any two wishes she desired and she demanded that Bharata be crowned and Rama banished to the forest for fourteen years. The king had no choice but to grant her wish and ordered Rama's banishment. Rama accepted the decree unquestioningly and when Sita, Rama's beautiful wife, heard he was to be banished, she begged to accompany him to his forest retreat. His brother, Laksmana, also went with him, leaving behind their riches to live a simple life.

When Bharata learned what his mother had done, he went after Rama to the forest and pleaded with him to come back and take his place as king, since the eldest must rule. However Rama refused to go against his father's command. Therefore Bharata took his brother's sandals and told him that he was going to place them on the throne as a symbol of his authority and that when the fourteen years of banishment were over he would return the kingdom to Rama. Rama was very impressed with Bharata's selflessness.

The years pass and Rama, Sita and Laksmana are very happy in the forest. Rama and Laksmana succeed in destroying the rakshasas (evil creatures) who were disturbing the sages in their meditations. One day a rakshasa princess called Surpanakha falls in love with Rama and tries to seduce him. Rama refuses her advances and Laksmana wounds her and drives her away. She flees to her brother, Ravana, the ten-headed ruler of the island kingdom of Lanka. She tells her brother, who has a weakness for beautiful women, about the beauty of Sita.

Ravana decides that he must have Sita and devises a plan to abduct her. He changes himself into a wandering holy man in order to find her in the forest. He sends a magical deer which Sita desires. Rama and Laksmana decide to go off to hunt the deer, but before leaving draw a protective circle around Sita, warning her that she will be safe as long as she doesn't step outside the circle. After they have left, Ravana appears as the wandering holy man and asks Sita for alms. The moment Sita steps outside the circle to give him food, Ravana grabs her and carries her off to his kingdom in Lanka.
When Rama returns, he is heartbroken to find the hut empty and Sita gone. He enlists the help of Hanuman, the monkey king, to help him find her.

Meanwhile in Lanka, Ravana cannot force Sita to be his wife and therefore puts her in his garden, where he tries to sweet-talk and threaten her, in an attempt to get her to marry him. Sita does not even look at him and thinks only of her beloved Rama. Hanuman has the ability to fly, since his father is the wind, and he flies to Lanka to find Sita in the garden. He comforts her by telling her that Rama will soon come to save her.

Rama defeating Ravana in battle

Sita is delighted to hear this, but Hanuman is caught and Ravana orders his men to wrap Hanuman's tail in cloth and set it on fire. With his tail on fire, Hanuman hops from roof to roof, setting Lanka alight. He then flies back to Rama to tell him where Sita is.

Rama, Laksmana, Hanuman and his monkey army build a causeway from the tip of India to Lanka and cross over. A mighty battle takes place and Rama kills several of Ravana's brothers. He then confronts the ten-headed Ravana and kills him. He frees Sita but does not accept her unreservedly: she has to prove her chastity, after having lived in the house of another man. When he asks her to undergo the test by fire, she agrees and proves her purity by remaining unscathed in the fire. They return to Ayodhya and Rama becomes king. His rule, Ram-rajya, is regarded as an ideal time, when everyone followed his or her dharma.

Characters and incidents in the Ramayana provide the ideals and wisdom of common life and help to bind the people of India, regardless of class and language. The obedience of Rama to his father's decree; the devotion and love of Sita; the brotherly faith of Laksmana; the loyalty and allegiance of Hanuman; and the great courage of Rama are all examples to be followed.

Two of the greatest Hindu festivals are directly linked to the Ramayana – Dussehra commemorates the siege of Lanka and Rama's victory over Ravana, while Diwali, the festival of lights, celebrates Rama and Sita's homecoming to their kingdom in Ayodha.

> **Seminar topic**
>
> *'Discuss what the Ramayana teaches about dharma'.*

Chapter 8

Ganesh

Ganesh is the son of Shiva and was appointed chief of the ganas (attendants) or lord of hosts by his father. There are many accounts of his birth and how he came to have the head of an elephant. According to one account, the goddess Parvati, the consort of Shiva, wanted a child. One day when Shiva had gone out to gather flowers on Mount Kailas, Parvati was alone in the house and went to have a bath. Before plunging into the water she rubbed her body with oil and from the dry skin that fell from her body she made a little boy and breathed life into him.

Parvati named the handsome boy Ganesh. She then asked him to guard the house and not to let anyone in until she had finished her bath. So Ganesh stood guard outside the house while Parvati took her bath. A few moments later Shiva returned, but he was refused entry by Ganesh, whom Parvati had already accepted as her son. Shiva did not know this and Ganesh did not know Shiva. He never suspected that it was his own father who wanted to enter the house. Shiva at first tried to persuade the little door-keeper and then threatened him, but Ganesh stood firm, faithfully keeping his mother's orders. Shiva was extremely angry and in his rage cut off the boy's head and forced his way into the house.

When Parvati came out of the bath and saw Ganesh lying murdered she was very angry with Shiva and insisted that he bring the boy back to life. Shiva promised to do so. He sent messengers to all parts of the earth with clear instructions to bring the head of the first living creature they found sleeping with its head turned towards the north. The first creature they found was a baby elephant, and Ganesh was given its head and brought back to life. This is also why Hindus do not sleep with their heads towards the north.

According to those who explain his appearance symbolically, Ganesh's trunk is bent to remove obstacles and his four arms represent the four categories (e.g. castes, varnas) into which people can be divided. This symbolises the several ways in which obstacles can be overcome to attain religious ends: the elephant tramples all in its path while the rat, his companion animal, creeps through narrow holes and cracks to achieve the same ends.

Ganesh is painted red in colour, although in some texts he is represented as a short fat man of yellow colour, with a protuberant belly, four hands and the head of an elephant with only one tusk. As a remover of difficulties and obstacles he is widely worshipped as a god of good luck, and prayers are offered to him on all auspicious occasions, at life-cycle rituals or at the beginning of a journey. As a god of learning and wisdom, he is adored by all seekers of knowledge.

Ganesh with his father Shiva and his mother Parvati

The Mother Goddess

Hindus worship the divine in female form. Every male deity has a female side which is called 'shakti'. While the male side tends to be passive in nature, shakti is usually an active energy. Many goddesses are worshipped in their own right and not merely as consorts of the gods. It is up to individual worshippers to decide which aspect of a deity they worship. Although 'devi' is the word for goddess, a more popular name is 'mata', which means mother.

Lakshmi

Lakshmi is the Goddess of wealth and prosperity, both material and spiritual, and is the wife or consort of the god Vishnu. In this role she plays the part of a model Hindu wife, obediently serving her husband as lord. In her many incarnations she appears as the wife of the Vishnu avatars e.g. when Vishnu became Rama, she was Sita. She sprang with other precious things from the foam of the ocean when it was churned up by the gods and demons for the recovery of the Amrata (the drink of immortality). She is also the embodiment of beauty, grace and charm.

The name Lakshmi comes from the Sanskrit word 'Laksya' meaning 'aim' or 'goal'. She is the household goddess of most Hindu families and is a great favourite amongst women and businessmen. Lakshmi is depicted as a beautiful woman of golden complexion, with four hands, sitting or standing on a full-bloomed lotus and holding a lotus bud, which stands for her beauty, purity and fertility. Her four hands represent the four ends of human life – dharma or righteousness; karma or desires; artha or wealth; and moksha liberation from the cycle of birth and death. Gold coins cascade from Lakshmi's hands, suggesting that she brings wealth to those who worship her. She always wears gold embroidered red clothes, the red symbolizing activity and the gold lining prosperity.

On the full moon following Durga Puja, Hindus worship Lakshmi ceremonially at home, pray for her blessings and invite neighbours to attend the puja. It is believed that on this full moon night the goddess herself visits the homes and replenishes the inhabitants with wealth. A special worship is also offered to Lakshmi on the Lakshmi-puja day during the Diwali festival.

Lakshmi

Chapter 8

Kali

Kali

Kali, as the Divine Mother, has a destructive and creative aspect. She is seen as the destroyer of evil spirits and the preserver of devotees. Kali is a consort of Shiva and has an awful, frightening appearance. The name Kali means 'black' and she is usually depicted as being very dark in appearance, usually naked, with long untidy hair, a skirt of severed arms, a necklace of freshly cut heads, earrings of children's corpses and bracelets of serpents. To add to her dreadful appearance she has long, sharp fangs, purple lips and red eyes. Her tusk-like teeth descend over her lower lip and her tongue lolls out. She has claw-like hands, with long nails, and her body is smeared with blood.

Kali is often shown standing on the body of her consort Shiva. In one of her ten arms she carries a sword and in another the head of the giant she slew. These objects symbolize both her creative and her destructive power in the unceasing cycle of life and death, creation and destruction. She laughs loudly and dances madly, usually in cremation grounds, and drinks blood.

In many temples in India goats are still sacrificed to Kali. She is very popular in West Bengal, South India and Kashmir, where she is regarded as a loving, if unpredictable, mother who shows infinite tenderness to her devotees.

> **For reflection**
>
> **Why do you think Hindu goddesses are important to Hindu women? What qualities do they show?**

Durga

Durga

Durga is a consort of the god Shiva and the warrior form of Parvati. She is also closely associated with Kali. Her name means 'inaccessible' or 'fort' and she was created by the male gods to destroy the buffalo-demon Mahisha. He had succeeded in defeating the gods in battle. The gods were angry at this and out of the energy of their anger a beautiful woman was created. Each god gave her his weapon – Shiva gave her his trident, Vishnu his chakra, the wind god his bow and arrows, and another god his sword and shield. Armed with these weapons she faced Mahisha and killed him. This victory is celebrated as Dussehra, which literally means 'the tenth day'. However the worship may take place over the nine preceding days – Navaratri.

Durga is usually depicted as a woman riding a lion, holding many weapons. In sculptures she is shown standing over the demon Mahisha as she kills him. She is often described as the mother of the world and with her prominent breasts is associated with fertility and crops and plants. Durga is very popular, especially in Bengal, where Durga Puja is the major autumn festival.

Chapter 8

> **For reflection**
>
> *Do you think Hinduism gives a positive image of women?*

Parvati

Parvati means 'daughter of the mountain', which refers to her birth from the Himalayan mountain range. She is the second consort of Shiva, the Hindu God of destruction and rejuvenation, and is the gentler aspect or representation of Durga. She symbolises power: some believe that she is the source of all power in this universe and that Shiva gets all his powers from her. Sometimes she is depicted as half of Shiva and as his consort she balances his passive and reclusive nature. She involves him more in the world and is the perfect wife and mother. Parvati is not worshipped as an independent goddess.

> **Seminar topic**
>
> *'It is clear that Hinduism is not a monotheistic religion.'*

Sita

Sita is held in esteem as an example of the virtues expected in a woman and wife. According to Hindu belief, Sita is an avatara of Lakshmi, who reincarnated herself as Sita. She is regarded as a daughter of the Earth Goddess because she was discovered in a furrow in a ploughed field. Sita is a principal character in the Ramayana with her husband Rama and it is her actions throughout her relationship with Rama that make her an example which every young girl in India is raised to follow.

Tasks

Writing tasks	a) Describe stories and practices associated with Durga.
	b) Explain the popularity of Kali in Hinduism.
Research task	Use the internet to find out more about different beliefs and practices associated with the gods and goddesses.

Glossary

Ramayana	Hindu epic scripture
Laksmana	Rama's brother
Hanuman	the monkey god
Ravana	ten-headed ruler of Lanka

Chapter 9

Festivals

Aim

After studying this chapter you will be able to show knowledge and understanding of why and how festivals are celebrated, and of the main features of specific Hindu festivals. You will also be able to show an understanding of how these festivals reflect Hindu belief and tradition, and have a critical perception of their value within Hinduism.

Why celebrate festivals?

Festivals play a very important part within Hinduism and almost certainly there are more festivals in this religious tradition than in any other. Once again, because of the nature of the religion, there is much variety in terms of region, tradition and movements.

There are many reasons why festivals are important:

a) They are a way of remembering and celebrating key events, historical or mythical, within the tradition e.g. Diwali in northern India celebrates the events of the Ramayana. They are also a way of celebrating important times of the year.

b) They create a special atmosphere, when people can forget worldly concerns and problems and focus on spiritual matters.

c) They are joyful and happy occasions which give people a lift, making them feel good about themselves and the world around them.

d) They are a way of confirming and strengthening a person's faith, which helps them perform their daily duties.

e) They remind worshippers of their duties and goals in life.

f) They create a sense of belonging by bringing the community together in celebration.

How festivals are celebrated

There are a variety of practices used within Hinduism to celebrate the different festivals, including:

- fasting and feasting
- sharing food (prasad)
- giving to charity
- visiting the temple
- visiting relatives
- praising God through dance, drama etc.
- making images of deities
- processing with temple deities
- wearing new clothes
- decorating house, street and temple.

Seminar topic

'Taking part in a festival does not show any commitment towards a religion.'

Chapter 9

Hindu Festivals

As already stated, Hinduism is a religion of festivals. These are some of the many celebrated within Hinduism:

FESTIVAL	MONTH	DEITY	STORY
SARASWATI PUJA	JANUARY	SARASVATI	SARASVATI CURSES BRAHMA
MAHA SHIVA RATRI	FEB/MARCH	SHIVA	STORIES OF SHIVA
HOLI	MARCH	VISHNU (NARASIMHA)	PRAHLADA AND NARASIMHA (AND HOLIKA)
RAMA NAVAMI	MARCH/APRIL	RAMA	RAMAYANA – (ESPECIALLY RAMA'S BIRTH)
HANUMAN JAYANTI	APRIL	HANUMAN	RAMAYANA – (ESPECIALLY LATER EPISODES)
RATHAYATRA	JUNE/JULY	JAGANNATHA	THE PROUD MERCHANT
RAKSHA BANDANA	AUGUST	–	INDRA WEARS A RAKHI
JANMASHTAMI	AUG/SEPT	KRISHNA	KRISHNA'S BIRTH/CHILDHOOD
GANESH CHATURTHI	AUG/SEPT	GANESH	HOW GANESH RECEIVED HIS HEAD
NAVARATRI / DURGA PUJA	SEPT/OCT	SHAKTI/PARVATI	DURGA KILLS MAHISHA/ THE RAMAYANA
DUSSEHRA	OCTOBER	RAMA	RAMAYANA
DIVALI	OCT/NOV	LAKSHMI/RAMA	STORIES OF LAKSHMI/RAMA

Holi

Holi is an annual Hindu spring festival. It takes place on two days in late March or early April. It is also called the festival of colour. There is more than one story about the origin of the festival: some believe it gets its name from Holika, who was the daughter of Hiranyakasipu, the king of demons. He demanded that people stop worshipping the gods and start praying to him. However his own little son, Prahlada, continued offering prayers to Vishnu. His father poisoned him, but the poison turned to nectar in his mouth; he was trampled by elephants but was unharmed; and although Hiranyakasipu tried various ways of killing him, he failed. He then ordered Holika to kill Prahlada and since, she had the ability to walk through fire unharmed, she picked up the child and walked into the fire with him. However Prahlada prayed to Vishnu for protection and was saved. Holika burnt to death, because she did not know that her power only worked if she walked into the fire alone. The burning of Holika is celebrated at Holi.

On the first day of Holi a bonfire is lit at night, to signify the burning of Holika. There is also a practice of throwing cow dung into the fire and shouting obscenities at it, as if shouting at Holika herself, which all suggests that the festival is associated with this particular story.

Throwing coloured powder to celebrate the festival of Holi

Other Hindus, however, celebrate Holi in memory of Krishna. As a youth, Krishna used to play all sorts of pranks on the cowgirls or gopis. One prank was to throw coloured powder all over them, and during the second day of Holi people go around until after noon throwing colours, powder and water at each other, and meet and have fun. Images of Krishna and his consort Radha are carried through the streets.

Whatever its origins, Holi is a very joyful festival and officially ushers in Spring, the season of love. Therefore rules are relaxed to further the fun.

Navaratri

Navaratri is a Hindu festival of worship and dance. The word Navaratri literally means 'nine nights' and the festival is celebrated twice each year, at the spring and autumn equinox, although the major festival has always been in the autumn. The festival is also called Durga Puja. The story which forms the background to this festival is the Ramayana. Rama lost not only his kingdom but also his wife, through deception. In order to gain enough strength to conquer Ravana, Rama turned to Durga for help. It is this that is remembered during Navaratri.

Most of the celebrations take place in the evening. The main feature is the dancing around the shrine to Durga, which has been built especially for the occasion.

An enactment of Durga and her shrine, in procession to celebrate Navaratri/Durga Puja

Chapter 9

Many Hindus also fast, eating only one meal of fruit and sweet foods made from milk each day. Prayers are offered for the protection of health and property and Navaratri is considered to be a good time for the starting of a new venture. Durga is also the divine Mother and, if possible, married daughters return home to their mothers.

Some Hindus divide Navaratri into sections of three days, in order to celebrate different aspects of the divine Mother. The first three days celebrate her power as Durga to destroy impurities, vices and defects. The next three days celebrate her as Lakshmi, the giver of sprirituaI wealth, who can give unlimited wealth to her worshippers. The final three days celebrate her wisdom as Saraswati. In order to have all round success in life, the blessing of all three aspects of the divine mother are needed.

Dussehra

This festival is celebrated at the end of Navaratri ('nine nights') — the meaning of the word 'das' is 'ten'. The festival remembers that it was on this day that Rama, with the help of his brother Laksmana and Hanuman, the monkey king, succeeded in rescuing his wife from the demon king, Ravana. It was on this 'glorious tenth' that Rama, having received strength from Durga, defeated the ten-headed demon, Ravana. Good triumphed over evil and Ravana failed to destroy the loyalty of Laksmana and faithfulness of Sita. The story celebrates friendship.

In Delhi a great celebration, Ram Lila, is held, when a large model of Ravana is burnt as the climax to a firework display. In India presents are given, especially to children.

Also on this day the spirit of the deity leaves the statue of Durga, created for the Navaratri festival, and the murti is carried in a joyous procession to the river and immersed in the water. As it sinks, Hindus believe that it takes with it all unhappiness and misfortune. The festival is also a special time for puja in the home and temple, and wives worship their husbands as Sita did with Rama.

This festival is a time when Hindus have a chance to forget their differences and any arguments which have split families or communities during the year. It is a festival which reminds Hindus of God's love and care, that good is greater than evil, and that everyone should be faithful and friendly to each other.

For reflection

What does taking part in a festival mean to you and your family?

Divali

Divali is the Hindu festival of lights. It is one of the most popular and eagerly awaited festivals of India and is celebrated for five consecutive days in October/November.

Divali marks the victory of good over evil, brightness over darkness. However, Hindus don't all celebrate Divali for the same reasons. In northern India Divali marks the return of Rama, his wife Sita and brother Laksmana to Ayodhya, after killing the demon king, Ravana. It is believed that the people lit oil lamps along the way to light their path in the darkness and that they were welcomed with rows of coloured lights.

In southern India Divali commemorates the killing of Narakasura (an evil demon who created havoc) by Krishna's wife Sathyabhama. In another version he was killed by Krishna himself. It is also said that Kali was born at this time and that the goddess Lakshmi, wife of Vishnu, visits each house, which has been cleaned and brightly lit, to bring gifts and prosperity during the coming year.

The celebration of Divali varies from region to region, but the main features are as follows:

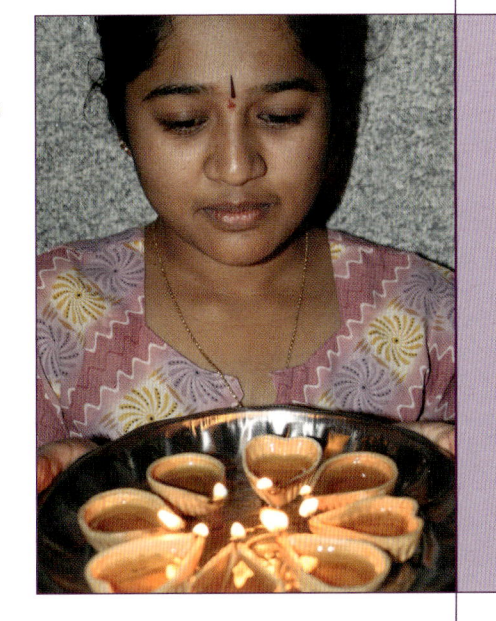

- homes, temples, offices etc. are decorated with coloured lights or clay Divali lamps (divas);
- families meet together;
- boys are often given parties by their sisters;
- new clothes are worn and houses are cleaned and decorated;
- girls make intricate designs, called rangoli patterns, in coloured chalk in front of their houses;
- animals are washed and decorated, have bells placed on their necks and are given special food to eat;
- accounts are settled and a new business year begins with no one in debt;
- everyone tries to turn over a new leaf;
- in India, dance groups tour the country dancing for money;
- presents are exchanged and Divali Cards are sent to friends and relatives at home and abroad;
- in India, girls make a Divali lamp and if they live by a river set their lights afloat on a small raft in the darkness. If the lamp stays alight for as long as they can see it, they will have good luck during the coming year.

Celebrating Divali

Ganesh Chaturthi

Ganesh Chaturthi lasts for ten days and celebrates the birthday of Ganesh, who is a very popular deity for worship. He is seen as a remover of obstacles and is prayed to particularly when people are beginning a new enterprise or starting a new buisness. Life-like clay models of Ganesh are made two to three months before the festival begins. On the first day of the festival they are placed on raised platforms in homes or in decorated outdoor tents, for people to see and worship. The priest then invokes life into the image with the chanting of mantras. Special prasad and food are offered –

Models of Ganesh, carried in procession to celebrate Ganesh Chaturthi

Chapter 9

sweetmeat, coconut, red flowers, sheaves of grass, vermilion and turmeric.

Aarti (a ritualistic puja with hymns) is performed twice a day. Most people in the community attend the evening aarti. As they sing hymns from the Rig Veda, everyone holds rice in their hands, which they later shower on Ganesh. Each ceremony concludes with the sharing of sweets, as it is believed that Ganesh himself liked them very much.

On the eleventh day the image is carried through the streets in a procession full of dancing and singing, to be immersed in a river or the sea. This symbolises bidding farewell to Ganesh as he begins his journey home to Kailash, taking the misfortune of the people with him. Everyone joins in this final procession, shouting 'Ganapathi Bappa Morya, Purchya Varshi Laukariya' which means 'O father Ganesha, come again early next year'. After the final offering of coconuts, flowers and camphor has been made, the image is carried to the river and immersed in the water.

Tasks

Writing tasks	a) Explain the main features of the Diwali festival in Hinduism. b) 'Festivals are the most important part of Hinduism.' Assess this view.
Research task	There are many festivals in Hinduism and many different practices within the festivals. Use the internet to find other festivals associated with the gods and goddesses of Hinduism and prepare a Powerpoint presentation on one of them for the rest of the class.
Presentation	Using Powerpoint, present one Hindu festival to the rest of the class, explaining its main features and the beliefs it is based on.

Glossary

Holi	the festival of colour
Navaratri	the festival of worship and dance, also called Durga Puja
Dussehra	celebration of Rama's viictory over Ravana
Divali	the festival of lights
Ganesh Chaturthi	celebration of the birthday of Ganesh

Chapter 10

Worship

Aim

After studying this chapter you will have knowledge and understanding of different types of worship within Hinduism and of specific puja associated with certain deities. You will also be able to give your own reaction to this type of worship and support it with reasoned comment.

In Hinduism worship can include a wide range of practices and activities, including dance and drama. Some practices are performed individually, some congregationally and many can be both. All can also be performed at home, as well as in the temple.

There are many types of worship, but the following ten are the most common:
1. Puja – ritual worship, especially of the deity;
2. Aarti – the greeting ceremony;
3. Bhajan or Kirtan – hymns and chants (often during arti);
4. Darshan – an audience with a deity or holy person;
5. Prasad – offering and eating sacred food;
6. Pravachan – talk or lecture on the scriptures;
7. Havan – the sacred fire ceremony;
8. Japa/meditation/prayer;
9. Parikrama/Pradakshina – circumambulation;
10. Seva – active service to the deity, holy people etc.

Hindu worship displays a number of distinct features :
(a) Since the presence of the Divine can be seen in a variety of ways within Hinduism, the focus of worship can also vary and can include focus on the Supreme, the various gods and goddesses, the guru etc.
(b) Much of the worship is done individually but in many traditions, including Hinduism in the UK, communal worship plays a central role.
(c) Most Hindu worship takes place outside the temple, most often in the home.
(d) In Hinduism there are no specific days of worship although particular deities are associated with particular days – Shiva is honoured on Monday and Hanuman on Tuesday. In the UK, however, Sunday has become the most important day, as most Hindus work during the week.
(e) The time when Hindus worship is also important. The hours on either side of dawn are considered very important for worship. In India many temples begin their first public ceremony between four and six in the morning. Evening worship is also popular.
(f) There is much warmth, joy and affection within Hindu worship, with God being regarded as a close friend or loved one.
(g) Hindu worship is more spontaneous and less liturgical than some western religions. People come and go freely during ceremonies.

Chapter 10

Puja being performed in India

Puja

Puja is a Sanskrit word which loosely translated means 'reverence' or 'worship' and refers to the worship Hindus perform daily, especially of the sacred image – Murti. Worship of the Murti is central to Hinduism and helps many Hindus develop and express their relationship with God. Murti are more than meditational aids or representations of the different aspects of God: to many Hindus the Murti is considered to be God or the deity it represents. Some Hindu groups consider the Murti to be a form of avatar.

Puja can be performed in the temple or at home. It is an act which shows devotion to God or the chosen deity. Puja can also be performed individually or in a group, in silence or accompanied by prayers. Most Hindus believe puja should be performed daily, although some believe that it should be done twice a day. Puja is also performed on a variety of special occasions, such as Durga Puja and Lakshmi Puja.

Puja can be performed for anything the performer considers to be the concept of God such as a Murti of Vishnu or a Shiva linga. Sometimes a puja is done for the benefit of certain people, for whom priests or relatives ask blessings. Because of the nature of Hinduism, the practical details of how to perform puja vary considerably. Many believe that it should only be done after a shower or bath and before breakfast, to ensure full concentration. Most puja usually include bathing and dressing the deity and the offering of various items to the deity such as water, perfume and flowers, and often lighting a candle or incense. It often ends with the offering of vegetarian food and is immediately followed by the aarti ceremony.

For reflection

Can puja be described as a religious experience? Why?

Most puja generally include a minimum of 16 devotional acts, for example:
1. The spirit of god is invited to enter the Murti by sprinkling rice grains on it and touching the eyes and heart of the Murti with a blade of grass dipped in ghee.
2. Rice grains are spread in a copper dish below the Murti and offered to god.
3. Water is offered to wash the feet of the Murti, which are touched with a wet flower.

A shrine at a home in north Wales

4   Fresh water is offered as a drink to the Murti.
5   The Murti is bathed symbolically with water and a honey-yogurt mixture.
6   Clothes are offered to the Murti and a red cloth is draped around the neck and shoulders of the deity. Holy thread is offered and draped around the Murti.
7   Sandalwood paste and red and yellow powder is put on the Murti's forehead.
8   Flowers are arranged around the Murti.
9   Incense is lit and waved before the deity.
10 Light in the form of a ghee lamp is waved before the deity.
11 Food is offered to the deity.
12 Fruit is offered to the deity.
13 Circumambulation in a clockwise direction in front of the Murti.
14 Flowers and prayers offered to the deity.

Love and devotion are the main characteristics of puja.

Durga Puja

Today Durga Puja is generally a community festival and lasts for nine days. It is also known as Navaratri – the festival of nights (see Chapter 9). On the first day a small bed of mud is prepared in the puja room of the house and barley seeds are sown on it. On the tenth day these shoots are about three to five inches in length. After the puja these seedlings are pulled out and given to devotees as a blessing from god.

During the first three days Durga is worshipped in her manifestations as Kumari, Parvati and Kali. During the next three days Durga is worshipped as Lakshmi, goddess of peace and prosperity. On the fifth day it is traditional to gather and display all literature available in the house, light a lamp or 'diya' to invoke Saraswati, goddess of knowledge and art. On the seventh and eighth days Saraswati is worshipped to gain spiritual knowledge. On the eighth day yagna (holy fire) is performed. Ghee (clarified butter), kheer (rice pudding) and sesame seeds are offered to Durga.

Chapter 10

Worshipping the goddess Durga at the festival of Durga Puja / Navaratri

Seminar topic

'Hindus worship the image rather than the god within it.'

Durga puja concludes with the Kanya puja, where nine young girls, representing the nine forms of Goddess Durga, are worshipped. Their feet are washed as a mark of respect for the Goddess and they are then offered new clothes as gifts by the worshipper.

Lakshmi Ganesh Puja

Lakshmi Ganesh Puja is one of the most important features of the Divali festival (see Chapter 9). Although unrelated in the Hindu pantheon, when placed side by side, Lakshmi and Ganesh hold out promise of a year of fulfilment, free from problems.

During the puja the image of Lakshmi is placed on the left and the image of Ganesh on the right. Lakshmi is the goddess of wealth and prosperity and is full of beauty, grace and charm. She is usually shown seated on a lotus with gold coins. Ganesh is the lord of wisdom and the remover of obstacles. Hindus believe that he must be kept happy to ensure smooth passage to success: this is why he is worshipped with Lakshmi during Divali. While Lakshmi is worshipped for wealth and prosperity, Ganesh is worshipped first to ensure that any obstacles to obtaining that wealth and prosperity are removed.

Puja takes place by placing the images on raised platforms, making various offerings of sandal paste, saffron paste, perfume, haldi, kumkum, flowers (especially marigolds), and leaves of Bel (wood apple tree). Incense sticks and dhoop are lit and offerings of sweets, coconut, fruit and tambul are made. At the end of puja aarti is performed.

Kali Puja

This puja is performed on the night of Kartik Amavasya, which falls in October/November. The main purpose of the puja is to seek the help of the goddess for protection. Kali Puja is performed essentially to seek protection against drought and is performed only at midnight.

In her worst form, Kali is portrayed as dancing in cremation grounds and drinking blood. Many of her devotees therefore go to cremation grounds to meditate and thousands of goats are still sacrificed to her, especially in the Kalighat temple in Calcutta. However Kali is not the goddess of death. Her followers worship her as the divine Mother, who releases humankind from samsara. That is why she is associated with cremation grounds – it is only through accepting death that a person can hope to be reborn and ultimately achieve Moksha.

Seminar topic
'Puja is done out of want, not need.'

Aarti

Aarti is the most popular ceremony within Hinduism, often performed in temples six or seven times a day. It is an offering of love and devotion to the deity and is also a greeting ceremony to the Murti and to gurus, holy people and other representations of the divine. Aarti is often called the 'ceremony of lights' because a tray with five lights, an arti tray, is waved before images of the deities. Sometimes lights are held in front of respected people from other religions, such as Jesus. Aarti can also be performed with a single lamp.

The priest or worshipper offers various articles by moving them in clockwise circles before the deity. At the same time he or she rings a small hand bell, while meditating on the form of the deity. During the ceremony, which normally lasts from five to thirty minutes, the worshipper offers incense, a flower, water, a five wick lamp, a lamp with camphor and other items. The ceremony is often started and ended by the blowing of a conch shell.

Performing aarti at home in north Wales

Chapter 10

During the ceremony the aarti tray is passed around the congregation, who pass their fingers over the flame and then touch their forehead and hair. The offered flowers are also passed around worshippers and the water is sprinkled over their heads. This symbolises the receiving of divine blessing, protection and power. Each person then receives Prasad, which is a mixture of dried fruits, nuts and sugar crystals. This represents the gifts given by the deity to the worshipper and is a symbol of the deity's love for them.

Aarti is usually accompanied by singing (Bhajan/Kirtan) and out of respect worshippers usually stand for the entire ceremony.

Other forms of worship

Apart from puja and aarti there are many other forms of worship within Hinduism.

Bhajan/Kirtan – Bhajan is the singing of hymns of praise in small groups or by the entire congregation. Kirtan is the repeated reciting of mantras to the sound of musical instruments. Bhajan and Kirtan are particularly important to bhakti movements. Common instruments used are drums, hand cymbals and the harmonium.

Darshan – literally means 'seeing' but is better translated as 'audience', when Hindus present themselves before the deity in the temple or before a holy person to receive their blessings. They bow their heads, fold their hands and bring an offering of money or produce, such as fruit or grain. They may also offer prayers. Afterwards they will sip some charanamrita (holy water collected after bathing the murti) or accept some prasad (sanctified food).

Prasad – Visitors to a temple finish their Darshan by accepting prasad (sacred food) offered to the deity. These left-overs are considered to purify the body, mind and soul and give spiritual merit. Some Hindus will only eat prasad and will offer all their meals to the household deity before eating it themselves.

Havan – 'fire sacrifice' or 'sacrificial fire'. This usually takes place on special occasions and rituals, such as initiation and marriage. However some Hindus practice it daily and offer grain and ghee through the fire, while chanting various mantras.

Japa and meditation – Japa is the silent recitation of a mantra, usually performed on a mala, a string of 108 beads. A popular form of meditation is the reciting of the Gagatri mantra, usually done by Brahmin at dawn, noon and dusk. Japa and other forms of meditation are thought to purify the heart of selfish desires and to invoke the love of God.

Seva – service to the deity is considered a form of worship. This could include cleaning, cutting vegetables etc. Service to holy people is also considered as a way of winning god's blessing, as well as service to holy places e.g. helping pilgrims.

Tasks

Writing tasks	a) Explain the main features of Durga Puja.
	b) 'Daily puja is just a meaningless routine, more habit than real devotion.' Assess this view.
Research task	Use the internet to find other practices found in Hindu worship. Discuss them with the rest of the class.

Glossary

puja	ritual worship, especially of the deity
aarti	greeting ceremony
prasad	offering and eating sacred food
Murti	sacred image

Bibliography

Fowler, Jeaneane, *Hinduism – Beliefs and Practices*, Sussex Academic Press, 1997

Kanitkar, V.P., and Cole, W. Owen, *Teach Yourself Hinduism*, Hodder & Stoughton, 2003

Kanitkar, V.P., *Hindu Festivals and Sacraments*, Published privately, 1984

Cole, W. Owen, *Meeting Hinduism*, Longman, Harlow, 1987

Flood, Gavin, *An Introduction to Hinduism*, Cambridge University Press, 1996

www.hindunet.org

www.hinduism.today.com

www.hindu.org